Lecture Notes in Computer Science 15764

Founding Editors

Juris Hartmanis
Gerhard Goos

AF848241

Series Editors

Elisa Bertino, USA
Wen Gao, China
Bernhard Steffen, Germany
Moti Yung, USA

Advanced Research in Computing and Software Science

Subline of Lecture Notes in Computer Science

Subline Series Editors

Giorgio Ausiello, *University of Rome 'La Sapienza', Italy*
Vladimiro Sassone, *University of Southampton, UK*

Subline Advisory Board

Susanne Albers, *TU Munich, Germany*
Benjamin C. Pierce, *University of Pennsylvania, USA*
Bernhard Steffen, *University of Dortmund, Germany*
Deng Xiaotie, *Peking University, Beijing, China*
Jeannette M. Wing, *Microsoft Research, Redmond, WA, USA*

More information about this series at https://link.springer.com/bookseries/558

Arnold Beckmann · Isabel Oitavem ·
Florin Manea
Editors

Crossroads of Computability and Logic: Insights, Inspirations, and Innovations

21st Conference on Computability in Europe, CiE 2025
Lisbon, Portugal, July 14–18, 2025
Proceedings

Springer

Editors
Arnold Beckmann
Swansea University
Swansea, UK

Isabel Oitavem
Universidade Nova de Lisboa
Lisboa, Portugal

Florin Manea
Georg-August Universität Göttingen
Göttingen, Germany

ISSN 0302-9743 ISSN 1611-3349 (electronic)
Lecture Notes in Computer Science
ISBN 978-3-031-95907-3 ISBN 978-3-031-95908-0 (eBook)
https://doi.org/10.1007/978-3-031-95908-0

© The Editor(s) (if applicable) and The Author(s), under exclusive license
to Springer Nature Switzerland AG 2025

This work is subject to copyright. All rights are solely and exclusively licensed by the Publisher, whether the whole or part of the material is concerned, specifically the rights of translation, reprinting, reuse of illustrations, recitation, broadcasting, reproduction on microfilms or in any other physical way, and transmission or information storage and retrieval, electronic adaptation, computer software, or by similar or dissimilar methodology now known or hereafter developed.
The use of general descriptive names, registered names, trademarks, service marks, etc. in this publication does not imply, even in the absence of a specific statement, that such names are exempt from the relevant protective laws and regulations and therefore free for general use.
The publisher, the authors and the editors are safe to assume that the advice and information in this book are believed to be true and accurate at the date of publication. Neither the publisher nor the authors or the editors give a warranty, expressed or implied, with respect to the material contained herein or for any errors or omissions that may have been made. The publisher remains neutral with regard to jurisdictional claims in published maps and institutional affiliations.

This Springer imprint is published by the registered company Springer Nature Switzerland AG
The registered company address is: Gewerbestrasse 11, 6330 Cham, Switzerland

If disposing of this product, please recycle the paper.

Preface

Computability in Europe 2025:
Crossroads of Computability and Logic:
Insights, Inspirations, and Innovations
Lisbon, Portugal
14–18 July 2025

The conference *Computability in Europe* (CiE) is organized yearly under the auspices of the Association CiE, a European association of mathematicians, logicians, computer scientists, philosophers, physicists, biologists, historians and others interested in new developments in computability and their underlying significance for the real world. CiE promotes the development of computability-related science, ranging over mathematics, computer science and applications in various natural and engineering sciences, such as physics and biology, as well as related fields, such as philosophy and history of computing. CiE 2025 had as its motto *Crossroads of Computability and Logic: Insights, Inspirations, and Innovations*, drawing on the numerous interconnections between computability research and broader logical methodologies, considering both well-established perspectives as well as recent innovations.

CiE 2025 was the 21st conference in the series, and the second to take place in Portugal — while the first took place in Ponta Delgada in 2010, this occurrence was hosted at the Science Faculty of the University of Lisbon in the city of Lisbon. The previous CiE conferences have been held in Amsterdam (The Netherlands) in 2005, Swansea (Wales) in 2006, Siena (Italy) in 2007, Athens (Greece) in 2008, Heidelberg (Germany) in 2009, Ponta Delgada (Portugal) in 2010, Sofia (Bulgaria) in 2011, Cambridge (UK) in 2012, Milan (Italy) in 2013, Budapest (Hungary) in 2014, Bucharest (Romania) in 2015, Paris (France) in 2016, Turku (Finland) in 2017, Kiel (Germany) in 2018, Durham (UK) in 2019, Salerno (Italy – virtually) in 2020, Ghent (Belgium – virtually) in 2021, Swansea (UK) in 2022, Batumi (Georgia) in 2023, and Amsterdam (The Netherlands) in 2024. Currently, the annual CiE conference is the largest international meeting focused on computability-theoretic issues. The proceedings containing the best submitted papers, as well as extended abstracts of invited, tutorial, and special session speakers, for all these meetings are published in the Springer series *Lecture Notes in Computer Science.*

Structure and Program of the Conference

The Program Committee invited six speakers to give plenary lectures at CiE 2025: Ugo Dal Lago (University of Bologna, Italy), Daniel Graça (University of Algarve, Portugal), Ekaterina Komendantskaya (University of Southampton, UK), Ng Keng Meng (Nanyang Technological University, Singapore), Paulo Oliva (Queen Mary University of London, UK) and Ana Sokolova (University of Salzburg, Austria). The conference also had two plenary tutorials, presented by Maria Paola Bonacina (University of Verona, Italy) and Igor Carboni Oliveira (University of Warwick, UK).

In addition, the conference had six special sessions:

Computable Analysis and Topology
Organizers: Takayuki Kihara (Nagoya University, Japan) and Elvira Mayordomo (University of Zaragoza, Spain)
Speakers: Djamel Amir (Paris-Saclay University, France), Emmanuel Rauzy (University of the Bundeswehr Munich, Germany), Holger Thies (Kyoto University, Japan) and Cécilia Pradic (Swansea University, UK).

Computability Aspects of Descriptive Set Theory
Organizers: Luca San Mauro (University of Bari, Italy) and Dino Rossegger (Vienna University of Technology, Austria)
Speakers: Antonio Montalban (University of California, Berkeley, USA), Noah Schweber (Proof School, San Francisco, USA), Dan Turetsky (Victoria University of Wellington, New Zealand) and Benjamin Siskind (Vienna University of Technology, Austria).

Human-Centered AI: Foundational, Historical, and Computational Perspectives
Organizers: Melissa Antonelli (University of Helsinki, Finland), Jean-Baptiste Joinet (Jean Moulin Lyon 3 University, France) and Mattia Petrolo (University of Lisbon, Portugal)
Speakers: Juan Manuel Duran (Delft University of Technology, The Netherlands), Paolo Pistone (Claude Bernard University Lyon 1, France), Carina Prunkl (Utrecht University, The Netherlands) and João Marques Silva (ICREA, Spain).

Proof Complexity and SAT
Organizers: Maria Luisa Bonet (Polytechnic University of Catalonia, Spain) and Pavel Pudlák (Czech Academy of Sciences, Czech Republic)
Speakers: Albert Atserias (Technical University of Catalonia, Spain), Paul Beame (University of Washington, USA), Sam Buss (University of California, San Diego, USA) and Robert Robere (McGill University, Canada).

Proof Theory: Pure and Applied
Organizers: Anton Freund (University of Würzburg, Germany) and Pedro Pinto (Technical University of Darmstadt, Germany)
Speakers: Horatiu Cheval (University of Bucharest, Romania), Azza Gaysin (University of Passau, Germany), Morenikeji Neri (University of Bath, UK) and Shuwei Wang (University of Leeds, UK).

Quantum Computing
Organizers: Georg Moser (University of Innsbruck) and Romain Péchoux (University of Lorraine, France)
Speakers: Alejandro Diaz-Caro (Inria, France & National University of Quilmes, Argentine), Robin Kaarsgaard (University of Southern Denmark, Denmark), Subhasree Patro (Eindhoven University of Technology, The Netherlands) and Simon Perdrix (Inria, France).

Speakers in these special sessions were selected by the respective special session organizers and were invited to contribute papers to this volume.

The members of the Program Committee of CiE 2025 selected 27 of the 49 non-invited submitted papers for publication in this volume and for presentation at the conference. Each paper received at least three single-blind reviews by the Program Committee and their subreviewers. In addition to the accepted contributed papers, this volume contains eight invited papers. The production of the volume would have been impossible without the diligent work of our expert referees, both Program Committee members and subreviewers. We would like to thank all of them for their excellent work.

Springer generously funded a Best Paper Award. The winner of this award was the paper *On the computational power of C-random strings* by Alexey Milovanov.

All authors who contributed to this conference were encouraged to submit significantly extended versions of their papers, with additional unpublished research content, to *Computability: The Journal of the Association CiE.*

The Steering Committee of the conference series CiE is concerned about the representation of women researchers in the field of computability. In order to increase women's participation, the series started the *Women in Computability* (WiC) programme in 2007.

In 2016, after the new constitution of the Association CiE allowed for the possibility of creating special interest groups, a special interest group named *Women in Computability* was established. The WiC program includes a workshop, the annual WiC dinner, the mentorship program and a grant program for young women researchers. The Women in Computability workshop continued in 2025, coordinated by Johanna Franklin.

The organizers of CiE 2025 would like to acknowledge and thank the following entities for their financial support (in alphabetical order): the *Association Computability in Europe* (ACiE), the *Association for Symbolic Logic* (ASL), the *Fundação para a Ciência e a Tecnologia, I.P.* (FCT) — national funds under the scope of the projects UIDB/00297/2020 and UIDP/00297/2020 (Center for Mathematics and Applications), UID/04561/2025 (Center for Mathematical Studies), UIDB/04106/2025 (Center for Research & Development in Mathematics and Applications), UID/00013 (Centro de Matemática da Universidade do Minho), UID/000408/2025 (LASIGE Research Unit), UID/5008 (Instituto de Telecomunicações) — the *International Center for Mathematics*, the *International Union of History and Philosophy of Science and Technology - DLMPST*, and the *Kurt Gödel Society.*

We extend our gratitude to all members of the CiE 2025 Organizing Committee for their work towards making the conference a successful event.

May 2025

Arnold Beckmann
Isabel Oitavem
Florin Manea

Organization

Program Committee Chairs

Arnold Beckmann	Swansea University, UK
Isabel Oitavem	NOVA University Lisbon, Portugal

Steering Committee Chair

Florin Manea	University of Göttingen, Germany

Organizing Committee Chair

Gilda Ferreira	Universidade Aberta, Portugal

Program Committee Members

Christel Baier	Dresden University of Technology, Germany
Arnold Beckmann	Swansea University, UK
Laurent Bienvenu	University of Bordeaux, France
Giuseppa Castiglione	University of Palermo, Italy
Benjamin Doerr	École Polytechnique, France
Fernando Ferreira	University of Lisbon, Portugal
Gilda Ferreira	Universidade Aberta, Portugal
Ekaterina Fokina	Vienna University of Technology, Austria
Lorenzo Galeotti	Amsterdam University College, The Netherlands
Jun Le Goh	National University of Singapore, Singapore
Angeliki Koutsoukou-Argyraki	University of Cambridge, UK
Ludovic Levy Patey	Mathematics Institute of Jussieu-Paris Rive Gauche, France
Zsuzsanna Lipták	University of Verona, Italy
Florin Manea	University of Göttingen, Germany
Simone Martini	University of Bologna, Italy
Russell Miller	City University of New York, USA
Georg Moser	University of Innsbruck, Austria
Isabel Oitavem	NOVA University Lisbon, Portugal

Paolo Pistone	Claude Bernard University Lyon 1, France
Romain Péchoux	University of Lorraine, France
Dino Rossegger	Vienna University of Technology, Austria
Stefan Vatev	Sofia University, Bulgaria
Damien Woods	Maynooth University, Ireland

Steering Committee Members

Veronica Becher	University of Buenos Aires, Argentina
Liesbeth De Mol	University of Lille, France
Gianluca Della Vedova	University of Milano-Bicocca, Italy
Benedikt Löwe	University of Hamburg, Germany & University of Cambridge, UK
Florin Manea	University of Göttingen, Germany
Klaus Meer	Brandenburg University of Technology Cottbus–Senftenberg, Germany
Russell Miller	City University of New York, USA
Mariya Soskova	University of Wisconsin-Madison, USA

Organizing Committee Members

Bruno Dinis	University of Évora & CIMA, Portugal
José Espírito Santo	University of Minho & CMAT, Portugal
Gilda Ferreira	Universidade Aberta & CEMS.UL, Portugal
Manuel Martins	University of Aveiro & CIDMA, Portugal
Isabel Oitavem	NOVA University Lisbon & NOVAMath, Portugal
Pedro Quaresma	University of Coimbra & CISUC, Portugal
João Rasga	University of Lisbon & IT, Portugal
Alexandra Sofia Rodrigues	NOVA University Lisbon & CICS.NOVA, Portugal
Paulo Guilherme Santos	Polytechnic University of Lisbon & CEMS.UL, Portugal
André Souto	University of Lisbon & LASIGE, Portugal
Maria Manuel Torres	University of Lisbon & CEMS.UL, Portugal

Administrative Staff

Rodrigo Marques	CEMS.UL, Portugal
Paula Nunes	FCiências.ID, Portugal

Abstracts of Invited Talks

Rigorous Explainability in Machine Learning – Logic to the Rescue

Joao Marques-Silva[1,2]

[1] ICREA, Catalunya, Spain
jpms@icrea.cat
[2] Univ. Lleida, Spain

Explainable artificial intelligence (XAI) is widely regarded as a mainstay of trustworthy AI. Recent years witnessed massive efforts towards devising trustable solutions for XAI. Unfortunately, most of these solutions are based on non-symbolic methods, and as a result, offer no guarantees of rigor. A more recent alternative is logic-based XAI. Logic-based explanations offer the strongest guarantees of rigor, given the machine learning (ML) model. However, logic-based XAI is not without drawbacks, including the complexity of logical reasoning.

This talk was organized into two main parts. The first part surveyed the emerging field of logic-based XAI, covering the core concepts, some of the most visible successes, but also some of the existing challenges. This part covered work reported in several publications [4, 7, 1, 5, 8]. Furthermore, the second part of the talk provided a brief glimpse into how logic-based XAI has served to uncover some of the most persisting misconceptions in non-symbolic XAI, integrating work from several publications [3, 6, 10, 2, 9].

References

1. Cooper, M.C., Marques-Silva, J.: Tractability of explaining classifier decisions. Artif. Intell. **316**, 103841 (2023)
2. Huang, X., Marques-Silva, J.: On the failings of Shapley values for explainability. Int. J. Approx. Reason. **171**, 109112 (2024)
3. Ignatiev, A.: Towards trustable explainable AI. In: IJCAI, pp. 5154–5158 (2020)
4. Ignatiev, A., Narodytska, N., Marques-Silva, J.: Abduction-based explanations for machine learning models. In: AAAI, pp. 1511–1519 (2019)
5. Izza, Y., Huang, X., Ignatiev, A., Narodytska, N., Cooper, M.C., Marques-Silva, J.: On computing probabilistic abductive explanations. Int. J. Approx. Reason. **159**, 108939 (2023)
6. Izza, Y., Ignatiev, A., Marques-Silva, J.: On tackling explanation redundancy in decision trees. J. Artif. Intell. Res. **75**, 261–321 (2022)
7. Marques-Silva, J.: Logic-based explainability in machine learning. In: Reasoning Web, pp. 24–104 (2022)

8. Marques-Silva, J.: Logic-based explainability: past, present and future. In: ISoLA, pp. 181–204 (2024)
9. Marques-Silva, J., Huang, X.: Explainability is not a game. Commun. ACM **67**(7), 66–75 (2024)
10. Marques-Silva, J., Ignatiev, A.: No silver bullet: interpretable ML models must be explained. Front. Artif. Intell. **6** (2023)

Probabilistic Kolmogorov Complexity

Igor C. Oliveira
Department of Computer Science, University of Warwick

Understanding the extent to which data can be compressed has implications across numerous areas of computer science, including data transmission, storage, and machine learning. The (in)compressibility of a string serves as a measure of its complexity. But how difficult is it to determine a string's compressibility? This question lies at the heart of an active research program known as meta-complexity—the study of the computational complexity of complexity itself.

In this tutorial, we present various results connecting compression and meta-complexity to fundamental questions in learning theory, cryptography, algorithms, and average-case complexity. Central to many recent advances is a novel notion of time-bounded Kolmogorov complexity, where a string is represented probabilistically rather than deterministically. We highlight the impact of this perspective and discuss a few recent advances where probabilistic Kolmogorov complexity plays an important role.

References

1. Carmosino, M.L., Impagliazzo, R., Kabanets, V., Kolokolova, A.: Learning algorithms from natural proofs. In: Conference on Computational Complexity (CCC), pp. 10:1–10:24 (2016)
2. Chen, L., Lu, Z., Oliveira, I.C., Ren, H., Santhanam, R.: Polynomial-time pseudodeterministic construction of primes. In: Symposium on Foundations of Computer Science (FOCS), pp. 1261–1270 (2023)
3. Goldberg, H., Kabanets, V., Lu, Z., Oliveira, I.C.: Probabilistic Kolmogorov complexity with applications to average-case complexity. In: Computational Complexity Conference (CCC), pp. 16:1–16:60 (2022)
4. Hirahara, S.: Average-case hardness of NP from exponential worst-case hardness assumptions. In: Symposium on Theory of Computing (STOC), pp. 292–302 (2021)
5. Hu, J., Lu, Z., Oliveira, I.C.: Meta-complexity of nondeterministic Kolmogorov complexity. In: Preprint (2025)
6. Liu, Y., Pass, R.: On one-way functions and Kolmogorov complexity. In: Symposium on Foundations of Computer Science (FOCS), pp. 1243–1254 (2020)
7. Lu, Z., Oliveira, I.C.: Theory and applications of probabilistic Kolmogorov complexity. Bull. EATCS **137** (2022). http://bulletin.eatcs.org/index.php/beatcs/article/view/700

8. Lu, Z., Oliveira, I.C., Zimand, M.: Optimal coding theorems in time-bounded Kolmogorov complexity. In: International Colloquium on Automata, Languages, and Programming (ICALP), pp. 92:1–92:14 (2022). https://doi.org/10.4230/LIPIcs.ICALP.2022.92
9. Oliveira, I.C.: Randomness and intractability in Kolmogorov complexity. In: International Colloquium on Automata, Languages, and Programming (ICALP), pp. 32:1–32:14 (2019)

Analysing Gödel's *L* in Realisability Models of CZF

Shuwei Wang

University of Leeds, England, UK

In this talk, we present a survey of some recent developments on the properties of Gödel's constructible universe L in the intuitionistic theory CZF. Here, L can be defined in the same way as the classical counterpart but its structure will be much more complicated since, most importantly, intuitionistic set theories do not prove that the ordinals are linearly ordered. The implications of this on L were first discussed by Robert Lubarsky [1] and more recently by Matthews and Rathjen [2].

In [2], it is shown that CZF cannot prove that L itself is a model of CZF — specifically, CZF cannot prove $L \models$Exponentiation, by reducing the problem through a variant of realisability-with-truth argument to the (non-)provability of the existence of a gap ordinal in KP($\mathcal{P}$). This shares some similarity to the classical proof-theoretic argument in [3] that KP($\mathcal{P}$)$+V = L$ is strictly stronger than KP($\mathcal{P}$) in consistency strength. However, we will explain that the non-linear nature of intuitionistic ordinals means that no similar argument can be made for the consistency strength of CZF$+V = L$, as we recently showed that, by constructing a non-classical ordinal through recursion-theoretic methods, CZF$+V = L$ can be realised in the type-theoretic structure $\mathrm{ML}_1\mathrm{V}$ and thus is equi-consistent with CZF.

References

1. Lubarsky, R.S.: Intuitionistic L. In: Crossley, J.N., Remmel, J.B., Shore, R.A., Sweedler, M.E. (eds.) Logical Methods: In Honor of Anil Nerode's Sixtieth Birthday, pp. 555–571. Birkhäuser, Boston (1993). https://doi.org/10.1007/978-1-4612-0325-4_18
2. Matthews, R., Rathjen, M.: Constructing the constructible universe constructively. Ann. Pure Appl. Logic **175**(3), 103392 (2024). https://doi.org/10.1016/j.apal.2023.103392
3. Rathjen, M.: Power Kripke–Platek set theory and the axiom of choice. J. Logic Comput. **30**(1), 447–457 (2020). https://doi.org/10.1093/logcom/exaa020

Contents

Invited Papers

Contributed Papers

Invited Papers

The CDSAT Method for Satisfiability Modulo Theories and Assignment: an Exposition

Maria Paola Bonacina(✉)

Dipartimento di Informatica, Università degli Studi di Verona, Verona, Italy
mariapaola.bonacina@univr.it
https://mariapaola.github.io/

Abstract. Most SMT instances involve symbols from more than one theory. The equality sharing method for combining theory satisfiability procedures has been a standard for over forty years. For the Boolean part, the equality sharing method is interfaced with the CDCL procedure for SAT. CDCL guides the search by learning lemmas from conflicts between clauses and candidate model. In the standard integration of CDCL and equality sharing, the conflict-driven reasoning remains propositional, even if conflict-driven theory procedures exist. The MCSAT method integrates CDCL with a single conflict-driven theory procedure. The CDSAT method generalizes CDCL, MCSAT, and equality sharing, by allowing the integration of multiple (conflict-driven or not) theory procedures.

Keywords: Conflict-driven satisfiability procedures · Satisfiability modulo theory and assignment · Combination of theories

1 Introduction

Automated reasoning (AR) is a foundation of computer science, since the quest for a formulation and a solution of the *Entscheidungsproblem* led Alan Turing to invent Turing machines in 1936. AR is also a foundation of artificial intelligence (AI), since machines that prove theorems have been a defining objective of AI since the Dartmouth Conference in 1956. AR is also an essential methodology for the analysis, verification, synthesis, and optimization of software. Logic proved to be the calculus of computation, as predicted by John McCarthy in 1963 [9]. AR is applied to proving verification/synthesis conditions, refining abstractions for model checking, generating tests for testing and examples for synthesis. Current goals in this field include correct-by-construction software, provable privacy, and verification of distributed protocols, distributed systems, and randomized algorithms. While generative AI as in chatbots can be used by anyone, AR is currently used mostly by computer scientists and mathematicians. The integration of AR and generative AI towards a better AI is a fascinating challenge (e.g., [3]).

© The Author(s), under exclusive license to Springer Nature Switzerland AG 2025
A. Beckmann et al. (Eds.): CiE 2025, LNCS 15764, pp. 3–18, 2025.
https://doi.org/10.1007/978-3-031-95908-0_1

Automated reasoning is concerned with the design and implementation of computer programs that solve problems formulated as validity or satisfiability queries in a logic or a theory. Validity queries are handled refutationally: a conjecture is shown valid by showing that its negation is unsatisfiable. AR procedures involve *inference* and *search*. In this paper we consider contexts where satisfiability is *decidable*, the problem is *quantifier-free*, and written as a *set of clauses*. We study *conflict-driven reasoning procedures* [4] that are at the heart of solvers for propositional satisfiability (SAT) and satisfiability modulo theories (SMT).

Conflict-driven reasoning procedures search for a model (to establish satisfiability) and perform inferences (to prove unsatisfiability), in such a way that search and inferences guide each other. The procedure searches for a model by proposing *assignments* to terms and *propagating* their consequences. Inferences are performed (almost) only when the procedure encounters a *conflict* or contradiction (e.g., one of the clauses to be satisfied is false in the current assignment). The inferences *explain* the conflict, resulting in the generation of a *lemma*. These inferences are steps towards a possible refutation. The lemma excludes the assignments that caused the conflict. Learning the lemma ensures that the model search will not incur in the same conflict, and the lemma is used to drive the search elsewhere. In a nutshell, the model search focuses the inferences on solving conflicts, and the inferences allow the search to escape dead-end's.

This paper gives an exposition of the CDSAT (*Conflict-Driven SATisfiability*) method [6–8]: to the best of our knowledge, CDSAT is currently the *most general conflict-driven reasoning procedure*. Its novel features include the following. CDSAT decides the satisfiability of a set of clauses modulo a *union of theories* and a possibly empty *initial assignment* (*satisfiability modulo theories and assignment*). The initial assignment may contain *both Boolean* (e.g., $L \leftarrow \mathsf{true}$) and *first-order* (e.g., $x \leftarrow 3$) assignments to terms occurring in the problem. The answer is sat, if there exists a satisfying assignment that includes the initial one, and unsat otherwise. CDSAT combines theories, not black-box theory procedures. The CDSAT transition system coordinates *theory modules* (theory inference systems), which collaborate as peers over a *trail* representing a candidate model by Boolean and first-order assignments. Since conflict-driveness is provided by CDSAT, the distinction between theory with conflict-driven procedure and theory without fades. CDSAT does not need the traditional architecture with the SAT solver at the center and the theory procedure(s) as satellite(s). Propositional logic is one of the theories, the *Boolean theory* denoted Bool.

The proceeding sections interleave outlines of conflict-driven reasoning procedures with examples, describing how CDSAT encompasses the other procedures. The CDSAT transition rules are described informally showing their behavior in the examples. A complete presentation with formal definitions, results, and proofs can be found in [6,8]. More references can be found in [4–6,8].

2 From CDCL to CDSAT

The CDCL (*Conflict-Driven Clause Learning*) procedure [22,23] for SAT (the problem of deciding the satisfiability of a set S of propositional clauses) evolved from the DPLL (Davis-Putnam-Logemann-Loveland) procedure [10], which in turn evolved[1] from the DP (Davis-Putnam) procedure [11]. CDCL was called DPLL or DPLL-CDCL until it was agreed that the two procedures deserve distinct names. The DP procedure featured rules incorporating *resolution* (clauses $C \vee L$ and $D \vee \overline{L}$ resolve upon complementary literals L and $\overline{L}$ to generate resolvent $C \vee D$), its special case *unit resolution* (L and $D \vee \overline{L}$ resolve to generate D), and *subsumption* (L subsumes $C \vee L$, and D subsumes $D \vee \overline{L}$). DPLL replaced resolution with *splitting*: consider assignment $L \leftarrow \mathsf{true}$ (written L), propagate its consequences, and if a conflict arises, backtrack to $L \leftarrow \mathsf{false}$ or $\overline{L} \leftarrow \mathsf{true}$ (written $\overline{L}$). DPLL demphasized inference in favor of backtracking search over partial models, represented as a *trail* of *Boolean assignments*. Propagation was *unit propagation* viewed as unit resolution and subsumption: if L is on the trail, clause $C \vee L$ is deleted, and clause $D \vee \overline{L}$ is replaced by D. A conflict arises if S contains $\{L, \overline{L}\}$. A splitting is done if no propagation is possible. The trail is organized as a *stack*: backtracking consists of popping L and pushing $\overline{L}$.

CDCL replaced chronological backtracking with *backjumping* and splitting with *decision*: guess L, provided neither L nor $\overline{L}$ is on the trail. If chronological backtracking is abandoned, $\overline{L}$ is not necessarily considered right after L, and a tentative assignment is a decision, rather than a case in a case analysis. Propagation replaced unit resolution and unit subsumption with detection of *implied literals* and *conflict clauses*. If $(C \vee L) \in S$ and the trail contains $\overline{Q}$ for every literal Q in C, literal L is *implied* and added to the trail with *justification* $C \vee L$. If $C \in S$ and the trail contains $\overline{Q}$ for every literal Q in C, clause C is in *conflict*. Every decision opens a new *level* on the trail, still organized as a stack. All implied literals belong to the level of the most recent decision, except those implied literals that are unit input clauses, which belong to level 0.

CDCL brought resolution back, applying it only to *explain* a conflict, by resolving a conflict clause containing L with the justification of literal $\overline{L}$ on the trail. One of the resolvents is retained as a lemma, and used to direct the backjumping to a level of the trail where the lemma can be satisfied and the search can restart. The choice of how much resolution to do and which resolvent to learn is a matter of heuristic. The popular *1UIP heuristic* prescribes to explain the conflict by resolution until it generates an *assertion clause*, namely a conflict clause C such that only one of its literals, the *assertion literal* L, is falsified by the current level of the trail. After learning C, the procedure jumps back to the smallest level where L is undefined, all other literals in C are false, and hence the conflict is solved by placing L on the trail with C as justification.

In CDSAT clause C is viewed as an abbreviation of the assignment $C \leftarrow \mathsf{true}$. There is no need to separate input set S and trail, and all input assignments sit on the trail at level 0. Decisions are performed by a transition rule named Decide. A decision is written as $_?L$ to record that it is a guess. CDSAT requires that every decision be *acceptable*. A Boolean decision $_?L$ is acceptable if neither L nor $\overline{L}$ is on the trail. The notion of implied literal is generalized to that of *justified assignment*: if there is a clause $C \vee L$ and the trail contains $\overline{Q}$ for every literal Q

[1] A historical account can be found in [21].

in C, the justified assignment ${}_{J}{\vdash}L$ is added to the trail, with $J = \{C \vee L, \neg C\}$ as *justification*. Propagations are performed by a transition rule named Deduce. A justified assignment ${}_{J}{\vdash}L$ belongs to the level of its justification J, which is the maximum among the levels of the elements of J. Input assignments are justified assignments with empty justification: input clause C is written as ${}_{\emptyset}{\vdash}C$. The trail is not necessarily a stack, as ${}_{J}{\vdash}L$ can be added after assignments of level higher than that of J. In this case, ${}_{J}{\vdash}L$ is called a *late propagation*.

In CDSAT the notion of conflict clause is replaced by that of *conflicting assignment*, or *conflict* tout court, meaning an unsatisfiable set E of assignments. Resolve transitions *explain* E by replacing an element ${}_{J}{\vdash}L$ of E with its justification J. Explanation continues until E contains only one literal whose level m is maximum in E. This is a generalization of 1UIP, because level m is not necessarily the current one. A LearnBackjump transition *learns* a clause ${}_{J}{\vdash}C$, where $C = L_1 \vee \ldots \vee L_k$ is the negation of a Boolean subset $H = \{\overline{L}_1, \ldots, \overline{L}_k\}$ of conflict E, and J is the rest of E, that is, $E = J \uplus H$. Indeed, if $S \cup (J \uplus H) \models \bot$, then $S \cup J \models C$. A conjunction, or set, of literals is called a *cube*. Then LearnBackjump jumps back to a level that is strictly smaller than that of H, in order to quit the conflict, and greater than or equal to that of J, so that ${}_{J}{\vdash}C$ can be placed on the trail. When LearnBackjump follows 1UIP, clause C is the first assertion clause and the destination level is the one prescribed by 1UIP. LearnBackjump can apply also other heuristics. Example 1 illustrates CDCL with 1UIP as done by CDSAT. The only active module is that of theory Bool.

Example 1. Let $S = \{\overline{A} \vee B,\ \overline{A} \vee C \vee E,\ \overline{B} \vee D,\ \overline{C} \vee \overline{D}, A \vee \overline{B} \vee E,\ B \vee \overline{C},\ F \vee \overline{E}\}$ be a subset of the input, where A, B, C, D, E, and F are propositional variables [18]. The trail contains ${}_{\emptyset}{\vdash}\overline{A} \vee B$ and all other input clauses at level 0. Suppose that in order to satisfy other clauses a Decide transition adds ${}_{?}\overline{F}$ to the trail. Say the trail already had $n-1$ levels prior to this decision, so that ${}_{?}\overline{F}$ opens level n. The Bool-module infers by unit propagation $\{F \vee \overline{E}, {}_{?}\overline{F}\} \vdash_{\mathsf{Bool}} \overline{E}$ and a Deduce transition adds ${}_{J}{\vdash}\overline{E}$ to the trail with $J = \{F \vee \overline{E}, {}_{?}\overline{F}\}$. The level of ${}_{J}{\vdash}\overline{E}$ is n, because the level of J is n, since the level of ${}_{?}\overline{F}$ is n and that of $F \vee \overline{E}$ is 0. Suppose that two other Decide transitions create levels $n+1$ and $n+2$, and a third one adds ${}_{?}A$ opening level $n+3$. More unit propagations and Deduce transitions add to the trail ${}_{H}{\vdash}B$ with $H = \{\overline{A} \vee B, {}_{?}A\}$, ${}_{I}{\vdash}C$ with $I = \{\overline{A} \vee C \vee E, {}_{J}{\vdash}\overline{E}, {}_{?}A\}$, and ${}_{K}{\vdash}D$ with $K = \{\overline{B} \vee D, {}_{H}{\vdash}B\}$. All these justified assignments have level $n+3$.

The unit propagation $\{\overline{C} \vee \overline{D},\ {}_{I}{\vdash}C\} \vdash_{\mathsf{Bool}} \overline{D}$ reveals a conflict, as ${}_{K}{\vdash}D$ is on the trail. The conflict is $E_0 = \{\overline{C} \vee \overline{D},\ {}_{I}{\vdash}C,\ {}_{K}{\vdash}D\}$. Clause $\overline{C} \vee \overline{D}$ is the conflict clause, but not an assertion clause. E_0 contains two literals of maximum level, namely ${}_{I}{\vdash}C$ and ${}_{K}{\vdash}D$ (both have level $n+3$). Resolve explains E_0 by replacing one of these two literals, say D, with its justification. The resulting conflict is $E_1 = \{\overline{C} \vee \overline{D},\ {}_{I}{\vdash}C,\ \overline{B} \vee D,\ {}_{H}{\vdash}B\}$. This unfolding of the conflict corresponds to the resolution step that resolves $\overline{C} \vee \overline{D}$ and $\overline{B} \vee D$ to yield $\overline{B} \vee \overline{C}$. The literal resolved upon D is the one explained away when going from E_0 to E_1. The new conflict clause $\overline{B} \vee \overline{C}$ is the negation of the cube $\{{}_{H}{\vdash}B,\ {}_{I}{\vdash}C\}$ in E_1.

Since E_1 does not have a unique literal of highest level (both ${}_{I}{\vdash}C$ and ${}_{H}{\vdash}B$ belong to level $n+3$), Resolve explains E_1 by replacing ${}_{I}{\vdash}C$ with I, yielding

$E_2 = \{\overline{C} \vee \overline{D},\ \overline{A} \vee C \vee E,\ {}_{J\vdash}\overline{E},\ {}_{?}A,\ \overline{B} \vee D,\ {}_{H\vdash}B\}$. This corresponds to resolving $\overline{B} \vee \overline{C}$ and $\overline{A} \vee C \vee E$ upon C to yield conflict clause $\overline{B} \vee \overline{A} \vee E$, negation of cube $\{{}_{H\vdash}B,\ {}_{?}A,\ {}_{J\vdash}\overline{E}\}$ in E_2. Since E_2 does not have a unique literal of highest level (both ${}_{?}A$ and ${}_{H\vdash}B$ belong to level $n+3$), Resolve explains E_2 by replacing ${}_{H\vdash}B$ with H, yielding $E_3 = \{\overline{C} \vee \overline{D},\ \overline{A} \vee C \vee E,\ {}_{J\vdash}\overline{E},\ {}_{?}A,\ \overline{B} \vee D,\ \overline{A} \vee B\}$. This corresponds to resolving $\overline{B} \vee \overline{A} \vee E$ and $\overline{A} \vee B$ upon B to yield conflict clause $\overline{A} \vee E$, negation of cube $\{{}_{?}A,\ {}_{J\vdash}\overline{E}\}$ in E_3.

As ${}_{?}A$ is the only literal of maximum level in E_3 (and $\overline{A} \vee E$ is the first assertion clause), 1UIP prescribes to learn $\overline{A} \vee E$ (we learn that no model of S can contain A and $\overline{E}$), and backjump to the smallest level where $\overline{A}$ is undefined and E is still false. This level is n. LearnBackjump takes us back to level n, clearing levels $n+1$, $n+2$ and $n+3$, and adding lemma ${}_{G\vdash}\overline{A} \vee E$ to the trail, where $G = \{\overline{C} \vee \overline{D},\ \overline{A} \vee C \vee E,\ \overline{B} \vee D,\ \overline{A} \vee B\}$. Level n contains ${}_{?}\overline{F}$, ${}_{J\vdash}\overline{E}$, and ${}_{G\vdash}\overline{A} \vee E$. Deduce performs unit propagation adding to the trail ${}_{M\vdash}\overline{A}$ with $M = \{\overline{A} \vee E, {}_{J\vdash}\overline{E}\}$, ${}_{N\vdash}\overline{B}$ with $N = \{A \vee \overline{B} \vee E,\ {}_{M\vdash}\overline{A},\ {}_{J\vdash}\overline{E}\}$, and ${}_{P\vdash}\overline{C}$ with $P = \{B \vee \overline{C}, {}_{N\vdash}\overline{B}\}$. The trail contains a model of S, namely $\{\overline{E},\ \overline{A},\ \overline{B},\ \overline{C}\}$, and the search can continue to take care of the other input clauses.

Several authors approached the SMT problem (deciding the satisfiability of a set of clauses in a first-order theory $\mathcal{T}$) by integrating in CDCL a decision procedure for the $\mathcal{T}$-satisfiability of a set of $\mathcal{T}$-literals. This integration was systematized in the DPLL($\mathcal{T}$) framework [26], later renamed CDCL($\mathcal{T}$). In this framework, CDCL works on a propositional abstraction of the first-order problem, where $\mathcal{T}$-atoms are replaced with propositional variables. The $\mathcal{T}$-satisfiability procedure detects *theory conflicts* and contributes *theory propagations*. Suppose that literals $L_1, \ldots, L_n$ are on the trail. If the $\mathcal{T}$-procedure detects that $\{L_1, \ldots, L_n\}$ is $\mathcal{T}$-unsatisfiable, it signals a $\mathcal{T}$-conflict with conflict clause $C = \overline{L}_1 \vee \ldots \vee \overline{L}_n$. If the $\mathcal{T}$-procedure detects that $\{L_1, \ldots, L_n\}$ entails L in $\mathcal{T}$, where L appears in the input problem, L is added to the trail with $C \vee L$ as justification. The model search, the reasoning about disjunction, and the explanation of conflicts remain propositional and are left to the CDCL procedure.

3 From the Equality Sharing Method to CDSAT

In many problems the theory $\mathcal{T}$ is the union of a few theories $\mathcal{T}_1, \ldots, \mathcal{T}_n$. The *equality sharing* or *Nelson-Oppen* scheme combines procedures for the satisfiability of sets of literals in the member theories to yield a procedure for the $\mathcal{T}$-satisfiability of a set S of $\mathcal{T}$-literals [25] (see [9, Ch. 10] and [5, Sect. 3] for a modern presentation). The theories are assumed to be *disjoint*, meaning that equality is the only shared symbol, *stably infinite*, meaning that every $\mathcal{T}_i$ admits models of infinite cardinality, and endowed with $\mathcal{T}_i$-satisfiability procedures $\mathcal{P}_i$ for $i = 1, \ldots, n$. Since the theories are disjoint, they only need to agree on equalities between *shared terms* and on the cardinalities of shared sorts. The latter point is solved by the stable infiniteness assumption.

Since the input literals mix symbols from the signatures of the theories, each theory $\mathcal{T}_i$ treats as a variable every term whose top symbol belongs to another theory. The equality sharing method features a *separation* phase where new variable symbols are introduced, and the set S of literals is separated into sets $S_1, \ldots, S_n$ sharing only equality and variables.

Example 2. Consider literal $f(2, y) \simeq f(x, y)$, where f is a function in the theory of Equality with Uninterpreted Functions (EUF or UF) [9, Ch. 9], constant symbol 2 comes from a fragment of arithmetic, say LIA for linear integer arithmetic, and x and y are variables. For EUF, term 2 is a variable, because 2 is a foreign symbol for EUF. For LIA, terms $f(2, y)$ and $f(x, y)$ are variables, because f is a foreign symbol for LIA. The set of *shared terms* is $\mathcal{V}_{\mathsf{sh}}(S) = \{f(2, y),\ 2,\ f(x, y)\}$. Separation yields $S_{\mathsf{EUF}} = \{w_1 \simeq f(w_2, y), w_3 \simeq f(x, y), w_1 \simeq w_3\}$, $S_{\mathsf{LIA}} = \{w_2 \simeq 2,\ w_1 \simeq w_3\}$, and a set of shared terms containing only variables: $\mathcal{V}_{\mathsf{sh}}(S) = \{w_1,\ w_2,\ w_3\}$.

The equality sharing method reduces a $\mathcal{T}$-satisfiability problem to $\mathcal{T}_i$-satisfiability problems. The reduction rests on whether the theories can agree on a *partition* of $\mathcal{V}_{\mathsf{sh}}(S)$, where two terms are equal iff they are in the same equivalence class. A partition is expressed as a set of equalities and disequalities called an *arrangement*: arrangement α contains $u \simeq v$, if u and v are in the same equivalence class, and $u \not\simeq v$ otherwise. Set S is $\mathcal{T}$-satisfiable iff there exists an arrangement α of $\mathcal{V}_{\mathsf{sh}}(S)$ such that $S_i \wedge \alpha$ is $\mathcal{T}_i$-satisfiable for $i = 1, \ldots, n$.

In order to determine whether an arrangement exists, each $\mathcal{P}_i$ deduces from S_i equalities between shared variables. Such equalities are shared by all $\mathcal{P}_i$'s and each $\mathcal{P}_i$ may use them to deduce more equalities. If a procedure $\mathcal{P}_i$ derives a contradiction in this process, no successful arrangement exists, and the $\mathcal{T}$-procedure returns unsat. Otherwise, let $\mathcal{E}$ be the final set of deduced and shared equalities, meaning that no more can be deduced by any of the $\mathcal{P}_i$'s. Set $\mathcal{E}$ represents the arrangement $\alpha_{\mathcal{E}}$ that contains $u \simeq v$, if $(u \simeq v) \in \mathcal{E}$, and $u \not\simeq v$ otherwise. Since no more equalities can be deduced, $S_i \wedge \alpha_{\mathcal{E}}$ is $\mathcal{T}_i$-satisfiable for $i = 1, \ldots, n$, and the $\mathcal{T}$-procedure returns sat.

Having procedure $\mathcal{P}_i$ deduce equalities between shared variables suffices if theory $\mathcal{T}_i$ is *convex*. A theory is *convex* if whenever a conjunctive formula (in our case a set of literals) implies a disjunction of equalities, one of the equalities is also implied. Otherwise, procedure $\mathcal{P}_i$ must be capable of deducing disjunctions of equalities between shared variables. Whenever a disjunction $\bigvee_{j=1}^{m} u_j \simeq v_j$ is deduced, the $\mathcal{T}$-procedure should split it and call itself recursively on each subproblem obtained by adding one of the disjuncts to the current set of shared equalities. In practice, the reasoning about disjunction is entrusted to the CDCL procedure (e.g., [1]). In the extension of the CDCL($\mathcal{T}$) framework to the case where the $\mathcal{T}$-procedure is an equality-sharing combination [1,19], the $\mathcal{T}$-procedure is allowed to send to CDCL (the propositional abstraction of) any disjunction $\bigvee_{j=1}^{m} u_j \simeq v_j$ that a procedure $\mathcal{P}_i$ may derive. The CDCL procedure treats such a disjunction as a clause to be satisfied. The $\mathcal{T}$-procedure considers the case where $u_j \simeq v_j$ is shared, when CDCL puts (the propositional literal standing for) $u_j \simeq v_j$ on the trail. Therefore, the only new (i.e., non-input) literals that can be $\mathcal{T}$-propagated are propositional abstractions of equalities between shared variables.

The equality sharing method is *not conflict-driven*. Rather, it resembles a *saturation* process, as it expects each theory procedure to be capable of deducing *all* (disjunctions of) equalities between shared variables that follow in the theory from its subproblem. Indeed, also *superposition* can emulate the equality sharing method, under suitable conditions and if superposition is a decision procedure for each component theory (see [5, Sect. 7] for an overview of superposition-based results and references). In equality sharing the theory procedures are combined as *black-boxes*, which communicate only by broadcasting (disjunctions of) equalities between shared variables. In the integration of equality sharing in the CDCL($\mathcal{T}$) framework, the broadcasting remains hidden within the $\mathcal{T}$-procedure.

CDSAT does not apply separation, so that the shared terms are not necessarily variables. CDSAT lets each theory module place on the trail its decisions and the results of its inferences. If CDSAT emulates equality sharing, the arrangement is built on the shared trail. The theory module inferences are not used only to deduce equalities between shared terms, but also to detect conflicts as shown in the next example. Regardless of whether Bool is a member of the union, every theory has the sort prop of the Boolean values.

Example 3. Consider the set of literals $\{x \leq y,\ y \leq (x + g(x)),\ P(h(x) - h(y)),\ \neg P(0),\ g(x) \simeq 0\}$, involving theories LIA and EUF [5]. It is customary to remove free predicate symbols by introducing new free function symbols and a new free constant symbol with an arbitrary name, say $\bullet$. The resulting set $S = \{x \leq y,\ y \leq (x + g(x)),\ f(h(x) - h(y)) \simeq \bullet,\ f(0) \not\simeq \bullet,\ g(x) \simeq 0\}$ is equisatisfiable.[2] The set of shared terms is $\mathcal{V}_{\mathsf{sh}}(S) = \{x, y, g(x), h(x), h(y), h(x) - h(y), 0\}$.

Suppose a Decide transition puts ${}_{?}(x \not\simeq y)$ on the trail opening level 1. The LIA-module infers $\{y \leq x + g(x),\ g(x) \simeq 0\} \vdash_{\mathsf{LIA}} y \leq x$ and a Deduce transition posts ${}_{J\vdash}(y \leq x)$ on the trail with $J = \{y \leq x + g(x),\ g(x) \simeq 0\}$. This is a *late propagation* as ${}_{J\vdash}(y \leq x)$ belongs to level 0 and the trail contains level 1. The LIA-module infers by antisymmetry $\{x \leq y,\ {}_{J\vdash}(y \leq x)\} \vdash_{\mathsf{LIA}} x \simeq y$, revealing conflict $E_0 = \{{}_{?}(x \not\simeq y),\ x \leq y,\ {}_{J\vdash}(y \leq x)\}$. Since ${}_{?}(x \not\simeq y)$ is the only assignment of highest level in E, a LearnBackjump transition takes us back to level 0 adding to the trail lemma ${}_{H\vdash}(x \simeq y)$ with $H = \{x \leq y,\ {}_{J\vdash}(y \leq x)\}$.

The EUF-module infers by congruence ${}_{H\vdash}(x \simeq y) \vdash_{\mathsf{EUF}} h(x) \simeq h(y)$, and a Deduce transition posts ${}_{I\vdash}(h(x) \simeq h(y))$ on the trail with $I = \{{}_{H\vdash}(x \simeq y)\}$. The LIA-module infers ${}_{I\vdash}(h(x) \simeq h(y)) \vdash_{\mathsf{LIA}} h(x) - h(y) \simeq 0$, and a Deduce transition posts ${}_{K\vdash}(h(x) - h(y) \simeq 0)$ on the trail with $K = \{{}_{I\vdash}(h(x) \simeq h(y))\}$. Then the EUF-module infers $\{f(h(x) - h(y)) \simeq \bullet,\ {}_{K\vdash}(h(x) - h(y) \simeq 0)\} \vdash_{\mathsf{EUF}} f(0) \simeq \bullet$, which is the complement of $f(0) \not\simeq \bullet$ on the trail. Since conflict $E_1 = \{f(h(x) - h(y)) \simeq \bullet,\ {}_{K\vdash}(h(x) - h(y) \simeq 0),\ f(0) \not\simeq \bullet\}$ is on level 0 (there is nowhere to backjump to), a Fail transition returns unsat.

[2] Assuming trivial models of cardinality 1 are excluded.

If the procedure $\mathcal{P}_i$ for a component theory $\mathcal{T}_i$ builds a candidate model $\mathcal{M}_i$, the equality sharing method can be implemented in CDCL($\mathcal{T}$) by *model-based theory combination* (MBTC) [12]. This variant of equality sharing allows procedure $\mathcal{P}_i$ to share equalities that are true in $\mathcal{M}_i$, even if they are not $\mathcal{T}_i$-entailed by S_i and the already shared equalities. Such an equality is placed on the CDCL($\mathcal{T}$) trail as a decision. If it turns out to cause a conflict, it gets retracted by the CDCL backjumping. CDSAT generalizes the concept of MBTC, because in CDSAT every theory $\mathcal{T}_i$ can place a decision on the shared trail by a Decide transition. In CDCL($\mathcal{T}$) with MBTC a $\mathcal{T}_i$-model $\mathcal{M}_i$ remains private to procedure $\mathcal{P}_i$. In CDSAT the $\mathcal{T}_i$-module share $\mathcal{M}_i$ on the trail, as all theory modules share assignments publicly on the trail.

Another variant of the equality sharing method is *polite theory combination*, where a non-stably-infinite theory can be combined with a polite theory. A polite theory satisfies stronger requirements about the cardinalities of models (e.g., [5, Sect. 4–6]). CDSAT has a different approach to the issue of agreement on the cardinalities of shared sorts. Theories are not required to be stably infinite. It suffices that there exists a *leading theory* that has all the sorts in the union of theories, and all the information about their cardinalities (e.g., at-most-m constraints) [8, Sect. 4.5]. This is because in the proof of completeness of CDSAT an agreement among all the $\mathcal{T}_i$'s is reached by having each $\mathcal{T}_i$ agree with the leading theory. It is not even necessary that the leading theory is one of the $\mathcal{T}_i$'s. If all the theories in the union are stably infinite, the leading theory is a fictional theory $\mathcal{T}_{\mathbb{N}}$, whose models interpret all sorts except prop as having the cardinality of the set $\mathbb{N}$ of the natural numbers [6, Sect. 8].

4 From GCDCL and MCSAT to CDSAT

The success of CDCL for SAT led to several conflict-driven reasoning procedures (e.g., for linear rational arithmetic [24], linear integer arithmetic [16], non-linear arithmetic [17]). In these procedures, the candidate model is a $\mathcal{T}$-model (for the respective fragment $\mathcal{T}$), assignments include *first-order assignments* (e.g., $x \leftarrow 2$), propagation consists of inexpensive theory deductions, such as term evaluation (e.g., if the trail contains assignment $x \leftarrow 2$ for a free variable x, then the term $x + 1$ evaluates to 3), and the inference rules for conflict explanation are relatively more expensive theory inferences.

Let $\mathcal{T}$ be linear rational arithmetic (LRA). The input is a set S of LRA-clauses, whose satisfiability is to be determined. An LRA-clause is a disjunction of literals of the form $t_1 \lessdot t_2$, where t_1 and t_2 are LRA-terms, and $\lessdot \in \{<, \leq\}$. An LRA-term is either a rational constant c or a sum $c_1 \cdot x_1 + \ldots + c_n \cdot x_n$, where c_i is a rational constant and x_i is a rational variable for $i = 1, \ldots n$. Since the ordering on the rational numbers is total, $\overline{(t_1 < t_2)}$ and $\overline{(t_1 \leq t_2)}$ can be replaced by $t_2 \leq t_1$ and $t_2 < t_1$, respectively. An equality $t_1 \simeq t_2$ can be rewritten as $t_1 \leq t_2$ and $t_2 \leq t_1$. If variable x appears in a literal with positive (negative) coefficient, the literal can be rearranged into an upper bound $x \lessdot t$ (lower bound $t \lessdot x$), where x does not occur in t.

Consider for example the *Generalized DPLL* (GDPLL) procedure [24], here renamed GCDCL for consistency. This procedure explains conflicts by the *shadow rule*, which is a generalization of *Fourier-Motzkin (FM) resolution* [18,20] from literals to clauses. The shadow rule collapses to FM-resolution if both premises are unit clauses. Given premises $t_1 \lessdot_1 x$ and $x \lessdot_2 t_2$, FM-resolution generates $t_1 \lessdot_3 t_2$, where $\lessdot_1, \lessdot_2, \lessdot_3 \in \{<, \leq\}$, and $\lessdot_3$ is $<$ if either $\lessdot_1$ or $\lessdot_2$ is $<$ and $\leq$ otherwise. For example, given clauses $(b < d) \vee (c < d)$ and $d < a$, the shadow rule generates clause $(b < a) \vee (c < a)$. If FM-resolution is applied systematically to eliminate one variable at a time as in the original Fourier-Motzkin algorithm [20], termination is guaranteed, but the algorithm is very inefficient (e.g., [18] and [8, Sect. 4.4]). The GCDCL procedure applies the shadow rule only to explain conflicts, and it ensures termination by additional restrictions. A fixed total ordering $\prec_{\mathsf{LRA}}$ on rational variables is assumed, and the shadow rule is applied only if the variable resolved upon is $\prec_{\mathsf{LRA}}$-maximal in both premises.

Abstract requirements for a conflict-driven procedure for the $\mathcal{T}$-satisfiability of $\mathcal{T}$-clauses for an arbitrary theory $\mathcal{T}$ were sketched in [24]: (1) embed reasoning about disjunction into theory reasoning, by generalizing to clauses a theory reasoning rule for literals; (2) apply the generalized rule only to explain conflicts; (3) devise additional restrictions to ensure termination.

The problem of how to generalize CDCL from propositional clauses to $\mathcal{T}$-clauses was reformulated as the problem of how to integrate in CDCL a conflict-driven $\mathcal{T}$-procedure for sets of $\mathcal{T}$-literals [13]. In this reformulation the reasoning about disjunction remains entrusted to CDCL, so that inference rules for $\mathcal{T}$-clauses are not necessary, and a $\mathcal{T}$-procedure for sets of $\mathcal{T}$-literals suffices. The reformulated problem was solved by the MCSAT method, where MCSAT stands for *Model Constructing SATisfiability* [13,15]. Unlike in CDCL($\mathcal{T}$), in MCSAT the $\mathcal{T}$-procedure also is conflict-driven, has access to the trail, proposes assignments to first-order terms, computes propagations, and explains $\mathcal{T}$-conflicts by $\mathcal{T}$-inferences that can generate *new* (i.e., non-input) literals. For termination, it suffices to ensure that the new literals come from a *finite basis* [13]. As observed in [24], in CDCL($\mathcal{T}$) the $\mathcal{T}$-procedure only derives clauses that are $\mathcal{T}$-valid. A conflict-driven $\mathcal{T}$-procedure (and MCSAT) can derive clauses that are logical consequences of the Boolean assignment on the trail (including the input), but are false under the first-order assignment on the trail and hence exclude it.

The discussion in [24] posed the open problem of *how to generalize conflict-driven reasoning to a generic combination of theories.* MCSAT is not a combination calculus, and the open problem was solved by CDSAT, generalizing MCSAT. In CDSAT, multiple theory modules have access to the trail, propose assignments, compute propagations, and explain conflicts by inferences that can generate *new* (i.e., non-input) literals. The modules have *finite local bases*, from which a *finite global basis* for termination is constructed [6,8].

For CDSAT a set of $\mathcal{T}$-clauses is a problem in a union of theories, namely $\mathcal{T}$ and Bool. In the next example, the LRA-module in CDSAT applies FM-resolution to explain LRA-conflicts. The LRA-module restricts FM-resolution by the restriction proposed in [24] for the shadow rule: assume a *fixed total ordering* $\prec_{\mathsf{LRA}}$ *on rational variables*, and allow FM-resolution only if the *variable resolved upon is* $\prec_{\mathsf{LRA}}$*-maximal in both premises*. This restriction is used to obtain a *finite basis* for the LRA-module [8, Sect. 4.4]. The next example portrays also acceptability of first-order assignments. A first-order decision $_?(t \leftarrow \mathfrak{c})$ is *acceptable* if the trail does not contain an assignment to term t, and $t \leftarrow \mathfrak{c}$ does not trigger an inference $J \cup \{t \leftarrow \mathfrak{c}\} \vdash \overline{L}$ for L a literal on the trail and J a subset of the trail. This condition excludes a first-order assignment that triggers an immediate conflict from which nothing can be learned.

Example 4. Consider the clause set $S = \{x < y,\ x < z,\ (y < w) \vee (z < w),\ w < x\}$, where x, y, z and w are rational variables [24]. Let $x \prec_{\mathsf{LRA}} y \prec_{\mathsf{LRA}} z \prec_{\mathsf{LRA}} w$ be the ordering on variables. Suppose that the CDSAT derivation starts with a series of three decisions by the LRA-module. A first Decide transition adds $_?(x \leftarrow 0)$ to the trail (level 1). A second Decide transition adds $_?(y \leftarrow 1)$ on level 2. Note that $_?(y \leftarrow 0)$ is *not acceptable*, as it would enable the LRA-evaluation inference $\{x \leftarrow 0,\ y \leftarrow 0\} \vdash_{\mathsf{LRA}} \overline{(x < y)}$ when $x < y$ is on the trail. A third Decide transition adds $_?(z \leftarrow 1)$ to the trail (level 3), whereas $_?(z \leftarrow 0)$ is not acceptable with $x < z$ and $_?(x \leftarrow 0)$ on the trail. At this point there is no acceptable value for w, because any rational value would create an immediate conflict with either $\{x \leftarrow 0,\ w < x\}$ or $\{y \leftarrow 1,\ z \leftarrow 1,\ (y < w) \vee (z < w)\}$.

The system needs to perform inferences to explain this LRA-conflict. Since clause $(y < w) \vee (z < w)$ is involved, a case analysis is necessary. Suppose that the Bool-module places $_?(y < w)$ on the trail (level 4) with another Decide transition. At this point the LRA-module can explain the conflict by FM-resolution. The FM-resolution inference $\{_?(y < w),\ _{\emptyset\vdash}(w < x)\} \vdash_{\mathsf{LRA}} y < x$ is allowed as w is the $\prec_{\mathsf{LRA}}$-maximal variable in both $y < w$ and $w < x$. A Deduce transition adds $_{J\vdash}(y < x)$ to level 4 of the trail with $J = \{_?(y < w),\ _{\emptyset\vdash}(w < x)\}$. The FM-resolution inference $\{_{\emptyset\vdash}(x < y),\ _{J\vdash}(y < x)\} \vdash_{\mathsf{LRA}} x < x$ is also allowed as y is the $\prec_{\mathsf{LRA}}$-maximal variable in both $x < y$ and $y < x$. A Deduce transition adds $_{I\vdash}(x < x)$ to level 4 of the trail with $I = \{_{\emptyset\vdash}(x < y),\ _{J\vdash}(y < x)\}$. Now $E_0 = \{_{I\vdash}(x < x)\}$ is an LRA-conflict. Two Resolve transitions unfold E_0 into $E_1 = \{_{\emptyset\vdash}(x < y),\ _{J\vdash}(y < x)\}$ and E_1 into $E_2 = \{_{\emptyset\vdash}(x < y),\ _?(y < w),\ _{\emptyset\vdash}(w < x)\}$. At this point a LearnBackjump transition allows the system to learn $\overline{y < w}$ as $_{H\vdash}(\overline{y < w})$, with $H = \{_{\emptyset\vdash}(x < y),\ _{\emptyset\vdash}(w < x)\}$, and jump back to level 0, which is the smallest level where $\overline{y < w}$ is undefined.

The unit propagation $\{_{H\vdash}(\overline{y < w}),\ _{\emptyset\vdash}((y < w) \vee (z < w))\} \vdash_{\mathsf{Bool}} z < w$ supports a Deduce transition that adds $_{G\vdash}(z < w)$ to level 0 with justification $G = \{_{H\vdash}(\overline{y < w}),\ _{\emptyset\vdash}((y < w) \vee (z < w))\}$. The shadow rule to work with clause $(y < w) \vee (z < w)$ is unnecessary: it suffices to let the Bool-module break apart the non-unit clause by decision and unit propagation, so that the LRA-module can reason about literals with FM-resolution. The FM-resolution step $\{_{G\vdash}(z < w),\ _{\emptyset\vdash}(w < x)\} \vdash_{\mathsf{LRA}} z < x$ is allowed as w is the $\prec_{\mathsf{LRA}}$-maximal variable in both $z < w$ and $w < x$. A Deduce transition adds $_{K\vdash}(z < x)$ to the trail (at level 0) with $K = \{_{G\vdash}(z < w),\ _{\emptyset\vdash}(w < x)\}$. The FM-resolution step $\{_{\emptyset\vdash}(x < z),\ _{K\vdash}(z < x)\} \vdash_{\mathsf{LRA}} x < x$ is allowed as z is the $\prec_{\mathsf{LRA}}$-maximal variable in both $x < z$ and $z < x$. The Deduce transition adding $_{M\vdash}(x < x)$ (at level 0), with $M = \{_{\emptyset\vdash}(x < z),\ _{K\vdash}(z < x)\}$, exposes LRA-conflict $E_3 = \{_{M\vdash}(x < x)\}$. Since the level of E_3 is 0, a Fail transition returns unsat.

The Deduce transition in CDSAT covers both propagation and conflict explanation. This allows CDSAT to apply theory inferences (e.g., congruence in Example 3 and FM-resolution in Example 4) more liberally than MCSAT. This flexibility spares the procedure from having to make decisions and create artificial conflicts only to enable inferences needed to discover unsatisfiability.

5 More on CDSAT

In a first-order assignment $t \leftarrow \mathfrak{c}$ (e.g., $x \leftarrow 1/2$, $f(y) \leftarrow 1$), the left side is a first-order (i.e., non-Boolean) term, and the right side is a *value* of the same sort. Values are constant symbols that a *conservative theory extension* $\mathcal{T}_i^+$ adds to the signature of theory $\mathcal{T}_i$ in order to name the elements of the domains of a model of $\mathcal{T}_i$ (one domain per sort). For example, if a theory has the sort of the integers, the signature of the theory can be extended with infinitely many constant symbols to name the integers. Since all the theories in a union are assumed to have the sort prop, the Boolean values true and false are added to every theory's signature. An extension that adds only true and false is called *trivial.* CDSAT keeps terms and values separate, because values are not terms. A term appears only on the left of an assignment, whereas a value appears only on the right. An assignment $t \leftarrow \mathfrak{c}$ cannot be replaced by the equality $t \simeq \mathfrak{c}$, because such an equality cannot be written, as $\mathfrak{c}$ is not a constant symbol in the original signature.

The notion of *theory view* defines what each theory $\mathcal{T}_i$ sees of the shared trail. Suppose that the theory of arrays Arr and LIA share the sort of the integers (e.g., Arr interprets indices as non-negative integers). If the LIA-module decides $A_1 = {}_?(x \leftarrow 3)$ and the Arr-module decides $A_2 = {}_?(y \leftarrow 3)$, the 3 in A_1 is a LIA-value and the 3 in A_2 is a Arr-value. If the Arr-module also decides $A_3 = {}_?(z \leftarrow 4)$, the LIA-*view* of the trail contains $\{A_1,\ x \simeq y,\ x \not\simeq z, y \not\simeq z\}$, and the Arr-*view* of the trail contains $\{A_2,\ A_3,\ x \simeq y,\ x \not\simeq z,\ y \not\simeq z\}$. In general, the $\mathcal{T}_i$-view contains the assignments of $\mathcal{T}_i$-values and all equalities and disequalities that can be gleaned from first-order assignments of a $\mathcal{T}_i$-sort, even if made by another theory. A Boolean assignment belongs to every theory view.

In addition to Decide, Deduce, Resolve, LearnBackjump, and Fail, the CDSAT transition system comprises the rules UndoClear and UndoDecide, which solve conflicts due to first-order assignments. Rule UndoClear intervenes when the assignment of highest level in the conflict is a first-order decision. Rule LearnBackjump cannot apply, because a first-order assignment does not have a complement that we can learn. Rule UndoClear applies instead, as shown in the next example, which also portrays the independent concept of *forced decision.*

Example 5. Let the input contains $\{2x+y \simeq 1,\ 2x+2y \simeq 1\}$. Suppose that the LRA-module ventures decision ${}_?(x \leftarrow 0)$, which is placed on level 1 by a Decide transition. The inference $\{2x + y \simeq 1,\ 2x + 2y \simeq 1\} \vdash_{\mathsf{LRA}} y \simeq 0$ supports a Deduce transition that adds $_{J\vdash}(y \simeq 0)$ to the trail with $J = \{2x + y \simeq$

1, $2x+2y \simeq 1\}$. The latter justified assignment belongs to level 0, and hence it is a late propagation. The LRA-inference $\{_?(x \leftarrow 0),\ _{J\vdash}(y \simeq 0)\} \vdash_{\mathsf{LRA}} 2x+y \not\simeq 1$ allows the LRA-module to detect the conflict $E = \{_?(x \leftarrow 0),\ _{J\vdash}(y \simeq 0),\ 2x + y \simeq 1\}$. Assignment $_?(x \leftarrow 0)$ has the highest level in the conflict. An UndoClear transition undoes $_?(x \leftarrow 0)$ and clears level 1, so that the trail only contains level 0. At this point $_?(x \leftarrow 1/2)$ is a *forced decision*, because 1/2 is the only acceptable value for x.

UndoClear incorporates backtracking from the level of the bad decision to the previous level. The state has changed due to a *late propagation*. It is certain that UndoClear fires after a late propagation, because the bad decision was acceptable prior to the late propagation and it causes a conflict afterwards.

If a term t of a sort s occurs on the trail (including as a subterm of another term), and theory $\mathcal{T}_i$ has values of sort s, then term t is said to be *relevant* to $\mathcal{T}_i$, because $\mathcal{T}_i$ can assign a value to t. Since all theories have the Boolean values, Boolean terms are relevant to every theory. Suppose that $\mathcal{T}_i$ does not have values of a sort s. If terms t_1 and t_2 of sort s occur on the trail, the equality $t_1 \simeq t_2$ is *relevant* to $\mathcal{T}_i$, even if $t_1 \simeq t_2$ itself does not appear on the trail. Indeed, $\mathcal{T}_i$ can assign a Boolean value to $t_1 \simeq t_2$. For a decision to be acceptable, the term being assigned must be *relevant* to the theory whose module is making the decision.

First-order assignments have impact on the *equality inferences*. In addition to the inferences based on reflexivity, symmetry, and transitivity, the equality inferences include inferences of the form $\{t_1 \leftarrow \mathfrak{c}, t_2 \leftarrow \mathfrak{c}\} \vdash t_1 \simeq t_2$, where t_1 and t_2 are terms of sort s and $\mathfrak{c}$ is a value of sort s, and $\{t_1 \leftarrow \mathfrak{c}_1, t_2 \leftarrow \mathfrak{c}_2\} \vdash t_1 \not\simeq t_2$, where $\mathfrak{c}_1$ and $\mathfrak{c}_2$ are distinct values of sort s. While the EUF-module (see Example 3) is responsible for inferences based on instances of the congruence axiom scheme for equality, all theory modules can perform equality inferences. When CDSAT emulates equality sharing, equalities derived from first-order assignments can contribute to build an arrangement.

The next example covers relevance, equality inferences, and the UndoDecide transition rule. The latter applies when the assignment of highest level in the conflict is a Boolean justified assignment L whose justification contains a first-order decision from the same level. UndoDecide undoes the first-order decision, backtracks, and puts $\overline{L}$ on the trail. A first-order assignment does not have a complement, but its Boolean consequence does. In this case Resolve is forbidden, because unfolding the conflict and undoing the first-order decision would lead us back to a previous state without learning anything.

Example 6. Assume that the input set S contains the following clauses:
C_1: $(i \not\simeq j) \vee (\mathsf{select}(\mathsf{store}(a,i,v),j) < \mathsf{select}(a,j))$,
C_2: $(\mathsf{select}(a,j) - \mathsf{select}(a,k)) \simeq 0$, and
C_3: $(\mathsf{select}(\mathsf{store}(a,i,v),j) \not< \mathsf{select}(a,j)) \vee (\mathsf{select}(a,j) + \mathsf{select}(a,k) \simeq v)$.
The theories are Bool, LRA for 0, subtraction, and the ordering, and the theory Arr of arrays, for the function symbols select and store. The select function takes an array and an index and returns the value of the array at that index. Thus,

$\mathsf{select}(a, j)$ is the term that corresponds to the $a[j]$ expression in programming languages. The store function takes an array, an index, and a value, and returns the array resulting from modifying the given array by writing the given value at the given index. Suppose that array indices are interpreted as integers, which means theory Arr has the sort of the integers. Suppose also that Arr has integer values. Say that the CDSAT derivation starts with two Decide transitions: the first one places $_?(i \leftarrow 0)$ on level 1; the second one places $_?(j \leftarrow 0)$ on level 2. Both are acceptable, because i and j are relevant to Arr.

A Deduce transition supported by equality inference $\{i \leftarrow 0, j \leftarrow 0\} \vdash_{\mathsf{Arr}} i \simeq j$ adds to level 2 assignment A_1: $_{J\vdash}(i \simeq j)$ with $J = \{_?(i \leftarrow 0), _?(j \leftarrow 0)\}$. Now unit propagation infers $\{A_1,\ C_1\} \vdash_{\mathsf{Bool}} \mathsf{select}(\mathsf{store}(a, i, v), j) < \mathsf{select}(a, j)$. A Deduce transition puts assignment A_2: $_{I\vdash}(\mathsf{select}(\mathsf{store}(a, i, v), j) < \mathsf{select}(a, j))$ on level 2 with $I = \{A_1,\ C_1\}$. Next, unit propagation infers $\{A_2,\ C_3\} \vdash_{\mathsf{Bool}} \mathsf{select}(a, j) + \mathsf{select}(a, k) \simeq v$, so that assignment A_3: $_{H\vdash}(\mathsf{select}(a, j) + \mathsf{select}(a, k) \simeq v)$ with $H = \{A_2,\ C_3\}$ is added to level 2 by a Deduce transition.

An axiom of the theory of arrays says that $i \simeq j \rightarrow \mathsf{select}(\mathsf{store}(a, i, v), j) \simeq v$ for all indices i and j, arrays a, and values v. Thus, the Arr-module can infer $\{A_1, A_2\} \vdash_{\mathsf{Arr}} v < \mathsf{select}(a, j)$. A Deduce transition adds A_4: $_{G\vdash}(v < \mathsf{select}(a, j))$ to level 2 with $G = \{A_1, A_2\}$. Suppose that the LRA-module infers $\{A_3,\ C_2\} \vdash_{\mathsf{LRA}} \mathsf{select}(a, j) \simeq v/2$. A Deduce transition adds A_5: $_{M\vdash}(\mathsf{select}(a, j) \simeq v/2)$ to level 2 with $M = \{A_3,\ C_2\}$. At this point, there is an LRA-conflict $E_0 = \{A_4,\ A_5\}$. A series of four Resolve transition unfolds the conflict as follows:

$E_1 = \{A_4,\ A_3,\ C_2\}$, $E_2 = \{A_1,\ A_2,\ A_3,\ C_2\}$,
$E_3 = \{A_1,\ A_2,\ C_3,\ C_2\}$, $E_4 = \{A_1,\ C_1,\ C_3,\ C_2\}$.

The justified assignment of highest level in E_4 is A_1, which has level 2, so that 2 is also the level of conflict E_4. The justification of A_1 is J, which contains the first-order decision $_?(j \leftarrow 0)$, whose level is also 2. Replacing A_1 with $J = \{_?(i \leftarrow 0), _?(j \leftarrow 0)\}$ and then undoing $_?(j \leftarrow 0)$ would not help, because the system would not learn that i and j cannot be equal. The UndoDecide transition rule fires in this kind of situation: it undoes $_?(j \leftarrow 0)$ clearing level 2 and going back to level 1, and then it opens a fresh level 2 with the decision $_?(i \not\simeq j)$. Now clause C_1 is satisfied. Three decisions by the LRA-module satisfy also clauses C_2 and C_3: for example, $_?(\mathsf{select}(a, j) \leftarrow 1)$ (level 3), $_?(\mathsf{select}(a, k) \leftarrow 1)$ (level 4), and $_?(v \leftarrow 2)$ (level 5). Note that $_?(\mathsf{select}(a, k) \leftarrow 1)$ is a *forced decision*, because no other value is acceptable for $\mathsf{select}(a, k)$ given clause C_2 and $_?(\mathsf{select}(a, j) \leftarrow 1)$.

Suppose that theory Arr does not have values for array indices. Then terms i and j are not relevant to Arr, and the derivation cannot start with decisions that are first-order assignments to i and j. However, since i and j occur on the trail, the equality $i \simeq j$ is relevant to Arr, and the derivation can start with a Boolean decision $_?(i \simeq j)$ on level 1. The same inferences as before lead to a conflict $\{_?(i \simeq j),\ C_1,\ C_3,\ C_2\}$. A LearnBackjump transition solves this conflict jumping back to level 0 and learning $_{N\vdash}(i \not\simeq j)$ with $N = \{C_1,\ C_3,\ C_2\}$. The satisfiability of the clauses can be detected as before.

6 Current and Future Work on CDSAT

CDSAT as in [6,8] is for a union of disjoint theories. We extended CDSAT to unions of theories that share predicate symbols other than equality [7]. We applied this extension to the theory of *arrays with abstract domain.*[3] Arrays are interpreted as updatable functions. Arrays with abstract domain are interpreted as updatable partial functions, that are defined only on those array indices that are *admissible.* The domain is abstract, because the array indices are not necessarily integers. We are currently working on theories of *maps and vectors with abstract domain.* Since admissibility is defined by a shared predicate symbol, these theories require the extension of CDSAT to the predicate-sharing case.

A main direction for future work is the implementation of CDSAT. In turn this requires working on the design of the search plans and the architecture for a CDSAT-based solver. The purpose of a search plan is to order the operations of the theory modules. CDSAT leaves much flexibility in terms of both search plan and architecture. The CDCL($\mathcal{T}$) paradigm abstracted existing architectures with the SAT solver at the center, and the theory solver as a satellite, also in the case where $\mathcal{T}$ is a union of theories. CDSAT neither requires nor favors an architecture with the SAT solver at the center, because it treats all theories as peers. The concept of an architecture without SAT solver at the center was very preliminarly explored in the context of MCSAT [2]. A prototype implementation of CDSAT in the Rust programming language is ongoing [14].

References

1. Barrett, C., Nieuwenhuis, R., Oliveras, A., Tinelli, C.: Splitting on demand in SAT modulo theories. In: Hermann, M., Voronkov, A. (eds.) LPAR 2006. LNCS (LNAI), vol. 4246, pp. 512–526. Springer, Heidelberg (2006). https://doi.org/10.1007/11916277_35
2. Bobot, F., Graham-Lengrand, S., Marre, B., Bury, G.: Centralizing equality reasoning in MCSAT. In: D'Silva, V., Dimitrova, R. (eds.) Proceedings of SMT-16 (2018)
3. Bonacina, M.P.: Automated reasoning for explainable artificial intelligence. In: Reger, G., Treytel, D. (eds.) Proceedings of ARCADE-1. EPiC Series in Computing, vol. 51, pp. 24–28. EasyChair (2017). https://doi.org/10.29007/4b7h
4. Bonacina, M.P.: On conflict-driven reasoning. In: Shankar, N., Dutertre, B. (eds.) Proceedings of 6th Workshop on Automated Formal Methods (AFM). Kalpa Publications, vol. 5, pp. 31–49. EasyChair (2018). https://doi.org/10.29007/spwm
5. Bonacina, M.P., Fontaine, P., Ringeissen, C., Tinelli, C.: Theory combination: beyond equality sharing. In: Lutz, C., Sattler, U., Tinelli, C., Turhan, A.-Y., Wolter, F. (eds.) Description Logic, Theory Combination, and All That. LNCS, vol. 11560, pp. 57–89. Springer, Cham (2019). https://doi.org/10.1007/978-3-030-22102-7_3
6. Bonacina, M.P., Graham-Lengrand, S., Shankar, N.: Conflict-driven satisfiability for theory combination: transition system and completeness. J. Autom. Reason.

3 Named theory of *arrays with abstract length* in [7].

64(3), 579–609 (2020). https://doi.org/10.1007/s10817-018-09510-y. Conference version at CADE 2017

7. Bonacina, M.P., Graham-Lengrand, S., Shankar, N.: CDSAT for nondisjoint theories with shared predicates: arrays with abstract length. In: Hyvärinen, A., Déharbe, D. (eds.) Proceedings of SMT-20. CEUR Proceedings, vol. 3185, pp. 18–37. CEUR WS-org, Aachen (2022), https://ceur-ws.org/Vol-3185/
8. Bonacina, M.P., Graham-Lengrand, S., Shankar, N.: Conflict-driven satisfiability for theory combination: lemmas, modules, and proofs. J. Autom. Reason. **66**(1), 43–91 (2022). https://doi.org/10.1007/s10817-021-09606-y. Conference version at CPP 2018
9. Bradley, A.R., Manna, Z.: The calculus of computation - decision procedures with applications to verification. Springer, Berlin (2007)
10. Davis, M., Logemann, G., Loveland, D.: A machine program for theorem-proving. Commun. ACM **5**(7), 394–397 (1962). https://doi.org/10.1145/368273.368557
11. Davis, M., Putnam, H.: A computing procedure for quantification theory. J. ACM **7**, 201–215 (1960). https://doi.org/10.1145/321033.321034
12. de Moura, L., Bjørner, N.: Model-based theory combination. In: Krstić, S., Oliveras, A. (eds.) Proc. SMT-5 (2007). ENTCS, vol. 198(2), pp. 37–49. Elsevier, Amsterdam (2008). https://doi.org/10.1016/j.entcs.2008.04.079
13. de Moura, L., Jovanović, D.: A model-constructing satisfiability calculus. In: Giacobazzi, R., Berdine, J., Mastroeni, I. (eds.) VMCAI 2013. LNCS, vol. 7737, pp. 1–12. Springer, Heidelberg (2013). https://doi.org/10.1007/978-3-642-35873-9_1
14. Denis, X.: A baby CDSAT-based verified solver written in Rust (2025). https://github.com/xldenis/cdsat, Accessed 13 Jan 2025
15. Jovanović, D., Barrett, C., de Moura, L.: The design and implementation of the model-constructing satisfiability calculus. In: Jobstman, B., Ray, S. (eds.) Proceedings of FMCAD-13, ACM and IEEE (2013). https://doi.org/10.1109/FMCAD.2013.7027033
16. Jovanović, D., de Moura, L.: Cutting to the chase: solving linear integer arithmetic. J. Autom. Reason. **51**(79–108), 2011 (2013). https://doi.org/10.1007/s10817-013-9281-x. Conference version at CADE 2011
17. Jovanović, D., de Moura, L.: Solving non-linear arithmetic. In: Gramlich, B., Miller, D., Sattler, U. (eds.) IJCAR 2012. LNCS (LNAI), vol. 7364, pp. 339–354. Springer, Heidelberg (2012). https://doi.org/10.1007/978-3-642-31365-3_27
18. Kroening, D., Strichman, O.: Decision procedures - an algorithmic point of view. Texts in Theoretical Computer Science, Springer, Berlin (2008)
19. Krstić, S., Goel, A.: Architecting solvers for SAT modulo theories: Nelson-Oppen with DPLL. In: Konev, B., Wolter, F. (eds.) FroCoS 2007. LNCS (LNAI), vol. 4720, pp. 1–27. Springer, Heidelberg (2007). https://doi.org/10.1007/978-3-540-74621-8_1
20. Lassez, J.L., Maher, M.J.: On Fourier's algorithm for linear arithmetic constraints. J. Autom. Reason. **9**, 373–379 (1992). https://doi.org/10.1007/BF00245296
21. Loveland, D., Sabharwal, A., Selman, B.: DPLL: the core of modern satisfiability solvers. In: Omodeo, E.G., Policriti, A. (eds.) Martin Davis on Computability, Computational Logic, and Mathematical Foundations. OCL, vol. 10, pp. 315–335. Springer, Cham (2016). https://doi.org/10.1007/978-3-319-41842-1_12
22. Marques Silva, J.P., Lynce, I., Malik, S.: Conflict-driven clause learning SAT solvers. In: Biere, A., Heule, M., Van Maaren, H., Walsh, T. (eds.) Handbook of Satisfiability (2nd Edition), Frontiers in Artificial Intelligence and Applications, vol. 336, pp. 133–182. IOS Press, Amsterdam (2021). https://doi.org/10.3233/FAIA200987

23. Marques Silva, J.P., Sakallah, K.A.: GRASP: a search algorithm for propositional satisfiability. IEEE Trans. Comput. **48**(5), 506–521 (1999). https://doi.org/10.1109/12.769433
24. McMillan, K.L., Kuehlmann, A., Sagiv, M.: Generalizing DPLL to richer logics. In: Bouajjani, A., Maler, O. (eds.) CAV 2009. LNCS, vol. 5643, pp. 462–476. Springer, Heidelberg (2009). https://doi.org/10.1007/978-3-642-02658-4_35
25. Nelson, G., Oppen, D.C.: Simplification by cooperating decision procedures. ACM Trans. Prog. Lang. Syst. **1**(2), 245–257 (1979). https://doi.org/10.1145/357073.357079
26. Nieuwenhuis, R., Oliveras, A., Tinelli, C.: Solving SAT and SAT modulo theories: from an abstract Davis-Putnam-Logemann-Loveland procedure to DPLL(T). J. ACM **53**(6), 937–977 (2006). https://doi.org/10.1145/1217856.1217859

Effective Second Countability in Computable Analysis

Vasco Brattka[1,2] and Emmanuel Rauzy[1(✉)]

[1] Fakultät für Informatik, Universität der Bundeswehr München, Werner-Heisenberg-Weg 39, 85577 Neubiberg, Germany
Vasco.Brattka@cca-net.de, emmanuel.rauzy.14@normalesup.org
[2] Department of Mathematics and Applied Mathematics, University of Cape Town, Private Bag X3, Rondebosch 7701, South Africa

Abstract. We compare Schröder's approach to computable topology with that of Weihrauch and Grubba. To do this, we introduce different notions of "effectively second countable" represented spaces, and prove that the most restrictive of these is precisely what was considered by Weihrauch and Grubba under the name "computable topological space". We then show that this strong notion of effective second countability is necessary in Schröder's Effective Metrization Theorem, but that some classical statements can be effectivized thanks to weaker hypotheses, we give as an example the statement "A second countable space is separable".

Keywords: Computable analysis · Represented spaces · Computable topology

1 Introduction

Choosing a good framework and correct definitions is one of the most important and sometimes most challenging problems in computable mathematics. Indeed, it is often the case that once a good framework has been found, many results that seemed non-trivial before can easily be proven. To see that this is the case in computable analysis, one can for instance look at early articles of Markov [14] and Lacombe [13], which, translated in today's vocabulary, end up having little content, and compare them to Pauly's 2016 survey paper [18] which deals with a great many notions and results in a mere 22 pages.

We are concerned here with computable topology as studied in Weihrauch's German school of computable analysis. We thus follow a "Type 2 approach": the basic notion of computable function is set on Baire space, and it is transferred from Baire space to other spaces with cardinality that of the continuum thanks to partial surjections which are called *representations*.

Our starting point is that within Weihrauch's school of computable analysis two approaches to studying computable topology coexist.

© The Author(s), under exclusive license to Springer Nature Switzerland AG 2025
A. Beckmann et al. (Eds.): CiE 2025, LNCS 15764, pp. 19–33, 2025.
https://doi.org/10.1007/978-3-031-95908-0_2

Namely:

- The approach devised by Matthias Schröder in his PhD thesis [21,22], which uses the representation of the Sierpiński space to define a representation of open sets on any represented space equipped with the final topology of its representation.
- The approach that Weihrauch developed with Grubba in the paper Elementary Computable Topology [29], which relies on a numbered basis $(B_i)_{i\in\mathbb{N}}$ for a space X to define a representations of X and a representation of the open sets of X.

Ideas very similar to those of Schröder were developed independently by Paul Taylor [26] and Martín Escardó [7] outside of the framework of represented sets.

It is clear that Schröder's approach is more general than Weihrauch's, in that it applies to spaces that need not be second countable. Yet the exact relationship between these two approaches had, until now, not been clarified.

In the present work, we explain the following:

- The approaches of Schröder and Weihrauch to computable topologies are compatible.
- The approach of Schröder is strictly more general than Weihrauch's, *even for second countable spaces*: not only does it encompass more topological spaces, but even on the topological spaces that fall under both approaches, phenomena purely related to computability occur in Schröder's approach that are not accounted for in Weihrauch's approach.
- Weihrauch and Grubba's notion of "computable topological space" in fact corresponds to a strong form of *effective second countability,* which is in most circumstances the *correct effective version of second countability.*
- Yet, even if this notion of effective second countability is in most cases the correct one, other weaker notions remain relevant, because the "effectivization" of some theorems can already be obtained thanks to those weaker notions.

There is no formal way of proving that an effective notion is the *correct* effective equivalent to a classical notion. However in many instances enough informal evidence is gathered so that no doubts remain.

We proceed as follows to justify that the Weihrauch-Grubba notion of "computable topological space" leads to the correct notion of effectively second countable space:

- We introduce five notions of "computable basis" of a represented space, by gathering different definitions that can be found in the literature, and by generalizing them so that they apply to represented bases instead of numbered bases.
- We discuss all the implications between these notions, implications of the form "a basis being computable in this sense implies it being computable in that sense". We note that these implication relations are not linearly ordered, in particular there are two notions of bases which are maximal and do not imply each other.

- Each notion of computable basis that we have introduced then yields a notion of "effective second countability". We prove that the notions of effective second countability that we thus obtain are linearly ordered by implication, and that the most restrictive notion of effective second countability corresponds precisely to Weihrauch and Grubba's "computable topological spaces".

Another way to understand the present article is as follows. One of the earliest representations ever studied is the *standard representation* introduced by Kreitz and Weihrauch in [10]. Given a second countable T_0 space X equipped with a countable subbasis $(B_i)_{i\in\mathbb{N}}$, this representation is given by

$$\rho(f) = x \iff \mathrm{Im}(f) = \{n \in \mathbb{N},\, x \in B_n\}.$$

Note that every totally numbered subbasis for X induces a standard representation, and thus each second countable space admits multiple standard representations. However, Kreitz and Weihrauch have shown that, up to continuous equivalence of representations, the choice of a standard representation is inconsequential:

Theorem 1. ([10]). *All standard representations of a second countable space are continuously equivalent. Every admissible representation of a second countable space is continuously equivalent to a standard representation.*

It is clear that not all standard representations of a second countable space need to be computably equivalent. What we want to emphasize here is that the effective analogue of the second point in the above theorem also fails: a computably admissible representation of a second countable space does not have to be computably equivalent to a standard representation. The representations that are computably equivalent to a standard representation are exactly those that give rise to what we call computably second countable spaces. In this paper, we show that the notion of a computably second countable represented space is extremely robust, as it emerges from a wide variety of different approaches. But we also show that among representations that are not computably equivalent to a standard representation, there is a range of weaker notions of effective second countability that can be, in different contexts, relevant.

The present paper constitutes an extended abstract which does not contain proofs. Proofs will appear in the full article [5]. In Sect. 2, we gather some general definitions about represented spaces and admissible representations. In Sect. 3, we give the definitions of the different notions of computable bases that are compared in the present work. In Sect. 4, we state our main results. Finally, in Sect. 5, we provide two examples of represented spaces that satisfy weak forms of effective second countability.

2 Preliminaries

The background we require is mostly contained in [18]. See also [23]. For two represented spaces (X, ρ) and (Y, τ), denote by $[\rho \to \tau]$ the representation of functions from (X, ρ) to (Y, τ) that are continuously realizable. Denote by $c_{\mathbb{S}}$ the usual representation of Sierpiński space. Then for any represented space $\mathbf{X} = (X, \rho)$, $[\rho \to c_{\mathbb{S}}]$ is a representation of the final topology of ρ, which we call the *Sierpiński representation* associated to ρ. We denote by $\mathcal{O}(\mathbf{X})$ the represented set thus obtained.

2.1 Overt Sets

Let $\mathbf{X} = (X, \rho)$ be a represented space. We denote by $\mathcal{V}(\mathbf{X})$ the represented space of overt subsets of X. Its underlying set is the set of closed subsets of X, equipped with the representation ψ^+ defined by:

$$\psi^+(p) = A \iff [[\rho \to c_{\mathbb{S}}] \to c_{\mathbb{S}}](p) = \{O \in \mathcal{O}(\mathbf{X}), O \cap A \neq \emptyset\}.$$

The problem Overt Choice was introduced in [6]. Hoyrup introduced in [9] the problem $\mathbf{\Pi}^0_2$ Overt Choice. The following problem, Local Overt Choice[1], is similar to this last problem, but we consider only sets that are the intersection of a closed set given by overt information with an open set (given as a characteristic function).

Definition 2. Let $\mathbf{X} = (X, \rho)$ be a represented space. *Local Overt Choice* is the following problem:

$$\begin{aligned} \mathrm{OVC}_X :\subseteq \mathcal{V}(\mathbf{X}) \times \mathcal{O}(\mathbf{X}) &\rightrightarrows \mathbf{X} \\ (V, U) &\mapsto V \cap U. \end{aligned}$$

with $\mathrm{dom}(\mathrm{OVC}_X) = \{(V, U) \mid V \cap U \neq \emptyset\}$.

2.2 Computably Sequential Embedding

If $\mathbf{X} = (X, \rho)$ is a represented space and $Y \subseteq X$, there are two natural representations of a topology on Y. First, we can consider the Sierpiński representation of $\mathcal{O}(\mathbf{X})$, $[\rho \to c_{\mathbb{S}}]$, and consider its trace on Y. We define a representation τ by:

$$\tau(p) = O \iff [\rho \to c_{\mathbb{S}}](p) \cap Y = O.$$

We can also consider first the restriction of ρ to Y, denoted $\rho_{|Y}$, and then take the associated Sierpiński representation, which is thus $[\rho_{|Y} \to c_{\mathbb{S}}]$. The following notion is due to Bauer, who introduced it in the context of synthetic topology under the name *intrinsic subset* [3]. We explain below why we choose a different terminology.

Definition 3. We say that Y is a *computably sequential subset* of $\mathbf{X}$ if the two representations defined above are computably equivalent: $\tau \equiv [\rho_{|Y} \to c_{\mathbb{S}}]$.

[1] The relevance of this problem to the present article was brought to our attention by Arno Pauly.

(In particular, they should be representations of the same topology on Y.) The name of the above notion is explained as follows: by results of Schröder [22], we know that the map

$$\begin{aligned} \mathcal{O}(\mathbf{X}) &\to \mathcal{O}(\mathbf{Y}) \\ U &\mapsto U \cap Y \end{aligned}$$

is surjective exactly when the subset topology on Y is sequential. Accordingly, we get a notion of computably sequential subset when this map is computably a surjection.

2.3 Computably Admissible Representation

Consider a represented space $\mathbf{X} = (X, \rho)$. If $\mathbf{X}$ is T_0, there is an embedding of $\mathbf{X}$ into $\mathcal{O}\mathcal{O}(\mathbf{X})$, defined by

$$x \mapsto \{U \in \mathcal{O}(\mathbf{X}), x \in U\}$$

A result of Schröder [22] says that a representation is admissible if and only if this embedding has a continuously realizable right inverse $\mathcal{O}\mathcal{O}(\mathbf{X}) \to \mathbf{X}$. This result leads to the following definition

Definition 4. (Schröder, [22]). A T_0 represented space $\mathbf{X}$ is *computably admissible* if the embedding of $\mathbf{X}$ into $\mathcal{O}\mathcal{O}(\mathbf{X})$ has a computable right inverse.

Computable admissibility can be understood as an effective equivalent of the T_0 property. Indeed, a topological space is T_0 when points are uniquely determined by the open sets to which they belong. A represented space is computably admissible when the previous equivalence holds computably: the name of a point can be translated to the name of a program that recognizes the open sets to which it belongs, and vice versa. We thus designate computably admissible spaces by using the abbreviation CT_0.

3 Computable Represented Bases

We consider five notions of effective bases for a represented space $\mathbf{X} = (X, \rho)$. Let $\mathfrak{B}$ be a subset of $\mathcal{O}(X)$, and β a representation of $\mathfrak{B}$.

3.1 Semi-effective Basis

Definition 5. We say that $(\mathfrak{B}, \beta)$ is a *semi-effective basis* if the map $\mathfrak{B} \hookrightarrow \mathcal{O}(\mathbf{X})$ is computable and if $\mathfrak{B}$ is a basis of $\mathcal{O}(\mathbf{X})$.

Thus the elements of $\mathfrak{B}$ are uniformly open, but the assumption that $\mathfrak{B}$ is a basis is purely a classical one.

3.2 Nogina Basis

Definition 6. We say that $(\mathfrak{B}, \beta)$ is a *Nogina basis* if it is a semi-effective basis and if furthermore the following multi-function is computable:

$$\begin{aligned} N :\subseteq X \times \mathcal{O}(X) &\rightrightarrows \mathfrak{B} \\ (x, O) &\mapsto \{B \in \mathfrak{B}, x \in B \subseteq O\} \end{aligned}$$

Here, $\mathrm{dom}(N) = \{(x, O), x \in O\} \subseteq X \times \mathcal{O}(X)$.

This notion of basis is closely related to [16,17], this explains our choice of name. It is similar to [24], however Spreen uses more restrictive conditions based on a *strong inclusion relation*. This notion is called "pointwise basis" in [2]. It was also used in [8], in the computably enumerable version, under the name "effective countable base".

The following is straightforward.

Proposition 7. *If $(\mathfrak{B}, \beta)$ is a Nogina basis, then the following multifunction is computable:*

$$\begin{aligned} N' :\subseteq X \times \mathfrak{B} \times \mathfrak{B} &\rightrightarrows \mathfrak{B} \\ (x, B_1, B_2) &\mapsto \{B \in \mathfrak{B}, x \in B \subseteq B_1 \cap B_2\} \end{aligned}$$

Here, $\mathrm{dom}(N') = \{(x, B_1, B_2), x \in B_1 \cap B_2\}$.

3.3 Lacombe Basis

Consider a semi-effective basis $(\mathfrak{B}, \beta)$ for $\mathbf{X} = (X, \rho)$. The following fact is well known (see for instance [18], Corollary 10.2, or [4]).

Lemma 8. *The computable injection $i : \mathfrak{B} \hookrightarrow \mathcal{O}(\boldsymbol{X})$ yields a computable function*

$$j : \begin{cases} \mathcal{V}(\mathfrak{B}) & \to \mathcal{O}(\boldsymbol{X}) \\ A \subseteq \mathfrak{B} & \mapsto \bigcup_{b \in A} i(b) \end{cases}$$

Definition 9. We say that the semi-effective basis $(\mathfrak{B}, \beta)$ is a **Lacombe basis** for $\mathbf{X}$ if the map $j : \mathcal{V}(\mathfrak{B}) \to \mathcal{O}(\mathbf{X})$ is onto, and if it has a computable multivalued right inverse: a computable $\psi : \mathcal{O}(\mathbf{X}) \rightrightarrows \mathcal{V}(\mathfrak{B})$ such that $j \circ \psi = \mathrm{id}_{\mathcal{O}(\mathbf{X})}$.

In other words, open sets of $\mathbf{X}$ can uniformly be written as overt unions of basic sets.

The name Lacombe basis comes from Lachlan [11] and Moschovakis [15] in reference to [12], where Lacombe introduced the idea that the computably open sets would be computable unions of basic open sets. This notion is called "pointfree basis" in [2]. It was also considered in [4] or in [6].

Note the following easy proposition:

Proposition 10. *For any represented space, $\mathcal{O}(\boldsymbol{X})$ is a Lacombe basis of $\boldsymbol{X}$.*

3.4 Representation Basis

Let X be a set. Consider a representation β of a certain set $\mathfrak{B}$ of subsets of X: $\beta :\subseteq \mathbb{N}^{\mathbb{N}} \to \mathfrak{B} \subseteq \mathcal{P}(X)$.

Denote by $\mathcal{N}_x^{\mathfrak{B}} = \{B \in \mathfrak{B},\, x \in B\}$. Suppose that :

1. for all x the set $\mathcal{N}_x^{\mathfrak{B}} \subseteq \mathfrak{B}$ is open in the final topology of β.
2. $\mathcal{N}_x^{\mathfrak{B}} \neq \mathcal{N}_y^{\mathfrak{B}}$ for $x \neq y$.

Under those hypotheses, the representation β of $\mathfrak{B}$ induces a representation β^* of X, defined as follows:

$$\beta^*(p) = x \iff [\beta \to c_{\mathbb{S}}](p) = \mathcal{N}_x^{\mathfrak{B}}.$$

In words: the β^*-name of a point x is a Sierpiński name of the open set $\mathcal{N}_x^{\mathfrak{B}}$.

Definition 11. We say that a semi-effective basis $(\mathfrak{B}, \beta)$ is a *representation basis for* (X, ρ) if the above hypotheses 1. and 2. are satisfied and $\rho \equiv \beta^*$.

As a consequence of results of Schröder [22], we get:

Proposition 12. *A represented space has a representation basis if and only if it is computably admissible (in which case $\mathcal{O}(\mathbf{X})$ is a representation basis).*

3.5 Enumeration Basis

The following notion is a generalization of the "standard representation" considered by Weihrauch, Kreitz and Grubba in [10,27,29]. The idea that a proper generalization of the standard representation uses a type of formal inclusion relation goes back to Spreen [25], see also [19].

The following construction applies only to first countable spaces, and note that a represented space which is first countable is automatically second countable. But the represented basis we consider does not have to be countable.

Consider a represented basis $(\mathfrak{B}, \beta)$. Consider a relation $\prec$ on $\mathrm{dom}(\beta)$ which is a *c.e. strong inclusion relation* [24,25]:

- The relation $\prec$ is transitive and semi-decidable: $\prec : \mathrm{dom}(\beta) \times \mathrm{dom}(\beta) \to \mathbb{S}$ is computable.
- For all p, q in $\mathrm{dom}(\beta)$, $p \prec q \implies \beta(p) \subseteq \beta(q)$.

Furthermore, suppose that every point of X admits a *formal neighborhood basis*:

$$\forall x \in X, \forall b_1 \in \mathrm{dom}(\beta),\, x \in \beta(b_1) \implies \exists b_2 \in \mathrm{dom}(\beta),\, x \in \beta(b_2) \& b_2 \prec b_1.$$

If $\prec$ is reflexive, points will automatically all admit formal neighborhood bases. Finally, we suppose that the topology generated by $\mathfrak{B}$ (as a classical basis) is T_0.

We then define a representation $\beta^{\prec}$ of X by the following:

$$\beta^{\prec}(\langle u_0, u_1, \ldots \rangle) = x \iff \begin{cases} \forall n \in \mathbb{N},\, u_n \in \mathrm{dom}(\beta) \,\&\, x \in \beta(u_n), \\ \forall b \in \mathrm{dom}(\beta),\, x \in \beta(b) \implies \exists n \in \mathbb{N},\, u_n \prec b. \end{cases}$$

(Here $\langle \cdot \rangle : \prod_{i \in \mathbb{N}} \mathbb{N}^{\mathbb{N}} \to \mathbb{N}^{\mathbb{N}}$ is a countable tupling function.)

Definition 13. We say that $(\mathfrak{B}, \beta)$ is an *enumeration basis for* (X, ρ) if $\rho \equiv \beta^{\prec}$.

A useful and easy to check property of representations coming from enumeration bases is that they are open:

Proposition 14. *The representation* $\beta^{\prec}$ *is an open map.*

3.6 Effective Second Countability

To each notion of effective basis introduced above, we associate a notion of effective second countability, following the following definition scheme:

Definition 15. A represented space $\mathbf{X}$ is *[Nogina, Lacombe, etc.] second countable* if it admits a totally numbered [Nogina, Lacombe, etc.] basis, i.e., if there exists a sequence $(B_i)_{i \in \mathbb{N}}$ which forms a [Nogina, Lacombe, etc.] basis of $\mathbf{X}$.

Theorem 16 shows that some implications between notions of second countability hold, that are not true for general represented bases.

4 Results

We finally state our results. Complete proofs will be found in [5].

Theorem 16. *All the implications and non-implications between the different notions of effective bases and of effective second countability appear on Figs. 1 and 2. The question mark indicates the only implication that is not known to be strict.*

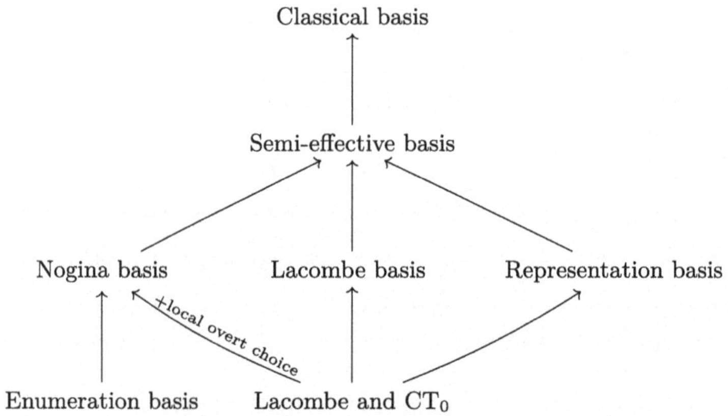

Fig. 1. Notions of bases

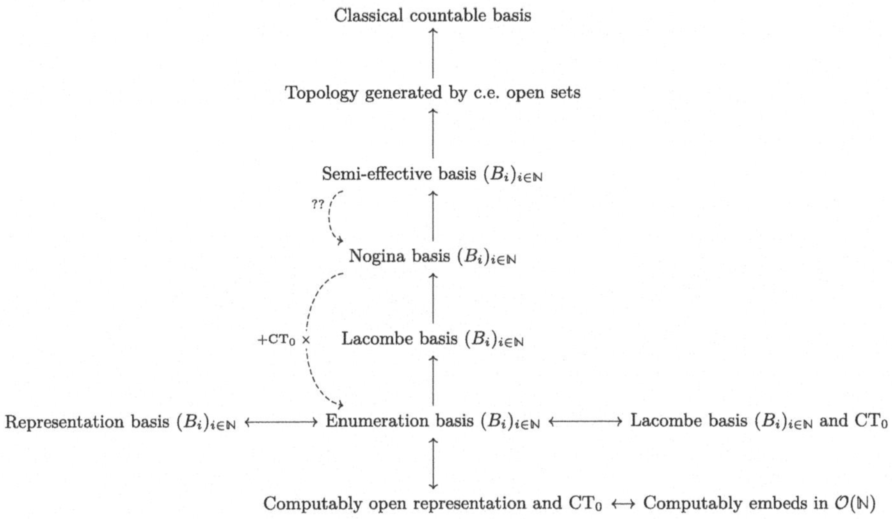

Fig. 2. Notions of effective second countability

Most of the implications given by this theorem are straightforward. The most interesting implication is the one which says that a representation basis $(B_i)_{i\in\mathbb{N}}$ always gives rise to a Lacombe basis. This is precisely the implication that shows that the Weihrauch-Grubba approach is compatible with Schröder's approach. An important part of the proof of Theorem 16 is the construction of counter-examples to separate different forms of effective second-countability. In Sect. 5, we include two examples which seem to us to be the most illustrative ones.

Definition 17. A represented space is called *computably second countable* if it satisfies one of the five equivalent characterizations that appear at the bottom of Fig. 2.

The following theorem proves that when the Weihrauch and Schröder approaches agree on a represented space, they also agree on all its subsets. Its proof relies on Theorem 16.

Theorem 18. *Let (X, ρ) be a computably second countable represented space, let $Y \subseteq X$ be a subset of X, and equip it with the induced representation $\rho_{|Y}$. Then $(Y, \rho_{|Y})$ is also computably second countable, and Y is a computably sequential subset of (X, ρ).*

When considering a classical statement of the form

If a topological set X is second countable and A, then B,

we can ask which form of effective second countability is required to effectivize it. We give two examples, one where a strong form is necessary, the other one where a weak form is sufficient.

Theorem 19. (Schröder-Urysohn Effective Metrization). *The following are equivalent for a represented space* $\mathbf{X}$*:*

1. $\mathbf{X}$ *computably embeds into the Hilbert cube,*
2. $\mathbf{X}$ *is computably second countable and strongly computably regular.*

The term "strongly computably regular" is due to Weihrauch [28], it designates a notion introduced by Schröder in [20]. The original theorem [20] was not expressed as an equivalence. The above theorem and its proof are very close to a result of [1], but there is a compactness assumption on $\mathbf{X}$ which we remove. The computable second countability hypothesis is necessary in this theorem.

Finally we study the statement:

A second countable space is separable.

A represented space $\mathbf{X}$ is called *computably separable* if it admits a dense and computable sequence, i.e., if there is a computable function $f : \mathbb{N} \to \mathbf{X}$ with dense image. The most obvious way of effectivizing the above statement relies on computable Open Choice, where Open Choice is the multi-function defined as follows:

$$\begin{aligned} OC : \mathcal{O}(X) \setminus \{\emptyset\} &\rightrightarrows X \\ O &\mapsto O. \end{aligned}$$

We show that this is too strong an assumption:

Theorem 20. *A represented space has computable Open Choice if and only if it is computably separable.*

The proof of this result relies on the fact that $\{X\}$ is dense in the Scott topology on $\mathcal{O}(X)$. We show that this holds also at the level of names (every open set has a name which is the limit of a sequence of names of X). Then, Open Choice applied to names of X will provide the desired computable and dense sequence. The correct hypothesis to render the above statement effective relies on Non-total Open Choice, Open Choice restricted to open sets that have non-empty complement.

Theorem 21. *A represented space* $\mathbf{X}$ *which is overt, has computable Non-total Open Choice and is semi-effectively second countable with a basis excluding* X *is computably separable.*

Yet having computable Non-total Open Choice does not imply effective separability.

The positive part of the above theorem is obvious. The non-trivial part is showing that a space can have computable Non-total Open Choice without being computably separable, i.e., without having computable Open Choice.

5 Two Examples of Spaces that Satisfy Weak Forms of Effective Second Countability

What may be the most interesting part of the proof of Theorem 16 is the construction of examples of represented spaces that satisfy different forms of effective second countability. We include here two examples of such spaces.

5.1 First Example

We first construct a CT_0 represented space whose c.e. open sets form a (necessarily countable) basis of the topology, but such that no computable sequence of open sets can form a basis of the topology: this space is not semi-effectively second countable.

We consider an effective version of a well known example of a sequential but not Frechet–Urysohn space. This example was studied by Schröder in [22] as an easy example of an admissibly represented space which is not second countable.

We first describe the classical example. Consider the following set X:

$$X = \mathbb{N}^2 \cup (\{\infty\} \times \mathbb{N}) \cup \{(\infty, \infty)\}.$$

For m_0, n_0 in $\mathbb{N}$ and $f : \mathbb{N} \to \mathbb{N}$, denote

$$D_{n_0,m_0} = \{(n, m_0), n \geq n_0, n \in \mathbb{N} \cup \{\infty\}\};$$

$$E_{m_0,f} = \{(\infty, \infty)\} \cup \bigcup_{m \geq m_0} D_{f(m),m}$$

Define also:

$$\mathfrak{B}_1 = \{\{(n, m)\}, n, m \in \mathbb{N}\},$$
$$\mathfrak{B}_2 = \{D_{n_0,m_0}, n_0, m_0 \in \mathbb{N}\},$$
$$\mathfrak{B}_3 = \{E_{m_0,f}, m_0 \in \mathbb{N}, f \in \mathbb{N}^{\mathbb{N}}\}.$$

Put $\mathfrak{B} = \mathfrak{B}_1 \cup \mathfrak{B}_2 \cup \mathfrak{B}_3$, and consider the topology on X generated by $\mathfrak{B}$.

Note that the set $\mathfrak{B}_3$ is not countable. And in X, while (∞, ∞) is adherent to $\mathbb{N}^2$, no sequence of points of $\mathbb{N}^2$ converges to (∞, ∞), because this would require that its first component should grow faster than *any* function $f : \mathbb{N} \to \mathbb{N}$.

Here we consider an effective version of this construction: we replace $\mathfrak{B}_3$ by $\mathfrak{B}_3^+$:

$$\mathfrak{B}_3^+ = \{E_{m_0,f}, m_0 \in \mathbb{N}, f \in \mathbb{N}^{\mathbb{N}} \text{ is a total computable function}\}$$

Denote $\mathfrak{B}^+ = \mathfrak{B}_1 \cup \mathfrak{B}_2 \cup \mathfrak{B}_3^+$. We define a numbering β of $\mathfrak{B}^+$ by

$$\beta(\langle 1, n, m \rangle) = \{(n, m)\},$$

$$\beta(\langle 2, n, m \rangle) = D_{n,m},$$

$$\beta(\langle 3, m, i \rangle) = E_{m,\varphi_i} \text{ if } \varphi_i \text{ is a total function.}$$

To the numbered basis $(\mathfrak{B}^+, \beta)$ we associate the representation ρ of X which is the basis representation β^*, as defined in Sect. 3.4. This guarantees that (X, ρ) is CT_0.

We will now prove that this space is the desired example.

Say that a point x is *effectively adherent* to a set A if the following multi-function is computable:

$$\begin{aligned} \Theta_{x,A} :\subseteq \mathcal{O}(X) &\rightrightarrows X \\ O &\mapsto A \cap O, \end{aligned}$$

with $\mathrm{dom}(\Theta_{x,A}) = \{O \in \mathcal{O}(X),\ x \in O\}$. Notice the following easy result:

Proposition 22. *If a represented space (X, ρ) is semi-effectively second countable, then any point that is effectively adherent to a set is the limit of a computable sequence of elements of this set.*

Lemma 23. *In X, the point (∞, ∞), while being effectively adherent to $\mathbb{N}^2$, is not the limit of a computable sequence of points of $\mathbb{N}^2$.*

Proof. We first show that (∞, ∞) is effectively adherent to $\mathbb{N}^2$. But this is obvious: any neighborhood of (∞, ∞) contains points of $\mathbb{N}^2$, one of these can be found by exhaustive search.

Now we show that (∞, ∞) is not the limit of a computable sequence of points of $\mathbb{N}^2$.

This follows immediately from the fact that if a sequence $((u_n, v_n))_{n\in\mathbb{N}} \subseteq \mathbb{N}^2$ converges to (∞, ∞), then we should have that for every computable function f and every set $E_{m_0,f}$, $((u_n, v_n))_{n\in\mathbb{N}}$ eventually belongs to $E_{m_0,f}$. This implies that eventually $u_n \geq f(v_n)$ for every computable function f. This is not possible for a computable sequence. □

5.2 Second Example

We now construct a CT_0 represented space which is very close to being computably second countable: indeed, it is CT_0 and Nogina second countable, yet not computably second countable. (See Fig. 2 to see where this lies in the hierarchy.)

We define a representation β of a basis for the discrete topology on $\mathbb{N}$ by the following:

$$\beta(k^\omega) = \{k\},\ k \in \mathbb{N},$$

$$\beta(\mathbb{1}_K) = \mathrm{Tot},$$

where $\mathbb{1}_K$ is the characteristic function of the halting set and Tot the set of codes of total computable functions. Thus the above is almost the obvious basis for the discrete topology on $\mathbb{N}$, but we artificially add a $\emptyset'$-computable name to Tot.

Let ρ denote the representation β^* induced by β (see Sect. 3.4). Let $\tilde{\mathbb{N}}$ be the represented space $(\mathbb{N}, \rho)$. It is easily seen to be CT_0.

Notice that $\rho \leq \mathrm{id}_{\mathbb{N}}$ (where $\mathrm{id}_{\mathbb{N}}$ is the identity numbering of $\mathbb{N}$). The ρ-name of a point n can be thought of as a pair, consisting of n together with a method to determine whether or not n belongs to Tot when given access to an oracle for the halting set K.

One immediately checks that the basis $\{\{k\}, k \in \mathbb{N}\}$ is a Nogina basis for $\tilde{\mathbb{N}}$. Thus it is Nogina second countable.

We now show that $\tilde{\mathbb{N}}$ does not have a totally numbered Lacombe basis.

We first consider another basis $(B_i)_{i\in\mathbb{N}}$ for the discrete topology on $\mathbb{N}$, given by $B_0 = \text{Tot}$, $B_i = \{i-1\}$ for $i > 0$. Consider the associated basis representation $\hat{\rho}$. The represented space $(\mathbb{N}, \hat{\rho})$ is, by construction, computably second countable. We then use the fact that a totally numbered representation basis automatically yields a Lacombe basis (this is one of the positive implications of Theorem 16): the Sierpiński representation $[\hat{\rho} \to c_{\mathbb{S}}]$ is equivalent to the "union representation" associated to $(B_i)_{i\in\mathbb{N}}$: open sets can uniformly be written as countable unions of basic sets. Denote by $\cup B$ this last representation.

The map $i \mapsto B_i$ is a numbering whose image is identical to the image of β, it can be seen as a representation which we denote by B. We then have $\beta \leq B$, and, because the map $\tau \mapsto [\tau \to c_{\mathbb{S}}]$ is order reversing, it follows that $\hat{\rho} \leq \rho$ and $[\rho \to c_{\mathbb{S}}] \leq [\hat{\rho} \to c_{\mathbb{S}}]$.

Suppose that there exists a Lacombe basis $(\mathfrak{B}_i)_{i\in\mathbb{N}}$ for $\tilde{\mathbb{N}}$, and denote by $\cup\mathfrak{B}$ the representation of open sets associated to countable unions of elements of $\mathfrak{B}$. We get:

$$\mathfrak{B} \leq \cup\mathfrak{B} \equiv [\rho \to c_{\mathbb{S}}] \leq [\hat{\rho} \to c_{\mathbb{S}}] \equiv \cup B.$$

By construction of ρ, Tot has an $\emptyset'$-computable $[\rho \to c_{\mathbb{S}}]$-name, and thus it also admits an $\emptyset'$-computable $\cup\mathfrak{B}$-name, which we call p_{Tot}. Translating this name to $\cup B$, we get a $\emptyset'$-computable name of Tot for $\cup B$. But it is clear that a $\cup B$-name of Tot either contains explicitly the B-name 0 for Tot, or it is not $\emptyset'$-computable. Thus the $\emptyset'$-computable $\cup B$-name of Tot contains 0. This implies that a finite prefix w_{Tot} of p_{Tot} is mapped, via the realizer of the translation $\cup\mathfrak{B} \leq \cup B$, to a finite prefix of a $\cup B$-name that contains 0. In turn, the prefix w_{Tot} can be extended to a computable $\cup\mathfrak{B}$-name of Tot, for instance the sequence $(w_{\text{Tot}})^\omega$ must be a valid $\cup\mathfrak{B}$-name for Tot.

And thus we get that Tot is actually a c.e. open set for ρ.

We conclude by proving that this is impossible.

For each $i \in \mathbb{N}$, consider a code t_i for the program that, on input k, runs $\varphi_i(i)$ for k computations steps, and halts if this computation does not stop during those k steps. Otherwise, it loops indefinitely.

There is a computable function Φ that transforms t_i into a ρ-name of t_i. Indeed, because, by construction, $t_i \in \text{Tot} \iff i \notin K$, there is an algorithm which, given the β-name $\not\Vdash_K$ of Tot, decides whether or not $t_i \in \text{Tot}$.

And thus if Tot were c.e. open for ρ, there would exist a program which, given i, would stop if and only if $t_i \in \text{Tot}$. Via the above equivalence this would give a program to semi-decide the complement of K. This completes the proof of the fact that the represented space $\tilde{\mathbb{N}}$ does not have a totally numbered Lacombe basis.

Acknowledgments. The first author was funded by the German Research Foundation (DFG, Deutsche Forschungsgemeinschaft) - project number 554999067 and by the National Research Foundation of South Africa (NRF) - grant number 151597.

The second author is funded by an Alexander von Humboldt fellowship.

References

1. Amir, D.E., Hoyrup, M.: Strong computable type. Computability **12**(3), 227–269 (2023). https://doi.org/10.3233/com-220430
2. Bauer, A.: The realizability approach to computable analysis and topology. Ph.D. thesis, School of Computer Science, Carnegie Mellon University (2000)
3. Bauer, A.: Spreen spaces and the synthetic Kreisel-Lacombe-Shoenfield-Tseitin theorem. J. Logic Anal. **17** (2025). https://doi.org/10.4115/jla.2025.17.fds3
4. Bauer, A., Lešnik, D.: Metric spaces in synthetic topology. Ann. Pure Appl. Logic **163**(2), 87–100 (2012). https://doi.org/10.1016/j.apal.2011.06.017
5. Brattka, V., Rauzy, E.: Effective bases and notions of effective second countability in computable analysis (In preparation)
6. de Brecht, M., Pauly, A., Schröder, M.: Overt choice. Computability **9**(3–4), 169–191 (2020). https://doi.org/10.3233/com-190253
7. Escardó, M.: Synthetic topology. Electron. Notes Theor. Comput. Sci. **87**, 21–156 (2004). https://doi.org/10.1016/j.entcs.2004.09.017
8. Gregoriades, V., Kispéter, T., Pauly, A.: A comparison of concepts from computable analysis and effective descriptive set theory. Math. Struct. Comput. Sci. **27**(8), 1414–1436 (2016). https://doi.org/10.1017/s0960129516000128
9. Hoyrup, M.: Notes on overt choice. Computability **12**(4), 351–369 (2023)
10. Kreitz, C., Weihrauch, K.: Theory of representations. Theoret. Comput. Sci. **38**, 35–53 (1985). https://doi.org/10.1016/0304-3975(85)90208-7
11. Lachlan, A.H.: Effective operations in a general setting. J. Symb. Log. **29**(4), 163–178 (1964). https://doi.org/10.2307/2270370
12. Lacombe, D.: Les ensembles récursivement ouverts ou fermés, et leurs applications à l'analyse récursive. Comptes rendus hebdomadaires des séances de l'Académie des Sciences (Paris) **245**, 1040–1043 (1957)
13. Lacombe, D.: Quelques procédés de définition en topologie récursive. In: Heyting, A. (ed.) Constructivity in mathematics. In: Proceedings of the colloquium held at Amsterdam, pp. 129–158. North-Holland Publishing Company (1957)
14. Markov, A.A.: On constructive functions, chap. In: Twelve Papers on Logic and Differential Equations, pp. 163–195. American Mathematical Society Translations (1963). https://doi.org/10.1090/trans2/029/05
15. Moschovakis, Y.: Recursive metric spaces. Fundam. Math. **55**(3), 215–238 (1964). https://doi.org/10.4064/fm-55-3-215-238
16. Nogina, E.Y.: Effectively topological spaces. Dokl. Akad. Nauk SSSR **169**, 28–31 (1966)
17. Nogina, E.Y.: Relations between certain classes of effectively topological spaces. Math. Notes Acad. Sci. USSR **5**(4), 288–294 (1969). https://doi.org/10.1007/bf01410800
18. Pauly, A.: On the topological aspects of the theory of represented spaces. Computability **5**(2), 159–180 (2016). https://doi.org/10.3233/com-150049
19. Rauzy, E.: Multi-representation associated to the numbering of a subbasis and formal inclusion relation. arXiv:2311.15861 (2023)

20. Schröder, M.: Effective metrization of regular spaces. In: Computability and Complexity in Analysis (1998)
21. Schröder, M.: Admissible representations of limit spaces. In: Blanck, J., Brattka, V., Hertling, P. (eds.) CCA 2000. LNCS, vol. 2064, pp. 273–295. Springer, Heidelberg (2001). https://doi.org/10.1007/3-540-45335-0_16
22. Schröder, M.: Admissible Representations for Continuous Computations. Ph.D. thesis, Fachbereich Informatik, FernUniversität Hagen (2002)
23. Schröder, M.: Admissibly represented spaces and Qcb-spaces. In: Handbook of Computability and Complexity in Analysis. TAC, pp. 305–346. Springer, Cham (2021). https://doi.org/10.1007/978-3-030-59234-9_9
24. Spreen, D.: On effective topological spaces. J. Symbolic Logic **63**(1), 185–221 (1998)
25. Spreen, D.: Representations versus numberings: on the relationship of two computability notions. Theoret. Comput. Sci. **262**(1–2), 473–499 (2001). https://doi.org/10.1016/s0304-3975(00)00319-4
26. Taylor, P.: Foundations for computable topology. In: Sommaruga, G. (ed.) Foundational Theories of Classical and Constructive Mathematics. No. 76 in Western Ontario Series in Philosophy of Science, Springer-Verlag, January 2011, PaulTaylor.EU/ASD/foufct
27. Weihrauch, K.: Computable Analysis. TTCSAES, Springer, Heidelberg (2000). https://doi.org/10.1007/978-3-642-56999-9
28. Weihrauch, K.: Computably regular topological spaces. Logical Methods Comput. Sci. **9**(3), (2013). https://doi.org/10.2168/lmcs-9(3:5)2013
29. Weihrauch, K., Grubba, T.: Elementary computable topology. J. Univers. Comput. Sci. **15**, 1381–1422 (2009)

Towards a Computational Quantum Logic
An Overview of an Ongoing Research Program

Alejandro Díaz-Caro[1,2(✉)]

[1] Université de Lorraine, CNRS, Inria, LORIA, 54000 Nancy, France
alejandro.diaz-caro@inria.fr
[2] Universidad Nacional de Quilmes, Bernal, Buenos Aires, Argentina

Abstract. This invited paper presents an overview of an ongoing research program aimed at extending the Curry-Howard-Lambek correspondence to quantum computation. We explore two key frameworks that provide both logical and computational foundations for quantum programming languages.

The first framework, the Lambda-S calculus, extends the lambda calculus by incorporating quantum superposition, enforcing linearity, and ensuring unitarity, to model quantum control. Its categorical semantics establishes a structured connection between classical and quantum computation through an adjunction between Cartesian closed categiries and additive symmetric monoidal closed categories.

The second framework, the $\mathcal{L}^{\mathbb{C}}$ calculus, introduces a proof language for intuitionistic linear logic augmented with sum and scalar operations. This enables the formal encoding of quantum superpositions and measurements, leading to a computational model grounded in categorical structures with biproducts.

These approaches suggest a fundamental duality between quantum computation and linear logic, highlighting structural correspondences between logical proofs and quantum programs. We discuss ongoing developments, including extensions to polymorphism, categorical and realizability models, as well as the integration of the modality !, which further solidify the connection between logic and quantum programming languages.

Keywords: Quantum Computing · Lambda calculus · Linear Logi · Categorical Semantics

1 Motivation

The term *quantum logic* was first introduced by Garrett Birkhoff and John von Neumann nearly a century ago [3] to describe a formal system inspired by the structure of quantum theory. Unlike classical logic, which is based on Boolean algebra, quantum logic replaces the distributive law with a weaker condition, leading to an orthocomplemented lattice structure, that is, a bounded lattice in

© The Author(s), under exclusive license to Springer Nature Switzerland AG 2025
A. Beckmann et al. (Eds.): CiE 2025, LNCS 15764, pp. 34–46, 2025.
https://doi.org/10.1007/978-3-031-95908-0_3

which each element has a unique orthocomplement satisfying involution, order-reversing, and complement laws. This formulation aligns with the mathematical properties of quantum mechanics, where propositions correspond to projections in a Hilbert space.

While quantum logic has been explored as a foundational framework for reasoning about quantum mechanics, its connection to computation remains underdeveloped. In contrast, the interplay between intuitionistic logic, typed lambda calculus, and Cartesian closed categories has been extensively studied, with the Curry-Howard-Lambek correspondence [4,16,19,24] revealing deep structural connections among these domains.

To extend this correspondence to quantum computation, a logical foundation that captures its structure is needed. This paper aims to provide an overview of a long-term research program focused on developing a *computational* quantum logic as a foundation for quantum programming languages. The approach is based on two key frameworks: the Lambda-S calculus [9,10], an extension of the lambda calculus for quantum computing, and the $\mathcal{L}^{\mathcal{S}}$ calculus [6], a proof language for intuitionistic linear logic whose proof terms enable the construction of quantum programs. The strategy follows a two-step process: first, deriving logic from computation (via the Lambda-S calculus), and then deriving computation from logic (via the $\mathcal{L}^{\mathcal{S}}$ calculus), ultimately aiming to converge in the middle.

2 The Lambda-S Calculus: From Computing to Logic

2.1 The *Lineal* Origin

Lambda-S [9] is a quantum lambda calculus designed to handle quantum superposition and control while preserving key computational properties such as strong normalization and subject reduction. This calculus extend the lambda calculus by incorporating algebraic linearity and type-based constraints to enforce quantum mechanical principles, particularly the no-cloning theorem.

This approach originates from the algebraic lambda calculus *Lineal* [1]. The primary goal of Lineal was to extend the untyped lambda calculus to effectively express quantum programs, following a simple yet powerful idea: in the lambda calculus, programs and data are indistinguishable. Thus, if quantum computing allows for superpositions of data, the lambda calculus must reflect this by allowing superpositions of programs. This idea was formalized in Lineal, a lambda calculus where program superposition is a first-class citizen. Specifically, if t and r are lambda terms, then so is $\alpha \cdot t + \beta \cdot r$, with $\alpha, \beta \in \mathbb{C}$. For example, if t and r are the encodings of true and false, respectively, then $\alpha \cdot t + \beta \cdot r$ is the encoding of the vector $\begin{pmatrix}\alpha\\ \beta\end{pmatrix}$, which represents a superposition of the two values.

The name Lineal derives from the Spanish word for linear, as its reduction strategy enforces linearity through the following rule:

$$(\lambda x.t)(\alpha \cdot v + \beta \cdot w) \longrightarrow \alpha \cdot (\lambda x.t)v + \beta \cdot (\lambda x.t)w \qquad (1)$$

where t is any term, and v and w are *basis terms*, meaning they are standard values, so they are not linear combinations of other terms. This reduction strategy, known as call-by-base [2], ensures that lambda terms such as $\lambda x.t$ are not explicitly required to be linear but are instead forced to behave linearly through the reduction process. Since quantum gates are unitary transformations, and thus are inherently linear, Lineal naturally encodes *measurement-free* quantum programs.

Indeed, the quantum measurement is not linear. To illustrate this, consider an orthonormal basis $\mathcal{B} = \{\boldsymbol{v}, \boldsymbol{w}\}$ of $\mathbb{C}^2$ and a superposition $\alpha\boldsymbol{v} + \beta\boldsymbol{w}$ with $|\alpha|^2 + |\beta|^2 = 1$. Measuring this state with respect to $\mathcal{B}$ collapses the state to $\boldsymbol{v}$ with probability $|\alpha|^2$ and to $\boldsymbol{w}$ with probability $|\beta|^2$. This *probabilistic collapse* is inherently non-linear, and since Lineal enforces linearity in its reduction strategy, it cannot directly encode quantum measurements.

This limitation motivated the development of the Lambda-S calculus. The key observation was that if a term $\lambda x.t$ includes a measurement of its argument, it must consume the entire superposition passed as an argument rather than distributing it according to rule (1). Lambda-S addresses this by enforcing linearity in the reduction strategy while allowing terms to consume full superpositions in cases where linearity is preserved through other means. Inspired by Intuitionistic Linear Logic, Lambda-S ensures that if $\lambda x.t$ consumes its whole argument, it uses it exactly once. This is achieved by combining the enforced linearity of the reduction strategy with the linearity constraints of the type system [9].

To accomplish this, Lambda-S introduces a type constructor S(A), where a simple type A represents a set of basis terms and S(A) denotes its span. This leads to two distinct β-reductions based on the type of the variable:

- If x has type S(A), then $(\lambda x^{S(A)}.t)r$ reduces in a call-by-name fashion, with the type system ensuring that t uses x only once.
- If $A \neq S(A')$, then $(\lambda x^A.t)r$ reduces in call-by-base mode.

Since the calculus is Church-style, reduction dependent on typing amounts to considering two abstractions.

A later refinement, Lambda-S_1 [12], further enforces norm-preserving constraints on superpositions and guarantees unitarity. This refinement was developed building upon a realizability model proposed earlier [10]. This realizability model addresses the long-standing issue of preserving quantum unitarity in quantum control lambda calculi.

2.2 Categorical Semantics

Lambda-S has been given a categorical semantics [11] through an adjunction between a Cartesian closed category $(\mathcal{S}, \times, I)$ and an additive symmetric monoidal closed category $(\mathcal{V}, \otimes, 1)$ (Fig. 1).

A concrete example of this adjunction is given by the category **Set** of sets and functions as $\mathcal{S}$, and the category **Vec** of vector spaces and linear maps as $\mathcal{V}$ [13]. In this case, F is the span functor, while G is the forgetful functor.

$$(\mathcal{S}, \times, I) \underset{(G,n)}{\overset{(F,m)}{\rightleftarrows}} (\mathcal{V}, \otimes, 1) \quad \bot$$

Fig. 1. The adjunction between the categories $\mathcal{S}$ and $\mathcal{V}$

In this setting, the type constructor S is interpreted by the monad GF, which first transforms a set into a vector space and then forgets its vector space structure, treating it again as a set.

This implies that, on the one hand, we have tight control over the Cartesian structure of the model (e.g., duplication), while on the other hand, the world of superpositions is, in a sense, embedded within the classical world—determined externally by classical rules until explicitly explored. This is represented by the following composition of maps:

$$GF(A) \times GF(A) \xrightarrow{n} G(F(A) \otimes F(A)) \xrightarrow{G(m)} GF(A \times A)$$

which allows operations within a monoidal structure, explicitly enabling algebraic manipulation, before returning to the Cartesian product. Intuitively, using **Set** and **Vec** as concrete examples of a Cartesian and a monoidal category, respectively, this composition corresponds to taking the Cartesian product of the spans of two sets (seen as sets), transforming it into their tensor product (in the linear world), and then returning to the span of the Cartesian product of the original sets. The passage through the tensor space allows for the necessary algebraic manipulation. This contrasts with linear logic, where the ! modality prevents algebraic manipulation; that is, in linear logic, $(!A) \otimes (!A)$ remains a product within a monoidal category.

The categorical semantics of Lambda-S_1 [12] is structured around a similar adjunction but diverges from Lambda-S by ensuring that all terms maintain a unitary norm.

Both calculi serve as foundations for quantum programming languages and categorical models of quantum computation, bridging classical and quantum computational paradigms through rigorous mathematical structures.

2.3 Extensions and Future Directions

A recent development in this line of research has been to extend the superpositions beyond the computational basis—which is essentially what we assume when choosing a measurement basis of non-superposed values—to include a diagonal basis [22]. In this work, the terms $|0\rangle$ and $|1\rangle$ are constants in the language, but also $|+\rangle = \frac{1}{\sqrt{2}} \cdot |0\rangle + \frac{1}{\sqrt{2}} \cdot |1\rangle$ and $|-\rangle = \frac{1}{\sqrt{2}} \cdot |0\rangle + \frac{-1}{\sqrt{2}} \cdot |1\rangle$. This serves as a proof of concept, demonstrating the ability to track duplicable basis terms across multiple bases.

A more general extension of Lambda-S_1 to arbitrary bases, following a realizability model, is currently under development.

The line of research on Lambda-S has demonstrated how to extend the lambda calculus to quantum computation. In terms of proof theory, it provides a proof language for intuitionistic logic with a modality S, indicating what cannot be duplicated.

As mentioned earlier, Intuitionistic Linear Logic is interpreted in a monoidal closed category, with an adjunction to a Cartesian closed category to account for duplicable data. This is the opposite approach to that of Lambda-S.

This duality strongly suggests that the structure of quantum computation is the dual of linear logic and that the Curry-Howard-Lambek correspondence can be extended to quantum computation—where this dual of linear logic serves as the logical counterpart, and Lambda-S as the computational counterpart.

3 The $\mathcal{L}^{\mathcal{S}}$ Calculus: From Logic to Computing

3.1 The *Sup* Connective

In the previous section, we introduced Lambda-S by extending the untyped lambda calculus with modifications necessary to handle quantum computation. This led to an intuitionistic logic with a novel modality S marking superpositions and a categorical model structurally related to linear logic.

The next step was to explore the opposite direction: starting with a well-established logic and modifying it to produce a quantum programming language within its proof system. Our first concern was to understand how the non-determinism from the quantum measurement could fit into a logical framework.

Our starting observation was that a quantum superposition cannot be represented by a conjunction, as its eliminations are deterministic: one can project either the first component or the second, but the choice is fixed. Similarly, a quantum superposition cannot be represented by a disjunction, as its introductions require only one of the components to be present, and thus, its elimination can only consider that one. This led us to introduce a new connective, which adopts the introduction rules of conjunction and the elimination rules of disjunction. This new connective, called sup and denoted $\odot$, requires both components to be present to be introduced, but its elimination allows for the non-deterministic choice of one of the components.

$$\frac{\Gamma \vdash A \quad \Gamma \vdash B}{\Gamma \vdash A \odot B}\ \odot_I \qquad\qquad \frac{\Gamma \vdash A \odot B \quad \Gamma, A \vdash C \quad \Gamma, B \vdash C}{\Gamma \vdash C}\ \odot_E$$

This way, the cut elimination process for the $\odot$ connective is not deterministic, since

$$\frac{\dfrac{\dfrac{\pi_A}{\Gamma \vdash A} \quad \dfrac{\pi_B}{\Gamma \vdash B}}{\Gamma \vdash A \odot B}\ \odot_I \quad \dfrac{\pi_C^A}{\Gamma, A \vdash C} \quad \dfrac{\pi_C^B}{\Gamma, B \vdash C}}{\Gamma \vdash C}\ \odot_E$$

can be reduced to either by replacing the proof π_A in the proof π_C^A or the proof π_B in the proof π_C^B.

This connective was introduced in [5], where we explored its properties and its application to encode quantum superpositions and measurements. While bits are usually encoded by $\top \vee \top$, in this logic, we encode qubits by $\top \odot \top$, and the non-determinism present in the elimination rule models the probabilistic nature of quantum measurements. However, we went further and demonstrated how to encode matrices and vectors using the $\odot$ connective by introducing two additional rules: one for scalars and another for sums of terms.

$$\frac{\Gamma \vdash A \quad \Gamma \vdash A}{\Gamma \vdash A}\ \text{Sum} \qquad\qquad \frac{\Gamma \vdash A}{\Gamma \vdash A}\ \text{Scalar}$$

The Sum rule enables a deterministic cut-elimination for the $\odot$ connective by summing the results of both paths. The Scalar rule allows us to transform non-deterministic choices into probabilistic ones. While these two rules are trivally admissible in the logic, they may stop cut-elimination from occurring. Indeed, they could appear in the middle of an introduction and an elimination, as in the following example

$$\frac{\dfrac{\dfrac{\dfrac{\pi_A}{\Gamma \vdash A} \quad \dfrac{\pi_B}{\Gamma \vdash B}}{\Gamma \vdash A \odot B}\ \odot_I}{\Gamma \vdash A \odot B}\ \text{Scalar} \quad \dfrac{\pi_C^A}{\Gamma, A \vdash C} \quad \dfrac{\pi_C^B}{\Gamma, B \vdash C}}{\Gamma \vdash C}\ \odot_E$$

Thus, we also introduced all the necessary commuting rules to ensure the introduction property, that is, guaranteeing that the cut-elimination process does not get stuck. In this case, for example, we introduced a rule commuting the rule Scalar with the rule $\odot_I$ as follows

$$\frac{\dfrac{\dfrac{\pi_A}{\Gamma \vdash A} \quad \dfrac{\pi_B}{\Gamma \vdash B}}{\Gamma \vdash A \odot B}\ \odot_I}{\Gamma \vdash A \odot B}\ \text{Scalar} \quad \longrightarrow \quad \frac{\dfrac{\dfrac{\pi_A}{\Gamma \vdash A}}{\Gamma \vdash A}\ \text{Scalar} \quad \dfrac{\dfrac{\pi_B}{\Gamma \vdash B}}{\Gamma \vdash B}\ \text{Scalar}}{\Gamma \vdash A \odot B}\ \odot_I \tag{2}$$

The proof language is then extended with a new kind of pair as proof terms for the $\odot$ connective, along with sums of proofs and scalar multiplication, much in the way of Lineal. Additionally, in order to represent vectors, we define every scalar $\alpha \in \mathcal{S}$—the set of scalars (usually taken as $\mathbb{C}$ to represent quantum computing)—as a proof of $\top$.

$$\frac{\Gamma \vdash t : A \quad \Gamma \vdash r : B}{\Gamma \vdash \langle t, r\rangle^{\odot} : A \odot B}\ \odot_I \qquad \frac{\Gamma \vdash t : A \odot B \quad \Gamma, x : A \vdash r : C \quad \Gamma, y : B \vdash s : C}{\Gamma \vdash \mathsf{match}^{\odot}(t, x.r, y.s) : C}\ \odot_E$$

$$\frac{\Gamma \vdash t : A \quad \Gamma \vdash r : A}{\Gamma \vdash t + r : A}\ \text{Sum} \qquad \frac{\Gamma \vdash t : A}{\Gamma \vdash \alpha \cdot t : A}\ \text{Scalar}(\alpha) \qquad \frac{}{\Gamma \vdash \star_\alpha : \top}\ \top_I(\alpha)$$

Thus, the commuting rule of Eq. (2) corresponds to the following rewriting rule of the proof-language

$$\alpha.\langle t, r\rangle^{\odot} \longrightarrow \langle \alpha \cdot t, \alpha \cdot r\rangle^{\odot}$$

Contrary to Lineal, where the linearity is enforced in the reduction strategy, we did not forced linearity in the proof-terms with an strategy in the $\odot$-calculus. Instead, we showed how to encode matrices and vectors with these contructions.

First, we defined the proposition $\mathcal{Q}^{\otimes n}$ by induction on n as follows: $\mathcal{Q}^{\otimes 0} = \top$ and $\mathcal{Q}^{\otimes(n+1)} = \mathcal{Q}^{\otimes n} \odot \mathcal{Q}^{\otimes n}$. Then, we have a one-to-one correspondance between the closed irreducible proofs of $\mathcal{Q}^{\otimes n}$ and the vectors of $\mathbb{C}^{2^n}$, associating to each closed irreductible proof-term t of $\mathcal{Q}^{\otimes n}$ a vector $\underline{t}$ of $\mathbb{C}^{2^n}$ as follows:

- If $t = \star_\alpha$, we let $\underline{t} = (\, \alpha \,)$.
- If $t = \langle t_1, t_2 \rangle^\odot$, we let $\underline{t}$ be the vector with two blocks $\underline{t_1}$ and $\underline{t_2}$: $\underline{t} = \begin{pmatrix} \underline{t_1} \\ \underline{t_2} \end{pmatrix}$.

In the same way, to each vector $\boldsymbol{v}$ of $\mathbb{C}^{2^n}$, we associate a closed irreducible proof $\overline{\boldsymbol{v}}$ of $\mathcal{Q}^{\otimes n}$.

- If $\boldsymbol{v} = (\, \alpha \,)$, we let $\overline{\boldsymbol{v}} = \star_\alpha$.
- If $\boldsymbol{v} = \left(\begin{smallmatrix} \boldsymbol{v}_1 \\ \boldsymbol{v}_2 \end{smallmatrix}\right)$, where $\boldsymbol{v}_1$ and $\boldsymbol{v}_2$ are the two blocks of size $\frac{n}{2}$, we let $\overline{\boldsymbol{v}} = \langle \overline{\boldsymbol{v}_1}, \overline{\boldsymbol{v}_2} \rangle^\odot$.

The interesting part is that the Sum and Scalar rules act as a sum of vectors and a scalar multiplication:

Theorem 3.1 (Sum and Scalar Multiplication [5, Props. 5.6 and 5.7]). *If t, t_1, and t_2 are closed proofs of $\mathcal{Q}^{\otimes n}$, and $\alpha \in \mathbb{C}$, we have*

$$\underline{t_1 + t_2} = \underline{t_1} + \underline{t_2}, \qquad \textit{and} \qquad \underline{\alpha \cdot t} = \alpha \underline{t}.$$

◁

Moreover, we can encode matrices as stated by the following theorem.

Theorem 3.2 (Matrices [5, Thm. 5.8]). *Let M be a matrix with 2^m columns and 2^n lines, then there exists a closed proof t of $\mathcal{Q}^{\otimes m} \multimap \mathcal{Q}^{\otimes n}$ such that, for all vectors $\boldsymbol{v} \in \mathbb{C}^{2^m}$, $\underline{t\overline{\boldsymbol{v}}} = M\boldsymbol{v}$.* ◁

3.2 The Sum and Scalar Rules in Linear Logic: The $\mathcal{L}^{\mathcal{S}}$ Calculus

In [6], we extended the study of the Sup-calculus to the context of linear logic, introducing the $\mathcal{L}\odot^{\mathcal{S}}$ calculus. Additionally, we studied a fragment without $\odot$ that retains the Sum and Scalar rules, which we referred to as the $\mathcal{L}^{\mathcal{S}}$ calculus. In all these calculi, $\mathcal{S}$ is a set of scalars, usually, a semiring.

This framework integrates the ideas from the previous section into intuitionistic multiplicative-additive linear logic (IMALL), establishing a formal proof language for this logical system. The inclusion of scalars and sums ensures that proofs in $\mathcal{L}^{\mathcal{S}}$ represent linear maps, and we provided a syntactic proof of this property. In particular, we proved the converse of Theorem 3.2, showing that every proof-term of $\mathcal{Q}^{\otimes m} \multimap \mathcal{Q}^{\otimes n}$ represents a linear map from $\mathbb{C}^{2^m}$ to $\mathbb{C}^{2^n}$. In fact, the statement of the theorem is more general, as it holds for any proposition as domain.

Theorem 3.3 (Linearity [6, Thm. 4.10]). *Let* $\vdash \lambda x.t : A \multimap \mathcal{Q}^{\otimes n}$, $\vdash r : A$, $\vdash s : A$, *and* $\alpha, \beta \in \mathbb{C}$. *Then*

$$(\lambda x.t)(\alpha \cdot r + \beta \cdot s) \longrightarrow^* v \qquad \Longleftrightarrow \qquad \alpha \cdot (\lambda x.t)r + \beta \cdot (\lambda x.t)s \longrightarrow^* v \quad \triangleleft$$

Moreover, this result also extends to any proposition in the codomain using an observational equivalence $\sim$.

Corollary 3.4 (Generalized linearity [6, Cor. 4.11]). *Let* $\vdash \lambda x.t : A \multimap B$, $\vdash r : A$, $\vdash s : A$, *and* $\alpha, \beta \in \mathbb{C}$. *Then the following observational equivalence holds*

$$(\lambda x.t)(\alpha \cdot r + \beta \cdot s) \qquad \sim \qquad \alpha \cdot (\lambda x.t)r + \beta \cdot (\lambda x.t)s \quad \triangleleft$$

3.3 Categorical Semantics

The Linear Case. The $\mathcal{L}\odot^{\mathcal{S}}$ calculus has been assigned a categorical semantics in a recent draft [15]. The core idea of introducing a connective that behaves as both a conjunction and a disjunction, while also incorporating sums and scalars within a linear logic framework, aligns naturally with an additive symmetric monoidal closed category. Such a category, equipped with biproducts, provides a suitable setting for interpreting this conjunction-disjunction connective, as well as the sum and scalar operations.

Introducing a probabilistic operator into a linear language is not straightforward. To aid intuition, consider the concrete category **Vec** of vector spaces and linear maps over $\mathbb{C}$. Our interpretation cannot use the Powerset Monad, as it is usual to express non-deterministic effects [21], since this monad is not valid in this category. Such approach would consist in using the Cartesian product of the several non-deterministic paths, and gather them together into a set. The problem is that the needed map $A \times A \xrightarrow{\xi} \mathcal{P}A$ defined by $\xi(a_1, a_2) = \{a_1, a_2\}$ is not linear and so it is not in the category.

Our approach was instead inspired by the density matrix quantum formalism (see, for example, [23, Section 2.4]), wherein we consider the linear combination of results as a representation of a probability distribution. Let t be a term reducing with probability p to t_1 and a probability of q to t_2, with $p + q = 1$. We interpret t as $\nabla_{pq}(t_1, t_2) = \hat{p} \cdot t_1 + \hat{q} \cdot t_2$, where, if $\hat{p}$ is the mapping that multiplies its argument by p, then ∇_{pq} is defined as $[\hat{p}, \hat{q}]$ (Fig. 2).

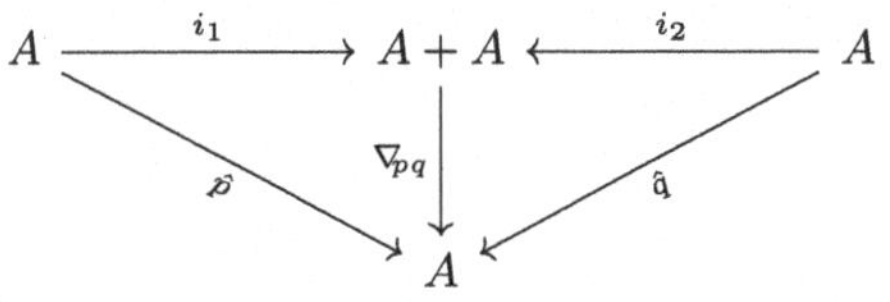

Fig. 2. The map ∇_{pq} as $[\hat{p}, \hat{q}]$

This approach is closed to that used for PCF$^{\mathcal{R}}$ [18]. However, in an abstract categorical setting, this means that we needed a category with biproducts, so we can interpret ∇_{pq} as $\nabla \circ (\hat{p} \oplus \hat{q})$, where $\hat{p}$ and $\hat{q}$ are appropriated maps $A \to A$. For those scalar maps, we considered the category to also be monoidal, so we can count with the semiring of scalars $\mathsf{Hom}(1,1)$ [17], where 1 is the tensor unit. Then, we defined a monomorphism $\mathcal{S} \to \mathsf{Hom}(I,I)$ to interpret the scalars, which ensures that two proof-terms interpreted by the same map are structurally equivalent.

In a category with biproducts, that is, where the product and the coproduct coincide, we have the diagram from Fig. 3. Continuing with **Vec** as a concrete example, we have $i_1(a) = (a, \mathbf{0})$, $i_2(b) = (\mathbf{0}, b)$, and $[g_1, g_2](a,b) = g_1(a) + g_2(b)$.

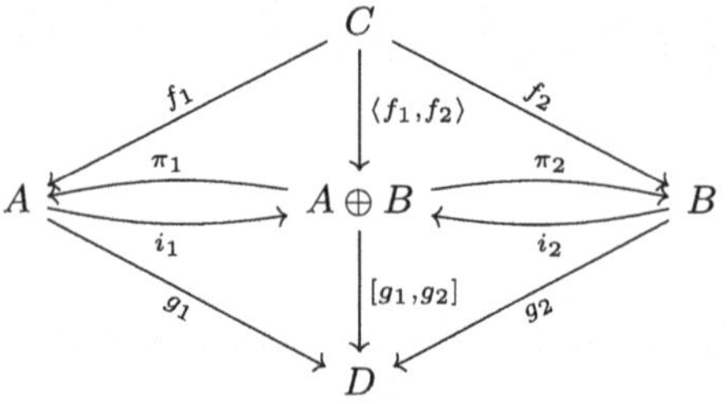

Fig. 3. The biproduct diagram

In this category, the additive conjunction and the additive disjunction from Linear Logic, with introduction rules given by:

$$\frac{\Gamma \vdash t : A \quad \Gamma \vdash r : B}{\Gamma \vdash \langle t, r \rangle : A \,\&\, B}\,\&_i \qquad \frac{\Gamma \vdash t : A}{\Gamma \vdash \mathsf{inl}(t) : A \oplus B}\,\oplus_i \qquad \frac{\Gamma \vdash t : B}{\Gamma \vdash \mathsf{inr}(t) : A \oplus B}\,\oplus_i$$

are both interpreted by the direct sum. The sum symbol is then interpreted as the sum of vectors, and thus, $\mathsf{inl}(t) + \mathsf{inr}(r)$ is encoded as the map $[i_1 \circ t, i_2 \circ r] \circ \Delta$.

$$\begin{array}{ccccccc} \Gamma & \xrightarrow{\Delta} & \Gamma \oplus \Gamma & \xrightarrow{[i_1 \circ t, i_2 \circ r]} & A \oplus B & & \\ g & \longmapsto & (g,g) & \longmapsto & (t(g), \mathbf{0}) + (\mathbf{0}, r(g)) & = & (t(g), r(g)) \end{array}$$

Thus, $\mathsf{inl}(t) + \mathsf{inr}(r)$ is interpreted in the same way as $\langle t, r \rangle$. Indeed, $\langle t, r \rangle$, as usual, is interpreted as $(t \oplus r) \circ \Delta$.

$$\begin{array}{ccccc} \Gamma & \xrightarrow{\Delta} & \Gamma \oplus \Gamma & \xrightarrow{t \oplus r} & A \oplus B \\ g & \longmapsto & (g,g) & \longmapsto & (t(g), r(g)) \end{array}$$

3.4 Sup Without Sup

The remark that $\mathsf{inl}(t) + \mathsf{inr}(r)$ is interpreted in the same way as $\langle t, r \rangle$ was first made in [7], where we proposed to drop the $\odot$ connective by adding a new rule for disjunction introduction:

$$\frac{\Gamma \vdash t : A \quad \Gamma \vdash r : B}{\Gamma \vdash \mathsf{inlr}(t, r) : A \vee B} \vee_{i3}$$

This rule allows us to encode quantum computing without the need of a new connective. We showed in that draft how to make sense of this new rule in propositional logic, and in linear logic, how to encode quantum computing, but also how to use inlr as a general method to solve commuting cuts, when there is the introduction of a connective blocking a cut.

The Non-Linear Case. In another recent draft [14] we consider the non-linear case of the Sum and Scalar rules, and provided a categorical characterisation for it. Indeed, the categorical characterisation for the linear case from Sect. 3.3 is meaningful only in the context of Linear Logic, where we can rely on categories such as **Vec**, which possess biproducts. However, in the context of propositional logic, the product and the coproduct are usually different, and the interpretation of the sum of disjunctions cannot be the same as the interpretation of the conjunction.

In [14], we address this problem by proposing models for Sum and Scalar rules within the framework of propositional logic, including disjunctions, as well as all the other logical connectives.

Let us focus first in the sum operator. Since we were in the context of propositional logic, we started from the category **Set**, and sought for the simplest way to enrich its objects with an operation so as to interpret the sum operator on it. Thus, we first considered the simpler algebraic structure, the category **Mag** of magmas, whose objects are magmas and whose arrows are magma homomorphisms. However, this category is not Cartesian closed, since the evaluation map is not an homomorphism, and so it is not a good candidate. Thus, we proposed studying the category $\mathbf{Mag_{Set}}$, whose objects are magmas but whose arrows are functions from the **Set** category. Naturally, it is Cartesian closed, since it is just the category **Set** whose objects have an operation to which we do not require any specific property. In addition to being Cartesian closed, not having homomorphisms as arrows turns out to be necessary. Indeed, without commutativity, and associativity, the sum operator is not preserved by proof-terms, as shown in the following example. Let f be the following proof-term.

$$\vdash \lambda x.(\pi_1 x + \pi_2 x) : (A \wedge A) \Rightarrow A$$

Then, $f\Big(\langle a_1, a_2 \rangle + \langle a_3, a_4 \rangle\Big) = f\langle a_1 + a_3, a_2 + a_4 \rangle = (a_1 + a_3) + (a_2 + a_4)$, while $f\langle a_1, a_2 \rangle + f\langle a_3, a_4 \rangle = (a_1 + a_2) + (a_3 + a_4)$.

The magmas provide the operation used to interpret the sum symbol. To our surprise, disjunction in this calculus is not interpreted by a coproduct, as our construction does not yield a coproduct unless the arrows in the category are homomorphisms. Our proposal was to interpret disjunction $A \vee B$ as the union of the coproduct and the product in **Set**, mimicking the idea of a biproduct from the Linear Logic setting, that is, $A \oplus B = (A \uplus \mathrm{B}) \cup (A \times B)$.

When adding also scalars, we needed to consider a category $\mathbf{AMag}^{\mathcal{S}}_{\mathbf{Set}}$ of *action magmas*, that is, magmas with an action map. The surprising aspect in both cases, $\mathbf{Mag}_{\mathbf{Set}}$ and $\mathbf{AMag}^{\mathcal{S}}_{\mathbf{Set}}$, is that we do not require algebraic properties in the categories beyond equipping each object with an operation, or with an operation and an action map. The disjunction is interpreted as $A \oplus B$, with its magma operator and its action map performing most of the work, grouping the left side to the left and the right side to the right, without requiring any associativity or commutativity in the underlying sets.

3.5 Extensions and Future Directions

In [8], we extended the $\mathcal{L}^{\mathcal{S}}$ calculus to ILL by adding the modality !, resulting in the $\mathcal{L}^{\mathcal{S}}_!$ calculus. Its model is currently under development for a forthcoming journal version of this paper. As expected, we employ the same adjunction as for Lambda-S (cf. Fig. 1), but interpret the calculus in a monoidal closed category and use the comonad FG to interpret the ! modality (see Fig. 4).

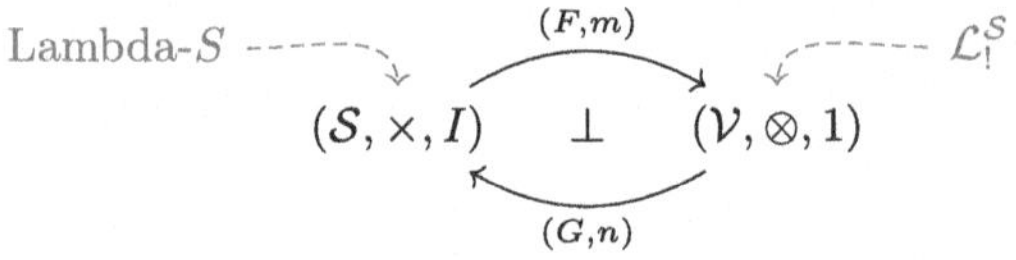

Fig. 4. The duality between Lambda-S and $\mathcal{L}^{\mathcal{S}}$

In the same paper, we also added polymorphism to the calculus, and we are currently studying the interaction between polymorphism and the sums and scalar products in this setting, following the ideas of [20].

The formal connection between the Lambda-S calculus and the $\mathcal{L}^{\mathcal{S}}_!$ calculus remains an open research question. We are currently working on a formal proof of the equivalence between the two calculi and on the development of a realizability model for the $\mathcal{L}^{\mathcal{S}}$ calculus, integrating ideas from [10].

4 Conclusion

In this invited paper, we provided an overview of a research program with the goal of extending the Curry-Howard-Lambek correspondence to quantum computation. We introduced the central ideas of two key frameworks. The Lambda-S calculus is a quantum lambda calculus that incorporates quantum superpositions of lambda terms while enforcing linearity to address quantum control. The $\mathcal{L}^{\mathcal{S}}$

calculus, in turn, is a proof language for intuitionistic linear logic, enhanced with novel rules to handle quantum superpositions and scalar operations.

The Lambda-S calculus incorporates quantum mechanical principles within the lambda calculus framework, providing a structured approach to encoding quantum programs. In parallel, the $\mathcal{L}^{\mathcal{S}}$ calculus extends intuitionistic linear logic with scalar and sum operations, providing a logical foundation for quantum computation and enabling categorical models that reveal deep structural connections between logic and quantum computation.

These calculi bridge classical and quantum computational paradigms and suggest a fundamental duality between quantum computation and linear logic. Future work includes further investigation of these foundational relationships, refinement and generalisation of the calculi, and the development of comprehensive categorical models. Ultimately, these frameworks provide a promising formal foundation for quantum programming languages and offer deeper theoretical insight into quantum computation.

Acknowledgement. This work is supported by the European Union through the MSCA SE project QCOMICAL (Grant Agreement ID: 101182520), by the Plan France 2030 through the PEPR integrated project EPiQ (ANR-22-PETQ-0007), and by the Uruguayan CSIC grant 22520220100073UD.

References

1. Arrighi, P., Dowek, G.: Lineal: a linear-algebraic λ-calculus. Logical Methods Comput. Sci. **13**(1:8), 1–33 (2017). https://doi.org/10.23638/LMCS-13(1:8)2017
2. Assaf, A., Díaz-Caro, A., Perdrix, S., Tasson, C., Valiron, B.: Call-by-value, call-by-name and the vectorial behaviour of the algebraic λ-calculus. Logical Methods Comput. Sci. **10**(4:8), 1–40 (2014). https://doi.org/10.2168/LMCS-10(4:8)2014
3. Birkhoff, G., von Neumann, J.: The logic of quantum mechanics. Ann. Math. **37**(4), 823–843 (1936). https://doi.org/10.2307/1968621
4. Crole, R.: Categories for Types. Cambridge University Press (1993). https://doi.org/10.1017/CBO9781139172707
5. Díaz-Caro, A., Dowek, G.: A new connective in natural deduction, and its application to quantum computing. Theoret. Comput. Sci. **957**, 113840 (2023). https://doi.org/10.1016/j.tcs.2023.113840
6. Díaz-Caro, A., Dowek, G.: A linear linear lambda-calculus. Math. Struct. Comput. Sci. **34**, 1103–1137 (2024). https://doi.org/10.1017/S0960129524000197
7. Díaz-Caro, A., Dowek, G.: A new introduction rule for disjunction. arXiv:2502.19172 (2025)
8. Díaz-Caro, A., Dowek, G., Ivnisky, M., Malherbe, O.: A linear proof language for second-order intuitionistic linear logic. In: Metcalfe, G., Studer, T., de Queiroz, R. (eds.) Logic, Language, Information and Computation (WoLLIC 2024), LNCS, vol. 14672, pp. 18–35. Springer (2024). https://doi.org/10.1007/978-3-031-62687-6_2
9. Díaz-Caro, A., Dowek, G., Rinaldi, J.: Two linearities for quantum computing in the lambda calculus. Biosystems **186**, 104012 (2019). https://doi.org/10.1016/j.biosystems.2019.104012

10. Díaz-Caro, A., Guillermo, M., Miquel, A., Valiron, B.: Realizability in the unitary sphere. In: Proceedings of the 34th Annual ACM/IEEE Symposium on Logic in Computer Science (LICS 2019), pp. 1–13 (2019). https://doi.org/10.1109/LICS.2019.8785834
11. Díaz-Caro, A., Malherbe, O.: A categorical construction for the computational definition of vector spaces. Appl. Categ. Struct. **28**(5), 807–844 (2020). https://doi.org/10.1007/s10485-020-09598-7
12. Díaz-Caro, A., Malherbe, O.: Quantum control in the unitary sphere: Lambda-S_1 and its categorical model. Logical Methods Comput. Sci. **18**(3:32) (2022). https://doi.org/10.46298/lmcs-18(3:32)2022
13. Díaz-Caro, A., Malherbe, O.: A concrete model for a linear algebraic lambda calculus. Math. Struct. Comput. Sci. **34**(1), 1–44 (2023). https://doi.org/10.1017/s0960129523000361
14. Díaz-Caro, A., Malherbe, O.: Parallel and algebraic lambda-calculi in intuitionistic propositional logic. arXiv:2408.16102 (2024)
15. Díaz-Caro, A., Malherbe, O.: The sup connective in IMALL: a categorical semantics. arXiv:2405.02142 (2024)
16. Howard, W.A.: The formulae-as-types notion of construction. In: Curry, H., B., H., Roger, S.J., Jonathan, P. (eds.) To H. B. Curry: Essays on Combinatory Logic, Lambda Calculus, and Formalism, pp. 479–490. Academic Press (1980), Hitherto unpublished note of 1969
17. Kelly, G.M., Laplaza, M.L.: Coherence for compact closed categories. J. Pure Appl. Algebra **19**, 193–213 (1980). https://doi.org/10.1016/0022-4049(80)90101-2
18. Laird, J., Manzonetto, G., McCusker, G., Pagani, M.: Weighted relational models of typed lambda-calculi. In: Proceedings of the 28rd Annual ACM/IEEE Symposium on Logic in Computer Science (LICS 2013), pp. 301–310 (2013). https://doi.org/10.1109/LICS.2013.36
19. Lambek, J.: From lambda calculus to cartesian closed categories. In: Curry, H., B., H., Roger, S.J., Jonathan, P. (eds.) To H. B. Curry: Essays on Combinatory Logic, Lambda Calculus, and Formalism, pp. 375–402. Academic Press (1980)
20. Maneggia, P.: Models of linear polymorphism. Ph.D. thesis, The University of Birmingham, February 2004
21. Moggi, E.: Notions of computation and monads. Inf. Comput. **93**(1), 55–92 (1991). https://doi.org/10.1016/0890-5401(91)90052-4
22. Monzón, N.: Extension of Lambda-S for different measurement bases. Master's thesis, Universidad Argentina de la Empresa, March 2025
23. Nielsen, M., Chuang, I.: Quantum Computation and Quantum Information. Cambridge University Press, 10th years anniversary edn. (2010). https://doi.org/10.1017/CBO9780511976667
24. Sørensen, M.H., Urzyczyn, P.: Lectures on the Curry-Howard Isomorphism. Elsevier (2006). https://doi.org/10.1016/S0049-237X(06)80499-4

Represented Spaces of Represented Spaces

Johanna Franklin[1], Eike Neumann[2], Arno Pauly[2], Cécilia Pradic[2(✉)], and Manlio Valenti[2]

[1] Hofstra University, Hempstead, USA
[2] Swansea University, Swansea, UK
c.pradic@swansea.ac.uk

Abstract. We investigate spaces of spaces in the category of represented spaces and type-2 computable maps. Concretely, such spaces of spaces are given by bundles whose bases are names for spaces. We give natural examples of such bundles for Polish and compact Polish spaces and show they are essentially equivalent. We then propose definitions of genericity and computable categoricity, according to which the Cantor space is computably categorical and generic as a compact Polish space. We also show that the degree of categoricity of $\mathcal{S}^1$ is the Weihrauch degree lim.

Keywords: Computable topology · Categoricity

1 Introduction

We want to discuss notions of represented spaces of spaces so that we may formalize some natural questions such as

- what is a generic X-space? (for X ranging over various adjectives such as Polish, QuasiPolish, compact Polish...)
- is a space computably categorical? If not, what is the (recursion-theoretic) complexity of computing an isomorphism?

But first, we need to explain what a space of spaces is. Describing a space of spaces involves coding names for spaces as a space and giving an interpretation for them consisting of a bundle over the space of names. We are doing this in the category ReprSp of represented spaces and type-2 computable maps.

Concretely, such bundles are simply computable maps $\mathrm{El}_A : A_\bullet \to A$ between represented spaces $A_\bullet$ and A. The points of A are codes for the spaces we want to represent and $A_\bullet$ is a sum of those spaces. A point $x \in A$ represents the fiber $\mathrm{El}_A^{-1}(x)$, so the map El_A should be thought of as a projection

$$\begin{array}{rcl} \sum_{x \in A} \mathrm{El}_A^{-1}(x) \;\cong\; A_\bullet & \longrightarrow & A \\ (\text{a code } x \text{ for a space } S_x \cong \mathrm{El}_A^{-1}(x), \text{a point of } S_x) & \longmapsto & x \end{array}$$

© The Author(s), under exclusive license to Springer Nature Switzerland AG 2025
A. Beckmann et al. (Eds.): CiE 2025, LNCS 15764, pp. 47–61, 2025.
https://doi.org/10.1007/978-3-031-95908-0_4

This in particular induces a map $\mathrm{El}_A^{-1} : A \to \mathsf{ReprSp}$, but also carries additional information about how the individual spaces $\mathrm{El}_A^{-1}(x)$ cohere together in $A_\bullet$, which allows us to talk about, say, maps that take as input a name x for a non-empty space and return a point of $\mathrm{El}_A^{-1}(x)$, *computably in* x. This is a standard approach to internalizing the notion of families in categories with pullbacks [11, 19].

Example 1. One bundle representing finite-dimensional Euclidean spaces is the first projection of vectors of real numbers onto their dimensions $\mathbb{R}^* \to \mathbb{N}$, where $\mathbb{R}^*$ is represented by $\delta_{\mathbb{R}^*}(\langle n, x\rangle) = \langle \delta_{\mathbb{N}}(n), \delta_{\mathbb{R}^n}(x)\rangle$, with $\delta_{\mathbb{N}}, \delta_{\mathbb{R}^n}$ being standard representations for $\mathbb{N}$ and $\mathbb{R}^n$.

Example 2. A hyperspace $\mathcal{H}$ over some space X is a represented space whose points are subsets of X. An example is the hyperspace $\mathcal{A}(X)$ of closed subsets of X.

Any hyperspace over X induces a bundle $\mathcal{H}_\bullet \xrightarrow{\pi_1} \mathcal{H}$ where $\mathcal{H}_\bullet$ is the subspace of $\mathcal{H} \times X$ consisting of those pairs (A, x) such that $x \in A$.

We first investigate bundles that represent Polish spaces and compact Polish spaces. There are several natural options, which we show to be equivalent in a suitable sense (Sect. 2). We then define what it means for a space represented in such bundles to be computably categorical and, more generally, what the Weihrauch degree of categoricity of a space is. We then show that $\{0,1\}^{\mathbb{N}}$ is computably categorical in the compact Polish spaces and that the degree of categoricity of the circle $\mathcal{S}^1$ is lim (Sect. 3). Finally we define what it means for a space to be generic in a bundle and show that $\{0,1\}^{\mathbb{N}}$ is Π^0_2-generic in the compact Polish spaces (Sect. 4).

Related Work. An early occurrence of the idea of a represented space of (certain) represented spaces is found in [17], where the space of Polish spaces is introduced. A representation of QuasiPolish spaces as an internal category of represented spaces was investigated by de Brecht [3], who shows that much of the usual structure over the category of QuasiPolish spaces does computably internalize.

The space $\mathcal{C}([0,1])$ is not computably categorical as a Banach space [14, Corollary 4.3] (up to isometry). An early contribution to the question of computable categoricity is [13], where it is e.g. shown that $\{0,1\}^{\mathbb{N}}$ is computably categorical in the sense of isometry.

Rather than looking at represented spaces of spaces, another approach is to investigate numberings of computable spaces of that kind. This approach is e.g. taken in [5].

Degrees of categoricity of various Lebesgue spaces have been extensively investigated [7,12].

2 Spaces of Spaces as Bundles

2.1 Equivalence of Bundles

As previously mentioned, a bundle is simply any map $\mathrm{El}_A : A_\bullet \to A$ that we want to interpret as a space of spaces.

Given two bundles El_A and El_B, we can define what it means for El_A to be essentially equivalent to El_B by using the notion of reindexing and pullbacks.

A pullback square is a commuting square as found in the diagram below, satisfying the following universal property: if there are morphisms α, β as depicted such that $f \circ \beta = g \circ \alpha$, then there is a unique γ such that $\alpha = k \circ \gamma$ and $\beta = h \circ \gamma$.

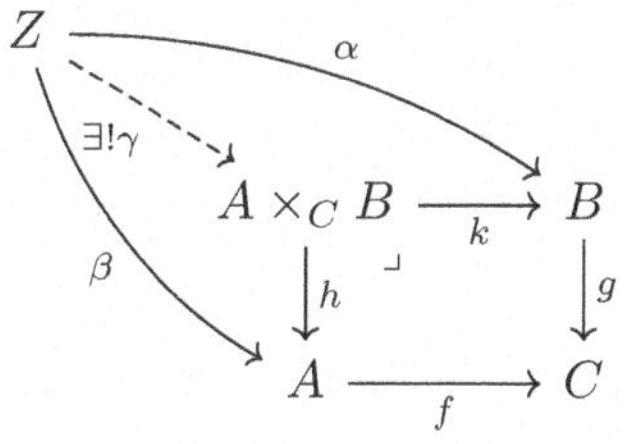

In this diagram $A \times_C B$ (together with the projections h and k) is a pullback of f and g. We use the notation around the top-left corner of the commuting square to indicate we intend this to be a pullback in a diagram. Pullbacks are only determined up to unique isomorphisms. Concretely, pullbacks in all categories we are interested in here can be built by taking the cartesian products of the domains of the maps f and g and restricting to the sets of pairs (a, b) with $f(a) = g(b)$.

We say that El_A is a pullback of El_B by $f : A \to B$ when there is a map $f_\bullet : A_\bullet \to B_\bullet$ such that the square below is a pullback.

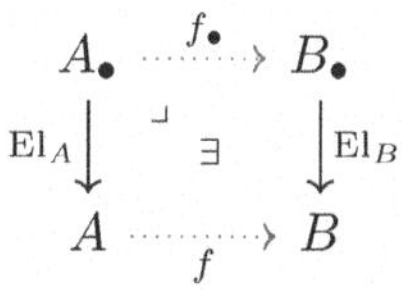

Essentially, f maps A-codes of spaces to B-codes of spaces and $f_\bullet$ gives us maps $\mathrm{El}_A^{-1}(a) \to \mathrm{El}_B^{-1}(f(a))$, uniformly in a. The square being a pullback moreover ensures that those maps are actually homeomorphisms whose inverses are also computable from codes $a \in A$. To define equivalence of two such bundles, we do not care if A and B are not isomorphic as long as El_A and El_B describe the same spaces and we can effectively translate a name for a space in A to a name for the same space in B and vice-versa. Hence we take the following definition for equivalence of bundles.

Definition 1. *We call two bundles (weakly) equivalent when they are pullbacks of one another.*

2.2 Intensional vs Extensional Equivalence of Bundles

Sometimes we do not want to work extensionally and work with multi-valued maps. However, the notion of bundle in the category of multi-valued computable functions does not do what we want, so we are doing something else to deal with that difficulty: we replace a represented space by the subspace of Baire space consisting of the names of its points. These spaces can actually be characterized up to homeomorphism as the regular projective spaces of the category of represented spaces.

Definition 2. *Let us call a computable map between represented spaces a* quotient map *if it is has a computable multi-valued right-inverse.*

A projective reindexing of a bundle $\mathrm{El}_B : B_\bullet \to B$ *is a bundle* $\mathrm{El}_A : A_\bullet \to A$ *which is a pullback of* El_B *along a quotient map* $e : A \twoheadrightarrow B$ *such that* A *is the domain of a representation of* B.

By our discussion above, any bundle has at least one projective reindexing determined by the representation map of A and a choice of pullbacks. We avoid defining this as "the" projective redindexing as we want our constructions to be closed under bundle equivalence.

Lemma 1. *Any two projective reindexings of a bundle are equivalent.*

Definition 3. *We say that two bundles are* intensionally equivalent *if any of their respective projective reindexings are equivalent.*

2.3 Bundles of All Polish Spaces and Compact Polish Spaces

We now consider several bundles meant to represent all (non-empty) Polish spaces.

We can code Polish spaces in the subspace PM of $\mathbb{R}^{\mathbb{N}^2}$ consisting of the pseudometrics over $\mathbb{N}$. Then the bundle is defined as $\mathrm{PM}_\bullet \xrightarrow{\pi_1} \mathrm{PM}$ where $\mathrm{PM}_\bullet$ is the obvious quotient of the subspace of $\mathrm{PM} \times \mathbb{N}^\mathbb{N}$ consisting of those pairs (d, s) such that s is a fast-converging sequence with respect to d.

The other alternatives are induced by hyperspaces as per Example 2. We write $\mathcal{A}(X)$, $\mathcal{V}(X)$ and $\Pi^0_2(X)$ for the hyperspaces of closed, overt and Π^0_2 subspaces of X (see [9,15] for definitions). Given a hyperspace $\mathcal{H}$ over some set, we write $\mathcal{H}_+$ for its restriction to non-empty subspaces. Given $\mathcal{H}$ and $\mathcal{H}'$ over the same set, we write $\mathcal{H} \wedge \mathcal{H}'$ for the join hyperspace determined by the subspace of $\mathcal{H} \times \mathcal{H}'$ of pairs (A, A).

Theorem 1. *The bundle* $\mathrm{PM}_\bullet \xrightarrow{\pi_1} \mathrm{PM}$ *is intensionally equivalent to each of the bundles generated by* $(\Pi^0_2 \wedge \mathcal{V})([0,1]^\omega)_+$, $(\mathcal{A} \wedge \mathcal{V})(\mathbb{R}^\omega)_+$ *and* $\mathcal{V}(\mathbb{R}^\omega)_+$.

Lemma 2. *For a computable metric space* $\mathbf{X}$ *the computable map* $f \mapsto f^{-1}(\{0\}) : \mathcal{C}(\mathbf{X}, [0,1]) \to \mathcal{A}(\mathbf{X})$ *has a computable multivalued right-inverse.*

Corollary 1. *For a computable metric space* $\mathbf{X}$ *the computable map* $f \mapsto f^{-1}((0,1]^\omega) : \mathcal{C}(\mathbf{X}, [0,1]^\omega) \to \Pi^0_2(\mathbf{X})$ *has a computable multivalued right-inverse.*

Proof (Theorem 1). We show that some of the bundles in the theorem statement, or projective reindexings thereof, are pullbacks of one another so that we end up with a cycle of pullbacks witnessing that those bundles are pairwise intensionally equivalent.

For each direction, we only describe the maps between the bases of the bundles, leaving the reader to complete the top part of the pullbacks in the obvious way.

- PM is a pullback of $(\Pi^0_2 \wedge \mathcal{V}_+)([0,1]^\omega)$:
Without loss of generality, we can assume that the pseudometrics are bounded by 1. We then map a pseudometric $d : \mathbb{N} \times \mathbb{N} \to [0,1]$ to

$$\{x \in [0,1]^\omega \mid \forall k\ \exists m\ \forall i \leq k\ |x_i - d(i,m)| < 2^{-k}\}$$

This immediately yields a Π^0_2-name for the set, and we obtain the overt information by observing that a basic open set $U_0 \times \ldots \times U_\ell \times [0,1]^\omega$ intersects the set if and only if $\exists m \in \mathbb{N}\ \forall i \leq \ell\ d(i,m) \in U_i$.
- Projective reindexings of $(\Pi^0_2 \wedge \mathcal{V})([0,1]^\omega)_+$ are pullbacks of $(\mathcal{A} \wedge \mathcal{V}_+)(\mathbb{R}^\omega)_+$:
Given some $A \in \Pi^0_2([0,1]^\omega)$, we can compute some $f : [0,1]^\omega \to [0,1]^\omega$ such that $A = f^{-1}((0,1]^\omega)$ by Corollary 1. This means that $\frac{1}{f} : A \to \mathbb{R}^\omega$ is well-defined, where the multiplicative inverse is taken component-wise, and we can compute $\frac{1}{f}$ from A. We then map A to $D_f := \left\{\left(x, \frac{1}{f}(x)\right) \mid x \in A\right\} \subseteq \mathbb{R}^\omega \times \mathbb{R}^\omega$. Since $(x,y) \notin D_f \Leftrightarrow \left(y \geq 1^\omega \wedge f(x) = \frac{1}{y}\right)$, we obtain $D_f \in \mathcal{A}(\mathbb{R}^\omega)$. Moreover, if we have $A \in \mathcal{V}([0,1]^\omega)$, we can obtain $D_F \in \mathcal{V}(\mathbb{R}^\omega)$, since overt sets are closed under continuous images.
- $(\mathcal{A} \wedge \mathcal{V})(\mathbb{R}^\omega)_+$ is a pullback of $\mathcal{V}(\mathbb{R}^\omega)_+$:
We can just forget the closed information.
- Projective reindexings of $\mathcal{V}_+(\mathbb{R}^\omega)$ are pullbacks of PM:
We can use overt choice to find a dense sequence $(x_n)_n$ in the given overt set, and then obtain the pseudometric $d(n,m) = d_{\mathbb{R}^\omega}(x_n, x_m)$.

Given a pseudometric d over $\mathbb{N}$, a witness of total boundedness is a function $t : \mathbb{N} \to \mathbb{N}$ such that for every $k \in \mathbb{N}$ and $x \in \mathbb{N}$, there exists $y < t(k)$ such that $d(x,y) < 2^{-k}$. Call $\mathrm{TBPM} \subseteq \mathbb{N}^{\mathbb{N}^2} \times \mathbb{N}^{\mathbb{N}}$ the subset consisting of pseudometrics and witnesses of total boundedness and $\mathrm{TBPM}_\bullet \xrightarrow{\pi_1} \mathrm{TBPM}$ the bundle obtained by pulling back $\mathrm{PM}_\bullet \xrightarrow{\pi_1} \mathrm{PM}$ along the first projection $\mathrm{TBPM} \to \mathrm{PM}$.

Theorem 2. *The bundle* $\mathrm{TBPM}_\bullet \xrightarrow{\pi_1} \mathrm{TBPM}$ *is intensionally equivalent to the bundle induced by* $(\mathcal{K} \wedge \mathcal{V})([0,1]^\omega)_+$.

Proof. To show that $\mathrm{TBPM}_\bullet \xrightarrow{\pi_1} \mathrm{TBPM}$ is a pullback of the bundle induced by $(\mathcal{K} \wedge \mathcal{V})([0,1]^\omega)_+$, we can use the same approach as in the proof of Theorem 1 by mapping the pair (d,t) of a pseudometric and a witness of total boundedness to the set

$$\{x \in [0,1]^\omega \mid \forall k\ \exists m \leq t(k)\ \forall i \leq k\ |x_i - d(i,m)| < 2^{-k}\}$$

This set is obviously effectively closed in $[0,1]^\omega$, so it is in particular effectively compact since $[0,1]^\omega$ is effectively compact. Note that, by the definition of witness of total boundedness, this set is the same as the set defined in the proof of Theorem 1; we can therefore get the overt information in the same way.

To show that any projective reindexing of $(\mathcal{K} \wedge \mathcal{V})([0,1]^\omega)_+$ is a pullback of $\text{TBPM}_\bullet \xrightarrow{\pi_1} \text{TBPM}$, we can use overt choice as before to find a dense sequence $(x_n)_n$ with which to build a pseudometric over $\mathbb{N}$. The corresponding witness of total boundedness can then be reconstructed using the compact information.

3 Computable Categoricity

We can also quantify the hardness of computing isomorphisms.

Definition 4. *Given a represented space S, the degree of categoricity $\text{CCat}(S, \text{El}_A)$ of S in a bundle $\text{El}_A : A_\bullet \to A$ is given by the following Weihrauch problem:*

- ***Input:*** *$(a, b) \in A^2$ such that S, $\text{El}_A^{-1}(a)$ and $\text{El}_A^{-1}(b)$ are homeomorphic*
- ***Output:*** *a homeomorphism witnessing $\text{El}_A^{-1}(a) \cong \text{El}^{-1}(b)$*

We say that S is computably categorical when $\text{CCat}(S, \text{El}_A) \leq_W \text{id}$. We omit El_A when it is clear from context.

For any S and bundle El_A, note that $\text{CCat}(S, \text{El}_A)$ is at most Σ^1_1.

Proposition 1. *For any space S and two intensionally equivalent bundles El_A and El_B, $\text{CCat}(S, \text{El}_A) \equiv_W \text{CCat}(S, \text{El}_B)$.*

Proof. Trivial.

Theorem 3. *$\{0,1\}^{\mathbb{N}}$ is computably categorical as a compact Polish space.*

Proof (Sketch). Given a set $X \in (\mathcal{K} \wedge \mathcal{V})([0,1]^\omega)$ that is homeomorphic to Cantor space, we can search for a cover of X by two basic open sets U_1, U_2, such that each U_i intersects X and the closures of U_1 and U_2 are disjoint. This search must succeed since X is homeomorphic to Cantor space. Now, the closures of U_1 and U_2 must again be homeomorphic to Cantor space, so that we can iterate this process, which allows us to construct an explicit isomorphism with the standard Cantor space $\{0,1\}^{\mathbb{N}}$.

The Degree of Categoricity of the Circle

As a somewhat more complicated example, we will investigate the degree of categoricity of the circle $\mathcal{S}^1$. We establish a slightly more general result, using the notation $(\mathcal{K} \wedge \mathcal{V})([0,1]^\omega)|_{\text{TC}}$ to denote the restriction of $(\mathcal{K} \wedge \mathcal{V})([0,1]^\omega)$ to sets which are classically homeomorphic to $\mathcal{S}^1$.

Theorem 4. *The following operations are Weihrauch equivalent:*

1. lim.
2. $\text{Image}^{-1} : (\mathcal{K} \wedge \mathcal{V})([0,1]^\omega)|_{\text{TC}} \rightrightarrows \mathcal{C}(\mathcal{S}_1, [0,1]^\omega)$ *mapping S to some $f : \mathcal{S}_1 \to [0,1]^\omega$ with $f[\mathcal{S}_1] = S$.*

3. Trace : $(\mathcal{K} \wedge \mathcal{V})([0,1]^\omega)|_{\mathrm{TC}} \rightrightarrows \mathcal{C}(\mathcal{S}_1, [0,1]^\omega)$ *mapping* S *to some injective* $f : \mathcal{S}_1 \to [0,1]^\omega$ *with* $f[\mathcal{S}_1] = S$.

Corollary 2. $\mathrm{CCat}(\mathcal{S}^1) \equiv_W \lim$.

Proof. A computable bijection between computably compact computably Hausdorff spaces is already a computable isomorphism.

The proof of the theorem follows later, after we have gathered a few crucial lemmas.

Definition 5. *For continuous* $f : S^1 \to [0,1]^\omega$ *and two disjoint open balls* $B_1, B_2 \in \mathcal{O}([0,1]^\omega)$*, we say that* f *oscillates between* B_1 *and* B_2 *at least* n *times if and only if there are* $t_1^1 < t_1^2 < t_2^1 \ldots < t_n^1 < t_n^2 \in [0,1]$ *such that* $f(t_i^j) \in B_j$ *(we view* S^1 *as a quotient of* $[0,1]$ *here). We say that* f *oscillates at most* n *times if it does not oscillate at least* $n+1$ *times.*

Lemma 3. *Given* $f \in \mathcal{C}(S^1, [0,1]^\omega)$*,* $x_1, x_2 \in [0,1]^\omega$ *and* $r_1, r_2 \in \mathbb{R}$ *with* $d(x_1, x_2) > r_1 + r_2$ *we can compute some* $n \in \mathbb{N}$ *such that* f *oscillates at most* n *times between* $B_1(x_1, r_1)$ *and* $B_2(x_2, r_2)$.

Proof. For some sufficiently large $n \in \mathbb{N}$, there must exist $t_1^1 < t_1^2 < t_2^1 \ldots < t_n^1 < t_n^2 \in S^1$ such that $f([t_i^1, t_i^2]) \subseteq \overline{B}(x_1, r_1)^C$, $f([t_i^2, t_{i+1}^2]) \subseteq \overline{B}(x_1, r_1)^C$ (with modular arithmetic on the indices). By overtness of S^1 and the availability of the compact-open representation on $\mathcal{C}(S^1, [0,1]^\omega)$, this is recognizable. Any such n is a valid answer, so we can just search for one that works.

Lemma 4. $\lim \leq_W \mathrm{Image}^{-1}$.

Proof. We actually show that $\widehat{\text{UpperBound}} \leq \mathrm{Image}^{-1}$ for UpperBound $:\subseteq \mathcal{O}(\mathbb{N}) \rightrightarrows \mathbb{N}$. The set we produce initially is a canonic unit circle, and we can provide its name in $(\mathcal{K} \wedge \mathcal{V})([0,1]^\omega)$ as a sequence of "tubes" shrinking with some known speed to it. For each of the ω-many UpperBound-instances we need to handle, we assign two open balls intersecting the unit circle, such that all balls are disjoint and the two balls for the same instance are next to each other. The situation at the site for a single UpperBound-instance is depicted in Fig. 1.

Once a new, larger value n gets enumerated into an UpperBound-instance we modify our current circle in a way that forces any function spanning it to oscillate at least n times between the two balls assigned to that instance as shown in Fig. 1.

For each site, we only need to do finitely many updates. This ensures that we actually end up with a topological circle. From any function $f : S^1 \to [0,1]^\omega$ having our circle as an image, we can extract upper bounds for how often it is oscillating at each side by Lemma 3, and these bounds are valid answers to the UpperBound-instances.

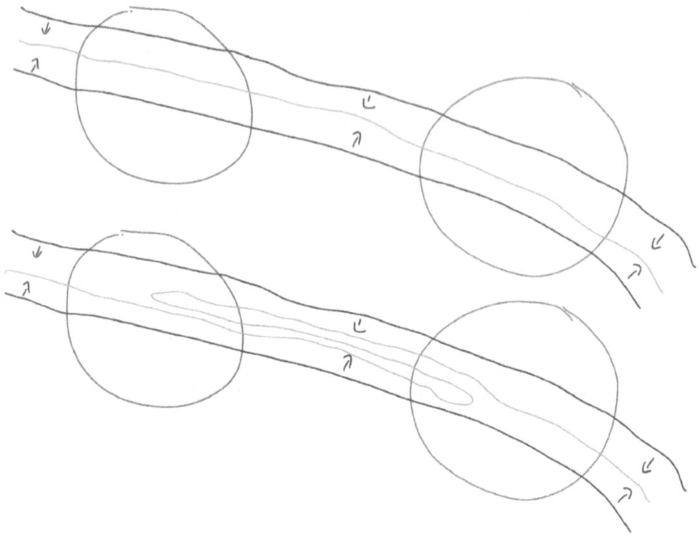

Fig. 1. The initial stage of the construction for Lemma 4 is shown at the top, the first update is depicted below. The tube going across the picture is the current approximation (which keeps shrinking). The actual circle we have in mind is depicted by the line lying in the tube and the two disjoint balls corresponding to one UpperBound-instance are depicted lying across the tube.

Recall that a *circular chain* is a sequence $B_0, \ldots, B_n$ of open balls such that B_i and B_{i+1} intersect, as do B_0 and B_n, and any other pair of balls is formally disjoint (i.e. the distance between their centres is greater than the sum of their radii). We will say that a circular chain $C_0, \ldots, C_m$ *refines* the circular chain $B_0, \ldots, B_n$ *without backtracking* if there is a partition $\{0, \ldots, m\} = \bigcup_{j \leq n} I_j$ into **intervals** such that $\forall i \in I_j\ \overline{C}_i \subseteq B_j$. A circular chain $B_0, \ldots, B_n$ *strictly covers* a set S if $S \subseteq \bigcup_{i \leq n} B_n$ and $\forall i \leq n\ S \cap B_n \neq \emptyset$. A circular chain has scale ε if each radius of a ball occurring in it is smaller than ε.

Lemma 5. *Let S be a topological circle. From a sequence of circular chains $(\mathfrak{B}^i)_{i \in \mathbb{N}}$ all strictly covering S such that $\mathfrak{B}^{i+1}$ has scale 2^{-i} and refines $\mathfrak{B}^i$ without backtracking, we can compute a homeomorphism $f : S^1 \to S$.*

Lemma 6. *Let S be a topological circle and consider a circular chain $(B_0, \ldots, B_n)$ strictly covering S. Either there are circular chains strictly covering S and refining $(B_0, \ldots, B_n)$ of arbitrarily small positive scale, or there exists a circular chain $(C_0, \ldots, C_m)$ strictly covering S such that $\forall j \leq m\ \exists i \leq n \overline{C}_j \subseteq B_i$ such that there are $j_0 < j_1 < j_2 < j_3$ and i with $\overline{C}_{j_0}, \overline{C}_{j_2} \subseteq B_i \setminus \overline{B}_{i+1}^C$ and $\overline{C}_{j_1}, \overline{C}_{j_3} \subseteq B_{i+1} \setminus \overline{B}_i^C$.*

Proof (Proof of Theorem 4). The reduction from (1) to (2) is the statement of Lemma 4. The reduction from (2) to (3) is trivial. To see that lim suffices to actually compute a homeomorphism, we consider the tree of circular chains strictly covering our circle where all balls have rational centres and radii ordered by refinement without backtracking on half the scale. By Lemma 5, having an infinite path through this tree suffices to find the desired homeomorphism. The basic parent-child relation in our tree is computably enumerable, and moreover, by invoking Lemma 6 we can recognize all dead ends. Thus, lim suffices to compute an infinite path.

4 Genericity

For $a \in A$, let $X_a = \mathrm{El}^{-1}(\{a\})$. We say that $a, b \in A$ are *homeomorphic* if $X_a \simeq X_b$. Morally, homeomorphic elements are names for the same object.

We call $U \subseteq A$ closed under homeo(morphism) if for all $a \in U$ and $b \in A$, if $X_b \simeq X_a$ then $b \in U$. We call $U \subseteq A$ *dense up to homeo(morphism)* if it intersects all open subsets of A which are closed under homeo.

Definition 6. *Let* $\mathrm{El}_A\colon A_\bullet \to A$ *be a bundle. Let* $\mathcal{C}$ *be a class of sets.*

Consider a space X *with* $X \simeq X_a$ *for some* $a \in A$. *The space* X *is called* $\mathcal{C}$*-generic as an element of* A *if for all* $C \subseteq A$ *belonging to* $\mathcal{C}$ *which are closed under homeomorphism and dense up to homeomorphism, we have* $a \in C$.

Morally, in the definition of genericity, we consider the represented space of homeomorphism types over a given space of spaces. A generic space is one whose homeomorphism type is contained in every dense set of class $\mathcal{C}$ in the space of homeomorphism types. This definition ensures that intensionally equivalent bundles have the same generic spaces.

We call a class $\mathcal{C}$ of sets a *pointclass* if it has the following properties:

1. The class $\mathcal{C}$ is closed under preimages of continuous maps.
2. If $e\colon \widetilde{A} \to A$ is a quotient map, then a set $C \subseteq A$ belongs to $\mathcal{C}$ if and only if $C = e\left(\widetilde{C}\right)$ for some $\widetilde{C} \in \mathcal{C}$ with $\widetilde{C} = e^{-1}\left(e\left(\widetilde{C}\right)\right)$.

Thus, a subset C of a represented space X is a member of a pointclass $\mathcal{C}$ if and only if it is the image of a set of names $\widetilde{C} \in \mathcal{C}$ under the representation, where $\widetilde{C}$ is extensional in the sense that a name of $x \in X$ belongs to $\widetilde{C}$ if and only if all names of x belong to $\widetilde{C}$. For a motivation of this definition, see [8,10,16].

Proposition 2. *Let* $\mathcal{C}$ *be a pointclass. Let* $\mathrm{El}_A\colon A_\bullet \to A$ *and* $\mathrm{El}_B\colon B_\bullet \to B$ *be equivalent bundles. Then a space* X *is* $\mathcal{C}$*-generic as an element of* A *if and only if it is* $\mathcal{C}$*-generic as an element of* B.

Proof. Assume that X is $\mathcal{C}$-generic as an element of A. We show that X is $\mathcal{C}$-generic as an element of B. By symmetry, this establishes the claim.

By assumption we have a diagram

$$\begin{array}{ccccc} A_\bullet & \xrightarrow{f_\bullet} & B_\bullet & \xrightarrow{g_\bullet} & A_\bullet \\ \downarrow{\scriptstyle \mathrm{El}_A} & & \downarrow{\scriptstyle \mathrm{El}_B} & & \downarrow{\scriptstyle \mathrm{El}_A} \\ A & \xrightarrow{f} & B & \xrightarrow{g} & A \end{array}$$

where all the squares are pullbacks.

Let $a \in A$ with $X \simeq X_a$. Let $\mathcal{C} \ni C \subseteq B$ be closed under homeo and dense up to homeo. We claim that $f^{-1}(C)$ is closed under homeo and dense up to homeo. Then $a \in f^{-1}(C)$ by assumption, so that $f(a) \in C$ with $X \simeq X_{f(a)}$.

Let a_0, a_1 be homeomorphic elements of A with $f(a_0) \in C$. Then $f(a_1)$ is homeomorphic to $f(a_0)$ so that $f(a_1) \in C$. Hence, $f^{-1}(C)$ is closed under homeo.

Now, let $U \in \mathcal{O}(A)$ be an open set which is closed under homeo. Then the set $g^{-1}(U)$ is closed under homeo. It hence intersects C. Let $b \in C \cap g^{-1}(U)$. Then $g(b) \in U$. We claim that $g(b) \in f^{-1}(C)$. Indeed, $X_{f \circ g(b)} \simeq X_b$, so that $f \circ g(b) \in C$ since C is closed under homeo.

Proposition 3. *Let $\mathcal{C}$ be a pointclass. Let $\mathrm{El}_A\colon A_\bullet \to A$ be a bundle. Let $\mathrm{El}_{\widetilde{A}}\colon \widetilde{A}_\bullet \to \widetilde{A}$ be a projective reindexing of A. Then a space X is $\mathcal{C}$-generic as an element of A if and only if it is $\mathcal{C}$-generic as an element of $\widetilde{A}$.*

Proof. Assume that X is $\mathcal{C}$-generic as an element of A. Pick $a \in A$ with $X \simeq X_a$. Pick some $\widetilde{a} \in \widetilde{A}$ with $e_A(\widetilde{a}) = a$. Then $X \simeq X_{\widetilde{a}}$. Let $\mathcal{C} \ni \widetilde{C} \subseteq \widetilde{A}$ be closed under homeo and dense up to homeo. Since $\mathcal{C}$ is closed under homeo it particularly satisfies $\mathcal{C} = e_A^{-1}(e_A(\mathcal{C}))$. Hence, since $\mathcal{C}$ is a pointclass, we have $e_A(\mathcal{C}) \in \mathcal{C}$. It is easy to see that $e_A(\mathcal{C})$ is closed under homeo and dense up to homeo. It follows that $e_A(\mathcal{C})$ contains a. Hence $\mathcal{C} = e_A^{-1}(e_A(\mathcal{C}))$ contains $\widetilde{a}$.

Conversely, assume that X is $\mathcal{C}$-generic as an element of $\widetilde{A}$. Pick $\widetilde{a} \in \widetilde{A}$ with $e_A(\widetilde{a}) = a$. Let C be closed under homeo and dense up to homeo. Then the set $e_A^{-1}(C)$ is closed under homeo and dense up to homeo. It hence contains $\widetilde{a}$, so that $C = e_A(e_A^{-1}(C))$ contains a.

Corollary 3. *Let $\mathcal{C}$ be a pointclass. Let $\mathrm{El}_A\colon A_\bullet \to A$ and $\mathrm{El}_B\colon B_\bullet \to B$ be intensionally equivalent bundles. Then a space X is $\mathcal{C}$-generic as an element of A if and only if it is $\mathcal{C}$-generic as an element of B.*

We work with density up to homeo rather than density, because the former is better behaved under pullbacks. For sufficiently nice bundles, there is a close connection between density and density up to homeo. Say that $\mathrm{El}\colon A_\bullet \to A$ is an *open representation* if the map

$$[\cdot]\colon A \to \mathcal{V}(A),\ a \mapsto \{b \in \mathrm{A} \mid X_a \simeq X_b\}$$

is computable.

Proposition 4. *Let $\mathrm{El}_A\colon A_\bullet \to A$ be an open representation. Let $X \subseteq A$ be closed under homeo. Then X is dense up to homeo if and only if it is dense.*

Proof. Let $U \in \mathcal{O}(A)$ be an open set. Since El_A is an open representation, the set

$$V = \{a \in A \mid [a] \cap U \neq \emptyset\}$$

is open. Since X is dense up to homeo, there exists $a \in V \cap X$. By construction, there exists $b \in U$ with $X_a \simeq X_b$. Since X is closed under homeo we have $b \in X$.

Our bundles for Polish spaces and compact Polish spaces are open representations. We illustrate this with two examples

Proposition 5. *The bundle* $\mathrm{PM}_\bullet \to \mathrm{PM}$ *is an open representation.*

Proof (Sketch). Basic open sets in PM yield constraints on a pseudometric of the form

$$d(i_1, j_1) \in [a_1, b_1] \wedge \cdots \wedge d(i_m, j_m) \in [a_m, b_m].$$

All these constraints reveal about a space up to homeomorphism is that the space has at least k distinct points for some $k \geq 1$.

Hence, we can compute the homeomorphism class of such a space as an overt set essentially as follows: given a basic open set that requires the space to have k distinct point, accept the set if and only if our space has at least k distinct points.

Proposition 6. *The bundle* $(\mathcal{V} \wedge \mathcal{K})([0,1]^\omega)_\bullet \to (\mathcal{V} \wedge \mathcal{K})([0,1]^\omega)$ *is an open representation.*

Proof (Sketch). Basic open sets in $(\mathcal{V} \wedge \mathcal{K})([0,1]^\omega)$ correspond to minimal covers of a set by balls with rational centres and radii. All these constraints reveal about a space up to homeomorphism is that the space has at least k connected components for some $k \geq 1$.

Hence, we can compute the homeomorphism class of a space as an overt set essentially as follows: given a basic open set that requires the space to have k connected components, accept the set if and only if our space has at least k connected components.

We now turn to the problem of classifying generic compact Polish spaces. In view of Corollary 3 and Theorem 2 we are justified in calling a space $\mathcal{C}$-generic as a compact Polish space if it is generic as an element of any of our intensionally equivalent representations of compact Polish spaces. It follows from the proof of Proposition 6 that $\mathcal{O}$-genericity is rather weak: a compact Polish space is $\mathcal{O}$-generic if and only if it has infinitely many connected components. We next show that Cantor space is up to homeomorphism the only Π^0_2-generic compact Polish space.

Proposition 7. *Being homeomorphic to Cantor space is a* Π^0_2*-property of compact Polish spaces that is closed under homeo and dense up to homeo.*

Proof. We encode compact Polish spaces by the bundle with base $(\mathcal{V} \wedge \mathcal{K})([0,1]^\omega)$.

Let $(x_n)_n$ be a computable dense sequence in $[0,1]^\omega$. We denote by $B(x,r) \subseteq [0,1]^\omega$ the open ball of radius r and by $\overline{B}(x,r)$ the corresponding closed ball. Observe that the closed ball is the closure of the open ball. Consider the following Π^0_2-property:

$$\begin{aligned}
&\forall n \in \mathbb{N}.\forall r \in \mathbb{Q}_{>0}.\exists m \in \mathbb{N}.\exists s \in \mathbb{Q}_{>0}.\exists m' \in \mathbb{N}.\exists s \in \mathbb{Q}_{>0}.\\
&\quad \Big(B(x_n,r) \cap X \neq \emptyset \rightarrow \big(\overline{B}(x_n,r) \cap X\big) \subseteq B(x_m,s) \cup B(x_{m'},s')\\
&\qquad\qquad \wedge B(x_m,s) \cap X \neq \emptyset\\
&\qquad\qquad \wedge B(x_{m'},s') \cap X \neq \emptyset\\
&\qquad\qquad \wedge \overline{B}(x_m,s) \cap \overline{B}(x_{m'},s') \cap X = \emptyset\Big).
\end{aligned}$$

Less formally, this property is stating that for every non-empty open rational ball, the corresponding closed ball can be covered by two non-empty open balls such that the corresponding closed balls are disjoint. It is easy to see that a set X has the above property if and only if it is homeomorphic to Cantor space. That this property is closed under homeo is obvious. It is dense up to homeo, since every open set that is closed under homeo contains a copy of Cantor space.

Proposition 8. *Every Π^0_2-subset of $(\mathcal{V} \wedge \mathcal{K})([0,1]^\omega)$ that is closed under homeo and dense up to homeo contains a copy (and hence every copy) of Cantor space.*

Proof. Proposition 7 establishes that the set $C_{\{0,1\}^\mathbb{N}}$ of all spaces homeomorphic to Cantor space is a dense Π^0_2-subset of $(\mathcal{V} \wedge \mathcal{K})([0,1]^\omega)$. Let C be a Π^0_2-subset of $(\mathcal{V} \wedge \mathcal{K})([0,1]^\omega)$ which is dense up to homeo and closed under homeo. Then by Proposition 4, C is dense. By the Baire category theorem, $C \cap C_{\{0,1\}^\mathbb{N}}$ is dense. If C does not contain a copy of Cantor space, then $C \cap C_{\{0,1\}^\mathbb{N}} = \emptyset$, contradiction!

Propositions 7 and 8 together establish:

Theorem 5. *Cantor space is, up to homeomorphism, the unique Π^0_2-generic compact Polish space.*

5 Further Questions

Further classes of represented spaces where finding suitable representations as bundles would be of immense interest include the QuasiPolish and the coPolish spaces. For the former, much of the groundwork was laid in [6], where it was shown that we can represent QuasiPolish spaces as spaces of ideals of preorders. Subsequent work by de Brecht ([2–4]) showed that many of the constructions we would want to perform on QuasiPolish spaces are indeed effective using this encoding.

The setting for coPolish spaces is less well understood. In particular, it is not even completely clear what a "computable coPolish space" should be. The primary objective here is to check to what extend the various characterizations (such as those provided in [18]) of coPolish spaces still work in the effective setting. The investigation needed here will require a representation of the QuasiPolish spaces, as we would want the representation of a coPolish space $\mathbf{X}$ to enable us to compute the QuasiPolish space $\mathcal{C}(\mathbf{X}, \mathbf{Y})$ for any given QuasiPolish space $\mathbf{Y}$.

Brattka discusses how to define computability of separable and some non-separable Banach spaces in [1], with a key insight being that if we represent separable Banach spaces as Polish spaces with additional structure, their duals naturally carry a coPolish topology. By expressing this in our framework, we would then have a general framework to study Banach spaces and their genericity, computable categoricity and so on.

Acknowledgments. This project was started during the BIRS meeting "A Convergence of Computable Structure Theory, Analysis, and Randomness" in 2023 and furthered during the Dagstuhl meeting 25131 "Weihrauch Complexity: Structuring the Realm of Non-Computability" in 2025, and we are grateful to the participants of those meetings for fruitful discussions.

The authors grant to Swansea University a non-exclusive, irrevocable, sub-licensable, worldwide license to make the accepted manuscript available on its institutional repository.

Appendix: proof of Lemma 1

Before moving on to the proof of Lemma 1, we first explain how the notions of Sect. 2.2 are actual general categorical notions. Then we use that to prove Lemma 1 via abstract nonsense. We hope this demonstrates in particular that the approach we use for represented spaces of represented spaces could be also done mutatis mutandi for other similar settings such as multirepresented spaces of multirepresented spaces.

Definition 7. *In a category, a* regular epimorphism *is a coequalizer of some pair of morphisms.*

In less categorical terms, regular epimorphisms are those surjections that arise from quotient maps. A characterization of regular epimorphisms in (multi)-represented spaces is that they have a multi-valued inverse; this is what we called *quotient maps* in Definition 2.

Example 3. The injections $2 \rightarrowtail \mathbb{S}$ are epimorphic but not regular.

Definition 8. *An object X of a category is called* (regular) projective *when it has the left lifting property against regular epimorphisms, that is, whenever we have a regular epimorphism $e : Z \twoheadrightarrow Y$ and a map $f : X \to Y$, there exists some $g : X \to Z$ such that $f = e \circ g$.*

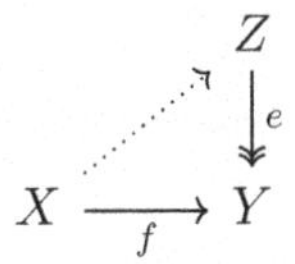

A (regular) projective cover *of an object X is a projective object $\widetilde{X}$ together with a regular epimorphism $\widetilde{X} \twoheadrightarrow X$*

In represented spaces, regular projective objects coincide with subspaces of $\mathbb{N}^{\mathbb{N}}$. It is also the case that a representation map $\delta_X :\subseteq \mathbb{N}^{\mathbb{N}} \twoheadrightarrow X$ determines a regular projective cover of the space X by its domain. Hence Definition 2 says that a projective reindexing is really a pullback along a regular projective cover.

Remark 1. In represented spaces, for any projective cover $c : \widetilde{X} \twoheadrightarrow X$ and multivalued function $f : X \rightrightarrows Y$, a computable realizer for f allows to build a computable map $\widetilde{f} : \widetilde{X} \to Y$ such that $\widetilde{f}(z) \in f(c(z))$ for every $z \in \widetilde{X}$.

Lemma 1. *Any two projective reindexings of a bundle are equivalent.*

Proof. Let $\mathrm{El}_A : A_\bullet \to A$ be a bundle and $\mathrm{El}_B, \mathrm{El}_C$ be projective reindexings, so we have regular epimorphisms e_B and e_C as in the following diagram.

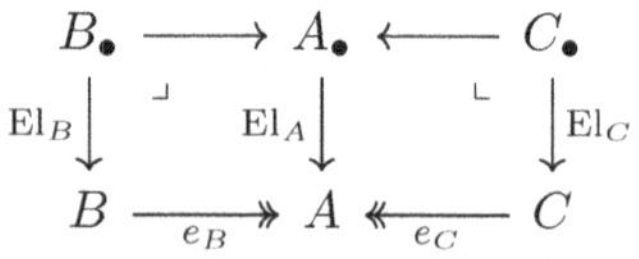

To conclude, it suffices to show that El_B embeds into El_C. Since B is regular projective and e_C is a regular epimorphism, there is a map $f : B \to C$ such that $e_C \circ f = e_B$. Then, using the fact that the right square is a pullback, we know there exists $f_\bullet$ such that the following diagram commute.

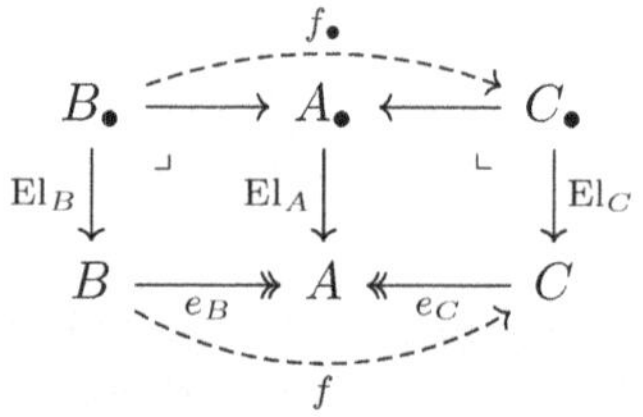

Then it remains to show that $f, f_\bullet, \mathrm{El}_B$ and El_C form a pullback square; this follows easily from the fact that the left square in the original diagram is a pullback.

References

1. Brattka, V.: Computability on non-separable Banach spaces and Landau's theorem. In: Crosilla, L., Schuster, P.M. (eds.) From sets and types to topology and analysis - Towards practicable foundations for constructive mathematics, Oxford logic guides, vol. 48. Oxford University Press (2005)

2. de Brecht, M.: Some notes on spaces of ideals and computable topology. In: Anselmo, M., Della Vedova, G., Manea, F., Pauly, A. (eds.) Beyond the Horizon of Computability, pp. 26–37. Springer, Cham (2020)
3. de Brecht, M.: The category of quasi-Polish spaces as a represented space. In: 68 (2021)
4. de Brecht, M.: Constructing the space of valuations of a quasi-Polish space as a space of ideals. In: Manea, F., Simpson, A. (eds.) 30th EACSL Annual Conference on Computer Science Logic, CSL 2022, February 14-19, 2022, Göttingen, Germany (Virtual Conference). LIPIcs, vol. 216, pp. 9:1–9:10. Schloss Dagstuhl - Leibniz-Zentrum für Informatik (2022). https://doi.org/10.4230/LIPICS.CSL.2022.9
5. de Brecht, M., Kihara, T., Selivanov, V.: Ideal presentations and numberings of some classes of effective quasi-Polish spaces (2023)
6. de Brecht, M., Pauly, A., Schröder, M.: Overt choice. Computability (2020)
7. Brown, T., McNicholl, T.H.: Analytic computable structure theory and L^p-spaces part 2. Arch. Math. Logic, 427–443 (2019). https://doi.org/10.1007/s00153-019-00697-4
8. Callard, A., Hoyrup, M.: Descriptive complexity on non-Polish spaces. In: Paul, C., Bläser, M. (eds.) STACS 2020, March 10-13, 2020, Montpellier, France. LIPIcs, vol. 154, pp. 8:1–8:16 (2020). https://doi.org/10.4230/LIPICS.STACS.2020.8
9. Gregoriades, V., Kispéter, T., Pauly, A.: A comparison of concepts from computable analysis and effective descriptive set theory. Math. Struct. Comput. Sci. **27**(8), 1414–1436 (2017). https://doi.org/10.1017/S0960129516000128
10. Hoyrup, M.: Descriptive complexity on non-Polish spaces II. In: Czumaj, A., Dawar, A., Merelli, E. (eds.) 47th International Colloquium on Automata, Languages, and Programming, ICALP 2020, July 8-11, 2020, Saarbrücken, Germany (Virtual Conference). LIPIcs, vol. 168, pp. 132:1–132:17. Schloss Dagstuhl - Leibniz-Zentrum für Informatik (2020). https://doi.org/10.4230/LIPICS.ICALP.2020.132
11. Jacobs, B.: Categorical Logic and Type Theory. Studies in Logic and the Foundations of Mathematics, Elsevier (2001)
12. McNicholl, T.H.: Computable copies of ℓ^p. Computability **6**(4), 391–408 (2017). https://doi.org/10.3233/com-160065
13. Melnikov, A.G.: Computably isometric spaces. J. Symb. Log. **78**(4), 1055–1085 (2013). https://doi.org/10.2178/jsl.7804030
14. Melnikov, A.G., Ng, K.M.: Computable structures and operations on the space of continuous functions. Fundamenta Mathematicae **233**(2), 101–141 (2016). http://eudml.org/doc/283338
15. Pauly, A.: On the topological aspects of the theory of represented spaces. Computability **5**(2), 159–180 (2016). https://doi.org/10.3233/COM-150049
16. Pauly, A., de Brecht, M.: Descriptive set theory in the category of represented spaces. In: LICS, pp. 438–449. IEEE Computer Society (2015). https://doi.org/10.1109/LICS.2015.48
17. Rettinger, R., Weihrauch, K.: Products of effective topological spaces and a uniformly computable Tychonoff theorem. Logical Methods Comput. Sci. **9**(4) (2013). https://doi.org/10.2168/LMCS-9(4:14)2013
18. Schröder, M.: Spaces allowing type-2 complexity theory revisited. Math. Log. Q. **50**(4–5), 443–459 (2004). https://doi.org/10.1002/MALQ.200310111
19. Streicher, T.: Fibred categories à la Jean Bénabou. arXiv preprint arXiv:1801.02927 (2018)

Computation with Real Numbers and Continuous-Time Dynamical Systems

Daniel S. Graça[1,2(✉)]

[1] Universidade do Algarve, C. Gambelas, 8005-139 Faro, Portugal
dgraca@ualg.pt

[2] Center for Research and Development in Mathematics and Applications (CIDMA), Department of Mathematics, University of Aveiro, 3810-193 Aveiro, Portugal

Abstract. In this paper we review some results about the interconnections between computation with real numbers and continuous dynamical systems. In particular, we take two complementary approaches: (i) to use standard computational models or theories such as Turing machines or computable analysis to understand which properties of continuous dynamical systems can be computed and (ii) to use continuous dynamical systems directly as models of computation and study their computational power. We will be particularly interested in continuous dynamical systems defined with analytic ordinary differential equations and, in particular, in dynamical systems defined with polynomial ordinary differential equations.

Keywords: Dynamical systems · Computation with real numbers · Computable Analysis · General Purpose Analog Computer · Polynomial ordinary differential equations

1 Introduction

Dynamical systems are mathematical models which can be used in a wide range of applications. Due to their versatility, dynamical systems have been the focus of abundant research, but much remains to be understood about them. Since dynamical systems are able to model many phenomena, it is not surprising that even very simple dynamical systems, such as chaotic systems, can display complex behavior. A common practical way to gain insights into the underlying complex behavior arising in the context of dynamical systems is to use computer simulations. For example, the pioneering work of Edward Lorenz [41] where he numerically studied a weather model, introduced notions such as sensitive dependence on initial conditions (the so-called butterfly effect) and a butterfly shaped attractor currently known as Lorenz attractor (see Fig. 1).

Given that computer simulations of dynamical systems typically rely on floating-point numbers, the accuracy of those simulations can be put into question, especially when trying to understand the long-term behavior of dynamical

© The Author(s), under exclusive license to Springer Nature Switzerland AG 2025
A. Beckmann et al. (Eds.): CiE 2025, LNCS 15764, pp. 62–75, 2025.
https://doi.org/10.1007/978-3-031-95908-0_5

systems. For example, in his well-known list of open problems for the next century, the Fields medalist Stephen Smale, asked whether the Lorenz attractor exists on the system originally studied by Edward Lorenz [53]. This question is pertinent because, although numerical simulations suggested the existence of an attractor such as the one found by Lorenz to which the dynamics converge, the use of floating-point numbers in the computations coupled with sensitive dependence on initial conditions prevented any definitive conclusion. Finally, in [54], using normal form theory and rigorous numerical analysis, Warwick Tucker provided an affirmative solution to Smale's problem.

The above situation suggests that it is important to understand in which contexts one can reliably compute some property concerning a dynamical system. This is especially relevant nowadays, since computers are used in an increasingly number of applications, many of them involving real numbers. In a first step it is natural to study those questions from a computability perspective, to understand which of the properties/features can be computed, independently of the computational resources being used. In a second step, after the situation from a computability perspective is clearer, it makes sense to refine those results at a computational complexity level, where the computational resources are taken into account.

Here we survey several results in that direction in Sect. 3. It should be noted that this review is far from exhaustive, as other results exist such as in [15,16,24,25,44], among many others which establish connections between dynamical systems and the theory of computation. See the recent surveys available in the book [13], such as [12,30,50], or [35] for a more complete discussion of results and references.

On the other hand, with the advent of novel computational models such as quantum computation, which may have a significant technological impact, it is interesting to understand what is the computational power of various dynamical systems. It is well known that continuous dynamical systems can be used directly to compute, and even achieve super-Turing power in certain cases (see e.g. [2,6], among many other references). Moreover, the analog computers in use before the explosive growth of digital computers, known as differential analyzers [18], can be modeled as polynomial ordinary differential equations and be used to compute. Hence, it would be interesting to understand which are the computational capabilities of such types of analog computers/dynamical systems. Some results in that direction are reviewed in Sect. 4 for a particular class of continuous-time dynamical systems.

2 Preliminaries

In this section we review some basic notions that will be needed throughout the paper, namely about dynamical systems and computable analysis. We assume that the reader is familiar with the classical theory of computability and computational complexity (see e.g. [52] for more details).

2.1 Dynamical Systems

In essence, a dynamical system consists on a set of points, called the state space and on a deterministic evolution rule which specifies how points in the state space, known as states, evolve along time. Here we will only consider dynamical systems where the state space is a subset of $\mathbb{R}^n$ and the time set is $\mathbb{T} = \mathbb{Z}$ or $\mathbb{T} = \mathbb{R}$. More formally (see [34, p. 141]):

Definition 1. *A smooth dynamical system on $S \subseteq \mathbb{R}^n$ over the time set $\mathbb{T}$ is a pair $(\mathbb{T}, S, \phi)$, where $\phi : \mathbb{T}\times S \to S$ is a continuously differentiable function satisfying (for simplicity, we write $\phi_t(x) = \phi(t, x)$):*

1. *$\phi_0 : S \to S$ is the identity map, i.e. $\phi_0(x) = x$ for all $x \in S$;*
2. *One has $\phi_t \circ \phi_s(x) = \phi_{t+s}(x)$ for all $x, s \in \mathbb{T}$.*

When the time set $\mathbb{T}$ is $\mathbb{Z}$ we are in the presence of a discrete-time dynamical system (sometimes known simply as a discrete dynamical system). In that case, we just need to study the iteration of the map $F = \phi_1$ to understand the behavior of ϕ_t for $t \geq 0$. Indeed, Definition 1 implies that

$$\phi_t(x) = \underbrace{\phi_1 \circ \ldots \circ \phi_1}_{t \text{ times}}(x) = \underbrace{F \circ \ldots \circ F}_{t \text{ times}}(x) = F^t(x).$$

When the time set is $\mathbb{R}$, we are in the presence of a continuous-time dynamical system (sometimes known simply as a continuous dynamical system). In that case, we can define a vector field $f : \mathbb{R}^n \to \mathbb{R}^n$ given by

$$f(x) = \left.\frac{d}{dt}\right|_{t=0} \phi_t(x)$$

and to understand the behavior of ϕ_t for $t \geq 0$ we just need to understand the behavior of the solutions of the ordinary differential equation (ODE)

$$x' = f(x) \tag{1}$$

Notice that the ODE above is autonomous, i.e. the vector field f does not depend on the time t (a non-autonomous ODE $x' = f(t, x)$, where $x = (x_1, \cdots, x_n) \in \mathbb{R}^n$ can be converted into an autonomous ODE by increasing the dimension by one and taking $x'_{n+1} = 1$ and $x_{n+1}(0) = 0$, since this yields $x_{n+1}(t) = t$). The trajectory of a dynamical system $(\mathbb{T}, S, \phi)$, starting on the initial point $x \in S$, is the function $\phi(\cdot, x) : \mathbb{T} \to S \subseteq \mathbb{R}^n$, while an orbit of the dynamical system is a set $\{\phi_t(x) : t \in \mathbb{T}\}$ for some $x \in S$ (sometimes these two terms are used interchangeably in the literature). We may focus on the forward (backward) orbit/trajectory by just considering values of $t \geq 0$ ($t \leq 0$, respectively). Unless otherwise stated, we will only consider forward orbits/trajectories and call these simply as orbits/trajectories. We also note that Definition 1 implies that ϕ_t has a C^1 inverse given by ϕ_{-t} since $\phi_t \circ \phi_{-t}(x) = \phi_{t+(-t)}(x) = \phi_0(x) = x$.

A common goal of dynamical systems theory is to understand the long-term (asymptotic) behavior of a system and the types of "steady states" that it can

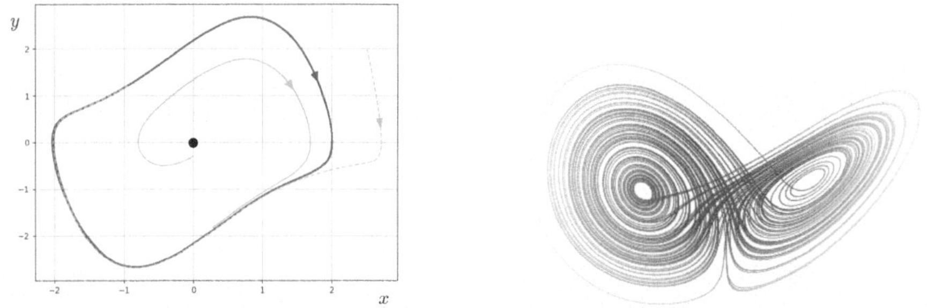

Fig. 1. The left figure represents the van der Pol system. It features a periodic orbit, represented in blue and an equilibrium point at the origin, which is unstable (it is called a source) since any trajectory starting nearby the origin will move away from the origin. The right image depicts the Lorenz attractor, a type of strange attractor. (Color figure online)

reach. Those steady states can take several forms. They can consist of equilibrium points, when $\phi_t(x) = x$ for all $t \in \mathbb{T}$, of periodic orbits, when $\phi_t(x) = x$ for some $t > 0$, or may take other more complex shapes such as strange attractors of which the Lorenz attractor is an example (see Fig. 1 for graphical depictions). In particular, a limit cycle is an isolated periodic orbit. An attractor is typically seen as the set of steady states to which the system evolves. Depending on the approach taken, one may define attractors in different ways (see e.g. [17, Chapter 1] or [49, Section 2.1]). The notion we take in this paper is the following. An attractor is a steady state that has a neighborhood with the property that any trajectory starting in this neighborhood converges to the steady state. We also say that this steady state is attracting. It is also required that the attractor cannot be decomposed into proper subsets which are themselves attractors. An attracting equilibrium point is called a sink. The set of all points which converge to an attractor is called the basin of attraction of that attractor.

2.2 Computable Analysis

To compute over continuous structures such as $\mathbb{R}^n$ we need to extend the classical theory of computation, since the latter is focused on discrete structures. This can be achieved using computable analysis, which originates in Turing's seminal paper [55]. Roughly, an element is computable if it can be approximated by a computable fast converging sequence of simple elements, in the sense that each of the elements of the sequence should have a finite description. For example, computable real numbers can be defined as those which are the limit of a computable fast converging sequence of rational numbers.

Definition 2. *A rational-valued function $\phi : \mathbb{N} \to \mathbb{Q}$ is called an oracle (or a name) for a real number $x \in \mathbb{R}$ if it satisfies the condition $|\phi(n) - x| \leq 2^{-n}$ for all $n \in \mathbb{N}$.*

Definition 3. *A real number is computable if it admits a computable oracle.*

We can also similarly define in a straightforward manner an oracle $\phi : \mathbb{N} \to \mathbb{Q}^m$ for an element $x \in \mathbb{R}^m$ and thus define computable elements in $\mathbb{R}^m$. The set of computable real numbers includes all the rationals numbers and all the usual numbers used in analysis such as $\pi, e, \sqrt{2}$, etc.

We can also compute with functions defined over the real numbers.

Definition 4. *Let $f : \mathbb{R}^m \to \mathbb{R}$ be a function.*

1. *The function f is computable if there is an oracle Turing machine that, when provided an oracle for an element $x \in \mathbb{R}^m$ provides as output an oracle for $f(x)$. In other words, given an oracle for x and an input $n \in \mathbb{N}$, the Turing machine outputs a rational q satisfying $|q - f(x)| \leq 2^{-n}$.*
2. *Assuming that f is C^k, the function f is C^k-computable if there is an oracle Turing machine that, when provided with an oracle for x and an input $n \in \mathbb{N}$, the Turing machine outputs $q = (q_0, q_1, \ldots, q_k)$, where $q_i \in \mathbb{Q}^m$ satisfies $\left\|q_i - D^i f(x)\right\| \leq 2^{-n}$ for all $0 \leq i \leq k$.*

Again, we can define in a similar way (C^k)-computable functions $f : \mathbb{R}^m \to \mathbb{R}^n$ by considering computability component wise. We note that we could have first defined names for real-valued functions, as done for real numbers, and then take as computable function those functions with computable names. For example, a name for a function $f : \mathbb{R}^m \to \mathbb{R}^n$ is a list of all pairs of open rational boxes (I, J) such that $f(\overline{I}) \subseteq J$, where $\overline{I}$ denotes the closure of the open set I. However, we think that the (equivalent) Definition 4 is more natural.

We can also define other computable structures such as open and closed sets. Roughly, a (planar) set $S \subseteq \mathbb{R}^2$ is computable if it can be rigorously drawn in a computer screen, using pixels, with arbitrary magnification (similar ideas apply for higher dimensional sets). See [14] for more details (again, we could use a computability approach using names for computable sets, but we do not follow that path here. The interested reader can check e.g. [14]).

Definition 5. *A closed subset $A \subseteq \mathbb{R}^m$ is computable if it is empty or if the distance function $d_A : \mathbb{R}^m \to \mathbb{R}$ given by*

$$d_A(x) = \inf_{y \in A} \|x - y\|$$

is computable. An open subset $U \subseteq \mathbb{R}^m$ is computable if its complement (which is a closed set) is computable.

3 Computing over Continuous-Time Dynamical Systems

In this section we consider the computability of several aspects of continuous-time dynamical systems, seen as ODEs (see Sect. 2.1).

3.1 Computability of Ordinary Differential Equations

Consider the initial-value problem (IVP)

$$\begin{cases} x' = f(x) \\ x(t_0) = x_0 \end{cases} \tag{2}$$

associated to the ODE (1), where $f : \mathbb{R}^n \to \mathbb{R}^n$. A first problem is to determine the state of the IVP at a given time. This can be achieved by understanding in which conditions the solution of (2) is computable. It is well known that, by Peano's existence theorem, (2) has at least a solution if f is continuous. However, when f is continuous, it was shown by Aberth in [1] (a different proof was provided in [47]) that (2) may have no computable solution when f is computable and when both t_0 and x_0 are 0. This result was later shown to hold even if f is assumed to be polynomial-time computable [39] (see [40] for the definition of a real function computable in polynomial time).

Theorem 1. *There is a polynomial-time computable function* $f : [0,1] \times [-1,1] \to \mathbb{R}$ *such that the initial-value problem* $x' = f(t,x)$, $x(0) = 0$, *does not have a computable solution on* $[0,\delta]$ *for any* $\delta > 0$.

If we additionally require that f satisfies a Lipschitz condition, i.e. if there is some $K > 0$ (the Lipschitz constant) such that

$$\|f(x) - f(y)\| \leq K \|x - y\| \tag{3}$$

for all x, y, then by the Picard-Lindelöf theorem the solution of (2) exists and is unique. The proof of this theorem, for example using Picard's scheme of successive approximations, is constructive and thus allows us to compute the solution of (2) when f, t_0, and x_0 are all computable, as noted in [40]. Note that if f is C^1 over $\mathbb{R}^n$, then it is locally Lipschitz, i.e. for any $x \in \mathbb{R}^n$ there is a neighborhood $U \subseteq \mathbb{R}^n$ such that for any U there is a (local) Lipschitz constant such that (3) holds. This was exploited in [33] to show that the solution of (2) is always computable when f is C^1-computable over $\mathbb{R}^n$, since the information about the derivative of f allows us to compute local Lipschitz constants, which are then used to locally compute solutions of (3) that are then sequentially extended.

But what happens when f is continuous and does not satisfy the Lipschitz condition (3)? It is also well known that in this setting, although (2) has a guaranteed solution by Peano's existence theorem, it may not be unique. In turns out that uniqueness is a crucial ingredient, alongside continuity, to ensure computability of the solution, as shown in a theorem of [23] (see also [9] for a similar result involving transfinite recursion). In that paper (and also in the previous results presented in [33]) the related problem of computing the maximal interval of existence of the solution of (2) is also studied.

We recall that given an IVP (2), the solution of (2) is defined in a maximal interval of existence (α, β). We recall (see e.g. [45, pp. 89–91]) that a maximal interval of existence (α, β) has the property that either $\beta = +\infty$ ($\alpha = -\infty$) or the solution of (2) gets unbounded as t approaches β (α, respectively).

Note that the type of restriction that applies to β may be different from the condition that applies to α. For example, the IVP

$$\begin{cases} x' = -x^2 \\ x(1) = 1 \end{cases}$$

has a unique solution, given by $x(t) = 1/t$, with the associated maximal interval of existence $(0, +\infty)$.

Theorem 2 ([23]). *Let $f : \mathbb{R}^n \to \mathbb{R}^n$ be continuous. Assume that the initial-value problem (2) has a unique solution, defined on the maximal interval of existence (α, β). Then:*

1. *The operator which maps (f, t_0, x_0) to the solution of (2) is computable. In particular, if f, t_0, x_0 are computable, then so is the solution x of (2).*
2. *The operator which maps (f, t_0, x_0) to the maximal interval (α, β) is semi-computable.*

Some clarifications are in order in the above theorem. In item 1, when we state that the operator is computable that means that there is an oracle Turing machine that, when provided an oracle for (f, t_0, x_0), it will output an oracle for the solution of (2). For item 2, semi-computability means that we can compute a sequence of rational numbers in (α, β) which converges to β (also another which converges to α), without knowing which is the rate of convergence, i.e. the information is partial and hence the name semi-computability.

In Theorem 2 we could ask whether the maximal interval of existence (α, β) is computable when f, t_0, x_0 are computable. The answer is negative, even if f is well behaved (analytic), as shown in the following result from [33] (see also [48] for more results related to the maximal interval of existence).

Theorem 3 ([33]). *There is an analytic and computable function $f : \mathbb{R} \to \mathbb{R}$ such that the unique solution of (2) with the initial condition $x(0) = 0$ is defined in a non-computable maximal interval of existence.*

We note that several results about the computational complexity of obtaining the solution of the IVP (2) are known. For example, if D is the unit ball centered on the origin, then the initial-value problem (2) can have arbitrarily high computational complexity, even if f is polynomial-time computable, . This follows from a result of [40], based on another result from [42]. However, again if we impose a right-Lipschitz condition on f, i.e. if f satisfies

$$\|f(t, x) - f(t, y)\| \leq K \|x - y\|$$

for all t, x, y, then the complexity of the IVP associated to $x' = f(t, x)$ drops significantly. Ko noted in [40] that if f is polynomial-time (right-)Lipschitz, then the solution of the IVP can be computed in polynomial space, thus providing an upper bound for the complexity. In [36], Kawamura shows a lower bound by showing that there is a polynomial-time computable f such that the solution of the IVP is PSPACE-complete. See also [37,38,43], for some related results about the computational complexity of ODEs on compact domains.

3.2 Computability of Properties Related to Asymptotic Behavior

A question that we may to wish to answer about the asymptotic behavior of a continuous-time dynamical system is, given an ODE (1), whether we can (uniformly) compute its simplest steady states, i.e. whether we can uniformly compute all the equilibrium points of (1). Since equilibrium points of (1) correspond to zeros of f, this is equivalent to uniformly determine the zeros of a function f. However, it is well known (see e.g. [14, Theorem 7.10]) that in general one cannot uniformly compute the set of all zeros of the function f. A natural question is then to tell under which conditions those zeros become uniformly computable. The next result from [31] shows that, in the closed unit ball, it is sufficient that Df is invertible at all zeros of f and that no zero lies on the boundary of the unit ball, to ensure that the zeros of f can be uniformly computed.

Note that x_e is an equilibrium point of (1) if $f(x_e) = 0$. Furthermore, the equilibrium point is called hyperbolic if none of the eigenvalues of $Df(x_e)$ has zero real part. We remark that if x_e is an hyperbolic equilibrium point, then $Df(x_0)$ is invertible.

It is possible to tell what type of equilibrium an hyperbolic equilibrium point is from its eigenvalues. Namely, if all eigenvalues of $Df(x_e)$ have negative real part, the equilibrium point is attracting and is a sink. If all eigenvalues of $Df(x_e)$ have positive real part, the equilibrium point is repelling, as the origin in the van der Pol system represented in Fig. 1, and is called a source. An intermediate case, where some eigenvalues have positive real part and the others have negative real part originate a saddle point. In a saddle point, some directions are attracting, and other directions are repelling (technically they are attracting when $t \to +\infty$). See e.g. [45] for more details and precise definitions.

Theorem 4 ([31]). *Let $\mathcal{Z}$ be the class of C^1 functions $f : K \to \mathbb{R}^m$ defined on the closed unit ball $K = \overline{B(0,1)} \subseteq \mathbb{R}^m$ with the property that: (i) no zero of f lies on the boundary of $\overline{B(0,1)}$ and (ii) all zeros of f are invertible, i.e. if $f(x_0) = 0$, then $Df(x_0)$ is invertible. Then the operator which assigns to each $f \in \mathcal{Z}$ its set of zeros and the exact number of zeros is uniformly computable.*

This above theorem can be somewhat generalized in the plane for the next type of steady states—periodic orbits. A well-known result in dynamical systems states that, on the unit ball (disc) K in $\mathbb{R}^2$, and assuming the set of vector fields $\mathcal{X}(\mathcal{K})$ with the property that the flow points inwards on the boundary of K, there is an open and dense set $\mathcal{S} \subseteq \mathcal{X}(\mathcal{K})$, called the set of structurally stable systems, such that the set of all steady states of (1) for $f \in \mathcal{S}$ consists only of a finite number of hyperbolic equilibrium points and hyperbolic periodic orbits (one can define hyperbolic periodic orbits as in [45]). The following result is from [32].

Theorem 5. *There is an open and dense set $\mathcal{S} \subseteq \mathcal{X}(\mathcal{K})$, where $\mathcal{S}$ consists of C^1 functions from $K = \overline{B(0,1)} \subseteq \mathbb{R}^2$ to $\mathbb{R}^2$, such that the operator $\Theta : \mathcal{S} \to K$ which maps each $f \in \mathcal{S}$ into its set of steady states, which consists only of equilibrium points and of periodic orbits, is computable. Furthermore, the map which maps*

$f \in \mathcal{S}$ into its set of periodic orbits and into the number of periodic orbits is also computable.

We have up to now considered equilibrium points and periodic orbits. What about the case of strange attractors such as the Lorenz attractor? The dynamical properties of higher dimensional dynamical systems are in general not well understood, but the following result from [29] shows a positive result.

Theorem 6. *For any geometric Lorenz flow defined on $\mathbb{R}^3$, if the data defining the system is computable, then so is its associated Lorenz attractor. Moreover, the physical measure supported on this attractor is a computable probability measure.*

4 Computing with Continuous-Time Dynamical Systems

As mentioned in the introduction, it is possible to compute directly with dynamical systems. See [12] for a recent survey on this topic. In this section our focus will be computing with dynamical systems of the form

$$y' = p(y) \tag{4}$$

where p is a (vectorial) function where each component is formed by a polynomial.

4.1 Computing with Polynomial ODEs and Shannon's General Purpose Analog Computer

As shown in [28] (see [46,51] for earlier results in that direction), the class of ODEs (4) corresponds to Shannon's General Purpose Analog Computer (GPAC) [51] which is a model of computation intended to capture computations done by the analog computers in use in the first half of the twentieth century, the differential analyzers [18]. The GPAC consists of circuits built with some basic units which perform operations such as generating constants and adding or multiplying their inputs (see Fig. 2). A crucial unit is the integrator, which allows one to solve ODEs. The other units allow us to compute polynomials thus providing the connection to polynomial ODEs of the form (4). However, not all connections are allowed, as some circuits are not meaningful (they might not have solutions or admit multiple solutions [28,51]). For this reason, some restrictions on the format of the circuit is needed to ensure that a unique solution exists [28], but without being too limiting to ensure that those circuits capture exactly all polynomial ODEs (4).

To compute with a GPAC, we may compute in "real time" as originally done in [51]. An example of how to compute the exponential function is provided in Fig. 2. In that case when the input t changes, the output e^t is immediately updated and in this sense the computation is done in real time. Using a similar approach, we can compute with polynomial ODEs (GPACs) polynomials, trigonometric functions, their compositions, inverses, etc.

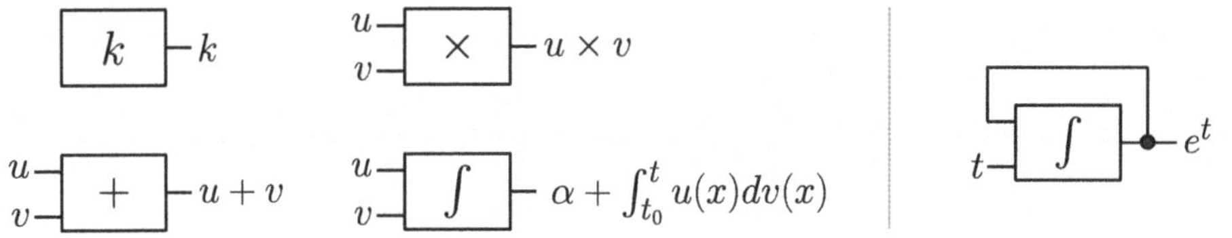

Fig. 2. The basic units used in a GPAC are represented on the left. The right image represents a GPAC computing the exponential function (in that case one takes for the integrator $t_0 = 0$ and $\alpha = 1$).

In particular, it was shown in [27] that, with a proper encoding of configuration of a Turing machine as real numbers, one can simulate an arbitrary Turing machine in real time with a polynomial ODE.

Theorem 7 ([27]). *Given an arbitrary Turing machine M, it can be simulated in real time by a polynomial ODE (4).*

We notice that more recently it was shown that even partial differential equations inspired from hydrodynamics can also simulate Turing machine, see e.g. [20–22].

4.2 Connections with Computable Analysis and the Classical Theory of Computation

As we have mentioned above, almost all standard functions from analysis are computable in real time by polynomial ODEs. Some notable functions which cannot be computed in real time by a polynomial ODE [51] are the gamma function Γ and the Riemann zeta function ζ, given by

$$\Gamma(x) = \int_0^{+\infty} t^{x-1}e^{-t}dt, \;\; \zeta(x) = \sum_{n=1}^{+\infty} \frac{1}{n^x}$$

which are computable under the computable analysis framework. However, as noted in [19], the notions of computation used in computable analysis, where more and more time is allowed to get increasingly better approximations of the output, is different from the much more restrictive view taken with real-time computation with the GPAC. In [8] it was shown that allowing more computation time for better approximations of the output with a GPAC, according to the definition provided below, polynomial ODEs and computable analysis actually compute the same set of functions.

Definition 6. *A function $f : \mathbb{R}^n \to \mathbb{R}^k$ is computable by a polynomial ODE (4) if there are two functions g, ε formed by components of the solution of (4) with initial condition $x(0) = x_0$ satisfying the following properties: given any $y_1, \ldots, y_n \in \mathbb{R}$, if we set the first n components of the initial condition*

x_0 of (4) to these values, i.e. $x_{0,1} = y_1, \ldots, x_{0,n} = y_n$ (the other components $x_{0,n+1}, \ldots, x_{0,m} \in \mathbb{R}$ are fixed) we get that

$$\|g(t) - f(y_1, \ldots, y_n)\| \leq \varepsilon(t) \text{ and } \varepsilon(t) \to 0 \text{ as } t \to +\infty.$$

This result was further expanded for polynomial time computable functions in [11]. See also [3] for a similar characterization involving discrete ordinary differential equations and [7] for some explanations on how one can program with ordinary differential equations. The following theorem is from [8].

Theorem 8. *Let $a, b \in \mathbb{R}$ be computable numbers (in the computable analysis sense). Then a function $f : [a, b] \to \mathbb{R}$ is computable (in the computable analysis sense) if and only if is computable by a polynomial ODE. Furthermore the ODE is computable as well as the fixed parameters in the initial condition.*

Surprisingly, we can also use polynomial ODEs to characterize classical computational complexity classes. In [11] it was shown that one can characterize the classical class P of languages accepted in polynomial time using polynomial ODEs. For this we have to introduce a measure of complexity for ODEs. This is done in those references using the length of the solution curve. See also [4] for a similar characterization involving discrete ordinary differential equations. Similarly, one can characterize $EXPTIME$, the Grzegorczyk hierarchy, the primitive recursive functions, or $PSPACE$ with polynomial ODEs [10,26]. The class $FPSPACE$ can also be characterized using ODEs [5]. As an example of these results, we provide the following theorem from [11], which characterizes the class P with polynomial ODEs.

Theorem 9. *Let $x_1 \geq 1$ define the accepting region of a polynomial ODE and $x_1 \leq -1$ the rejecting region. Then there is an encoding of words into $\mathbb{R}^m$ such that for any language $L \in P$, there is an ODE (4) such that its solution starting on the encoding of $w \in L$ ($w \notin L$) will reach the accepting (rejecting) region in polynomial length on the size of $|w|$ and vice versa.*

Acknowledgement. This work is supported by CIDMA under the Portuguese Foundation for Science and Technology (FCT, https://ror.org/00snfqn58) Multi-Annual Financing Program for R&D Units.

References

1. Aberth, O.: The failure in computable analysis of a classical existence theorem for differential equations. Proc. Am. Math. Soc. **30**, 151–156 (1971)
2. Asarin, E., Maler, O.: Achilles and the tortoise climbing up the arithmetical hierarchy. J. Comput. Syst. Sci. **57**(3), 389–398 (1998)
3. Blanc, M., Bournez, O.: A characterisation of functions computable in polynomial time and space over the reals with discrete ordinary differential equations: Simulation of turing machines with analytic discrete odes. In: Leroux, J., Lombardy, S., Peleg, D. (eds.) 48th International Symposium on Mathematical Foundations of Computer Science, MFCS 2023. LIPIcs, vol. 272, pp. 21:1–21:15. Dagstuhl-Leibniz-Zentrum für Informatik (2023)

4. Blanc, M., Bournez, O.: Quantifiying the robustness of dynamical systems. relating time and space to length and precision. In: Murano, A., Silva, A. (eds.) 32nd EACSL Annual Conference on Computer Science Logic (CSL 2024). Leibniz International Proceedings in Informatics (LIPIcs), vol. 288, pp. 17:1–17:20. Schloss Dagstuhl – Leibniz-Zentrum für Informatik (2024)
5. Blanc, M., Bournez, O.: The complexity of computing in continuous time: space complexity is precision. In: Leibniz International Proceedings in Informatics (LIPIcs). vol. 51st International Colloquium on Automata, Languages, and Programming (ICALP 2024). Schloss Dagstuhl – Leibniz-Zentrum für Informatik, Tallinn, Estonia (2024). https://doi.org/10.4230/LIPIcs.ICALP.2024.129
6. Bournez, O.: Achilles and the Tortoise climbing up the hyper-arithmetical hierarchy. Theoret. Comput. Sci. **210**(1), 21–71 (1999)
7. Bournez, O.: Programming with ordinary differential equations: Some first steps towards a programming language. In: Berger, U., Franklin, J.N.Y., Manea, F., Pauly, A. (eds.) Revolutions and Revelations in Computability - 18th Conference on Computability in Europe, CiE 2022, Swansea, UK, 11–15 July 2022, Proceedings. Lecture Notes in Computer Science, vol. 13359, pp. 39–51. Springer (2022). https://doi.org/10.1007/978-3-031-08740-0_4
8. Bournez, O., Campagnolo, M.L., Graça, D.S., Hainry, E.: Polynomial differential equations compute all real computable functions on computable compact intervals. J. Complex. **23**(3), 317–335 (2007)
9. Bournez, O., Gozzi, R.: Solving discontinuous initial value problems with unique solutions is equivalent to computing over the transfinite. In: Beyersdorff, O., Kanté, M.M., Kupferman, O., Lokshtanov, D. (eds.) 41st International Symposium on Theoretical Aspects of Computer Science (STACS 2024). Leibniz International Proceedings in Informatics (LIPIcs), vol. 289, pp. 20:1–20:19. Schloss Dagstuhl – Leibniz-Zentrum für Informatik, Dagstuhl, Germany (2024). https://doi.org/10.4230/LIPIcs.STACS.2024.20, https://drops.dagstuhl.de/entities/document/10.4230/LIPIcs.STACS.2024.20
10. Bournez, O., Gozzi, R., Graça, D.S., Pouly, A.: A continuous characterization of PSPACE using polynomial ordinary differential equations. J. Complex. **77**, 101755 (2023)
11. Bournez, O., Graça, D.S., Pouly, A.: Polynomial time corresponds to solutions of polynomial ordinary differential equations of polynomial length – the general purpose analog computer and computable analysis are two efficiently equivalent models of computations. In: Chatzigiannakis, I., Mitzenmacher, M., Rabani, Y., Sangiorgi, D. (eds.) Proceedings of 43rd International Colloquium on Automata, Languages and Programming (ICALP 2016). Leibniz International Proceedings in Informatics (LIPIcs), vol. 55, pp. 109:1–109:15. Schloss Dagstuhl-Leibniz-Zentrum fuer Informatik (2016)
12. Bournez, O., Pouly, A.: A survey on analog models of computation. In: Brattka, V., Hertling, P. (eds.) Handbook of Computability and Complexity in Analysis, pp. 173–226. Springer International Publishing, Cham (2021)
13. Brattka, V., Hertling, P. (eds.): Handbook of Computability and Complexity in Analysis. Springer (2021)
14. Brattka, V., Hertling, P., Weihrauch, K.: A tutorial on computable analysis. In: Cooper, S.B., , Löwe, B., Sorbi, A. (eds.) New Computational Paradigms: Changing Conceptions of What is Computable, pp. 425–491. Springer (2008)
15. Braverman, M., Grigo, A., Rojas, C.: Noise vs computational intractability in dynamics. In: Proceedings of the 3rd Innovations in Theoretical Computer Sci-

ence Conference, ITCS 2012, pp. 128–141. Association for Computing Machinery (2012). https://doi.org/10.1145/2090236.2090247, https://doi.org/10.1145/2090236.2090247
16. Braverman, M., Yampolsky, M.: Non-computable Julia sets. J. Am. Math. Soc. **19**(3), 551–578 (2006)
17. Buescu, J.: Exotic Attractors: from Liapunov Stability to Riddled Basins. Birkhäuser (1997)
18. Bush, V.: The differential analyzer. A new machine for solving differential equations. J. Franklin Inst. **212**, 447–488 (1931)
19. Campagnolo, M.L., Moore, C., Costa, J.F.: Iteration, inequalities, and differentiability in analog computers. J. Complex. **16**(4), 642–660 (2000)
20. Cardona, R., Miranda, E., Peralta-Salas, D., Presas, F.: Constructing turing complete euler flows in dimension 3. Proc. Natl. Acad. Sci. **118**(19) (2021)
21. Cardona, R., Miranda, E., Peralta-Salas, D., Presas, F.: Universality of euler flows and flexibility of reeb embeddings. Adv. Math. **428**, 109142 (2023)
22. Cardona, R., Miranda, E., Peralta-Salas, D.: Turing universality of the incompressible Euler equations and a conjecture of moore. Int. Math. Res. Notices (2021)
23. Collins, P., Graça, D.S.: Effective computability of solutions of ordinary differential equations the thousand monkeys approach. In: Brattka, V., Dillhage, R., Grubba, T., Klutsch, A. (eds.) Proceedings of 5th International Conference on Computability and Complexity in Analysis (CCA 2008), Electronic Notes in Theoretical Computer Science, vol. 221, pp. 103–114. Elsevier (2008)
24. Galatolo, S., Hoyrup, M., Rojas, C.: Statistical properties of dynamical systems - simulation and abstract computation. Chaos, Solitons Fractals **45**, 1–14 (2012)
25. Gangloff, S., Herrera, A., Rojas, C., Sablik, M.: Computability of topological entropy: from general systems to transformations on cantor sets and the interval. Discrete Contin. Dynam. Systems **40**(7), 4259–4286 (2020)
26. Gozzi, R., Graça, D.: Characterizing time computational complexity classes with polynomial differential equations. Computability **12**(1), 23–57 (2023)
27. Graça, D.S., Campagnolo, M.L., Buescu, J.: Computability with polynomial differential equations. Adv. Appl. Math. **40**(3), 330–349 (2008)
28. Graça, D.S., Costa, J.F.: Analog computers and recursive functions over the reals. J. Complex. **19**(5), 644–664 (2003)
29. Graça, D.S., Rojas, C., Zhong, N.: Computing geometric Lorenz attractors with arbitrary precision. Trans. Am. Math. Soc. **370**, 2955–2970 (2018)
30. Graça, D.S., Zhong, N.: Computability of differential equations. In: Brattka, V., Hertling, P. (eds.) Handbook of Computability and Complexity in Analysis, pp. 71–99. Springer International Publishing, Cham (2021)
31. Graça, D.S., Zhong, N.: The set of hyperbolic equilibria and of invertible zeros on the unit ball is computable. Theoret. Comput. Sci. **895**, 48–54 (2021)
32. Graça, D.S., Zhong, N.: Computing the exact number of periodic orbits for planar flows. Trans. Am. Math. Soc. **375**, 5491–5538 (2022)
33. Graça, D.S., Zhong, N., Buescu, J.: Computability, noncomputability and undecidability of maximal intervals of IVPs. Trans. Am. Math. Soc. **361**(6), 2913–2927 (2009)
34. Hirsch, M.W., Smale, S., Devaney, R.: Differential Equations, Dynamical Systems, and an Introduction to Chaos. Academic Press (2004)
35. Hoyrup, M., Rute, J.: Computable measure theory and algorithmic randomness. In: Brattka, V., Hertling, P. (eds.) Handbook of Computability and Complexity in Analysis, pp. 227–270. Springer International Publishing, Cham (2021)

36. Kawamura, A.: Lipschitz continuous ordinary differential equations are polynomial-space complete. Comput. Complex. **19**(2), 305–332 (2010)
37. Kawamura, A., Cook, S.: Complexity theory for operators in analysis. ACM Trans. Comput. Theor. **42**(2), 5 (2012)
38. Kawamura, A., Ota, H., Rösnick, C., Ziegler, M.: Computational complexity of smooth differential equations. Logical Methods Comput. Sci. **10**(1:6), 1–15 (2014)
39. Ko, K.I.: On the computational complexity of ordinary differential equations. Inf. Control **58**, 157–194 (1983)
40. Ko, K.I.: Complexity Theory of Real Functions. Birkhäuser (1991)
41. Lorenz, E.N.: Deterministic non-periodic flow. J. Atmos. Sci. **20**, 130–141 (1963)
42. Miller, W.: Recursive function theory and numerical analysis. J. Comput. Syst. Sci. **4**, 465–472 (1970)
43. Müller, N.T.: Uniform computational complexity of Taylor series. In: Ottmann, T. (ed.) ICALP 1987. LNCS, vol. 267, pp. 435–444. Springer, Heidelberg (1987). https://doi.org/10.1007/3-540-18088-5_37
44. Mycka, J., Costa, J.F.: A new conceptual framework for analog computation. Theoret. Comput. Sci. **374**(1–3), 277–290 (2007)
45. Perko, L.: Differential Equations and Dynamical Systems. Springer, 3rd edn. (2001)
46. Pour-El, M.B.: Abstract computability and its relations to the general purpose analog computer. Trans. Am. Math. Soc. **199**, 1–28 (1974)
47. Pour-El, M.B., Richards, J.I.: A computable ordinary differential equation which possesses no computable solution. Ann. Math. Logic **17**, 61–90 (1979)
48. Rettinger, R., Weihrauch, K., Zhong, N.: Topological complexity of blowup problems. J. Univ. Comput. Sci. **15**(6), 1301–1316 (2009)
49. Rojas, C., Sablik, M.: On the algorithmic descriptive complexity of attractors in topological dynamics. Ergodic Theor. Dyn. Syst. 1–28 (2025). https://doi.org/10.1017/etds.2024.135
50. Rojas, C., Yampolsky, M.: Computable geometric complex analysis and complex dynamics. In: Brattka, V., Hertling, P. (eds.) Handbook of Computability and Complexity in Analysis, pp. 143–172. Springer International Publishing, Cham (2021)
51. Shannon, C.E.: Mathematical theory of the differential analyzer. J. Math. Phys. **20**, 337–354 (1941)
52. Sipser, M.: Introduction to the Theory of Computation. Cengage Learning, 3rd edn. (2012)
53. Smale, S.: Mathematical problems for the next century. Math. Intell. **20**(2), 7–15 (1998). https://doi.org/10.1007/BF03025291
54. Tucker, W.: A rigorous ODE solver and Smale's 14th problem. Found. Comput. Math. **2**(1), 53–117 (2002)
55. Turing, A.M.: On computable numbers, with an application to the entscheidungsproblem. Proc. London Math. Soc. **s2-42**(1), 230–265 (1937)

Proof-Carrying Neuro-Symbolic Code

Ekaterina Komendantskaya[1,2](✉)

[1] Southampton University, Southampton, UK
ek1u23@soton.ac.uk
[2] Heriot-Watt University, Edinburgh, UK

Abstract. This invited paper introduces the concept of "proof-carrying neuro-symbolic code" and explains its meaning and value, from both the "neural" and the "symbolic" perspectives. The talk outlines the first successes and challenges that this new area of research faces.

Keywords: Neural Networks · Cyber-Physical System Verification · Programming Languages · Neuro-Symbolic Programs

1 Neuro-Symbolic Proofs and Programs

Proof Carrying Code is a long tradition within programming language research, broadly referring to methods that interleave verification with executable code, thus avoiding the inevitable discrepancies that arise when the code and the proofs are handled in different languages. Although the term was coined by Necula [50] almost three decades ago, with time, it grew to encompass any languages that are powerful enough to handle both the coding and the proving. Examples are dependently-typed (Agda, Idris, Coq/Rocq) and refinement-typed (F*, Liquid Haskell) languages. In the last decades, both families of languages have seen substantial successes in the proof-carrying code direction [35,44].

An equally impressive, though very different in nature, revolution has happened elsewhere in computer science: machine learning methods grew in quality, diversity and quickly proliferated to different applications. Machine learning attracted attention of programming language researchers, and there have been recent advances in functional algorithms for automated differentiation, probabilistic programming, as well as *neuro-symbolic programming* and *neuro-symbolic theorem proving*. The latter two aim to develop, respectively:

- methods of merging machine learning code and standard (symbolic) code [17]; and
- methods automating theorem proving with the help of machine learning deployed at the stage of proof search or lemma conjecturing [34,42,63].

The natural question to ask is: *Does the concept of proof-carrying code bear any value, or indeed meaning, in the age of the neuro-symbolic paradigm shift?* This invited talk will answer both questions in the positive.

© The Author(s), under exclusive license to Springer Nature Switzerland AG 2025
A. Beckmann et al. (Eds.): CiE 2025, LNCS 15764, pp. 76–87, 2025.
https://doi.org/10.1007/978-3-031-95908-0_6

2 What is Proof-Carrying Neuro-Symbolic Code?

Consider a purely symbolic program $s(\cdot)$, whose completion requires computing a complex, unknown function $\mathcal{H} : \mathcal{P} \to \mathcal{R}$ that maps objects in the *problem input space* $\mathcal{P}$ to those in the *problem output space* $\mathcal{R}$. Given an embedding function $e : \mathcal{P} \to \mathbb{R}^m$ and an unembedding function $u : \mathbb{R}^n \to \mathcal{R}$, we can approximate $\mathcal{H}$ by training a neural network $f : \mathbb{R}^m \to \mathbb{R}^n$ such that $u \circ f \circ e \approx \mathcal{H}$. We refer to $u \circ f \circ e$ as the *solution*, and refer to $\mathbb{R}^m$ and $\mathbb{R}^n$ as the *embedding input space* and *embedding output space* respectively. Unlike objects in the problem space, the vectors in the embedding space are often not directly interpretable. The complete *neuro-symbolic program* is then modelled as $s(u \circ f \circ e)$. Examples of u and e would be the normalization of inputs, resizing operations for images, or data augmentation operations that are commonplace in machine learning pipelines. Most common examples of neuro-symbolic programs come from the area of *cyber-physical system* modelling, as such systems often implement machine learning algorithms in order to process sensor data. We will illustrate this class of applications using a simple example (Example 1 and Fig. 1).

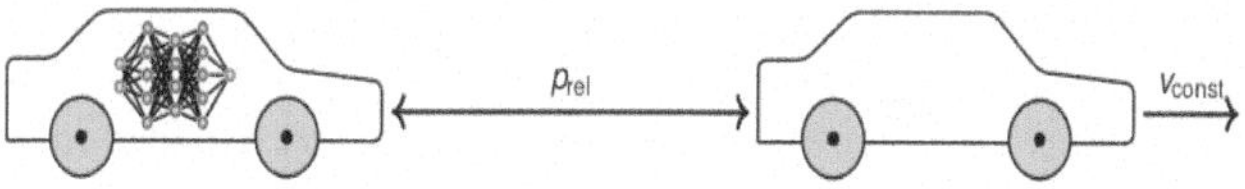

$$p_{rel} > \frac{v_{rel}^2}{2B} \to \left[a_{rel} := f(v_{rel}, p_{rel}); t := 0; \{p'_{rel} = v_{rel}, v'_{rel} = -a_{rel}, t' = 1\}\right] p_{rel} > 0$$

Fig. 1. Example of a cyber-physical system with a neural controller (given by the left-most car) written in the syntax of the Differentiable Dynamic Logic [29,53,58].

What might one want to prove about such a neuro-symbolic program? The end goal may be to prove that $s(u \circ f \circ e)$ satisfies a property Ψ, which we will refer to as the *program property*. The natural way to proceed is to establish a *solution property* Φ and a *network property* Ξ such that the proof of Ψ is decomposable into the following three lemmas:

$$\Xi(f) \tag{1}$$

$$\forall g : \Xi(g) \Rightarrow \Phi(u \circ g \circ e) \tag{2}$$

$$\forall h : \Phi(h) \Rightarrow \Psi(s(h)) \tag{3}$$

i.e. Lemma 1 proves that the network f obeys the network property Ξ, then Lemma 2 proves that this implies $u \circ$ f$\circ e$, the neural network lifted to the problem space, obeys the solution property Φ, and finally Lemma 3 proves that this implies $s(u \circ f \circ e)$, the neuro-symbolic program, obeys the program property Ψ.

In conclusion, writing proof-carrying neuro-symbolic code amounts to writing a program $s(u \circ f \circ e)$ and completing a proof as in Eqs. (1), (2) and (3).

Example 1 (A Cyber-Physical System with a Neural Network Controller). Fig. 1 shows to cars moving. The second car is "autonomous", i.e. it has a neural network f for controlling the (de)acceleration a_{rel} (we use red colour to indicate that it is potentially unsafe). The high-level safety specification boils down to asserting that, *assuming the front car has constant speed, the back car can be trusted to maintain a safe distance.* The relative distance p_{rel} and velocity v_{rel} are modelled by differential equations. We highlight the dynamical (continuous) parts of the system with blue colour. The relative distance p_{rel} between the two cars is a function of velocity of the autonomous car ($p'_{rel} = v_{rel}$), and the latter is a function of the car's relative acceleration: $v'_{rel} = -a_{rel}$. This continuous dynamics has to interact with the discrete actions of the autonomous car; e.g., "set the acceleration to value $f(v_{rel}, p_{rel})$", written as $a_{rel} := f(v_{rel}, p_{rel})$, or "start following the car in front of you at time 0" ($t := 0$). Traditionally, the simplest way to control this car was to set $a_{rel} := -B$, for some breaking constant B. The neural network f replaces this classical discrete controller. In the classical example, the safety specification is given by pre- and post-conditions: assuming the autonomous car satisfies $p_{rel} > \frac{v_{rel}^2}{2B}$, there will never be a collision: $p_{rel} > 0$. Suppose a neural controller f is trained on previous data, the input-output pairs where the input is given by sensor readings v_{rel}, p_{rel} and the output by $-B$.[1] For training prposes, the data is normalised as floating point numbers between some fixed interval, e.g. $[-1, 1]$, thus $f : [-1, 1]^2 \rightarrow [-1, 1]$. In order for the whole cyber-physical system to be proven safe, the neural network f must be verified to satisfy the following property (subject to normalisation of input and outputs of f) : $p_{rel} > \frac{v_{rel}^2}{2B} \rightarrow f(p_{rel}, v_{rel}) \leq -B$. This will then provably imply that the desired postcondition $p_{rel} > 0$ holds.

The purpose of *proof-carrying neuro-symbolic code* is to propose languages or programming environments where implementation of such a system, as well as all necessary proofs of its safety, can be carried out in a sound and consistent way. This has proven to be a challenge, see [18] for a detailed discussion of the problems raised in the literature.

3 Why Proof-Carrying Neuro-Symbolic Code?

Arguably, there are good reasons for both machine learning and programming language communities to invest efforts into the development of proof-carrying neuro-symbolic code.

Machine Learning Perspective. Traditionally, statistical machine learning has distinguished its methods from "algorithm-driven" programming: the consensus has been that machine learning is deployed when there is example input-output data but no general algorithm for computing outputs from inputs. Thus,

[1] This training scenario is somewhat convoluted, as we are assuming that the neural network is trained to simulate the classical controller which in this toy example was given by $a := -B$, but the reader should imagine that in a real-life examples the training data comes from a less trivial distribution.

neural networks are commonly seen as programs that emerge from data via training, without direct human guidance on how to perform the computation. This unfortunate dichotomy has led to a divide between programming language and machine learning research that is still awaiting resolution.

The first hint that this dichotomy is not as fundamental as was thought came from the machine learning community itself. The famous paper by Szegedy et al. [56] pointed out the "intriguing" problem that even the most accurate neural networks fail to satisfy the property of *robustness*, i.e. small perturbations of their inputs should result in small changes to their output. Szegedy's key example concerned imperceptible perturbations of pixels in an image that can sway the neural network's classification decisions. This lack of robustness can have safety and security implications: for example, an autonomous car's vision unit may fail to recognise pedestrians on the road. For that reason, the problem attracted significant attention [14] but remains unresolved to this day. Partial solutions often deploy methods of *adversarial training*—i.e., training based on computing *adversarial attacks*—which augment the training set with the worst-case perturbations of the input data points with respect to the output loss of the neural network [41].

The robustness of neural networks actually yields a formal specification [15]. Given a neural network $f : \mathbb{R}^m \to \mathbb{R}^n$, f is *robust around* $\hat{x} \in \mathbb{R}^m$, if

$$\forall x, \|\hat{x} - x\| \leq \epsilon \implies \|f(\hat{x}) - f(x)\| \leq \delta, \tag{4}$$

where $\epsilon, \delta \in \mathbb{R}$ are small constants and $\|.\|$ computes a vector distance. From the programming language perspective, robustness can be seen as a refinement type that refines input and output types of f, cf. [40]. At the same time, robustness is an example of a *desirable property* that neural networks cannot learn from data alone: note the quantification over vectors x that do not belong to the data set. This challenges the classical dichotomy between algorithm-driven and data-driven programming, demonstrating the inevitability of property specification in both cases – and the need for solutions that are *proof-carrying*.

Proofs and Programs Perspective. It is clear that neuro-symbolic systems generally, and cyber-physical systems in particular, are the next holy grail for formal verification. This gives one reason for proof and program communities to engage in designing the principles of proof-carrying neuro-symbolic code. The less obvious reason, that will be demonstrated in the next section, is that, despite seemingly exotic nature of the neuro-symbolic set up, some of the best methods developed in the programming language community in the past few decades find unexpectedly direct applications in this new domain.

4 Challenges in Developing Proof-Carrying Neuro-Symbolic Code

4.1 Challenge 1. Multi-backend Interfaces

The languages rich enough to host proof-carrying code generally (e.g. Agda, Rocq or Idris), and prove higher-level properties such as (3) in particular, do not tend to be good at proving low-level *reachability properties* of functions over real vector spaces (cf. $\Xi(f)$ in Eq. (1) or Eq. (4) in Sect. 3). There exists a range of specialised solvers that can prove properties such as $\Xi(f)$, e.g. Marabou [38,62], $\alpha\beta$-CROWN [60] and PyRAT [43]. These solvers have a specialised syntax to handle proof search for properties such as $\Xi(f)$. But this syntax is not general enough to serve as a proof-carrying code for general properties such $\Psi(s(h))$.

In many ways, the argument replicates the one that has already been made by programming languages such as F* and Liquid Haskell that successfully argued for delegation of low-level proofs to the SMT solvers. The language **Vehicle** [21–23] was the first to adapt this argument to neural network solvers. The adaptation introduced a few interesting ideas concerning type checking and compilation fit for the neuro-symbolic set up [19,20].

Vehicle provides an interface and a DSL (*Domain-Specific Language*) in which properties such as (2) can be written, type checked, and compiled to both low-level solvers that can prove the property (1) and high-level languages that can express the property (3). At the time of writing, however, **Vehicle** has only interface to Marabou for part (1) and to Agda for part (3) of the pipeline shown in Sect. 2. For practical applications, the range of solvers and theorem provers to which **Vehicle** compiles to needs to grow. Crucially, it must involve higher-order provers that are better geared to work with features that are valuable in cyber-physical system verification, for example:

- proving properties involving systems of ODEs (*Ordinary Differential Equations*), which is possible in KeymaeraX [29,53] (cf. Example 1);
  or
- proving with probabilistic properties, which is possible in e.g. Rocq Mathematical Components library [57] that covers substantial fragments of Measure & Probability Theory [2,4–6].

4.2 Challenge 2. Formal Proof Certificates

If we are to use neural network solvers as part of our compilers, as **Vehicle** or **CAISAR** [32] do, we need to be sure of their correctness. Neural network solvers are variants of established proof search methods, such as SMT-solving with bound tightening (Marabou [38,62]), interval bound propagation ($\alpha\beta$-CROWN [60]), or abstract interpretation (AI2 [31]). Unfortunately, even complete neural network solvers are prone to implementation bugs and numerical imprecision that, in turn, might compromise their soundness and can be maliciously exploited, as was shown in e.g. [18,37,64].

One could consider verifying neural network solvers directly. However, as Wu et al. exemplify, a mature neural network verifier is a complex multi-platform library; its direct verification would be close to infeasible [61]. A similar problem was faced by the SMT solvers [10,11,49]. A solution was found: to produce (and then check) the proof evidence, instead of checking the verifier in its entirety. The software that checks the proof evidence is called the *proof checker*.

This idea only works on two conditions: the proof checker should be significantly simpler than the original verifier and it should yield strong guarantees of code correctness. For Marabou, the first half of this research agenda was accomplished by Isac et al. [36] with proof checking for Marabou being reduced to the application of the *Farkas lemma* [59], a well-known solvability theorem for a finite system of linear equations, along with a tree structure that reflects Marabou's verification procedure. As the proof checker in [36] was implemented in C++, that code itself raised the question of trust (e.g., in the precision of floating point calculations or guarantees for implementation soundness). It thus had no formal guarantee of correctness.

Therefore, the next step was to establish the necessary guarantees of the checker's correctness, while adhering to the principles of *proof-carrying code* [50] and avoiding any potential discrepancy between implementation and formalization; and ensuring that the checker itself did not suffer from floating point imprecision. The industrial theorem prover Imandra [52] was chosen to for this purpose: it supports arbitrary precision real arithmetic and can prove properties of programs written in a subset of OCaml. It strikes a balance between interactive and automated proving: i.e., it features strong automation while admitting user interaction with the prover via tactics. In Imandra, Desmartin et al. [25] obtained the checker implementation that can be safely executed *and* proven correct in the same language. The result was the first *certified* proof checker for neural network verification.

To develop this initial success further, one needs to incorporate similar implementations into languages such as **Vehicle**, and also extend the range of certified neural network solvers. The new generation of neural network solvers that produce probabilistic, stochastic and/or dynamic guarantees [1,45,46,51] are about to encounter the very same obstacle of missing proof certification. However, this time, the mathematical sophistication of formal proofs involved in turning their output into *proof certificates* will require libraries and languages formalising measure/probability theory or proofs involving differential equations. A good hosting language for formalising probabilistic proof certificates would be an interactive theorem prover with strong Probability theory libraries, such as the Mathematical Components library [2,4–6,57]) in Rocq.

4.3 Challenge 3. Compilation of Verification Properties to Machine Learning Objective Functions

No implementation of proof-carrying neuro-symbolic programs could be complete without acknowledging that, at the heart of the matter lies the question whether the object we verify – the neural network f – actually adheres to the verification conditions we wish to verify in Eq. (1). In Sect. 3 we outlined that this problem boils down to deploying methods that can optimise f to satisfy the desired properties. Initially, for the robustness properties such as the one given in Eq. (4), various *adversarial training* methods were suggested [27,33,41]. Generally, the weakest form of property-based training boils down to translating a specification written in a subset of first-order logic into a *loss function*, that serves directly as an optimisation objective within the implementation of a training algorithm. Thus, the training algorithm optimises the neural network to satisfy the desired property. This translation method is known under the name of *differentiable logic* (or DL) [27,28,54]. **Vehicle** implements DL as one of its backends [21] and serves as a prototype of a compiler for neural network property specification languages.

The remaining challenge in this domain is to establish a coherent theory of the differentiable logics, seen as a sub-family of quantitative logics. Quantitative logics, i.e. logics that have semantics in interval domains instead of the Booleans have been studied for decades, and date back to the ideas of Kleene, Gödel, and Łukasiewicz at the start of the 20th century [48]. They also have been shown to relate to the family of substructural logics, known as *superintuitionistic* logics [30]. Fuzzy logics [48], and the logics of the Lawvere quantale [9,26] are important examples of quantitative logics. To illustrate, consider a toy syntax with atomic propositions and conjunction, such as

$$\Phi \ni \phi := A \mid \phi \wedge \phi \tag{5}$$

where A is interpreted in a domain $D \subseteq [-\infty, \infty]$. D varies among logics and restricts the interpretation of connectives. For example, Gödel logic has an interpretation on $[0, 1]$ with conjunctions interpreted as min, and the Lawvere quantale logics would interpret this syntax in the interval $[0, \infty]$ and take multiplication to model conjunction.

Recently, there was a surge of interest in quantitative logics, stimulated by the growing interest in *safer machine learning* [24]. Nevertheless, there is one fundamental problem that quantitative logics face in this domain. Many specifications of interest for machine learning are first-order, yet the majority of quantitative logic results concern propositional syntax [9,48,55]. For example, language in Definition (5) is propositional, while property in Eq. (4) is first-order. Generalizing some sound and complete propositional quantitative logics to first-order logic often comes at a cost of either completeness or function continuity. For example, among several known fuzzy logics [48,55], the only first-order extension that is sound and complete involves the Gödel logic, that interprets conjunction as min, disjunction as a max, and universal and existential quantifiers via the infimum and supremum [8]. However, connectives of this logic are not continuous and therefore not suitable for gradient-descent algorithms.

Recently, a promising solution was proposed by Capucci [13], interpreting quantifiers as *p-means*, a generalization of *p*-norms over a probability space [12]. This new semantics gives hope that the open problem of finding a suitable approach to quantification in quantitative logics will find its resolution, and we can soon find a logic that is sound and complete relative to this new quantitative semantics.

Finally, in the context of using quantitative logics for compilation of specifications of neuro-symbolic programs into the machine learning backends, there comes a question of proving their soundness in a theorem prover, in order to obtain a form of a certified machine learning compiler in the style of [35,44].

It has been shown that verifying soundness of quantitative logics (as well as their so-called "geometric properties" assumed by machine learning algorithms: continuity, shadow-lifting, etc.) results in laborious proofs that are error prone [3,16,55]. To overcome these challenges, a rigorous formalization of propositional semantics for quantitative (differentiable) logics has been proposed in [3]. Extending this formalisation to first-order logics is a non-trivial challenge that is yet to be overcome [7,47]. Moreover, the new semantics proposed by Capucci presents a particular challenge for formal verification, since, unlike the previous formalizations of quantitative logics [3], it now also involves results from real analysis and probability. Most notably, it involves formalisation of measure spaces, probability spaces, and Lebesgue integrals, as well as the use of results such as Jensen's and Hölder's inequalities [39] that require extensions to the existining Mathematical Components Libraries [57].

5 Conclusions

In this paper, we argue for introducing the term *proof-carrying neuro-symbolic code* to refer to implementation of neuro-symbolic programs with strong safety guarantees expressed in the code and maintained by a provably sound compiler. Using three prominent examples of:

- a multi-backend neuro-symbolic compiler;
- neural network proof certificate checking; and
- formalisation of properties of symbolic loss functions,

we illustrated how these novel methods fit with the existing programming language traditions, and how they depart from them or re-interpret them.

This overview is based on many discussions with my co-authors and colleagues, whose papers were used and cited throughout this paper in order to illustrate more technical points. The interested reader is invited to follow up on these references to learn more.

Acknowledgments. The author acknowledges the partial support of the EPSRC grant AISEC: AI Secure and Explainable by Construction (EP/T026960/1), and support of ARIA: Mathematics for Safe AI grant. She thanks Matthew Daggitt for providing a nice notation for expressing the essence of neuro-symbolic code with proofs in Sect. 2.

References

1. Abate, A., Giacobbe, M., Roy, D.: Stochastic omega-regular verification and control with supermartingales. In: Gurfinkel, A., Ganesh, V. (eds.) Computer Aided Verification - 36th International Conference, CAV 2024, Montreal, QC, Canada, 24–27 July 2024, Proceedings, Part III, LNCS, vol. 14683, pp. 395–419. Springer (2024). https://doi.org/10.1007/978-3-031-65633-0_18

2. Affeldt, R., et al.: Analysis library compatible with mathematical components. https://github.com/math-comp/analysis (2017), latest version: 1.8.0 (2024-12-19)
3. Affeldt, R., Bruni, A., Komendantskaya, E., Ślusarz, N., Stark, K.: Taming differentiable logics with coq Formalisation. In: Bertot, Y., Kutsia, T., Norrish, M. (eds.) 15th International Conference on Interactive Theorem Proving (ITP 2024). Leibniz International Proceedings in Informatics (LIPIcs), vol. 309, pp. 4:1–4:19. Schloss Dagstuhl – Leibniz-Zentrum für Informatik, Dagstuhl, Germany (2024). https://doi.org/10.4230/LIPIcs.ITP.2024.4, https://drops.dagstuhl.de/entities/document/10.4230/LIPIcs.ITP.2024.4
4. Affeldt, R., Bruni, A., Roux, P., Saikawa, T.: Yet another formal theory of probabilities (with an application to random sampling). In: 30th International Conference on Types for Proofs and Programs TYPES 2024–Abstracts, p. 11 (2024)
5. Affeldt, R., Cohen, C.: Measure construction by extension in dependent type theory with application to integration. J. Autom. Reason. **67**(3), 28 (2023). https://doi.org/10.1007/S10817-023-09671-5
6. Affeldt, R., Stone, Z.: A comprehensive overview of the lebesgue differentiation theorem in coq. In: Bertot, Y., Kutsia, T., Norrish, M. (eds.) 15th International Conference on Interactive Theorem Proving, ITP 2024, September 9-14, 2024, Tbilisi, Georgia. LIPIcs, vol. 309, pp. 5:1–5:19. Schloss Dagstuhl - Leibniz-Zentrum für Informatik (2024). https://doi.org/10.4230/LIPICS.ITP.2024.5
7. Atkey, R., Capucci, M., Komendantskaya, E., Mardare, R.: Quantitative predicate logic as a foundation for verified ml. ARIA grant (2024)
8. Baaz, M., Preining, N., Zach, R.: First-order gödel logics. Ann. Pure Appl. Logic **147**(1), 23–47 (2007). https://doi.org/10.1016/j.apal.2007.03.001, https://www.sciencedirect.com/science/article/pii/S016800720700019X
9. Bacci, G., Mardare, R., Panangaden, P., Plotkin, G.: Propositional logics for the lawvere quantale. Electron. Notes Theor. Inf. Comput. Sci. **3** (2023)
10. Barbosa, H., et al.: Flexible proof production in an industrial-strength SMT solver. In: Proceedings of 11th International Joint Conference on Automated Reasoning (IJCAR), pp. 15–35 (2022)
11. Barrett, C., de Moura, L., Fontaine, P.: Proofs in satisfiability modulo theories. All about Proofs, Proofs for All **55**(1), 23–44 (2015)
12. Bourgain, J.: New classes of Lp-spaces, vol. 889. Springer (2006)
13. Capucci, M.: On quantifiers for quantitative reasoning. arXiv preprint arXiv:2406.04936 (2024)
14. Carlini, N.: A complete list of all (arxiv) adversarial example papers (2019)
15. Casadio, M., et al.: Neural network robustness as a verification property: a principled case study. In: Computer Aided Verification (CAV 2022), LNCS, Springer (2022)
16. Casadio, M., et al.: Neural network robustness as a verification property: a principled case study. In: International Conference on Computer Aided Verification, pp. 219–231. Springer (2022)
17. Chaudhuri, S., Ellis, K., Polozov, O., Singh, R., Solar-Lezama, A., Yue, Y.: Neurosymbolic programming. Found. Trends® Program. Lang. **7**(3), 158–243 (2021). https://doi.org/10.1561/2500000049, http://dx.doi.org/10.1561/2500000049
18. Cordeiro, L., et al.: Neural network verification is a programming language challenge. In: European Symposium on Programming Languages, ESOP 2025 (2025)
19. Daggitt, M.L., Atkey, R., Kokke, W., Komendantskaya, E., Arnaboldi, L.: Compiling higher-order specifications to smt solvers: how to deal with rejection constructively. In: Proceedings of the 12th ACM SIGPLAN International Conference on Certified Programs and Proofs, pp. 102–120 (2023)

20. Daggitt, M.L., Kokke, W., Atkey, R.: Efficient compilation of expressive problem space specifications to neural network solvers (2024). https://arxiv.org/abs/2402.01353
21. Daggitt, M.L., Kokke, W., Atkey, R., Slusarz, N., Arnaboldi, L., Komendantskaya, E.: Vehicle: bridging the embedding gap in the verification of neuro-symbolic programs. CoRR **abs/2401.06379** (2024). https://doi.org/10.48550/ARXIV.2401.06379
22. Daggitt, M.L., Kokke, W., Atkey, R., Slusarz, N., Arnaboldi, L., Komendantskaya, E.: Vehicle: bridging the embedding gap in the verification of neuro-symbolic programs. arXiv preprint arXiv:2401.06379 (2024)
23. Daggitt, M.L., et al.: The vehicle tutorial: neural network verification with vehicle. In: Narodytska, N., Amir, G., Katz, G., Isac, O. (eds.) Proceedings of the 6th Workshop on Formal Methods for ML-Enabled Autonomous Systems, FoMLAS@CAV 2023, Paris, France, 17–18 July 2023, Kalpa Publications in Computing, vol. 16, pp. 1–5. EasyChair (2023). https://doi.org/10.29007/5S2X
24. Dalrymple, D.: Safeguarded AI: constructing guaranteed safety (2024), programme Thesis
25. Desmartin, R., Isac, O., Komendantskaya, E., Stark, K., Passmore, G.O., Katz, G.: A certified proof checker for deep neural network verification in imandra. In: Submitted to ITP 2025 (2025). https://doi.org/10.48550/arXiv.2405.10611
26. Fischer, M., Balunovic, M., Drachsler-Cohen, D., Gehr, T., Zhang, C., Vechev, M.: Dl2: training and querying neural networks with logic. In: International Conference on Machine Learning, pp. 1931–1941. PMLR (2019)
27. Fischer, M., Balunovic, M., Drachsler-Cohen, D., Gehr, T., Zhang, C., Vechev, M.T.: DL2: training and querying neural networks with logic. In: Chaudhuri, K., Salakhutdinov, R. (eds.) Proceedings of the 36th International Conference on Machine Learning, ICML 2019, 9–15 June 2019, Long Beach, California, USA. Proceedings of Machine Learning Research, vol. 97, pp. 1931–1941. PMLR (2019). http://proceedings.mlr.press/v97/fischer19a.html
28. Flinkow, T., Pearlmutter, B.A., Monahan, R.: Comparing differentiable logics for learning with logical constraints (2024). https://arxiv.org/abs/2407.03847
29. Fulton, N., Mitsch, S., Quesel, J.-D., Völp, M., Platzer, A.: KeYmaera X: an axiomatic tactical theorem prover for hybrid systems. In: Felty, A.P., Middeldorp, A. (eds.) CADE 2015. LNCS (LNAI), vol. 9195, pp. 527–538. Springer, Cham (2015). https://doi.org/10.1007/978-3-319-21401-6_36
30. Galatos, N., Jipsen, P., Ono, T.K.H.: Residuated lattices: an algebraic glimpse at substructural logics. Elsevier (2007)
31. Gehr, T., Mirman, M., Drachsler-Cohen, D., Tsankov, P., Chaudhuri, S., Vechev, M.: AI2: safety and robustness certification of neural networks with abstract interpretation. In: Proceedings of 39th IEEE Symposium on Security and Privacy (SP), pp. 3–18 (2018)
32. Girard-Satabin, J., Alberti, M., Bobot, F., Chihani, Z., Lemesle, A.: Caisar: a platform for characterizing artificial intelligence safety and robustness. arXiv preprint arXiv:2206.03044 (2022)
33. Giunchiglia, E., Stoian, M.C., Lukasiewicz, T.: Deep learning with logical constraints. In: Thirty-First International Joint Conference on Artificial Intelligence (IJCAI-22), pp. 5478–5485. International Joint Conferences on Artificial Intelligence Organization, July 2022. https://doi.org/10.24963/ijcai.2022/767, survey Track

34. Heras, J., Komendantskaya, E., Johansson, M., Maclean, E.: Proof-pattern recognition and lemma discovery in ACL2. In: McMillan, K., Middeldorp, A., Voronkov, A. (eds.) LPAR 2013. LNCS, vol. 8312, pp. 389–406. Springer, Heidelberg (2013). https://doi.org/10.1007/978-3-642-45221-5_27
35. Ho, S., Fromherz, A., Protzenko, J.: Modularity, code specialization, and zero-cost abstractions for program verification. Proc. ACM Program. Lang. **7**(ICFP) (2023). https://doi.org/10.1145/3607844
36. Isac, O., Barrett, C., Zhang, M., Katz, G.: Neural network verification with proof production. In: Proceedings of 22nd International Conference on Formal Methods in Computer-Aided Design (FMCAD), pp. 38–48 (2022)
37. Jia, K., Rinard, M.: Exploiting verified neural networks via floating point numerical error. In: Proceedings of 28th International Static Analysis Symposium (SAS), pp. 191–205 (2021)
38. Katz, G., et al.: The marabou framework for verification and analysis of deep neural networks, pp. 443–452, July 2019
39. Kazarinoff, N.D.: Analytic inequalities. Courier Corporation (2014)
40. Kokke, W., Komendantskaya, E., Kienitz, D., Atkey, R., Aspinall, D.: Neural networks, secure by construction. In: Oliveira, B. (ed.) APLAS 2020. LNCS, vol. 12470, pp. 67–85. Springer, Cham (2020). https://doi.org/10.1007/978-3-030-64437-6_4
41. Kolter, Z., Madry, A.: Adversarial robustness—theory and practice. NeurIPS 2018 tutorial (2018). https://adversarial-ml-tutorial.org/
42. Komendantskaya, E., Heras, J., Grov, G.: Machine learning in proof general: interfacing interfaces. In: Kaliszyk, C., Lüth, C. (eds.) Proceedings 10th International Workshop On User Interfaces for Theorem Provers, UITP 2012, Bremen, Germany, July 11th, 2012. EPTCS, vol. 118, pp. 15–41 (2012). https://doi.org/10.4204/EPTCS.118.2
43. Lemesle, A., Lehmann, J., Gall, T.L.: Neural network verification with pyrat (2024). https://arxiv.org/abs/2410.23903
44. Leroy, X.: Formal verification of a realistic compiler. Commun. ACM **52**(7), 107–115 (2009)
45. Mandal, U., et al.: Formally verifying deep reinforcement learning controllers with lyapunov barrier certificates (2024). https://arxiv.org/abs/2405.14058
46. Manginas, V., et al.: A scalable approach to probabilistic neuro-symbolic verification (2025). https://arxiv.org/abs/2502.03274
47. Marulanda-Giraldo, J.M., Komendantskaya, E., Bruni, A., Affeldt, R., Capucci, M.: Quantifiers for quantitative logics in rocq: a new project description (2025), a draft
48. Metcalfe, G., Olivetti, N., Gabbay, D.M.: Proof theory for fuzzy logics, vol. 36. Springer Science & Business Media (2008)
49. de Moura, L., Bjørner, N.: Satisfiability modulo theories: introduction and applications. Commun. ACM **54**(9), 69–77 (2011)
50. Necula, G.: Proof-carrying code. In: Proceedings of 24th Symposium on Principles of Programming Languages (POPL), pp. 106–119 (1997)
51. Neustroev, G., Giacobbe, M., Lukina, A.: Neural continuous-time supermartingale certificates. CoRR **abs/2412.17432** (2024). https://doi.org/10.48550/ARXIV.2412.17432
52. Passmore, G., et al.: The Imandra automated reasoning system (System Description). In: Proceedings of 10th International Joint Conference Automated Reasoning (IJCAR), pp. 464–471 (2020)

53. Platzer, A.: Logical Foundations of Cyber-Physical Systems. Springer, Cham (2018). https://doi.org/10.1007/978-3-319-63588-0
54. Slusarz, N., Komendantskaya, E., Daggitt, M.L., Stewart, R.J., Stark, K.: Logic of differentiable logics: Towards a uniform semantics of DL. In: Piskac, R., Voronkov, A. (eds.) LPAR 2023: Proceedings of 24th International Conference on Logic for Programming, Artificial Intelligence and Reasoning, Manizales, Colombia, 4-9th June 2023. EPiC Series in Computing, vol. 94, pp. 473–493. EasyChair (2023). https://doi.org/10.29007/C1NT
55. Slusarz, N., Komendantskaya, E., Daggitt, M.L., Stewart, R.J., Stark, K.: Logic of differentiable logics: towards a uniform semantics of DL. In: Piskac, R., Voronkov, A. (eds.) LPAR 2023: Proceedings of 24th International Conference on Logic for Programming, Artificial Intelligence and Reasoning, Manizales, Colombia, 4–9th June 2023. EPiC Series in Computing, vol. 94, pp. 473–493. EasyChair (2023). https://doi.org/10.29007/C1NT
56. Szegedy, C., et al.: Intriguing properties of neural networks (2014)
57. Team, M.C.: Mathematical components library. https://github.com/math-comp/math-comp (2007)
58. Teuber, S., Mitsch, S., Platzer, A.: Provably safe neural network controllers via differential dynamic logic. In: NeurIPS 2024 (2024). http://papers.nips.cc/paper_files/paper/2024/hash/031b5fd7d847f69ed33378a9a1117b4b-Abstract-Conference.html
59. Vanderbei, R.: Linear programming: foundations and extensions. J. Oper. Res. Soc. (1996)
60. Wang, S., et al.: Beta-crown: efficient bound propagation with per-neuron split constraints for neural network robustness verification. Adv. Neural. Inf. Process. Syst. **34**, 29909–29921 (2021)
61. Wu, H., et al.: Marabou 2.0: a versatile formal analyzer of neural networks. In: Proceedings of 36th International Conference on Computer Aided Verification (CAV) (2024)
62. Wu, H., et al.: Marabou 2.0: a versatile formal analyzer of neural networks. In: Computer Aided Verification (CAV) (2024)
63. Yang, K., et al.: Formal mathematical reasoning: a new frontier in AI. CoRR **abs/2412.16075** (2024). https://doi.org/10.48550/ARXIV.2412.16075
64. Zombori, D., Bánhelyi, B., Csendes, T., Megyeri, I., Jelasity, M.: Fooling a complete neural network verifier. In: International Conference on Learning Representations (2021). https://api.semanticscholar.org/CorpusID:235613486

Uniform Functional Interpretations

Paulo Oliva(✉)

Queen Mary University of London, London, UK
p.oliva@qmul.ac.uk
https://www.eecs.qmul.ac.uk/~pbo/

Abstract. We study a variant of Gödel's *dialectica* (functional) interpretation whereby quantifiers are treated *uniformly*, while atomic formulas are assumed to carry computational content. We argue that this leads to a more flexible approach to functional interpretations, aligning the *dialectica* interpretation with Cohen's method of forcing – following recent work of Alexander Miquel on implicative structures. We also discuss how this could lead to an alternative way of dealing with abstract spaces in *proof mining*. The interpretation is first presented in the setting of *classical* first-order logic, but is subsequently factorised as a composition of Krivine's negative translation and a uniform *intuitionistic* interpretation. For any (multi-sorted) first-order theory, a concrete functional interpretation can be obtained once a choice of *base interpretation* for the predicate symbols is fixed.

Keywords: functional interpretation · dialectica interpretation · realizability · forcing

1 A Bit of History

The study of the connections between formal *mathematical proofs* and *computable functions* started with the Brouwer-Heyting-Kolmogorov (BHK) interpretation [57] (informally) and Kleene's realizability [27,56] (formally). It later also led to the Curry-Howard correspondence between proofs in minimal logic and simply-typed λ-terms: Formulas correspond to types, and proofs of A correspond to λ-terms of type A. Kleene realizability [27] focuses on Heyting (intuitionistic) arithmetic **HA** and computable functions. It associates each closed formula A of **HA** to a new unary formula $n \mathbf{r} A$, read as n *realizes* A, such that from a proof of A one can "read-off" a code n of a computable function such that $n \mathbf{r} A$. Therefore, Kleene realizability can be seen as mapping formulas A to sets of codes

$$A \qquad \mapsto \qquad A^{\mathbf{r}} :\equiv \{n \mid n \mathbf{r} A\}$$

such that from a proof of A we can obtain an element of the set $A^{\mathbf{r}}$. Intuitively, realizers are witnesses for the (positive) existential quantifiers in the formula A.

Around the same time (although only published later), Gödel was also developing an alternative computational interpretation of arithmetic, the *dialectica*

© The Author(s), under exclusive license to Springer Nature Switzerland AG 2025
A. Beckmann et al. (Eds.): CiE 2025, LNCS 15764, pp. 88–103, 2025.
https://doi.org/10.1007/978-3-031-95908-0_7

(or functional) interpretation [2,24]. According to Gödel, formulas A are interpreted as *binary relations* $A_D(\boldsymbol{x}; \boldsymbol{y})$ between *realizers* (or *witnesses*) $\boldsymbol{x}$ and *co-realizers* (or *counter-witnesses*) $\boldsymbol{y}$, where $\boldsymbol{x}$ and $\boldsymbol{y}$ are both tuples. While in Kleene realizability one focuses simply on the existential quantifiers of A, in Gödel's *dialectica* interpretation both existential and universal quantifiers are considered – leading to a more refined computational interpretation of A. One can still think of A as a set of realizers, by saying that the tuple $\boldsymbol{x}$ is a "proper" realizer if it defeats any tuple of "co-realizers" $\boldsymbol{y}$, i.e.

$$A \quad \mapsto \quad A^D :\equiv \{\boldsymbol{x} \mid \forall \boldsymbol{y} A_D(\boldsymbol{x}; \boldsymbol{y})\}.$$

One important difference is that, while Kleene worked with codes of computable functions, Gödel developed an extension of the primitive recursive functions to arbitrary (finite) types, so that in Gödel's interpretation the elements in $\boldsymbol{x}$ and $\boldsymbol{y}$ range over primitive recursive functionals[1]. But just as in Kleene's realizability, Gödel showed that if A is a theorem of **HA**, then from any proof of A one can extract a tuple of functionals $\boldsymbol{t}$ such that $\boldsymbol{t} \in A^D$, i.e. $\forall \boldsymbol{y} A_D(\boldsymbol{t}; \boldsymbol{y})$. Again, a theorem A of **HA** corresponds to a non-empty set A^D.

One advantage of Gödel's finer treatment of quantifiers is that one can more easily combine the *dialectica* interpretation with a negative (double negation) translation [20,21,23,37,42], leading to an interpretation of classical (Peano) arithmetic **PA**. In 1962, Spector [44,54] also extended Gödel's interpretation of arithmetic to classical analysis, using a new form of recursion on well-founded trees called *bar recursion*. Recently, some surprising connections between bar recursion, game theory, category theory and topology have been discovered [10–13].

An interesting alternative presentation of Gödel's dialectica interpretation of **PA** is given by Shoenfield [53], where a double negation translation (Krivine's negative translation) and the intuitionistic interpretation are combined into a single step [55]. This note is heavily inspired by Shoenfield's approach. In particular, we will use Shoenfield's formalisation of first-order logic.

Although Gödel's *dialectica* interpretation allows us to "extract" primitive recursive functionals from any proof in Peano arithmetic, Gödel's original motivation for developing the interpretation was to give a relative consistency proof for **PA** (relative to the primitive recursive functionals). To make the reduction as "finitistic" as possible, Gödel completely eliminated quantifiers, in favour of functionals of finite type, so $A_D(\boldsymbol{x}; \boldsymbol{y})$ is a quantifier-free formula. However, this means that to interpret the contraction axiom $A \to A \wedge A$, Gödel must assume that quantifier-free formulas are decidable. This restriction was eliminated by Diller-Nahm [8], where they showed that one no longer needs the decidability restriction if one interprets the contraction axioms via finite sets. In here we also follow Diller-Nahm's approach so that we do not need to require that atomic formulas are decidable.

[1] Kreisel's *modified realizability* can be seen as a variant of Kleene realizability where realizers are also primitive recursive functionals [38,39], as in Gödel's *dialectica* interpretation.

The *dialectica* interpretation has also been studied in the setting of Categorical Logic, by Valéria de Paiva [51], where Gödel's and Diller-Nahm's interpretations are presented in terms of categorical constructions. Following a suggestion from Girard, de Paiva also includes in the study interpretations of *propositional linear logic* – which was later extended to *first-order linear logic* by Shirahata [52]. Shirahata's and de Paiva's work has been the basis for most of my work (together with various collaborators) on realizability and functional interpretations of intuitionistic and linear logic, culminating in a unifying approach to both realizability and functional interpretations [9,18,22,25,45–49].

The most impressive applications of Gödel's functional interpretation come, without doubt, from Kohlenbach's programme of *proof mining* [28,29,31,32, 35,36]. The main tool used for proof mining is Kohlenbach's *monotone functional interpretation*, a systematic combination of the *dialectica* interpretation and Howard-Bezem majorizability relation [6,26]. The use of majorizability not only also removes the need for decidability of atomic formulas (characteristic functions are easily majorizable), but it guarantees the extraction of highly uniform computational bounds, with various breakthrough results in Applied Proof Theory, in particular, in the fields of Approximation Theory [30,36], Fixed-Point Theory [33], and Probability Theory [1,3,34].

In this note we aim to reflect on the fundamental structure of Gödel's (Diller-Nahm variant) functional interpretation, by applying the interpretation to first-order logic, where

(i) quantifiers are interpreted *completely uniformly*, while
(ii) atomic formulas are assumed to carry *arbitrary computational content.*

We present this as an alternative way of understanding how the functional interpretation propagates computational content both within formulas (from atomic formulas to more complex ones) and proofs (from axioms to theorems). We will show that nothing is lost in this alternative approach, on the contrary, it seems to lead to a much deeper and general understanding of the interpretation, in particular the way in which *uniformities* (independence of the witness on certain parameters) show up in proof mining.

The functional interpretation presented here (Definition 2 and Theorem 1) builds on a recent joint work with Bruno Dinis [9] on the *parametrised functional interpretation*, where we have first considered the uniform treatment of quantifiers combined with a computational interpretation of atomic formulas. The interpretation is also similar to the *herbrandized functional interpretation* proposed by Fernando Ferreira and Gilda Ferreira in [14], except that in here quantifiers are treated uniformly.

2 The Uniform Functional Interpretation

We will follow Shoenfield [53], and work with first-order logic and first-order theories built from atomic formulas, the logical connectives $A \vee B$ and $\neg A$, and the universal quantifier $\forall x A$. All the other logical connectives and the existential quantifier are definable from those, e.g. $A \to B$ is an abbreviation for $\neg A \vee B$. Let $\mathbf{S}$ be a first-order theory which we wish to interpret (we will call it the *source theory*) and $\mathbf{T}$ the (*target*) theory in which we are interpreting $\mathbf{S}$. Following Gödel's approach, we want to associate to each formula $A(\boldsymbol{x})$ of $\mathbf{S}$ a binary relation $\{A(\boldsymbol{x})\}^{\boldsymbol{a}}_{\boldsymbol{b}}$ in $\mathbf{T}$, where $\boldsymbol{a}$ and $\boldsymbol{b}$ are tuples of fresh free-variables[2]. Therefore, if $\boldsymbol{x}$ are the free-variables of $A(\boldsymbol{x})$, then $\boldsymbol{x}, \boldsymbol{a}$ and $\boldsymbol{b}$ are the free-variables of $\{A(\boldsymbol{x})\}^{\boldsymbol{a}}_{\boldsymbol{b}}$. Our interpretation will assume that we are given a "base" interpretation of atomic formulas:

Definition 1 (Base interpretation of S into T). *A* base interpretation *of* $\mathbf{S}$ *associates to each n-ary predicate symbol* $P(\boldsymbol{x})$ *of the language of* $\mathbf{S}$ *a new relation* $\boldsymbol{x} \lhd^P_{\boldsymbol{b}} \boldsymbol{a}$ *in* $\mathbf{T}$, *between* $\boldsymbol{x}, \boldsymbol{a}$ *and* $\boldsymbol{b}$.

In the above, any of $\boldsymbol{x}, \boldsymbol{a}$ and/or $\boldsymbol{b}$ could be the empty tuple. In the formula $\boldsymbol{x} \lhd^P_{\boldsymbol{b}} \boldsymbol{a}$ we should think of $\boldsymbol{a}$ as evidence or witnesses that $\boldsymbol{x}$ has property P, and $\boldsymbol{b}$ a counter-evidence or counter-witness. We will interpret the validity of $P(\boldsymbol{x})$ as

$$\exists \boldsymbol{a} \forall \boldsymbol{b} (\boldsymbol{x} \lhd^P_{\boldsymbol{b}} \boldsymbol{a}). \tag{1}$$

For example, consider the predicate $\mathbb{R}(x)$ saying that x is a real number. Several possible base interpretations of $\mathbb{R}(x)$ are possible. Let us consider four of them:
1. Cauchy sequence interpretation. We could take sequence of rationals $\alpha \colon \mathbb{N} \to \mathbb{Q}$ as the possible witnesses for x being in $\mathbb{R}$. This corresponds to the base interpretation

$$x \lhd^{\mathbb{R}} \alpha \;:\equiv\; \forall n^{\mathbb{N}} (|\alpha(n) - x| \leq 2^{-n}) \tag{2}$$

so that $x \lhd^{\mathbb{R}} \alpha$ says that α is a (fast converging) Cauchy sequence representation of $x \colon \mathbb{R}$.
2. Cauchy sequence interpretation (pointwise). The interpretation above can also be made "pointwise" by allowing $\mathbb{R}(x)$ to be refuted via some counter-witness n. More precisely, we can take a sequence of rationals $\alpha \colon \mathbb{N} \to \mathbb{Q}$ as the possible witnesses for $\mathbb{R}(x)$, as above, but also a number $n \colon \mathbb{N}$ as a counter-witness that α (fast) converges to x. This leads to the base interpretation

$$x \lhd^{\mathbb{R}}_n \alpha \;:\equiv\; |\alpha(n) - x| \leq 2^{-n} \tag{3}$$

so that α is a sequence of rationals that converges to x with rate 2^{-n} only if no n refutes $x \lhd^{\mathbb{R}}_n \alpha$.
3. Bounding interpretation. In the above interpretations the witness α aims to completely capture the real number x. Alternatively, we can work with much weaker interpretation where the witnesses for $\mathbb{R}(x)$ only provides partial information about x. For instance, for n a natural number, we can define

$$x \lhd^{\mathbb{R}} n \;:\equiv\; |x| \leq n. \tag{4}$$

[2] In Gödel's terminology one would write $\{A(\boldsymbol{x})\}^{\boldsymbol{a}}_{\boldsymbol{b}}$ as $(A(\boldsymbol{x}))_D(\boldsymbol{a}; \boldsymbol{b})$. We prefer the notation $\{A(\boldsymbol{x})\}^{\boldsymbol{a}}_{\boldsymbol{b}}$ as it more clearly highlights the opposing roles of $\boldsymbol{a}$ and $\boldsymbol{b}$.

This interpretation is related to the *bounded functional interpretation* [15,16].
4. Rational intervals interpretation. One could say that the bounding interpretation above is too course. We can adjust it to make it more precise by working with rational intervals. More precisely, let $p, q \colon \mathbb{Q}$, and define

$$x \lhd^{\mathbb{R}} p, q \;:\equiv\; x \in [p, q]. \tag{5}$$

In this case, a pair of rationals p, q witnesses that x is real number if x belongs to the interval $[p, q]$. This choice is closely related to the *confined interpretation* studied in [19].
See [17] for a more detailed discussion on various choices of base interpretations.

Definition 2 (Uniform functional interpretation). *Let a base interpretation of* $\mathbf{S}$ *into* $\mathbf{T}$ *be given. For each formula* A *of* $\mathbf{S}$*, possibly with free-variables, associate a formula* $\{A\}^{\boldsymbol{a}}_{\boldsymbol{b}}$ *of* $\mathbf{T}$*, by induction on* A*, as follows. For atomic formulas we let:*

$$\begin{aligned} \{s = t\} &:\equiv s = t \\ \{P(\boldsymbol{x})\}^{\boldsymbol{V}}_{\boldsymbol{p}} &:\equiv \exists \boldsymbol{a} \in \boldsymbol{V} \forall \boldsymbol{b} \in \boldsymbol{pa}(\boldsymbol{x} \lhd^{P}_{\boldsymbol{b}} \boldsymbol{a}) \end{aligned}$$

where $\boldsymbol{V}$ *ranges over tuples of finite sets, and* $\boldsymbol{pa}$ *denotes the tuple* $p_1(a_1, \ldots, a_m), \ldots, p_n(a_1, \ldots, a_m)$*, when* $\boldsymbol{p} = p_1, \ldots, p_n$ *and* $\boldsymbol{a} = a_1, \ldots, a_m$*. Suppose* $\{A\}^{\boldsymbol{a}}_{\boldsymbol{b}}$ *and* $\{B\}^{\boldsymbol{c}}_{\boldsymbol{d}}$ *are the interpretations of* A *and* B*, respectively*[3]*. We define*

$$\begin{aligned} \{\neg A\}^{\boldsymbol{V}}_{\boldsymbol{p}} &:\equiv \exists \boldsymbol{b} \in \boldsymbol{V} \neg \{A\}^{\boldsymbol{pb}}_{\boldsymbol{b}} \\ \{A \vee B\}^{\boldsymbol{a},\boldsymbol{c}}_{\boldsymbol{b},\boldsymbol{d}} &:\equiv \{A\}^{\boldsymbol{a}}_{\boldsymbol{b}} \vee \{B\}^{\boldsymbol{c}}_{\boldsymbol{d}} \\ \{\forall x A\}^{\boldsymbol{a}}_{\boldsymbol{b}} &:\equiv \forall x \{A\}^{\boldsymbol{a}}_{\boldsymbol{b}}. \end{aligned}$$

Remark 1. The interpretation of atomic formulas might require some justification. One argument for why this is the correct definition is that, as we will see in Sect. 4, it allows for the decomposition of the interpretation of classical logic into an interpretation of intuitionistic logic with a negative translation (as in [50,55]). Another justification is that when trying to write $\exists \boldsymbol{a} \forall \boldsymbol{b}(\boldsymbol{x} \lhd^{P}_{\boldsymbol{b}} \boldsymbol{a})$ as a $\forall\exists$-formula, we can first "herbrandize" the universal quantification (leaving a bounded $\boldsymbol{b}$)

$$\exists \boldsymbol{a} \forall \boldsymbol{b}(\boldsymbol{x} \lhd^{P}_{\boldsymbol{b}} \boldsymbol{a}) \quad \Rightarrow \quad \forall \boldsymbol{p} \exists \boldsymbol{a} \forall \boldsymbol{b} \in \boldsymbol{pa}(\boldsymbol{x} \lhd^{P}_{\boldsymbol{b}} \boldsymbol{a}) \tag{6}$$

followed by splitting of the unbounded quantification on $\boldsymbol{a}$ into a bounding set $\boldsymbol{V}$ and a bounded quantification:

$$\forall \boldsymbol{p} \exists \boldsymbol{a} \forall \boldsymbol{b} \in \boldsymbol{pa}(\boldsymbol{x} \lhd^{P}_{\boldsymbol{b}} \boldsymbol{a}) \quad \Rightarrow \quad \forall \boldsymbol{p} \exists \boldsymbol{V} \underbrace{\exists \boldsymbol{a} \in \boldsymbol{V} \forall \boldsymbol{b} \in \boldsymbol{pa}(\boldsymbol{x} \lhd^{P}_{\boldsymbol{b}} \boldsymbol{a})}_{\{P(\boldsymbol{x})\}^{\boldsymbol{V}}_{\boldsymbol{p}}}. \tag{7}$$

[3] The arities of the tuples $\boldsymbol{a}$ and $\boldsymbol{b}$ in $\{A\}^{\boldsymbol{a}}_{\boldsymbol{b}}$ are completely determined by the logical structure of A and the arities of the tuples in the base interpretation of the predicate symbols in A.

Remark 2. We will see that an alternative interpretation of $\forall x A$, namely

$$\{\forall x A\}_{\boldsymbol{b}}^{\boldsymbol{V}} \quad :\equiv \exists \boldsymbol{a} \in \boldsymbol{V} \forall x \{A\}_{\boldsymbol{b}}^{\boldsymbol{a}}. \tag{8}$$

also works and leads to an equivalent interpretation. This (more complex) definition will be useful for our decomposition result in Sect. 4.

Remark 3. It is common for functional interpretations to interpret a formula A as a new formula $\exists \boldsymbol{a} \forall \boldsymbol{b} A_I(\boldsymbol{a}; \boldsymbol{b})$, where $A_I(\boldsymbol{a}; \boldsymbol{b})$ is either *quantifier-free* or *bounded*. This is not the case in $\{A\}_{\boldsymbol{b}}^{\boldsymbol{a}}$, which might contain unbounded quantifications (see clause for $\forall x A$). Nevertheless, we can still think of these unbounded quantifications as "computationally empty", since the quantification over x is not qualified. The computational content will come from a qualification of x, for instance, $\forall x(\mathbb{R}(x) \rightarrow A(x))$, so witnesses for $A(x)$ can still be computed from witnesses of $\mathbb{R}(x)$, *uniformly in* x.

Remark 4. Since $A \rightarrow B$ is an abbreviation for $\neg A \vee B$, it is easy to see that

$$\{A \rightarrow B\}_{\boldsymbol{p},\boldsymbol{d}}^{\boldsymbol{V},\boldsymbol{c}} \quad \equiv \forall \boldsymbol{b} \in \boldsymbol{V} \{A\}_{\boldsymbol{b}}^{\boldsymbol{p}\boldsymbol{b}} \rightarrow \{B\}_{\boldsymbol{d}}^{\boldsymbol{c}},$$

The interpretation above is very similar to the *herbrandized functional interpretation* presented in [14], except in the treatment of atomic formulas (which we consider to have computational content) and the quantifiers (which we treat uniformly). The two interpretations deal with disjunction and negation in the same way.

Following [17], we assume an *intensional* treatment of equality $s = t$, i.e. equality is not assumed to carry computational content. This is not the case for *extensional* equality, which could carry witnesses. For instance, extensional equality on real numbers would naturally be interpreted as

$$\{x =_{\mathbb{R}} y\}_k :\equiv |x - y| \leq 2^{-k}. \tag{9}$$

so that $x =_{\mathbb{R}} y$ corresponds to $\forall k^{\mathbb{N}}(|x - y| \leq 2^{-k})$.

Finally, note that Definition 2 makes use of function application and quantifications over finite sets. For now, all we need to assume is that the target theory $\mathbf{T}$ is expressive enough to capture the above definition. Formally speaking, we are working with λ-terms containing constructs for working with finite sets, in the spirit of the star calculus of [14]. As we will see in the next section, we will also need to assume that the terms of $\mathbf{S}$ are terms of $\mathbf{T}$, and that $\mathbf{T}$ contains at least one constant symbol.

2.1 Related Work on Uniform Quantifiers

The uniform interpretation of quantifiers presented above is inspired by Krivine's interpretation of second-order arithmetic [41], where quantifiers are interpreted as intersections (and hence, uniformly).

Ulrich Berger has also considered uniform quantifiers in [5], in the context of modified realizability, where the main motivation was to selectively choose which quantifiers to witness, and hence extract programs that only focused on the important parts of a proof.

In [4], modified realizability and functional interpretations are introduced to interpret nonstandard analysis. In these interpretations, quantifications over standard elements carry computational content as usual, but quantifications over all elements (standard and non-standard) are also treated uniformly.

2.2 Soundness of Interpretation

Intuitively, the uniform functional interpretation $A \mapsto \{A\}_{\boldsymbol{b}}^{\boldsymbol{a}}$ of Definition 2 is associating a formula A of first-order logic with a binary relation $\{A\}_{\boldsymbol{b}}^{\boldsymbol{a}}$ between two tuples $\boldsymbol{a}$ and $\boldsymbol{b}$ in such a way that the validity of A corresponds to the validity of the formula

$$\forall \boldsymbol{b} \exists \boldsymbol{a} \{A\}_{\boldsymbol{b}}^{\boldsymbol{a}}. \tag{10}$$

We will show that if A is provable then $\exists \boldsymbol{p} \forall \boldsymbol{b} \{A\}_{\boldsymbol{b}}^{\boldsymbol{pb}}$ will be witnessed by a tuple of closed λ-terms $\boldsymbol{t}$, i.e. $\mathbf{T}$ will prove $\forall \boldsymbol{b} \{A\}_{\boldsymbol{b}}^{\boldsymbol{tb}}$.

Definition 3 (T-interpretable formulas). *Let $\{A\}_{\boldsymbol{b}}^{\boldsymbol{a}}$ be the interpretation of A. The formula A is said to be* $\mathbf{T}$-interpretable *if $\forall \boldsymbol{b} \{A\}_{\boldsymbol{b}}^{\boldsymbol{tb}}$ is provable in $\mathbf{T}$, for some tuple of closed λ-terms $\boldsymbol{t}$.*

Note that in the above definition A does not need to be a sentence, i.e. A can contain free-variables $\boldsymbol{x}$. Nevertheless, the witnessing terms $\boldsymbol{t}$ will be closed terms due to the uniform treatment of quantifiers. Let us also remark that the variables $\boldsymbol{a}$ in $\{A\}_{\boldsymbol{b}}^{\boldsymbol{a}}$ range over finite sets, and that the following monotonicity property holds:

Proposition 1 (Monotonicity property). *If $\{A\}_{\boldsymbol{b}}^{\boldsymbol{a}}$ and $\boldsymbol{a} \subseteq \boldsymbol{a}'$ then $\{A\}_{\boldsymbol{b}}^{\boldsymbol{a}'}$. In particular, we have*

(i) $\{A\}_{\boldsymbol{b}}^{\boldsymbol{a}}$ implies $\{A\}_{\boldsymbol{b}}^{\boldsymbol{a} \cup \boldsymbol{a}'}$,
(ii) $\exists \boldsymbol{b} \in \boldsymbol{V} \{B\}_{\boldsymbol{d}}^{\boldsymbol{pb}}$ implies $\{B\}_{\boldsymbol{d}}^{\bigcup_{\boldsymbol{b} \in \boldsymbol{V}} \boldsymbol{pb}}$.

We can then prove our main result:

Theorem 1 (Soundness Theorem). *For any first-order theory* $\mathbf{S}$, *if its non-logical axioms are* $\mathbf{T}$*-interpretable then all of its theorems are* $\mathbf{T}$*-interpretable.*

Proof. Let us show that for every theorem A of $\mathbf{S}$ we have a λ-expression $\boldsymbol{t}[\boldsymbol{b}]$, with free-variables $\boldsymbol{b}$, such that $\mathbf{T}$ proves $\{A\}_{\boldsymbol{b}}^{\boldsymbol{t}[\boldsymbol{b}]}$. We do this by induction on proofs, with the interpretability of the non-logical axioms as an assumption. The equality axioms are also trivially interpretable due to the uniform interpretation of equality. Assume A and B are interpreted as $\{A\}_{\boldsymbol{b}}^{\boldsymbol{a}}$ and $\{B\}_{\boldsymbol{d}}^{\boldsymbol{c}}$.
Propositional axiom. Consider an instance $A \vee \neg A$ of the propositional axiom schema. Its interpretation $\{A \vee \neg A\}_{\boldsymbol{b},\boldsymbol{p}}^{\boldsymbol{a},\boldsymbol{V}}$ is

$$\{A\}_{\boldsymbol{b}}^{\boldsymbol{a}} \vee \exists \boldsymbol{b}' \in \boldsymbol{V} \neg \{A\}_{\boldsymbol{b}'}^{\boldsymbol{pb}'}.$$

Given $\boldsymbol{b}, \boldsymbol{p}$, let $\boldsymbol{V} := \{\boldsymbol{b}\}$ and $\boldsymbol{a} := \boldsymbol{p}\boldsymbol{b}$. Clearly we have $\{A\}_{\boldsymbol{b}}^{\boldsymbol{p}\boldsymbol{b}} \vee \exists \boldsymbol{b}' \in \{\boldsymbol{b}\} \neg \{A\}_{\boldsymbol{b}'}^{\boldsymbol{p}\boldsymbol{b}'}$.
Substitution axiom. Consider an instance $\neg \forall x A(x) \vee A(s)$ of the substitution axiom, where s is a term of the source theory $\mathbf{S}$. Its interpretation is

$$\{\neg \forall x A(x) \vee A(s)\}_{\boldsymbol{p},\boldsymbol{b}}^{\boldsymbol{V},\boldsymbol{a}} \equiv \forall \boldsymbol{b}' \in \boldsymbol{V} \forall x \{A(x)\}_{\boldsymbol{b}'}^{\boldsymbol{p}\boldsymbol{b}'} \to \{A(s)\}_{\boldsymbol{b}}^{\boldsymbol{a}}.$$

Given $\boldsymbol{p}$ and $\boldsymbol{b}$, we can take $\boldsymbol{V} := \{\boldsymbol{b}\}$ and $\boldsymbol{a} := \boldsymbol{p}\boldsymbol{b}$. With the more complex interpretation (8) the assumption would be $\forall \boldsymbol{b}' \in \boldsymbol{V} \exists \boldsymbol{a} \in \boldsymbol{p}\boldsymbol{b}' \forall x \{A(x)\}_{\boldsymbol{b}'}^{\boldsymbol{a}}$ and we can take $\boldsymbol{V} := \{\boldsymbol{b}\}$ and $\boldsymbol{a} := \bigcup \boldsymbol{p}\boldsymbol{b}$, since $\boldsymbol{p}\boldsymbol{b}$ will be a set of sets. In this case we also need to use the monotonicity property.
Weakening rule (infer $B \vee A$ from A). Suppose $\boldsymbol{t}[\boldsymbol{b}]$ is a tuple of λ-terms such that $\{A\}_{\boldsymbol{b}}^{\boldsymbol{t}[\boldsymbol{b}]}$. Then, by the weakening rule, $\{A\}_{\boldsymbol{b}}^{\boldsymbol{t}[\boldsymbol{b}]} \vee \{B\}_{\boldsymbol{d}}^{\boldsymbol{c}}$, for any tuple of constants $\boldsymbol{c}$ of appropriate arity[4].
Contraction rule (infer A from $A \vee A$). Suppose $\boldsymbol{t}_1[\boldsymbol{b}_1, \boldsymbol{b}_2], \boldsymbol{t}_2[\boldsymbol{b}_1, \boldsymbol{b}_2]$ are such that that

$$\{A \vee A\}_{\boldsymbol{b}_1,\boldsymbol{b}_2}^{\boldsymbol{t}_1[\boldsymbol{b}_1,\boldsymbol{b}_2],\boldsymbol{t}_2[\boldsymbol{b}_1,\boldsymbol{b}_2]}$$

i.e. $\{A\}_{\boldsymbol{b}_1}^{\boldsymbol{t}_1[\boldsymbol{b}_1,\boldsymbol{b}_2]} \vee \{A\}_{\boldsymbol{b}_2}^{\boldsymbol{t}_2[\boldsymbol{b}_1,\boldsymbol{b}_2]}$. In particular, $\{A\}_{\boldsymbol{b}}^{\boldsymbol{t}_1[\boldsymbol{b},\boldsymbol{b}]} \vee \{A\}_{\boldsymbol{b}}^{\boldsymbol{t}_2[\boldsymbol{b},\boldsymbol{b}]}$. By the monotonicity property (Proposition 1 (i)) and an application of the contraction rule we can conclude $\{A\}_{\boldsymbol{b}}^{\boldsymbol{t}_1[\boldsymbol{b},\boldsymbol{b}] \cup \boldsymbol{t}_2[\boldsymbol{b},\boldsymbol{b}]}$.
Associativity rule (infer $(A \vee B) \vee C$ from $A \vee (B \vee C)$). Trivial.
Cut rule. Suppose we have witnessing terms for $A \vee B$ and $A \to C$, i.e. $\boldsymbol{r}[\boldsymbol{b}, \boldsymbol{d}], \boldsymbol{s}[\boldsymbol{b}, \boldsymbol{d}]$ such that

$$\{A\}_{\boldsymbol{b}}^{\boldsymbol{r}[\boldsymbol{b},\boldsymbol{d}]} \vee \{B\}_{\boldsymbol{d}}^{\boldsymbol{s}[\boldsymbol{b},\boldsymbol{d}]} \tag{11}$$

and $\boldsymbol{V}[\boldsymbol{p}, \boldsymbol{f}], \boldsymbol{t}[\boldsymbol{p}, \boldsymbol{f}]$ such that

$$\forall \boldsymbol{b} \in \boldsymbol{V}[\boldsymbol{p}, \boldsymbol{f}] \{A\}_{\boldsymbol{b}}^{\boldsymbol{p}\boldsymbol{b}} \to \{C\}_{\boldsymbol{f}}^{\boldsymbol{t}[\boldsymbol{p},\boldsymbol{f}]}. \tag{12}$$

Let us show that

$$\{B\}_{\boldsymbol{d}}^{\bigcup_{\boldsymbol{b} \in \boldsymbol{V}[\lambda \boldsymbol{b}.\boldsymbol{r}[\boldsymbol{b},\boldsymbol{d}],\boldsymbol{f}]} \boldsymbol{s}[\boldsymbol{b},\boldsymbol{d}]} \vee \{C\}_{\boldsymbol{f}}^{\boldsymbol{t}[\lambda \boldsymbol{b}.\boldsymbol{r}[\boldsymbol{b},\boldsymbol{d}],\boldsymbol{f}]}.$$

Assume $\boldsymbol{d}$ and $\boldsymbol{f}$ are fixed. From (11), restricting the quantification over $\boldsymbol{b}$ to $\boldsymbol{V}[\lambda \boldsymbol{b}.\boldsymbol{r}[\boldsymbol{b}, \boldsymbol{d}], \boldsymbol{f}]$, we get

$$\forall \boldsymbol{b} \in \boldsymbol{V}[\lambda \boldsymbol{b}.\boldsymbol{r}[\boldsymbol{b}, \boldsymbol{d}], \boldsymbol{f}] \{A\}_{\boldsymbol{b}}^{\boldsymbol{r}[\boldsymbol{b},\boldsymbol{d}]} \vee \exists \boldsymbol{b} \in \boldsymbol{V}[\lambda \boldsymbol{b}.\boldsymbol{r}[\boldsymbol{b}, \boldsymbol{d}], \boldsymbol{f}] \{B\}_{\boldsymbol{d}}^{\boldsymbol{s}[\boldsymbol{b},\boldsymbol{d}]}).$$

By the monotonicity property (Proposition 1 (ii)) we have

$$\forall \boldsymbol{b} \in \boldsymbol{V}[\lambda \boldsymbol{b}.\boldsymbol{r}[\boldsymbol{b}, \boldsymbol{d}], \boldsymbol{f}] \{A\}_{\boldsymbol{b}}^{\boldsymbol{r}[\boldsymbol{b},\boldsymbol{d}]} \vee \{B\}_{\boldsymbol{d}}^{\bigcup_{\boldsymbol{b} \in \boldsymbol{V}[\lambda \boldsymbol{b}.\boldsymbol{r}[\boldsymbol{b},\boldsymbol{d}],\boldsymbol{f}]} \boldsymbol{s}[\boldsymbol{b},\boldsymbol{d}]}).$$

On the other hand, from (12), taking $\boldsymbol{p} := \lambda \boldsymbol{b}.\boldsymbol{r}[\boldsymbol{b}, \boldsymbol{d}]$ we get

$$\forall \boldsymbol{b} \in \boldsymbol{V}[\lambda \boldsymbol{b}.\boldsymbol{r}[\boldsymbol{b}, \boldsymbol{d}], \boldsymbol{f}] \{A\}_{\boldsymbol{b}}^{\boldsymbol{r}[\boldsymbol{b},\boldsymbol{d}]} \to \{C\}_{\boldsymbol{f}}^{\boldsymbol{t}[\lambda \boldsymbol{b}.\boldsymbol{r}[\boldsymbol{b},\boldsymbol{d}],\boldsymbol{f}]}.$$

[4] We must assume that the target theory $\mathbf{T}$ has at least one constant symbol.

From which we can conclude $\{B\}_{\boldsymbol{d}}^{\bigcup_{b \in \boldsymbol{v}[\lambda b.\boldsymbol{r}[b,\boldsymbol{d}],\boldsymbol{f}]} \boldsymbol{s}[b,\boldsymbol{d}]} \vee \{C\}_{\boldsymbol{f}}^{\boldsymbol{t}[\lambda b.\boldsymbol{r}[b,\boldsymbol{d}],\boldsymbol{f}]}$.
$\forall$-*introduction rule* (infer $A \vee \forall x B$ from $A \vee B$, where x is not free in A). Suppose we have terms $\boldsymbol{t}$ and $\boldsymbol{r}$ such that $\{A\}_{\boldsymbol{b}}^{\boldsymbol{t}[\boldsymbol{b},\boldsymbol{d}]} \vee \{B\}_{\boldsymbol{d}}^{\boldsymbol{r}[\boldsymbol{b},\boldsymbol{d}]}$. By the $\forall$-introduction rule, $\{A\}_{\boldsymbol{b}}^{\boldsymbol{t}[\boldsymbol{b},\boldsymbol{d}]} \vee \forall x \{B\}_{\boldsymbol{d}}^{\boldsymbol{r}[\boldsymbol{b},\boldsymbol{d}]}$. Hence, we can conclude either $\{A\}_{\boldsymbol{b}}^{\boldsymbol{t}[\boldsymbol{b},\boldsymbol{d}]} \vee \{\forall x B\}_{\boldsymbol{d}}^{\boldsymbol{r}[\boldsymbol{b},\boldsymbol{d}]}$ or $\{A\}_{\boldsymbol{b}}^{\boldsymbol{t}[\boldsymbol{b},\boldsymbol{d}]} \vee \{\forall x B\}_{\boldsymbol{d}}^{\{\boldsymbol{r}[\boldsymbol{b},\boldsymbol{d}]\}}$, depending on the way we interpret $\forall x A$. □

3 Motivation

Starting with Gödel's seminal *dialectica* (functional) interpretation [24], the focus of different new functional interpretations has mainly been on the way that (usually typed) quantifiers $\forall x^X A(x)$ and $\exists x^X A(x)$ are interpreted. So, why shift the focus from quantifiers (now treated uniformly) to atomic formulas (now carrying computational content)? We offer several arguments in favour of this approach.

3.1 Nothing is Lost

Since we can think of quantifications over a sort or type X

$$\forall x^X A(x) \qquad \text{and} \qquad \exists x^X A(x) \tag{13}$$

as 'unqualified' quantifications relativized to elements having property X

$$\forall x(X(x) \to A(x)) \qquad \text{and} \qquad \exists x(X(x) \wedge A(x)) \tag{14}$$

nothing is lost by focusing on the computational interpretation of atomic formulas. For instance, in arithmetic one would normally say that a witness for $\exists n^{\mathbb{N}} A(n)$ is a number n and a witness for $A(n)$. But if we have $\mathbb{N}(n)$ as an atomic formula, and choose the base interpretation:

$$n \lhd^{\mathbb{N}} k \;:\equiv\; n = k, \tag{15}$$

then a witness for $\exists n(\mathbb{N}(n) \wedge A(n))$ is also a k such that $\exists n(n = k \wedge A(n))$, which is just $A(k)$. The axioms of arithmetic are then stated with quantifications relativized to $\mathbb{N}(n)$, e.g. the induction axiom schema would be

$$A(0) \wedge \forall n^{\mathbb{N}}(A(n) \to A(n+1)) \to \forall n^{\mathbb{N}} A(n)$$

where $\forall n^{\mathbb{N}} A(n)$ is an abbreviation for $\forall n(\mathbb{N}(n) \to A(n))$. It is easy to check that the above is a sound base interpretation, i.e. the non-logical axioms are $\mathbf{HA}^\omega$-interpretable. This is essentially the Diller-Nahm's variant [8] of Gödel's dialectica interpretation [24]. Most other functional interpretations can be recovered in this way (see [9,17]).

3.2 Some Might be Gained

In Sect. 2 we purposely did not mention *types*. If the base interpretation $\boldsymbol{x} \lhd_{\boldsymbol{b}}^{P} \boldsymbol{a}$, is such that the tuples $\boldsymbol{a}$ and $\boldsymbol{b}$ have fixed types (e.g. in the language of finite types), then it is easy to see that this property also holds for all $\{A\}_{\boldsymbol{b}}^{\boldsymbol{a}}$ – i.e. the types of $\boldsymbol{a}$ and $\boldsymbol{b}$ will be determined by the logical structure of A. On the other hand, if we are working in an untyped language, then the λ-terms constructed in Theorem 1 can also be viewed as untyped λ-terms.

Alexander Miquel [43] has recently developed a novel theory of *implicative structures* with the main goal of unifying Krivine's realizability [41] of second-order arithmetic with Cohen's method of forcing in set theory [7].

Definition 4 ([43]). *An* implicative structure $(\mathcal{A}, \preccurlyeq, \bigwedge, \to)$ *is a complete meet-semilattice* $(\mathcal{A}, \preccurlyeq, \bigwedge)$ *together with a binary operation* $(a, b) \mapsto (a \to b)$*, called the* implication of $\mathcal{A}$*, satisfying the following two conditions:*

1. *Implication is anti-monotonic w.r.t. its first operand and monotonic w.r.t. its second operand*
$$a' \preccurlyeq a, b \preccurlyeq b' \quad \textit{entails} \quad (a \to b) \preccurlyeq (a' \to b').$$
2. *Implication commutes with arbitrary meets on its second operand, i.e. for all* $B \subseteq \mathcal{A}$
$$\bigwedge_{b \in B} (a \to b) = a \to \bigwedge_{b \in B} b.$$

As described in [43], application and λ-abstraction can be defined in any implicative structure[5] as

$$t(s) :\equiv \bigwedge\{x \in \mathcal{A} \mid t \preccurlyeq (s \to x)\} \qquad \lambda x.t[x] :\equiv \bigwedge\{x \to t[x] \mid x \in \mathcal{A}\} \quad (16)$$

where $t[x] \in \mathcal{A}$ for each $x \in \mathcal{A}$. Therefore, for a fixed implicative structure $(\mathcal{A}, \preccurlyeq, \to)$, we can view the witnessing terms constructed in the proof of Theorem 1 as elements of $\mathcal{A}$. In particular, when $(\mathcal{A}, \preccurlyeq, \to)$ is derived from a complete Heyting algebra $(\mathcal{A}, \preccurlyeq, \wedge)$, by defining

$$a \to b \ := \ \max\{c \in \mathcal{A} \mid (c \wedge a) \preccurlyeq b\},$$

we have that function application pa becomes the meet $p \wedge a$. Moreover, once we have a lattice structure on the realizers, we no longer need finite sets, but could just work with the lattice order, replacing unions with joins. In this case, the two clauses in Definition 2 involving function applications become statements about the lattice structure:

$$\begin{aligned} \{P(\boldsymbol{x})\}_{\boldsymbol{u}}^{\boldsymbol{v}} &:\equiv \exists \boldsymbol{a} \preccurlyeq \boldsymbol{v} \forall \boldsymbol{b} \preccurlyeq (\boldsymbol{u} \wedge \boldsymbol{a})(\boldsymbol{x} \lhd_{\boldsymbol{b}}^{P} \boldsymbol{a}) \\ \{\neg A\}_{\boldsymbol{a}}^{\boldsymbol{v}} &:\equiv \exists \boldsymbol{b} \preccurlyeq \boldsymbol{v} \neg \{A\}_{\boldsymbol{b}}^{\boldsymbol{a} \wedge \boldsymbol{b}}. \end{aligned}$$

[5] See [43] for more details, including proofs that the application and abstraction as defined above satisfy the main properties required from a λ-calculus.

One can think of the above in two ways: (1) as a generalisation of the functional interpretation where the witnesses for A are elements of a complete Heyting algebra, or (2) a relational (pointwise) version of forcing, where $\{A\}_{\boldsymbol{b}}^{\boldsymbol{a}}$ could also be written as $\boldsymbol{a} \Vdash_{\boldsymbol{b}} A$, read as "$\boldsymbol{a}$ *forces* A *relative to* $\boldsymbol{b}$".

3.3 Alternative Interpretation of Abstract Spaces

Most of the recent successes in proof mining rely on meta-theorems that involve extensions of $\mathbf{PA}^\omega$ with so-called *abstract spaces*, introduced in [31]. Abstract spaces are a way of formalising mathematical structures such as normed spaces, W-hyperbolic spaces and CAT(0)-spaces as new base types $X, Y, \ldots$ together with axiomatisations that ensure these new base types behave as the corresponding spaces. Quantifications over these spaces, e.g. $\exists x^X A(x)$, are dealt with via Kohlenbach's *monotone functional interpretation*. By considering X as a new base type, the *dialectica* interpretation of a statement involving quantifications over X would require the witnessing functionals to be built from the new type X. By extending the majorizability relation in a way that elements of X can be majorized by computational types (e.g. $\mathbb{N}$ or $\mathbb{N} \to \mathbb{N}$), the bounding witnesses are then standard computable functionals.

For instance, in the case of *normed spaces* X, majorizability is defined as

$$x \leq^*_X n \;:\equiv\; \|x\| \leq n \tag{17}$$

where n is a natural number. The *dialectica* interpretation of a statement such as

$$\exists x^X P(x) \to \exists y^X Q(y), \tag{18}$$

assuming $P(x)$ and $Q(x)$ are quantifier-free, would be

$$\exists f^{X \to X} \forall x^X (P(x) \to Q(fx)).$$

Therefore, the *monotone functional interpretation* of (18) would ask for a term $t \colon \mathbb{N} \to \mathbb{N}$ such that

$$\exists f^{X \to X} (\underbrace{\forall x^X, n^{\mathbb{N}} (\|x\| \leq n \to \|fx\| \leq tn)}_{t \text{ majorizes } f} \wedge \underbrace{\forall x^X (P(x) \to Q(fx))}_{f \text{ witnesses (18)}}). \tag{19}$$

The base interpretation of atomic formulas as discussed above presents an alternative approach: We can view quantifications over X as relativizations, e.g. $\exists x^X A(x)$ stands for $\exists x (X(x) \wedge A(x))$, where $\exists x$ can then be treated uniformly and the atomic formula $X(x)$ can be given a *base interpretation* as

$$x \lhd^X n \;:\equiv\; \|x\| \leq n. \tag{20}$$

By doing this, instead of applying majorizability at the end, after the *dialectica* interpretation has been performed, one can instead inductively work with the functional interpretation and the boundedness of elements of X together.

Considering again the simple example (18), the uniform interpretation built on the base interpretation (20) would ask for a term t such that

$$\forall n^{\mathbb{N}}(\exists x^X(\|x\| \leq n \wedge P(x)) \to \exists y^X(\|y\| \leq tn \wedge Q(y))). \tag{21}$$

By axiom of choice, the above is equivalent to

$$\forall n^{\mathbb{N}}\exists f^{X\to X}\forall x^X \underbrace{((\|x\| \leq n \wedge P(x)) \to (\|fx\| \leq tn \wedge Q(fx)))}_{\text{majorizability and witnessing combined}}. \tag{22}$$

By comparing (19) and (22) one can see, already at this simple case of $\exists \to \exists$, that the two interpretations will have very different properties and hence might lead to finer results in different scenarios. For instance, in (22) the realizer t is not really needed to majorize f, but only needs to do so for values x that satisfy $P(x)$. Moreover, the function f in (19) always maps P's to Q's, whereas in (22) we can only conclude that a sequence of f's exists, where each f works for some given n and $\|x\| \leq n$. In an upcoming joint work with Fernando Ferreira [17], we investigate this alternative approach in more detail, including a comparison with the *bounded functional interpretation* approach to proof mining.

4 Decomposing the Classical Uniform Interpretation

Let us conclude by showing that the uniform functional interpretation of classical first-order logic presented here (Definition 2) can be seen as a composition of a negative translation and a (uniform) functional interpretation of intuitionistic first-order logic. This is similar to what has been done in [50] (Krivine classical realizability) and [55] (Shoenfield interpretation). We will work informally on the $\{\neg, \wedge, \exists\}$-fragment of intuitionistic first-order logic.

Definition 5 (Krivine negative translation [40]**).** *For any formula A in the $\{\neg, \vee, \forall\}$-fragment of classical first-order logic, let $A^{\neg}$ be a formula in the $\{\neg, \wedge, \exists\}$-fragment of intuitionistic first-order logic, defined as follows:*

$$\begin{aligned} (P(\boldsymbol{x}))^{\neg} &:\equiv \neg P(\boldsymbol{x}) & \qquad (\neg A)^{\neg} &:\equiv \neg A^{\neg} \\ (A \vee B)^{\neg} &:\equiv A^{\neg} \wedge B^{\neg} & (\forall x A)^{\neg} &:\equiv \exists x A^{\neg}. \end{aligned}$$

Finally, let $A^{\neg\neg} :\equiv \neg A^{\neg}$.

Proposition 2 ([40])**.** *If A is provable classically then $A^{\neg\neg}$ is provable intuitionistically.*

Assuming a base interpretation of atomic formulas (Definition 1), we can also give a uniform functional interpretation of *intuitionistic* first-order logic as follows, where we now assume that the source theory $\mathbf{S}$ is an intuitionistic theory.

Definition 6 (Intuitionistic uniform functional interpretation, see [9]). *Let a base interpretation of* **S** *into* **T** *be given. For each formula A of* **S**, *possibly with free-variables, associate a formula $\{A\}^{\boldsymbol{a}}_{\boldsymbol{b}}$ of* **T**, *by induction on A, as follows. For atomic formulas we let:*

$$\begin{aligned} |s=t| &:\equiv s=t \\ |P(\boldsymbol{x})|^{\boldsymbol{a}}_{\boldsymbol{b}} &:\equiv \boldsymbol{x} \lhd^{P}_{\boldsymbol{b}} \boldsymbol{a}. \end{aligned}$$

Suppose $|A|^{\boldsymbol{a}}_{\boldsymbol{b}}$ and $|B|^{\boldsymbol{c}}_{\boldsymbol{d}}$ are the interpretations of A and B, respectively. We define

$$\begin{aligned} |\neg A|^{\boldsymbol{p}}_{\boldsymbol{a}} &:\equiv \neg \forall \boldsymbol{b} \in \boldsymbol{pa} |A|^{\boldsymbol{a}}_{\boldsymbol{b}} \\ |A \wedge B|^{\boldsymbol{a},\boldsymbol{c}}_{\boldsymbol{b},\boldsymbol{d}} &:\equiv |A|^{\boldsymbol{a}}_{\boldsymbol{b}} \wedge |B|^{\boldsymbol{c}}_{\boldsymbol{d}} \\ |\exists x A|^{\boldsymbol{a}}_{\boldsymbol{V}} &:\equiv \exists x \forall \boldsymbol{b} \in \boldsymbol{V} |A|^{\boldsymbol{a}}_{\boldsymbol{b}}. \end{aligned}$$

The above can be seen as a variant of the Diller-Nahm interpretation [8] where the existential quantifier is treated uniformly, and the atomic formulas are given computational content. It is one of the interpretations considered in [9], except that we allow here for a more general interpretation of atomic formulas (atomic formulas are also permitted to carry "negative" computational content).

Lemma 1. $\{A\}^{\boldsymbol{V}}_{\boldsymbol{p}} \Leftrightarrow \exists \boldsymbol{a} \in \boldsymbol{V} \neg |A^{\neg}|^{\boldsymbol{p}}_{\boldsymbol{a}}$.

Proof. Induction on A, using alternative interpretation (8) of $\forall x A$. □

We can now see that the classical uniform functional interpretation of Definition 2 is indeed a composition of the intuitionistic interpretation (Definition 6) with Krivine's negative translation (Definition 5).

Proposition 3. $\{A\}^{\boldsymbol{\phi p}}_{\boldsymbol{p}} \Leftrightarrow |A^{\neg\neg}|^{\boldsymbol{\phi}}_{\boldsymbol{p}}$.

Proof. By induction on the structure of A. Let us consider two cases. When $A = P(\boldsymbol{x})$ we have

$$\begin{aligned} \{P(\boldsymbol{x})\}^{\boldsymbol{\phi p}}_{\boldsymbol{p}} &\equiv \exists \boldsymbol{a} \in \boldsymbol{\phi p} \forall \boldsymbol{b} \in \boldsymbol{pa} (\boldsymbol{x} \lhd^{P}_{\boldsymbol{b}} \boldsymbol{a}) \\ &\Leftrightarrow \neg \forall \boldsymbol{a} \in \boldsymbol{\phi p} \neg \forall \boldsymbol{b} \in \boldsymbol{pa} |P(\boldsymbol{x})|^{\boldsymbol{a}}_{\boldsymbol{b}} \\ &\equiv \neg \forall \boldsymbol{a} \in \boldsymbol{\phi p} |\neg P(\boldsymbol{x})|^{\boldsymbol{p}}_{\boldsymbol{a}} \\ &\equiv |(P(\boldsymbol{x}))^{\neg\neg}|^{\boldsymbol{\phi}}_{\boldsymbol{p}}. \end{aligned}$$

For the universal quantifier $\forall x A$, using interpretation (8), we have

$$\begin{aligned} \{\forall x A\}^{\boldsymbol{\phi p}}_{\boldsymbol{p}} &\Leftrightarrow \exists \boldsymbol{a} \in \boldsymbol{\phi p} \forall x \{A\}^{\boldsymbol{a}}_{\boldsymbol{p}} \\ &\overset{\text{L.1}}{\Leftrightarrow} \exists \boldsymbol{a} \in \boldsymbol{\phi p} \forall x \exists \boldsymbol{a}' \in \boldsymbol{a} \neg |A^{\neg}|^{\boldsymbol{p}}_{\boldsymbol{a}'} \\ &\Leftrightarrow \neg \forall \boldsymbol{a} \in \boldsymbol{\phi p} \exists x \forall \boldsymbol{a}' \in \boldsymbol{a} |A^{\neg}|^{\boldsymbol{p}}_{\boldsymbol{a}'} \\ &\equiv \neg \forall \boldsymbol{a} \in \boldsymbol{\phi p} |\exists x A^{\neg}|^{\boldsymbol{p}}_{\boldsymbol{a}} \\ &\equiv |(\forall x A)^{\neg\neg}|^{\boldsymbol{\phi}}_{\boldsymbol{p}}. \end{aligned}$$

The other two cases (disjunction and negation) are similar. □

Acknowledgment. The work presented here is based on recent joint work with Bruno Dinis [9], and on several discussions I have had with Fernando Ferreira over these past few months. Fernando and I are currently working on an application of the uniform interpretation to the setting of finite types [17], including a novel interpretation of the function types. I would also like to thank the anonymous referees for several interesting comments and questions.

References

1. Arthan, R., Oliva, P.: On the Borel-Cantelli lemmas, the Erdös-Rényi theorem, and the Kochen-stone theorem. J. Logic Anal. **13**, 1–23 (2021)
2. Avigad, J., Feferman, S.: Gödel's functional ("Dialectica") interpretation. In: Buss, S.R. (ed.), Handbook of proof theory, volume 137 of Studies in Logic and the Foundations of Mathematics, pp. 337–405. North Holland, Amsterdam (1998)
3. Avigad, J., Gerhardy, P., Towsner, H.: Local stability of ergodic averages. Trans. Am. Math. Soc. **362**, 261–288 (2010)
4. van den Berg, B., Briseid, E., Safarik, P.: A functional interpretation for nonstandard arithmetic. Ann. Pure Appl. Logic **163**(12), 1962–1994 (2012)
5. Berger, U.: Uniform Heyting arithmetic. Ann. Pure Appl. Logic **133**, 125–148 (2005)
6. Bezem, M.: Strongly majorizable functionals of finite type: a model for bar recursion containing discontinuous functionals. J. Symbolic Logic **50**, 652–660 (1985)
7. Cohen, P.: Set Theory and the Continuum Hypothesis. Benjamin, W. A (1966)
8. Diller, J., Nahm, W.: Eine Variant zur Dialectica-interpretation der Heyting Arithmetik endlicher Typen. Arch. Math. Logik Grundlagenforsch **16**, 49–66 (1974)
9. Dinis, B., Oliva, P.: A parametrised functional interpretation of Heyting arithmetic. Ann. Pure Appl. Logic **172**(4) (2021)
10. Escardó, M.H., Oliva, P.: Computational interpretations of analysis via products of selection functions. In: Ferreira, F., Lowe, B., Mayordomo, E., Gomes, L.M.: (eds.), Computability in Europe 2010, LNCS 6158, pp. 141–150. Springer (2010)
11. Escardó, M.H., Oliva, P.: Selection functions, bar recursion, and backward induction. Math. Struct. Comput. Sci. **20**(2), 127–168 (2010)
12. Escardó, M.H., Oliva, P.: The Herbrand functional interpretation of the double negation shift. J. Symbolic Logic **82**(2), 590–607 (2017)
13. Escardó, M.H., Oliva, P.: Sequential games and optimal strategies. Roy. Soc. Proc. A **467**, 1519–1545 (2011)
14. Ferreira, F., Ferreira, G.: A herbrandized functional interpretation of classical first-order logic. Arch. Math. Logic **56**(5), 523–539 (2017)
15. Ferreira, F., Oliva, P.: Bounded functional interpretation. Ann. Pure Appl. Logic **135**, 73–112 (2005)
16. Ferreira, F., Oliva, P.: Bounded functional interpretation in feasible analysis. Ann. Pure Appl. Logic **145**, 115–129 (2007)
17. Ferreira, F., Oliva, P.: The uniform functional interpretation with informative types. In: Preparation (2025)
18. Ferreira, G., Oliva, P.: Functional interpretations of intuitionistic linear logic. In: Grädel, E., Kahle, R. (eds.), Proceedings of CSL, vol. 5771 of LNCS, pp. 3–19. Springer (2009)
19. Ferreira, G., Oliva, P.: Confined modified realizability. Math. Log. Q. **56**(1), 13–28 (2010)

20. Ferreira, G., Oliva, P.: On various negative translations. In: Proceedings of Classical Logic and Computation 2010, volume 47 of Electronic Proceedings in Theoretical Computer Science, pp. 21–33. EPTCS (2011)
21. Ferreira, G., Oliva, P.: On the relation between various negative translations. Logic, Constr. Comput. Ontos-Verlag Math. Logic Ser. **3**, 227–258 (2012)
22. Gaspar, J., Oliva, P.: Proof interpretations with truth. Math. Log. Q. **56**(6), 591–610 (2010)
23. Gödel, K.: Zur intuitionistischen Arithmetik und Zahlentheorie. Ergebnisse eines Mathematischen Kolloquiums **4**, 34–38 (1933)
24. Gödel, K.: Über eine bisher noch nicht benützte Erweiterung des finiten Standpunktes. Dialectica **12**, 280–287 (1958)
25. Hernest, M.-D., Oliva, P.: Hybrid functional interpretations. Proc. CiE, LNCS **5028**, 251–260 (2008)
26. Howard, W.A.: Hereditarily majorizable functionals of finite type. In: Troelstra, A.S. (ed.) Metamathematical investigation of intuitionistic Arithmetic and Analysis. Lecture Notes in Mathematics, vol. 344, pp. 454–461. Springer, Berlin (1973)
27. Kleene, S.C.: On the interpretation of intuitionistic number theory. J. Symbolic Logic **10**, 109–124 (1945)
28. Kohlenbach, U.: Theory of majorizable and continuous functionals and their use for the extraction of bounds from non-constructive proofs: effective moduli of uniqueness for best approximations from ineffective proofs of uniqueness (German). PhD thesis, Frankfurt, pp. xxii+278 (1990)
29. Kohlenbach, U.: Effective bounds from ineffective proofs in analysis: an application of functional interpretation and majorization. J. Symbolic Logic **57**, 1239–1273 (1992)
30. Kohlenbach, U.: New effective moduli of uniqueness and uniform a-priori estimates for constants of strong unicity by logical analysis of known proofs in best approximation theory. Numer. Funct. Anal. Optim. **14**, 581–606 (1993)
31. Kohlenbach, U.: Some logical metatheorems with applications in functional analysis. Trans. Amer. Math. Soc. **357**(1), 89–128 (2005)
32. Kohlenbach, U.: Applied Proof Theory: Proof Interpretations and their Use in Mathematics. Monographs in Mathematics. Springer (2008)
33. Kohlenbach, U., Leustean, L.: The approximate fixed point property in product spaces. Nonlinear Anal. **66**, 806–818 (2007)
34. Kohlenbach, U., Leustean, L.: A quantitative mean ergodic theorem for uniformly convex banach spaces. Ergodic Theor. Dyn. Syst. **29**, 1907–1915 (2009)
35. Kohlenbach, U., Oliva, P.: Proof mining: a systematic way of analysing proofs in mathematics. Proc. Steklov Inst. Math. **242**, 136–164 (2003)
36. Kohlenbach, U., Oliva, P.: Proof mining in L_1-approximation. Ann. Pure Appl. Logic **121**, 1–38 (2003)
37. Kolmogorov, A.N.: On the principle of the excluded middle (russian). Mat. Sb. **32**, 646–667 (1925)
38. Kreisel, G.: Interpretation of analysis by means of constructive functionals of finite types. In: Heyting, A. (ed.) Constructivity in Mathematics, pp. 101–128. North Holland, Amsterdam (1959)
39. Kreisel, G.: On weak completeness of intuitionistic predicate logic. J. Symbolic Logic **27**, 139–158 (1962)
40. Krivine, J.: Opérateurs de mise en mémoire et traduction de gödel. Arch. Math. Logic **30**(4), 241–267 (1990)
41. Krivine, J.: Dependent choice, 'quote' and the clock. Th. Comp. Sci. **308**, 259–276 (2003)

42. Kuroda, S.: Intuitionistische Untersuchungen der formalistischen Logik. Nagoya Math. J. **3**, 35–47 (1951)
43. Miquel, A.: Implicative algebras: a new foundation for realizability and forcing. Math. Struct. Comput. Sci. **30**(5), 458–510 (2020)
44. Oliva, P.: Understanding and using Spector's bar recursive interpretation of classical analysis. In: Beckmann, A., Berger, U., Löwe, B., Tucker, J.V. (eds.), Proceedings of CiE 2006, LNCS 3988, pp. 423–234. Springer (2006)
45. Oliva, P.: Unifying functional interpretations. Notre Dame J. Formal Logic **47**(2), 263–290 (2006)
46. Oliva, P.: Modified realizability interpretation of classical linear logic. In: Proceedings of the Twenty Second Annual IEEE Symposium on Logic in Computer Science LICS 2007, IEEE Press (2007)
47. Oliva, P.: An analysis of Gödel's dialectica interpretation via linear logic. Dialectica **62**(2), 269–290 (2008)
48. Oliva, P.: Functional interpretations of linear and intuitionistic logic. Inf. Comput. **208**(5), 565–577 (2010)
49. Oliva, P.: Hybrid functional interpretations of linear and intuitionistic logic. J. Log. Comput. **22**(2), 305–328 (2012)
50. Oliva, P., Streicher, T.: On Krivine's realizability interpretation of classical second-order arithmetic. Fund. Inform. **84**(2), 207–220 (2008)
51. de Paiva, V.: The Dialectica categories. In: Gray, J.W., Scedrov, A. (eds.) Proceedings of Categories in Computer Science and Logic, Boulder, CO, 1987, vol. 92, pp. 47–62. Contemporary Mathematics, American Mathematical Society (1989)
52. Shirahata, M.: The Dialectica interpretation of first-order classical linear logic. Theor. Appl. Categories **17**(4), 49–79 (2006)
53. Shoenfield, J.R.: Mathematical Logic. Addison-Wesley Publishing Company (1967)
54. Spector, C.: Provably recursive functionals of analysis: a consistency proof of analysis by an extension of principles in current intuitionistic mathematics. In: Dekker, F. (ed.) Recursive Function Theory: Proc. Symposia in Pure Mathematics, volume 5, pp. 1–27. American Mathematical Society, Providence, Rhode Island (1962)
55. Streicher, T., Kohlenbach, U.: Shoenfield is Gödel after Krivine. Math. Log. Q. **53**, 176–179 (2007)
56. Troelstra, A.S.: Realizability. In: Buss, S.R. (ed.) Handbook of proof theory. volume 137, pp. 408–473. North Holland, Amsterdam (1998)
57. Troelstra, A.S.: History of constructivism in the 20th century. In: Set Theory, Arithmetic, and Foundations of Mathematics: Theorems, Philosophies, vol. 36, pp. 150 – 179. Cambridge University Press (2011)

Computable Analysis for Extraction of Certified Programs and Its Applications

Holger Thies(✉)

Kyoto University, Kyoto, Japan
thies.holger.5c@kyoto-u.ac.jp

Abstract. We present recent progress on the formalization of exact real computation within a dependent type theory and its implementation as the Coq library cAERN. A key feature of cAERN is the extraction of certified Haskell programs for exact real computation from proofs in the system. The extracted programs can be used to compute rigorous numerical approximations with arbitrary precision. We introduce the main aspects of the formalization and present some recent extensions, specifically focusing on the problem of solving initial value problems for systems of first-order ordinary differential equations.

Keywords: Exact Real Computation · Program Extraction · Computable Analysis · Formal Verification

1 Introduction

In the Type-2 Theory of Effectivity (TTE) approach to computable analysis [27], real numbers and other uncountable structures are treated by encoding them as infinite sequences of natural numbers, called *representations*. Computability of operations is shown by defining computable *realizers*, i.e., algorithms transforming infinite sequences encoding the inputs to infinite sequences encoding the outputs. For example, the standard way to represent real numbers is the *Cauchy representation* ρ_{Cauchy}, encoding real numbers by rapidly converging sequences of rationals. Basic operations such as addition, subtraction, multiplication, and division (where defined) can be shown to be computable functions with respect to the Cauchy representation. Although TTE relies on computation over infinite sequences, it provides a realistic model of computation in the sense that it can be used as the theoretical foundation of computer programs over exact representations of real numbers, referred to as *exact real computation*. Implementations of exact real computation (for instance [1,16,22]) typically hide details about concrete representations by adding an abstract data-type for real numbers, along with operations to manipulate such objects exactly. Internally, the implementation still needs to track errors and, for instance, increase the working precision

H. Thies—Supported by JSPS KAKENHI Grant Numbers JP20K19744, JP23H03346 and JP24K20735.

© The Author(s), under exclusive license to Springer Nature Switzerland AG 2025
A. Beckmann et al. (Eds.): CiE 2025, LNCS 15764, pp. 104–115, 2025.
https://doi.org/10.1007/978-3-031-95908-0_8

when necessary. However, the details of this internal implementation should not be visible to the user. Consequently, it should be possible to argue about the correctness of exact real computation programs without explicit reference to the underlying representation.

Indeed, in [7] it is shown that the following list of properties suffices to uniquely describe the standard computability structure on the reals up to computable isomorphism:

1. The constants $0, 1$ are computable.
2. Addition, Subtraction, and Multiplication are computable.
3. Division x/y as partial function defined whenever $y \neq 0$ is computable.
4. The partial order relation $x < y$ defined whenever $x \neq y$ is computable.
5. The limit operation defined on fast Cauchy sequences, i.e. sequences converging with a rate 2^{-n}, is computable.

In [17] a formalization of real numbers and other types relevant for exact real computation inspired by this axiomatization is given in a simple dependent type theory. Furthermore, the theory has been implemented in the Coq proof assistant, as the cAERN library [14]. In cAERN, axiomatically defined types (roughly) correspond to (multi-)represented spaces in TTE and functions over them to continuously/computably realizable functions. This is made precise by showing soundness of the formalization with regards of the realizability interpretation from computable analysis. Furthermore, the Coq implementation allows to extract efficient Haskell programs, built on top of the exact real computation library AERN [16], providing a practical tool for certified exact real computation.

In recent ongoing work, we extended the library towards several different domains, such as non-determinism and operations on complex numbers [13], subsets of Euclidean spaces [15] and a more general theory of spaces of subsets of Polish spaces [12]. Most recently, we have studied the problem of solving *initial value problems* (IVPs) for systems of first-order ordinary differential equations (ODEs) and extended cAERN to solution operators for *one-dimensional polynomial ODEs* using a method based on power series expansions [23].

In this paper, we present key features of the formal development, give an overview over recent developments and, as an application, present some novel results regarding the formalization of ordinary differential equations. More precisely, we present how to generalize the one-dimensional polynomial solver to *analytic* ODEs of *arbitrary dimension*, significantly broadening its possible applications. Additionally, we formulate the results in a more abstract way, mostly using purely algebraic methods without making explicit use of the real number structure, only requiring few properties of the real number structure in cAERN. We think that this makes it more feasible to adapt the results to other formal developments of real numbers in proof assistants.

1.1 Preliminaries and Notation

As in previous work, we assume to work over a simple constructive dependent type theory inspired by the calculus of inductive constructions (CIC) [5,6], the type-theory underlying the Coq proof assistant. We assume two distinct universes: a universe Type of types and a universe Prop of (classical) propositions. We further assume the existence of basic inductive types B for Booleans, N for the natural numbers Z, and for each type X a type $[\mathsf{X}]$ of finite lists of elements of type X, all defined in the usual way. For a list $a : [X]$ of length d, we denote its elements by $a_1, a_2, \ldots, a_d$ and when working with concrete lists sometimes denote them in the form $[a_1, \ldots, a_d]$.

The type of reals R is defined axiomatically as a complete Archimedean ordered field with the above mentioned computability properties. That is, we assume that it contains two distinct constants 0 and 1, the standard arithmetical operators $+, \cdot, /$, a semi-decidable comparison operator $<$, and a limit operator lim. Division x/y is restricted to require a proof that x is nonzero and the limit operator requires a proof that the given sequence converges rapidly. We frequently use other computable operations such as the absolute value $|\cdot|$, the max and min functions, roots and powers, etc., all of which can be defined within the axiomatization. For each $d : \mathsf{N}$ the type R^d for d-dimensional Euclidean space is defined as the type of d-element lists of R, and (point-wise) arithmetic operations and the maximum norm $\|\cdot\|$ are defined in the obvious way.

Occasionally, we also reference results from classical mathematics without a specific formalization in mind, intended to serve as a motivation for the formalization. To clearly distinguish statements formalized in our system from statements in the classical meta-theory, we use $\mathbb{R}, \mathbb{N}, \mathbb{Z}, \ldots$ instead of $\mathsf{R}, \mathsf{N}, \mathsf{Z}, \ldots$ for the corresponding spaces and set theoretic notion (e.g., membership $\in$) instead of type-theoretical notation.

2 Realizability and Program Extraction

As a formal model for our theory, we use the category of assemblies (or multi-representations) over Kleene's second algebra $\mathsf{Asm}(\mathcal{K}_2)$ [26]. An assembly $\mathbf{X}$ is a pair of a set $|\mathbf{X}|$ and a binary relation $\Vdash_{\mathbf{X}} \subseteq \mathbb{N}^{\mathbb{N}} \times |\mathbf{X}|$ such that $\forall (x \in |\mathbf{X}|). \exists (\varphi \in \mathbb{N}^{\mathbb{N}}). \varphi \Vdash_{\mathbf{X}} x$. When $\varphi \Vdash_{\mathbf{X}} x$, we say φ is a realizer of x. Given two assemblies $\mathbf{X}, \mathbf{Y}$, a mapping $f : |\mathbf{X}| \to |\mathbf{Y}|$ is said computable when there is a type-2 machine that for all $x \in |\mathbf{X}|$, transforms each realizer of x to a realizer of $f(x)$. The category $\mathsf{Asm}(\mathcal{K}_2)$ denotes the category of assemblies and computable mappings between them. A type in our type theory is interpreted as an assembly and a term is interpreted as a computable mapping. The soundness of an axiom can thus be validated by an existence of a computable point in the underlying set of the assembly. For example we can consider the computational structure of the reals using the Cauchy assembly $\mathbf{R}_{\mathrm{Cauchy}}$ where $\varphi \in \mathbb{N}^{\mathbb{N}}$ is a realizer of $x \in \mathbb{R}$ if and only if φ encodes a sequence of rationals (in turn appropriately encoded by a sequence of integers) converging rapidly towards x.

Moreover, basic types, their operations, and standard type constructions have well-defined interpretations in the category of assemblies (see e.g. [2]). Classical propositions in Prop can be realized trivially, effectively erasing non-computational content and we can extend the realizability interpretation to axiomatically defined types (see [17, Section 5] for details). As a result, every term in our system naturally provides a computational interpretation as a program, where real numbers are represented in a way suitable for exact real computation.

In the Coq implementation we go one step further and, instead of mapping operations on real numbers to some specific algorithm on infinite sequences, we map them directly to a higher-order type for exact real numbers provided by the Haskell library AERN. Concretely, this is achieved by using Coq's code extraction mechanisms combined with some automatic post-processing of the extracted code to realize axioms by custom implementations in AERN. That is, in the extracted programs e.g. the type R, and its axiomatically defined basic operations such as arithmetic or limits, are directly mapped to corresponding type `CREAL` and its operations in AERN. Although this requires to trust the implementation of basic operations in the AERN library, it leads to significantly more efficient programs as demonstrated by the examples in [17, Section 8.4]. Additionally, it has the advantage that the extracted code can be easily integrated with non-verified components of AERN.

3 Non-Determinism and Continuity

A central result in computable analysis is that any computable function is continuous. However, while inevitable, these continuous semantics can pose difficulties in the practice of exact real computation. For example, no total function from the reals to the booleans is computable, and partiality, corresponding to non-terminating programs, is usually not desirable and challenging to work with in the formal verification setting.

Exact real computation software therefore typically implements primitive operations to construct *non-deterministic* (or *multivalued*) functions, which for a given input have a set of possible outcomes, each of them considered a valid result of the computation. Formally, this means a realizer of a multivalued function needs to map a name of an input to a name of one of the possible function values. Usually, this kind of non-determinism arises from the fact that the same real number can be represented by different infinite sequences, and thus, while being deterministic when regarded as program on these sequences, the program is non-deterministic as a program on real numbers.

To include this kind of non-determinism in our theory, it is modeled by a monad such that for any type $X : \texttt{Type}$, we have a new type $\mathsf{M}X : \texttt{Type}$ that stands for the type of non-deterministic computations in X, characterized by relating it to the classical non-empty power-set monad [13]. This allows us to regard a term of the form $\mathsf{M}X$ as the result of a non-deterministic computation in X. Note that in AERN real valued functions are intrinsically non-deterministic, thus in the extracted code M simply becomes the identity monad.

The fact that every computable function is continuous itself can sometimes be useful for formal verification. Thus, we would like to formalize this continuity property within our system. In constructive mathematics and computable analysis, statements of the form that every function is continuous in some sense, are typically known as *continuity principles*. Introducing continuity principles in type-theoretical settings is quite subtle, and introducing such principles incorrectly can lead to inconsistencies within the system [28].

While in constructive mathematics, typically continuity of functions on Baire space is considered, we choose a different approach which is more natural in our setting. That is, we formalize the continuity principle directly for our axiomatic type of real numbers. As with other aspects of our formalization, the idea of the formulation of the continuity principle in cAERN is based on practical implementations of exact real computation. Such implementations often internally represent real numbers by converging sequences of intervals with dyadic rational endpoints and allow to output a rational approximation of the result of computations with an arbitrarily precise prescribed error-bound. Thus, for a real-valued function $f : \mathsf{R} \to \mathsf{R}$, consider the operation of generating a rational approximation d of $f(x)$ with error at most 2^{-n}. To generate this approximation, the software internally replaces x with a stream of rational approximations converging to x. Thus, as it needs to generate the output in finite time, it can only have read approximations of x up to some finite precision 2^{-m}. By inspecting this internal computation process, we effectively get a modulus of continuity for f at x. In summary, we get the following continuity principle

$$\forall(x : \mathsf{R}).\ \forall(m : \mathsf{N}).\ \mathsf{M}\Sigma(n : \mathsf{N}).\ \forall(y : \mathsf{N}).\ |x - y| < 2^{-n} \to |(f\ x) - (f\ y)| < 2^{-m}.$$

for real valued functions $f : \mathsf{R} \to \mathsf{R}$, i.e. for any point, we can *non-deterministically* find a value for the modulus of continuity of f at x. Note that the non-determinism is essential here, as the same real number x can be represented by different converging intervals, and thus the output depends on the internal representation.

As shown in [15], this form of continuity is valid in the realizability interpretation and can be extracted to a modulus of continuity function implemented in the AERN library. Furthermore, it can be extended to spaces other than the real numbers in a natural way, which turns out to be useful to prove the equivalence of various representations of subsets of metric spaces [12].

4 Polynomials and Analytic Functions

As a more practical example of the formalization, let us introduce some types to compute efficiently with certain classes of real functions.

First, consider multivariate real polynomials as they are fundamental in many areas of computational mathematics. A polynomial of the form

$$p(x_1, \ldots, x_d) = \sum_{(i_1,\ldots,i_d)} a_{i_1,\ldots,i_d} x_1^{i_1} \ldots x_d^{i_d}$$

where $a_{i_1,\ldots,i_d}$ are the real-valued coefficients, can be encoded simply by a (nested) list of its coefficients. However, to smoothly generalize to arbitrary dimension in the formal setting, it is beneficial to consider polynomials over a more general type of coefficients.

In algebra, polynomials are typically defined over a ring, as the structure supports the required fundamental operations, such as addition and multiplication. Furthermore, the ring structure extends naturally to polynomials, i.e. the set of polynomials over a ring forms again a ring, where the operations of addition and multiplication are defined pointwise on the coefficients.

However, in our setting, as we consider structures where equality is not decidable, we cannot extend the ring structure to polynomials while keeping natural and computable definitions of the ring operations. For example, for real polynomial p, q we would need to define subtraction in a way such that $p - p$ always reduces to the (unique) zero polynomial, and thus effectively would need to check that each of the coefficients of p and q are additive inverses to each other.[1]

The slightly weaker notion of a semiring therefore proves to be more suitable in our setting. A *semiring* (also called *rig*) is an algebraic structure similar to a ring, but without the requirement of additive inverses. That is, a semiring consists of a set equipped with two associative and commutative binary operations, addition and multiplication (denoted by $+$ and $\cdot$), such that addition and multiplication have identity elements 0_R and 1_R, multiplication distributes over addition, and 0_R annihilates multiplication.

We define the type of polynomials over a semiring $\mathcal{R}$, denoted by $\mathcal{R}[x]$, as a list $[a_0, \ldots, a_d] : [\mathcal{R}]$ of its coefficients. Standard polynomial operations, such as addition, multiplication, composition, and subtraction (in case $\mathcal{R}$ is a ring), are defined naturally. Furthermore we can define an evaluation function $\mathcal{R}[x] \to \mathcal{R}$, e.g. by the straight-forward evaluation of the finite sum, or more efficient methods like the Horner scheme.

Arithmetic operations can be defined on $\mathcal{R}[x]$ such that they respect the semiring structure of $\mathcal{R}$, ensuring that $\mathcal{R}[x]$ (with the empty list $[]$ and the single-element list $[1]$ as identity for addition and multiplication, respectively) itself forms a semiring. Note, however, that even when $\mathcal{R}$ is a ring, we can not necessarily show that $\mathcal{R}[x]$ forms a ring in our setting.

Multivariate polynomials can then be defined inductively as polynomials with polynomial coefficients. That is, we define the type $\mathcal{R}[x_1, \ldots, x_d]$ of multivariate polynomials over the semiring $\mathcal{R}$ inductively via $\mathcal{R}[x_1, \ldots, x_{d+1}] := \mathcal{R}[x_2, \ldots, x_{d+1}][x_1]$ with multivariate evaluation $p(x_1, \ldots, x_d)$ and multivariate composition $p(q_1, \ldots, q_d) : \mathcal{R}[x_1, \ldots, x_r]$ of polynomials $p : \mathcal{R}[x_1, \ldots, x_d]$ and $q_1, \ldots, q_d : \mathcal{R}[x_1, \ldots, x_r]$ defined recursively. We typically denote polynomials by letters $p, q, \ldots$ and use standard mathematical notation such as $p + q$, pq, $p \circ q$, $p(x)$ for the operations.

[1] Alternatively, we could identify two lists when they represent the same real-valued function. However, as Coq does not support quotient types, this would either require the introduction of additional axioms or to use a setoid structure instead of the standard Leibniz equality. As we wanted to avoid introducing this additional complexity in the formalization, choosing the semi-ring structure appears to be a more practical solution.

A useful property of real-valued polynomials is that they are smooth, i.e., that they are infinitely many times differentiable. Now, the derivative of a polynomial can be defined purely algebraically, without using analytic methods such as limits. We define a differentiation operator $\partial : \mathcal{R}[x] \to \mathcal{R}[x]$ by $\partial[a_0, \ldots, a_d] := [a_1, 2a_2, \ldots, da_d]$ where a natural number $i : \mathsf{N}$ is embedded into the semiring $\mathcal{R}$ via the canonical map summing up the multiplicative identity 1_R exactly i times. It is not difficult to verify that this operator satisfies the standard properties of the derivative, i.e., linearity $\partial(p+q) = \partial(p) + \partial(q)$, the Leibniz Rule $\partial(pq) = \partial(p)q + p\partial(q)$ and the chain rule $\partial(p \circ q) = (\partial p \circ q) \cdot \partial q$.

Furthermore, for any $d : \mathsf{N}$, we define the partial derivatives $\partial_1, \ldots, \partial_d : \mathcal{R}[x_1, \ldots, x_d] \to \mathcal{R}[x_1, \ldots, x_d]$ inductively by $\partial_1 := \partial$, and $\partial_{i+1}[a_0, \ldots, a_d] := [\partial_i a_0, \ldots, \partial_i a_d]$ for $k = 1, \ldots, d-1$. Using linearity and the Leibniz rule, we can extend the chain rule to the multivariate case: For $p : R[x_1, \ldots, x_r]$, $q_1, \ldots, q_r : R[x_1, \ldots, x_d]$ and $k = 1, \ldots, d$, we get $\partial_k p(q_1, \ldots, q_r) = \sum_{i=1}^{r} (\partial_i p)(q_1, \ldots, q_r)\partial_k q_i$.

4.1 Polynomial Approximations of Real Functions

In numerical computations, real functions are frequently approximated by polynomials as they support operations such as evaluation, differentiation, and integration and allow their efficient implementation.

In validated numerics [11], a rigorous approximation of a function over a compact domain is sometimes called a *polynomial model* or *Taylor model* [3,19]. Formally, a polynomial model for a real-valued function $f : B \subseteq \mathbb{R}^d \to \mathbb{R}$ on a compact box $B = [a_1, b_1] \times \cdots \times [a_d, b_d]$ is a multivariate polynomial p together with a guaranteed error bound Δ such that $\forall (x \in B).\ \|f(x) - p(x)\| \leq \Delta$. Arithmetic and other mathematical operations, such as, for instance, integration, can be defined by performing the corresponding operation on the underlying polynomials and then combining the error bounds in a systematic way. This allows to replace real-valued functions by polynomial models in calculations while maintaining rigorous error bounds.

We define the type $\mathtt{PM}_d$ of d-variate *polynomial models* as a triple $(D, p, \Delta) : \mathsf{I}^d \times \mathsf{R}[X^d] \times \mathsf{R}$ where I is a type of real-valued intervals defined as the pair of its end-points. The triple is meant to represent the approximation of some real valued function, where D gives the domain where the approximation is valid, p is a multivariate polynomial and Δ is a rigorous error bound for the difference between the polynomial and the function over D. Arithmetic operations on polynomial models are defined such that the resulting model accurately represents the intended function over the domain.

In the setting of exact real computation, we are particularly interested in the case where we can systematically make the error bound Δ arbitrarily small. That is, instead of single polynomial models, we would like to uniformly generate sequences of polynomial models with $\Delta \to 0$ effectively. For analytic functions there is a canonical way to generate such sequences, simply by taking initial segments of their infinite power series. That is, we generalize polynomials to infinite series $\sum_{n=0}^{\infty} a_n x^n$, called *formal power series*. We define the type $\mathcal{R}[[x]]$

of formal power series over a semiring $\mathcal{R}$ as the type of infinite sequence $a : \mathsf{N} \to \mathcal{R}$. Similar to polynomials, addition and multiplication can be defined on $\mathcal{R}[[x]]$ as coefficient-wise addition and convolution, equipping $\mathcal{R}[[x]]$ itself with a semiring structure and thus allowing multivariate formal power series to be defined analogously to the polynomial case. Furthermore, partial derivatives can be generalized to formal power series in the obvious way. Formal power series, however, do not need to converge for a given $x : \mathcal{R}$, and even when they do, their value might not be computable [21]. Thus, there is no evaluation operator on formal power series. However, enriching the formal power series with some additional information, we can extend them to a type for analytic functions.

A function $f : \mathbb{R}^d \to \mathbb{R}$ is said to be analytic at a point $x_0 \in \mathbb{R}^d$ if it can be locally expressed by a power series, i.e. there is a neighborhood of x_0 such that $f(x) = \sum_{\alpha \in \mathbb{N}^d} a_\alpha (x - x_0)^\alpha$ where a is a multivariate power series. Such a function is smooth at x_0 and the power series is given by $a_\alpha = \frac{1}{\alpha!} D^\alpha(x_0)$. Furthermore, there exist constants $M, r : \mathbb{R}$ such that $|a_\alpha| \leq M r^{-\alpha}$. These contants are useful to make a tail estimate bounding the difference between the finite sum $\sum_{n=N}^{\infty} a_\alpha (x - x_0)^\alpha$ and the infinite power series whenever $\|x - x_0\| < r$. Thus, given M, r and the infinite sequence, it is possible to uniformly define an evaluation operator for the power series in a neighborhood of x_0. We define the type PS of *bounded power series* as the type of triples $(a, M, r) : \mathsf{R}[x_1, \ldots, x_d] \times \mathsf{R} \times \mathsf{R}$ satisfying the above bound. It can furthermore be shown that on a fixed interval, the error term decreases exponentially in the order of the approximation N, and it is possible to define addition, multiplication and taking derivatives on bounded power series. Details of these constructions can e.g. be found in [10].

5 Ordinary Differential Equations

Let us now consider the problem of solving initial value problems (IVPs) for Ordinary Differential Equations (ODEs). An IVP for a system of ODEs consists of a set of equations describing the evolution of unknown functions along with initial conditions, i.e., equations of the form

$$\dot{\vec{y}}(t) = \vec{f}(t, \vec{y}(t)), \quad \vec{y}(t_0) = \vec{y_0} \tag{1}$$

where $\vec{y}(t) = (y_1(t), \ldots, y_d(t))$ is a vector of unknown functions depending on time t, $\vec{f}(t, \vec{y}(t)) = (f_1(t, y_1(t), \ldots, y_d(t)), \ldots, f_d(y, y_1(t), \ldots, y_d(t))$ is a vector of given functions describing the evolution of the system (also called the *right-hand side* of the IVP), $\vec{y_0} = (y_{0,1}, \ldots y_{0,d})$ is a vector of initial conditions at $t = t_0$, and $\dot{\vec{y}}(t) = (\dot{y}_1(t), \ldots, \dot{y}_d(t))$ denotes the derivative of $\vec{y}(t)$ with respect to the time variable t. We call $\vec{y}(t)$ the *solution* of the IVP (1).

There are IVPs that do not have a solution, and even for those that do, the solution is not always unique, in which case it might not have any computable solution even when all inputs are computable [24]. On the other hand, if the ODE has a unique solution, and $\vec{f}$ and $\vec{y_0}$ are computable, then so is the solution function [20]. Nevertheless, computing the solution may not be tractable in terms of time complexity, even for Lipschitz-continuous right-hand side [9]. Implementations in exact real computation (e.g. [4,10,25]) therefore often focus on the case where the right-hand side functions are polynomials or at least analytic, which has been shown to offer better computational complexity properties [21].

Here, we also consider the case where the right-hand side functions are analytic at $\vec{y_0}$. Analytic ODEs have a unique solution in a neighborhood of t_0 which is again analytic. Before we define the problem formally, let us note that any system (1) can be transformed into an autonomous system (i.e. one where the right-hand side functions do not explicitly depend on the variable t) by introducing the auxiliary function $y_{d+1}(t) = t$ and thus it suffices to consider only such systems. As autonomous systems do not depend explicitly on the time t, we can further assume that $t_0 = 0$.

5.1 Formal Power Series Solution

We formalize a d-dimensional autonomous analytic IVP as a tuple of a d-element list $[p_1, p_2, \ldots, p_d]$ of analytic right-hand side functions (encoded as formal power series with bounds on the coefficients) and a d-element list $[y_{0,1}, y_{0,2}, \ldots, y_{0,d}] : [\mathsf{R}]$ of initial values. Computing the power series of the solution from the power series of the right hand side functions can be formulated purely in algebraic terms. That is, our goal is to compute a sequence $y_1, \ldots, y_d : \mathsf{R}[[t]]$ satisfying $\partial y_i = p_i(y_1, \ldots, y_d)$. Applying the chain rule to get the derivative of the right-hand side and substituting the expressions for the first derivative, we get the expression

$$\partial^2 y_i = \sum_{j=1}^{d} \dot{y_j}(\partial_j p_i)(y_1, \ldots, y_d) = \sum_{j=1}^{d} p_j(y_1, \ldots, y_d) \cdot (\partial_j p_i)(y_1, \ldots, y_d)$$

for the second derivative.

As all operations are defined on formal power series, this reduces again to a formal power series, meaning we can apply this recursively to get higher derivatives $\delta^k y_i$. More precisely, we inductively define a sequence $p_i^{[n]} : \mathsf{R}[[x_1, \ldots, x_d]]$ of multivariate formal power series such that $y_i(t) = \sum_{m=0}^{\infty} p_i^{[m]}(y_0)(t - t_0)^m.$, thus giving us an expression of the formal power series of y at t_0.

It remains to show that this power series converges and that we can evaluate it effectively. Note that even in case that the right-hand side functions are all polynomial, only in trivial cases the solution will again be polynomial and thus we need to evaluate an infinite series, requiring an error bound when approximating the function by a finite initial segment of the power series. By Taylor's theorem we get that for any t close enough to t_0,

$$y_i(t) = \sum_{m=0}^{k} p_i^{[m]}(y_0)(t - t_0)^m + R_k(t)$$

with remainder term given by

$$R_k(t) = \frac{1}{(n+1)!} y_i^{(k+1)}(\xi)(t - t_0)^{k+1} = p_i^{[k+1]}(y_1(\xi), \dots, y_d(\xi))(t - t_0)^{k+1} \quad (2)$$

for some $\xi \in [t_0, t]$. Now assume that $B, r_0 \in \mathsf{R}$ satisfy $\|p_i(y)\| \leq B$ for all $i = 1, \dots, d$ and $y \in \mathsf{R}$ with $\|y - y_0\| \leq r_0$. Such a bound B can always be found for any r_0 when p_i are polynomials, for example simply by inspecting its coefficients or by using methods from interval arithmetic. Moreover, for a bounded power series with series bound (M, r), B can be explicitly computed from the series bound for any $r_0 < r$. It follows that the solutions $y_i(t)$ exists for $t - t_0 < \min(r_0, \frac{B}{r_0})$ and are bounded by B. Thus, choosing t in this range, we can bound the remainder term (2). Moreover, from analyticity of the solution it follows that the error decreases exponentially in k. Thus, given the IVP (1), we can compute an interval $[t_0, t_1]$ such that for any $t \in [t_0, t_1]$, we can approximate the solution $y(t)$ up to any desired output precision and combining this with the limit operator, we can compute $y(t) : \mathsf{R}$ exactly. Although the time interval is typically small, it can be used to find a new initial value $y(t_1)$, and thus extending the solution by iteratively calling the solution operator, leading to a single-step method for solving the IVP over a larger interval.

5.2 Extending to Polynomial Models

Although the method requires the right-hand side functions to be analytic, it can be extended to more general functions simply by replacing them with polynomial approximations. Assume we are given the IVP (1) with right-hand side $f_1, \dots, f_d : \mathbb{R}^d \to \mathbb{R}$ not necessarily polynomial or analytic and denote its solution by y_f. We get a rigorous approximation of the solution by applying it on polynomial models, that is, assume $(p_1, \Delta_1), \dots, (p_d, \Delta_d)$ are polynomial models for $f_1, \dots, f_d$ on a box of radius r centered at y_0 and denote by $y_p(t)$ the solution of the IVP where the right-hand side functions are replaced by $p_1, \dots, p_d$. Further, let $\Delta := \max_i \Delta_i$ and $B : \mathsf{R}$ a bound on $p_1, \dots, p_d$ on the interval. Then for $t \leq \min(r, t_0 + \frac{r}{B+\Delta})$ both $y_f(t)$ and $y_p(t)$ stay inside the domain of the polynomial model, thus $\|f(y(t)) - p(y(t))\| \leq \Delta$. By Grönwall's Lemma it follows $|y_f(t) - y_p(t)| \leq \frac{\Delta}{L}(e^{L(t-t_0)} - 1)$ where L is a Lipschitz constant for $p_1, \dots, p_d$. Consequently, by choosing t sufficiently small, the solution to the approximate system can be made arbitrarily close to that of the actual system, providing a rigorous approximations.

6 Conclusion

We presented recent progress on formal verification of computable analysis and program extraction within the cAERN library. We summarized its theoretical foundations and some of the main features implemented in recent work. Building upon this theory, we developed a formal framework for solving initial value problems for ordinary differential equations in the setting of exact real computation.

The new framework presented in this paper allows a more abstract formalization, making most results independent from the concrete real number type and in fact mostly independent of the cAERN library itself. Future work includes extending the framework to partial differential equations [25], formalizing more efficient algorithms, and comparing our approach to other formalizations in proof assistants such as [8,18].

References

1. Balluchi, A., Casagrande, A., Collins, P., Ferrari, A., Villa, T., Sangiovanni-Vincentelli, A.L.: Ariadne: a framework for reachability analysis of hybrid automata. In: Proceedings of 17th International Symposium on Mathematical Theory of Networks and Systems, Kyoto (2006)
2. Birkedal, L.: Developing theories of types and computability via realizability. Carnegie-Mellon Univ. Pittsburgh PA School of Computer Science, Technical report (1999)
3. Brauße, F., Korovina, M., Müller, N.: Using taylor models in exact real arithmetic. In: Revised Selected Papers of the 6th International Conference on Mathematical Aspects of Computer and Information Sciences, vol. 9582, MACIS 2015, pp. 474–488, Berlin, Heidelberg (2015)
4. Brauße, F., Korovina, M., Th Müller, N.: Towards using exact real arithmetic for initial value problems. In: Perspectives of System Informatics: 10th International Andrei Ershov Informatics Conference, PSI 2015, in Memory of Helmut Veith, Kazan and Innopolis, Russia, 24–27 August 2015, Revised Selected Papers 10, pp. 61–74. Springer (2016)
5. Coquand, T., Huet, G.P.: The calculus of constructions. Inf. Comput. **76**(2/3), 95–120 (1988)
6. Coquand, T., Paulin, C.: Inductively defined types. In: Martin-Löf, P., Mints, G., (eds.), COLOG-88, International Conference on Computer Logic, Tallinn, USSR, December 1988, Proceedings, volume 417 of *LNCS*, pp. 50–66. Springer (1988)
7. Hertling, P.: A real number structure that is effectively categorical. Math. Log. Q. **45**, 147–182 (1999)
8. Immler, F.: A Verified ODE Solver and Smale's 14th Problem (Ein Verifizierter GDGL-Löser und Smales 14. Problem). PhD thesis, Technical University of Munich, Germany (2018)
9. Kawamura, A.: Lipschitz continuous ordinary differential equations are polynomial-space complete. Comput. Complex. **19**(2), 305–332 (2010)
10. Kawamura, A., Steinberg, F., Thies, H.: Parameterized complexity for uniform operators on multidimensional analytic functions and ODE solving. In: International Workshop on Logic, Language, Information, and Computation, pp. 223–236. Springer (2018)
11. Baker Kearfott, R.: Interval computations: Introduction, uses, and resources. Euromath Bulletin **2**(1), 95–112 (1996)
12. Konecný, M., Park, S., Thies, H.: Formalizing hyperspaces and operations on subsets of polish spaces over abstract exact real numbers. *CoRR*, abs/2410.13508 (2024)
13. Konečný, M., Park, S., Thies, H.: Certified computation of nondeterministic limits. In: NASA Formal Methods Symposium, pp. 771–789. Springer (2022)

14. Konečný, M., Park, S., Thies, H.: cAERN: axiomatic reals and certified efficient exact real computation. https://github.com/holgerthies/coq-aern (2024)
15. Konečný, M., Park, S., Thies, H.: Formalizing hyperspaces for extracting efficient exact real computation. In: Leroux, J., Lombardy, S., Peleg, D., (eds.), 48th International Symposium on Mathematical Foundations of Computer Science (MFCS 2023), volume 272 of Leibniz International Proceedings in Informatics (LIPIcs), pp. 59:1–59:16, Dagstuhl, Germany (2023). Schloss Dagstuhl – Leibniz-Zentrum für Informatik
16. Konečný, M.: aern2-real: a haskell library for exact real number computation. https://hackage.haskell.org/package/aern2-real (2021)
17. Konečný, M., Park, S., Thies, H.: Extracting efficient exact real number computation from proofs in constructive type theory. J. Logic Comput. p. exae066, October 2024
18. Makarov, E., Spitters, B.: The picard algorithm for ordinary differential equations in coq. In: Blazy, S., Christine, P.-M., Pichardie, D. (eds.), Interactive Theorem Proving, pp. 463–468, Berlin, Heidelberg (2013)
19. Makino, K., Berz, M.: Taylor models and other validated functional inclusion methods. Int. J. Pure Appl. Math. **6**, 239–316 (2003)
20. Miller, W.: Recursive function theory and numerical analysis. J. Comput. Syst. Sci. **4**(5), 465–472 (1970)
21. Th Müller, N.: Constructive aspects of analytic functions. In: Proceedings of Workshop on Computability and Complexity in Analysis, vol. 190, pp. 105–114 (1995)
22. Th Müller, N.: The iRRAM: exact arithmetic in C++. In: International Workshop on Computability and Complexity in Analysis, pp. 222–252. Springer (2000)
23. Park, S., Thies, H.: A coq formalization of taylor models and power series for solving ordinary differential equations. In: Bertot, Y., Kutsia, T., Norrish, M., (eds.), 15th International Conference on Interactive Theorem Proving, ITP 2024, 9–14 September 2024, Tbilisi, Georgia, volume 309 of *LIPIcs*, pp. 30:1–30:19. Schloss Dagstuhl - Leibniz-Zentrum für Informatik (2024)
24. Pour-el, M.B., Richards, I.: A computable ordinary differential equation which possesses no computable solution. Ann. Math. Logic **17**(1–2), 61–90 (1979)
25. Selivanova, S., Steinberg, F., Thies, H., Ziegler, M.: Exact real computation of solution operators for linear analytic systems of partial differential equations. In: Computer Algebra in Scientific Computing: 23rd International Workshop, CASC 2021, Sochi, Russia, 13–17 September 2021, Proceedings 23, pp. 370–390. Springer (2021)
26. Van Oosten, J.: Realizability: an introduction to its categorical side. Elsevier (2008)
27. Weihrauch, K.: Computable Analysis. Springer, Berlin (2000)
28. Xu, C.: A continuous computational interpretation of type theories. PhD thesis, University of Birmingham (2015)

Contributed Papers

The Existence of Primitive Recursive Structures

Pavel E. Alaev[1,2(✉)]

[1] Novosibirsk State University, Novosibirsk 630090, Russia
[2] Sobolev Institute of Mathematics, pr. Koptuga 4, Novosibirsk 630090, Russia
alaev@math.nsc.ru

Abstract. We suggest a general, relatively simple criterion for the existence of a primitive recursive (p.r.) or punctual presentation of a given structure. As an example, we prove that every computable Abelian group with a computably enumerable linear basis has a p.r. presentation. A series of other examples is found.

Keywords: computability · computable structure · punctual structure · primitive recursive structure · Abelian group · partial ordering · Boolean algebra

1 Introduction

Let $\mathcal{A} = (A, L^{\mathcal{A}})$ be a structure of a finite signature L. We say that $\mathcal{A}$ is a *primitive recursive (p.r.) structure* if $A \subseteq \omega$ is a primitive recursive set and all operations and relations from $L^{\mathcal{A}}$ are primitive recursive. This definition can be extended to the case of an infinite signature, the exact formulation is given below. In addition, we can introduce another stronger notion. A structure $\mathcal{A}$ is called *punctual* if it is primitive recursive and its universe A is an initial segment of ω. A long series of recent articles is devoted to the study of such structures. Very brief historical notes are given below. In particular, a series of facts is proved in [1] that every computable structure in a class K always has an isomorphic punctual presentation. We note that in [1] it was called a *fully primitive recursive presentation*. A general question (main question) was also formulated there: *when does a computable structure have a punctual presentation*? Recall that a structure $\mathcal{A} = (A, L^{\mathcal{A}})$ is called *computable* if $A \subseteq \omega$ is a computable set and all operations and predicates from $L^{\mathcal{A}}$ are computable.

This article is devoted to answering this question. We formulate a general criterion that says when a structure $\mathcal{A}$ has a primitive recursive or punctual presentation. Then we give some examples of applying this criterion, which shows that it can be easily used.

Let a class K be in the following list: structures with equivalence, linear orderings, torsion-free Abelian groups, or torsion Abelian p-groups, where p is a prime number. It is proved in [1] that every computable structure in K has a

© The Author(s), under exclusive license to Springer Nature Switzerland AG 2025
A. Beckmann et al. (Eds.): CiE 2025, LNCS 15764, pp. 119–132, 2025.
https://doi.org/10.1007/978-3-031-95908-0_9

punctual presentation. The proofs of these facts are based on specific constructions, where we define an increasing chain of finite structures with a number of special properties, sometimes quite cumbersome. Using a general criterion, we suggest a very simple proof for each of these facts.

In addition, we get one more general fact: if $\mathcal{A}$ is a computable Abelian group with a computably enumerable (c.e.) linear basis, which is not torsion, then $\mathcal{A}$ has a punctual presentation.

A structure is called *locally finite* if each of its finitely generated substructure is finite. It is proved in [2] that every computable locally finite structure of a finite signature has an isomorphic presentation computable in polynomial time (P-*computable*). Hence, it also has a primitive recursive presentation. Using a general criterion, we give a very simple proof of the latter fact.

The general definition of a p.r. structure can be found already in [3], where it is formulated in the language of numbered structures. It was also considered in [4]. Some basic results can be found in [5]. Several new results on this topic are proved in [6].

The form of the criterions formulated in Theorems 1–4 is close to a criterion for the existence of P-computable structures which is found in [7,8]. The latter one requires rather complicated and cumbersome calculations, while the criterion in this paper has a quite simple and straightforward proof. However, it turned out to be a handy tool for applications.

Despite the fact that the criterion works well in many special cases, it contains a quantifier over an infinite set, and therefore has the form of a Σ^1_1-statement. This does not allow to call it simple in general. It is shown in [9] that this is not a coincidence, since the index set of computable structures having a P-computable presentation is Σ^1_1-complete. The criteria of this kind cannot be simple.

Perhaps the interest in structures with a full universe arose due to [10], where it is proved that every infinite computable linear order has a P-computable presentation with universe $\{0,1\}^*$, consisting of all words in the alphabet $\{0,1\}$. The size limitations do not allow us to provide any complete overview of research on this topic here.

2 A Criterion for Primitive Recursive Structures

The book [11] can be used as a source of basic facts on computability theory. Let $A \subseteq \omega$. We say that a function $f : A \to \omega$ is *primitive recursive* (*p.r.*) if there exists a p.r.f. $f_0 : \omega \to \omega$ such that $f = f_0|_A$. Let $\mathcal{A}, \mathcal{B}$ be two structures of a signature L with p.r. universes $A, B \subseteq \omega$. We say that they are *primitive recursively isomorphic*, $\mathcal{A} \cong_{\text{pr}} \mathcal{B}$, if there exists an isomorphism $f : \mathcal{A} \to \mathcal{B}$ such that f and f^{-1} are p.r.f. This means identity from the point of view of the theory of primitive recursive structures.

In the proofs below, we use the following simple fact. If $A \subseteq \omega$ is an infinite p.r. set and $f : \omega \to A$ is an injective p.r.f., then there is a bijection $g : \omega \to A$ such that g and g^{-1} are p.r.f. Hence, we have the following

Note 1. *If $\mathcal{A}$ is a p.r. structure with universe $A \subseteq \omega$, and there exists a p.r. injection $f : \omega \to A$, then $\mathcal{A} \cong_{\mathrm{pr}} \mathcal{B}$, where $\mathcal{B}$ is a punctual structure.*

It easily follows that every p.r. structure $\mathcal{A}$ of a finite signature, which is not locally finite, has a punctual presentation $\mathcal{B} \cong_{\mathrm{pr}} \mathcal{A}$. The proof of these facts can be found, for example, in [12]. A significant difference between p.r. and punctual structures appears only for locally finite structures.

If we want to consider a p.r. structure of an infinite signature $L = (L_P, L_F, L_C, \mu)$, consisting of predicates, functions, constants, and an arity function $\mu : L_P \cup L_F \to \omega \backslash \{0\}$, then we need to choose an injective numbering $\nu : I \to L_P \cup L_F \cup L_C$, where $I \subseteq \omega$. In this case, $\mathcal{A} = (A, L^{\mathcal{A}})$ is a *uniformly p.r. structure* of the numbered signature (L, ν) if $A \subseteq \omega$ is a p.r. set and the elements of $L^{\mathcal{A}}$ are primitive recursive uniformly in their numbers. This means that there exists a p.r.f. $F(i, x)$ such that if, for example, $\nu(i) = P_i$ is an n-place predicate, then $P_i^{\mathcal{A}}(x_1, \ldots, x_n) = F(i, \langle x_1, \ldots, x_n \rangle_\omega)$. The definition for functions and constants is similar. Here $\langle x_1, \ldots, x_n \rangle_\omega$ denotes the number of a tuple $\langle x_1, \ldots, x_n \rangle$ in a natural numbering of all tuples in $\omega^{<\omega}$. By $\langle x_1, \ldots, x_n \rangle_n$ we denote the number of this tuple in a numbering of all tuples in ω^n.

Below, we will always assume that (L, ν) is a numbered p.r. signature. This means that the sets $\nu^{-1}(L_P)$, $\nu^{-1}(L_F)$, $\nu^{-1}(L_C)$ and the function $\mu \circ \nu$ are p.r. Uniformly p.r. structures are simply called *p.r. structures* below.

If (L, ν) is a p.r. numbered signature, then it generates a natural Gödel numbering $\gamma : T_L \to \omega$, where T_L is the set of all terms in L. We will assume that all variables in terms from T_L are in the fixed sequence $v_1, v_2, \ldots$ The notation $t = t(v_1, \ldots, v_n)$ for a term t means that all variables in t are included in the set $\{v_1, v_2, \ldots, v_n\}$. Since the theorems below can depend on the definition of γ, we may specify that

1) $\gamma(v_i) = \langle 1, i \rangle_2$;
2) $\gamma(c) = \langle 2, i \rangle_2$ if $\nu(i) = c \in L_C$;
3) $\gamma(f(t_1, \ldots, t_k)) = \langle 3, \langle i, \langle \gamma(t_1), \ldots, \gamma(t_k) \rangle_\omega \rangle_2 \rangle_2$ if $\nu(i) = f \in L_F$.

The case when the signature does not include predicates is more simple.

Theorem 1. *Let $\mathcal{A}$ be a structure of a p.r. numbered signature (L, ν) without predicates. Then $\mathcal{A}$ has an isomorphic p.r. presentation if and only if the following objects exist:*

a) a sequence $\{a_n\}_{n \geqslant 1}$ generating $\mathcal{A}$;

b) a computable function $e : \omega \to \omega$;

c) a p.r.f. $T(s, y, z)$ with the following property: if $n \geqslant 0$ and y, z are the numbers of terms $t_1(v_1, \ldots, v_n)$ and $t_2(v_1, \ldots, v_n)$ in L, then

$$\mathcal{A} \models t_1(a_1, \ldots, a_n) = t_2(a_1, \ldots, a_n) \Leftrightarrow T(e(n), y, z) = 1.$$

Proof. ($\Rightarrow$): let $\mathcal{A} = (A, L^{\mathcal{A}})$ be a p.r. structure. Then A is a p.r. set. If $A = \{a_1, \ldots, a_n\}$ is a finite set, then we may assume that $a_i = a_n$ for $i > n$, and the proof remains true. We suppose that A is infinite. Then there exists a strictly monotonic surjective computable function $g : \omega \to A$. Let $a_n = g(n-1)$ for $n \geqslant 1$.

There is a p.r.f. $h(s,x) = h_s(x)$ with the following property: for every $x \in \omega$, there exists $s_0 \in \omega$ such that $h_s(x) = 0$ for $s < s_0$ and $h_s(x) = g(x) + 1$ for $s \geqslant s_0$. We define $e(n) = \mu s\,[\forall x < n \;\; h_s(x) > 0\,]$, it is a computable function.

It is easy to verify that the values of terms in $\mathcal{A}$ can be computed from their numbers by a p.r.f. More exactly, if $x = \langle a_1, \ldots, a_n \rangle_\omega$ and y is the number of a term $t(v_1, \ldots, v_n)$, then its value $t^{\mathcal{A}}(a_1, \ldots, a_n) = g(x,y)$, where g is a p.r.f. It can be easily constructed by recursion on y using the properties of Gödel numberings.

Now we construct the p.r.f. $T(s,y,z)$. If y, z are not the numbers of terms in T_L, then $T(s,y,z) = 0$. Suppose that y, z are the numbers of terms t_1, t_2. We find minimal $n \geqslant 0$ such that $t_i = t_i(v_1, \ldots, v_n)$ for $i = 1, 2$. If they are constant terms, then $n = 0$, otherwise it is the largest number of a variable in t_1 or t_2.

We check that $h_s(x) > 0$ for all $x < n$. If this fails, $T(s,y,z) = 0$. Suppose that the condition is true and $a_i = h_s(i-1) - 1$ for $1 \leqslant i \leqslant n$. Then $a_i = g(i-1)$ are elements of A, and, having y, we can find the value of the term $t_1^{\mathcal{A}}(a_1, \ldots, a_n) = b_1$. Then we find $t_2^{\mathcal{A}}(a_1, \ldots, a_n) = b_2$ and define $T(s,y,z) = 1 \Leftrightarrow b_1 = b_2$.

($\Leftarrow$): suppose that we have $\{a_n\}_{n \geqslant 1}$, $e(x)$, and $T(s,y,z)$ with the properties from a) – c). We add new constants $a_1, a_2, \ldots$ to L, obtaining the signature $L' = L \cup \{a_n \mid n \geqslant 1\}$. Choose a natural numbering ν' for L', coordinated with ν. For example, $\nu'(2i) = \nu(i)$ and $\nu'(2i+1) = a_i$. It generates a Gödel numbering $\gamma' : T_{L'} \to \omega$.

Since $e(x)$ is computable, there exists a p.r.f. $h(s,x) = h_s(x)$ with the following property: for every $x \in \omega$, there exists $s_0 \in \omega$ such that $h_s(x) = 0$ for $s < s_0$ and $h_s(x) = e(x) + 1$ for $s \geqslant s_0$. Let $\mu(x) = \mu s\,[\,h_s(x) > 0\,]$. Then $e(x) = h_{\mu(x)}(x) - 1$.

We suppose that t' is a constant term in the signature L'. Then $t' = t(a_1, \ldots, a_n)$, where $t(v_1, \ldots, v_n)$ is a term in L. To find t, we need to replace a_i by v_i in t'. Let $n = n(t')$ be the smallest number with the specified property. This means that the constant a_n is in t', or that t' is a term in L and $n = 0$.

Let $B_0 = \{\langle y, \mu(n) \rangle_2 \mid y = \gamma'(t')$, where t' is a constant term in L' and $n = n(t')\}$. We show that B_0 is a p.r. set. Consider $x = \langle y, m \rangle_2$. If y is not the number of a constant term, then $x \notin B_0$. Suppose $y = \gamma'(t')$. The number $n = n(t')$ can be found from y by a p.r.f. Now $m = \mu(n) \Leftrightarrow h_m(n) > 0$ and $h_s(n) = 0$ for each $s < m$, which is a p.r. condition.

Since B_0 is essentially a free algebra, we can naturally define the operations from L on it. Let $f \in L_F$ be an r-place function and $x_1, \ldots, x_r \in B_0$. Then $x_i = \langle \gamma'(t'_i), m_i \rangle_2$ for $i \leqslant r$. Let $t' = f(t'_1, \ldots, t'_r)$. Clearly, $n(t') = \max_{i \leqslant r}\{n(t'_i)\}$, and we can find $j \leqslant r$ such that $n(t') = n(t'_j)$. Define $f(x_1, \ldots, x_r) = \langle \gamma'(t'), m_j \rangle_2$. We obtain a p.r. operation on B_0, defined uniformly in l such that $f = \nu(l)$. If c is a constant from L, then its value in B_0 is $\langle \gamma'(c), \mu(0) \rangle_2$. We have constructed a p.r. structure $\mathcal{B}_0$ of the signature L.

Now we define a natural map $g : B_0 \to A$. If $x = \langle \gamma'(t'), m \rangle_2 \in B_0$ then $t' = t(a_1, \ldots, a_n)$, where $t(v_1, \ldots, v_n)$ is a term in L. Let $g(x) = t^{\mathcal{A}}(a_1, \ldots, a_n)$, where a_i denote the elements of $\mathcal{A}$. We obtain a surjective homomorphism $g : \mathcal{B}_0 \to \mathcal{A}$.

We define $E = \{\langle x_1, x_2\rangle \in B_0^2 \mid g(x_1) = g(x_2)\}$. It is a congruence relation on $\mathcal{B}_0$. Let $[x] = \{y \in B_0 \mid xEy\}$ be the equivalence class of x. The natural mapping $g^* : \mathcal{B}_0/E \to \mathcal{A}$, $g^*([x]) = g(x)$, is an isomorphism. It remains to understand that E is primitive recursive.

We consider $x_1 = \langle \gamma'(t_1'), m_1\rangle_2$ and $x_2 = \langle \gamma'(t_2'), m_2\rangle_2$. Suppose that $n = n(t_1') \geqslant n(t_2')$, $t_1' = t_1(a_1, \dots, a_n)$, and $t_2' = t_2(a_1, \dots, a_n)$, where $t_i(v_1, \dots, v_n)$ are terms in L. Then $g(x_1) = g(x_2) \Leftrightarrow \mathcal{A} \models t_1(a_1, \dots, a_n) = t_2(a_1, \dots, a_n) \Leftrightarrow T(e(n), y, z) = 1$, where y, z are the numbers of $t_1(\bar{v})$ and $t_2(\bar{v})$. In this case, $e(n) = h_{\mu(n)}(n) - 1$ and $\mu(n) = m_1$. The above procedure describes a p.r. condition.

Next, we pass in the standard way from the quotient structure $\mathcal{B}_0/E$ to a usual p.r. structure, defining $B = \{x \in B_0 \mid \forall y < x \quad \neg yEx\}$ and the natural interpretation of L on B, for which $g : B \to A$ is an isomorphism. □

If the structure $\mathcal{A}$ is finitely generated and $\bar{a} = a_1, \dots, a_n$ are its generators, then this criterion says that $\mathcal{A}$ has an isomorphic p.r. presentation if and only if the set $\{\langle t_1(\bar{v}), t_2(\bar{v})\rangle \in T_L^2 \mid \mathcal{A} \models t_1(\bar{a}) = t_2(\bar{a})\}$ is p.r. This is a simple and well-known fact noted in [3].

In the proof of this theorem, we construct a p.r. structure $\mathcal{B}$ and an isomorphism $g : \mathcal{B} \to \mathcal{A}$. In addition, the sequence $\{g^{-1}(a_n)\}_{n \geqslant 1}$ is computable, and the set $\{g^{-1}(a_n) \mid n \geqslant 1\}$ is p.r. This fact imposes certain limitations on the capabilities of this construction. If the structure $\mathcal{A}$ is computable and $\{a_n\}_{n\in\omega}$ is a computable sequence in $\mathcal{A}$, then the isomorphism g is also computable.

Now we generalize Theorem 1 to the case of a signature with predicates.

Theorem 2. *Let $\mathcal{A}$ be a structure of a p.r. numbered signature (L, ν). Then $\mathcal{A}$ has an isomorphic p.r. presentation if and only if the following objects exist:*

a) a sequence $\{a_n\}_{n \geqslant 1}$ generating $\mathcal{A}$;

b) a computable function $e : \omega \to \omega$;

c) a p.r.f. $T(s, y, z)$ with the following property: if $n \geqslant 0$ and y, z are the numbers of terms $t_1(v_1, \dots, v_n)$ and $t_2(v_1, \dots, v_n)$ in L, then
$\mathcal{A} \models t_1(a_1, \dots, a_n) = t_2(a_1, \dots, a_n) \Leftrightarrow T(e(n), y, z) = 1$;

d) a p.r.f. $T^*(s, i, y)$ with the following property: if $n \geqslant 0$, $\nu(i) = P_i$ is an r-place predicate, and $y = \langle y_1, \dots, y_r\rangle_\omega$, where y_j are the numbers of terms $t_j(v_1, \dots, v_n)$ in L for $j \leqslant r$, then
$\mathcal{A} \models P_i(t_1(a_1, \dots, a_n), \dots, t_r(a_1, \dots, a_n)) \Leftrightarrow T^*(e(n), i, y) = 1$.

Proof. ($\Rightarrow$): we slightly modify the proof of Theorem 1. Let $\mathcal{A} = (A, L^A)$ be a p.r. structure. We define the function g, put $a_n = g(n-1)$, and define $e(n)$ in the same way. The functions $h_s(y)$ and $T(s, y, z)$ are defined literally as in Theorem 1.

Now we describe $T^*(s, i, y)$. Check that $P_i = \nu(i)$ is a predicate from L. Let r be its arity. Check that $y = \langle y_1, \dots, y_r\rangle_\omega$, where y_j is the number of a term t_j for $j \leqslant r$. Find the minimal $n \geqslant 0$ such that $t_j = t_j(v_1, \dots, v_n)$. Check that $h_s(x) > 0$ for all $x < n$. If at least one condition fails, $T^*(s, i, y) = 0$. Below we suppose that everything holds.

We define $a_i = h_s(i-1) - 1$ for $1 \leqslant i \leqslant n$. They are elements of $\mathcal{A}$, we can find the values of terms $\langle b_1, \dots, b_r\rangle_\omega = \langle t_1(\bar{a}), \dots, t_r(\bar{a})\rangle_\omega$, where $\bar{a} = a_1, \dots, a_n$, and then find $P_i(b_1, \dots, b_r)$, using the definition of a uniformly p.r. structure. The last value is $T^*(s, i, y)$.

($\Leftarrow$): literally repeating Theorem 1, we define L', ν', γ', $h_s(x)$, $\mu(x)$, $n(t')$, the p.r. set B_0 and the function $g : B_0 \to A$. Next, we define the operations and constants from L on B_0 in the same way, as on a free algebra, obtaining a series of p.r.f.

Now let P be a r-place predicate from L_P and $x_1, \dots, x_r \in B_0$. Suppose that $x_j = \langle\gamma'(t_j'), m_j\rangle_2$ for $j \leqslant r$, and $n = \max_{j \leqslant r}\{n(t_j')\}$. Then $t_j' = t_j(a_1, \dots, a_n)$, where $t_j(v_1, \dots, v_n)$ are terms in L. We define the predicate P by B_0 so that

$$P^{B_0}(x_1, \dots, x_r) \Leftrightarrow \mathcal{A} \models P(g(x_1), \dots, g(x_r)) \Leftrightarrow \mathcal{A} \models P(t_1(\bar{a}), \dots, t_r(\bar{a})),$$

where $\bar{a} = a_1, \dots, a_n$.

Let $n = n(t_j')$. Then $m_j = \mu(n)$ and $e(n) = h_{m_j}(n) - 1$. We have found $e(n)$ using a p.r. algorithm. Next, we find $y = \langle\gamma(t_1), \dots, \gamma(t_r)\rangle_\omega$. If $P = \nu(i)$ then $\mathcal{B}_0 \models P(x_1, \dots, x_r) \Leftrightarrow T^*(e(n), i, y) = 1$. We see that P is a p.r. predicate on $\mathcal{B}_0$ uniformly in i.

Now we repeat Theorem 1 again. The relation $E = \{\langle x_1, x_2\rangle \in B_0^2 \mid g(x_1) = g(x_2)\}$ is a p.r. congruence on $\mathcal{B}_0$. The quotient structure $\mathcal{B}_0/E$ has the natural p.r. presentation, whose elements are the smallest members of equivalence classes. □

In the proof of implication ($\Rightarrow$) here, we do not use the fact that the universe A is a p.r. set. It can be even computably enumerable. It is important that the operations and predicates are primitive recursive. This explains the following

Corollary 1. *If $\mathcal{A}$ is a p.r. structure and $E \subseteq A$ is a nonempty c.e. set, then the substructure $\mathcal{A}\langle E\rangle$ generated by E also has an isomorphic p.r. presentation.*

Proof. There is a p.r.f. $h : \omega \to E$ such that $\operatorname{ran}(h) = E$. We define $a_n = h(n-1)$ for $n \geqslant 1$ and $e(n) = 0$. Then $\{a_n\}_{n \geqslant 1}$ generates $\mathcal{A}\langle E\rangle$, and items c) and d) in Theorem 2 are met, since we can compute the values of terms $t^{\mathcal{A}}(h(0), \dots, h(n-1))$ in $\mathcal{A}$ uniformly in n and the number of t using a p.r.f. By Theorem 2, $\mathcal{A}\langle E\rangle$ has a p.r. presentation. □

Corollary 2. *Let $\mathcal{A}$ be a computable locally finite structure of a finite signature L. Then $\mathcal{A}$ has an isomorphic p.r. presentation.*

Proof. If $\mathcal{A}$ is finite, then there is nothing to prove. We may assume that $A = \omega$. We use Theorem 2 and put $a_n = n - 1$ for $n \geqslant 1$.

We define $e(n)$. Let A_n be the substructure of $\mathcal{A}$ generated by $\{a_1, \dots, a_n\}$ for $n \geqslant 1$. We can find the number of the finite set A_n with a computable function, passing over all terms $t(v_1, \dots, v_n)$ in L, computing $t^{\mathcal{A}}(a_1, \dots, a_n)$, and waiting for the moment when the set becomes closed under all operations from L. We put the number of A_n in $e(n)$, and add the diagram for all operations and relations from $\mathcal{A}$ on A_n.

For example, if f is an r-place operation from L, then we define

$D_{f,n} = \{\langle x_1, \dots, x_r, x_{r+1}\rangle \in A_n^{r+1} \mid f^{\mathcal{A}}(x_1, \dots, x_r) = x_{r+1}\}$.

The diagram $D_{P,n}$ for a predicate $P \in L$ is defined similarly. The function $e(n)$ is the number of the tuple consisting of the number of A_n and the numbers of all these sets.

To construct $T(s, y, z)$, we assume that s is the number of the diagram of some substructure with a finite universe $A' \subseteq A$. Then we can compute the values of terms $t(a_1, \dots, a_n)$ by recursion on the number of t. It is a p.r. operation. The same is true for the function $T^*(s, i, y)$, where $i \in I$ and I is a finite set. □

The corollaries will be false if we replace p.r. structures by punctual structures. In a sense, the theory of p.r. structures is simpler and more natural than the theory of punctual structures. A counterexample can be found in [1]: in Theorem 3.3, a computable Abelian group $\mathcal{A}_S = \bigoplus\limits_{p \in S} \mathbb{Z}_p$ is constructed, where S is a c.e. set of prime numbers, which has no punctual presentation. Here $\mathbb{Z}_p$ denotes the cyclic group of order p. The group $\mathcal{A}_S$ is locally finite, as a torsion Abelian group, and has a p.r. presentation.

In addition, it is easy to construct a natural punctual presentation for the group $\mathcal{A} = \bigoplus_{p \in P} \mathbb{Z}_p$, where P is the set of all prime numbers and the generating elements of components $\mathbb{Z}_p$ form a computable set. Then $\mathcal{A}_S$ can be isomorphically embedded to $\mathcal{A}$ as a substructure generated by a c.e. set.

We give another example of applying p.r. structures. Basic information about Boolean algebras can be found in [13]. By $\mathcal{B}(L)$ we denote the Boolean algebra generated by a linear order L. If $\mathcal{A}$ is a Boolean algebra and $a \in A$, then $\widehat{a}$ denotes the algebra with the universe $\{x \in A \mid x \leqslant a\}$ and the order induced from $\mathcal{A}$.

Let $\mathcal{A} = (A, +, \cdot, -, 0, 1)$ be an infinite computable Boolean algebra. It is locally finite and has a p.r. presentation. If $\mathcal{A}$ is not atomic, then it contains an atomless element $a \neq 0$, and $\mathcal{A} \cong \widehat{a} \times \widehat{-a}$, where $\widehat{a} \cong \mathcal{B}(\eta)$. If $\mathcal{A}$ is atomic, F is its Fréchet ideal, and $\mathcal{A}/F$ contains an atom, then $\mathcal{A} \cong \widehat{a} \times \widehat{-a}$, where $\widehat{a} \cong \mathcal{B}(\omega)$. Otherwise $\mathcal{A} \cong \mathcal{B}(\omega \times \eta)$. The algebras $\mathcal{B}(\eta)$, $\mathcal{B}(\omega)$, and $\mathcal{B}(\omega \times \eta)$ have natural punctual presentations. Therefore, $\mathcal{A}$ can be presented as $\mathcal{A}_1 \times \mathcal{A}_2$, where $\mathcal{A}_1$ is an infinite punctual algebra and $\mathcal{A}_2$ is p.r. In this case, $\mathcal{A}_1 \times \mathcal{A}_2$ is a p.r. structure containing an injective p.r. sequence. By Note 1, we have that $\mathcal{A}$ also has a punctual presentation, that has already been proved in Theorem 3.1 of [1] with a rather impressive construction.

3 A Criterion for Punctual Structures

Now we transfer the above criterion to punctual structures. Again, we first consider the case of a signature without predicates. In this theorem, the generators of $\mathcal{A}$ are divided into two parts: good b_n, for which the problem of the equality of terms is uniformly p.r., and bad a_n, for which we need a hint $e(n)$ to recognize the equality. It is important that there are infinitely many good generators.

Theorem 3. *Let $\mathcal{A}$ be a countable structure of a p.r. numbered signature (L, ν) without predicates. Then $\mathcal{A}$ has an isomorphic punctual presentation if and only if the following objects exist:*

a) *an injective infinite sequence $\{b_n\}_{n\geqslant 1}$ in $\mathcal{A}$;*

b) *a sequence $\{a_n\}_{n\geqslant 1}$ in $\mathcal{A}$ such that $\{b_n\}_{n\geqslant 1} \cup \{a_n\}_{n\geqslant 1}$ generate $\mathcal{A}$;*

c) *a computable function $e : \omega \to \omega$;*

d) *a p.r.f. $T(s, n, y, z)$ with the following property: if $n, k \geqslant 0$ and y, z are the numbers of terms $t_1(v_1, \dots, v_{n+k})$ and $t_2(v_1, \dots, v_{n+k})$ in L, then*

$\mathcal{A} \models t_1(a_1, \dots, a_n, b_1, \dots, b_k) = t_2(a_1, \dots, a_n, b_1, \dots, b_k) \Leftrightarrow T(e(n), n, y, z) = 1.$

Proof. ($\Rightarrow$): let $\mathcal{A}$ be a punctual structure of the signature L with universe ω. We define $b_n = a_n = n - 1$ for $n \geqslant 1$ and $e(n) = 0$. Since the values of terms are computed in $\mathcal{A}$ uniformly in their numbers, $T(s, n, y, z)$ can be constructed as a p.r.f. without using s.

($\Leftarrow$): we suppose that the sequences $\{a_n\}_{n\geqslant 1}$, $\{b_n\}_{n\geqslant 1}$, $e(x)$, and the function $T(s, n, y, z)$ are given. We slightly modify the proof of Theorem 1. Again, we find the p.r.f. $h(s, x) = h_s(x)$ with the following property: for every $x \in \omega$, there exists $s_0 \in \omega$ such that $h_s(x) = 0$ for $s < s_0$ and $h_s(x) = e(x) + 1$ for $s \geqslant s_0$. Let $\mu(x) = \mu s\, [\, h_s(x) > 0\,]$. Then $e(x) = h_{\mu(x)}(x) - 1$.

Let $L' = L \cup \{a_n, b_n \mid n \geqslant 1\}$, where a_n, b_n are new constants. We choose a natural numbering ν' for L' coordinated with ν. It generates a Gödel numbering $\gamma' : T_{L'} \to \omega$.

We consider a constant term t' in L'. Then $t' = t(a_1, \dots, a_n, b_1, \dots, b_k)$, where $t(v_1, \dots, v_{n+k})$ is a term in L. Let $n = n(t')$ be the minimal number with the specified property. This means that a_n is in t', and a_i is not included in t' for $i > n$. If $n(t') = 0$ then a_i is not included in t' for all $i \geqslant 1$.

We define $B_0 = \{\langle y, \mu(n)\rangle_2 \mid y = \gamma'(t'),\ t'$ is a constant term in L', and $n = n(t')\}$. Just as in Theorem 1, we can verify that B_0 is a p.r. set.

Considering B_0 as a free algebra, we define operations from L on it. Let $f \in L_F$ be an r-place operation and $x_1, \dots, x_r \in B_0$. Then $x_j = \langle \gamma'(t'_j), m_j\rangle_2$ for $j \leqslant r$. Let $t' = f(t'_1, \dots, t'_r)$. As in Theorem 1, $n(t') = \max_{j\leqslant r}\{n(t'_j)\}$. Similarly reasoning, we see that f is a p.r.f. on B_0. The values of constants are defined in the same way, and we get a p.r. structure $\mathcal{B}_0$.

The mapping $g : B_0 \to \mathcal{A}$ is constructed similarly. If $x = \langle \gamma'(t'), m\rangle_2 \in B_0$ then $g(x) = t^{\mathcal{A}}(a_1, \dots, a_n, b_1, \dots, b_k)$, where $t' = t(a_1, \dots, a_n, b_1, \dots, b_k)$ and $t(v_1, \dots, v_{n+k})$ is a term in L. This is a surjective homomorphism.

We define $E = \ker(g) = \{\langle x_1, x_2\rangle \in B_0^2 \mid g(x_1) = g(x_2)\}$. It is a congruence on $\mathcal{B}_0$, and the natural mapping $g^* : \mathcal{B}_0/E \to \mathcal{A}$, $g^*([x]) = g(x)$, is an isomorphism. We prove that E is p.r. Let $x_1 = \langle \gamma'(t'_1), m_1\rangle_2$ and $x_2 = \langle \gamma'(t'_2), m_2\rangle_2$. Suppose that $n = n(t'_1) \geqslant n(t'_2)$, $t'_1 = t_1(a_1, \dots, a_n, b_1, \dots, b_k)$, and $t'_2 = t_2(a_1, \dots, a_n, b_1, \dots, b_k)$. Then $g(x_1) = g(x_2) \Leftrightarrow T(e(n), n, y, z) = 1$, where y, z are the numbers of terms $t_1(v_1, \dots, v_{n+k})$ and $t_2(v_1, \dots, v_{n+k})$. Since $e(n) = h_{\mu(n)}(n) - 1$ and $\mu(n) = m_1$, this is a p.r. condition.

Next, we proceed verbatim as in Theorem 1: $B = \{x \in B_0 \mid \forall y < x\ \neg yEx\}$, we transfer predicates and functions in L to B, and obtain a p.r. structure $\mathcal{B}$ and an isomorphism $g : \mathcal{B} \to \mathcal{A}$.

In addition, the mapping $h_0 : \omega\backslash\{0\} \to B_0$, $h_0(i) = \langle\gamma'(b_i), \mu(0)\rangle_2$, is a p.r. injection. Replacing $h_0(i)$ by $h(i) \in B$ such that $h_0(i)Eh(i)$, we again get a p.r. injection, since $b_i \neq b_j$ in $\mathcal{A}$ for $i \neq j$. By Note 1, $\mathcal{B}$ has a punctual presentation. □

As for Theorem 2, now we change the formulation for the case of a signature with predicates.

Theorem 4. *Let $\mathcal{A}$ be a countable structure of a p.r. numbered signature (L, ν). Then $\mathcal{A}$ has an isomorphic punctual presentation if and only if the following objects exist:*

a) *an injective infinite sequence $\{b_n\}_{n\geqslant 1}$ in $\mathcal{A}$;*

b) *a sequence $\{a_n\}_{n\geqslant 1}$ in $\mathcal{A}$ such that $\{b_n\}_{n\geqslant 1} \cup \{a_n\}_{n\geqslant 1}$ generate $\mathcal{A}$;*

c) *a computable function $e : \omega \to \omega$;*

d) *a p.r.f. $T(s, n, y, z)$ with the following property: if $n, k \geqslant 0$ and y, z are the numbers of terms $t_1(v_1, \ldots, v_{n+k})$ and $t_2(v_1, \ldots, v_{n+k})$ in L, then*

$\mathcal{A} \models t_1(a_1, \ldots, a_n, b_1, \ldots, b_k) = t_2(a_1, \ldots, a_n, b_1, \ldots, b_k) \Leftrightarrow T(e(n), n, y, z) = 1$.

e) a p.r.f. $T^*(s, i, n, y)$ with the following property: if $n, k \geqslant 0$, $\nu(i) = P_i$ is an r-place predicate, and $y = \langle y_1, \ldots, y_r\rangle_\omega$, where y_j are the numbers of terms $t_j(v_1, \ldots, v_{n+k})$ in L for $j \leqslant r$, then

$\mathcal{A} \models P_i(t_1(a_1, \ldots, a_n, b_1, \ldots, b_k), \ldots, t_r(a_1, \ldots, a_n, b_1, \ldots, b_k)) \Leftrightarrow T^*(e(n), i, n, y) = 1$.

Proof. We need to change the proof of Theorem 3 in the same way as we have transformed the proof of Theorem 1 into Theorem 2 by adding p.r. predicates P_i to the structure $\mathcal{B}_0$. We leave this proof as an exercise. □

We give a series of examples of using Theorems 1–4. A poset $(A, \leqslant)$ is called *directed* (upward) if for every $x, y \in A$, there exists $z \in A$ such that $x, y \leqslant z$. A torsion Abelian group $\mathcal{A}$ is called a *p-group*, where p is a prime number, if the order of any of its elements is equal to p^n for some $n \geqslant 0$. Items (1), (3), and (4) below are proved in [1]. Here we offer a very simple and elementary proof for them. It does not require any sketches or metaphors, and does not use step-by-step constructions.

Theorem 5. *Every computable structure from the following classes has a punctual presentation:*

(1) *a structure with equivalence $\mathcal{A} = (A, E)$;*

(2) *a directed poset $\mathcal{A} = (A, \leqslant)$ without greatest element;*

(3) *a linear ordering $\mathcal{A} = (A, \leqslant)$;*

(4) *an Abelian p-group $\mathcal{A} = (A, +, -, 0)$, where p is a prime number.*

Proof. (1): let $\mathcal{A} = (A, E)$ be a computable structure with an equivalence relation E. We define $B = \{x \in A \mid \forall y < x\ \neg xEy\}$, choosing the minimal element in every equivalence class. If B is finite, then the number of classes is finite, and a punctual presentation for $\mathcal{A}$ looks trivial. Suppose B is infinite. This it is a computable set, and there exists a computable bijection $f : \omega \to B$. Let

$b_n = f(n-1)$ for $n \geqslant 1$. If $A \backslash B$ is finite then $\mathcal{A}$ is again trivial. We suppose that $A \backslash B$ is infinite. There is a computable bijection $g : \omega \to A \backslash B$, let $a_n = g(n-1)$ for $n \geqslant 1$.

If $a \in A$ then there is unique $b_i \in B$ such that aEb_i. Define $h(a) = i$ and $e(n) = \langle h(a_1), \dots, h(a_n) \rangle_\omega$. We check the conditions in Theorem 4. The signature $L = \{E\}$ has no functions, and every term $t(v_1, \dots, v_{n+k})$ is equal to v_j. Therefore, in d) we need to check conditions of the form $a_i = b_j$, $b_i = b_j$, or $a_i = a_j$. The first one is always false, and the rest are equivalent to $i = j$.

In e), we need to check conditions of the form $E(b_i, b_j)$, $E(a_i, b_j)$, or $E(a_i, a_j)$. Here $E(b_i, b_j) \Leftrightarrow i = j$ and $E(a_i, b_j) \Leftrightarrow h(a_i) = j$. Knowing $e(n)$, we find $h(a_i)$ and get a p.r. condition. Similarly $E(a_i, a_j) \Leftrightarrow h(i) = h(j)$. By Theorem 4, $\mathcal{A}$ has a punctual presentation.

(2): we suppose that A is infinite and $f : \omega \to A$ is a computable bijection. Using the hypothesis, we can construct a computable sequence $\{b_n\}_{n \geqslant 1}$ such that $b_n < b_{n+1}$ and $f(x) \leqslant b_n$ for all $x \leqslant n$. Let $a_n = f(n-1)$ for $n \geqslant 1$. Then $a_n \leqslant b_n$. We define $e(n)$ as the diagram of the set $\{a_i, b_i \mid i \leqslant n\}$. Namely, let $D_n = \{\langle i, j \rangle_2 \mid i, j \leqslant n \text{ and } a_i \leqslant a_j\}$, $D'_n = \{\langle i, j \rangle_2 \mid i, j \leqslant n \text{ and } b_i \leqslant a_j\}$ and $D''_n = \{\langle i, j \rangle_2 \mid i, j \leqslant n \text{ and } a_i \leqslant b_j\}$. Then $e(n)$ is the triple of the numbers of D_n, D'_n, D''_n.

We check the conditions in Theorem 4. Since $x = y$ is reduced to $x \leqslant y$ and $y \leqslant x$, it suffices to check the conditions $a_i \leqslant b_t$, $b_t \leqslant a_i$, $b_t \leqslant b_s$, and $a_i \leqslant a_j$, where $i, j \leqslant n$ and $t, s \geqslant 1$. Here $b_t \leqslant b_s \Leftrightarrow t \leqslant s$, $a_i \leqslant b_t$ for $t > n$, $b_t \leqslant a_i$ is false for $t > n$, and if $t \leqslant n$ then all other answers are contained in $e(n)$.

(3): we follow the proof scheme of Theorem 3.1 in [1]. We may assume that $\mathcal{A}$ is infinite and has greatest and least elements, since they can always be added. We suppose that b is a left limit point in $\mathcal{A}$. This means that b is not least, and for every $x < b$, there is y such that $x < y < b$. Then $\mathcal{A}_1 = (-\infty, b)$ is a directed poset without greatest element, and by (2) $\mathcal{A}_1$ has a punctual presentation. Since $\mathcal{A}_2 = [b, +\infty)$ has a p.r. presentation as a locally finite structure, we have that $\mathcal{A}_1 + \mathcal{A}_2$ has a p.r. presentation with an injective p.r. subsequence. The proof for a right limit point is similar.

Now we suppose that $\mathcal{A}$ contains an interval $[x, y] \cong \omega + \omega^*$. Then $\mathcal{A} \cong (-\infty, x) + [x, y] + (y, +\infty)$, $(-\infty, x)$ and $(y, +\infty)$ have p.r. presentations, $[x, y]$ has a natural punctual presentation, and $\mathcal{A}$ also has a punctual presentation.

If $\mathcal{A}$ has no limit points, then every element, except the largest, has the next one, and every non-least element has the preceding one. If there are no intervals like $\omega + \omega^*$, then $\mathcal{A} \cong \omega + (\omega^* + \omega) \times \eta + \omega^*$ and has a natural punctual presentation.

(4): let $B = \{x \in A \mid px = 0\}$. It is a computable subgroup in $\mathcal{A}$, and the multiplication $\lambda, x \mapsto \lambda x$ defines on it a vector space over the finite field F_p with p elements. If B is finite then $\mathcal{A} \cong F \times \mathbb{Z}^n_{p^\infty}$, where F is a finite group, as noted in [1]. In this case $\mathcal{A}$ has a natural punctual presentation. We suppose that B is infinite.

It is not difficult to construct a computable basis $E = \{b_n\}_{n \geqslant 1}$ in the vector space B. If $f : \omega \to B$ is a computable bijection, then we can define E as the set

$\{f(x) \mid f(x)$ does not lie in the linear shell of $f(0), f(1), \ldots, f(x-1)\}$. If $x \in B$ then we can compute an uniquely defined decomposition $x = \lambda_1 b_1 + \ldots + \lambda_n b_n$, where $\lambda_i \in F_p$ and $\lambda_n \neq 0$ for $n \neq 1$.

We suppose that $g : \omega \to A$ is a computable bijection. Let $a_n = g(n-1)$ for $n \geqslant 1$. For every $n \geqslant 1$, we can compute $l_n \geqslant 1$, the order of a_n. We put the number $\langle l_1, \ldots, l_n \rangle_\omega$ in $e(n)$. Next, we go through all tuples $\langle t_1, \ldots, t_n \rangle$, where $0 \leqslant t_i < l_i$ for $i \leqslant n$, find $b = t_1 a_1 + \ldots + t_n a_n$ and check the condition $b \in B$.

If it holds then we find a decomposition $b = \lambda_1 b_1 + \ldots + \lambda_s b_s$ and add the pair $\langle \langle t_1, \ldots, t_n \rangle_\omega, \langle \lambda_1, \ldots, \lambda_s \rangle_\omega \rangle$ to $e(n)$. We check the conditions in Theorem 3, since there are no predicates in the signature.

In the class of Abelian groups, every term $t(v_1, \ldots, v_n)$ in the signature $L = (+, -, 0)$ can be reduced to the form $t_1 v_1 + \ldots + t_n v_n$, where $t_i \in \mathbb{Z}$ for $i \leqslant n$, using a p.r. procedure. In addition, the condition $t_1(\bar{v}) = t_2(\bar{v})$ is equivalent to $t_1(\bar{v}) - t_2(\bar{v}) = 0$. Therefore, in d) we need only to check conditions of the form $t(\bar{a}, \bar{b}) = 0$, where $\bar{a} = a_1, \ldots, a_n$, $\bar{b} = b_1, \ldots, b_k$, and $t(v_1, \ldots, v_{n+k}) = t_1 v_1 + \ldots + t_n v_n + s_1 v_{n+1} + \ldots + s_k v_{n+k}$, $t_i, s_j \in \mathbb{Z}$. If we know $l_i \geqslant 1$ such that $l_i a_i = 0$, then can always assume that $0 \leqslant t_i < l_i$ and $0 \leqslant s_j < p$. All such combinations are described in $e(n)$: if $t_1 a_1 + \ldots + t_n a_n \notin B$, then the equality $t(\bar{a}, \bar{b}) = 0$ is impossible. If $t_1 a_1 + \ldots + t_n a_n = \lambda_1 b_1 + \ldots + \lambda_s b_s$, then $t(\bar{a}, \bar{b}) = 0 \Leftrightarrow \lambda_1 b_1 + \ldots + \lambda_s b_s + s_1 b_1 + \ldots + s_k b_k = 0$, and this equality can be easily verified in the linear space B. By Theorem 3, $\mathcal{A}$ has a punctual presentation with a p.r. basis in B. □

Corollary 3. *Let $\mathcal{A} = (A, +, -, 0)$ be a computable torsion Abelian group that has no punctual presentation. Then $\mathcal{A} \cong \bigoplus_{p \in P} \mathcal{A}_p$, where P is the set of all primes, and $\mathcal{A}_p$ is a finite sum of cyclic groups of the form $\mathbb{Z}_{p^n}$, $n \geqslant 1$.*

Proof. Let p be a prime number and $A_p = \{x \in A \mid \exists n \geqslant 1 \;\; p^n x = 0\}$. Then A_p is a c.e. subgroup in $\mathcal{A}$, uniformly in $p \in P$. It is well known that $\mathcal{A} = \bigoplus_{p \in P} \mathcal{A}_p$. Therefore, $\mathcal{A} = \mathcal{A}' \oplus \mathcal{A}''$, where $\mathcal{A}' = \mathcal{A}_p$ and $\mathcal{A}'' = \bigoplus_{q \in P \setminus \{p\}} \mathcal{A}_q$ are two c.e. subgroups. If $x \in A$ then there is a unique decomposition $x = x' + x''$, where $x' \in A'$ and $x'' \in A''$. Hence, the subgroups A', A'' are computable. If $\mathcal{A}'$ is infinite, then it has a punctual presentation, $\mathcal{A}''$ has a p.r. presentation, and their product also has a punctual presentation.

We have that every A_p is a finite Abelian p-group. It is known that it is isomorphic to a direct sum of cyclic p-groups. □

Let $\mathcal{A}$ be an Abelian group. A set $E \subseteq A$ is *linearly independent* if for every different elements $e_1, \ldots, e_n \in E$ and for every $t_i \in \mathbb{Z}$, the equality $t_1 e_1 + \ldots + t_n e_n = 0$ implies that $t_i = 0$ for all $i \leqslant n$. A set $E \subseteq A$ is a *linear basis* in $\mathcal{A}$ if it is a maximal linearly independent set. A family $\{e_i\}_{i \in I}$ is a linear basis in $\mathcal{A}$ if the set $E = \{e_i \mid i \in I\}$ is a linear basis and $e_i \neq e_j$ for $i \neq j$.

Lemma 1. *Let $\mathcal{A} = (A, +, -, 0)$ be a computable Abelian group and let $\{e_i\}_{i \in I}$ be a computable sequence in $\mathcal{A}$, which is a linear basis, where I is an initial segment in $\omega \setminus \{0\}$. Then $E = \{e_i \mid i \in I\}$ and the subgroup $\mathcal{A}\langle E \rangle$ generated by E are computable.*

Proof. We suppose that $x \in A \backslash E$. Then the set $E \cup \{x\}$ is not independent, and there are $m, t_1, \dots, t_n \in \mathbb{Z}$ and different $e_1, \dots, e_n \in E$ such that

$$mx = t_1 e_1 + \ldots + t_n e_n,$$

where $m \neq 0$. We can compute such a decomposition from x. If $x \in E$ then $x = e_n$ for some $n \in I$.

We show that $x \in \mathcal{A}\langle E\rangle \Leftrightarrow m \mid t_i$ for all $i \leqslant n$ and $x = \frac{t_1}{m} e_1 + \ldots + \frac{t_n}{m} e_n$. The implication ($\Leftarrow$) is obvious. We prove ($\Rightarrow$). Suppose $x = t'_1 e_1 + \ldots + t'_{n'} e_{n'}$. For simplicity, we may assume that $n' = n$. Then $mx = mt'_1 e_1 + \ldots + mt'_n e_n$ and

$$(t_1 - mt'_1)e_1 + \ldots + (t_n - mt'_n)e_n = 0.$$

Hence, $t_i - mt'_i = 0$ for $i \leqslant n$.

We see that the set $\mathcal{A}\langle E\rangle$ is computable. If $x \notin \mathcal{A}\langle E\rangle$ then $x \notin E$. If $x \in \mathcal{A}\langle E\rangle$ then there is a unique decomposition $x = t_1 e_1 + \ldots + t_n e_n$, where $t_i \in \mathbb{Z}$ and $t_n \neq 0$ for $n \neq 1$, and we can check the condition $x \in E$. □

In particular, if a computable Abelian group $\mathcal{A}$ has a c.e. linear basis, then it is computable.

Theorem 6. *Let $\mathcal{A} = (A, +, -, 0)$ be a computable Abelian group that is not torsion. Suppose that $\mathcal{A}$ has a c.e. linear basis. Then $\mathcal{A}$ has an isomorphic punctual presentation.*

Proof. We suppose that $\{e_i\}_{i \in I}$ is a c.e. linear basis in $\mathcal{A}$, where I is an initial segment in $\omega \backslash \{0\}$. Let $E = \{e_i \mid i \in I\}$ and let $B = \mathcal{A}\langle E\rangle$ be the subgroup generated by E. As noted in Lemma 1, it is computable. We choose a computable bijection $g : \omega \to A$. Let $a_n = g(n-1)$ for $n \geqslant 1$. We use Theorem 1.

We fix $n \geqslant 1$ and define $e(n)$. As noted in the proof of Lemma 1, for each $i \leqslant n$, there are numbers $k, m \geqslant 1$ and $u_{i1}, \dots, u_{ik} \in \mathbb{Z}$ such that

$$ma_i = u_{i1} e_1 + \ldots + u_{ik} e_k. \tag{1}$$

Since we can multiply (1) by any non-zero number and add the summand $0 \cdot e_{k+1}$, we may assume that m, k are the same for all $i \leqslant n$. We put k, m and the entire matrix $\{u_{ij}\}_{i \leqslant n, j \leqslant k}$ in $e(n)$.

Next, we go through all tuples $\langle r_1, \dots, r_n\rangle$, where $0 \leqslant r_i < m$ for $i \leqslant n$. Let $x = r_1 a_1 + \ldots + r_n a_n$. If $x \in B$ then there are $k' \geqslant 1$ and numbers $p_1, \dots, p_{k'} \in \mathbb{Z}$ such that $x = p_1 e_1 + \ldots + p_{k'} e_{k'}$. Again, we may assume that $k = k'$. If $x \in B$ then we also put the pair $\langle \langle r_1, \dots, r_n\rangle_\omega, \langle p_1, \dots, p_k\rangle_\omega\rangle$ in $e(n)$.

Now we check condition c) in Theorem 1. As noted in the proof of Theorem 5, a condition $t_1(\bar{v}) = t_2(\bar{v})$ can be converted to $t_1 v_1 + \ldots + t_n v_n = 0$ using a p.r.f., where $t_i \in \mathbb{Z}$ for $i \leqslant n$. We consider an arbitrary set $\langle t_1, \dots, t_n\rangle \in \mathbb{Z}^n$. For each $i \leqslant n$, we can find $q_i, r_i \in \mathbb{Z}$ such that $t_i = q_i m + r_i$ and $0 \leqslant r_i < m$. Using (1), we bring the condition $t_1 a_1 + \ldots + t_n a_n = 0$ to the form

$$r_1 a_1 + \ldots + r_n a_n = s_1 e_1 + \ldots + s_k e_k, \tag{2}$$

where $s_j \in \mathbb{Z}$ for $j \leqslant k$. If $x = r_1 a_1 + \ldots + r_n a_n \notin B$ then this equality is impossible. If $x = p_1 e_1 + \ldots + p_k e_k$ then (2) $\Leftrightarrow$ $s_j = p_j$ for all $j \leqslant k$, by uniqueness of decomposition in B. We have found a p.r. algorithm for c) using $e(n)$. By Theorem 1, $\mathcal{A}$ has a p.r. presentation.

By hypothesis, $\mathcal{A}$ has at least one element e_1 of infinite order. If $\mathcal{A}$ is a p.r. structure, then the mapping $t \mapsto te_1$ is a p.r. injection from ω into A, and $\mathcal{A}$ has a punctual presentation. □

An Abelian group $\mathcal{A}$ is called a *torsion-free group* if $nx = 0$, where $n \in \mathbb{Z}$ and $x \in \mathcal{A}$, implies that $n = 0$ or $x = 0$. If $\mathcal{A}$ is a torsion-free group and has a linear basis of cardinality λ, then $\mathcal{A}$ can be isomorphically embedded to $\mathbb{Q}^\lambda = \sum\limits_{i\in\lambda} \mathbb{Q}$. The cardinality of all bases is the same and is called the *rank* of the group $\mathcal{A}$.

If $\mathcal{A}$ is an arbitrary Abelian group, then $T = T(\mathcal{A}) = \{x \in A \mid nx = 0$ for some $n \geqslant 1\}$ is the *torsion part* of $\mathcal{A}$. Then $\mathcal{A}/T$ is a torsion-free group. It is easy to verify that $\{e_i\}_{i\in I}$ is a linear basis in $\mathcal{A} \Leftrightarrow \{e_i/T\}_{i\in I}$ is a linear basis in $\mathcal{A}/T$.

Corollary 4. *If $\mathcal{A}$ is a computable non-torsion Abelian group and the rank of the quotient group $\mathcal{A}/T$ is finite, then $\mathcal{A}$ has an isomorphic punctual presentation.*

Proof. A finite basis is a c.e. set. □

In [14], it was proved that every computable torsion-free Abelian group $\mathcal{A}$ has an isomorphic computable presentation with a computable basis. A more modern proof of this fact can be found in [15]. In particular, every computable torsion-free Abelian group $\mathcal{A}$ has a punctual presentation, which is also proved in [1]. This leads to the following natural

Question. *Is it true that every computable Abelian group $\mathcal{A} = (A, +, -, 0)$ has a p.r. presentation?*

We note that it suffices to get the positive answer to this question for groups $\mathcal{A} = (A, +, 0)$ in the signature without $\{-\}$. In [16], it is proved that every group $(A, \cdot, 1)$, not necessarily Abelian, that has a p.r. presentation, also has a p.r. presentation in the signature $\{\cdot, ^{-1}, 1\}$. For Abelian groups, this fact can be generalized to a very wide range of complexity classes, as noted in [17].

The author is sincerely grateful to the anonymous reviewers for their valuable remarks and suggestions.

Acknowledgement. The research was supported by grant of the Russian Science Foundation No. 23-11-00170, https://rscf.ru/project/23-11-00170.

Disclosure of Interests. The authors have no competing interests to declare that are relevant to the content of this article.

References

1. Kalimullin, I., Melnikov, A., Ng, K.M.: Algebraic structures computable without delay. Theoret. Comput. Sci. **674**, 73–98 (2017)

2. Alaev, P.E.: Structures computable in polynomial time. I. Algebra and Logic **55**(6), 421–435 (2016)
3. Maltsev, A.I.: Constructive algebras. I. Russian Math. Surv. **16**(3), 77–129 (1961)
4. Cenzer, D., Remmel, J.: Polynomial time versus recursive models. Ann. Pure Appl. Logic **54**(1), 17–58 (1991)
5. Bazhenov, N., Downey, R., Kalimullin, I., Melnikov, A.G.: Foundations of online structure theory. Bull. Symb. Log. **25**(2), 141–181 (2019)
6. Kalociński, D., San Mauro, L., Wrocławski, M.: Punctual presentability in certain classes of algebraic structures. In: 49th International Symposium on Mathematical Foundations of Computer Science (MFCS), LIPIcs, vol. 306, pp. 65:1-65:15 (2024)
7. Alaev, P.E.: A criterion for P-computability of structures. Algebra Logic **61**(5), 437–441 (2022)
8. Alaev P.E.: Description of structures computable in polynomial time. In: Submitted to Annals of Pure and Applied Logic
9. Bazhenov, N., Harrison-Trainor, M., Kalimullin, I., Melnikov, A., Ng, K.M.: Automatic and polynomial-time algebraic structures. J. Symb. Logic **84**(4), 1630–1669 (2019)
10. Grigorieff, S.: Every recursive linear ordering has a copy in DTIME(n). J. Symb. Log. **55**, 260–276 (1990)
11. Rogers, H.: Theory of Recursive Functions and Effective Computability. McGraw-Hill, New York (1967)
12. Alaev, P.E.: Categoricity for primitive recursive and polynomial Boolean algebras. Algebra Logic **57**(4), 251–274 (2018). https://doi.org/10.1007/s10469-018-9498-1
13. Goncharov, S.S.: Countable Boolean Algebras and Decidability. Consultants Bureau, Plenum Publishing Corporation, New York, London, and Moscow (1997)
14. Nurtazin, A.: Computable classes and algebraic criteria of autostability. Math. Inst. SB USSRAS, Novosibirsk, Summary of Scientific Schools (1974)
15. Harrison-Trainor, M., Melnikov, A., Montalban, A.: Independence in computable algebra. J. Algebra **443**, 441–468 (2015)
16. Alaev, P.E.: Complexity of the inversion operations in groups. Algebra Logic **62**(2), 103–118 (2023)
17. Alaev, P.E.: Inversion operations in algebraic structures. Computability **12**(4), 315–322 (2023)

Space-Bounded Online Kolmogorov Complexity is Additive

Bruno Bauwens[1(✉)] and Maria Marchenko[2]

[1] Faculty of Computer Science, HSE University, Moscow 101000, Russia
brbauwens@gmail.com
[2] University of Amsterdam, Amsterdam, The Netherlands

Abstract. The *even online Kolmogorov complexity* of a string $x = x_1x_2\cdots x_n$ is the minimal length of a program that for all $i \le n/2$, on input $x_1x_3\cdots x_{2i-1}$ outputs x_{2i}. The *odd* complexity is defined similarly. The sum of the odd and even complexities is called the *dialogue* complexity. In [4] it is proven that for all n, there exist n-bit x for which the dialogue complexity exceeds the Kolmogorov complexity by $n \log \frac{4}{3} + O(\log n)$. Let $\mathrm{C}^s(x)$ denote the Kolmogorov complexity with space bound s. Here, we prove that the space-bounded dialogue complexity with bound $s + 6n + O(1)$ is at most $\mathrm{C}^s(x) + O(\log(sn))$, where $n = |x|$.

Keywords: Online Kolmogorov complexity · Symmetry of information · Space-bounded computation

1 Introduction

The Kolmogorov complexity of a bitstring x is the minimal length of a program that prints x. It was first used by Solomonoff to define an ideal algorithm for general sequence prediction [19–21]. However, in real life situations, prediction often exploits side information. For example, one may aim to predict whether certain stocks go up or down relative to macro economic indicators. This corresponds to a complexity notion where one alternates between receiving information and generating output.

The *online* Kolmogorov complexity is a generalization of conditional complexity, introduced in [7]. Given two lists of strings $x_1, \ldots, x_k$ and $y_1, \ldots, y_k$, it is defined by the minimal length of a program that for all $j \le k$, on input $y_1, \ldots, y_j$ prints x_j. This notion has important applications in the foundations of reinforcement learning [11,12].[1]

[1] An agent and an environment exchange messages. The AIXI-agent defined in [11,12] models the environment using online a priori probability, which is a probabilistic version of complexity. The online coding theorem provides a close relation between the probabilistic and program complexity versions. A space bounded variant of this theorem is essential to prove the main result below.

The work is supported by the HSE University Basic Research Program.

© The Author(s), under exclusive license to Springer Nature Switzerland AG 2025
A. Beckmann et al. (Eds.): CiE 2025, LNCS 15764, pp. 133–142, 2025.
https://doi.org/10.1007/978-3-031-95908-0_10

A special case is the *even complexity* $\mathrm{C}_{\mathrm{ev}}(x)$, which is the task of predicting each next even bit in x given previous odd bits. The *odd* complexity $\mathrm{C}_{\mathrm{odd}}(x)$ is defined similarly. A time-bounded variant of this complexity is used in a recent combinatorial characterization of 2 cryptographic conjectures: the existence of public key cryptography and one way functions [1].

Kolmogorov complexity satisfies an additivity property: the information in a pair (x, y) of bitstrings equals the information in x and the information in y given x, thus

$$\mathrm{C}(x, y) = \mathrm{C}(x) + \mathrm{C}(y|x) + O(\log \mathrm{C}(x, y)). \qquad (*)$$

Note that for Shannon entropy, the chain rule is exact $H(X, Y) = H(X) + H(Y|X)$. The additional non constant big-O term is (among other things) a consequence of the uncomputability of the complexity function [6,8]. Remarkably, such a result does not hold for online complexity: the sum of odd and even complexities can be much larger than the complexity of a string. Again, this is a consequence of the uncomputability of complexity.

Theorem 1 ([4,16]). *There exists a sequence $\omega_1\omega_2\cdots$ such that for all n,*

$$(\mathrm{C}_{\mathrm{ev}} + \mathrm{C}_{\mathrm{odd}})(\omega_1 \cdots \omega_n) \geq n \log \tfrac{4}{3} + \mathrm{C}(\omega_1 \cdots \omega_n).$$

There is a close relation to Muchnik's paradox [16] in algorithmic randomness: if an infinite sequence is not random, one might neither blame the even nor the odd bits. Similar results for computable randomness and Hausdorff dimensions are given in [2,3].

In this paper, we study additivity properties for space bounded variants. Note that there is no hope to prove a time-bounded version of this result, because it would contradict the existence of one-way functions [10,14,15].

The Kolmogorov complexity $\mathrm{C}^s(x|y)$ *with space-bound* s, is the minimal length of a program that prints x on input y and uses at most s cells of memory [5,9], (see Sect. 2 for the full definition). It is well known that this complexity satisfies an additivity property. Note that the equality $(*)$ can be viewed as two inequalities. With space-bounded complexity, the smaller side of each inequality needs a larger space-bound. The easy direction is

$$\mathrm{C}^{s+2|x|+O(1)}(x, y) \leq \mathrm{C}^s(x) + \mathrm{C}^s(y|x) + O(\log \mathrm{C}^s(x)).$$

This is proven by simply running the program for x and for y given x. Also, the other inequality holds.

Theorem 2 ([5]). *For all strings x, y and number s,*

$$\mathrm{C}^{s'}(x) + \mathrm{C}^{s'}(y|x) \leq \mathrm{C}^s(x, y) + O(\log \mathrm{C}^s(x, y)),$$

where $s' = s + O(|xy|)$.

In the limit for large s, this inequality reduces to additivity for unbounded complexity. Only the weaker result from [14,15] is well known: if $s' = 2s +$

$O(|xy| + \log s)$, then $\mathrm{C}^{s'}(x) + \mathrm{C}^{s'}(y|x) \le \mathrm{C}^s(x, y) + O(\log s + \log \mathrm{C}^s(x, y))$. The proof follows the one for unbounded complexity.[2] Because of the $O(\log s)$ in the right side, the result (*) for unbounded complexity is not a corollary.

The main result of this paper is that additivity of online information does hold for space-bounded complexity. This is indeed possible, because space bounded complexity is computable (in small space), see [5, lemma 3] or proposition 4 below. Moreover, the gap $s' - s$ is very small.

Theorem 3. *For all x and s,*

$$\mathrm{C}_{\mathrm{ev}}^{s'}(x) + \mathrm{C}_{\mathrm{odd}}^{s'}(x) \le \mathrm{C}^s(x) + O\big(\log s + \log |x|\big),$$

where $s' = s + 6|x| + O(1)$.

Remark. It is not possible to drop the term $O(\log s)$, because otherwise, the limit for large s would contradict Theorem 1. Also, the proof gives a better precision of the big-O term at the right: $2d + 4 \log d + O(1)$ where $d = \mathrm{C}^s(s, \mathrm{C}^s(x), |x|)$.[3]

Proof Structure. In the next section we define odd and even online semimeasures. We fix such semimeasures P_{ev} and $\mathrm{P}_{\mathrm{odd}}$ that can be computed with $s + O(|x|)$ space and satisfy

$$\mathrm{C}^s(x) \ge \log \frac{1}{\mathrm{P}_{\mathrm{ev}}(x)} + \log \frac{1}{\mathrm{P}_{\mathrm{odd}}(x)}.$$

Let $k = \mathrm{C}^s(x)$. We prove a space-bounded version of the online coding theorem from [7]. This implies

$$\mathrm{C}_{\mathrm{odd}}^{s}(x) \le \log \frac{1}{\mathrm{P}_{\mathrm{odd}}(x)} + O(\mathrm{C}^s(s, |x|, k))$$

$$\mathrm{C}_{\mathrm{ev}}^{s}(x) \le \log \frac{1}{\mathrm{P}_{\mathrm{ev}}(x)} + O(\mathrm{C}^s(s, |x|, k)).$$

The theorem follows by combining the 3 inequalities.

---

2 There is a small caveat: the decompresser needs to iterate over pairs (x, y) with $\mathrm{C}^s(x, y) < k$, and hence, iterate through all halting programs that use at most s bits of memory. To check whether a program halts, one needs an additional s bits of memory. The main idea of the proof of Theorem 2 is to do this more efficiently using Sipser's trick from [18]: given a halting configuration of a machine, consider all computations that terminate in this state, and organize all states in a tree. Then do a DFS search starting in this halting configuration.

3 We believe it is possible to improve the $O(d)$ term to $O(\mathrm{C}_s(\mathrm{C}_s(s)))$ when using prefix or monotone complexity, and still $s' - s \le O(|x|)$. However, the proof seems significantly harder. We hope to publish it in an extended version of this paper.

2 Definitions

For background in Kolmogorov complexity, see [13,17]. Typically, space-bounds in Kolmogorov complexity are defined and used up to arbitrary constant factors, see for example [9,14,15]. In [5], a definition is given with additive precision $O(1)$. We use it here as well.

We consider Turing machines that contain the following.

- For each input, there is a separate one-directional read-only tape with end markers. (Both for the program and each auxiliary string in the definition of complexity.)
- A one-directional write-only output tape.
- Two binary stacks, with PUSH, POP and ISEMPTY requests.

Remarks.

- A computation uses *space* s if the sum of the sizes of the 2 stacks is always at most s.
- One may think of the 2 stacks as a standard work tape, where 1 stack represents the part at the left of the computation head and the other stack the right part. Note that to make this analogy work, one needs to think about a binary work tape with some "end marker" that can be moved.[4] This definition allows for cleaner statements and proofs of space-bounded additivity (with $s' = s + O(|xy|)$ without extra $O(\log s)$ term).
- In proofs, one may use a more general type of machine, which is equipped with extra binary work tapes that have end markers. Consider a machine with k such tapes. Also, consider a computation on which at most $t_1, \ldots, t_k$ bits of these tapes are used, and the total stack size is at most s. This computation is simulated on a machine of the above type using space $s + \sum_{i \le k} t_i + 2\lceil \log t_i \rceil$.

Definition 1. *Let* $x_1, y_1, \ldots, x_k, y_k$ *be strings. The* Kolmogorov complexity of tasks $x_1 {\leftarrow} y_1; \ldots; x_k {\leftarrow} y_k$ with space bound s *relative to machine* V *is*

$$\mathrm{C}_V^s(x_1 {\leftarrow} y_1; \ldots; x_k {\leftarrow} y_k) = \min \big\{ |p| : \forall j \le k \big[V(p, y_j) = x_j \text{ using space } s \big] \big\}.$$

Note that even for $s = +\infty$, this definition is more general than in [7], because only y_j is used to produce x_j, (not $y_1, \ldots, y_j$).

Proposition 1. *There exists a machine* U *with the following property. For each machine* V *there exists a constant* c *such that for all* s, k *and strings* $x_1, y_1, \ldots, x_k, y_k$,

$$\mathrm{C}_U^{s+c}(x_1 {\leftarrow} y_1; \ldots; x_k {\leftarrow} y_k) \le \mathrm{C}_V^s(x_1 {\leftarrow} y_1; \ldots; x_k {\leftarrow} y_k) + c.$$

---

4 Popping an element from the stack corresponds to shifting all elements on one side of the tape, and this operation can be done if an endmarker is available.

The proof follows the one of [5, proposition 1], where a universal machine is constructed in which a self-delimiting description of the simulated machine V is prepended to its program. This description is stored on an auxiliary tape. The constant c is the extra space needed by the auxiliary tape (in the simulation on the stacks).

We fix such a machine U and drop subscripts U in all complexities.

Definition 2. *The* conditional *complexity of string x given string y is* $\mathrm{C}^s(x|y) = \mathrm{C}^s(x \leftarrow y)$. *If y is empty, we write* $\mathrm{C}^s(x)$. *We use the prefix code* $\bar{y} = 0y_1 \cdots 0y_{|y|}1$ *and encode a tuple of strings* $(t_1, \ldots, t_e)$ *as* $\bar{t}_1\bar{t}_2 \cdots \bar{t}_{e-1}t_e$. *The conditional complexity with tuples is defined using this encoding, thus* $\mathrm{C}^s(y, z|x) = \mathrm{C}^s(\bar{y}z|x)$.

The even *complexity of x is*

$$\begin{aligned}\mathrm{C}^s_{\mathrm{ev}}(x) &= \min\left\{|p| : \forall j \le |x|/2\left[U(p, x_1x_3\cdots x_{2j-1}) = x_{2j} \text{ using space } s\right]\right\}\\ &= \mathrm{C}^s(x_2 \leftarrow x_1;\ x_4 \leftarrow x_1x_3;\ \ldots).\end{aligned}$$

The odd *complexity of x is defined similarly. For unbounded complexity, thus $s = +\infty$, we drop s, for example* $\mathrm{C}^\infty(x|y) = \mathrm{C}(x|y)$.

Remark. The *decision* variant of the Kolmogorov complexity of x, is the minimal length of a program that prints some string or infinite sequence that has x as prefix. It would be fair to compare $\mathrm{C}_{\mathrm{ev}}(x) + \mathrm{C}_{\mathrm{odd}}(x)$ to this variant, since the odd and even complexity do not require to indicate where x stops. However, the decision and plain complexities differ by at most $O(\log \mathrm{C}(|x|))$, and this is less than the precision $O(\log(s|x|))$ of the main result. Hence, there is no reason to deviate from the more standard notions.

3 Even and Odd Online Measures

Definition 3. *Let P be a mapping from strings to the real interval* $[0, 1]$.

- *P is a* semimeasure *if for all x,*

$$P(x) \ge P(x0) + P(x1).$$

- *P is an* even semimeasure *if for all x,*

$$\begin{aligned}\text{if } |x| \text{ is even } &\Rightarrow P(x) = P(x0) = P(x1)\\ \text{if } |x| \text{ is odd } &\Rightarrow P(x) \ge P(x0) + P(x1).\end{aligned}$$

- *P is an* odd semimeasure *if for all x,*

$$\begin{aligned}\text{if } |x| \text{ is odd } &\Rightarrow P(x) = P(x0) = P(x1)\\ \text{if } |x| \text{ is even } &\Rightarrow P(x) \ge P(x0) + P(x1).\end{aligned}$$

An online coding theorem for lower-semicomputable even semimeasures was proven in [7]. However, it has a logarithmic term in the right side. Below, we prove a variant for computable even semimeasures without this term.

Proposition 2. *For each computable even semimeasure* P_{ev} *there exists a constant* c *such that for all* x,

$$\mathrm{C}_{\mathrm{ev}}(x) \leq \log \frac{1}{\mathrm{P}_{\mathrm{ev}}(x)} + c.$$

The goal of this section is to prove a version of this theorem with a space-bound. We use the following.

Definition 4. *The complexity* $\mathrm{C}^{\tilde{s}}(F)$ *of* $F : \{0,1\}^* \to \mathbb{R}$ *with space bound* $\tilde{s} : \mathbb{N} \times \mathbb{N} \to \mathbb{N}$ *is the minimal length of a program on* U *that on input* (k, x) *uses space* $\tilde{s}(k, |x|)$ *and outputs an integer* N *such that* $|F(x) - N2^{-k}| \leq 2^{-k}$*, where* k *is given in binary. If no such program exists, then* $\mathrm{C}^{\tilde{s}}(F) = +\infty$.

Proposition 3. *Assume* $\tilde{s}$ *is non-decreasing in both arguments. There exists a* c *such that for every even semimeasure* P_{ev} *and string* x,

$$\mathrm{C}^{s'}_{\mathrm{ev}}(x) \leq \log \frac{1}{\mathrm{P}_{\mathrm{ev}}(x)} + O(\mathrm{C}^{\tilde{s}}(\mathrm{P}_{\mathrm{ev}})),$$

where $s' = \tilde{s}(\lceil \log \frac{8|x|}{\mathrm{P}_{\mathrm{ev}}(x)} \rceil, |x|) + |x| + O(\log |x|)$.

A similar result for $\mathrm{P}_{\mathrm{odd}}$ holds with a similar proof.

Proof (of Proposition 2). We first explain that any real even semimeasure can be transformed to a rational one that is at most a constant factor smaller. Indeed, we may assume that $\mathrm{P}_{\mathrm{ev}}(x) \geq 2^{-\lfloor |x|/2 \rfloor - 1}$, otherwise, we replace it by $2^{-\lfloor |x|/2 \rfloor - 1} + \frac{1}{2}\mathrm{P}_{\mathrm{ev}}(x)$. Then, it is sufficient to round the value down with precision $2^{-|x|}$ for even $|x|$, and rescale recursively. This decreases the semimeasure by at most a factor $\prod_{i=1}^{\infty}(1 - 2^{-i}) > 0$.

We now assume that P_{ev} is rational. Let the cumulative semimeasure $\mathrm{PC}(y)$ represent the measure of all z with $|z| = |y|$ such that $0.z < 0.y$, more precisely, let PC(empty str) $= 0$ and

$$\mathrm{PC}(yb) = \begin{cases} \mathrm{PC}(y) & \text{if } |y| \text{ is even or } b = 0, \\ \mathrm{PC}(y) + \mathrm{P}_{\mathrm{ev}}(y0) & \text{otherwise, thus } |y| \text{ is odd and } b = 1. \end{cases}$$

The decompresser depends on this function. For notational convenience, consider a decompresser in the definition of even complexity that also takes the previous answers as input. For example, in $\mathrm{C}_{\mathrm{ev}}(y)$, on input $y_1y_2y_3y_4y_5$ the program outputs y_6, (instead of input $y_1y_3y_5$).

Machine V *on input* p *and* y: *if* $2^{-|p|}p < \mathrm{PC}(y1)$ *output 0, otherwise output 1.*

Recall that P_{ev} is a computable rational function. Hence, also PC is a computable rational function, because $\mathrm{PC}(y)$ equals the sum of at most $|y|/2$ values $\mathrm{P}_{\mathrm{ev}}(\cdot)$. Thus V computes 2 rational numbers and outputs the result of their comparison.

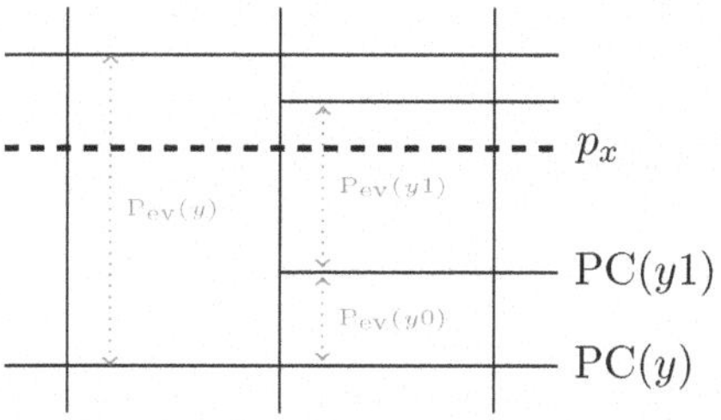

We verify that if y is a strict prefix of x with odd length and if

$$p2^{-|p|} \in \left[\mathrm{PC}(x), \mathrm{PC}(x) + \mathrm{P_{ev}}(x)\right),$$

then $V(p, y) = x_{|y|+1}$. This is quite logical, but we give the details anyway. Fix a prefix z of x, and let $b = x_{|z|+1}$. If $b = 0$, then

$$\left[\mathrm{PC}(zb), \mathrm{PC}(zb) + \mathrm{P_{ev}}(zb)\right) \subseteq \left[\mathrm{PC}(z), \mathrm{PC}(z) + \mathrm{P_{ev}}(z)\right).$$

This also holds for $b = 1$, by expanding the definition of PC($z1$). It also holds for $|z|$ of even length. For $b = x_{|y|+1}$, it holds that

$$p2^{-|p|} \in \left[\mathrm{PC}(x), \mathrm{PC}(x) + \mathrm{P_{ev}}(x)\right) \subseteq \left[\mathrm{PC}(yb), \mathrm{PC}(yb) + \mathrm{P_{ev}}(yb)\right).$$

We verify that $b = 0$ if and only if $2^{-|p|}p < \mathrm{PC}(y1)$:

- If $b = 0$, then $p2^{-|p|} < \mathrm{PC}(y0) + \mathrm{P_{ev}}(y1) = \mathrm{PC}(y1)$.
- If $b = 1$, then $p2^{-|p|} \geq \mathrm{PC}(y1)$.

This implies $V(p, y) = b$.

To finish the proof, consider any x with $\mathrm{P_{ev}}(x) > 0$. Let $k_x = \lceil \log \frac{1}{\mathrm{P_{ev}}(x)} \rceil$ and let

$$p_x = \lceil 2^{k_x} \cdot \mathrm{PC}(x) \rceil$$

be a number in binary with k_x bits (if needed, add leading zeros). Note that $2^{-k_x}p_x$ lies in the interval $\left[\mathrm{PC}(x), \mathrm{PC}(x) + \mathrm{P_{ev}}(x)\right)$, because $\mathrm{P_{ev}}(x) \geq 2^{-k_x}$. Thus,

$$\mathrm{C}_{\mathrm{ev},V}(x) \leq |p_x| = k_x = \left\lceil \log \frac{1}{\mathrm{P_{ev}}(x)} \right\rceil.$$

□

Proof (of Proposition 3*).* Now $\mathrm{P_{ev}}$ is a real valued function and computable with space bound $\tilde{s}$. The idea is to use the program $p'_x = p_x01$, which represents a number in the interval $\left[\mathrm{PC}(x), \mathrm{PC}(x) + \mathrm{P_{ev}}(x)\right)$ that is separated from the borders by at least $2^{-|p_x|-2}$. We use the same algorithm for V, but PC($y1$) is calculated with precision $2^{-|p_x|-2}$.

For this, all of the at most $|y|/2$ used values of $\mathrm{P_{ev}}$ are computed with $k_x + 2 + \lceil \log |y| \rceil$ bits precision, where $k_x = \lceil \log \mathrm{P_{ev}}(x) \rceil$. This requires space

$$\tilde{s}(k_x + 2 + \lceil \log |x| \rceil, |x|) + k_x + \log |x| + O(1).$$

□

4 Proof of Theorem 3

In [5, lemma 3] it is shown that on input x and s, one can compute $\mathrm{C}^s(x)$ in space $s + O(|x|)$. With a similar proof we obtain the following.

Proposition 4. *There exists a machine that on input x, k, s decides whether $\mathrm{C}^s(x) < k$ using space $s + 2|x| + O(\log |x|)$.*

The proof of Theorem 3 relies on defining an odd and an even semimeasure. They are derived from the following (standard) semimeasure. Fix s, k and n. Let

$$Q(y) = \frac{\left|\{z : |yz| = n \text{ and } \mathrm{C}^s(yz) < k\}\right|}{2^k}.$$

Remarks.

- At most $2^k - 1$ strings have complexity less than k, thus $Q(\text{empty}) < 1$.
- Q is indeed a semimeasure with $Q(x) = 0$ if $|x| > n$.
- If $\mathrm{C}^s(x) < k$ and $|x| = n$, then $-\log Q(x) \leq k$.
- The space needed to evaluate $Q(y)$ exactly is

$$(n + O(\log n)) + (s + 2n + O(\log n)),$$

where the first term, is for storing the counter z (in the definition of Q), and the second is the bound of Proposition 4 for checking $\mathrm{C}^s(yz) < k$.

For x of even length ℓ, the odd and even semimeasures are

$$\mathrm{P}_{\text{odd}}(x) = Q(x_1) \cdot \frac{Q(x_1x_2x_3)}{Q(x_1x_2)} \cdot \ldots \cdot \frac{Q(x_1 \cdots x_{\ell-2}x_{\ell-1})}{Q(x_1 \cdots x_{\ell-2})}$$
$$\mathrm{P}_{\text{ev}}(x) = \frac{Q(x_1x_2)}{Q(x_1)} \cdot \frac{Q(x_1x_2x_3x_4)}{Q(x_1x_2x_3)} \cdot \ldots \cdot \frac{Q(x_1 \cdots x_{\ell-1}x_\ell)}{Q(x_1 \cdots x_{\ell-1})},$$

and these semimeasures are then uniquely defined on strings of odd lengths.

Remarks.

- $Q(x) = \mathrm{P}_{\text{odd}}(x)\, \mathrm{P}_{\text{ev}}(x)$.
- For $\tilde{s}(k', n') = s + 3n' + k' + O(\log n')$, we prove that

$$\mathrm{C}^{\tilde{s}}(\mathrm{P}_{\text{odd}}) = \mathrm{C}^s(s, n, k) + O(1).$$

Indeed, if $|y| > n$, then $\mathrm{P}_{\text{odd}}(y) = 0$. Otherwise, P_{odd} is the product of $|y| = n'$ numbers, and each is computed and stored with precision $k' + \lceil \log n' \rceil$.

Now we finish the proof of the main theorem. For some x and s, consider Q with parameters $s = s$, $k = \mathrm{C}^s(x) + 1$ and $n = |x|$. We may assume $k \leq n$, otherwise the theorem is trivial. For $s' = \tilde{s}(n + 3 + \lceil \log n \rceil, n) + n + O(\log n)$,

$$\begin{aligned} \mathrm{C}^s(x) + 1 &\geq -\log Q(x) \\ &= -\log \Big(\mathrm{P}_{\text{odd}}(x)\ \mathrm{P}_{\text{ev}}(x) \Big) \\ &\geq \mathrm{C}^{s'}_{\text{ev}}(x) + \mathrm{C}^{s'}_{\text{odd}} - O(\mathrm{C}^s(s, n, k)). \end{aligned}$$

Since $s' \leq s + 6n + O(1)$, this finishes the proof of Theorem 3.

References

1. Ball, M., Liu, Y., Mazor, N., Pass, R.: Kolmogorov comes to cryptomania: on interactive Kolmogorov complexity and key-agreement. In: 2023 IEEE 64th Annual Symposium on Foundations of Computer Science (FOCS), pp. 458–483. IEEE (2023)
2. Barmpalias, G., Liu, L.: Irreducibility of enumerable betting strategies. arXiv preprint arXiv:2112.14416 (2021)
3. Barmpalias, G., Liu, L.: Aspects of muchnik's paradox in restricted betting. arXiv preprint arXiv:2201.07007 (2022)
4. Bauwens, B.: Asymmetry of the kolmogorov complexity of online predicting odd and even bits. In: 31st International Symposium on Theoretical Aspects of Computer Science, p. 125 (2014)
5. Bauwens, B., Gács, P., Romashchenko, A., Shen, A.: Inequalities for space-bounded Kolmogorov complexity. Computability **11**, 165–185 (2022)
6. Bauwens, B., Shen, A.: Complexity of complexity and maximal plain versus prefix-free Kolmogorov complexity. J. Symb. Log. **79**(2), 620–632 (2013)
7. Chernov, A., Shen, A., Vereshchagin, N., Vovk, V.: On-line probability, complexity and randomness. In: ALT 2008: Proceedings of the 19th International Conference on Algorithmic Learning Theory, pp. 138–153. Springer, Heidelberg (2008)
8. Gács, P.: On the symmetry of algorithmic information. Soviet Math. Dokl. **15**, 1477–1480 (1974)
9. Hartmanis, J.: Generalized kolmogorov complexity and the structure of feasible computations. In: 24th Annual Symposium on Foundations of Computer Science (SFCS 1983), pp. 439–445. IEEE (1983)
10. Hirahara, S., Ilango, R., Lu, Z., Nanashima, M., Oliveira, I.C.: A duality between one-way functions and average-case symmetry of information. In: Proceedings of the 55th Annual ACM Symposium on Theory of Computing, pp. 1039–1050 (2023). https://hdl.handle.net/1721.1/151049
11. Hutter, M.: Universal Artificial Intelligence: Sequential Decisions Based on Algorithmic Probability. Springer, Cham (2005)
12. Hutter, M., Quarel, D., Catt, E.: An Introduction to Universal Artificial Intelligence. CRC Press, Boca Raton (2024)
13. Li, M., Vitányi, P.M.B.: An Introduction to Kolmogorov Complexity and Its Applications. TCS, Springer, Cham (2019). https://doi.org/10.1007/978-3-030-11298-1_8
14. Longpré, L.: Resource bounded Kolmogorov complexity, a link between computational complexity and information theory. PhD thesis, Cornell University, 1986. PhD thesis. TR 86-776 (1986), 115 pp. Dept. of Computer Science, Cornell University, Ithaca, NY 14583. https://ecommons.cornell.edu/handle/1813/6616
15. Longpré, L., Mocas, S.: Symmetry of information and one-way functions. Inf. Process. Lett. **46**(2), 95–100 (1993)
16. Muchnik, A.A.: Algorithmic randomness and splitting of supermartingales. Probl. Inf. Transm. **45**(1), 54–64 (2009)
17. Shen, A., Uspensky, V.A., Vereshchagin, N.: Kolmogorov complexity and algorithmic randomness, vol. 220. American Mathematical Soc. (2017)
18. Sipser, M.: Halting space-bounded computations. Theoret. Comput. Sci. **10**(3), 335–338 (1980)

19. Solomonoff, R.J.: A preliminary report on a general theory of inductive inference (1960). available at http://raysolomonoff.com/publications/rayfeb60.pdf, revised version of this report was published in November 1960, *A Preliminary Report on a General Theory of Inductive Inference,* (Revision of Report V-131), Contract AF 49(639)-376, Report ZTB-138, Zator Co., Cambridge, Mass., Nov. 1960, available at http://raysolomonoff.com/publications/z138.pdf
20. Solomonoff, R.J.: A formal theory of inductive inference. Part I. Inf. Control **7**(1), 1–22 (1964)
21. Solomonoff, R.J.: A formal theory of inductive inference. Part II. Inf. Control **7**(2), 224–254 (1964)

On Learning Existentially Definable Subsets in a Computable Structure

Nikolay Bazhenov[1,2,3](✉) and Manat Mustafa[3]

[1] Novosibirsk State University, Novosibirsk, Russia
nickbazh@yandex.ru
[2] Innopolis University, Innopolis, Russia
[3] Nazarbayev University, Astana, Kazakhstan
manat.mustafa@nu.edu.kz

Abstract. The paper studies learnability from positive data for families of existentially definable subsets in a given computable structure $\mathcal{S}$. While provided larger and larger pieces of a subset V of the domain of $\mathcal{S}$, a learner tries to guess the Gödel number of an $\exists$-formula which defines V in $\mathcal{S}$. In this setting, we consider several classical learning criteria.

For a computable structure $\mathcal{S}$, we work with the class $\Sigma_1^0(\mathcal{S})$ containing all $\exists$-definable (without parameters) subsets of $\mathcal{S}$. We focus on some familiar classes of structures $\mathcal{S}$, including equivalence structures and Boolean algebras. We establish the following results. If the $\forall\exists$-theory of a structure $\mathcal{S}$ is decidable, then the notions of explanatory learnability and behaviorally correct learnability for families $\mathcal{F} \subseteq \Sigma_1^0(\mathcal{S})$ coincide. If $\mathcal{S}$ is a computable equivalence structure, then the notions of vacillatory learnability and behaviorally correct learnability for such families $\mathcal{F}$ coincide.

We build a computable equivalence structure $\mathcal{E}$ such that $\Sigma_1^0(\mathcal{E})$ is vacillatorily learnable but not explanatorily learnable. We prove that every computable Boolean algebra has a computable isomorphic copy $\mathcal{C}$ such that $\Sigma_1^0(\mathcal{C})$ is confidently learnable. We construct a computable Heyting algebra $\mathcal{H}$ such that for any computable copy $\mathcal{C}$ of $\mathcal{H}$, the family $\Sigma_1^0(\mathcal{C})$ is not behaviorally correctly learnable.

Keywords: Algorithmic learning theory · Inductive inference · Computable structure · Definable set · Equivalence structure · Boolean algebra · Heyting algebra

1 Introduction

The scope of the paper lies at the intersection of two major areas of the modern computability theory—computable structure theory and algorithmic learning theory. We study learnability properties for families of definable subsets in a computable structure $\mathcal{S}$.

Computable structure theory studies relations between algorithmic and structural properties of various mathematical structures [2,11]. The core methods of

© The Author(s), under exclusive license to Springer Nature Switzerland AG 2025
A. Beckmann et al. (Eds.): CiE 2025, LNCS 15764, pp. 143–158, 2025.
https://doi.org/10.1007/978-3-031-95908-0_11

this theory are based on Turing computable algorithms. Consider an algebraic structure $\mathcal{S} = (A; f_1, f_2, \dots, f_n)$—for example, one can think about the abelian group of integers $(\mathbb{Z}; +)$ or the field of rationals $(\mathbb{Q}; +, \times)$. The structure $\mathcal{S}$ is *computable* if A is a (Turing) computable set and the functions $f_1, f_2, \dots, f_n$ are also computable.

Algorithmic learning theory goes back to the classical works of Putnam [25] and Gold [14]. As formulated in the monograph [18], algorithmic learning theory "attempts to clarify the process by which a child or adult discovers systematic generalizations about her environment". The classical directions in this area focus on learning (or identification) of formal languages and learning of recursive functions (see [18] and the surveys [19,28]).

Stephan and Ventsov [27] initiated systematic studies of learnability for families of substructures in a given computable structure $\mathcal{S}$. Within their framework, a *learner* M obtains only positive data (or text) about substructures of $\mathcal{S}$. While provided larger and larger pieces of a substructure $\mathcal{T} \subseteq \mathcal{S}$ to be learned, M outputs *hypotheses*. These hypotheses are essentially the learner's attempts to guess the code of a program which enumerates the substructure $\mathcal{T}$. There are several familiar criteria for successful learning. For example, if the outputted sequence of hypotheses converges semantically to a correct code, then one gets *behaviorally correct learning* (or **BC**-learning, for short). In the case of syntactic convergence, one gets *explanatory learning* (or **Ex**-learning). The formal details about the considered learning criteria are given in Subsect. 2.1.

It turned out that in this setting, learnability has strong connections with the algebraic properties of a considered structure $\mathcal{S}$. For example, the family of all ideals of a computable ring $\mathcal{S}$ is **BC**-learnable (from text) if and only if the ring $\mathcal{S}$ is Noetherian [27]. Another illustrative (computability-theoretic) example comes from the standard computable presentation V_∞ of the countably-dimensional vector space over $\mathbb{Q}$. For a given computably enumerable (c.e.) subspace $V \subset V_\infty$, the family of all c.e. subspaces which contain V is **BC**-learnable if and only if the quotient space V_∞ / V has finite dimension [17].

Within the described direction, the following substructures have been studied extensively: ideals of rings [27], subgroups of groups [27], subspaces of vector spaces [17], closed sets in matroids [13], down-sets in quasi-orders [5], etc. We refer to [3,4,10] for other approaches to learning classes of computable structures. We note that within this area, learning for the following classes was studied: equivalence structures [12], Boolean algebras [4], linear orders [4].

In this paper, we continue investigations of learnability from positive data in a given computable structure $\mathcal{S}$ [5,17,27]. Most of the previous works used the standard universal numbering of all computably enumerable sets $(W_e)_{e \in \mathbb{N}}$ to form the *hypothesis space* for a learner M. That is, given a finite sequence σ (which is meant to encode a finite part of a substructure) containing elements from the domain of $\mathcal{S}$, the outputted hypothesis $M(\sigma) = i$ is interpreted as follows: the learner M guesses that the 'perceived' substructure $\mathcal{T} \subseteq \mathcal{S}$ has domain W_i.

In this paper, we slightly shift the focus of the learning paradigm. Intuitively speaking, our learner M wants to identify first-order definable subsets V of the domain of $\mathcal{S}$, and this goal is obtained by changing the hypothesis space of M. More formally, here we focus on the following learnability modification. Given a finite sequence σ, the outputted hypothesis $M(\sigma) = i$ is interpreted in the following way. M 'thinks' that the perceived set V is definable by an existential formula (without parameters) $\psi(x)$ in the structure $\mathcal{S}$, and M guesses that i is the Gödel number of the defining formula $\psi(x)$. (The formal definitions are given in Sect. 2.) In order to distinguish this interpretation of the hypothesis space, we refer to this approach as *learnability by indices of* Σ^0_1*-formulas.*

Roughly speaking, given the positive data about a set V, the learner M intends to identify the syntactic description of V in terms of first-order logic, inside the underlying algebraic structure $\mathcal{S}$. We focus on *existentially definable* subsets V of $\mathcal{S}$, since one can intuitively say that Σ^0_1-definability is a 'syntactic counterpart' of computable enumerability. We note that one can also work with arbitrary first-order formulas $\psi(x)$, with or without parameters from $\mathcal{S}$, but this general approach lies outside the scope of the paper.

For a structure $\mathcal{S}$, by $\Sigma^0_1(\mathcal{S})$ we denote the family of all Σ^0_1-definable (without parameters) subsets of the domain of $\mathcal{S}$.

The discussed 'syntactic' learnability approach lies in line with the theory of classifications introduced by Goncharov and Kogabaev [15]. In this theory, an algebraic structure $\mathcal{S}$ admits a computable Σ^0_1-classification if there exists a computable list $\mathbb{L} = (\psi_i(x))_{i \in \mathbb{N}}$ of Σ^0_1-formulas (in the signature of $\mathcal{S}$) such that $\mathbb{L}$ lists the class $\Sigma^0_1(\mathcal{S})$ without repetitions. The paper [15] provides the first example of a computable structure $\mathcal{S}$ such that the class $\Sigma^0_1(\mathcal{S})$ is infinite, but $\mathcal{S}$ does not admit a computable Σ^0_1-classification. Further results on Σ^0_n-classifications can be found in [1,8,15].

Similarly to the paper [6] (which studies learnability for c.e. families of sets realized by a given c.e. equivalence relation), here we focus on a few selected learning criteria. We list these criteria from strongest to weakest: confident (**Conf**), explanatory (**Ex**), vacillatory (**Vac**), and behaviorally correct (**BC**). (For example, being **Ex**-learnable always implies being **Vac**-learnable, but not vice versa.)

The main goal of this paper is to study how the considered criteria behave in the new setting of learnability by indices of Σ^0_1-formulas. The paper is arranged as follows. Section 2 contains the necessary preliminaries. In particular, Subsect. 2.1 introduces the formal definitions of the learnability notions which are used throughout the text.

In Sect. 3, we establish some basic results about our learnability paradigm. Proposition 1 gives a sufficient condition for non-**BC**-learnability (by indices of Σ^0_1-formulas) of the family $\Sigma^0_1(\mathcal{S})$. Corollary 2 shows that separations between the considered learning criteria can be obtained by choosing appropriate families $\mathcal{F} \subseteq \Sigma^0_1(\mathcal{N})$, where $\mathcal{N}$ is the standard model of Peano arithmetic. We also prove the following: if the $\forall\exists$-theory of a computable structure $\mathcal{S}$ is decidable, then the notions of **Ex**-learnability and **BC**-learnability for families $\mathcal{F} \subseteq \Sigma_1^0(\mathcal{S})$ coincide (Proposition 2).

A structure $\mathcal{A} = (A; E)$ is an *equivalence structure* if E is an equivalence relation on the set A. Section 4 is focused on computable equivalence structures $\mathcal{A}$. We prove that for Σ_1^0-definable subsets of an equivalence structure $\mathcal{A}$, the notions of **Vac**-learnability and **BC**-learnability coincide (Theorem 1 and Corollary 3). We construct a computable equivalence structure $\mathcal{E}$ such that the family $\Sigma_1^0(\mathcal{E})$ is **Vac**-learnable but not **Ex**-learnable (Theorem 2). We also give sufficient conditions for **Conf**-learnability of the family $\Sigma_1^0(\mathcal{A})$ (Propositions 4 and 5).

Section 5 works with Boolean algebras. We prove that every computable Boolean algebra $\mathcal{B}$ has a computable isomorphic copy $\mathcal{C}$ such that $\Sigma_1^0(\mathcal{C})$ is **Conf**-learnable (Theorem 3 and Corollary 4). We provide an example of non-**BC**-learnability for a Boolean algebra with distinguished atoms (Proposition 6). We also build a computable Heyting algebra $\mathcal{H}$ such that for any computable copy $\mathcal{C} \cong \mathcal{H}$, the family $\Sigma_1^0(\mathcal{C})$ is not **BC**-learnable (Theorem 4).

2 Preliminaries

For the preliminaries on computability theory, we refer to the monograph [22]. Background information on computable structure theory can be found in [2,11].

For a finite string σ, by $|\sigma|$ we denote the length of σ. By Λ we denote the empty string. For a set X, $\mathrm{card}(X)$ denotes the cardinality of X.

If $\mathcal{S}$ is an infinite computable structure, then (as is customary in computable structure theory) without loss of generality we assume that $\mathrm{dom}(\mathcal{S}) = \mathbb{N}$.

Definition 1. *Let $\mathcal{S}$ be a structure in a signature L, and let V be a subset of the domain of $\mathcal{S}$. The set V is* Σ_1^0-definable *(without parameters) in the structure $\mathcal{S}$ if there exists an existential L-formula—i.e., a formula of the form*

$$\delta(x) = \exists y_1 \exists y_2 \dots \exists y_k \theta(x, y_1, y_2, \dots, y_k), \text{ where } \theta \text{ is quantifier-free,}$$

such that $V = \{a \in \mathcal{S} : \mathcal{S} \models \delta(a)\}$. In what follows, this fact is denoted as $V = \delta[\mathcal{S}]$.

If $V = \delta[\mathcal{S}]$, then we say that δ is a Σ_1^0-defining formula *for the set V (inside the structure $\mathcal{S}$).*

For a signature L, by $\mathrm{Form}_\exists(L)$ we denote the set of all existential L-formulas $\delta(x)$. Recall that for an L-structure $\mathcal{S}$, by $\Sigma_1^0(\mathcal{S})$ we denote the family of all Σ_1^0-definable subsets of $\mathrm{dom}(\mathcal{S})$.

It is well-known that for a computable structure $\mathcal{S}$, every Σ_1^0-definable subset of $\mathrm{dom}(\mathcal{S})$ is a c.e. set (see, e.g., Sect. 3.15 in [2]).

2.1 Algorithmic Learning

Our notations mainly follow the conventions of [6,27] which, in turn, follow the Gold-style learning for families of c.e. sets [14,18]. In this paper, we consider only learnability from *positive data* (or text).

Fix a special pause symbol #. A *text* for a set $V \subseteq \mathbb{N}$ is an arbitrary function $t\colon \mathbb{N} \to \mathbb{N} \cup \{\#\}$ such that $\mathrm{range}(t) \setminus \{\#\} = V$. Informally speaking, a text t provides a list of all elements from V, with possible repetitions and with possible pauses (when $t(i) = \#$).

Given a finite sequence $a_0, a_1, \dots, a_n$ from $\mathbb{N} \cup \{\#\}$, a learner M (algorithmically) outputs a conjecture $e_n \in \mathbb{N}$. More formally, a *learner* M is a computable function acting from $(\mathbb{N} \cup \{\#\})^{<\mathbb{N}}$, the set of all finite sequences of natural numbers and #, into $\mathbb{N}$.

If t is a text and $n \in \mathbb{N}$, then by $t \upharpoonright n$ we denote the finite string $t(0)t(1)\dots t(n-1)$ from $(\mathbb{N} \cup \{\#\})^{<\mathbb{N}}$. If $\sigma \in (\mathbb{N} \cup \{\#\})^{<\mathbb{N}}$, then by $\sigma \subset t$ we denote the fact that σ is an initial segment of t (that is, $\sigma = t \upharpoonright |\sigma|$).

Let L be a computable signature. Consider a standard Gödel numbering γ which maps the set of existential formulas $\mathrm{Form}_{\exists}(L)$ into $\mathbb{N}$ (see, e.g., Sect. 1.4 of [11] for the formal definition of γ).

Convention 2. Since the set $\gamma(\mathrm{Form}_{\exists}(L))$ is infinite and computable, without loss of generality, we assume that γ is an effective *bijection* from $\mathrm{Form}_{\exists}(L)$ onto $\mathbb{N}$. For an index $i \in \mathbb{N}$, by $\delta_i(x)$ we denote the existential L-formula $\gamma^{-1}(i)$.

Now we are ready to formally discuss some classical learning criteria (tailored to our setting of Σ^0_1-definable subsets). Let $\mathcal{S}$ be a computable structure, and let V be a Σ^0_1-definable subset of $\mathrm{dom}(\mathcal{S})$.

(i) A learner M *behaviorally correctly learns* (**BC**-learns) the set V if for every text t of V, there exists $n_0 \in \mathbb{N}$ such that

$$(\forall n \geq n_0)(\delta_{M(t\upharpoonright n)}[\mathcal{S}] = V).$$

(ii) A learner M *explanatorily learns* (**Ex**-learns) the set V if for every text t of V, there exists $n_0 \in \mathbb{N}$ such that

$$(\forall n \geq n_0)(M(t \upharpoonright n) = M(t \upharpoonright n_0) \text{ and } \delta_{M(t\upharpoonright n_0)}[\mathcal{S}] = V).$$

(iii) A learner M *confidently learns* (**Conf**-learns) the set V if M **Ex**-learns V and, additionally, M converges to some conjecture on every text for *any* subset $A \subseteq \mathrm{dom}(\mathcal{S})$: that is, if t is an arbitrary text for an arbitrary A, then there exists k_0 such that $M(t \upharpoonright k) = M(t \upharpoonright k_0)$ for all $k \geq k_0$.

(iv) A learner M *vacillatorily learns* (**Vac**-learns) the set V if M **BC**-learns V, and for any text t of V, the set of outputted conjectures $\{M(t \upharpoonright n) : n \in \mathbb{N}\}$ is finite.

Let $\mathbf{N} \in \{\mathbf{Ex}, \mathbf{BC}, \mathbf{Conf}, \mathbf{Vac}\}$, i.e., $\mathbf{N}$ is one of the learning criteria given above. We say that a family $\mathcal{F} \subseteq \Sigma^0_1(\mathcal{S})$ is $\mathbf{N}$-*learnable* (by indices of Σ^0_1-formulas) if there exists a learner M which $\mathbf{N}$-learns each set $V \in \mathcal{F}$. Observe the following implications for our learnability notions:

$$\mathbf{Conf} \Rightarrow \mathbf{Ex} \Rightarrow \mathbf{Vac} \Rightarrow \mathbf{BC}.$$

Sometimes we will also use (standard Gold-style) learnability by indices of c.e. sets: for example, a learner M **BC**-learns a set $V \subseteq \mathbb{N}$ *by indices of c.e. sets* iff for every text t of V, there exists $n_0 \in \mathbb{N}$ such that $(\forall n \geq n_0)(W_{M(t\upharpoonright n)} = V)$.

3 Basic Facts About Learnability

First, we show that the technique of *locking sequences* can be applied in our setting. Note that locking sequences constitute a fundamental tool in the classical algorithmic learning theory for languages [18]. Let M be a learner, and let $V \in \Sigma^0_1(\mathcal{S})$ for a computable structure $\mathcal{S}$. We say that a string $\sigma \in (\mathbb{N} \cup \{\#\})^{<\mathbb{N}}$ is a *locking sequence for* M *on* V if $\mathrm{range}(\sigma)\backslash\{\#\} \subseteq V$ and the following holds:

$$(\forall \tau \in (\mathbb{N} \cup \{\#\})^{<\mathbb{N}})[\mathrm{range}(\tau)\backslash\{\#\} \subseteq V \;\rightarrow\; (M(\sigma\tau) = M(\sigma)\ \&\ \delta_{M(\sigma)}[\mathcal{S}] = V)].$$

A string σ is a **BC**-*locking sequence for* M *on* V if $\mathrm{range}(\sigma)\backslash\{\#\} \subseteq V$ and

$$(\forall \tau \in (\mathbb{N} \cup \{\#\})^{<\mathbb{N}})[\mathrm{range}(\tau)\backslash\{\#\} \subseteq V \;\rightarrow\; \delta_{M(\sigma\tau)}[\mathcal{S}] = V]. \tag{1}$$

Lemma 1. *Let* $\mathcal{S}$ *be a computable structure, and let* $V \in \Sigma^0_1(\mathcal{S})$. *If a learner* M **Ex**-*learns a set* V *by indices of* Σ^0_1-*formulas, then there exists a locking sequence for* M *on* V. *A similar fact is true for* **BC**-*learning and* **BC**-*locking sequences.*

Proof. Fix a learning criterion $\mathbf{N} \in \{\mathbf{Ex}, \mathbf{BC}\}$. Observe the following: given a number $i \in \mathbb{N}$, one can uniformly compute an index $f(i)$ such that the c.e. set $W_{f(i)}$ is equal to $\delta_i[\mathcal{S}]$. By the *s*–*m*–*n* Theorem, one may assume that the function f is injective. The observation implies the following fact: if a learner M **N**-learns a set V by indices of Σ^0_1-formulas, then we can effectively construct another learner $\widehat{M}$ which **N**-learns the set V by indices of c.e. sets (in the standard Gold-style paradigm [18]). (Indeed, here we define $\widehat{M}(\sigma) = f(M(\sigma))$.) Therefore, the existence of an appropriate locking sequence for M on V immediately follows from the existence of a locking sequence for $\widehat{M}$ on V (this is a classical result of [7]). □

Lemma 1 allows us to obtain an example of a non-learnable family $\Sigma^0_1(\mathcal{S})$.

Proposition 1. *Suppose that* $\mathcal{S}$ *is an infinite computable structure such that every finite subset of* $\mathrm{dom}(\mathcal{S})$ *is* Σ^0_1-*definable. Then the family* $\Sigma^0_1(\mathcal{S})$ *is not* **BC**-*learnable by indices of* Σ^0_1-*formulas.*

Proof. Our proof mirrors the classical example of a non-learnable family of languages, given in Theorem I.8 of [14]. Notice that the set $\mathbb{N} = \mathrm{dom}(\mathcal{S})$ is Σ^0_1-definable by the formula $\psi(x) = (x = x)$.

Towards a contradiction, assume that $\Sigma^0_1(\mathcal{S})$ is **BC**-learnable by a learner M. Choose a **BC**-locking sequence σ for M on the set $\mathbb{N}$. Then the sequence $t = \sigma\#\#\#\dots$ is a text for the finite Σ^0_1-definable set $V = \mathrm{range}(\sigma)\backslash\{\#\}$. On the other hand, by Eq. (1), we must have $\delta_{M(t\restriction n)}[\mathcal{S}] = \mathbb{N}$ for all $n > |\sigma|$. Therefore, the learner M does not learn the finite set V. This contradicts the choice of M. □

Corollary 1. *Consider the standard model* $\mathcal{N} = (\mathbb{N}; s, +, \times, 0)$ *of the Peano arithmetic. The family* $\Sigma_1^0(\mathcal{N})$ *is not* **BC***-learnable by indices of* Σ_1^0*-formulas.*

We also observe the following useful fact:

Lemma 2. *Let* $\mathbf{N} \in \{\mathbf{Conf}, \mathbf{Ex}, \mathbf{Vac}, \mathbf{BC}\}$*, and let* $\mathcal{S}$ *be a computable structure. Suppose that* $\mathcal{F} \subseteq \Sigma_1^0(\mathcal{S})$*. If the family* $\mathcal{F}$ *is* $\mathbf{N}$*-learnable, then* $\mathcal{F}$ *is* $\mathbf{N}$*-learnable by a primitive recursive learner.*

3.1 Separating the Learnability Notions

Informally speaking, working with the standard model of arithmetic $\mathcal{N} = (\mathbb{N}; s, +, \times, 0)$ allows us to transfer many classical learning results into the setting of $\Sigma_1^0(\mathcal{N})$. This is witnessed by the following result:

Lemma 3. *Let* $\mathbf{N} \in \{\mathbf{Conf}, \mathbf{Ex}, \mathbf{Vac}, \mathbf{BC}\}$*, and let* $\mathcal{F} \subseteq \Sigma_1^0(\mathcal{N})$*. Then the family* $\mathcal{F}$ *is* $\mathbf{N}$*-learnable by indices of* Σ_1^0*-formulas if and only if* $\mathcal{F}$ *is* $\mathbf{N}$*-learnable by indices of c.e. sets (in the classical sense of [18]).*

Proof. As in the proof of Lemma 1, one can fix a computable function $f(x)$ such that for a given Σ_1^0-formula $\delta_i(x)$, we have $W_{f(i)} = \delta_i[\mathcal{N}]$. If a learner M $\mathbf{N}$-learns a family $\mathcal{F} \subseteq \Sigma_1^0(\mathcal{N})$ by indices of Σ_1^0-formulas, then the new learner $\widehat{M} = f \circ M$ $\mathbf{N}$-learns $\mathcal{F}$ by indices of c.e. sets.

The celebrated solution of the Hilbert's Tenth Problem [20] implies that there exists a computable function $g(x)$ such that for a given c.e. set W_e, we have $\delta_{g(e)}[\mathcal{N}] = W_e$. Therefore, if a learner M $\mathbf{N}$-learns a family $\mathcal{F}$ by indices of c.e. sets, then the learner $M' = g \circ M$ $\mathbf{N}$-learns $\mathcal{F}$ by indices of Σ_1^0-formulas. □

Corollary 2. *For each of the following cases, there exists a family* $\mathcal{F} \subseteq \Sigma_1^0(\mathcal{N})$ *such that:*

1. $\mathcal{F}$ *is* **Ex***-learnable but not* **Conf***-learnable,*
2. $\mathcal{F}$ *is* **Vac***-learnable but not* **Ex***-learnable,*
3. $\mathcal{F}$ *is* **BC***-learnable but not* **Vac***-learnable.*

Proof. This fact immediately follows from Lemma 3 and the classical results on learning: (1) Proposition 4.6.2A of [23]; (2) Proposition 1 of [24]; (3) Proposition 3.10 of [9]. □

3.2 When Ex and BC Coincide

We give a general fact about learnability properties of families $\mathcal{F} \subseteq \Sigma_1^0(\mathcal{S})$.

Proposition 2. *Let* $\mathcal{S}$ *be a computable structure such that the* $\forall\exists$*-theory of* $\mathcal{S}$ *is decidable. Consider a family* $\mathcal{F} \subseteq \Sigma_1^0(\mathcal{S})$*. Then the family* $\mathcal{F}$ *is* **Ex***-learnable if and only if* $\mathcal{F}$ *is* **BC***-learnable.*

Proof. Suppose that a family $\mathcal{F} \subseteq \Sigma_1^0(\mathcal{S})$ is **BC**-learnable by a learner M. We will construct a learner $\widehat{M}$ which **Ex**-learns $\mathcal{F}$. We put $\widehat{M}(\Lambda) := M(\Lambda)$.

Suppose that $\tau = \sigma a$ for some $a \in \mathrm{dom}(\mathcal{S}) \cup \{\#\}$. Compute $i = \widehat{M}(\sigma)$ and $j = M(\tau)$. By using the decidability of the $\forall\exists$-theory of $\mathcal{S}$, we can effectively check whether the sets $\delta_i[\mathcal{S}]$ and $\delta_j[\mathcal{S}]$ are equal. More formally, we check the following $\forall\exists$-condition: $\mathcal{S} \models \forall x(\delta_i(x) \leftrightarrow \delta_j(x))$. If this condition is true, then we put $\widehat{M}(\tau) := \widehat{M}(\sigma)$. Otherwise, define $\widehat{M}(\tau) := M(\tau)$. Notice that in each of the two cases, we have $\delta_{\widehat{M}(\tau)}[\mathcal{S}] = \delta_{M(\tau)}[\mathcal{S}]$.

Let t be a text for a set $V \in \mathcal{F}$. Then there exists n_0 such that $\delta_{M(t\upharpoonright n)}[\mathcal{S}] = \delta_{M(t\upharpoonright n_0)}[\mathcal{S}] = V$ for all $n \geq n_0$. The definition of the learner $\widehat{M}$ implies that $\widehat{M}(t \upharpoonright n) = \widehat{M}(t \upharpoonright n_0)$ for $n \geq n_0$, and $\delta_{\widehat{M}(t\upharpoonright n_0)}[\mathcal{S}] = \delta_{M(t\upharpoonright n_0)}[\mathcal{S}]$. We conclude that $\widehat{M}$ **Ex**-learns the set V. □

4 Equivalence Structures

We fix the signature $L_{eq} = \{E^2\}$. By T_{eq} we denote the theory of the class of all equivalence structures (in the signature L_{eq}).

Let $\mathcal{A} = (A; E)$ be an equivalence structure. For an element $a \in A$, by $[a]_{\mathcal{A}}$ we denote the E-equivalence class of a.

An equivalence structure $\mathcal{A}$ is *unbounded* if for every $k \in \mathbb{N}$, there exists a finite equivalence class $[a]_{\mathcal{A}}$ such that $\mathrm{card}([a]_{\mathcal{A}}) > k$. This section is focused on the results about unbounded equivalence structures. The (simpler) case of bounded (i.e., not unbounded) equivalence structures is considered at the end of the section (see Proposition 5).

Convention 3. For $n \geq 1$, we introduce the following existential formula:

$$\xi_n(x) = \exists y_1 \exists y_2 \ldots \exists y_n [y_i \text{ are pairwise different, and for each } i \leq n, E(x, y_i)].$$

In other words, the formula $\xi_n(x)$ says that $\mathrm{card}([x]_{\mathcal{A}}) \geq n$.

So, for an equivalence structure $\mathcal{A}$, the set $\xi_n[\mathcal{A}]$ equals $\{x : \mathrm{card}([x]_{\mathcal{A}}) \geq n\}$. Notice the following: if $\mathcal{A}$ is unbounded, then for every $n \geq 1$, the set $\xi_n[\mathcal{A}]$ is non-empty. In addition, $\xi_{n+1}[\mathcal{A}] \subseteq \xi_n[\mathcal{A}]$.

Proposition 3 (Proposition 1 in [8]). *Let $\psi(x)$ be an existential formula in the signature L_{eq}.*

(a) If the formula $\psi(x)$ is consistent with T_{eq}, then there exists $n \geq 1$ such that for all unbounded equivalence structures $\mathcal{A}$, we have $\psi[\mathcal{A}] = \xi_n[\mathcal{A}]$.
(b) There exists an algorithm which, given $\psi(x)$, either indicates that ψ is inconsistent with T_{eq} (i.e., $\psi[\mathcal{A}] = \emptyset$), or outputs n such that $\psi[\mathcal{A}] = \xi_n[\mathcal{A}]$ for all unbounded $\mathcal{A}$.

Intuitively speaking, Proposition 3 implies the following: when learning non-empty sets $V \in \Sigma_1^0(\mathcal{A})$ (where $\mathcal{A}$ is an unbounded equivalence structure), one can work only with the formulas $\xi_n(x)$, $n \geq 1$, in place of arbitrary $\exists$-formulas $\delta_i(x)$ from Convention 2.

We emphasize the following: in general, in a given unbounded equivalence structure $\mathcal{A}$, the formulas ξ_n and ξ_m, where $n \neq m$, could define *the same set.* For example, if $\mathcal{A}$ does not contain equivalence classes of size 1, then we have $\xi_2[\mathcal{A}] = \xi_1[\mathcal{A}] = \mathrm{dom}(\mathcal{A})$.

For the sake of convenience, we introduce the following notation: $\xi_0(x) = (x \neq x)$, that is, the formula $\xi_0(x)$ always defines the empty set. In this section, we identify formulas ξ_n, $n \in \mathbb{N}$, with their Gödel numbers.

Our first result of this section establishes that for unbounded equivalence structures, the notions of **BC**-learnability and **Vac**-learnability coincide:

Theorem 1. *Let $\mathcal{A}$ be a computable unbounded equivalence structure. Consider a family $\mathcal{F} \subseteq \Sigma_1^0(\mathcal{A})$. Then the family $\mathcal{F}$ is* **Vac***-learnable if and only if $\mathcal{F}$ is* **BC***-learnable.*

Proof. Let M be a learner which **BC**-learns the family $\mathcal{F}$. We construct a new learner $\widehat{M}$.

Given $\sigma \in (\mathbb{N} \cup \{\#\})^{<\mathbb{N}}$, we compute $i = M(\sigma)$. By applying Proposition 3, we effectively find an index $n \in \mathbb{N}$ such that $\xi_n[\mathcal{A}] = \delta_i[\mathcal{A}]$. We define $\widehat{M}(\sigma)$ as (the Gödel number of) the formula ξ_n.

The construction of the learner $\widehat{M}$ immediately implies that $\widehat{M}$ **BC**-learns the family $\mathcal{F}$. If $\emptyset \in \mathcal{F}$, then it is clear that for the unique text $t_\emptyset = \#\#\#\dots$ for $\emptyset$, the learner $\widehat{M}$ outputs $\widehat{M}(t_\emptyset \upharpoonright n) = \xi_0$ for almost all $n \in \mathbb{N}$.

Suppose that a set $V \in \mathcal{F}$ is non-empty. Then by Proposition 3, there exists $k_0 \geq 1$ such that $V = \xi_{k_0}[\mathcal{A}]$. Fix numbers $k_2 > k_1 \geq k_0$ such that the structure $\mathcal{A}$ has a class of size k_1 and a class of size k_2. Observe that $V \supseteq \xi_{k_1}[\mathcal{A}] \supset \xi_m[\mathcal{A}]$ for all $m \geq k_2$. Therefore, for every $m \geq k_2$ we have $V \neq \xi_m[\mathcal{A}]$. This implies that for any text t of the set V, there exists $n_0 \in \mathbb{N}$ such that for all $n \geq n_0$, we have $\widehat{M}(t \upharpoonright n) = \xi_j$ for some $j < k_2$. We deduce that

$$\{\widehat{M}(t \upharpoonright n) : n \in \mathbb{N}\} \subseteq \{\widehat{M}(t \upharpoonright l) : l < n_0\} \cup \{\xi_j : j < k_2\},$$

and this set is finite. Hence, the learner $\widehat{M}$ **Vac**-learns the set V. □

Now we separate the notions of **Ex**-learnability and **Vac**-learnability for definable subsets of equivalence structures:

Theorem 2. *There exists a computable unbounded equivalence structure $\mathcal{E}$ such that the family $\Sigma_1^0(\mathcal{E})$ is* **Vac***-learnable but not* **Ex***-learnable.*

Proof. We will build a computable equivalence structure $\mathcal{E}$ such that for each $e \in \mathbb{N}$, the structure $\mathcal{E}$ has infinitely many classes of size $2e + 2$ (and $\mathcal{E}$ has at most one class of size $2e + 1$).

We fix a computable list $(M_e)_{e\in\mathbb{N}}$ of all primitive recursive learners. By Lemma 2, in order to establish non-**Ex**-learnability, it is sufficient to ensure that for every $e \in \mathbb{N}$, the learner M_e does not explanatorily learn the family $\Sigma_1^0(\mathcal{E})$. In addition, we will show that $\Sigma_1^0(\mathcal{E})$ is vacillatorily learnable.

For a given $e \in \mathbb{N}$ we will define an equivalence structure $\mathcal{A}_e$ on the domain $C_e = \{\langle e, x\rangle : x \in \mathbb{N}\}$. The structure $\mathcal{A}_e$ will be used to 'diagonalize' against the learner M_e. The desired equivalence structure $\mathcal{E}$ will be defined as the disjoint union of the structures $\mathcal{A}_e$, $e \in \mathbb{N}$.

Construction of the Structure $\mathcal{A}_e$. Every equivalence class of the structure $\mathcal{A}_e$ will either have size $2e+1$ or have size $2e+2$. The structure $\mathcal{A}_e$ is built in stages. At a stage s, we define a finite structure $\mathcal{A}^s$ and a finite string $\sigma^s \in \mathbb{N}^{<\mathbb{N}}$. The intuition behind σ^s is as follows: our intention is to produce a text $t = \bigcup_s \sigma^s$ such that range(t) is equal to either $\xi_{2e+1}[\mathcal{E}]$ or $\xi_{2e+2}[\mathcal{E}]$, and the learner M_e *does not converge* (to a correct hypothesis) on the text t.

We will ensure that at every stage s, the structure $\mathcal{A}^s$ contains precisely one class $[a]_{\mathcal{A}^s}$ of size $2e+1$.

We define $\mathcal{A}^0$ as the structure containing only one equivalence class $[\langle e, 0\rangle]_{\mathcal{A}^0} = \{\langle e, i\rangle : i < 2e+1\}$, and we choose σ^0 such that range$(\sigma^0) =$ dom$(\mathcal{A}^0)$.

At the end of each stage s, we also add into $\mathcal{A}^s$ a fresh equivalence class $[v]_{\mathcal{A}^s}$ of size $2e+2$ (this class is comprised by the least unused numbers from the set C_e). In addition, we find the least number d such that $d \notin$ range(σ^s) and $d \in \bigcup_{j>e} C_j$, and we update the current σ^s to a string $\tau \supset \sigma^s$ such that range$(\tau) =$ range$(\sigma^s) \cup [v]_{\mathcal{A}^s} \cup \{d\}$.

By applying Proposition 3, without loss of generality we may assume that for every string τ, the value $M_e(\tau)$ is equal to (the index of) some formula ξ_k, for $k \in \mathbb{N}$.

At the beginning of a stage $s+1$, we have one of the following three cases:

Case 1. Suppose that range(σ^s) contains an element $a \in C_e$ such that card$([a]_{\mathcal{A}^s}) = 2e+1$; and in addition, suppose that $M_e(\sigma^s) = \xi_{2e+1}$.

Then we add a fresh element $r \in C_e$ into the (unique) class $[a]_{\mathcal{A}^s}$. That is, now the class $[a]_{\mathcal{A}}$ gets size $2e+2$ inside $\mathcal{A}^{s+1}$. We also add into $\mathcal{A}^{s+1}$ a fresh class $[\widehat{a}]_{\mathcal{A}^{s+1}}$ of size $2e+1$.

We define σ^{s+1} as a string $\tau \supset \sigma^s$ such that range$(\tau) =$ range$(\sigma^s) \cup \{r\}$.

Case 2. Suppose that every $b \in$ range$(\sigma^s) \cap C_e$ satisfies card$([b]_{\mathcal{A}^s}) = 2e+2$; and also suppose that $M_e(\sigma^s) = \xi_{2e+2}$.

Then we choose σ^{s+1} as a string $\tau \supset \sigma^s$ such that range$(\tau) =$ range$(\sigma^s) \cup$ dom$(\mathcal{A}^s)$.

Case 3. If neither of Cases 1 or 2 is satisfied, then put $\sigma^{s+1} = \sigma^s$.

This concludes the construction of the structure $\mathcal{A}_e = \bigcup_s \mathcal{A}^s$. We define $\mathcal{E} = \bigsqcup_{e\in\mathbb{N}} \mathcal{A}_e$ (that is, $\mathcal{E}$ is the disjoint union of the structures $\mathcal{A}_e$). It is clear that $\mathcal{E}$ is a computable unbounded equivalence structure.

Verification. Towards a contradiction, assume that a primitive recursive learner M_e **Ex**-learns the family $\Sigma_1^0(\mathcal{E})$. Take the text $t = \bigcup_s \sigma^s$ from the construction of the structure $\mathcal{A}_e$. Then the properties of the construction imply the following: $\bigcup_{j>e} C_j \subset \text{range}(t)$; and if $\text{card}([x]_{\mathcal{E}}) = 2e+2$, then $[x]_{\mathcal{E}} \subset \text{range}(t)$. In other words, we have $\xi_{2e+2}[\mathcal{E}] \subseteq \text{range}(t)$. We note that one of the following three cases holds.

Case A. Suppose that there are infinitely many stages $s+1$ satisfying Case 1 of the construction. Then first, we show that there are infinitely many stages $s'+1$ satisfying Case 2.

Indeed, assume that s_0+1 is the last stage satisfying Case 2. Choose $s_1 > s_0$ such that s_1+1 satisfies Case 1. Then for every $s > s_1$ and every $b \in \text{range}(\sigma^s) \cap C_e$, we have $\text{card}([b]_{\mathcal{A}^s}) = 2e+2$. Hence, no stage $s+1$ for $s > s_1$ can satisfy Case 1, and this contradicts the original assumption.

Since each of Cases 1 and 2 is satisfied infinitely often, every $x \in \mathcal{A}_e$ satisfies $\text{card}([x]_{\mathcal{E}}) = 2e+2$, and hence, we have $\text{range}(t) = \xi_{2e+2}[\mathcal{E}] = \xi_{2e+1}[\mathcal{E}]$. On the other hand, there are infinitely many n with $M_e(t \upharpoonright n) = \xi_{2e+1}$, and there are infinitely many n' with $M_e(t \upharpoonright n') = \xi_{2e+2}$. Therefore, M_e does not **Ex**-learn the set $\xi_{2e+1}[\mathcal{E}] = \xi_{2e+2}[\mathcal{E}]$.

Case B. Suppose that there are only finitely many stages $s+1$ satisfying Case 1, but there are infinitely many stages $s'+1$ satisfying Case 2. Then $\mathcal{E}$ contains a (unique) class of size $2e+1$, and we have $\text{dom}(\mathcal{A}_e) \subset \text{range}(t) = \xi_{2e+1}[\mathcal{E}] \neq \xi_{2e+2}[\mathcal{E}]$. On the other hand, there are infinitely many n' satisfying $M_e(t \upharpoonright n') = \xi_{2e+2}$. Therefore, M_e does not even **BC**-learn the set $\xi_{2e+1}[\mathcal{E}]$.

Case C. Otherwise, almost every stage $s+1$ satisfies Case 3. In particular, this implies that $\mathcal{E}$ has a unique class $[a^*]_{\mathcal{E}}$ of size $2e+1$, and $\xi_{2e+1}[\mathcal{E}] \neq \xi_{2e+2}[\mathcal{E}]$.

First, assume that $\text{range}(t)$ contains a^*. Then t is a text for the set $\xi_{2e+1}[\mathcal{E}]$, and at almost every stage $s+1$, the set $\text{range}(\sigma^s)$ contains a^*. Since Case 1 is satisfied only finitely often, there exist infinitely many n such that $M_e(t \upharpoonright n) = \xi_{k_n}$ for some $k_n \neq 2e+1$. Hence, the set $\xi_{2e+1}[\mathcal{E}]$ is not **BC**-learnable by M_e.

Second, assume that $a^* \notin \text{range}(t)$. Then we have $\text{range}(t) = \xi_{2e+2}[\mathcal{E}]$. Since Case 2 is satisfied only finitely often, there exist infinitely many n such that $M_e(t \upharpoonright n) = \xi_{\ell_n}$ for some $\ell_n \neq 2e+2$. This implies that the set $\xi_{2e+2}[\mathcal{E}]$ is not **BC**-learnable by M_e.

In each of Cases A–C we deduced a contradiction with the assumption that M_e **Ex**-learns the family $\Sigma^0_1(\mathcal{E})$. Therefore, $\Sigma^0_1(\mathcal{E})$ is not **Ex**-learnable.

To finish the proof of Theorem 2, we construct a learner M' which **Vac**-learns the family $\Sigma^0_1(\mathcal{E})$. If $\text{range}(\sigma) \subseteq \{\#\}$, then we put $M'(\sigma) = \xi_0$.

Suppose that $\text{range}(\sigma) \cap \mathbb{N} \neq \emptyset$. Then we find the least index $e \in \mathbb{N}$ such that $\text{range}(\sigma) \cap C_e \neq \emptyset$. Consider all elements $b_0, b_1, \ldots, b_q$ from $\text{range}(\sigma) \cap C_e$. Compute the least number $u(\sigma)$ such that $b_0, b_1, \ldots, b_q \in \mathcal{A}^{u(\sigma)}$ (here the sequence $(\mathcal{A}^s)_{s\in\mathbb{N}}$ is taken from the construction of the structure $\mathcal{A}_e$).

Note that in this case (by the properties of the construction of $\mathcal{E}$) one can deduce the following: if $\sigma \subset t$ for some text t of a Σ^0_1-definable set V, then $b_0 \in (\xi_{2e+1}[\mathcal{E}] \backslash \xi_{2e+3}[\mathcal{E}]) \cap V$ and $V \neq \xi_m[\mathcal{E}]$ for all $m \geq 2e+3$. Indeed, since $b_0 \in C_e$, we have $\text{card}([b_0]_{\mathcal{E}}) \in \{2e+1, 2e+2\}$.

If there exists $j \leq q$ such that $\text{card}([b_j]_{\mathcal{A}^{u(\sigma)}}) = 2e+1$, then we put $M'(\sigma) = \xi_{2e+1}$. Otherwise, we set $M'(\sigma) = \xi_{2e+2}$.

We prove that for a non-empty Σ_1^0-definable subset V, the constructed learner M' **Vac**-learns the set V. Suppose that $V \in \{\xi_{2e+1}[\mathcal{E}], \xi_{2e+2}[\mathcal{E}]\}$ for some $e \in \mathbb{N}$, and let t be a text for V. We have the following two cases:

Case I. First, assume that $\xi_{2e+1}[\mathcal{E}] \neq \xi_{2e+2}[\mathcal{E}]$. Choose an element a^* such that $\mathrm{card}([a^*]_{\mathcal{E}}) = 2e+1$. Also choose an element b^* such that $b^* > a^*$ and $\mathrm{card}([b^*]_{\mathcal{E}}) = 2e+2$. Fix $n_0 \in \mathbb{N}$ such that $b^* \in \mathrm{range}(t \upharpoonright n_0)$ and n_0 satisfies the following additional condition: if $V = \xi_{2e+1}[\mathcal{E}]$, then $a^* \in \mathrm{range}(t \upharpoonright n_0)$. Then one can show that for every $n \geq n_0$, $M'(t \upharpoonright n)$ correctly guesses the Σ_1^0-defining formula for V (this is either ξ_{2e+1} or ξ_{2e+2}). Thus, M' **Ex**-learns the set V.

Case II. Now assume that $\xi_{2e+1}[\mathcal{E}] = \xi_{2e+2}[\mathcal{E}]$. Choose an arbitrary element b such that $\mathrm{card}([b]_{\mathcal{E}}) = 2e+2$. Fix n_0 such that $b \in \mathrm{range}(t \upharpoonright n_0)$. It is clear that for every $n \geq n_0$, we have $M'(t \upharpoonright n) \in \{\xi_{2e+1}, \xi_{2e+2}\}$. Hence, M' **Vac**-learns the set V.

We conclude that the family $\Sigma_1^0(\mathcal{E})$ is **Vac**-learnable but not **Ex**-learnable. Theorem 2 is proven. □

In conclusion of this section, we give two sufficient conditions for confident learnability:

Proposition 4. *Let $\mathcal{A}$ be a computable unbounded equivalence structure such that the existential diagram of $\mathcal{A}$ is decidable. Then the family $\Sigma_1^0(\mathcal{A})$ is* **Conf***-learnable.*

Proposition 5. *Let $\mathcal{A}$ be a computable bounded equivalence structure (that is, there exists $N \geq 1$ such that every finite equivalence class of $\mathcal{A}$ has size at most N). Then the family $\Sigma_1^0(\mathcal{A})$ is* **Conf***-learnable.*

For reasons of space, the proofs of Propositions 4 and 5 are omitted. The following fact is an immediate consequence of Theorem 1 and Proposition 5.

Corollary 3. *Let $\mathcal{A}$ be an arbitrary computable equivalence structure, and let $\mathcal{F} \subseteq \Sigma_1^0(\mathcal{A})$. Then the family $\mathcal{F}$ is* **Vac***-learnable if and only if $\mathcal{F}$ is* **BC***-learnable.*

In conclusion, we summarize the separation results of Sect. 4. Let $\mathcal{A}$ be a computable equivalence structure. The notions of **BC**-learnability and **Vac**-learnability for an $\mathcal{F} \subseteq \Sigma_1^0(\mathcal{A})$ coincide. In general, **Vac**-learnability and **Ex**-learnability are different notions. We leave open whether **Ex**-learnability is equivalent to **Conf**-learnability or not.

5 Boolean Algebras

Boolean algebras are viewed as structures in the signature $L_{BA} = \{\vee, \wedge, \overline{(\cdot)}, 0, 1\}$. For a Boolean algebra $\mathcal{B}$, its standard ordering $\leq_{\mathcal{B}}$ is recovered as follows: $x \leq_{\mathcal{B}} y$ if and only if $x \vee y = y$.

Recall that an element $a \in \mathcal{B}$ is an *atom* if a is a minimal non-zero element in $\mathcal{B}$, that is, $0 <_{\mathcal{B}} a$ and there is no b with $0 <_{\mathcal{B}} b <_{\mathcal{B}} a$. By $\mathrm{Atom}(\mathcal{B})$ we denote the set of atoms of $\mathcal{B}$.

Convention 4. For $k \geq 1$, we introduce an existential formula $\theta_k(x)$ which states that $x \neq 1_{\mathcal{B}}$ and there exist $y_1, y_2, \dots y_k$ such that: every y_i is not equal to $0_{\mathcal{B}}$, and $y_i \wedge y_j = 0_{\mathcal{B}}$ for $i \neq j$, and $y_1 \vee y_2 \vee \dots \vee y_k = x$. In other words, $\theta_k(x)$ says that the element $x \neq 1_{\mathcal{B}}$ can be represented as a disjoint union of at least k-many non-zero parts. We put $\rho_k(x) = \theta_k(\overline{x})$, where $\overline{x}$ is the complement of x.

For an infinite Boolean algebra $\mathcal{B}$ and any non-zero k, ℓ, notice the following inclusions: $\theta_{k+1}[\mathcal{B}] \subseteq \theta_k[\mathcal{B}]$, $\rho_{k+1}[\mathcal{B}] \subseteq \rho_k[\mathcal{B}]$, $\mathrm{dom}(\mathcal{B}) \backslash \theta_k[\mathcal{B}] \subseteq \rho_\ell[\mathcal{B}]$, and $\mathrm{dom}(\mathcal{B}) \backslash \rho_k[\mathcal{B}] \subseteq \theta_\ell[\mathcal{B}]$.

Recall that a set $X \subseteq \mathbb{N}$ is *immune* if X is infinite and X does not contain infinite c.e. subsets. Our first result of this section is the following:

Theorem 3. *Let $\mathcal{B}$ be a computable Boolean algebra such that its set of atoms is not immune. Then the family $\Sigma_1^0(\mathcal{B})$ is* **Conf**-*learnable.*

Remmel (Corollary 2.7 in [26]) proved that for any computable Boolean algebra $\mathcal{B}$ with infinite set of atoms, there exists a computable $\mathcal{C} \cong \mathcal{B}$ such that the set $\mathrm{Atom}(\mathcal{C})$ is not immune. Therefore, Theorem 3 implies the following:

Corollary 4. *For any computable Boolean algebra $\mathcal{B}$, there exists a computable structure $\mathcal{C}$ isomorphic to $\mathcal{B}$ such that $\Sigma_1^0(\mathcal{C})$ is* **Conf**-*learnable.*

Proof. (of Theorem 3). If the algebra $\mathcal{B}$ is finite, then it is clear that the family $\Sigma_1^0(\mathcal{B})$ is **Conf**-learnable (since $\Sigma_1^0(\mathcal{B})$ is a finite family of finite sets). Therefore, we assume that the algebra $\mathcal{B}$ is infinite.

Following Sect. 2.6 of [16], one can show that every non-empty $V \in \Sigma_1^0(\mathcal{B})$ can be represented as a finite union of the sets having one of the following forms: $\theta_k[\mathcal{B}]$ for $k \geq 1$; $\rho_\ell[\mathcal{B}]$ for $\ell \geq 1$; $\{0_{\mathcal{B}}\}$; or $\{1_{\mathcal{B}}\}$.

Without loss of generality, we may assume that the set $\mathrm{Atom}(\mathcal{B})$ is infinite (the case of finite $\mathrm{Atom}(\mathcal{B})$ could be treated in a similar way). Since the set $\mathrm{Atom}(\mathcal{B})$ is not immune, we can choose a computable sequence $(a_i)_{i \in \mathbb{N}}$ which lists some subset of $\mathrm{Atom}(\mathcal{B})$ without repetitions. For a non-zero k, we set $c_k = a_0 \vee a_1 \vee \dots \vee a_{k-1}$ and $d_k = \overline{c_k}$. Notice that $c_k \in \theta_k[\mathcal{B}] \backslash \theta_{k+1}[\mathcal{B}]$ and $d_k \in \rho_k[\mathcal{B}] \backslash \rho_{k+1}[\mathcal{B}]$.

We sketch how to construct a confident learner M. Let σ be a finite string such that $\mathrm{range}(\sigma) \cap \mathbb{N} \neq \emptyset$. Then $M(\sigma)$ outputs a finite disjunction of the following formulas:

- $\theta_k(x)$ for the least non-zero k such that $c_k \in \mathrm{range}(\sigma)$, if such k exists;
- $\rho_\ell(x)$ for the least non-zero ℓ such that $d_\ell \in \mathrm{range}(\sigma)$, if such ℓ exists;
- $(x = 0_{\mathcal{B}})$, if $0_{\mathcal{B}} \in \mathrm{range}(\sigma)$;
- $(x = 1_{\mathcal{B}})$, if $1_{\mathcal{B}} \in \mathrm{range}(\sigma)$;
- the formula $(x \neq x)$ defining $\emptyset$.

It is clear that the constructed M **Ex**-learns every non-empty set $V \in \Sigma_1^0(\mathcal{B})$. In addition, M converges on every text t, hence M confidently learns $\Sigma_1^0(\mathcal{B})$. Theorem 3 is proven. □

The next result considers Boolean algebras with distinguished atoms (viewed as structures in the extended signature $L_{BA} \cup \{\text{Atom}\}$). We show that learnability properties for this class change significantly, when compared with Theorem 3.

A Boolean algebra $\mathcal{B}$ is *atomic* if for every non-zero element $x \in \mathcal{B}$, there exists an atom a such that $a \leq_{\mathcal{B}} x$.

Proposition 6. *Let $\mathcal{B}$ be a computable, infinite atomic Boolean algebra. Then the family $\Sigma_1^0(\mathcal{B}, \text{Atom})$ is not* **BC***-learnable.*

We note that Proposition 6 can be viewed as follows: one cannot **BC**-learn all $\exists$-definable subsets in the structural jump of $\mathcal{B}$ (see Section X.3 of [21]). For reasons of space, the proof of Proposition 6 is omitted. Notice that Proposition 6, e.g., implies the following fact. If $\mathcal{B}$ is the Boolean algebra of finite/cofinite subsets of $\mathbb{N}$, then for any computable structure $\mathcal{C} \cong \mathcal{B}$, the family $\Sigma_1^0(\mathcal{C}, \text{Atom})$ cannot be **BC**-learnable. This contrasts with Corollary 4.

Such a behavior also extends to other familiar classes which contain the class of Boolean algebras. For example, we have the following:

Theorem 4. *There exists a computable Heyting algebra $\mathcal{H}$ such that for any computable structure $\mathcal{S} \cong \mathcal{H}$, the family $\Sigma_1^0(\mathcal{S})$ is not* **BC***-learnable.*

For reasons of space, the proof of Theorem 4 is omitted.

Acknowledgments. The work was supported by Nazarbayev University Faculty Development Competitive Research Grants 201223FD8823. The work of Bazhenov is supported by the Mathematical Center in Akademgorodok under the agreement No. 075-15-2025-349 with the Ministry of Science and Higher Education of the Russian Federation. We are grateful to the anonymous referees for their helpful comments.

Disclosure of Interests. The authors have no competing interests to declare that are relevant to the content of this article.

References

1. Aleksandrova, S., Bazhenov, N., Zubkov, M.: Complexity of Σ_n^0-classifications for definable subsets. Arch. Math. Log. **62**(1–2), 239–256 (2023). https://doi.org/10.1007/S00153-022-00842-6
2. Ash, C.J., Knight, J.: Computable Structures and the Hyperarithmetical Hierarchy, Stud. Logic Found. Math., vol. 144. Elsevier, Amsterdam (2000)
3. Bazhenov, N., Fokina, E., Rossegger, D., Soskova, A., Vatev, S.: Learning families of algebraic structures from text. In: Levy Patey, L., Pimentel, E., Galeotti, L., Manea, F. (eds.) CiE 2024. LNCS, vol. 14773, pp. 166–178. Springer, Cham (2024). https://doi.org/10.1007/978-3-031-64309-5_14
4. Bazhenov, N., Fokina, E., San Mauro, L.: Learning families of algebraic structures from informant. Inf. Comput. **275**, 104590 (2020). https://doi.org/10.1016/J.IC.2020.104590

5. Bazhenov, N., Mustafa, M.: On learning down-sets in quasi-orders, and ideals in Boolean algebras. Theory Comput. Syst. **69**(1), 1 (2025). https://doi.org/10.1007/S00224-024-10201-Y
6. Belanger, D., Gao, Z., Jain, S., Li, W., Stephan, F.: Learnability and positive equivalence relations. Inf. Comput. **295**, 104913 (2023). https://doi.org/10.1016/J.IC.2022.104913
7. Blum, L., Blum, M.: Toward a mathematical theory of inductive inference. Inf. Control **28**(2), 125–155 (1975). https://doi.org/10.1016/S0019-9958(75)90261-2
8. Boyadzhiyska, S., Lange, K., Raz, A., Scanlon, R., Wallbaum, J., Zhang, X.: Classifications of definable subsets. Algebra Log. **58**(5), 383–404 (2019). https://doi.org/10.1007/s10469-019-09559-7
9. Case, J.: The power of vacillation in language learning. SIAM J. Comput. **28**(6), 1941–1969 (1999). https://doi.org/10.1137/S0097539793249694
10. de Brecht, M., Yamamoto, A.: Topological properties of concept spaces (full version). Inf. Comput. **208**(4), 327–340 (2010). https://doi.org/10.1016/J.IC.2009.08.001
11. Ershov, Y.L., Goncharov, S.S.: Constructive Models. Kluwer Academic, New York (2000)
12. Fokina, E.B., Kötzing, T., San Mauro, L.: Limit learning equivalence structures. In: Garivier, A., Kale, S. (eds.) ALT 2019. Proceedings of Machine Learning Research, vol. 98, pp. 383–403 (2019)
13. Gao, Z., Stephan, F., Wu, G., Yamamoto, A.: Learning families of closed sets in matroids. In: Dinneen, M.J., Khoussainov, B., Nies, A. (eds.) WTCS 2012. LNCS, vol. 7160, pp. 120–139. Springer, Heidelberg (2012). https://doi.org/10.1007/978-3-642-27654-5_10
14. Gold, E.M.: Language identification in the limit. Inf. Control **10**(5), 447–474 (1967). https://doi.org/10.1016/S0019-9958(67)91165-5
15. Goncharov, S.S., Kogabaev, N.T.: On Σ_1^0-classification of relations on computable structures. Vestn. Novosib. Gos. Univ., Ser. Mat. Mekh. Inform. **8**(4), 23–32 (2008). https://www.mathnet.ru/eng/vngu304. in Russian
16. Goncharov, S.S.: Countable Boolean Algebras and Decidability. Consultants Bureau, New York (1997)
17. Harizanov, V.S., Stephan, F.: On the learnability of vector spaces. J. Comput. Syst. Sci. **73**(1), 109–122 (2007). https://doi.org/10.1016/j.jcss.2006.09.001
18. Jain, S., Osherson, D., Royer, J.S., Sharma, A.: Systems that Learn, 2nd edn. MIT Press, Cambridge (1999)
19. Lange, S., Zeugmann, T., Zilles, S.: Learning indexed families of recursive languages from positive data: a survey. Theor. Comput. Sci. **397**(1–3), 194–232 (2008). https://doi.org/10.1016/j.tcs.2008.02.030
20. Matiyasevich, Y.V.: Hilbert's Tenth Problem. MIT Press, Cambridge (1993)
21. Montalbán, A.: Computable Structure Theory: Within the Arithmetic. Cambridge University Press, Cambridge (2021)
22. Odifreddi, P.: Classical Recursion Theory, Stud. Logic Found. Math., vol. 125. Elsevier, Amsterdam (1992)
23. Osherson, D.N., Stob, M., Weinstein, S.: Systems that Learn. MIT Press, Cambridge (1986)
24. Osherson, D.N., Weinstein, S.: Criteria of language learning. Inf. Control **52**(2), 123–138 (1982). https://doi.org/10.1016/S0019-9958(82)80025-9
25. Putnam, H.: Trial and error predicates and the solution to a problem of Mostowski. J. Symb. Log. **30**(1), 49–57 (1965). https://doi.org/10.2307/2270581

26. Remmel, J.B.: Recursive isomorphism types of recursive Boolean algebras. J. Symb. Log. **46**(3), 572–594 (1981). https://doi.org/10.2307/2273757
27. Stephan, F., Ventsov, Y.: Learning algebraic structures from text. Theor. Comput. Sci. **268**(2), 221–273 (2001). https://doi.org/10.1016/S0304-3975(00)00272-3
28. Zeugmann, T., Zilles, S.: Learning recursive functions: a survey. Theor. Comput. Sci. **397**(1–3), 4–56 (2008). https://doi.org/10.1016/j.tcs.2008.02.021

Transfinite Structured Programming

Siddharth Bhaskar(✉)

University of Southern Denmark, 5230 Odense, Denmark
bhaskar@imada.sdu.dk

Abstract. There are barely any high-level imperative languages for implementing transfinite algorithms. We construct one by taking the standard model of structured programming—a language of *while-programs* built from assignment operations with selection, sequencing, and iteration, and equip it with a denotational semantics. We show while-programs are equivalent in expressive power to (unstructured) *register programs*, thus demonstrating a transfinite *structured program theorem*. We additionally provide a *native* operational semantics and prove its correctness. Finally, we show how to implement a transfinite insertion sort with exactly the same code one might use in the finitary setting. Our work shows how the basic theory of imperative programming lifts cleanly to the transfinite setting and suggests the outline of a theory of infinite data structures and transfinite algorithms.

Keywords: Transfinite computation · Structured programming · Programming language semantics

1 Introduction

If you want to implement a (finitary) algorithm in code, you have a vast array of different languages and programming paradigms to choose from. On the other hand, if you want to implement a transfinite algorithm, your choices of language are much fewer and further between. To program in an imperative style—where lines of code act on a program state—you are more or less limited to the unstructured assembly languages of infinite-time Turing or register machines.

Such was once the situation for finitary computation as well. But the 1950's and 60's saw the emergence of *high-level languages* which abstracted away from physical components of the machine. For example, physical memory locations storing actual bits were replaced by typed variables and unstructured control flow (like goto statements) was replaced by blocks of iterative control (like while-, do-while, and for-loops).

This structured programming paradigm is now widely considered *de rigueur* for its understandability and ease of maintenance in comparison with unstructured code, though it was subject to active debate at the time. Its theoretical basis is the *structured programming theorem,* which establishes that structured code is no less expressive than its unstructured counterpart [1].

© The Author(s), under exclusive license to Springer Nature Switzerland AG 2025
A. Beckmann et al. (Eds.): CiE 2025, LNCS 15764, pp. 159–174, 2025.
https://doi.org/10.1007/978-3-031-95908-0_12

Our Contributions. In the current paper, we retell this history in the setting of transfinite computation, where the running time can be some infinite ordinal. We define a transfinite programming language using the familiar elements of structured program control—sequencing (concatenation), selection (if-then-else branching), and iteration (while loops)—and give it a denotational semantics. We show how to compile to and from unstructured programs, thus establishing a transfinite structured programming theorem. We further show the robustness of the model by defining a native operational semantics, i.e., one that does not rely on compiling to another language.

Furthermore, we establish all of these results in the presence of arbitrary data. In so doing, we have to say what we mean by an infinitary or transfinite data type, which will simply turn out to be ordinary data equipped with a mechanism for taking limits. We demonstrate the utility of this definition by implementing insertion sort for well-ordered lists using *exactly the same code* one might reasonably use for finite lists. This suggests that the design philosophy of *data encapsulation*—keeping your code simple at the expense of your data structures—extends to the transfinite setting as well.

Related Work. The typical usage of transfinite recursion by set theorists may be considered a model of transfinite computation, though the programs involved are usually informal. The earliest self-conscious models of imperative transfinite computation are probably *infinite-time Turing machines* of Hamkins and Kidder [3], followed quickly by *ordinal register machines* of Koepke and others [4]. There is now a large and mature literature on these and similar machines characterizing their expressive power in descriptive set-theoretic terms. (See [9] for a broad survey of transfinite computation models and [2] for a textbook on ordinal computability.) Our work is directly inspired by this tradition even though our concerns are primarily with programming language theory and not computability.

Of particular note are the ordinal register programs of Koepke and Siders, cf. [5,6]. These *are* structured imperative program models of transfinite computation, and they use the same set of program constructors that we do, viz., selection, sequencing, and iteration. Our contributions are (1) defining a native semantics for such programs, which then allows to prove (2) a structured programming theorem, and moreover doing all of this in (3) the context of arbitrary data types, instead of registers which store ordinals in particular.

Notation and Preliminaries. We assume basic familiarity with ordinals and their arithmetic. By $\mathcal{ON}$ we mean the class of ordinals. Every ordinal is identified with the set of its predecessors. Therefore $2 = \{0, 1\}$, and we also identify it with the set of booleans $\{\mathrm{ff}, \mathrm{tt}\}$. By $f : X \rightharpoonup Y$ we mean that f is a partial function with domain X and codomain Y. We use $\equiv$ for syntactic equality of program terms.

For partial functions f, g of the same type $f \sqsubseteq g$ means g extends f, i.e., $(\forall x, y)\, f(x) = y \implies g(x) = y$, and $f = g$ means $f \sqsubseteq g \wedge g \sqsubseteq f$. Composition of partial functions is strict: $(f \circ g)(x) = y$ means $(\exists w)\, g(x) = w \wedge f(w) = y$.

A subset Y of an ordinal α is *cofinal* if Y is unbounded in α. If $(X, \leq)$ is a partial order and $Y \subseteq X$, then by $\inf Y$ we mean the $\leq$-greatest lower bound of Y, if it exists. If $(x_\beta : \beta < \alpha)$ is an α-indexed sequence of elements from X, then by a *tail* of this sequence, we mean $(x_\beta : \gamma < \beta < \alpha)$ for some $\gamma < \alpha$.

2 Programs and Their Denotations

In this section we introduce both *register programs* and *while-programs*, parameterized by a signature Σ, that take semantics over Σ-structures. These refer to unstructured and structured programs respectively. Our presentation in this section is influenced by [7].

Definition 1. *Let S be a set. Then by an S-*indexed set *we mean a collection of sets $\langle X_s : s \in S\rangle$.*

In other words, an S-indexed set is a function from S to the class of sets. We can similarly talk about S-indexed groups, S-indexed partial orders, etc.

Definition 2. *Let S be a set. Then an S-*sorted signature *is a collection of function and relation symbols such that*

- *each function symbol has type $s_0 \times \cdots \times s_{n-1} \to s_n$, and*
- *each relation symbol has type $s_0 \times \cdots \times s_{n-1}$,*

for some $n \in \mathbb{N}$ and $s_0, \ldots, s_n \in S$.

In other words, any S-sorted signature can be thought of as types indexed by product types and function types generated by S.

Definition 3. *Given an S-sorted signature Σ, a Σ-*structure *consists of*

- *for each $s \in S$, a set X_s (the* carrier *or* domain *of s) together with a way of taking limits of ordinal-indexed sequences in S, and*
- *interpretations of each symbol in Σ by a partial function or relation of the right type.*

While we talk about *sets* and *functions* above, this is force of habit: nothing really breaks if we allow proper class-sized structures.

Remark 1. Ordinarily, the carrier of a sort s is a pure set; the novelty here is the limit operator, which has type

$$\lim : \bigcup_{\alpha \in \mathcal{ON}} X_s^\alpha \rightharpoonup X_s,$$

where $\mathcal{ON}$ is the class of all ordinals and X_s^α is the set of α-indexed sequences over X_s. Note that limits may be undefined, hence the partial function type. This is a class function, but in practice, our limits will rely on something much tamer:

- Either X_s is a partial order and limits are given by limits inferior, i.e.,

$$\lim_{\alpha<\gamma} x_\alpha := \sup_{\alpha<\gamma} \inf_{\alpha<\beta<\gamma} x_\beta,$$

- or X_s is the function type $A \to B$, B already has a notion of limit, and we lift this to X_s pointwise, i.e.,

$$(\lim_{\alpha<\gamma} x_\alpha)(a) := \lim_{\alpha<\gamma} (x_\alpha(a)).$$

We shall follow the convention of putting structures in boldface. If Σ is an S-sorted signature and $\mathbf{M}$ is a Σ-structure, then by M_s for $s \in S$ we mean the carrier of s in $\mathbf{M}$. For a Σ-structure $\mathbf{M}$ and $f \in \Sigma$, the interpretation of f in $\mathbf{M}$ is written $[\![f]\!]^{\mathbf{M}}$.

2.1 Program Syntax

Let us fix an index set S and, for each $s \in S$, a countable stock of variables of type S.

Definition 4. *A* Σ-assignment statement *is an expression*

$$Y \leftarrow f(X_0, \ldots, X_{n-1}),$$

where $X_0, \ldots, X_{n-1}$ *are variables,* $f \in \Sigma$ *is a function symbol, and all types agree.*

Definition 5. *A* Σ-register program *is a finite, consecutively numbered list of lines, each of which is either*

- *a* Σ*-assignment statement, the line* halt, *or the line* skip,
- *a line of the form* $\mathsf{goto}\ \ell$, *or*
- *a line of the form* $\mathsf{if}\ B(X_0, \ldots, X_{n-1})\ \mathsf{then\ goto}\ \ell_1\ \mathsf{else\ goto}\ \ell_2$,

where ℓ, ℓ_1, ℓ_2 *range over line numbers and* B *ranges over relation symbols of* Σ.

Definition 6. *A* Σ-while program *is defined by the grammar*

$$T := \varepsilon \mid A \mid T_0; T_1 \mid \mathsf{if}\ B\ \mathsf{then}\ T_0\ \mathsf{else}\ T_1 \mid \mathsf{while}(B)\ \mathsf{do}\ T,$$

where ε *denotes the empty syntax tree,* A *ranges over* Σ*-assignment statements, and* B *abbreviates* $B(X_0, \ldots, X_{n-1})$ *and ranges over the relation symbols of* Σ.

Finally, a few auxiliary terms:

- The *support* of a while-program or register program is simply the (finite) set of variables that occur in it.
- If $\mathbf{V}$ is a set of Σ-variables and $\mathbf{M}$ is a Σ-structure then by $M^{\mathbf{V}}$ we mean the $\mathbf{V}$-indexed set

$$\langle M_{\mathrm{sort}(V)} : V \in \mathbf{V} \rangle.$$

- A *valuation* is a tuple $\langle x_V : V \in \mathbf{V}\rangle$ such that $x_V \in M_{\mathrm{sort}(V)}$. When $\mathbf{V}, \mathbf{M}$ are understood, we shall simply write x. (For those inclined to see it that way, $M^{\mathbf{V}}$ is a dependent function type and x is an inhabitant.)

Finally, let us give a semantics to assignment statements. Suppose α is the assignment

$$W \leftarrow f(U_0, \ldots, U_{n-1})$$

and $U_0, \ldots, U_{n-1}, W \in \mathbf{V}$. Suppose that x is a valuation. Then we define

$$[\![\alpha]\!](x) := y \tag{1}$$

where

$$y_V := \begin{cases} x_V & \text{if } V \neq W \\ [\![f]\!]^{\mathbf{M}}(x_{U_0}, \ldots, x_{U_{n-1}}) & \text{if } V = W \end{cases}$$

This is not saying anything more complicated than "compute f on the values stored by the variables $U_0, \ldots, U_{n-1}$ and store the result in W."

2.2 Register Program Semantics

Fix a Σ-register program p with support $\mathbf{V}$ and a Σ-structure $\mathbf{M}$. Then a *configuration* of p is a pair (ℓ, x), where ℓ is a line number of p and $x \in M^{\mathbf{V}}$. Lines of p act on configurations as follows:

- Assignment statements increment the line number and act on the valuation according to Eq. (1).
- The line skip increments the line number and leaves the valuation unchanged.
- The line goto ℓ changes the line number to ℓ and leaves the valuation unchanged.
- The line

  $$\text{if } B(U_0, \ldots, U_{n-1}) \text{ then goto } \ell_1 \text{ else goto } \ell_2$$

  leaves the valuation unchanged and changes the line number to ℓ_1 or ℓ_2 depending on whether $[\![B]\!]^{\mathbf{M}}(x_{U_0}, \ldots, x_{U_{n-1}})$ is true or false.

This is all routine; the important and novel bit for transfinite computation is how we compute the limit (ℓ_γ, x_γ) of a sequence $(\ell_\alpha, x_\alpha)_{\alpha<\gamma}$ of configurations, where γ is a limit ordinal. This is defined as

$$\ell_\gamma := \liminf_{\alpha<\gamma} \ell_\alpha, \quad (x_\gamma)_V := \lim_{\alpha<\gamma} (x_\alpha)_V, \tag{2}$$

where the limit inferior on the left is taken with respect to the ordering on line numbers on the left and the limit on the right is given by the carrier $M_{\mathrm{sort}(\mathbf{V})}$.

A *computation sequence* is an ordinal-indexed sequence of configurations such that successor stages are given by the action of the previous line on the previous configuration, and limit stages are defined by (2). We define

$$[\![p]\!] : M^{\mathbf{V}} \rightharpoonup M^{\mathbf{V}} \tag{3}$$

by

$$[\![p]\!](x) = y$$

if there exists a computation sequence starting at (ℓ_0, x) and ending at some (ℓ_Γ, y), where ℓ_0 is the first line of the program and ℓ_Γ labels some halt command.

Divergence, etc. Since programs are deterministic, Eq. (3) defines a partial function. If the unique maximal computation sequence starting at (ℓ_0, x) does not halt, it either *diverges,* i.e., has length the class of all ordinals, or is *stuck.* There are two types of the latter: those which are stuck because some limit is undefined, and those which are stuck because they tried to go to a line which does not exist.

Relationship to Ordinal Register Machines. An ordinal register machine is a register machine whose registers (program variables) store ordinals, either unbounded or bounded by a fixed ordinal β. In the latter case we might have *overflows* where the contents of the registers is unbounded in β; in this case, there are two semantics: the "strong," where the register resets to 0, and the "weak," where the limit is undefined.

Any of these situations can be modeled in the present formalism. The structure in question has domain either the class of ordinals or the set of ordinals bounded by β, with whatever primitive operations you want—usually a constant naming 0, successor, and an equality test. For the weak semantics of ordinals-in-β, limits are defined to be limits inferior with respect to the usual ordering on the ordinals. For the strong semantics, we need two copies of 0, one at the bottom as usual, and one at the top.

Admittedly the latter workaround is a bit hackish—really the correct domain is "ordinals modulo β," but then $<$ is no longer transitive, so limits are not given by limits inferior. We wonder whether there is a generalization of the notion of structure that handles it more elegantly.

2.3 While-Program Semantics

Fix a Σ-while program T with support $\mathbf{V}$ and a Σ-structure $\mathbf{M}$. We define its *denotation*

$$[\![T]\!] : M^{\mathbf{V}} \rightharpoonup M^{\mathbf{V}} \tag{4}$$

by induction on the construction of T as follows:

- If $T \equiv \varepsilon$ then $[\![T]\!]^{\mathbf{M}}$ is the identity function.
- If $T \equiv f$ for some $f \in \Sigma$, then $[\![T]\!]^{\mathbf{M}} = [\![f]\!]^{\mathbf{M}}$.
- If $T \equiv T_0; T_1$ then $[\![T]\!]^{\mathbf{M}} = [\![T_1]\!]^{\mathbf{M}} \circ [\![T_0]\!]^{\mathbf{M}}$.
- If $T \equiv \mathsf{if}\ B\ \mathsf{then}\ T_0\ \mathsf{else}\ T_1$ then $[\![T]\!]^{\mathbf{M}}(x) = y$ means

$$\left([\![B]\!]^{\mathbf{M}}(x) = \mathrm{tt} \wedge [\![T_0]\!]^{\mathbf{M}}(x) = y\right) \vee \left([\![B]\!]^{\mathbf{M}}(x) = \mathrm{ff} \wedge [\![T_1]\!]^{\mathbf{M}}(x) = y\right).$$

This is all identical to the finitary setting thus far. Now we come to the important case:

- If $T \equiv \mathsf{while}(B)\ \mathsf{do}\ T_0$, then $[\![T]\!]^{\mathbf{M}}(x) = y$ means that there exists an ordinal Γ and an ordinal-indexed sequence $(x_\alpha : \alpha \leq \Gamma)$ such that
  - $x_0 = x$ and $x_\Gamma = y$,
  - for all $\alpha < \Gamma$ $[\![B]\!]^{\mathbf{M}}(x_\alpha) = \mathrm{ff}$,
  - $[\![B]\!]^{\mathbf{M}}(x_\Gamma) = \mathrm{tt}$,
  - if $\alpha = \beta + 1$, then $x_\alpha = [\![T_0]\!]^{\mathbf{M}}(x_\beta)$, and
  - if α is limit, then $x_\alpha = \lim(x_\beta : \beta < \alpha)$.

Remark 2. This complicated definition of $[\![\mathsf{while}(B)\ \mathsf{do}\ T]\!]^{\mathbf{M}}(x) = y$ must be defended. In the finitary case, one can simply write

$$(\exists w \in M^{\mathbf{V}})([\![T]\!]^{\mathbf{M}}(x) = w \wedge [\![\mathsf{while}(B)\ \mathsf{do}\ T]\!]^{\mathbf{M}}(w) = y),$$

at least in the case that $[\![B]\!]^{\mathbf{M}}(x) = \mathrm{ff}$. This is an *elementary* (i.e., first-order) monotone inductive definition, and elegantly avoids the kind of list quantification we use above. It exploits the fact that for a finite ordinal n, if n is decomposed as $1 + m$, then necessarily $m < n$, so the induction is well-founded.

But if an infinite ordinal α is decomposed as $1+\beta$—or even as $\gamma+\beta$ for nonzero γ and β—there is no guarantee that $\beta < \alpha$. Whether there exists an elementary inductive definition of $[\![\mathsf{while}(B)\ \mathsf{do}\ T]\!]$ is an interesting open question. That said, the present definition is a legitimate inductive definition, thus justifying the phrase "denotational semantics," even though it is not first-order.

3 While-Programs vs. Register Programs

In finitary programming, we can compile a while-program into an equivalent register program, thereby "forgetting the structure." We can also compile a register program into a while program by essentially implementing the register program semantics in a single while-loop. In this section, we replicate both constructions in the transfinite setting. Together, this gives us a *structured programming theorem* which shows the extensional equivalence of these two models under mild assumptions on the data.

3.1 Compiling While-Programs to Register Programs

Let T be a while-program and n be a natural number. We define the register program $\boxed{T, n}$ as follows: $\boxed{\varepsilon, n}$ is the single-line program skip, $\boxed{X \leftarrow f(X), n}$ is the single-line program $X_n \leftarrow f(X_n)$, and the other cases are covered in Table 1.

For the purposes of exposition, it makes things easier to assume that T uses a single variable X, but this restriction is not essential in any way. Notice the support of $\boxed{T, n}$ is $\{X_n, X_{n+1}, \dots X_{n+m}\}$, where m is the depth of while-nesting in T.

Finally, given a while-program T, the *compiled register program* is $\boxed{T, 0}$ followed by halt. We omit the proof of the following theorem.

Table 1. Definition of $\boxed{T, n}$. In the rightmost column, ℓ_2 labels the first line after the entire block of code.

$\boxed{T_1; T_2, n}$	$\boxed{\text{if } B(X) \text{ then } T_1 \text{ else } T_2, n}$	$\boxed{\text{while}(B(X)) \text{ do } T, n}$
$\boxed{T_1, n}$	if $B(X_n)$ then goto ℓ_1 else goto ℓ_2	ℓ_0 : if $B(X_n)$ then goto ℓ_1 else goto ℓ_2
$\boxed{T_2, n}$	$\ell_1 : \boxed{T_1, n}$	$\ell_1 : X_{n+1} \leftarrow X_n$
	goto ℓ_3	$\boxed{T, n+1}$
	$\ell_2 : \boxed{T_2, n}$	$X_n \leftarrow X_{n+1}$
		goto ℓ_0

Theorem 1. *For any while-program T and values x and y, $[\![T]\!](x) = y$ iff its compiled register program with X_0 initialized to x (and the other variables initialized arbitrarily) halts with X_0 storing y.*

We don't label the variables of register programs as "input" and "output," but if we did, you can easily see how to compute exactly the partial function $[\![T]\!]$: just make X_0 the single input and output variable.

Remark 3. The compiling function in the finitary case is almost identical, but simpler—we don't need to make multiple copies of the same variable, and so we don't need the additional parameter n. What's going on here?

The answer lies in the semantics of while loops at limit stages. According to the semantics of while-programs, the value of the variable X in the limit is the limit of the values it takes *at the top of the loop.* If we failed to create a new copy of X in the body of the loop, its value in the limit would be the limit of the values it takes at *every line in the body of the loop.*

To remedy this discrepancy, we introduce a new copy of X, use it for computation during the body of the loop, and update the original when it's done. That way the original copy of X only changes right before the top of the loop. This construction echoes the usual way of implementing a *function call* in assembly language, albeit in a lightweight manner.

3.2 Compiling Register Programs to While-Programs

In this direction, we assume that the structure Σ is rich enough to interpret any finite linearly ordered list of lines. Limits are given by the limit inferior with respect to the line ordering, we should have constants naming each of these lines, and we need to be able to test equality on lines. This is a very mild assumption, but it is probably necessary.

Given a register program p, we construct a while-program with exactly the same set of variables, plus a variable L storing the line number. The program has the form

```
L ← 0
while(L is not halting) do
  if L = 0 then ... else
  if L = 1 then ... else
  ⋮
```

where what we do in each case depends on the particular line of p. For example, if line 1 were $Y \leftarrow f(X)$, then the relevant term of the while-program would be

$$\text{if } L = 1 \text{ then } Y \leftarrow f(X); L \leftarrow 2 \text{ else } \ldots;$$

if line 1 were if $B(X)$ then goto 4 else goto 8, then the same term would be

$$\text{if } L = 1 \text{ then if } B(X) \text{ then } L \leftarrow 4 \text{ else } L \leftarrow 8 \text{ else } \ldots$$

This is exactly the same code you would use in the finitary setting. For the purposes of correctness, we need to observe that the sequence of valuations we see at the top of the loop is exactly the computation sequence of the original register program. The correctness at successor stages depends on correctly implementing the body of the loop (which hopefully should be clear). At limit stages, the semantics of the register program and of the compiled while-program say exactly the same thing: take limits coordinate-wise (i.e., variable by variable), including defining L by the limit inferior. Since each variable changes at most once during the body of the loop, we don't have to worry about taking the limit over all configurations vs. those at the top of the loop like we did in the other direction, cf. Remark 3.

4 A Native Operational Semantics

The register program semantics of Eq. (3) is operational; if we pull it back along the function compiling while-programs to register programs, it certainly yields an operational semantics for while-programs. However, it is much more preferable to have operational semantics that deals directly with the syntax of the language, such as we provide in this section and prove correct in the next. A good semantics of a language is not only correct but informative: it should help us get a feel for the intensional behavior of its program control structures. Hiding the semantics behind a compiling function destroys this.

Fix a signature Σ, Σ-program P with support $\mathbf{V}$, and Σ-structure $\mathbf{M}$. By a *P-term* we mean an occurrence of a subprogram of P; we shall typically let T vary over these. (*Occurrence* meaning: we distinguish between identical terms located at different parts of P—this is crucial.)

Definition 7. *Let $\succ$ be the transitive closure of the union of the following relations on P-terms:*

- *If* $T \equiv \text{if } B \text{ then } T_0 \text{ else } T_1$, *then* $T_0 \succ T$ *and* $T_1 \succ T$.

- *If* $T \equiv \mathsf{while}(B)\ \mathsf{do}\ T_0$, *then* $T_0 \succ T$.
- *If* $T \equiv T_0; T_1$, *then* $T_0 \succ T_1 \succ T$.

Notice that this is almost the definition of the subterm relation, except that $T_0 \succ T_1$ in case $T \equiv T_0; T_1$.

Remark 4. $\succ$ is a partial order on P-terms, as it is the transitive closure of an acyclic relation. It is also *treelike*, i.e., the set of elements less than any fixed element forms a chain, and there is a global minimum in P itself.

Definition 8. *A* configuration *is a tuple of the form*

$$x : T_0\ T_1\ \ldots\ T_{n-1}$$

for some $n > 0$, *where* $x \in M^{\mathbf{V}}$ *and* $(T_i : i < n)$ *are* P*-terms such that* $T_0 \succ \cdots \succ T_{n-1}$.

Note that the length of any configuration is bounded by the length of the longest $\succ$-chain among P-terms.

In a finite operational semantics, computations are finite sequences of configurations, and the transition relation simply specifies when one configuration can follow another. In a transfinite semantics, computations are well-ordered sequences of configurations, and the transition relation additionally has to specify what happens at limit stages.

Definition 9 (Transition relation at successor stages). *We define the single-step relation* $\sigma \mapsto \sigma'$ *as follows. The definition is by cases on* $x : T_0$.

(i) If $T_0 \equiv f$ *for some* $f \in \Sigma$, *then*

$$x : f\ T_1\ \ldots\ T_{n-1} \mapsto y : T_1\ \ldots\ T_{n-1}$$

if $[\![f]\!]^{\mathbf{M}}(x) = y$.

(ii) If $T_0 \equiv S; S'$ *then*

$$x : S; S'\ T_1\ \ldots\ T_{n-1} \mapsto x : S\ S'\ T_1\ \ldots\ T_{n-1}.$$

(iii) If $T_0 \equiv \mathsf{if}\ B\ \mathsf{then}\ S\ \mathsf{else}\ S'$ *and* $[\![B]\!]^{\mathbf{M}}(x) = \mathrm{tt}$, *then*

$$x : (\mathsf{if}\ B\ \mathsf{then}\ S\ \mathsf{else}\ S')\ T_1\ \ldots\ T_{n-1} \mapsto x : S\ T_1\ \ldots\ T_{n-1}.$$

(iv) If $T_0 \equiv \mathsf{if}\ B\ \mathsf{then}\ S\ \mathsf{else}\ S'$ *and* $[\![B]\!]^{\mathbf{M}}(x) = \mathrm{ff}$, *then*

$$x : (\mathsf{if}\ B\ \mathsf{then}\ S\ \mathsf{else}\ S')\ T_1\ \ldots\ T_{n-1} \mapsto x : S'\ T_1\ \ldots\ T_{n-1}.$$

(v) If $T_0 \equiv \mathsf{while}(B)\ \mathsf{do}\ S$ *and* $[\![B]\!]^{\mathbf{M}}(x) = \mathrm{tt}$ *then*

$$x : (\mathsf{while}(B)\ \mathsf{do}\ S)\ T_1\ \ldots\ T_{n-1} \mapsto x : S\ (\mathsf{while}(B)\ \mathsf{do}\ S)\ T_1\ \ldots\ T_{n-1}.$$

(vi) If $T_0 \equiv \mathsf{while}(B)\ \mathsf{do}\ S$ *and* $[\![B]\!]^{\mathbf{M}}(x) = \mathrm{ff}$ *then*

$$x : (\mathsf{while}(B)\ \mathsf{do}\ S)\ T_1\ \ldots\ T_{n-1} \mapsto x : T_1\ \ldots\ T_{n-1}.$$

It is straightforward to check that $\mapsto$ *preserves the property of configurations that* P*-terms are listed in* $\succ$*-decreasing order.*

Now we come to our central definition.

Definition 10 (Transition relation at limit stages). *Suppose that* γ *is a limit ordinal and* $(\sigma_\alpha : \alpha < \gamma)$ *is a* γ*-indexed sequence of configurations. Write* σ_α *as*

$$x_\alpha : T_{0,\alpha}\ T_{1,\alpha}\ T_{2,\alpha}\ \cdots$$

with the convention that for sufficiently large n, $T_{n,\alpha}$ *is* blank. *Then define the limit configuration* σ_γ *by*

$$x_\gamma : T_{0,\gamma}\ T_{1,\gamma}\ T_{2,\gamma}\ \cdots$$

where

- *for each* $n < \omega$, $T_{n,\gamma} = \liminf(T_{n,\alpha} : \alpha < \gamma)$ *with respect to the order* $\succ$*, where blanks are regarded as* $\succ$*-minimal, and*
- $(x_\gamma)_V = \lim((x_\alpha)_V : \alpha \in I)$*, where* $I \subseteq \gamma$ *is the set of stages* i *at which* $T_{0,\gamma} \equiv T_{0,i}$ *and the limit is given by* $M_{\mathrm{sort}(\mathbf{V})}$.

Several remarks are in order. The limit of $(x_\alpha : \alpha \in I)$ may or may not exist; this of course depends on the structure **M**. But for each n, $T_{n,\gamma}$ is well-defined and $T_{0,\gamma} \succ T_{1,\gamma} \succ \ldots$. We state the result and omit its proof.

Finally, this limit rule cannot be given by a partial order on configurations. However, if we think about configurations as functions from ω to the set of program terms, the limit configuration is essentially a pointwise limit as discussed in Remark 1.

Definition 11. *A* computation sequence *is an ordinal-indexed sequence* $(\sigma_\alpha : \alpha < \gamma)$ *of configurations such that*

- *for any successor ordinal* $\alpha < \gamma$, σ_α *is related to* σ_β *according to Definition 9, where* $\alpha = \beta + 1$*, and*
- *for any limit ordinal* $\alpha < \gamma$, σ_α *is related to* $(\sigma_\beta : \beta < \alpha)$ *according to Definition 10.*

Lemma 1. *Suppose that* γ *is a limit ordinal and* $(\sigma_\alpha : \alpha < \gamma)$ *is a computation sequence. Then for each* $n < \omega$ *there is a unique* $\succ$*-minimal* P*-term* T *that occurs cofinally often in* $(T_{n,\alpha} : \alpha < \gamma)$.

Recall that $T_{n,\gamma}$ is defined as $\lim(T_{n,\alpha} : \alpha < \gamma)$. If, among all terms that occur cofinally often in an ordinal-indexed sequence, there is a unique $\succ$-minimal one, then that term must actually *be* the limit of the sequence. Hence, Lemma 1 shows us that $T_{n,\gamma}$ is actually that unique $\succ$-minimal term that occurs cofinally often in $(T_{n,\alpha})_{\alpha<\gamma}$. Next, we show that $T_{0,\gamma} \succ T_{1,\gamma} \succ \dots$.

Lemma 2. *Suppose that γ is a limit ordinal and $(\sigma_\alpha : \alpha < \gamma)$ is a computation sequence. Define $T_{n,\gamma}$ for $n < \omega$ according to Definition 10. Then, for each n,*

- *if $T_{n,\gamma}$ is blank, so is $T_{n+1,\gamma}$; otherwise,*
- *$T_{n,\gamma} \succ T_{n+1,\gamma}$, regarding a blank as $\succ$-minimal.*

Proof. Suppose $T_{n,\gamma}$ is blank. Then $T_{n,\alpha}$ is blank for cofinitely many $\alpha < \gamma$. Hence $T_{n+1,\alpha}$ is blank for the same α, and $T_{n+1,\gamma}$ is blank.

Suppose that $T_{n,\gamma}$ is an actual P-term, i.e., not blank. Then it occurs as $T_{n,\alpha}$ for cofinally many $\alpha < \gamma$, and hence $T_{n,\gamma} \succ T_{n+1,\alpha}$ for cofinally many α. Therefore $T_{n,\gamma} \succ \lim(T_{n+1,\alpha} : \alpha < \gamma)$, but the latter is simply $T_{n+1,\gamma}$.

Lemmata 1 and 2 show that the limit σ_γ of $(\sigma_\alpha : \alpha < \gamma)$ given by Definition 10 is a genuine configuration, as long as $\lim(x_\alpha : \alpha \in I)$ exists in $\mathbf{M}$. Moreover, it is unique.

5 Equivalence of Operation and Denotation

Suppose we have a computation sequence

$$x_0 : \boldsymbol{T}_0,\ x_1 : \boldsymbol{T}_1,\ \dots,$$

where $\boldsymbol{T}_\alpha = T_{0,\alpha}\ T_{1,\alpha}\ \dots\ T_{n_\alpha-1,\alpha}$. Suppose furthermore that $\boldsymbol{S}$ is a list $S_0\ \dots\ S_m$ of P-terms such that $S_0 \succ \dots \succ S_m$ and $T_{n_\alpha-1,\alpha} \succ S_0$. Then we would like to show that appending $\boldsymbol{S}$ to each configuration, viz.,

$$x_0 : \boldsymbol{T}_0\ \boldsymbol{S},\ x_1 : \boldsymbol{T}_1\ \boldsymbol{S},\ \dots,$$

is still a computation sequence. Note that it suffices to show that appending each configuration with a *single* term S preserves being a computation sequence. The proof is by induction on the length of the sequence. Successor ordinals pose no problem, as each of the rules in Definition 9 is preserved by appending S on both sides. The only essential difficulty is in the limit case, which is omitted in the interests of space.

It's also useful to know that we can go backwards: given a computation sequence where $\boldsymbol{S}$ appears on the right-hand side of every configuration, we can *delete* $\boldsymbol{S}$ everywhere. The verification of this fact is much more straightforward. We repeatedly appeal to the fact can we can append or delete a list of terms on the right of every configuration in a computation sequence in the (omitted) proofs of the following results.

Definition 12. *For configurations σ and τ, by $\sigma \rightsquigarrow \tau$ mean that there is a computation sequence starting in σ and ending in τ.*

Lemma 3. *Let $\boldsymbol{T} \succ S$ be P-terms and $x, y \in M^{\mathbf{V}}$. Suppose there is a computation sequence $(x: \ \boldsymbol{T}\ S) \rightsquigarrow (y: \ )$. Then it must have the form*

$$x : \boldsymbol{T}\ S, \ldots,\ w : S, \ldots,\ y :$$

for some $w \in M$, where S appears as the rightmost term in $(x : \boldsymbol{T}\ S) \rightsquigarrow (w : S)$ and not at all after $w : S$.

In particular, every computation sequence $(x : T_0 \ \ldots \ T_{n-1}) \rightsquigarrow (y : \ )$ can be decomposed as

$$(x : T_0 \ \ldots \ T_{n-1}) \rightsquigarrow (x_1 : T_1 \ \ldots \ T_{n-1}) \rightsquigarrow \ldots \rightsquigarrow (x_{n-1} : T_n) \rightsquigarrow (y : \ )$$

This fact is used in both directions of the equivalence

$$[\![T]\!]^{\mathbf{M}}(x) = y \iff (x : T) \rightsquigarrow (y : \ ),$$

for any P-term T and $x, y \in M^{\mathbf{V}}$. The proof in the forwards direction is by induction on the stage that (T, x, y) gets added to the monotone least-fixed point definition of Eq. 4, and the proof in the backwards direction is by induction on the length of the computation sequence $(x : T) \rightsquigarrow (y :)$.

6 Case Study: Insertion Sort

The *comparison sorting* problem is to compute the unique sorted permutation of a finite list of characters from a linearly ordered alphabet. The following is a reasonable imperative implementation of insertion sort. The *tail* of a list is everything but its first element and *cons*-ing an element onto a list simply prepends it to the front. We hope the other operations are self-explanatory:

```
while(nonempty(A)) do
  x ← A[0]; A ← tail(A);
  B ← cons(x, B); i ← 0;
  while(i < length(B)) do
    if B[i] < B[i + 1] then B ← swap(B, i, i + 1);
    i ← i + 1
```

In this section, we sketch how *exactly the same program* implements insertion sort over infinite lists. The first observation is that we want lists to be *well-ordered,* as we want to destruct A by repeatedly deleting its minimum and iterate through B by starting at 0 and taking successors. This also means that *the alphabet should be well-ordered,* for otherwise we cannot hope to sort them into an ordinal-indexed list. Let us therefore fix a well-ordered alphabet Δ.

The crucial question is how to define the internal limit operation on lists. *Not every reasonable-sounding definition works.* For example, suppose we formalized lists as ordinal-indexed sequences from Δ, i.e., functions of type $\alpha \to \Delta$ for some ordinal $\alpha \in \mathcal{ON}$. This is a dependent type; we can remove this dependence by thinking of all lists as functions $f : \mathcal{ON} \to \Delta \cup \{\uparrow\}$ that are undefined (equal to $\uparrow$) past some ordinal.

Suppose we were to define limits of lists pointwise, viz.,

$$(\lim_{\alpha<\gamma} f_\alpha)(x) := \liminf_{\alpha<\gamma} f_\alpha(x), \tag{5}$$

where the limit inferior on the right is defined with respect to the partial order on $\Delta \cup \{\uparrow\}$ that puts $\uparrow$ at the bottom and makes all elements of Δ pairwise incomparable. (This is the only reasonable choice. The partial order on Δ should be blind to the well-ordering, because in the comparison sorting model we may only gain information about the order by explicitly querying it.)

However, suppose $\Delta = \omega + 1$ and consider the sequence of lists

$$(\omega, 0, 1, 2, \dots),\ (0, \omega, 1, 2, \dots),\ (0, 1, \omega, 2, \dots),\ \dots \tag{6}$$

such as we might see when inserting ω into the front of the list. The *correct* limit is $(0, 1, 2, \dots, \omega)$. However, the limit given by Eq. (5) is $(0, 1, 2, 3, \dots)$—in other words, ω runs right off the map!

Back to the drawing board. Let's now think of lists as a well-ordering of some set $D \subseteq \Delta$. (We do lose a little—we can now only model lists *of distinct elements.* This is probably fixable.) Along the same lines as before, we formalize this as a function of type $\Delta \times \Delta \to 2 \cup \{\uparrow\}$, where $(x, y) \mapsto \mathrm{tt}$ means $x < y$, $(x, y) \mapsto \mathrm{ff}$ means $y < x$, and $(x, y) \mapsto \uparrow$ means either x or y is not in the domain of the well-order. Define limits pointwise as in Eq. (5). This means that

$$x <_\gamma y \iff (\forall^\infty \alpha < \gamma)\ x <_\alpha y,$$

where $\forall^\infty$ means "for sufficiently large." But notice that in this regime, the limit of the lists in Eq. (6) is $(0, 1, 2, \dots, \omega)$, as it should be! Furthermore,

- this limit also correctly computes the limit of a sequence of end-extensions (such as B in the outer loop) as the union of those extensions, and
- it correctly computes the limit of a sequence of restricting the partial order to smaller and smaller domains (such as A in the outer loop) as the intersection of those restrictions

So this works! We have the data type for variables A and B. We simply make x have type $\Delta \cup \{\uparrow\}$ with the same partial order as above, and i have type $\mathcal{ON}$ ordered by the usual ordering on the ordinals. One could now prove that, equipped with these data types, the code above correctly sorts a list.

7 Discussion

In this paper, we exhibited a model of structured transfinite programming and showed it to be *robust,* by both exhibiting a structured programming theorem and constructing a native operational semantics, all in the presence of arbitrary (infinitary) data. We collect some take-aways below:

Limits Inferior are Important. Almost every limit behavior in the literature of infinite time Turing and register machines is defined as some limit inferior (or equivalently, limit superior). Our work suggests that this is a surprisingly general framework, especially when lifted pointwise to function classes. It applies both to the limits we see in our operational semantics as well as insertion sort.

Choose the Correct Limiting Behavior. Two different data types that are apparently isomorphic in the finitary setting may be differentiated by their limiting behavior, and this can have consequences for the correctness of the algorithms they are used in.

Limiting Behavior Should be Encapsulated in the Data. In other words, a program implementing a transfinite algorithm should be identical to (or at least, no more complicated than) a program implementing its finitary counterpart. Rather, all of the work should go into defining the right limiting behavior.

These comments suggest that **developing a theory of infinitary data types** is an important open direction. Does the theory of algebraic data types and initial-algebra semantics of [7] extend to the transfinite setting? Another question with a category-theoretic flavor: can we **recover the native operational semantics canonically from the denotation** of a while-program, along with an automatic proof of correctness, along the lines of [8]? Finally, there is the question of **expressive power.** What can we compute with these while-programs, and what is their relationship to well-known definability classes in abstract model theory and the theory of inductive relations? What is the ordinal running time, as a measure of complexity?

References

1. Böhm, C., Jacopini, G.: Flow diagrams, turing machines and languages with only two formation rules. Commun. ACM **9**(5), 366–371 (1966). https://doi.org/10.1145/355592.365646
2. Carl, M.: Ordinal Computability: An Introduction to Infinitary Machines. De Gruyter, Boston (2019)
3. Hamkins, J.D., Lewis, A.: Infinite time turing machines. J. Symb. Log. **65**(2), 567–604 (2000)
4. Koepke, P.: Turing computations on ordinals. Bull. Symb. Log. **11**(3), 377–397 (2005)
5. Koepke, P., Siders, R.: Minimality considerations for ordinal computers modeling constructibility. Theor. Comput. Sci. **394**(3), 197–207 (2008). https://doi.org/10.1016/j.tcs.2007.12.012, https://www.sciencedirect.com/science/article/pii/S0304397507008808. from Gödel to Einstein: Computability between Logic and Physics
6. Koepke, P., Siders, R.C.: Register computations on ordinals. Arch. Math. Logic **47**, 529–548 (2008)

7. Meseguer, J., Goguen, J.A.: Initiality, induction, and computability. In: Nivat, M., Reynolds, J.C. (eds.) Algebraic Methods in Semantics. Cambridge University Press (1986)
8. Rutten, J.J.: A note on coinduction and weak bisimilarity for while programs. Technical report NLD (1998)
9. Welch, P.D.: Discrete Transfinite Computation, pp. 161–185. Springer, Cham (2015). https://doi.org/10.1007/978-3-319-22156-4_6

Computability of Real Functions with Oracle Pointer Machines Implies Real-Time Simulation of Chemical Reaction Networks

Saptarshi Biswas(✉)

Iowa State University, Ames, IA 50014, USA
sbiswas@iastate.edu

Abstract. According to Hartmanis and Stearns conjecture (1965), algebraic irrational numbers cannot be computed by a Turing Machine (TM) in real-time ($O(n)$). Brent (1976) demonstrated that Newton's method can find the square root of a floating-point number in $O(M(n))$ operations, with $M(n)$ being cost of multiplying two *n-bit* integers. Further, Schönhage (1979) showed that a Storage Modification Machine (SMM) can reduce the cost of multiplication to $O(n)$ operations, implying real-time computability of square-roots. In 2018, Huang et al. proved that a Deterministic Chemical Reaction Network (analog computation) could compute all algebraic irrational numbers in real-time. In this regard, this work explores the relationship between these two computing models. To answer this question, this work introduces the concept of Oracle SMM, which computes real functions more efficiently than the Oracle TM model established by Ko and Friedman (1982). In contrast to their results, this work shows that a real function $f \in \mathcal{C}[0,1]$, with the modulus of continuity $m(n)$, can be computed by an oracle SMM in $O(\lceil \log(m(n)) \rceil)$ rather than $O(n+1) + O(m(n+1))$ operations, where n represents the precision parameter. The proposed model further simulates any real-time CRN, i.e., find trajectory points of species concentration in $O(t)$ many operations, where t defines the precision parameter proportional to time. As a result, this work implies that an Oracle SMM in $O(t)$ operations can compute any algebraic irrational or transcendental number that a real-time CRN computes to a precision of *t-bits*.

Keywords: Storage Modification Machine · Pointer Machines · Chemical Reaction Network · Real-time computability · Computational complexity · Real functions

1 Introduction

In 2018, the investigation results of Huang et al. [17] opened the door for real-time computability of real numbers in analog computing through real-time chemical reaction networks. The work reported that a Chemical Reaction Network

© The Author(s), under exclusive license to Springer Nature Switzerland AG 2025
A. Beckmann et al. (Eds.): CiE 2025, LNCS 15764, pp. 175–190, 2025.
https://doi.org/10.1007/978-3-031-95908-0_13

(CRN) can compute algebraic irrational and several transcendental numbers in real-time (fast) by a Chemical Reaction Network (CRN). Such CRNs are termed Real-Time Chemical Reaction Networks (RTCRN), and the class of real numbers they compute is denoted as $\mathbb{R}_{RTCRN}$.

The idea of real-time computability of real numbers flared up through the paper of Hartmanis and Stearns in 1965 [15]. This paper conjectures that if a Turing Machine (TM) computes a number in real-time, it is rational or transcendental: algebraic irrational numbers cannot be computed by a TM in real-time. Before going any further, it is a good place for defining real-time computability.

Definition 1. *A number is said to be real-time computable if any Turing Machine can compute the number till $n-$bits of precision in $O(cn)$ (this is same as $O(n)$) many operations where c is a constant*[1]*. In other words, a number is said to be real-time computable if each of its bits can be computed in constant amortized time by a TM [11, 23].*

The Hartmanis-Stearns conjecture is an open problem with significant implications in the computability of real numbers. Huang et al. did not directly refute this conjecture; however, their findings on the real-time computability of real numbers align with the definition of real-time computability in the context of analog computing via chemical reaction networks (CRNs). Specifically, the trajectories of real-time CRNs converge exponentially to the numbers in $\mathbb{R}_{RTCRN}$, achieving a precision of t bits in a time t. This conjecture is essential to understanding the relationship between continuous analog computation and discrete computational models.

Additionally, exploring Brent's work (1976) [8] and Schönhage's Storage Modification Machines (SMM, 1979) [25] reveals interesting findings. Brent showed that the root of any number with fixed precision can be computed in $O(M(n))$ steps, where $M(n)$ is the multiplication cost of two n-bit integers. Schönhage demonstrated that two n-bit numbers can be multiplied using an SMM in $O(n)$ steps. This suggests that roots of fixed precision, a class of irrational numbers, can be computed in $O(n)$ operations. This leads to four important questions that inspired this work.

Question 1. Do Brent's method and RTCRN compute the same class of numbers in real-time?

Question 2. If an SMM and a CRN can find roots of finite precision numbers in real-time, do their computation paths intersect? In other words, given the sequences of numbers computed by the two models of computation, $\{x_i^{SMM}\}_{i\in\mathbb{N}}$ and $\{x_i^{CRN}\}_{i\in\mathbb{N}}$, while converging to the solution, i.e., $\exists n \in \mathbb{N}, \forall i \geq n, \exists m \in \mathbb{N}$ such that $x_i^{SMM} = x_m^{CRN}$, or vice versa.

[1] It is important to note that, as stated in [11], real-time is equivalent to linear time. Consequently, this paper uses the terms "linear time" and "real-time" interchangeably in certain sections. In these instances, it is meant to convey that operations are conducted in $O(n)$ steps, where n represents the precision parameter.

Question 3. Can an SMM simulate an RTCRN, and what is the complexity of this simulation?

Question 4. Is there a discrete model of computation that can compute a class of algebraic irrational numbers in real-time besides the TM?

However, before initiating any discussion or investigation of the above questions, let us focus on the literature that provided numerous facts and evidence that clarify the given problems. Let us begin with the following definitions.

Definition 2. *Consider two models of computation, A and B. A simulates B in real-time (denoted as $A \xrightarrow{r} B$) if, for every input x, A performs the same operations as B with only a constant factor of slowdown allowed. Specifically, if B performs n operations in times $t_1, t_2, \ldots, t_n$ and A in $t'_1, t'_2, \ldots, t'_n$, then for all $i, 1 \leq i \leq n$, it holds that $t'_i - t'_{i-1} \leq c(t_i - t_{i-1})$ [25].*

Definition 3. *Consider two models of computation, A and B. A simulates B in Nearly Linear Time (NLT) (denoted as $A \xrightarrow{NLT} B$) if, for every input x, whenever B performs n operations on x, A performs the same operations in at most $O(n\ p(\log n))$ steps, where $p()$ is a polynomial [14].*

The paper [25] presented a description of the RAM0 and RAM1 models, besides proposing SMM, based on the fundamental RAM models of computation [1]. The paper further demonstrated that the RAM0, RAM1, and SMM models could simulate each other in real-time and that SMMs can simulate a TM in real-time, as shown in Fig. 1(a). Additionally, a collection of Random Access and pointer machines can *effectively* simulate each other, meaning the simulation occurs in Nearly Linear Time [14]. These are Frugal Random access Turing Machine (RTM), Random Access Computers (RAC) and Frugal RAC, BTM [14], SMM and Kolmogorov-Uspenskii (KU) Machine [21]. Diagram 1(b) summarizes this relationship. It is also noted that KU machines can simulate SMMs in real-time [25].

To simulate an RTCRN using an SMM, one can create an SMM-based algorithm replicating each species' trajectory in the RTCRN. In 2019, Huang et al. recognized deterministic CRNs as a specific case of Shanon's General Purpose Analog Computer (GPAC) [17], revealing real-time equivalence between the two models [16]. Consequently, the concentration trajectories generated by the CRN match those produced by a GPAC. Polynomial Ordinary Differential Equations (pODEs) and CRNs exhibit strong Turing completeness [10,12]. Further details regarding GPACs can be found in [7,13,22].

Definition 4. *Any function generated by a General Purpose Analog Computer is called a GPAC-generable function. In other words, if a function happens to be one of the solutions of a pODE system representing a GPAC, then this solution is said to be a function generated by a GPAC. [3,6]*

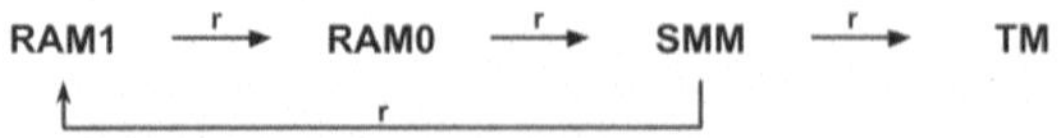

(a) Between SMM, RAM models, and TM

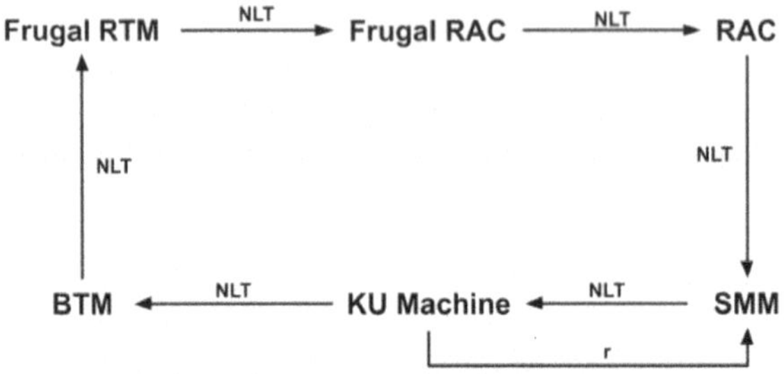

(b) Between Random Access and Pointer Machines

Fig. 1. Simulation relationship

This implies that the trajectories of species concentration of a deterministic CRN are real analytic functions which introduces another problem in the computability of real functions. Let us consider the following Initial Value Problem (IVP) defined by an Ordinary Differential Equation (ODE):

$$\frac{dy}{dt} = f(y), \qquad y(t=0) = y_0 \tag{1}$$

According to the existence and uniqueness theorem, if f is continuous, the solution of the Initial Value Problem (IVP) exists. If f is Lipschitz continuous, then the solution is unique. However, for polynomial time computable functions with a unique solution for (1), computing the solution can be arbitrarily complex [19]. Ko showed that an IVP is solvable in polynomial time if $P = PSPACE$ and f satisfy a weaker form of Lipschitz continuity. Kawamura refined this, indicating that an IVP with a polynomial time computable f Lipschitz continuous has a PSPACE-complete solution [18]. If f is analytic, its solution can be approximated in polynomial time [5].

Proposition 1. *A function f is a real analytic if a GPAC generates it. [4]*

Next, let us define *sparse* and *dense* computation.

Definition 5. *Consider a real function $f : \mathbb{R} \to \mathbb{R}$ and a sequence $\{x_i\}_{i \in \mathbb{N}}$ with $x_{i+1} > x_i$. This sequence can be represented as a k-ary tree, where each node at a level corresponds to points from $\{x_i\}$, with the root node at x_0. The height/nodes of the tree can be finite or infinite.*

1. *The function f is computed* ***sparsely*** *if an algorithm computes one point/node per level, creating a unique connected path from the root.*

2. *The function f is computed **densely** if all points/nodes at each level are computed.*

In this context, this work contributes in the following grounds:

1. Proposes Oracle Pointer Machine based on Oracle SMM.
2. Main findings:
   (a) An Oracle SMM can compute a function f that is continuous in $[0, 1]$ with a modulus of continuity $m(n)$, in $O(\lceil \log m(n) \rceil)$ steps.
   (b) The trajectory of an RTCRN can be computed sparsely using Oracle SMM in real time.
3. Answers questions 1–4.

The remaining paper is structured as follows: Sect. 2 presents related background. Section 3 details the Oracle Storage Modification Machine. Sections 4, 5, and 6 discuss key research results on the computability of real functions and real-time simulation of RTCRN. Section 7 examines the relationship between real numbers computable by Brent's method and those computable by RTCRNs. Finally, Sect. 8 wraps up this work.

2 Background

2.1 Deterministic Chemical Reaction Networks (CRN)

CRN (Chemical Reaction Network) is a mathematical model for solving computational problems, such as arithmetic and logical operations [9]. It involves designing a CRN to perform specific tasks, followed by mathematical or simulation analysis for verification. Although practical implementations were pursued with identified chemicals, DNA strand displacement systems eventually replaced this approach due to the challenges of finding suitable reactions and compounds [26].

Definition 6. *A Chemical Reaction Network (CRN) is a tuple in the form (S, R), where $S = \{X_0, \ldots, X_n\}$ is the set of species and R is the set of reactions involved in the network.*

Definition 7. *$\rho \in R$ are the reactions in the CRN over a finite set of species $C \subseteq S$ defined as the following tuple:*

$$\rho = (r, p, k) \in \mathbb{N}^C \times \mathbb{N}^C \times (0, \infty) \text{ where } r \neq p$$

The r and p are the reactants, and product vectors can be constructed from elements in S, representing the physical concentration of each species where $\dim(r) + \dim(p) = n$.

To generalize and ease the convention of the notation further for systems considering more than one reaction, the reactant vector, product vector, and rate constant for some reaction ρ is denoted by $r(\rho)$, $p(\rho)$ and $k(\rho)$ respectively. The net effect of a ρ is further expressed by

$$\Delta\rho(A) = p(\rho)(A) - r(\rho)(A), \ \textit{where}\ A \in C \tag{2}$$

The reaction rate at time t is defined by following the equation.

$$rate_\rho(t) = k(\rho) \prod_{A\in C} x(t)(A)^{r(\rho)(A)} \tag{3}$$

Finally, the total rate of change of species A is given by

$$\frac{dx(t)(A)}{dt} = \sum_{\rho\in R} \Delta\rho(A) \times rate_\rho(t) \tag{4}$$

Here, $x(t)(A)$ represents the quantity of A at time t.

Furthermore, some CRNs can perform real-time computation of real numbers. These are called Real-Time Chemical Reaction Networks (RTCRN). A Real-Time Chemical Reaction Network can be defined as follows.

Definition 8. *A number $\gamma \in \mathbb{R}$ is said to be Real-Time CRN computable, i.e., $\gamma \in \mathbb{R}_{RTCRN}$, if a CRN $N = (S, R)$ consists of a species $X \in S$ such that,*

1. ***Integrality:*** *Rate constants of all reactions should be integers greater than 0, i.e., $k(\rho) \in \mathbb{Z}^+$*
2. ***Boundedness:*** *For all species $X \in S$, their concentrations are bounded above with a constant $c > 0$ for any time $t \in [0, \infty)$, i.e., $x(t) \leq c$*
3. ***Real-time convergence:*** *If the CRN N is initialized with all species concentration 0 ($x(0) = 0$, $\forall X \in S$), then $\forall t \in [1, \infty)$ we have $|x(t) - |\gamma|| \leq 2^{-t}$*

2.2 Storage Modification Machine

In 1980, Schönhage introduced the Storage Modification Machine (SMM), a pointer machine model consisting of nodes and pointers to other nodes or memory locations. Computation occurs by rearranging these nodes and pointers or by tracing the paths formed by pointers in a defined structure called Δ-structure, where Δ represents the unique pointers any node can have. This work adheres to the instruction set outlined in [25].

Each Δ−structure is defined as $D = (X, a, p)$, through a center $a \in X$ from which it expands using the set of pointers Δ, where X denotes the set of nodes that are finite in number. The p is a set of transition functions described as $p = (p_d | d \in \Delta)$, where $p_d : X \to X$. On the other hand, recursive mapping $p^* : \Delta^* \to X$ is defined as,

$$p^*(\lambda) = a \qquad p^*(Wd) = p_d(p^*(W))$$

where $W \in \Delta^*$. All the other operations follow Schönhage's proposal in [25]. Lastly, if a pointer points to ϵ, it is not set to point to any node.

Δ–structure and B-Scale for numeric computation on SMM Like [25], the idea of utilizing a B-scale for B-ary arithmetic operations has been considered in the proposed method. The Δ–structure for the numbering system followed by the designed SMM model uses the set of pointers defined as $\Delta = \{S, T, P, Q, E, F, B, SP, SQ, SE\}$. In this regard, a B-scale structure can be represented, as shown in Fig. 2.

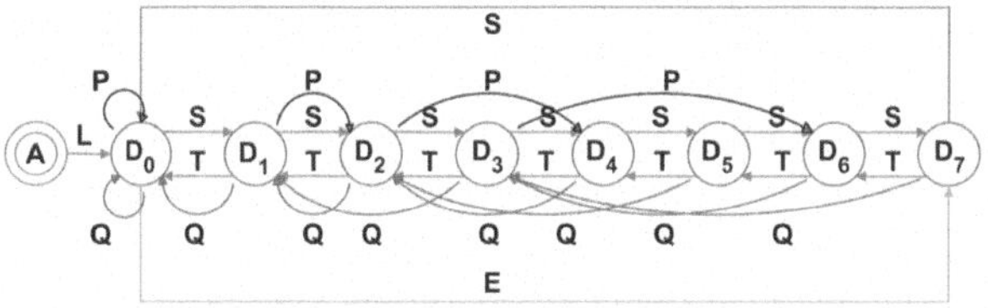

(a) Single B-scale structure for $B = 8$

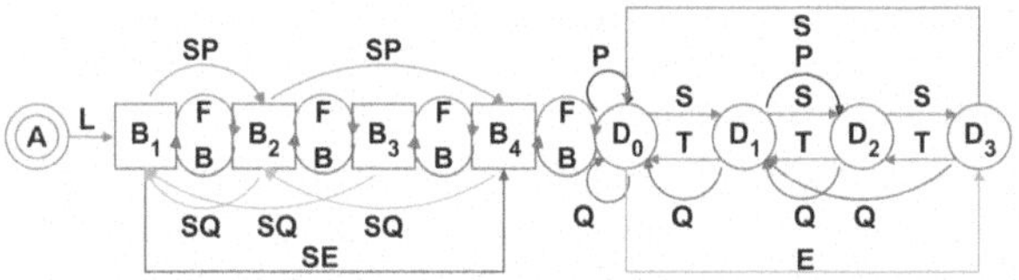

(b) Structure representing a number > 7 with chaining of B-scale blocks

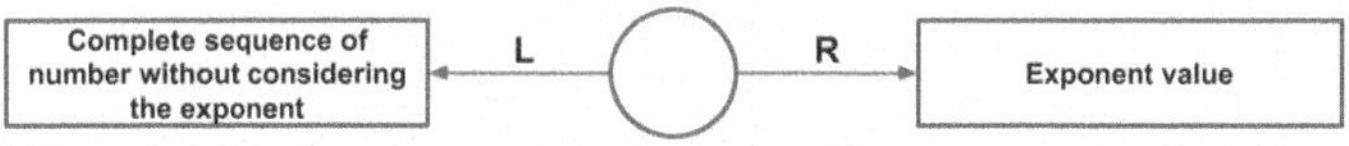

(c) Δ–structure representing a decimal number

Fig. 2. Δ–structure for representation of numbers with A as the center

The Δ–structure, as shown in Fig. 2(c), represents a number with the center A with the L pointer holding the structure for the number while the R pointer stores the counter value relevant to the decimal point placement. In other words, the R pointer holds a value r such that the actual number represented by the complete Δ–structure is $val(L) \times 10^{-r}$, where $val(L)$ is the value stored using the pointer L. Additional structural blocks can also be introduced to indicate the arithmetic sign of a number. Lastly, all the numeric values considered in this paper refer to binary numbers that are represented by the proposed Δ–structure (as shown in Fig. 2(a)) such that both numeric systems are equivalent and interconvertible. The Δ–structure represented in Fig. 2(a) is similar to the one proposed in [25] but differs only in the addition of the E and T pointers.

The numbers in each block, indicated by the L and R pointers, are formed using B-scale blocks (see Fig. 2(b)). The square block B_i, where $i \in \mathbb{N}$, represents a complete 8-node block, as shown in Fig. 2(a). The circular nodes at the end signify the nodes in the last block. This last block may or may not be complete, meaning it may contain fewer than 8 nodes (D_0 to D_i where $i < 7$). If it contains fewer than 8 nodes, it is referred to as a terminal incomplete block; otherwise, it is called a terminal complete block, which again represents a complete B-scale block, as depicted in Fig. 2(a). The SP and SQ pointers denote multiplication and division by two, respectively, using only the floor value for division, aiding in extracting the binary bits of the represented number. The F and B pointers connect consecutive blocks, while the SE pointer links the first node of the first complete block to the last node of the last complete block.

Inside a single B-scale block (Fig. 2(a)), the organization of pointers (P, Q, S, T, and E) mirrors that of the outer structure but points to nodes within the same block. Notably, the S pointer of the last node connects back to the first node of the block. This structure can represent the binary bits of any floating point or dyadic rational number. Note that a single B-scale block can encode 3 bits of information for $B = 8$. In Fig. 2(b), the terminal block represents 3 bits of information while each intermediate block B_i corresponds to a single bit of information. For e.g., the structure in Fig. 2(b) represents the binary number 100011 where the last 3 bits 011 is acquired from the terminal block. In contrast, the first $1 + \lceil \log n \rceil$ bits are acquired from the organization of the intermediate blocks B_i where $n = \max i \in \mathbb{N}$ such that B_i exists in the structure. However, this differs from how a Turing Machine encodes dyadic rational numbers, highlighting a difference in the computability of bits between the two models. The next definition clarifies the concept of an SMM for bit computation.

Definition 9. *Let there exist an SMM that holds the encoding of n bits within its $\Delta-$structure. In this regard, an SMM is said to compute n bits if it holds such a $\Delta-$structure for which a computing algorithm can extract all the n bits within $O(n)$ many operations. We further call such a $\Delta-$structure to be Real-Time extractable and the algorithm to be a Real-Time extractor.*

Considering the above definition we can conclude the following Lemma.

Lemma 1. *Any dyadic rational number represented in the form of Fig. 2(a) and 2(b) is Real-Time extractable, i.e., there exists a Real-Time extractor that can generate the bits encoded by the $\Delta-$structure in $O(n)$ time.*

3 Introduction to Oracle SMM

3.1 $\Delta-$structure of Oracle SMM

Figure 3(a) introduces the idea of the Oracle structure within the computational structure of SMM. Here, A is the external head, which accesses the underlying structure with the *Access* pointer. It leads to a node from where three different pointers originate: S, C, and P. These three pointers lead to computing $\Delta-$ *structure*, control/state structure, and Oracle structure. Each of those structures is explained below.

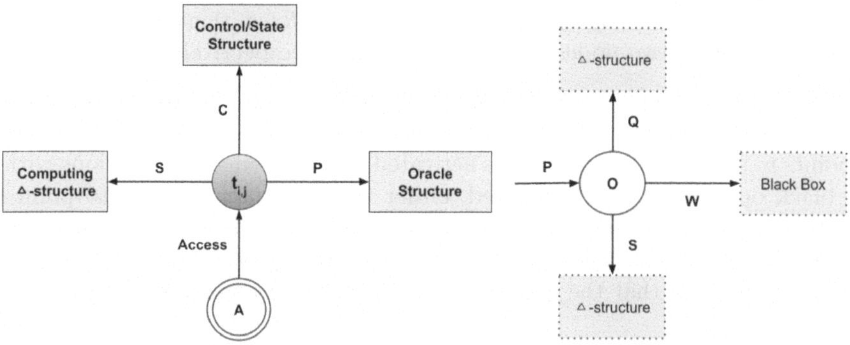

(a) Computing Model (b) Oracle Structure

Fig. 3. Oracle SMM

- **Computing Δ – structure:** This structure performs general computations of the SMM. It executes the main program for which the SMM is designed.
- **Control/State Structure:** This consists of a $\Delta - structure$ that keeps track of the number of computational steps performed by the main program
- **Oracle Structure for $\mathbf{t_{i,j}}$:** The oracle structure of the SMM is analogous to the one for Turing Machines. The underlying $\Delta - structure$ of the Oracle is shown in Fig. 3(b). Each $t_{i,j}$ (considering time nodes) is associated with an oracle structure that will accept a query and provide its answer. The significance of the notation $t_{i,j}$ will be discussed later in this paper.

3.2 Working Principles

Figure 3(b) provides the internal representation of the Oracle structure. The working behavior of an SMM oracle is the same as the one for Turing Machines. The node O serves as the access point of the structure. This node exhibits three pointers, namely, Q, W and S.

These pointers further point to the following structures:

1. **Q :** This pointer represents the query nodes. Unlike the Oracle TM, the Oracle of an Oracle SMM accepts queries in Δ-structures, but may limit its understanding to specific forms or predefined orientations. For example, it might accept inputs using a specific B-scale structure or follow certain encoding rules. Despite these limitations, this oracle model allows for more dynamic query representations compared to the TM model's binary string format, optimizing data representation and reducing read/write overhead.
2. **W :** This pointer leads to the Oracle's working black box, which reads the query from the Q pointer and provides a solution.
3. **S :** This pointer points to the structure representing the solution from the black box based on the query. It serves as the answer to the oracle computation, which the main structure can collect. For example, if the Oracle has

to report outcomes like *yes*, *no*, and *error/undefined*, then S could point to structures with two nodes, one node, and none, respectively.

The working principle of the Oracle structure is similar to the one of the Turing Machine model of computation. Once the oracle structure finds that the Q points to a valid structure, it gets activated and performs its operations within the black box. Once it is completed, either the S pointer is made to point to the structure representing the solution, or the solution structure is constructed in the S branch of the oracle structure. Finally, the Q pointer is set to point to ϕ again, indicating that the query has been removed from the Q branch.

3.3 Δ–structure Representation of Oracle-Set E, Encoding of $\langle n, k\rangle$ and $\langle d, n, e\rangle$

The oracle set E was originally presented as a single set constituting two different oracle sets E_1 and E_2 with elements indexed at odd and even positions by Ko and

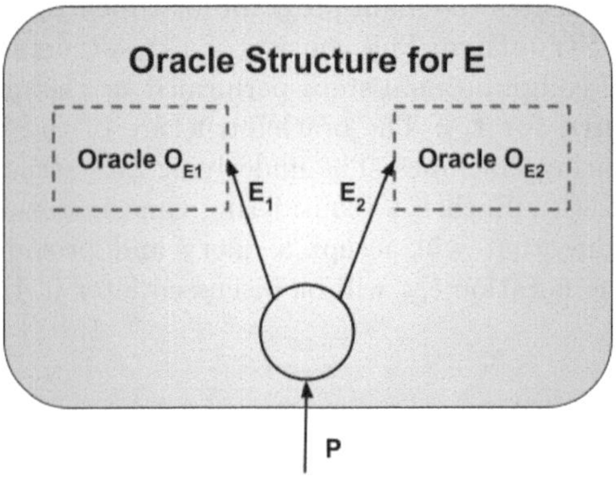

(a) Oracle set E

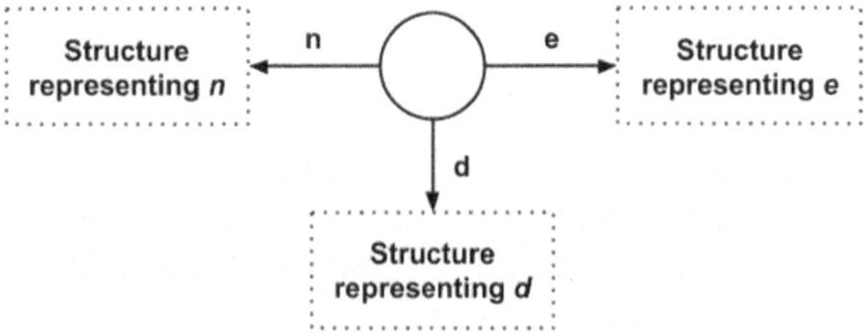

(b) $\langle d, n, e\rangle$

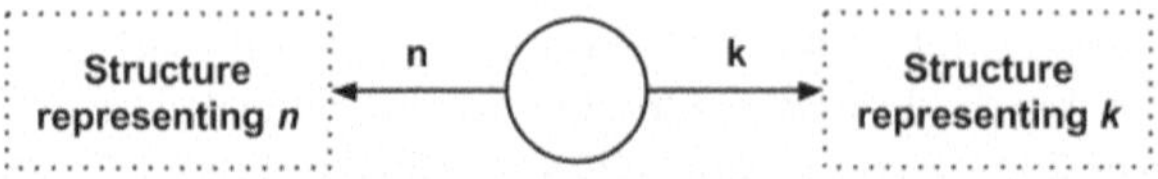

(c) $\langle n, k\rangle$

Fig. 4. Δ–structure representation

Friedman (1982) [20]. Analogously, in SMM, it can be represented as a collection of two different Δ−structures representing sets E_1 and E_2, which are connected by predefined pointers. This, as a whole, constitutes the oracle structure E. A computing Δ−structure can query either of the two oracle sets using the E_1 and E_2 pointers as shown in Fig. 4(a).

One of the primary differences of an Oracle SMM is that it accepts queries and returns its answer in the form of structures rather than a stream of bits. As a result, we can consider a standard format of the structural representation of the queries that our Oracle accepts. Figure 4(c) and 4(b) show one such structural representation of the query to the oracle E_1 and E_2, respectively, considered for this paper. This representation aligns with the encoding function π presented in [20]. The Δ−structure representing $\langle n, k\rangle$ consists of a central node that has n and k pointers pointing to structures representing the values of n and k respectively. The same idea holds for encoding $\langle d, n, e\rangle$ as shown in Fig. 4(b).

Note: It is assumed that the oracle sets O_{E1} and O_{E2} used in this paper accept structures of the form shown in 4(c) and 4(b) respectively as valid queries.

4 Binary Search with SMM and Search Δ−structure Using the Oracle Set E_2

This section introduces a lemma and a theorem that provide essential support for the main theorems of this paper.

Lemma 2. *Given a list of n sorted elements, there exists a Δ−structure for SMM through which it can perform binary search within $O(\log n)$ operations.*

Theorem 1. *Consider there is an oracle set $E_2 = \{\langle n, k\rangle | n, k \subset \mathbb{N}, k \leq m(n)\}$ where $m : \mathbb{N} \to \mathbb{N}$ is a modulus function. Then there exists a Δ−structure that for a given n can search for the $\sup\{k | k \leq m(n)$ and $\langle n, k\rangle \in E_2\}$ with at most $O(\lceil \log m(n)\rceil)$ many queries to the oracle set E_2. This further implies that the oracle SMM needs $O(\lceil \log m(n)\rceil)$ many operations to search for the modulus $m(n)$.*

5 Computing Trajectory Points of RTCRN in Bounded Time Domain

Two fundamental findings in [20] are of greater significance to understanding pointer machines' capability in computing real functions. These findings are listed below.

1. **Finding 1:** If a function f is continuous on $[0, 1]$, then there is a set-oracle E in which the function is partially recursive having a modulus of continuity $m(n)$ defined as a function $m : \mathbb{N} \to \mathbb{N}$. This further signifies, $\forall x, y \in [0, 1]$

$$|f(x) - f(y)| \leq 2^{-n}, \ \textit{whenever}\ |x - y| \leq 2^{-m(n)}$$

2. **Finding 2:** If $f \in \mathcal{C}[0,1]$ has a modulus of continuity $m(n)$, then there exists a set oracle E such that f is computable in time $O(n+1) + O(m(n+1))$. If m is polynomially bounded, then f is computable in polynomial time.

Consider that a Real Time CRN N exhibits trajectory functions defined as $T_X : [0,1] \to \mathbb{R}$ for each of the species $X \in S$. This trajectory function is said to be computable if there exists a computable function $\hat{f} : \mathbb{Q} \times \mathbb{N} \to \mathcal{D}_n$ such that,

$$|T(x) - \hat{f}(y,n)| \leq 2^{-n},\ \textit{whenever}\ |x-y| \leq 2^{-m(n)}\ \textit{and}\ x, y \in [0,1]$$

Here, $m(n)$ is the modulus of continuity. On the other hand, every trajectory function of a deterministic CRN is an analytic function on the positive real line. As a result, they are differentiable and continuous on every point in [0,1]. On applying Ko and Friedman's generalization and results from finding 1 and 2, we get the following theorems and corollary:

Theorem 2. *Suppose there is a Real-Time computable CRN with a trajectory function T_x in [0,1]. Then the function exhibits a linear modulus of continuity $m(n)$ such that $\forall x, y \in [0,1]$, we have*

$$|T(x) - T(y)| \leq 2^{-n},\ \textit{whenever}\ |x-y| \leq 2^{-m(n)}$$

Theorem 3. *Consider a Real-Time CRN with a trajectory function T_x in [0,1]. Then for any point $y \in [0,1]$, the $T_x(y)$ is computable by an oracle SMM in $O(n)$ operations using a set-oracle E.*

Proof. From Theorem 2, we know T_x has a linear modulus of continuity. Considering the $\Delta - structure$ representing the set-oracle E, one needs $O(\lceil \log m(n) \rceil) = O(\log n)$ operations to find out the modulus. Then considering a modified search tree $\Delta - structure$, one can find the value $e \in \mathcal{D}_n$ in $O(\log n) + O(\frac{2n}{3})$ many operations. This results in a total of $O(\log n) + O(\log n) + O(\frac{2n}{3})$ operations to approximate $T_x(y)$ to a precision of n bits for any point lying in $[0,1]$.

Corollary 1. *Consider any trajectory function T_x which is continuous, unbounded above, but whose gradient is bounded above by $2^{2^{q(n)}}$, where q is a polynomial. Then, the function is computable by an oracle SMM within polynomial time.*

Proof. Let the gradient be denoted by g, where $g \leq 2^{2^{q(n)}}$. Here, the gradient is defined as $g = \frac{T_x(y_1) - T_x(y_2)}{y_1 - y_2}, \forall y_1, y_2 \in domain(T_x)$. This is consistent with the original definition $g = \lim_{y_1 \to y_2} \frac{T_x(y_1) - T_x(y_2)}{y_1 - y_2}, \forall y_2 \in domain(T_x)$ [24]. Then we have

$$\log g + n \leq \log 2^{2^{q(n)}} + n \leq m(n) \Rightarrow 2^{q(n)} + n \leq m(n)$$

Now, adopting the same method used to prove Theorem 3 gives us the total number of operations required to compute T_x to be $O(\lceil \log 2^{q(n)} \rceil) + O(\log n) + O(\frac{2n}{3}) = O(q(n))$.

Note that CRNs with exponential species concentration trajectories exist, for e.g. $x(t) = e^{e^t}$, for which the above corollary will be applicable.

6 Oracle SMM for Computing the Continuous Trajectory of RTCRN in Unbounded Time Domain

This section discusses the consequence of using Oracle SMM in computing trajectory points of RTCRN. Note that trajectories of CRNs can be defined on the unbounded time domain. This section presents the following main theorem of this paper.

Theorem 4. *Given a trajectory function $\tau : \mathbb{R}^+ \cup \{0\} \rightarrow \mathbb{R}^+ \cup \{0\}$ defined such that $\tau(t) = x(t)$, where t represents time and $x(t)$ is the concentration of species X at time t of an RTCRN. Let $k \in \mathbb{N}$ and break each unit time interval into k equally spaced subintervals covering the whole unit interval. There exists an oracle SMM with a defined Δ-structure that can compute one trajectory point from each tree level along a unique computation path in $O(t)$ operations while following the convergent species concentration trajectory of the deterministic RTCRN.*

7 Relation Between Brent's Numbers and $\mathbb{R}_{RTCRN}$

Brent's method utilizes Newton's iteration method to compute the square root of a floating point number (numbers with $n-$bit fraction for some $n \in \mathbb{N}$) $c \geq 0$ [8]. This method uses the following recursive equation:

$$y_{j+1} = \frac{y_j + \frac{c}{y_j}}{2} \tag{5}$$

We will assume that the floating-point number c in Brent's algorithm has a finite degree of precision. It is important to note that such a number can be represented in the form $\frac{p}{q}$, where p and q are positive integers ($\mathbb{Z}^+$) and $q \neq 0$. This representation reflects the finite degree of precision of the floating-point number. From the results reported in [2,17] one can conclude the following corollary.

Corollary 2. *Let c be a floating point number for Brent's Newton iteration method to find $\sqrt{c}$. A deterministic Real-Time CRN exists for every such c that also converges to $\sqrt{c}$.*

The lemmas and theorems presented below illustrate the relationship between the two classes of numbers that can be computed using Brent's method and those computed through the RTCRN. In addition to the last section, these findings also represent one of the key contributions of this paper.

Lemma 3. *The trajectory points computed by Brent's method and Real-Time CRN may differ.*

Lemma 4. *Different computation paths exist for deterministic CRNs that converge to the same point for different initialization concentrations of the species.*

Theorem 5. *Let c be a floating point number for which Brent's method of Newton's iteration can find $\sqrt{c}$ in $O(M(n))$ operations, where $M(n)$ is the cost of performing multiplication of two $n-$ bit integers. Then, a deterministic Real-Time CRN exists that can compute all values generated by Newton's method at every iteration on the convergence path.*

It is important to note that this result can be generalized to other constants that can be computed iteratively, as discussed in [8]. However, the formulation of the ordinary differential equation (ODE) system representing the chemical reaction networks (CRNs) must adhere to the conditions outlined in definition 8.

8 Conclusion

The study of discrete models of computation has a long-standing history. However, recent decades have seen significant advancements that have brought analog models of computation into the spotlight. According to Hartmanis and Stearns's conjecture, algebraic irrational numbers are considered computable but not in real-time. Interestingly, research by Huang et al. has shown that algebraic irrational numbers can be computed in real-time using Deterministic Chemical Reaction Networks. In this context, the current work introduces the concepts of the Oracle Pointer Machine to examine its impact on simulating Real-Time Chemical Reaction Networks.

In pursuit of the primary goal of this work, the proposed computational models yielded intriguing insights into the computability of real functions. A significant finding in this area extends Ko's research. Our results indicate that for any function f that is continuous on the interval $[0, 1]$ with $m(n)$ being the modulus of continuity, the function can be computed sparsely using an Oracle Storage Modification Machine in $O(\lceil \log m(n) \rceil)$ operations.

Our investigation further revealed that the functions representing the trajectory of species concentration within RTCRNs exhibit a linear modulus of continuity. Consequently, these functions can be computed in sub-linear time using Oracle SMM. Additionally, some CRNs display a trajectory that follows an exponential modulus of continuity, allowing for computation in polynomial time by an Oracle SMM. This work presents the mathematical details surrounding these findings. Finally, the research results suggest that the computational trajectories of Brent's method and RTCRN for determining algebraic irrational numbers may differ. Specifically, Brent's method may not produce sequences of numbers that converge to the same algebraic irrational number as any RTCRN. Conversely, certain RTCRNs can replicate all the sequences generated by Brent's method.

Acknowledgments. This work has been supported in parts by the National Science Foundation grant number 1900716. The author is grateful to Dr. Jack Lutz for his support, insightful discussion, and valuable suggestions on this research. The author conveys his sincere thanks to Dr. Tamalika Ash and Dr. Aniket Banerjee for technical discussions.

Disclosure of Interests. The National Science Foundation, USA, has funded the paper's author.

References

1. Aho, A.V., Hopcroft, J.E.: The Design and Analysis of Computer Algorithms, 1st edn. Addison-Wesley Longman Publishing Co., Inc, USA (1974)
2. Anderson, D.F., Joshi, B.: Chemical mass-action systems as analog computers: implementing arithmetic computations at specified speed. Theoret. Comput. Sci. **1025**, 114983 (2025). https://doi.org/10.1016/j.tcs.2024.114983
3. Bournez, O.: Computability, complexity and programming with ordinary differential equations. In: Paul, C., Bläser, M. (eds.) 37th International Symposium on Theoretical Aspects of Computer Science (STACS 2020). Leibniz International Proceedings in Informatics (LIPIcs), vol. 154, pp. 3:1–3:13. Schloss Dagstuhl – Leibniz-Zentrum für Informatik, Dagstuhl, Germany (2020). https://doi.org/10.4230/LIPIcs.STACS.2020.3
4. Bournez, O., Campagnolo, M.L., Graça, D.S., Hainry, E.: Polynomial differential equations compute all real computable functions on computable compact intervals. J. Complex. **23**(3), 317–335 (2007). https://doi.org/10.1016/j.jco.2006.12.005
5. Bournez, O., Graça, D.S., Pouly, A.: Solving analytic differential equations in polynomial time over unbounded domains. In: Murlak, F., Sankowski, P. (eds.) Mathematical Foundations of Computer Science 2011, pp. 170–181. Springer, Berlin Heidelberg, Berlin, Heidelberg (2011)
6. Bournez, O., Graça, D., Pouly, A.: On the functions generated by the general purpose analog computer. Inf. Comput. **257**, 34–57 (2017). https://doi.org/10.1016/j.ic.2017.09.015
7. Bournez, O., Pouly, A.: A survey on analog models of computation, pp. 173–226. Springer, Cham (2021). https://doi.org/10.1007/978-3-030-59234-9_6
8. Brent, R.P.: Fast multiple-precision evaluation of elementary functions. J. ACM **23**(2), 242–251 (1976). https://doi.org/10.1145/321941.321944
9. Cook, M., Soloveichik, D., Winfree, E., Bruck, J.: Programmability of chemical reaction networks, pp. 543–584. Springer, Berlin Heidelberg (2009). https://doi.org/10.1007/978-3-540-88869-7_27
10. Fages, F., Le Guludec, G., Bournez, O., Pouly, A.: Strong turing completeness of continuous chemical reaction networks and compilation of mixed analog-digital programs. In: Feret, J., Koeppl, H. (eds.) Computational Methods in Systems Biology, pp. 108–127. Springer International Publishing, Cham (2017)
11. Fischer, P.C., Meyer, A.R., Rosenberg, A.L.: Time-restricted sequence generation. J. Comput. Syst. Sci. **4**(1), 50–73 (1970)
12. GraÇa, D.S., Zhong, N.: Computability of differential equations, pp. 71–99. Springer, Cham (2021). https://doi.org/10.1007/978-3-030-59234-9_3
13. Graça, D.S., Campagnolo, M.L., Buescu, J.: Computability with polynomial differential equations. Adv. Appl. Math. **40**(3), 330–349 (2008). https://doi.org/10.1016/j.aam.2007.02.003
14. Gurevich, Y., Shelah, S.: Nearly linear time. In: Meyer, A.R., Taitslin, M.A. (eds.) Logic at Botik '89, pp. 108–118. Springer, Berlin Heidelberg, Berlin, Heidelberg (1989)
15. Hartmanis, J., Stearns, R.E.: On the computational complexity of algorithms. Trans. Am. Math. Soc. **117**, 285–306 (1965)

16. Huang, X., Klinge, T.H., Lathrop, J.I.: Real-time equivalence of chemical reaction networks and analog computers. In: Thachuk, C., Liu, Y. (eds.) DNA Computing and Molecular Programming, pp. 37–53. Springer International Publishing, Cham (2019)
17. Huang, X., Klinge, T.H., Lathrop, J.I., Li, X., Lutz, J.H.: Real-time computability of real numbers by chemical reaction networks. Nat. Comput. **18**(1), 63–73 (2019). https://doi.org/10.1007/s11047-018-9706-x
18. Kawamura, A.: Lipschitz continuous ordinary differential equations are polynomial-space complete. In: 2009 24th Annual IEEE Conference on Computational Complexity, pp. 149–160 (2009). https://doi.org/10.1109/CCC.2009.34
19. Ko, K.I.: Complexity theory of real functions. Birkhauser Boston Inc., 1st edition edn. (1991). https://doi.org/10.1007/978-1-4684-6802-1
20. Ko, K.I., Friedman, H.: Computational complexity of real functions. Theoret. Comput. Sci. **20**(3), 323–352 (1982)
21. Kolmogorov, A.N., Uspenskii, V.A.: On the definition of an algorithm. Russ. Math. Surv. **13**, 3–28 (1958)
22. Pouly, A., Graça, D.S.: Computational complexity of solving polynomial differential equations over unbounded domains. Theoret. Comput. Sci. **626**, 67–82 (2016). https://doi.org/10.1016/j.tcs.2016.02.002
23. Rabin, M.O.: Real time computation. Israel J. Math. **1**, 203–211 (1963). https://doi.org/10.1007/BF02759719
24. Schey, H.: Div, Grad, Curl, and All that: An Informal Text on Vector Calculus. Norton, W.W (2005)
25. Schönhage, A.: Storage modification machines. SIAM J. Comput. **9**(3), 490–508 (1980). https://doi.org/10.1137/0209036
26. Soloveichik, D., Seelig, G., Winfree, E.: Dna as a universal substrate for chemical kinetics. Proc. Natl. Acad. Sci. **107**(12), 5393–5398 (2010). https://doi.org/10.1073/pnas.0909380107

Computability of Initial Value Problems

Vasco Brattka[1,2(✉)] and Hendrik Smischliaew[1]

[1] Fakultät Für Informatik, Universität der Bundeswehr München, Werner-Heisenberg-Weg 39, 85577 Neubiberg, Germany
Vasco.Brattka@cca-net.de, hendrik@hsmischliaew.de

[2] Department of Mathematics and Applied Mathematics, University of Cape Town, Private Bag X3, Rondebosch 7701, South Africa

Abstract. We demonstrate that techniques of Weihrauch complexity can be used to get easy and elegant proofs of known and new results on initial value problems. Our main result is that solving continuous initial value problems is Weihrauch equivalent to weak Kőnig's lemma, even if only solutions with maximal domains of existence are considered. This result simultaneously generalizes negative and positive results by Aberth and by Collins and Graça, respectively. It can also be seen as a uniform version of a Theorem of Simpson. Beyond known techniques we exploit for the proof that weak Kőnig's lemma is closed under infinite loops. One corollary of our main result is that solutions with maximal domain of existence of continuous initial value problems can be computed non-deterministically, and for computable instances there are always solutions that are low as points in the function space. Another corollary is that in the case that there is a fixed finite number of solutions, these solutions are all computable for computable instances and they can be found uniformly in a finite mind-change computation.

Keywords: Computable analysis · Weihrauch complexity · ordinary differential equations

1 Introduction

We consider *initial value problems* of the form

$$\begin{cases} y'(x) = f(x, y(x)) \\ y'(x_0) = y_0 \end{cases} \tag{1}$$

for continuous functions $f : U \to \mathbb{R}^n$ with $U \subseteq \mathbb{R} \times \mathbb{R}^n$ and $(x_0, y_0) \in U$. A *solution* of such a problem is a differentiable function $y : I \to \mathbb{R}^n$ that satisfies the equations in (1) on some interval $I \subseteq \mathbb{R}$ with $x_0 \in I$. Being a solution entails that $(x, y(x)) \in U$ for all $x \in I$. Any such solution y is automatically continuously differentiable. We say that $I = (a, b)$ with $a \in \mathbb{R} \cup \{-\infty\}$ and $b \in \mathbb{R} \cup \{\infty\}$ is a *maximal interval of existence* for a solution $y : I \to \mathbb{R}^n$, if y is a solution and no proper extension of y to a strictly larger interval of the above form is a solution.

© The Author(s), under exclusive license to Springer Nature Switzerland AG 2025
A. Beckmann et al. (Eds.): CiE 2025, LNCS 15764, pp. 191–206, 2025.
https://doi.org/10.1007/978-3-031-95908-0_14

There are two classical theorems that guarantee the existence of solutions of initial value problems, which are relevant in our context. The Picard-Lindelöf theorem guarantees the uniqueness of the solution on some small interval, in the case that f satisfies some Lipschitz condition [22, Theorem 2.2].

Theorem 1. (Picard-Lindelöf). *Let $U \subseteq \mathbb{R} \times \mathbb{R}^n$ be open with $(x_0, y_0) \in U$ and let $f : U \to \mathbb{R}^n, (t, s) \mapsto f(t, s)$ be continuous and locally Lipschitz continuous in the second argument $s \in \mathbb{R}^n$, uniformly with respect to the first argument $t \in \mathbb{R}$. Then the initial value problem (1) has a unique solution $y : [x_0 - \varepsilon, x_0 + \varepsilon] \to \mathbb{R}^n$ on some interval with $\varepsilon > 0$.*

Here local Lipschitz continuity in the second argument, uniformly with respect to the first argument, means that for any compact subset $K \subseteq U$ there is a Lipschitz constant in the second argument that works uniformly for each fixed first argument (see [22]). The theorem can be proved with the help of the Banach fixed-point theorem, applied to a *Picard operator* $T : D \to D$ (see (2) below) with a suitably chosen domain $D \subseteq \mathcal{C}([a, b], \mathbb{R}^n)$ of continuous functions. It is not too difficult to see that the fixed points of T correspond to the solutions of (1) for the respective interval $I = [a, b]$ (see [22, Section 2.2]).

In the case that f is only continuous and not necessarily Lipschitz continuous, it is still guaranteed that there are solutions, but not necessarily a unique one. This existence of solutions follows from the Peano theorem [22, Theorem 2.19].

Theorem 2. (Peano). *Let $U \subseteq \mathbb{R} \times \mathbb{R}^n$ be open with $(x_0, y_0) \in U$ and let $f : U \to \mathbb{R}^n$ be continuous. Then the initial value problem (1) has at least one solution $y : [x_0 - \varepsilon, x_0 + \varepsilon] \to \mathbb{R}^n$ on some interval with $\varepsilon > 0$.*

Again this result can be proved with the help of a suitable Picard operator as in (2) below, but in this case one needs to apply a different fixed-point theorem, such as the Schauder fixed-point theorem in order to obtain a fixed point. This explains why this result is less constructive than the Picard-Lindelöf theorem.

The study of initial value problems has a long tradition in computable analysis (for a survey see Graça and Zhong [12]). One of the earliest results is by Aberth [1], who proved that even for computable f there is not necessarily a computable solution in the non-unique case (Aberth's proof idea is the basis of the proof of Proposition 20).

Theorem 3. (Aberth). *There exists a computable function $f : [-1, 1]^2 \to \mathbb{R}$ such that the initial value problem (1) with $x_0 = y_0 = 0$ has no computable solution $y : I \to \mathbb{R}$, defined on some interval $I \subseteq \mathbb{R}$ with interior point $x_0 = 0$.*

This example was later strengthened by Pour-El and Richards [16], who proved that there is even such a counterexample that has no computable solution irrespectively of the chosen computable initial value.

On the positive side, it follows from a computable version of the Theorem of Picard-Lindelöf that for computable f that satisfies some suitable Lipschitz condition there is a unique computable solution of the initial value problem (1) for some interval. And more than this, even the solution on the maximal interval of existence $I = (a, b)$ is computable by a theorem of Graça, Zhong and Buescu [11, Theorem 3.1]. Ruohonen [17] showed in general that every uniquely determined solution for a computable f is computable. This result was extended to maximal domains of existence by Collins and Graça [9, Theorem 21].

Theorem 4. (Collins and Graça). *Let $f : U \to \mathbb{R}^n$ be a computable function on a c.e. open set $U \subseteq \mathbb{R} \times \mathbb{R}^n$ and let $(x_0, y_0) \in U$ be computable. Suppose that (1) has a unique solution $y : I \to \mathbb{R}^n$ with a maximal interval of existence $I = (a, b) \subseteq \mathbb{R}$ such that $x_0 \in I$. Then I is c.e. open and y is computable.*

More than this, the authors have shown that given (f, x_0, y_0), one can even uniformly compute (y, I).

In a different direction it has been proved in reverse mathematics by Simpson that the Peano existence theorem is equivalent to weak Kőnig's lemma over the base system of recursive comprehension RCA_0 [19] and [20, Theorem IV.8.1].

Theorem 5. (Simpson). *In second-order arithmetic (a suitable version of) the theorem of Peano is equivalent to weak Kőnig's lemma WKL_0 over RCA_0.*

One direction of the proof of this result is essentially based on the proof idea of Aberth's theorem (Theorem 3), whereas the other direction uses an appropriate version of the Schauder fixed point theorem.

Our goal here is to establish a similar result for Weihrauch complexity, which offers a computational way of classifying the computational content of mathematical problems (see [6] for a recent survey). Our proof incorporates ideas of Aberth, Simpson, and of Graça, Zhong and Buescu [11], but it also requires some new techniques and ideas, for instance regarding infinite loops. Our main result is the following Weihrauch complexity classification of the initial value problem.

Theorem 6. *The following are pairwise (strongly) Weihrauch equivalent:*

1. *The initial value problem IVP for the special case $x_0 = y_0 = 0$ and $n = 2$.*
2. *The initial value problem IVP.*
3. *The initial value problem for maximal domains of existence $\mathsf{IVP}_{\max}$.*
4. *Weak Kőnig's lemma WKL.*

Here IVP denotes the problem, given $(f, U, x_0, y_0) \in \mathcal{C}(U, \mathbb{R}^n) \times \mathcal{O}(\mathbb{R}^{n+1}) \times \mathbb{R}^{n+1}$, find $(y, I) \in \mathcal{C}(I, \mathbb{R}^n) \times \mathcal{O}(\mathbb{R})$ such that $y : I \to \mathbb{R}^n$ is a solution of (1). Here $\mathcal{C}(X, Y)$ denotes the space of continuous functions $f : X \to Y$ and $\mathcal{O}(X)$ the space of open subsets $U \subseteq X$, both represented in the standard way. The problem $\mathsf{IVP}_{\max}$ is defined analogously, except that I is additionally required to be a maximal domain of existence. And as usual, WKL denotes the problem, given an infinite binary tree T, find an infinite path $p \in [T]$ of T.

The equivalence of 2. and 4. in Theorem 6 can be seen as a uniform version of the theorem of Simpson (Theorem 5). However, we also obtain a version of the theorems of Aberth (Theorem 3) and of Collins and Graça (Theorem 4) as immediate corollaries of Theorem 6. And more than this, we can say something on solutions with maximal domain of existence in the general case.

Corollary 7. *Let $f : U \to \mathbb{R}^n$ be computable with a c.e. open set $U \subseteq \mathbb{R} \times \mathbb{R}^n$ and let $(x_0, y_0) \in U$ be computable. Then there exists a solution $y : I \to \mathbb{R}^n$ of (1) with a maximal domain of existence $I = (a, b) \subseteq \mathbb{R}$ such that y is low as a point in $\mathcal{C}(I, \mathbb{R}^n)$. And given (f, U, x_0, y_0), a solution (y, I) can be found uniformly in a non-deterministic way.*

We recall that a *low* point y is one that has a name $p \in \mathbb{N}^{\mathbb{N}}$ that is low, i.e., whose Turing jump is computable relative to the halting problem.[1]

These statements are immediate consequences of Theorem 6, as it is a property of the Weihrauch equivalence class of WKL that all problems in this class are non-deterministically computable and computable instances of such problems have low solutions [4, Corollary 7.13, Theorem 8.3]. Theorem 4 is a corollary of Theorem 6 since single-valued functions below WKL are automatically computable [4, Corollary 5.2]. Using further well-known results from Weihrauch complexity [14, Proposition 3.2] we can, for instance, obtain the following result, which is reminiscent of results by Hauck [13].

Corollary 8. *Let $f : U \to \mathbb{R}^n$ be computable with a c.e. open set $U \subseteq \mathbb{R} \times \mathbb{R}^n$ and computable $(x_0, y_0) \in U$. If there are only finitely many solutions $y : I \to \mathbb{R}^n$ of (1) with a maximal domain of existence $I = (a, b) \subseteq \mathbb{R}$, then all of these solutions y are computable, and all the maximal domains of existence are c.e. open. Given (f, U, x_0, y_0) with a fixed finite number of solutions, one such maximal solution (y, I) can be found uniformly with a finite mind-change computation.*

Here a problem f is *finite mind-change computable* (i.e., it can be computed with a finite number of mind changes on the output side) if and only if it is Weihrauch reducible to the limit on $\mathbb{N}$ (see [4]).

We close this section with a description of the further content of this article. In the next section we provide some basic definitions that are required to define the initial value problem formally, and we introduce some concepts from Weihrauch complexity. In Sect. 3 we prove that the initial value problem is Weihrauch reducible to weak Kőnig's lemma. In Sect. 4 we strengthen this result to the initial value problem with maximal domain of existence. In Sect. 5 we discuss the reduction in the opposite direction.[2]

[1] We note that a low point $y \in \mathcal{C}(I, \mathbb{R}^n)$ is not the same thing as a low function $y : I \to \mathbb{R}^n$ (see [2] for this distinction).

[2] The results presented in this article are based on the master's thesis of the second author [21], which was written under supervision of the first author.

2 Weihrauch Complexity and the Initial Value Problem

We introduce some concepts from computable analysis and Weihrauch complexity and we refer the reader to [7,23] for all concepts that have not been introduced here. We follow the representation based approach to computable analysis and we recall that a *representation* of a space X is a surjective partial map $\delta_X :\subseteq \mathbb{N}^{\mathbb{N}} \to X$. In this case (X, δ_X) is called a *represented space*. If we have two represented spaces (X, δ_X) and (Y, δ_Y), then we automatically have a representation $\delta_{\mathcal{C}(X,Y)}$ of the space of functions $f : X \to Y$ that have a continuous realizer. Here $F :\subseteq \mathbb{N}^{\mathbb{N}} \to \mathbb{N}^{\mathbb{N}}$ is called a *realizer* of some partial multivalued function $f :\subseteq X \rightrightarrows Y$, if

$$\delta_Y F(p) \in f\delta_X(p)$$

for all $p \in \operatorname{dom}(f\delta_X)$. In this situation we also write $F \vdash f$. It is well-known that there are universal functions $\mathsf{U} :\subseteq \mathbb{N}^{\mathbb{N}} \to \mathbb{N}^{\mathbb{N}}$ such that for every continuous $F :\subseteq \mathbb{N}^{\mathbb{N}} \to \mathbb{N}^{\mathbb{N}}$ there is some $q \in \mathbb{N}^{\mathbb{N}}$ such that $F(p) = \mathsf{U}\langle q, p\rangle$ for all $p \in \operatorname{dom}(F)$. Here $\langle . \rangle$ denotes some standard pairing function on Baire space $\mathbb{N}^{\mathbb{N}}$ (we use this notation for pairs as well as for the pairing of sequences). For short we write $\mathsf{U}_q(p) := \mathsf{U}\langle q, p\rangle$ for all $q, p \in \mathbb{N}^{\mathbb{N}}$. Now we obtain a representation $\delta_{\mathcal{C}(X,Y)}$ of the set $\mathcal{C}(X, Y)$ of total singlevalued functions $f : X \to Y$ with continuous realizers by

$$\delta_{\mathcal{C}(X,Y)}(q) = f :\iff \mathsf{U}_q \vdash f.$$

It is well-known that for admissibly represented T_0–spaces X, Y the function space $\mathcal{C}(X, Y)$ consists exactly of the usual continuous functions (see [7,23] for more details). The informal discussion of IVP in the introduction shows that we need representations of function spaces $\mathcal{C}(U, Y)$ for varying domains U. We use coproducts of the following form for this purpose.

Definition 9. (Coproduct function spaces). Let X, Y be represented spaces and let $(\mathcal{P}(X), \delta_{\mathcal{P}})$ be a represented space with $\mathcal{P}(X) \subseteq 2^X$. Then

$$\bigsqcup_{A \in \mathcal{P}(X)} \mathcal{C}(A, Y) := \{(f, A) : A \in \mathcal{P}(X) \text{ and } f \in \mathcal{C}(A, Y)\}$$

denotes the *coproduct function space* that we represent by δ, defined by

$$\delta\langle q, p\rangle = (f, A) :\iff \delta_{\mathcal{P}}(p) = A \text{ and } \mathsf{U}_q \vdash f$$

for all total singlevalued functions $f : A \to Y$ in $\mathcal{C}(A, Y)$ with $A \in \mathcal{P}(X)$.

We note that the representations of X and Y occur implicitly in the definition of $\vdash$. Typically, we will use for $\mathcal{P}(X)$ the set $\mathcal{O}(X)$ of open subsets of X. This set is represented via characteristic functions in $\mathcal{C}(X, \mathbb{S})$ to *Sierpiński space* $\mathbb{S} = \{0, 1\}$, which in turn is represented in the standard way. Also the *Euclidean space* $\mathbb{R}^n$ is represented in the standard way. The space of closed sets $\mathcal{A}_-(X)$ equipped with negative information is represented using complements of open sets in $\mathcal{O}(X)$. The space of compact subsets $\mathcal{K}_-(X)$ (of some computable metric space X) is represented via the *universal map*

$$\forall_K : \mathcal{O}(X) \to \mathbb{S}, \forall_K(U) = 1 :\iff K \subseteq U,$$

see [15,18] for details. The computable points in $\mathcal{O}(X)$, $\mathcal{A}_-(X)$ and $\mathcal{K}_-(X)$ are called *c.e. open*, *co-c.e. closed* and *co-c.e. compact* sets, respectively.

In general a *problem* is a multivalued function $f :\subseteq X \rightrightarrows Y$ on represented spaces X, Y that has a realizer. A typical example of a problem is *weak Kőnig's lemma*

$$\mathsf{WKL} :\subseteq \mathrm{Tr} \rightrightarrows 2^{\mathbb{N}}, T \mapsto [T]$$

that is defined for all infinite binary trees $T \in \mathrm{Tr}$ and maps those to the set $[T]$ of infinite paths. Here Tr denotes the set of binary trees represented via their characteristic functions. Hence, input and output space of WKL can be seen as subspaces of $\mathbb{N}^{\mathbb{N}}$. Another typical problem is *compact choice*

$$\mathsf{K}_X :\subseteq \mathcal{K}_-(X) \rightrightarrows X, K \mapsto K,$$

which maps any non-empty compact set $K \subseteq X$ to its points. The problem $\mathsf{LLPO} := \mathsf{K}_{\{0,1\}}$ is also know as *lesser limited problem of omniscience.*

We use the usual concept of Weihrauch reducibility (see [6] for a survey) in order to compare problems. Here $\mathrm{id} : \mathbb{N}^{\mathbb{N}} \to \mathbb{N}^{\mathbb{N}}$ denotes the identity on Baire space.

Definition 10. (Weihrauch reducibility). Let $f :\subseteq X \rightrightarrows Y$ and $g :\subseteq Z \rightrightarrows W$ be problems. We say that

1. f is *Weihrauch reducible* to g, in symbols $f \leq_{\mathrm{W}} g$, if there are computable $H, K :\subseteq \mathbb{N}^{\mathbb{N}} \to \mathbb{N}^{\mathbb{N}}$ such that $H\langle \mathrm{id}, GK\rangle \vdash f$, whenever $G \vdash g$ holds.
2. f is *strongly Weihrauch reducible* to g, in symbols $f \leq_{\mathrm{sW}} g$, if there are computable $H, K :\subseteq \mathbb{N}^{\mathbb{N}} \to \mathbb{N}^{\mathbb{N}}$ such that $HGK \vdash f$, whenever $G \vdash g$ holds.

As usual, we denote the corresponding equivalences by $\equiv_{\mathrm{W}}$ and $\equiv_{\mathrm{sW}}$, respectively. The following was essentially proved in [10] (see also [5, Theorem 8.5]).

Theorem 11. (Gherardi and Marcone). $\mathsf{K}_X \leq_{\mathrm{sW}} \mathsf{WKL}$ *for every computable metric space X.*

We recall that a *computable metric space* is a metric space (X, d) together with a dense sequence $\alpha : \mathbb{N} \to X$ such that $d \circ (\alpha \times \alpha) : \mathbb{N} \times \mathbb{N} \to \mathbb{R}$ is computable.

We can now define our versions of the initial value problem formally. For the remainder we assume that $n \geq 2$ is some fixed dimension. With a little more effort, we could also make the dimension variable in the coproduct.

Definition 12. (Initial value problem). By

$$\mathsf{IVP} :\subseteq \bigsqcup_{U \in \mathcal{O}(\mathbb{R}^{n+1})} \mathcal{C}(U, \mathbb{R}^n) \times \mathbb{R} \times \mathbb{R}^n \rightrightarrows \bigsqcup_{I=(a,b)\in\mathcal{O}(\mathbb{R})} \mathcal{C}(I, \mathbb{R}^n)$$

we denote the *initial value problem*, defined by

$$\mathsf{IVP}(f, U, x_0, y_0) := \{(y, I) : I = (a, b) \in \mathcal{O}(\mathbb{R}) \text{ and } y : I \to \mathbb{R}^n \text{ solves } (1)\}$$

with $\mathrm{dom}(\mathsf{IVP}) := \{(f, U, x_0, y_0) : (x_0, y_0) \in U\}$. By $\mathsf{IVP}_{\max}$ we denote the restriction of IVP in the image to such solutions $y : I \to \mathbb{R}^n$ for which I is a maximal domain of existence.

We also need a number of operators on problems that are commonly used in Weihrauch complexity, such as the *parallelization*

$$\widehat{f} :\subseteq X^{\mathbb{N}} \rightrightarrows Y^{\mathbb{N}}, (x_n)_{n\in\mathbb{N}} \mapsto \bigtimes_{n\in\mathbb{N}} f(x_n),$$

which is defined for every problem $f :\subseteq X \rightrightarrows Y$. We also define the concept of an *inverse limit* on problems, which can be seen as an infinite loop operation. For technical simplicity we define this for problems on Baire space.

Definition 13. (Inverse limit). Let $f :\subseteq \mathbb{N}^{\mathbb{N}} \rightrightarrows \mathbb{N}^{\mathbb{N}}$ be a problem. Then we define the *inverse limit* $f^\infty :\subseteq \mathbb{N}^{\mathbb{N}} \rightrightarrows \mathbb{N}^{\mathbb{N}}$ of f by

$$f^\infty(q_0) := \{\langle q_0, q_1, q_2, ...\rangle \in \mathbb{N}^{\mathbb{N}} : (\forall i)\ q_{i+1} \in \mathsf{U} \circ \langle \mathrm{id} \times f\rangle(q_i)\}$$

where $\mathrm{dom}(f^\infty)$ consists of all $q_0 \in \mathbb{N}^{\mathbb{N}}$ such that $A_0 := \{q_0\} \subseteq \mathrm{dom}(\mathsf{U} \circ \langle \mathrm{id} \times f\rangle)$ and $A_{i+1} := \mathsf{U} \circ \langle \mathrm{id} \times f\rangle(A_i) \subseteq \mathrm{dom}(\mathsf{U} \circ \langle \mathrm{id} \times f\rangle)$ for all $i \in \mathbb{N}$.

Using standard techniques we can extend this definition to represented spaces. The main result we need about the inverse limit is that weak Kőnig's lemma is closed under this operation, which was proved in [3]. We also use the fact that the parallelization of LLPO is equivalent to weak Kőnig's lemma [5, Theorem 8.2].

Proposition 14. $\mathsf{WKL}^\infty \equiv_{\mathrm{sW}} \mathsf{WKL} \equiv_{\mathrm{sW}} \widehat{\mathsf{LLPO}}$.

It has also been proved in [3] that $f \mapsto f^\infty$ is a monotone operation with respect to (strong) Weihrauch reducibility. Now we are well prepared to discuss the proof of our main result.

3 Reduction of IVP to WKL

In this section we prove that the initial value problem IVP is reducible to weak Kőnig's lemma WKL. The main idea is pretty simple: the solutions of IVP can be obtained as fixed points of the Picard operator (2) below for a suitable domain D. Most of the work goes into choosing an appropriate compact D. Finding a fixed point can then be achieved with the help of compact choice K_X for a suitable space X (see below).

We borrow some proof ideas from Simpson [19] and [20, Theorem IV.8.1]. The difference to the proof in reverse mathematics is that we do not need to re-prove the Peano theorem (see Theorem 2) in an effective form, but we can just use the classical theorem that already guarantees the existence of solutions and hence the existence of fixed points of the Picard operator. Even though computable versions of fixed-point theorems exist (see [8]), we do not need to use them here either. On the other hand, our proof has to be more uniform than the proof in reverse mathematics. For this purpose, we need some preliminary results about coproduct function spaces.

In the following we want to consider $X = \bigsqcup_I \mathcal{C}(I, \mathbb{R}^n)$ as a computable metric space. Here the coproduct is taken over all rational closed intervals $I = [a, b] \subseteq \mathbb{R}$ with $a, b \in \mathbb{Q}$ and $a < b$, represented as pairs $(a, b) \in \mathbb{Q}^2$. The following is easy to see and it is based on the standard construction of a metric for coproducts.

Lemma 15. (Coproduct function space). *The space* $X = \bigsqcup_I \mathcal{C}(I, \mathbb{R}^n)$ *is a computable metric space endowed with the metric* $d : X \times X \to \mathbb{R}$, *given by*

$$d((f, I), (g, J)) := \begin{cases} \sup_{x \in I} \frac{||f(x)-g(x)||}{1+||f(x)-g(x)||} & \text{if } I = J \\ 2 & \text{otherwise} \end{cases}$$

A suitable computable dense subset can be constructed with the help of rational polynomials. It is also clear that the injection into coproduct spaces is computable.

Lemma 16. (Injection into coproduct spaces). *The canonical injection maps*

$$\text{inj} : \mathcal{C}(J, \mathbb{R}^n) \to \bigsqcup_I \mathcal{C}(I, \mathbb{R}^n), f \mapsto (f, J)$$

are computable for every $J = [a, b]$ *with rational* $a < b$.

Now we are prepared to prove the main result of this section. As usual, we denote by $B(x, r) := \{y \in X : d(x, y) < r\}$ the *open ball* in a metric space (X, d) and by $\overline{A}$ the *closure* of a set $A \subseteq X$. On $\mathbb{R}^n$ we use the *maximum metric*.

Proposition 17. $\mathsf{IVP} \leq_{\mathrm{sW}} \mathsf{WKL}$.

This result even holds for a version of IVP whose outputs consist of functions $y : [a, b] \to \mathbb{R}$ defined on closed intervals $[a, b]$ with rational endpoints $a < b$.

Proof. Given $U \in \mathcal{O}(\mathbb{R}^{n+1})$, $f \in \mathcal{C}(U, \mathbb{R}^n)$ and $(x_0, y_0) \in U$ we can compute a rational $\delta > 0$ such that

$$K := \overline{B((x_0, y_0), \delta)} = [x_0 - \delta, x_0 + \delta] \times \overline{B(y_0, \delta)} \subseteq U.$$

This allows us to compute

$$M := \max_{(s,t) \in K} ||f(s, t)|| + 1$$

and $a, b \in \mathbb{Q}$ with

$$x_0 - \frac{\delta}{M} \leq a < x_0 < b \leq x_0 + \frac{\delta}{M}.$$

If we can find a solution $y : [a, b] \to \mathbb{R}^n$ of the initial value problem (1), then also the restriction $y : (a, b) \to \mathbb{R}^n$ to the open interval (a, b) is a solution. We let $I := [a, b]$. By [22, Theorem 2.19] there exists a solution $y : I \to \mathbb{R}^n$ of the initial value problem (1) for this particular I. Any such solution is in

$$D := \{y \in \mathcal{C}(I, \mathbb{R}^n) : y(x_0) = y_0 \text{ and } (\forall s, t \in I)\, ||y(s) - y(t)|| \leq M|s - t|\}$$

and any $y \in D$ satisfies $y(I) \subseteq \overline{B(y_0, \delta)}$. We consider the *Picard operator*

$$T : D \to D, y \mapsto \left(x \mapsto y_0 + \int_{x_0}^{x} f(t, y(t))\, \mathrm{d}t \right) \tag{2}$$

for this domain D. It is not too difficult to see that T is well-defined, i.e., $T(D) \subseteq D$. Given the input data (f, U, x_0, y_0), we can also compute a name of $T \in \mathcal{C}(D, D)$, as integration is computable by [23, Theorem 6.4.1]. The set

$$A := \{y \in D : T(y) = y\}$$

of fixed points of T contains exactly the solutions $y : I \to \mathbb{R}^n$ of the initial value problem (1) (see [22, Section 2.2]). We can compute a name of $A \in \mathcal{A}_-(D)$ as a closed set. We claim that we can also compute a name of $D \in \mathcal{K}_-(\mathcal{C}(I, \mathbb{R}^n))$ as a compact set. Then it follows by [15, Proposition 5.5 (4)] that we can also compute a name of $A \in \mathcal{K}_-(\mathcal{C}(I, \mathbb{R}^n))$ as a compact set. By Lemma 15 also the coproduct space $X = \bigsqcup_I \mathcal{C}(I, \mathbb{R}^n)$ is a computable metric space and the natural injection of $\mathcal{C}(I, \mathbb{R}^n)$ into X is computable by Lemma 16. Hence, it can be lifted to a computable injection from $\mathcal{K}_-(\mathcal{C}(I, \mathbb{R}^n))$ into $\mathcal{K}_-(X)$ by [15, Proposition 5.5 (6)]. Altogether, with Theorem 11 this yields the reduction

$$\mathsf{IVP} \leq_{\mathrm{sW}} \mathsf{K}_X \leq_{\mathrm{sW}} \mathsf{WKL}.$$

It remains to prove the claim on computability of D as a compact subset of $\mathcal{C}(I, \mathbb{R}^n)$. For technical simplicity, we describe the construction for the case $n = 1$. The general case can be treated similarly. Let $(q_i)_{i \in \mathbb{N}}$ be a computable enumeration of the rational numbers in $I = [a, b]$. We claim that the following map is a computable embedding relative to the input data

$$e : D \to [-1, 1]^{\mathbb{N}}, y \mapsto \left(\frac{1}{\delta}(y(q_i) - y_0) \right)_{i \in \mathbb{N}}.$$

Firstly, e is well-defined as $y \in D$ implies $y(I) \subseteq \overline{B(y_0, \delta)}$, and e is obviously computable relative to the input data. Since two $y_1, y_2 \in D$ with $y_1 \neq y_2$ must differ on a rational input, it follows that e is injective. Now we consider the set

$$C := \left\{ z \in [-1, 1]^{\mathbb{N}} : (\forall i, j)\ ||z_i - z_j|| \leq \frac{M}{\delta} \cdot |q_i - q_j| \right\}.$$

We claim that there is a function $g : C \to \mathcal{C}(I, \mathbb{R}^n)$ that is a left inverse to e and computable relative to the input data. Firstly, it is clear that $e(D) \subseteq C$. Given $z \in C$, there is a uniquely defined continuous function $y : I \to \mathbb{R}^n$ given by

$$y(q_i) := y_0 + \delta \cdot z_i$$

for all $i \in \mathbb{N}$. If we set $g(z) := y$, then it is clear that g is left inverse to e. We still need to prove that g is computable. The definition of y implies

$$||y(q_i) - y(q_j)|| \leq M \cdot |q_i - q_j|$$

for all $i, j \in \mathbb{N}$. Hence, in order to evaluate $g(z) = y$ up to precision 2^{-k}, it suffices to find some $j \in \mathbb{N}$ with $|x - q_j| < \frac{2^{-k}}{M}$. Then $||y(x) - y(q_j)|| \leq 2^{-k}$ follows. This also proves that e is an embedding and e and its partial inverse e^{-1} are both computable relative to the input data.

If we can now prove that $e(D)$ is co-c.e. closed in the compact metric space $[-1, 1]^{\mathbb{N}}$ relative to the input data, then it follows that $D = g \circ e(D)$ is a computable point in $\mathcal{K}_-(\mathcal{C}(I, \mathbb{R}^n))$ relative to the input data by [15, Proposition 5.5]. Now, if $z \in [-1, 1]^{\mathbb{N}} \setminus e(D)$ then either $z \notin C$, which can be recognized or $z \in C \setminus e(D)$, which means $g(z)(x_0) \neq y_0$, which can also be recognized. Altogether, this proves the claim. □

Instead of using the computable metric space X, we could also directly prove $\mathsf{IVP} \leq_{\mathrm{sW}} \mathsf{K}_{[-1,1]^{\mathbb{N}}}$.

As an immediate corollary of Proposition 17 we obtain the following result, which is a version of Corollary 7 for the non-maximal case.

Corollary 18. *Let $f : U \to \mathbb{R}^n$ be computable with a c.e. open set $U \subseteq \mathbb{R} \times \mathbb{R}^n$ and computable $(x_0, y_0) \in U$. Then there exists a solution $y : I \to \mathbb{R}^n$ of (1) with $I = [a, b]$ and rational $a < b$ such that y is low as a point in $\mathcal{C}(I, \mathbb{R}^n)$. And given (f, U, x_0, y_0), a solution (y, I) can be found uniformly in a non-deterministic way.*

4 Reduction of $\mathsf{IVP}_{\max}$ to WKL

In this section we want to strengthen Proposition 17 in the sense that we can even solve initial value problems for their maximal domains of existence with the help of WKL. For this strengthening we will actually use a refined version of Proposition 17 repeatedly. In fact, we will follow the proof idea of Graça, Zhong, and Buescu [11], which essentially is to apply $\mathsf{IVP}(f, U, x_0, y_0)$ repeatedly with values x_0, y_0 at the boundary of an already existing solution, in order to extend the domain of the solution step by step.

Proposition 19. $\mathsf{IVP}_{\max} \leq_{\mathrm{sW}} \mathsf{WKL}$.

Proof. We first refine the proof of Proposition 17 by making the choice of $\delta > 0$ and $a, b \in \mathbb{Q}$ more specific. Let (f, U, x_0, y_0) be the input given to IVP. We can assume, without loss of generality, that there are $c_m \in \mathbb{R}^{n+1}$ and $r_m > 0$ such that

$$U = \bigcup_{m \in \mathbb{N}} B(c_m, r_m) \text{ with } \overline{B(c_m, r_m)} \subseteq U$$

for all $m \in \mathbb{N}$. For each pair $\langle m, k \rangle \in \mathbb{N}$ at least one of the conditions

$$||c_m - (x_0, y_0)|| < r_m - 2^{-k} \text{ or } ||c_m - (x_0, y_0)|| > r_m - 2^{-k+1}$$

has to hold, and we check for each pair, which condition we can verify first, until we have found a pair $\langle m', k' \rangle$ for which we can recognize the first condition first.

Such a pair needs to exist, as k' can be made sufficiently large. This guarantees that the $\langle m', k'\rangle \in \mathbb{N}$ that we have found is smaller than or equal to the first $\langle m, k\rangle$ with

$$||c_m - (x_0, y_0)|| \le r_m - 2^{-k+1}. \tag{3}$$

We set $\delta := 2^{-k'}$ and we choose M and $a, b \in \mathbb{Q}$ as in the proof of Proposition 17, but with the additional property that

$$x_0 - a > \tfrac{1}{2} \cdot \tfrac{\delta}{M} \text{ and } b - x_0 > \tfrac{1}{2} \cdot \tfrac{\delta}{M}. \tag{4}$$

We obtain $\overline{B((x_0, y_0), \delta)} \subseteq B(c_{m'}, r_{m'})$ and hence $M \le M_{m'}$ for $M_{m'} := \max_{z \in \overline{B(c_{m'}, r_{m'})}} ||f(z)|| + 1$.

We consider a version of IVP that produces an output $y : [a, b] \to \mathbb{R}^n$ with the above additional requirements (3) and (4). The remainder of the proof of Proposition 17 still shows $\mathsf{IVP} \le_{\mathrm{sW}} \mathsf{WKL}$ for this version of IVP.

Now we use this specific version of IVP in an infinite loop inductively as follows. Given (f, U, x_0, y_0) as input to $\mathsf{IVP}_{\max}$, we choose $a_0 := b_0 := x_0$, $y_{a_0}(a_0) := y_{b_0}(b_0) := y_0$. Then we determine inductively in a loop for all $i \in \mathbb{N}$

- $(y_{a_{i+1}}, [a_{i+1}, .]) \in \mathsf{IVP}(f, U, a_i, y_{a_i}(a_i))$,
- $(y_{b_{i+1}}, [., b_{i+1}]) \in \mathsf{IVP}(f, U, b_i, y_{b_i}(b_i))$.

The infinite loop that determines these values can be realized with the help of $(\mathsf{IVP} \times \mathsf{IVP})^\infty$. The final result of this computation is $(y, (a, b))$ with $a := \inf_{i \in \mathbb{N}} a_i$, $b := \sup_{i \in \mathbb{N}} b_i$ and $y : (a, b) \to \mathbb{R}^n$ with

$$y(x) := \begin{cases} y_{a_{i+1}}(x) \text{ if } a_{i+1} \le x \le a_i \\ y_{b_{i+1}}(x) \text{ if } b_i \le x \le b_{i+1} \end{cases}.$$

The set $(a, b) \in \mathcal{O}(\mathbb{R})$ and the function $y \in \mathcal{C}((a, b), \mathbb{R}^n)$ can be computed relative to the previously determined objects, as $y_{a_{i+1}}(a_i) = y_{a_i}(a_i)$ and $y_{b_{i+1}}(b_i) = y_{b_i}(b_i)$. We claim that y is a maximal solution of the initial value problem (1). With the help of Propositions 17 and 14 this proves

$$\mathsf{IVP}_{\max} \le_{\mathrm{sW}} (\mathsf{IVP} \times \mathsf{IVP})^\infty \le_{\mathrm{sW}} \mathsf{WKL}^\infty \equiv_{\mathrm{sW}} \mathsf{WKL}.$$

We still need to prove the claim. Let us assume that the solution $y : (a, b) \to \mathbb{R}^n$ is not maximal and that there is some $b' > b$ such that y can be extended to a solution $y : (a, b') \to \mathbb{R}^n$. Then $(b, y(b)) \in U$ and there are $m, k, j \in \mathbb{N}$ with

$$B((b_j, y(b_j)), 2^{-k+1}) \subseteq B((b, y(b)), 2^{-k+2}) \subseteq B(c_m, r_m) \text{ and} \tag{5}$$

$$b_j + \tfrac{1}{2} \cdot \min_{\langle m', k'\rangle \le \langle m, k\rangle} \tfrac{2^{-k'}}{M_{m'}} > b, \tag{6}$$

since $\lim_{i \to \infty} b_i = b$. If we apply (3) and (4) to the pair $(b_j, y(b_j))$ (in place of (x_0, y_0)), then we obtain that the $\langle m', k'\rangle$ chosen by the algorithm for IVP

on input $(f, U, b_j, y(b_j))$ satisfies $\langle m', k' \rangle \leq \langle m, k \rangle$ by (3) and (5) and hence we obtain with $\delta = 2^{-k'}$

$$b_{j+1} > b_j + \tfrac{1}{2} \cdot \tfrac{\delta}{M_{m'}} \geq b_j + \tfrac{1}{2} \cdot \min_{\langle m', k' \rangle \leq \langle m, k \rangle} \tfrac{2^{-k'}}{M_{m'}} > b$$

by (4) and (6). But this is a contradiction to $b = \sup_{i \in \mathbb{N}} b_i$. Hence, there cannot be any extension $y : (a, b') \to \mathbb{R}^n$ of the solution $y : (a, b) \to \mathbb{R}^n$ with $b' > b$. Likewise, it follows that there cannot be any extension $y : (a', b) \to \mathbb{R}^n$ with $a' < a$. □

Theorem 4 and Corollaries 7 and 8 are immediate consequences of Proposition 19.

5 Reduction of WKL to IVP

In this section we will translate the proof idea of Aberth [1] into a proof of $\mathsf{WKL} \leq_{\mathrm{sW}} \mathsf{IVP}$. This idea has also been used by Simpson [20, Theorem IV.8.1]. However, the uniform version of this proof needs a little modification, which ensures that the required information can be reconstructed from a solution of the constructed initial value problem on some arbitrarily small domain.

Proposition 20. $\mathsf{WKL} \leq_{\mathrm{sW}} \mathsf{IVP}$.

Proof. By Proposition 14 it suffices to prove $\widehat{\mathsf{LLPO}} \leq_{\mathrm{sW}} \mathsf{IVP}$ and for this purpose we first describe a gadget that does the job for a single instance of LLPO. We can assume that $\mathsf{LLPO} :\subseteq \mathbb{N}^{\mathbb{N}} \rightrightarrows \{0,1\}$ is defined for all $p \in \mathbb{N}^{\mathbb{N}}$ with $\{0,1\} \not\subseteq \mathrm{range}(p)$ and all $i \in \{0,1\}$ by

$$i \in \mathsf{LLPO}(p) \iff i \notin \mathrm{range}(p).$$

Given $p \in \mathbb{N}^{\mathbb{N}}$ we construct a continuous function $g_p : \mathbb{R}^2 \to \mathbb{R}$ such that all solutions $y : [0,4] \to \mathbb{R}$ of the corresponding initial value problem $y'(x) = g_p(x, y(x))$ with $y(0) = 0$ satisfy $y(4) = 0$ and

$$y(2) > -1 \Longrightarrow 0 \in \mathsf{LLPO}(p) \tag{7}$$
$$y(2) < 1 \Longrightarrow 1 \in \mathsf{LLPO}(p). \tag{8}$$

The idea is to construct the function g_p such that the unique solutions are sufficiently positive or negative for $\mathsf{LLPO}(p) = \{0\}$ and $\mathsf{LLPO}(p) = \{1\}$, respectively, and such that there are positive and negative solutions for $\mathsf{LLPO}(p) = \{0,1\}$. Figure 1 illustrates the situation.

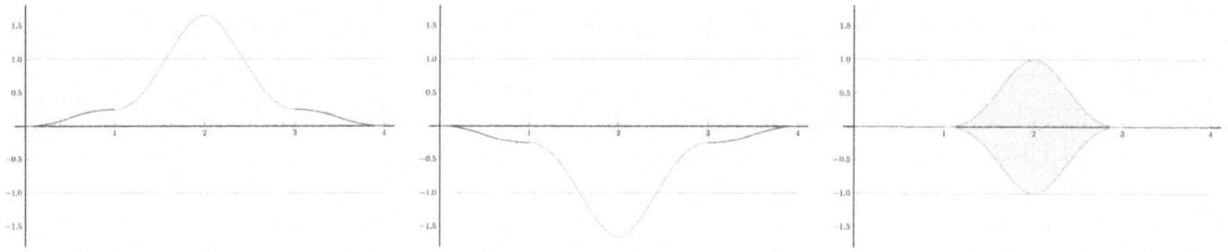

$\mathsf{LLPO}(p) = \{0\}$ $\mathsf{LLPO}(p) = \{1\}$ $\mathsf{LLPO}(p) = \{0, 1\}$

Fig. 1. Solutions y of the initial value problem $y'(x) = g_p(x, y(x))$ with $y(0) = 0$.

The function $g_p : \mathbb{R}^2 \to \mathbb{R}$ is defined piecewise by

$$g_p(x, y) := \begin{cases} h_p(x) & \text{if } x \in [0, 1] \\ s(x-1, y) & \text{if } x \in [1, 2] \\ -s(x-2, y) & \text{if } x \in [2, 3] \\ -h_p(x-3) & \text{if } x \in [3, 4] \\ 0 & \text{otherwise} \end{cases},$$

where h_p governs the blue parts and s the orange parts of the solutions in Fig. 1. The function $h_p : [0, 1] \to \mathbb{R}$ is given by

$$h_p(x) := \max(0, 1 - |2x - 1|) \cdot (T(p, 1) - T(p, 0))$$

with

$$T(p, i) := \begin{cases} 2^{-\min\{j \in \mathbb{N}: p(j) = i\}} & \text{if } i \in \text{range}(p) \\ 0 & \text{otherwise} \end{cases}.$$

That is h_p is a triangle function with a peak at $\frac{1}{2}$ and value 2^{-k} or -2^{-k}, respectively, depending on whether 1 or 0 first appears at position k in p, respectively, and hence $\mathsf{LLPO}(p) = \{0\}$ or $\mathsf{LLPO}(p) = \{1\}$, respectively. If neither of $0, 1$ appears in p, then h_p is constantly 0. The function $s : [0, 1] \times \mathbb{R} \to \mathbb{R}$ is given by

$$s(x, y) := 9x(1 - x) \cdot \text{sign}(y)|y|^{\frac{1}{3}}.$$

The initial value problem $y'(x) = s(x, y(x))$ with $y(0) = y_0$ has the solution

$$y(x) = \text{sign}(y_0) \left(x^2(3 - -2x) + |y_0|^{\frac{2}{3}} \right)^{\frac{3}{2}}$$

for $y_0 \neq 0$, and for $y_0 = 0$ and all $c \in [0, 1]$ one has the solutions

$$y(x) = \begin{cases} 0 & \text{if } x \leq c \\ \pm \left(x^2(3 - -2x) - c^2(3 - -2c) \right)^{\frac{3}{2}} & \text{if } c \leq x \leq 1 \end{cases}.$$

As explained in [1], these are all possible solutions of the given initial value problem for s. The solutions y for $y_0 > 0$ satisfy $y(1) > 1$ and the solutions for

$y_0 < 0$ satisfy $y(1) < -1$ and hence the solutions y for the initial value problem of g_p satisfy the properties given in (7) and (8).

In a second step we combine a countable number of the gadgets for a single instance of LLPO in a single initial value problem in order to reduce the problem $\widehat{\mathsf{LLPO}}$ to IVP. Given an input $p = \langle p_0, p_1, p_2, ...\rangle$ of $\widehat{\mathsf{LLPO}}$, we can compute the function $f : [-1,1] \times \mathbb{R} \to \mathbb{R}$ with

$$f(x,y) := \begin{cases} \sum_{m=\langle k,i\rangle \in \mathbb{N}} 2^{-(m+3)} g_{p_i}\left(2^{m+3}(x+2^{-m}), 2^{2(m+3)}y\right) & \text{if } x \leq 0 \\ f(-x,y) & \text{otherwise} \end{cases}.$$

We note that the construction ensures that the information on any value of $\mathsf{LLPO}(p_i)$ is included for infinitely many m in f, which is necessary as a solution $y : I \to \mathbb{R}$ of the initial value problem $y'(x) = f(x, y(x))$ with $y(0) = 0$ might only be known on some small interval $I = (a,b)$ around 0. We claim that any such solution y satisfies

$$y(-2^{-m} + 2^{-(m+2)}) > -2^{-(m+3)} \Longrightarrow 0 \in \mathsf{LLPO}(p_i) \tag{9}$$

$$y(-2^{-m} + 2^{-(m+2)}) < 2^{-(m+3)} \Longrightarrow 1 \in \mathsf{LLPO}(p_i) \tag{10}$$

for all $m = \langle k,i\rangle \in \mathbb{N}$ with $-2^{-m} + 2^{-(m+2)} \in I$. This enables us to compute a value $q \in \widehat{\mathsf{LLPO}}(p)$ from $\mathsf{IVP}(f, (-1,1) \times \mathbb{R}, 0, 0)$ and hence we obtain

$$\mathsf{WKL} \equiv_{\mathrm{sW}} \widehat{\mathsf{LLPO}} \leq_{\mathrm{sW}} \mathsf{IVP}.$$

It still remains to show how the claim follows from the implications given in (7) and (8). We consider $m = \langle k,i\rangle \in \mathbb{N}$ with $-2^{-m} + 2^{-(m+2)} \in I$ and solutions y of $y'(x) = f(x,y(x))$ with $y(0) = 0$ and solutions $\hat{y}$ of $\hat{y}'(\hat{x}) = g_{p_i}(\hat{x}, \hat{y}(\hat{x}))$ with $\hat{y}(0) = 0$ on the intervals $[-2^{-m}, -2^{-(m+1)}]$ and $[0,4]$, respectively. The transformation $\hat{x} = 2^{m+3}(x + 2^{-m})$ maps the interval $[-2^{-m}, -2^{-(m+1)}]$ onto the interval $[0,4]$. Together with the transformation $\hat{y} = 2^{2(m+3)}y$ we obtain

$$\frac{\mathrm{d}y}{\mathrm{d}x} = 2^{-2(m+3)} \cdot \frac{\mathrm{d}\hat{y}}{\mathrm{d}y} \cdot \frac{\mathrm{d}y}{\mathrm{d}x} \cdot \frac{\mathrm{d}x}{\mathrm{d}\hat{x}} \cdot 2^{m+3} = 2^{-(m+3)} \cdot \frac{\mathrm{d}\hat{y}}{\mathrm{d}\hat{x}} = 2^{-(m+3)} \cdot g_{p_i}(\hat{x}, \hat{y}(\hat{x}))$$

and hence

$$y(-2^{-m} + 2^{-(m+2)}) = 2^{-(m+3)}\hat{y}(2).$$

This shows that (7) and (8) imply (9) and (10), respectively. □

We note that this reduction requires IVP only in the special case $x_0 = y_0 = 0$.

Acknowledgments. The first author was funded by the German Research Foundation (DFG, Deutsche Forschungsgemeinschaft) – project number 554999067 and by the National Research Foundation of South Africa (NRF) – grant number 151597.

References

1. Aberth, O.: The failure in computable analysis of a classical existence theorem for differential equations. Proc. Am. Math. Soc. **30**, 151–156 (1971)
2. Brattka, V.: A Galois connection between turing jumps and limits. Logical Methods Comput. Sci. **14**(3:13), 1–37 (2018). https://doi.org/10.23638/LMCS-14(3:13)2018. https://lmcs.episciences.org/4794
3. Brattka, V.: Loops, inverse limits and non-determinism. arXiv arXiv:2501.17734 (2025)
4. Brattka, V., de Brecht, M., Pauly, A.: Closed choice and a uniform low basis theorem. Ann. Pure Appl. Logic **163**, 986–1008 (2012). https://doi.org/10.1016/j.apal.2011.12.020
5. Brattka, V., Gherardi, G.: Weihrauch degrees, omniscience principles and weak computability. J. Symbolic Logic **76**(1), 143–176 (2011). https://doi.org/10.2178/jsl/1294170993
6. Brattka, V., Gherardi, G., Pauly, A.: Weihrauch complexity in computable analysis. In: Brattka, V., Hertling, P. (eds.) Handbook of Computability and Complexity in Analysis, Theory and Applications of Computability, pp. 367–417. Springer, Cham (2021). https://doi.org/10.1007/978-3-030-59234-9_11
7. Brattka, V., Hertling, P. (eds.): Handbook of computability and complexity in analysis. Theory and Applications of Computability. Springer, Cham (2021). https://doi.org/10.1007/978-3-030-59234-9
8. Brattka, V., Le Roux, S., Miller, J.S., Pauly, A.: Connected choice and the brouwer fixed point theorem. J. Math. Log. **19**(1), 1–46 (2019). https://doi.org/10.1142/S0219061319500041
9. Collins, P., Graça, D.S.: Effective computability of solutions of differential inclusions: the ten thousand monkeys approach. J. Univ. Comput. Sci. **15**(6), 1162–1185 (2009). https://doi.org/10.3217/jucs-015-06-1206
10. Gherardi, G., Marcone, A.: How incomputable is the separable Hahn-Banach theorem? Notre Dame Journal of Formal Logic **50**(4), 393–425 (2009). https://doi.org/10.1215/00294527-2009-018
11. Graça, D.S., Zhong, N., Buescu, J.: Computability, noncomputability and undecidability of maximal intervals of IVPs. Trans. Am. Math. Soc. **361**(6), 2913–2927 (2009). https://doi.org/10.1090/S0002-9947-09-04929-0
12. Graça, D.S., Zhong, N.: Computability of differential equations. In: Brattka, V., Hertling, P. (eds.) Handbook of Computability and Complexity in Analysis, Theory and Applications of Computability, pp. 71–99. Springer, Cham (2021). https://doi.org/10.1007/978-3-030-59234-9_3
13. Hauck, J.: Ein Kriterium für die konstruktive Lösbarkeit der Differentialgleichung $y' = f(x, y)$. Zeitschrift für Mathematische Logik und Grundlagen der Mathematik **31**, 357–362 (1985)
14. Le Roux, S., Pauly, A.: Finite choice, convex choice and finding roots. Logical Methods Comput. Sci. **11**(4), 4:6, 31 (2015). https://doi.org/10.2168/LMCS-11(4:6)2015
15. Pauly, A.: On the topological aspects of the theory of represented spaces. Computability **5**(2), 159–180 (2016). https://doi.org/10.3233/COM-150049
16. Pour-El, M.B., Richards, J.I.: A computable ordinary differential equation which possesses no computable solution. Ann. Math. Logic **17**, 61–90 (1979)
17. Ruohonen, K.: An effective Cauchy-Peano existence theorem for unique solutions. Int. J. Found. Comput. Sci. **7**(2), 151–160 (1996). https://doi.org/10.1142/S0129054196000129

18. Schröder, M.: Admissibly represented spaces and Qcb-spaces. In: Brattka, V., Hertling, P. (eds.) Handbook of Computability and Complexity in Analysis, Theory and Applications of Computability, pp. 305–346. Springer, Cham (2021). https://doi.org/10.1007/978-3-030-59234-9_9
19. Simpson, S.: Which set existence axioms are needed to prove the Cauchy/Peano theorem for ordinary differential equations?. J. Symbolic Logic **49**, 783–802 (1984). https://doi.org/10.2307/2274131
20. Simpson, S.G.: Subsystems of Second Order Arithmetic, 2nd edn. Perspectives in Logic, Cambridge University Press (2009)
21. Smischliaew, H.A.: Berechenbarkeitseigenschaften von Differentialgleichungen. Master's thesis, Fakultät für Informatik, Universität der Bundeswehr München, Neubiberg (2024)
22. Teschl, G.: Ordinary differential equations and dynamical systems, Graduate Studies in Mathematics, vol. 140. American Mathematical Society, Providence, RI (2012). https://doi.org/10.1090/gsm/140
23. Weihrauch, K.: Computable Analysis. Springer, Berlin (2000)

Full Generalized Effective Reducibility

Merlin Carl(✉)

Institut für Mathematik, Europa-Universität Flensburg, Flensburg, Germany
merlin.carl@uni-flensburg.de

Abstract. We extend the concept of effective reducibility between statements of set theory with ordinal Turing machines (OTMs) explored in [2] for Π_2-statements to statements of arbitrary quantifier complexity in prenex normal form and use this to compare various fundamental set-theoretical principles, including the power set axiom, the separation scheme, the collection scheme and the replacement scheme, with respect to effective reducibility. This notion of reducibility is both different from classical truth and from the OTM-realizability of the corresponding implications. Along the way, we obtain a computational characterization of HOD as the class of sets that are OTM-computable relative to every effectivizer of Σ_2-separation.

Keywords: Ordinal Computability · Ordinal Turing Machines · Effective Reducibility · HOD · set theory

1 Introduction

A Π_2-statement $\forall x\exists y\varphi(x,y)$ (with φ a Δ_0-formula) can be regarded as *effective* if there is an effective method for constructing a corresponding y for each given x. In effective reducibility, one then asks whether, given access to any method effectivizing a statement ψ, one can effectivize φ relative to that method. Usually, "effective method" is interpreted as "Turing program" (see, e.g., Hirschfeldt, [10], pp. 22f).

In [1], we introduced a variant of effective reducibility based on Koepke's ordinal Turing machines (OTMs) rather than classical Turing machines; in [2], we conducted an investigation concerning the OTM-reducibility of various formulations of the axiom of choice. However, as sketched in section 6 of [2], the ideas behind effectivity and reducibility can be extended to arbitrary set-theoretic statements, provided they are given in prenex normal form. We also briefly mentioned that, in this setting, the separation scheme was reducible to the replacement scheme.

In the present work, we will begin a study of this extension of effective OTM-reducibility to statements of arbitrary quantifier complexity.

Another way to "effectivize" set-theoretical statements with OTMs is to consider notions of *realizability* based on OTMs; this path was pursued in [4]. These approaches agree in the sense that, for formulas in prenex normal form, OTM-effectivity is equivalent to uniform OTM-realizability ([4], Definition 18) with a

© The Author(s), under exclusive license to Springer Nature Switzerland AG 2025
A. Beckmann et al. (Eds.): CiE 2025, LNCS 15764, pp. 207–221, 2025.
https://doi.org/10.1007/978-3-031-95908-0_15

single ordinal parameter. However, the OTM-effective reducibility of φ to ψ differs considerably from the OTM-realizability of $\psi \rightarrow \varphi$: While an OTM-realizer must be able to convert OTM-*realizers* for ψ into OTM-realizers for φ, an OTM-effective reduction must achieve the same for *arbitary* effectivizers, which, from the point of view of OTMs, may be arbitrarily ineffective. As we will see below, OTM-effective reducibility is indeed strictly stronger than the OTM-realizability of the corresponding implication, for formulas in prenex normal form.

In particular, we will investigate interdependencies between the axioms of ZFC in terms of effective reducibility, showing, for example, that an effectivizer for the separation scheme makes all heriditarily ordinally definable sets computable (Theorem 15), that the power set axiom is effectively reducible to the Σ_3-replacement scheme (Theorem 17(4)) and that replacement is reducible to the conjunction of the separation scheme and the power set axiom (Lemma 18(1)), while a number of other reductions, such as the reducibility of Σ_3-replacement to separation alone (Lemma 18(4)), or the reducibility of the power set axiom to the statement that each cardinal has a successor (Theorem 17(3)), are independent of ZFC.

2 Basic Definition

We assume familiarity with Koepke's ordinal Turing machines (OTMs) and refer to [13] and [3] for details. In this paper, OTM-computations always allow for ordinal parameters, i.e., at the start of the computation, some ordinal on the scratch tape may be marked. Many of our results, however, do not depend on the presence or absence of parameters.

A definition for effectivizers for set-theoretical formulas in prenex normal form of arbitrary quantifier complexity was sketched in [2], section 6. We will now state it in a form that we regard as much clearer, which at the same time generalizes to game formulas (i.e., formulas given in prenex normal form with infinitely many quantifier alternations). We take for granted familiarity with the standard way of encoding arbitary sets as sets of ordinals as described, e.g., in [2].[1]

Definition 1. *Let $\varphi :\Leftrightarrow \forall x_0 \exists y_0 ... \forall x_n \exists y_n \psi(x_0, y_0, ..., x_n, y_n)$ be an $\in$-formula in prenex normal form, where ψ is quantifier-free. An effectivizer of φ is a class function $F : V \rightarrow V$ such that, for all $x_0, ..., x_n$, we have*[2]

$$\psi(x_0, F((x_0)), x_1, F((x_0, x_1)), ..., x_n, F((x_0, ..., x_n))).$$

An encoding $\hat{F}$ of F is a class function $\hat{F} : \mathfrak{P}(On) \rightarrow \mathfrak{P}(On)$ such that, for all sets x, whenever $c_x \subseteq On$ codes x, then $\hat{F}(c_x)$ codes $F(x)$.

If $\hat{F}$ is OTM-computable by a program P in some parameter $\rho \in On$, we call (P, ρ) an OTM-effectivizer of φ and say that φ is OTM-effective. If F is

[1] Also cf. [4] for basic results of OTM-computations with set codes.

[2] Note the double brackets in $F((x_0))$, which indicate that F is applied to the sequence consisting of the single element x_0, rather than to x_0 itself.

restricted to sequences of length $k \leq n$, we call F a k-effectivizer. Now, as in [2], a statement is called effective if some effectivizer of it has an OTM-computable encoding.

This definition extends in a straightforward way to schemes of statements: If $\Phi := \{\varphi_i : i \in \omega\}$ is a set of $\in$-statements,[3] *then an effectivizer for Φ is a class function $F : \omega \times V \to V$ such that, for all $i \in \omega$, the function $x \mapsto F(i, x)$ is an effectivizer for φ_i. As usual, we will write $\ulcorner\psi\urcorner$ for that element $i \in \omega$ such that $\psi = \varphi_i$. If φ is a formula and p is a set parameter, then $\ulcorner\varphi(p)\urcorner$ abbreviates $(\ulcorner\varphi\urcorner, p)$.*

Remark 1. Below, we will slightly abuse our terminology by using the term "effectivizers" to also refer to encodings of effectivizers.

Remark 2. The same definitions work for *game formulas*, i.e., infinite formulas of the form $\forall x_0 \exists y_0 \forall x_1 \exists y_1 ... \psi(x_0, y_0, ...)$ in which an infinitary quantifier-free formula ψ is preceded by a series of quantifiers of length ω. However, such formulas will not be considered in this paper.

Definition 2. *If φ and ψ are $\in$-statements in prenex normal form, then we refer to an ordered pair (F, G), consisting of an effectivizer F of φ and an effectivizer G of ψ, as an effectivizer of $\varphi \oplus \psi$.*

Remark 3. Note that this is to be read as a definition of the phrase "an effectivizer of $\varphi \oplus \psi$", not of the notation "$\varphi \oplus \psi$", which we only use within the context of effectivizers. Also, this notation is a mere matter of convenience. Effectivizers for $\varphi \oplus \psi$ are easily (and, in particular, effectively) converted into effectivizers for the prenex normal form of $\varphi \wedge \psi$ and vice versa.

The following now generalizes [2], Definition 3 (which, in turn, is a variant of classical effective reducibility for OTM-computability) to formulas of arbitrary quantifier complexity:

Definition 3. *Let φ, ψ be $\in$-formulas in prenex normal form. We say that φ is OTM-reducible to ψ, written $\varphi \leq_{OTM} \psi$, if and only if there are an OTM-program P and some $\rho \in On$ such that, for each encoded effectivizer F_ψ of ψ, we have $P^{F_\psi}(\rho)$ – i.e., P with access to F_ψ as an oracle – computes an encoded effectivizer for φ. If P computes a k-effectivizer for φ from every l-effectivizer for ψ for some $k, l \in \omega$, we write $\varphi \leq^{k,l}_{OTM} \psi$.*

This definition also extends to effectivizers for schemes in the obvious way.

Remark 4. In a ZFC framework, class functions must be definable; for the purpose of this paper, this is an unwanted restriction of possible effectivizers. In particular, reductions may become trivial when definable effectivizers do not exist or when they do exist, but, due to the lack of a global well-ordering, cannot be encoded. The way to formalize reducibility in ZFC is then the following: We add a new unary function symbol F to the language of ZFC and formulate

[3] Generalizations to uncountable schemes are easily possible, but will not be considered in this work.

the ZFC-axioms for this extended language (which modifies only separation and replacement); call the resulting theory ZFC(F). We then add an axiom $\mathcal{A}$ (or, in the case of a scheme, a scheme) expressing that F is an effectivizer for the statement ψ. We say that φ is ZFC-provably OTM-reducible to ψ if and only if, for some OTM-program P, the theory ZFC(F) + $\mathcal{A}$ proves that, for each encoding $\hat{F}$ of F, $P^{\hat{F}}$ computes an effectivizer for φ. Most of our results and arguments work equally well for both definitions; it is only in our treatment of AC below that we need to switch to ZFC-provable OTM-reducibility.

Remark 5. For several principles analyzed below, it is more natural to consider 1-effectivizers (and the respective reducibility) rather than full reducibility. For example, for separation and replacement, an effectivizer should deliver the separated set and the image of a set under a class function; however, a full effectivizer would also yield witnesses to (many) existential quantifiers nested in complicated formulas, and thus deliver far more information than it should.

3 Axioms of ZFC and Other Basic Statements

We will now consider the axiom (schemes) of ZFC, along with some basic consequences of ZFC, with respect to their OTM-effectivity and mutual OTM-reducibility.

Definition 4. *We will use the following abbreviations:*

- $\bot$ *denotes* $0 = 1$, *while* $\top$ *denotes* $0 = 0$.
- HP*: For each OTM-program P and each ordinal parameter ρ, there is $i \in \{0, 1\}$ such that either $P(\rho)\downarrow$ and $i = 1$ or $P(\rho)\uparrow$ and $i = 0$.*
- GreaterCard*: For each $\alpha \in On$, there is a cardinal $\kappa > \alpha$.*
- NextCard*: For each $\alpha \in On$, there is a next largest cardinal $\kappa > \alpha$; we will denote this cardinal as α^+.*
- Pot*: $\forall x \exists y \forall z (z \in y \leftrightarrow z \subseteq x)$*
- Sep$_\varphi$*: $\forall X, p \exists Y \forall z (z \in Y \leftrightarrow z \in X \wedge \varphi(z, p))$, for all formulas φ. By Σ_n-*Sep*, we denote the set of all formulas* Sep$_\varphi$ *where φ is Σ_n, for $n \in \omega$. The totality of all these formulas will be denoted by* Sep.
- Rep$_\varphi$*: $\forall X, p(\forall x \in X \exists! y \varphi(x, y, p) \rightarrow \exists Y \forall x \in X \exists y \in Y \varphi(x, y, p))$; by Σ_n-*Rep*, we denote the set of all formulas* Sep$_\varphi$*, where φ is Σ_n, for $n \in \omega$. The totality of all these formulas will be denoted by Rep.*
- Coll$_\varphi$*: $\forall X, p(\forall x \in X \exists y \varphi(x, y, p) \rightarrow \exists Y \forall x \in X \exists y \in Y \varphi(x, y, p))$; by Σ_n-*Coll*, we denote the set of all formulas* Coll$_\varphi$*, where φ is Σ_n, for $n \in \omega$; the totality of all these formulas will be denoted by* Coll.
- AC*: $\forall X (\emptyset \notin X \rightarrow \exists f : X \rightarrow \bigcup X \forall x \in X f(x) \in x)$.*
- *For $n \in \omega$, let $T_n(k, \rho)$ be a truth-predicate for Σ_n-formulas in the language of set theory. That is, for a natural enumeration $(\varphi_i : i \in \omega)$ of the Σ_n-formulas, $T_n(k, \rho)$ is true if and only if $\varphi_k(\rho)$.[4] This can be expressed by an*

[4] To simplify the notation, a finite sequence of parameters is encapsulated in a single parameter ρ, if φ has several free variables.

$\in$-formula $\exists x_0 \forall y_0 ... \exists x_n \forall y_n \psi_n^T(x_0, y_0, ..., x_n, y_n)$ with ψ^T quantifier-free; likewise, $\neg T_n$ can be expressed as $\forall v_0 \exists w_0 ... \exists v_n \forall w_n \neg \psi_n^T(v_0, w_0, ..., v_n, w_n)$, where $\{v_0, w_0, ..., v_n, w_n\}$ is disjoint from $\{x_0, y_0, ..., x_n, y_n\}$. Let Q, $Q_\neg$ be the quantifier sequences starting these formulas. By Σ_n-TND ("tertium non datur"), we denote the statement

$$\forall k \forall \rho \exists i \in \{0,1\} Q Q_\neg ((i = 1 \wedge \psi_n^T(k, \rho)) \vee (i = 0 \wedge \neg \psi_n^T(k, \rho))).$$

Remark 6. This list contains several axiom (schemes) of ZFC. It is easy to see that all ZFC-axioms not mentioned in this list are OTM-effective and can thus be disregarded in considerations about reducibility.

Effectivizers for Sep and Rep may not exist, since the straightforward way of defining them would involve a truth predicate for the universe; this would lead to an unwanted trivialization of reduction results. We can, however, still define halting computations relative to such effectivizers (and hence reductions) in the following way: Each halting computation will only require instances of Sep or Rep from some V_α. If we assume that there are arbitrarily large κ such that V_κ is an elementary submodel of V, then an encoding of such an effectivizer is always definable. The extension of ZFC by the statement that arbitrarily large such κ exist is known as Feferman set theory,[5] which is equiconsistent with ZFC (ibid.). We thus assume, in all results concerning Sep and Rep in this paper, Feferman set theory, which will be indicated by writing (Fef) at the beginning of the respective statement.

Remark 7. Clearly, HP is not OTM-effective, as an OTM-effectivizer would amount to a halting problem solver. It immediately follows that not even all logical tautologies are OTM-effective, so that every axiomatic system will, in classical logic, prove some non-OTM-effective statements, no matter the specific axioms. In particular, classical logic does not preserve effectiveness.

3.1 Realizability, Reducibility, and Truth

It is easy to see that every OTM-effective statement must be true, for the OTM-computability of witnesses implies a fortiori the existence of witnesses. We already saw that HP, although true on the basis of classical first-order logic, is not OTM-effective. Thus, we note:

Proposition 8. *OTM-effectivity is strictly stronger than classical truth.*

We note that, under $V = L$, truth is, however, equivalent to the existence of a *definable* (but not necessarily OTM-computable) effectivizer:

Proposition 9. *In ZFC+$V = L$, it is provable, for every $\in$-formula φ, that φ is equivalent to the existence of a definable effectivizer for φ.*

[5] See [6], p. 202.

Proof. This follows from the existence of definable Skolem functions in L, see, e.g., Schindler and Zeman [16].

We note that provability of the implication and OTM-realizability (defined in [4], Definition 17) are both different from OTM-effective reducibility:

Lemma 10. *1. There are $\in$-sentences φ and ψ such that $\psi \leq_{OTM} \varphi$, but $\varphi \rightarrow \psi$ is not provable in ZFC.*

2. There are $\in$-sentences φ and ψ such that $\varphi \rightarrow \psi$ is provable in ZFC, but $\psi \not\leq_{OTM} \varphi$.

3. The same, in fact, holds for every consistent $\in$-theory that is sufficiently strong to formalize the relevant statements; in particular, this includes extensions of KP.

4. There are statements φ and ψ in prenex normal form such that $\varphi \rightarrow \psi$ is OTM-realizable, but $\psi \not\leq_{OTM} \varphi$.

5. If φ and ψ are statements in prenex normal form such that $\psi \leq_{OTM} \varphi$, then $\varphi \rightarrow \psi$ is (uniformly) OTM-realizable.

Proof. 1. Let φ be HP and let ψ be $\perp$. Then $\psi \leq_{\text{OTM}} \varphi$ is vacuously true (since HP has no effectivizers), but, since φ is a logical tautology, $\varphi \rightarrow \psi$ is not provable in ZFC (unless ZFC is inconsistent).

2. An obvious example is $\top \rightarrow$HP.
3. The same arguments work in the more general case.
4. Let φ be NextCard (see Definition 4) and ψ be $\perp$. By [4], Lemma 41, NextCard is not OTM-realizable (since OTM-programs cannot output objects whose cardinality is larger than the length of their input and the size of their parameter), so $\varphi \rightarrow \psi$ is trivially OTM-realizable. But NextCard has the (obvious) effectivizer $\kappa \mapsto \kappa^+$, so we cannot have $\psi \leq_{\text{OTM}} \varphi$.
5. (Sketch) Let r be an OTM-realizer for φ. Then r induces, in an OTM-computable way, an OTM-computable effectivizer F for φ (note that here, the assumption that φ is in prenex normal form is crucially used). By definition of $\psi \leq_{\text{OTM}} \varphi$, we can compute an effectivizer G for ψ relative to F. By combining the reduction with the realizer for φ, we obtain a realizer for ψ.

4 Some Reducibility Results

Lemma 11. *1. For all $n \in \omega$, there is an OTM-program which, relative to any effectivizer $F_{n\text{-}Sep}$ for Σ_n-separation, decides truth of Σ_n-formulas.*

2. (Fef) There is an OTM-program which, relative to any effectivizer F_{Sep} for Sep, decides truth of arbitary $\in$-formulas.

Proof. Given a Σ_n-formula φ and a parameter $\boldsymbol{p}$, obtain $F_{n-Sep}(\ulcorner\varphi(\boldsymbol{p})\urcorner, \{\emptyset\})$, then return "true" if and only if the result is non-empty, and "false" otherwise.

Lemma 12. *$\leq_{OTM}$ is a reflexive, transitive order on the set of $\in$-sentences with minimal element $\top$ and maximal element $\perp$. Moreover, for any $\varphi <_{OTM} \perp$ (in prenex normal form), there is an $n \in \omega$ such that $\varphi <_{OTM} \Sigma_n$-TND.*

Proof. Trivial.

Proposition 13. *1. GreaterCard$\leq_{OTM}$NextCard.*
2. NextCard$\equiv_{OTM}$GreaterCard is independent of ZFC.
3. HP$\leq_{OTM}$GreaterCard.
4. If $V = L$, then HP$\equiv_{OTM}$GreaterCard
(In particular, NextCard and GreaterCard are not OTM-effective.)[6]

Proof. 1. Since the cardinal successor of an ordinal α is in particular a greater cardinal than α, any effectivizer for NextCard is also one for GreaterCard.

2. By (1), it only remains to consider NextCard$\leq_{\text{OTM}}$GreaterCard.
This is true in L: Let F_{gc} be an effectivizer for GreaterCard, and let α be an ordinal. Then $F((\alpha))$ will be a cardinal greater than α. Using standard techniques for OTMs (cf., e.g., [3], Lemma 3.5.3), we can then compute a code for $L_{F((\alpha))}$ from $F(\alpha)$. By basic fine-structure theory, an ordinal $\beta < F(\alpha)$ will be a cardinal in $L_{F((\alpha))}$ if and only if it is a cardinal in L. We can thus search through $L_{F((\alpha))}$, using the ability of OTMs to evaluate set-theoretic truth in a given structure (cf., e.g., Koepke, [13], Lemma 6.1 or [3], Theorem 2.3.28) to check whether $L_{F((\alpha))}$ contains a cardinal κ above α. If so, we return κ as the next cardinal after α; otherwise, we return $F((\alpha))$.
We now show how to obtain a model of ZFC which will satisfy the statement NextCard$\not\leq_{\text{OTM}}$GreaterCard. First, use Cohen-forcing to obtain a real number $g \subseteq \omega$ that is Cohen-generic over L. The idea is now to code g into the cardinal structure of M. To this end, consider a forcing extension M of L in which, for all $k \in \omega$, $(\aleph_{2k+1})^M = ((\aleph_{2k}^M)^+)^{L}$[7] if and only if $k \in g$, and otherwise, $(\aleph_{2k+1})^M = ((\aleph_k^M)^{++})^L$, while $((\aleph_{2k+1}^M)^+)^M = ((\aleph_{2k+1}^M)^+)^L$ for all $k \in \omega$. That is, the cardinal successor of $\aleph_{2k}^M$ in M is the next L-cardinal after $\aleph_{2k}^M$ if $k \in g$, and otherwise, it is the second-next, while all other cardinals remain as in L. Note that this implies $\aleph_k^M \leq \aleph_{2k}^L$ for all $k \in \omega$. Such an extension of L can, e.g., be obtained by Easton-forcing (see, e.g., [14], p. 262f). Let us assume for a contradiction that, in M, (P, ρ) effectively reduces NextCard to GreaterCard. Now, let F be the class function which sends an ordinal $\alpha \in [\aleph_k^L, \aleph_{k+1}^L)$ to $(\aleph_{2k+2})^L$ for $k \in \omega$ and all other ordinals α to $(\alpha^+)^L$. Then, in M, F is an effectivizer of GreaterCard. By definition, F is definable in L. Thus, if G is the effectivizer of NextCard obtained via (P, ρ) from F, then G is definable in L as well. Then, using the definition of NextCard, we can define in L the sequence $s := (\aleph_k^M : k \in \omega)$. But then, relative to s, g is definable (by definition of s), so we have $g \in L$, a contradiction. Hence $M \models$NextCard$\not\leq_{\text{OTM}}$GreaterCard, as desired.

3. It is well-known (see, e.g., the proof of [3], Lemma 8.6.3) that, for an OTM-program P, a parameter $\rho \in \text{On}$ and an infinite cardinal $\kappa > \rho$, $P(\rho)$ will either halt in less than κ many steps or it will not halt at all. Thus, given an

---

6 This statement, which amounts to the undecidability of the class of cardinals on an OTM, is well-known in the literature; the first explicit mention we know of is in Dawson [5].

7 Note that $((\aleph_{2k}^M)^+)^L$ is the cardinal successor of $\aleph_{2k}^M$ in L.

effectivizer F for GreaterCard, it suffices to run $P(\rho)$ for $\max\{F((\rho)), F((\omega))\}$ many steps to determine whether it will halt.

4. If $V = L$, consider the OTM-program $P_{\text{card-check}}$ that, in the input α, enumerates L,[8] searches for a bijection that collapses α to some smaller ordinal and halts once such an ordinal is found. This program will terminate if and only if α is a not cardinal in L. Thus, given access to an effectivizer of HP, an OTM can identify L-cardinals. Now, given some ordinal α, we can simply count upwards from α, checking each arising ordinal for being an L-cardinal, and halting once we have found one.

Lemma 14. *Let $n \in \omega$, and let a, b be sets. Suppose that P is an OTM-program such that, for every 1-effectivizer F of Σ_n-Sep and every code c_a for a, $P^F(c_a)$ computes a code for b. Then $b \in HOD(a)$.*

Proof. For the sake of simplicity, we only consider the unrelativized case where $a = \emptyset$. We first show that Σ_n-Sep has a definable encoded 1-effectivizer, for every $n \in \omega$. Assume that the conditions of the lemma are satisfied. We pick a particularly simply encoded effectivizer F, where, given codes for a set X and a parameter $\mathbf{p}$, the set $\bar{X} := \{x \in X : \varphi(x, \mathbf{p})\}$ is encoded by taking the elements of $\text{tc}(\bar{X})$ in the same order in which they appear in the encoding of X. We show by transfinite induction that, for any OTM-program P and any ordinal α, the partial computation of P^F of length α is contained in HOD. First, note that all such computations are clearly ordinal definable in the parameter α and thus elements of OD. It thus suffices to see that the *elements* of such a computation sequence belong to HOD. Any such element is an ordered pair $(\xi, (t, s, \rho))$, where ξ is an ordinal less than or equal to α (indicating the point in time), t is a map (or, if several tapes are used, a finite tuple of such) from some ordinal to $\{0, 1\}$ (indicating the tape content at time ξ), s is a natural number (indicating the inner state at time ξ) and ρ is an ordinal (indicating the head position at time ξ). The only component that is not obviously an element of HOD is t. However, $(\xi, (t, s, \rho))$ is clearly ordinal definable from the computation of length α in the parameter ξ, and t is definable as the first component of the second component of $(\xi, (t, s, \rho))$, so t belongs to OD; since the elements of t are ordered pairs of ordinals, it follows that $t \in \text{HOD}$, as required.

Concerning full separation, we obtain the following:

Theorem 15. *(Fef) The class S of sets that are OTM-computable relative to every 1-effectivizer of Sep is HOD.*

Proof. By using the same argument as for Lemma 14 within (Fef), we see that $S \subseteq \text{HOD}$.

Now let $x \in \text{HOD}$. Since $\text{HOD} \models \text{ZFC}$, HOD will also contain a set $c_x \subseteq \alpha \in \text{On}$ which codes x, for some $\alpha \in \text{On}$. By definition of HOD, there is an $\in$-formula φ and some $\rho \in \text{On}$ such that c_x is the unique y with $\varphi(y, \rho)$. Let F be an effectivizer for Sep. Now, using the parameters α and ρ, apply Sep to α with the formula $\psi(\iota) :\Leftrightarrow \exists y(\varphi(y, \rho) \wedge \iota \in y)$. The result will be c_x. Thus, a code for x is OTM-computable relative to F.

[8] Such a program exists by the work of Koepke; see, e.g., [13].

Remark 16. 1. In fact, the same argument, combined with the well-known fact that every ordinal definable set is already ordinal definable with a Σ_2-definition (see, e.g., [8]) yields that one can replace Sep with Σ_2-Sep in Theorem 15 (which will no longer require Fef).

2. It is somewhat suggestive that, while plain OTM-computability with ordinal parameters yields all sets in L (Koepke, [13]), OTM-computability relative to an effectivizer for separation yields HOD, which can be obtained by replacing first-order logic with second-order logic in the definition of L (see [12]). We do not know a direct connection between the two results, nor whether other of the "alternative Ls" described in [12] can be given a computable characterization.

Theorem 17. *1. NextCard$\leq_{OTM}$Pot*

2. If $V = L$ holds, then NextCard$\equiv_{OTM}$Pot.

3. In general, Pot$\leq_{OTM}$NextCard is independent of ZFC.

4. Pot$\leq^{1,1}_{OTM}$ Σ_3-Rep.[9] Moreover, we have $\varphi \leq^{n,1}_{OTM}$Rep for every true $\in$-formula whose prenex normal form can be written in the form $\forall x_0 \exists! y_0 ... \forall x_n \exists! y_n \psi(x_0, y_0, ..., x_n, y_n, p)$.

5. NextCard$\leq_{OTM}$ Σ_1-Sep.

6. For all $n \geq 1$, Pot$\leq^{1,1}_{OTM}$ Σ_n-Sep is independent of ZFC. (Fef) Pot$\leq^{1,1}_{OTM}$Sep is independent of ZFC.

Proof. 1. Let F be an effectivizer for Pot, and let α be an ordinal. Then we obtain the cardinal successor of α from F as follows: We have $F((\alpha)) = \mathfrak{P}(\alpha)$. An OTM can check whether a set of ordinals codes a well-ordering.[10] We use this to compute the set of all codes for well-orderings in $\mathfrak{P}(\alpha)$ – which will contain codes for all ordinals below α^+, and from this set, compute a code for their sum, which will be α^+.

2. One direction follows from (1). We consider the other direction: In L, we have $\mathfrak{P}(\alpha) = \mathfrak{P}^{L[\alpha^+]}(\alpha)$. Hence, we can compute $(\alpha^+)^L$ using an effectivizer for NextCard, then use that to compute a code for $L_{(\alpha^+)^L}$ and then extract all subsets of α contained in it.

3. By (2), Pot$\leq_{\mathrm{OTM}}$NextCard is consistent relative to ZFC. For the other direction, let $L[g]$ be a generic extension of L by one Cohen real $g \subseteq \omega$. Suppose for a contradiction that P is an OTM-program that, in the parameter $\rho \in \mathrm{On}$, reduces Pot to NextCard in $L[g]$. Let F be an effectivizer of NextCard in $L[g]$. Then F is definable in $L[g]$, and, since $L[g]$ has the same cardinals as L, F is also definable in L. By absoluteness of computations, the computation of $P^F(\omega, \rho)$ is the same in L and in $L[g]$. It follows that $g \in \mathfrak{P}^{L[g]}(\omega) \in L$, a contradiction.

[9] In fact, we have Pot$\leq_{\mathrm{OTM}}$ Π_2-Rep, for the obvious definition of the latter. Note, however, that this cannot be simplified further, since we require our formulas to be in prenex normal form, and expressing Pot requires bounded quantifiers on top of the initial unbounded universal quantifier.

[10] See, e.g., [3], Theorem 2.3.25; the argument is essentially the same given by Hamkins and Lewis for subsets of ω, see [9], Theorem 2.2.

4. Let F_{Rep} be an effectivizer for the replacement scheme. Let φ be a formula with a prenex normal form of the given shape. For any $k \leq n$, let $\varphi^k(x_0, y_0, ..., x_k, y_k, p)$ be the formula
$$\forall x_{k+1} \exists y_{k+1} ... \forall x_n \exists y_n \psi(x_0, y_0, ..., x_n, y_n, p).$$
Then, for any $k \leq n$ and any tuple $(a_0, ..., a_k)$, let $b_0, ..., b_{k-1}$ be the unique witnesses fort he first $(k-1)$ existential quantifiers when $x_0, ..., x_{k-1}$ are instantiated with $a_0, ..., a_{k-1}$; by induction, these can be obtained from F_{Rep}. Then
$$F_{\text{Rep}}(\ulcorner \varphi^k \urcorner, p, \{(a_0, b_0, ..., a_{k-1}, b_{k-1}, a_k)\})$$
will yield (a code for) the respective $\{y_k\}$, from which y_k can be computed. This reduces φ to Rep.
This argument then applies to Pot, noting that $y = \mathfrak{P}(x)$ can be expressed by a Π_2-formula in prenex normal form.
5. Using the fact that an effectivizer for Σ_n-Sep allows us to evaluate the truth of Σ_n-statements (and the fact that "α is not an ordinal" is Σ_1), we can, for a given $\alpha \in \text{On}$, successively count through the ordinals, checking for each ordinal β whether $\beta > \alpha$ and β is a cardinal. The first ordinal for which this holds true is NextCard(α).
6. First suppose that $V = L$. For a given (encoded) set x, run the OTM-program that enumerates L until some $L_\alpha \ni x$ is found. Then use (5) to obtain the cardinal successor κ of α in L from an effectivizer for Σ_1-Sep. Then compute $\mathfrak{P}^{L_\kappa}(x)$, which will be $\mathfrak{P}^L(x)$. Thus, in L, we have Pot $\leq^{1,1}_{\text{OTM}} \Sigma_1$-Sep.
On the other hand, let V be such that $V \neq \text{HOD}$.[11] Then there must be some set $x \in \text{HOD}$ such that $\mathfrak{P}(x) \neq \mathfrak{P}^{\text{HOD}}(x)$ (for example, there must be some minimal ordinal α such that $V_\alpha \notin \text{HOD}$). Now, if P was a program reducing Pot to Σ_n-Sep for some $n \in \omega$, then, for any effectivizer F of Σ_n-Sep, we would have the result of the computation $P^F(x) \in \text{HOD}$ by Lemma 14. But then, we would have $\mathfrak{P}(x) \in \text{HOD}$, a contradiction. Thus, it is consistent with ZFC that Pot$\not\leq_{\text{OTM}} \Sigma_n$-Sep, for all $n \in \omega$.

Lemma 18. *1. Sep$\leq_{OTM}$Rep*[12]
2. Rep$\leq_{OTM}$Sep$\oplus$Pot
3. Rep$\equiv_{OTM}$Coll; moreover, Σ_n-Rep$\leq_{OTM}$ Σ_n-Coll, for all $n \in \omega$.
4. Σ_n-Coll$\leq_{OTM}$ Σ_n-Sep$\oplus$Pot, for all $n \in \omega$.
5. Coll$\leq_{OTM}$Sep$\oplus$Pot
6. If $V = L$, then Σ_n-Rep$\equiv_{OTM}$ Σ_n-Sep, for all $n \in \omega$. (Fef) Thus, also Rep$\equiv_{OTM}$Sep.
7. Σ_3-Rep$\leq^{1,1}_{OTM}$Sep is independent of ZFC.

[11] For example, one can pick $V := L[x]$, where x is Cohen-generic over L. Since Cohen forcing is weakly homogenous (and definable without parameters in L), we then have $\text{HOD}^{L[x]} \subseteq \text{HOD}^L = L$ (see, e.g., [17], p. 2).
[12] This result was briefly mentioned in [2], section 6.

Proof. 1. Given an $\in$-formula $\varphi(x, \boldsymbol{z})$, a (coded) parameter $\boldsymbol{p}$, and a (code for a) set X, and an effectivizer F for the replacement scheme, let $\psi(x, y, \boldsymbol{z})$ be the $\in$-formula $(\varphi(x, \boldsymbol{z}) \wedge y = x) \vee (\neg\varphi(x, \boldsymbol{z}) \wedge y = X)$, use F to obtain $Y' := F(\ulcorner\psi\urcorner, \boldsymbol{z}, X)$ and then compute Y from Y' by deleting X. Since OTMs can test whether two sets of ordinals code the same set, the final step can be done on an OTM.

2. Let F_{Sep} and F_{Pot} be effectivizers for the separation scheme and the power set axiom, respectively, let X be a set and let $\varphi(x, y, \boldsymbol{z})$ be an $\in$-formula and $\boldsymbol{p}$ a parameter such that φ is functional on X (i.e., such that $\forall x \in X \exists! y \varphi(x, y, \boldsymbol{z})$) for $\boldsymbol{z} = \boldsymbol{p}$. By Proposition 11, we can obtain a truth predicate from F_{Sep}. Now use F_{Pot} iteratively, starting with $\emptyset$, applying F_{Pot} at sucessor levels and forming unions at limit levels (forming unions is computable on an OTM) to successively obtain the stages V_ι of the von Neumann hierarchy.[13] Use the truth predicate to test, for each of these levels, whether it satisfies $\forall x \in X(\exists y \varphi(x, y, \boldsymbol{p}) \rightarrow \exists y \in V_\iota \varphi(x, y, \boldsymbol{p}))$. By assumption, such a level will eventually be found. Now use $F_{\text{Sep}}(\ulcorner\exists x \varphi(x, y, \boldsymbol{p}\urcorner, V_\iota)$ to obtain the desired set.

3. For all $n \in \omega$, Σ_n-Rep$\leq_{\text{OTM}}$ Σ_n-Coll, and hence also Rep$\leq_{\text{OTM}}$Coll, are trivial, as every instance of replacement is also an instance of collection.
For the other direction, we use an effective version of the usual argument that replacement implies collection in ZF: Let an $\in$-formula φ, a parameter $\boldsymbol{p}$ and a set X be given such that $\forall x \in X \exists y \varphi(x, y, \boldsymbol{p})$. Consider the formula $\varphi'(x, \alpha, Y, \boldsymbol{p})$, which states that $\alpha \in \text{On}$ is minimal with the property that $\exists y \in V_\alpha \varphi(x, y, \boldsymbol{p})$ and that Y is the set of all $y \in V_\alpha$ that satisfy $\varphi(x, y, \boldsymbol{p})$. Then $\varphi''(x, Y, \boldsymbol{p}) :\Leftrightarrow \exists \alpha \varphi'(x, \alpha, Y, \boldsymbol{p})$ defines a function from X to V. Applying an effectivizer for F_{Rep} to the set X for φ'' in the parameter $\boldsymbol{p}$, we obtain a set Z such that $\bigcup Z$ will be the set required to exist by the given instance of Coll.

4. We use a similar idea as in (3), using an additional idea, which is basically an effective version of Scott's trick: Let a Σ_n-formula φ, a parameter $\boldsymbol{p}$ and a set X be given. For each $x \in X$, we again use F_{Pot} to enumerate the universe until we arrive at some V_α such that $\exists y \in V_\alpha \varphi(x, y, \boldsymbol{p})$ (which can be identified by searching through each occurring V-level and testing each element using $F_{\Sigma_n\text{-Sep}}$). Now, again using $F_{\Sigma_n\text{-Sep}}$, compute the set $S_x := \{y \in V_\alpha : \varphi(x, y, \boldsymbol{p}\}$ and store the result. Once this has been done for all $x \in X$, compute $\bigcup_{x \in X} S_x$, which will be as desired.

5. This is an immediate consequence of (2), (3) and the transitivity of $\leq_{\text{OTM}}$.

6. Let $n \in \omega$, $\boldsymbol{p}$ a parameter, and let $\varphi(x, y, z)$ be a Σ_n-formula which is functional for $z = \boldsymbol{p}$. Moreover, let an effectivizer F for Σ_n-Sep be given, along with a (code for a) set X. Recall that, using F, we can evaluate the truth of Σ_n-formulas. The required OTM-program now works by enumerating L, and, for each $Y \in L$, performing the following:

[13] OTMs with access to a power set operator ("Power-OTMs") are equivalent to the (full) Set Register Machines introduced and studied by Passmann, [15]; their ability to enumerate the V-hierarchy is essentially [15], Proposition 3.10.

(a) Running through X, and, for each $x \in X$, running through Y and searching for some $y \in Y$ such that $\varphi(x, y, \boldsymbol{p})$. If, for some $x \in X$, no such y is found, the search through L is continued.
(b) Running through Y, and, for each $y \in Y$, running through X and searching for some $x \in X$ such that $\varphi(x, y, \boldsymbol{p})$. If, for some $y \in Y$, no such x is found, the search through L is continued.

Since L is a model of replacement, the search will eventually reveal a set Y for which both checks are successful; this will then be returned as the output. Since the procedure works uniformly in φ, it also witnesses Rep$\leq_{\text{OTM}}$Sep in general, provided the relevant computations are defined.

7. The relative consistency follows from (6). For the converse, we have already seen that Pot$\leq_{\text{OTM}}$ Σ_3-Rep. If additionally Σ_3-Rep$\leq_{\text{OTM}}$ Σ_n-Sep for any $n \in \omega$, then it follows by transitivity of $\leq_{\text{OTM}}$ that Pot$\leq_{\text{OTM}}$ Σ_n-Sep. Thus, any model of ZFC where the latter fails is also one where Σ_3-Rep$\not\leq_{\text{OTM}}$ Σ_n-Sep, for all $n \in \omega$. But the existence of such models was already seen in the proof of Theorem 17(6).

Given that Rep$\leq_{\text{OTM}}$Sep$\oplus$Pot, one may ask how far Pot can be weakened in this statement. We note that NextCard is not enough:

Lemma 19. *It is consistent with ZFC that:*

1. *Pot$\not\leq^{1,1}_{OTM}$Sep$\oplus$NextCard*
2. *Σ_3-Rep$\not\leq^{1,1}_{OTM}$Sep$\oplus$NextCard*

Proof. Let $V = L[g]$, where $g \subseteq \omega$ is Cohen-generic over L. Then it follows that NextCard=NextCard^L. As in Theorem 15 above, it follows that every set computable by an OTM relative to every Sep$\oplus$NextCard-effectivizer is an element of $\text{HOD}^{L[g]} = L$. However, we have $g \notin L$, but $g \in \mathfrak{P}(\omega)$, so $\mathfrak{P}(\omega)$ is not OTM-computable from Sep$\oplus$NextCard. Thus, in $L[g]$, Pot$\not\leq_{\text{OTM}}$Sep$\oplus$NextCard, for, if P and ρ were a program and a parameter witnessing the reduction, then we could compute $\mathfrak{P}(\omega)$ from Sep$\oplus$NextCard by first computing ω and then applying Pot to it. Since Pot$\leq_{\text{OTM}}$ Σ_3-Rep, it also follows that Σ_3-Rep$\not\leq_{\text{OTM}}$Sep$\oplus$NextCard in $L[g]$.

4.1 The Axiom of Choice

Finally, we have a look at the axiom of choice. The mutual effective reducibility of various versions of the axiom of choice was already considered in [2], where it was shown (ibid., Proposition 14) that the usual equivalent formulations of this axiom are also mutually OTM-reducible to each other. One might think that AC$\leq_{\text{OTM}}$Pot for the following reason: If X is a set of non-empty sets, then $\mathfrak{P}(X \times \bigcup X)$ will contain a choice function for X. Thus, an OTM with access to the power set operator can, uniformly in X, compute a set S containing a choice function for X; and then one can search through S, identify a choice function and write it to the output. Unfortunately, S will, except in trivial cases, contain not only one, but many choice functions, and choosing one of these is as hard as choosing from the elements of X in the first place.

Definition 5. *A definable effectivizer for an $\in$-formula φ is an effectivizer for φ that is definable as a class function in the language of ZFC (i.e., using the binary relation symbol $\in$ alone) and such that ZFC proves that F is in fact an effectivizer for φ.*

Lemma 20. *Let φ and ψ be $\in$-formulas. Suppose that ψ has a definable effectivizer, and that φ is ZFC-provably OTM-reducible to ψ. Then φ has a definable effectivizer.*

Proof. (Note that the statement is not immediate, since the program that implements the reduction requires an encoded effectivizer for ψ, rather than merely an effectivizer. In the absence of a definable global well-ordering, this may fail.)

Suppose that P is an OTM-program that ZFC-provably reduces φ to ψ, let F be a definable effectivizer for ψ, and let M be a model of ZFC. Suppose that δ_F is an $\in$-formula and $\boldsymbol{p} \in M$ is a parameter such that $F = \{(a,b) \in M : \delta_F(a,b,\boldsymbol{p})\}$. Over M, consider the class forcing $(\mathbb{P}, \preceq)$ for obtaining a global well-ordering of M.[14] Let $M[G]$ be the extension. Then $M[G]$ is a model of ZFC. Moreover, since this class forcing does not add any new sets[15] (see [11], Example 3.2 and Lemma 3.5), the defining formula δ_F of F yields the same function in the parameter $\boldsymbol{p}$ in $M[G]$ as in M. However, in $M[G]$, we can use the global well-ordering to encode F as $\hat{F}$. Now, for any $x \in M$ and any code c_x for x, $P^{\hat{F}}(c_x)$ will by assumption terminate and output a code c_y for some set y such that y depends only on x (and not on c_x) and such that the function mapping x to y is an effectivizer for φ. Since the computation $P^{\hat{F}}(c_x)$ halts, it can only use set-many instances of $\hat{F}$. Thus, there is some $\hat{f} \in M$ such that $P^{\hat{f}}(c_x)$ halts without requiring instances of $\hat{F}$ other than those encoded in $\hat{f}$ with output c_y.

We can thus define (in M) an effectivizer F_φ for φ as follows: $F_\varphi(x) = y$ if and only if, for every code c_x for x and every encoded function $\hat{f}$ which satisfies $\forall a, b(f(a) = b \rightarrow \delta_F(a,b,\boldsymbol{p}))$, $P^{\hat{f}}(c_x)$ halts without requiring values of the oracle function not encoded in $\hat{f}$ and its output is a code for y. Note that these encodings $\hat{f}$ are guaranteed to exist for every set-sized function f because AC holds in M.)

This suffices to clarify the ZFC-provable OTM-reduction relations between AC and the other non-effective axioms of ZFC:

Theorem 21. *1. AC is not ZFC-provably $(1,1)$-OTM-reducible to Σ_n-Sep, for all $n \in \omega$.*
2. AC is not ZFC-provably $(1,1)$-OTM-reducible to Σ_n-Rep, for all $n \in \omega$.
3. AC is not ZFC-provably OTM-reducible to Pot.

Proof. Since Pot, Σ_n-Sep and Σ_n-Rep all have definable effectivizers, AC being ZFC-provably OTM-reducible to any of these would, by Lemma 20, imply the

[14] That is, $\mathbb{P}$ consists of all well-orderings in M and $\preceq$ is the end-extension relation. See, e.g., [7].

[15] We thank Philipp Schlicht for pointing us to this reference.

existence of a definable global choice function. Thus, any transitive class model M of ZFC that does not have a definable global choice function[16] is a counterexample to the reduction.

Remark 22. We remark, however, that AC$\leq_{\mathrm{OTM}}$ φ is consistent with ZFC for every formula φ, since, if $V = L$, then AC is in fact OTM-effective ([2], Proposition 3) and thus trivially reducible to any formula.

Proposition 23. *None of Pot, Σ_1-Sep and Σ_1-Rep is OTM-reducible to AC.*

Proof. Pot$\not\leq_{\mathrm{OTM}}$AC follows from [2], Lemma 7, combined with the fact that effectivizers for Pot increase cardinalities of the transitive closure of sets, while effectivizers for AC do not.

If we had Σ_1 $\leq_{\mathrm{OTM}}$AC in any model M of ZFC, we would, in particular, have it in L^M (since L^M is Σ_1-definable in M); however, this would in fact make Σ_1-Sep OTM-effective for L^M (since L^M has an OTM-computable global well-ordering). But this cannot be, for HP$\leq_{\mathrm{OTM}}$ Σ_1-Sep.

A fortiori, since Σ_1-Sep$\leq_{\mathrm{OTM}}$ Σ_1-Rep, we also have Σ_1-Rep$\not\leq_{\mathrm{OTM}}$AC.

5 Conclusion and Further Work

The above investigations can, of course, be extended to further set-theoretical statements.

We may also ask about the overall structure of the ordering induced on the set of set-theoretical formulas by $\leq_{\mathrm{OTM}}$. Is it linear? Is it dense? Which elements e (if any) can be "decomposed" in the sense that $e \equiv_{\mathrm{OTM}} a \oplus b$ where $a, b <_{\mathrm{OTM}} e$?

The statement HP can be regarded as an analogue of the Turing jump in the context of OTM-effective reducibility. In general, degree-theoretical questions can be asked about set-theoretical statements. We regard the following one as particularly interesting:

Question 24. Is there a "natural" set-theoretical statement φ such that $\top <_{\mathrm{OTM}} \varphi <_{\mathrm{OTM}} HP$?

As already mentioned above, the notion of OTM-reducibility can also be applied to formulas with infinitely many quantifier alternations. In particular, one can define an effectivizer for the axiom of determinacy, AD, as a map $F : \mathfrak{P}(\omega^\omega) \times \omega^{<\omega} \to \omega$ such that, for all $X \subseteq \omega^\omega$, $s \mapsto F(X, s)$ is a winning strategy for X.

Question 25. Which of the ZF-provable consequences of AD are OTM-reducible to AD, or to the formula expressing the existence of a winning strategy for the game used in the proof of the respective implication?

Acknowledgements. We thank our *five* referees for many valuable comments on the presentation, and for suggesting the simplifications of the proofs of Theorems 15 and 17(3), which are now implemented in this paper.

[16] By [8], this is the case in any model of ZFC+V $\neq$ HOD, such as a Cohen extension of L.

References

1. Carl, M.: Generalized effective reducibility. In: Beckmann, A., et al. (eds.) Pursuit of the Universal. 12th Conference on Computability in Europe (2016)
2. Carl, M.: Effectivity and reducibility with ordinal turing machines. Computability **10**(4) (2021)
3. Carl, M.: Ordinal Computability. An Introduction to Infinitary Machines. De Gruyter, Berlin (2019)
4. Carl, M., Galeotti, L., Passmann, R.: Realisability for infinitary intuitionistic set theory. Ann. Pure Appl. Log. **174**(6) (2023)
5. Dawson, B.: Ordinal time turing computation. Ph.D. thesis, University of Bristol (2010)
6. Feferman, S.: Set-theoretical foundations of category theory. In: MacLane, S. (ed.) Reports of the Midwest Category Seminar III. Springer, Heidelberg (1969). file://C:/Users/DOZMATCAM/Downloads/BFb0059139.pdf
7. Gitman, V.: Forcing to add proper classes to a model of GBC: the technicalities. Blog Post. https://victoriagitman.github.io/research/2013/10/12/forcing-to-add-proper-classes-to-a-model-of-gbc-the-technicalities.html. Accessed 11 Apr 2025
8. Hamkins, J.: Complexity of definable global choice functions. https://mathoverflow.net/users/1946/joel-david-hamkins, https://mathoverflow.net/q/472862. Accessed 08 June 2024
9. Hamkins, J., Lewis, A.: Infinite time turing machines. J. Symb. Logic **65**(2) (2000)
10. Hirschfeldt, D.: Slicing the Truth. On the Computable and Reverse Mathematics of Combinatorial Principles. World Scientific Publishing (2015)
11. Holy, P., Krapf, R., Schlicht, P.: Sufficient conditions for the forcing theorem, and turning proper classes into sets. Fundam. Math. **246** (2019)
12. Kennedy, J., Magidor, M., Väänänen, J.: Inner models from extended logics. J. Math. Log. **21**(2) (2021)
13. Koepke, P.: Turing computations on ordinals. Bull. Symb. Log. **11**(3) (2005)
14. Kunen, K.: Set Theory. An Introduction to Independence Proofs. Elsevier, Amsterdam (1980)
15. Passmann, R.: The first-order logic of CZF is intuitionistic first-order logic. J. Symb. Log. **89**(1) (2022)
16. Schindler, R., Zeman, M.: Fine structure. In: Foreman, M., Kanamori, A. (eds.) Handbook of Set Theory, pp. 605–656. Springer (2010)
17. Woodin, H., Davis, J., Rodríguez, D.: The HOD dichotomy. In: Cummings, J., et al. (eds.) Appalachian Set Theory: 2006-2012. London Mathematical Society Lecture Note Series. Cambridge University Press (2012)

On S-Degrees of Some Representations of Irrational Numbers

Ivan Georgiev[1] and Lars Kristiansen[2,3](✉)

[1] Department of Mathematical Logic and its Applications, Faculty of Mathematics and Informatics, Sofia University "St. Kliment Ohridski", Sofia, Bulgaria
ivandg@fmi.uni-sofia.bg
[2] Department of Mathematics, University of Oslo, Oslo, Norway
larsk@math.uio.no
[3] Department of Informatics, University of Oslo, Oslo, Norway

Abstract. We study degrees of representations of irrational numbers. We explore a novel general idea to produce new representations using difference with respect to base expansions.

Keywords: representations of irrational numbers · subrecursive functions and degrees · computational complexity · computable analysis

1 Introduction

Through the years, mathematicians have developed various ways to represent irrational numbers. Their work has not been in vain: We need a wide variety of representations as a representation that works well in one context, might not work very well, or even fail spectacularly, in another context. No representation seems to be superior in the sense that it works well in every context.

Our daily life standard representation of irrational numbers is base-10 expansions ($\pi = 3.14159\ldots$). Base-2 expansions ($\pi = 11.001001\ldots$) will of course in many respects work as well as base-10 expansions, but not in every respect, e.g., they are visually hard to read. But neither base-10 nor base-2 expansions do work at all in computable analysis and exact real number computations. In computable analysis Weihrauch's representation by nested intervals is frequently used, see [12]. Clock makers in the 19th century found it convenient to use continued fractions and other representations related to the Stern-Brocot tree, see [1]; nested intervals would not have been of much use to them. In studies of the foundation of mathematics Cauchy sequences and Dedekind cuts tend to be useful and convenient representations of irrational numbers.

Intuitively, a representation R_1 is *subrecursive in* another representation R_2, written $R_1 \preceq_S R_2$, if we can convert an R_2-representation of an irrational $\alpha \in (0,1)$ into an R_1-representation of α without undertaking unbounded search. The S-degrees are the equivalence classes induced on the representations of the irrational numbers by the reducibility relation $\preceq_S$. We use $\mathcal{S}$ to denote the corresponding degree structure.

© The Author(s), under exclusive license to Springer Nature Switzerland AG 2025
A. Beckmann et al. (Eds.): CiE 2025, LNCS 15764, pp. 222–236, 2025.
https://doi.org/10.1007/978-3-031-95908-0_16

The study of S-degrees was initiated in Kristiansen [8,9] and followed up in, e.g., Georgiev et al. [6], Kristiansen [10], Georgiev [4,5] and Hiroshima & Kawamura [7]. A full formal definition of the degree structure $\mathcal{S}$ appeared for the first time in Ben-Amram et al. [2], and Ben-Amram & Kristiansen [3] have proven that $\mathcal{S}$ is a lattice with a maximum and a minimum element: The maximum degree is the degree of the representation by continued fractions; the minimum degree is the degree of the representation by Weihrauch intersections (the representation by nested intervals belongs to this degree).

The relationship between the degrees of a number of prominent representations was established in the early papers [8] and [9]: Let $\mathbf{0}$ denote the degree of the representation by Weihrauch intersections, and let $\mathbf{1}$ denote the degree of the representation by continued fractions (thus we have $\mathbf{0} \leq \mathbf{a} \leq \mathbf{1}$ for any degree $\mathbf{a}$), furthermore, let $\mathbf{a}_C$ denote the degree of the representation by Cauchy sequences; $\mathbf{a}_{Eb}$ denote the degree of the representation by base-b expansions (for any natural number $b \geq 2$); $\mathbf{a}_D$ denote the degree of the representation by Dedekind cuts. Then, (for any $b \geq 2$) we have

$$\mathbf{0} < \mathbf{a}_C < \mathbf{a}_{Eb} < \mathbf{a}_D < \mathbf{1}$$

moreover, we have $\mathbf{a}_{Ea} \leq \mathbf{a}_{Eb}$ iff every prime that divides a also divides b (thus, we have, e.g., $\mathbf{a}_{E2} < \mathbf{a}_{E10}$ and $\mathbf{a}_{E2} \not\leq \mathbf{a}_{E3}$ and $\mathbf{a}_{E3} \not\leq \mathbf{a}_{E2}$). In the more recent paper [6], it was proved that the degrees $\mathbf{a}_L$ and $\mathbf{a}_R$ of the representations by left and right, respectively, best approximations are incomparable to each other ($\mathbf{a}_L \not\leq \mathbf{a}_R$ and $\mathbf{a}_R \not\leq \mathbf{a}_L$), and moreover, lie strictly between $\mathbf{a}_D$ and $\mathbf{1}$ (left and right best approximations are representations related to the Stern-Brocot tree). It was also proved that $\mathbf{1}$ is the least upper bound of the two degrees. For more on the degree structure and the degrees of a number of natural representations, see [2].

In the current paper we study some new and pretty natural representations which are due to the authors. These representations yield interesting degrees. A class of representations, which we refer to as *difference representations*, yields degrees that lie closer to the maximum degree $\mathbf{1}$ than any other natural degree known so far: We find degrees $\mathbf{b}_L$ and $\mathbf{b}_R$ such that $\mathbf{a}_L < \mathbf{b}_L < \mathbf{1}$ and $\mathbf{a}_R < \mathbf{b}_R < \mathbf{1}$ (these two degrees will be incomparable as $\mathbf{1}$ is the least upper bound of $\mathbf{a}_L$ and $\mathbf{a}_R$). We also study the representation by expansions in base-3 with digits $\{-1, 0, 1\}$, and how the degree of this representation relates to degrees of the standard base-b expansions with digits $\{0, 1, \ldots, b-1\}$.

2 Preliminaries

The *Dedekind cut* of an irrational number α is the function $\alpha : \mathbb{Q} \rightarrow \{0, 1\}$ where $\alpha(q) = 0$ if $q < \alpha$; and $\alpha(q) = 1$ if $q > \alpha$. Each irrational number has a unique Dedekind cut, and we will identify an irrational number α with its Dedekind cut $\alpha : \mathbb{Q} \rightarrow \{0, 1\}$. Furthermore, we will work with (parameterized function) oracle Turing machines, and we use Φ_M^f to denote the function computed by the Turing machine M using the function f as an oracle.

Definition 1. *Let R be a class of functions with countable domain and countable co-domain. Then, R is a* representation (of the irrational numbers) *if*

1. *There exists an oracle Turing machine M with the following property: For every $f \in R$ there exists an irrational α in the interval $(0,1)$ such that $\alpha = \Phi_M^f$. When $\alpha = \Phi_M^f$, we say that f* represents *α and that f is an* R-representation *of α.*
2. *There exists an oracle Turing machine N with the following property: For every irrational α in the interval $(0,1)$ there exists an R-representation f of α such that $f = \Phi_N^\alpha$.*

We say that an oracle Turing machine M converts an R_1-representation into an R_2-representation if for any $f \in R_1$ representing α there exists $g \in R_2$ representing α such that $g = \Phi_M^f$.

To put it more informally (and perhaps more simply), a class of functions is a representation if it is uniformly computably equivalent to the class of Dedekind cuts of irrationals in the interval $(0,1)$. See the introductory sections of [2] and [3] for examples and further explanations of Definition 1.

The transition relation of a Turing machine M with a function-oracle f is defined as for standard Turing machines, except for the query state q_q: If M enters state q_q with the word x on its query tape, then (i) the contents of the query tape are instantaneously changed to $f(x)$, (ii) the query-tape head is reset to the origin, while other heads do not move, and (iii) M moves to state q_a. The *time complexity* of a function-oracle machine is counted as for standard Turing machines, with the transition between q_q and q_a taking $|f(x)|$ time steps (where $|f(x)|$ denotes the length of the representation of $f(x)$).

A function $t : \mathbb{N} \longrightarrow \mathbb{N}$ is a *time bound* if (i) $n \leq t(n)$, (ii) $t(n) \leq t(n+1)$ and (iii) t is time-constructible: there is a single-tape Turing machine that, on input 1^n, computes the binary expansion of $t(n)$ in $\Theta(t(n))$ steps (all natural running-time bounds on Turing machines will be time bounds).

Definition 2. *Let t be a time bound, and let R be a representation. Then, $O(t)_R$ denotes the class of all irrational α in the interval $(0,1)$ such that at least one R-representation of α is computable by a Turing machine running in time $O(t(n))$ (where n is the length of the input).*

Let R_1 and R_2 be representations. The relation $R_1 \preceq_S R_2$ holds if for any time-bound t there exists a time-bound s such that $O(t)_{R_2} \subseteq O(s)_{R_1}$.

Intuitively, if an oracle Turing machine can convert an R_2-representation into an R_1-representation without undertaking unbounded search, the relation $R_1 \preceq_S R_2$ will hold. Algorithms (conversions, computations, etc.) that do not perform unbounded search will be referred to as *subrecursive* algorithms (conversions, computations, etc.), in general, the word *subrecursive* signifies *absence of unbounded search.* This terminology might not be standard, but it will be very convenient.

If $R_1 \preceq_S R_2$, we will say that the representation R_1 is *subrecursive in* or *S-reducible to* the representation R_2. The relation $R_1 \equiv_S R_2$ holds iff $R_1 \preceq_S R_2$ and

$R_2 \preceq_S R_1$. If $R_1 \equiv_S R_2$, we will say that the representation R_1 is *subrecursively equivalent* or *S-equivalent* to the representation R_2. If $R_1 \not\preceq_S R_2$ and $R_2 \not\preceq_S R_1$, we will say that the representation R_1 is *subrecursively incomparable* or *S-incomparable* to the representation R_2. The relation $R_1 \prec_S R_2$ holds iff $R_1 \preceq_S R_2$ and $R_2 \not\preceq_S R_1$.

We define the *S-degree* of the representation R, denoted $\deg_S(R)$, as the equivalence class given by $\deg_S(R) = \{Q \mid Q \equiv_S R\}$. The set of all S-degrees, denoted $\mathcal{S}$, is given by $\mathcal{S} = \{\deg_S(R) \mid R \text{ is a representation}\}$.

We will use the theory of honest functions when we need to prove that the relation $\preceq_S$ does not hold.

We use $\mathcal{E}(\psi)$ to denote the class of functions *elementary in* ψ, that is, the class of functions generated from the initial functions ψ, 2^x, max, 0, S (successor), I_i^n (projections) by composition and bounded primitive recursion. We use $\mathcal{E}$ to denote the class of functions elementary in the constant function 0, that is, $\mathcal{E} = \mathcal{E}(0)$ (thus, $\mathcal{E}$ is the class of Kalmár elementary functions), and we refer to the functions in $\mathcal{E}$ as the *elementary functions*.

We may say that an algorithm (computation, conversion, operation, etc.) is *elementary*. Formally, this means that the algorithm (computation, conversion, operation, etc.) can be implemented by a Turing machine running in time $O(f)$ for some $f \in \mathcal{E}$. If a function is computed by an elementary algorithm, then the function is of course elementary.

A function $f : \mathbb{N} \to \mathbb{N}$ is *honest* if (1) $f(x) \leq f(x+1)$, (2) $f(x) \geq 2^x$ and (3) f has elementary graph, that is, the relation $f(x) = y$ is elementary. For any honest function f, we define *the jump of* f, written f', by $f'(x) = f^{x+1}(x)$. An honest function f enjoys the following properties: (i) for any $\psi \in \mathcal{E}(f)$, there exists $k \in \mathbb{N}$ such that $\psi(\vec{x}) \leq f^k(\max(\vec{x}))$; (ii) the jump f' is an honest function; (iii) for any unary function $\psi \in \mathcal{E}(f)$, we have $\psi(x) < f'(x)$ for all sufficiently large x; (iv) $f' \notin \mathcal{E}(f)$. For proofs and more on honest functions, see [11] or [6].

For any class of functions $\mathcal{F}$ and any representation R, let $(\mathcal{F})_R$ denote the set of irrational numbers with an R-representation in $\mathcal{F}$. The next theorem yields a very useful method to prove $R_1 \not\preceq_S R_2$: *Take an arbitrary honest function f, and use f to construct an irrational $\alpha \in (0,1)$ such that $\alpha \in (\mathcal{E})_{R_2}$ and $\alpha \notin (\mathcal{E}(f))_{R_1}$.*

Theorem 3. *Let R_1 and R_2 be representations. If for any honest f there exists α such that $\alpha \in (\mathcal{E})_{R_2}$ and $\alpha \notin (\mathcal{E}(f))_{R_1}$, then $R_1 \not\preceq_S R_2$.*

Proof (Sketch). The theorem is the contrapositive statement of the following claim.

(Claim). If $R_1 \preceq_S R_2$, then there exists honest f such that $(\mathcal{E})_{R_2} \subseteq (\mathcal{E}(f))_{R_1}$.

In order to prove the claim, assume $R_1 \preceq_S R_2$. Then, by Definition 2, for any time-bound t there will exist a time-bound s such that $O(t)_{R_2} \subseteq O(s)_{R_1}$. Thus, it should be fairly easy to see (at least for those familiar with honest functions) that we have $(\mathcal{E})_{R_2} \subseteq (\mathcal{E}(f))_{R_1}$ for some sufficiently fast-increasing honest function f (for any time-bound s there exists an honest f such that $s(x) < f(x)$). □

3 Representations

In this section we discuss a number of representations. First we discuss the standard base-b expansions. Any real we consider will be in the interval $[0, 1)$.

Any natural number $b \geq 2$ is called a *base*, and the numbers in $\{0, 1, \ldots, b-1\}$ will be called *base-b digits*. For any base-b digit $\mathtt{D}$, we define the *complement digit* $\overline{\mathtt{D}}$ by $\overline{\mathtt{D}} = b - 1 - \mathtt{D}$. For any finite sequence $\mathtt{D}_1, \mathtt{D}_2, \ldots, \mathtt{D}_n$ of base-b digits, we use $(0.\mathtt{D}_1\mathtt{D}_2 \ldots \mathtt{D}_n)_b$ to denote rational number $\sum_{k=1}^{n} \mathtt{D}_k b^{-k}$. For an infinite sequence $\mathtt{D}_1, \mathtt{D}_2, \ldots$ of base-b digits, we will say that $(0.\mathtt{D}_1\mathtt{D}_2 \ldots)_b$ is the *base-b expansion* of the real number α if we have

$$(0.\mathtt{D}_1\mathtt{D}_2 \ldots \mathtt{D}_n)_b \leq \alpha < (0.\mathtt{D}_1\mathtt{D}_2 \ldots \mathtt{D}_n)_b + b^{-n} \tag{1}$$

for all $n \geq 1$. We identify the base-b expansion of α with the function $E_b^\alpha : \mathbb{N} \to \{0, 1, \ldots, b-1\}$ given by $E_b^\alpha(0) = 0$ and $E_b^\alpha(n) = \mathtt{D}_n$ (for $n \geq 1$).

Since the second inequality of (1) is strict, every real number $\alpha \in [0, 1)$ has a unique base-b expansion (and thus E_b^α is well defined). According to this definition, a base-b expansion cannot end with an infinite sequence of $\overline{\mathtt{0}}$s. Thus, for a rational number $q \in [0, 1)$ we have two alternatives:

(1) The base-b expansion of q ends with an infinite sequence of $\mathtt{0}$s. In this case we say that q has a *finite* base-b expansion. This happens precisely when q can be written in the form $\frac{m}{b^n}$ for some $m, n \in \mathbb{N}$; the least such n is called *the length* of the base-b expansion.
(2) The base-b expansion of q is infinite and periodic, that is, it is of the form $0.\mathtt{D}_1\mathtt{D}_2 \ldots \mathtt{D}_s(\mathtt{D}_{s+1} \ldots \mathtt{D}_n)^\omega$. The number n can be chosen strictly less than the denominator of q. The period must contain at least one digit different from $\mathtt{0}$ (otherwise q has a finite base-b expansion) and at least one digit different from $\overline{\mathtt{0}}$ (according to our definition of base-b expansions).

Let $\mathrm{dig}_b : \mathbb{Q} \times \mathbb{N} \to \{0, 1, \ldots, b-1\}$ be the function such that $\mathrm{dig}_b(q, n)$ equals the nth digit in the base-b expansion of q. Clearly, the function dig_b is elementary (for every fixed b). Moreover, we can check elementarily whether a rational q has finite or infinite base-b expansion, and if the expansion is infinite, and thus of the form $0.\mathtt{D}_1\mathtt{D}_2 \ldots \mathtt{D}_s(\mathtt{D}_{s+1} \ldots \mathtt{D}_n)^\omega$, then s and n can be computed elementarily in q. For convenience, we will write $E_b^q(n)$ in place of $\mathrm{dig}_b(q, n)$.

We can compare two distinct real numbers using the lexicographic ordering of their base-b expansions. More precisely, $\beta < \alpha$ iff $E_b^\beta(n) < E_b^\alpha(n)$, where n is the first position in which the base-b expansions of β and α are different. In general, we will need unbounded search to determine the witnessing n, but in the particular case when α is irrational and q has a finite base-b expansion of length n, we have $q < \alpha$ iff $q \leq \sum_{k=1}^{n} E_b^\alpha(k) b^{-k}$.

Oracle Turing machines can convert Dedekind cuts of irrationals into base-b expansions and base-b expansions of irrationals into Dedekind cuts (the latter requires unbounded search), and thus, the class of functions $\mathcal{E}_b$, given by $\mathcal{E}_b = \{E_b^\alpha \mid \alpha \text{ is an irrational in the interval } (0, 1)\}$, is a representation according to Definition 1.

Next we will discuss the representation by *signed digits base-3 expansions.* The representation is similar to the representation by standard base-3 expansions, but we use the digits $\mathtt{-1}, \mathtt{0}, \mathtt{1}$ in place of the digits $\mathtt{0}, \mathtt{1}, \mathtt{2}$. For convenience, we use $\widehat{\mathtt{1}}$ to denote $\mathtt{-1}$.

The numbers $\widehat{\mathtt{1}}, \mathtt{0}, \mathtt{1}$ will be called the *signed digits* of base 3, or shortly *base-3s digits.* For $M \in \{\mathtt{0}, \mathtt{1}\}$ and a finite sequence $\mathtt{D}_1, \mathtt{D}_2, \dots, \mathtt{D}_n$ of base-3s digits, $(M.\mathtt{D}_1\mathtt{D}_2 \dots \mathtt{D}_n)_{3s}$ denotes the rational number $M + \sum_{k=1}^{n} \mathtt{D}_k 3^{-k}$. For $M \in \{\mathtt{0}, \mathtt{1}\}$ and an infinite sequence $\mathtt{D}_1, \mathtt{D}_2, \dots$ of base-3s digits, we say that $(M.\mathtt{D}_1\mathtt{D}_2 \dots)_{3s}$ is the *base-3s expansion* of the real number α if we have

$$(M.\mathtt{D}_1\mathtt{D}_2 \dots \mathtt{D}_n)_3 - \frac{1}{2}3^{-n} < \alpha \leq (M.\mathtt{D}_1\mathtt{D}_2 \dots \mathtt{D}_n)_3 + \frac{1}{2}3^{-n} \tag{2}$$

for all $n \in \mathbb{N}$. Note that the first equality of (2) is strict, and hence, every real number $\alpha \in [0, 1)$ has a unique base-3s expansion. A base-3s expansion can end with an infinite sequences of $\mathtt{1}$s, but not with an infinite sequence of $\widehat{\mathtt{1}}$s. E.g., $\mathtt{0.111111}\dots$ is the base-3s expansion of $1/2$, while $\mathtt{1.}\widehat{\mathtt{111111}}\dots$ is not considered to be a valid base-3s expansion. We identify α's base-3s expansion $(M.\mathtt{D}_1\mathtt{D}_2 \dots)_{3s}$ with the function $E^{\alpha}_{3s} : \mathbb{N} \to \{\widehat{\mathtt{1}}, \mathtt{0}, \mathtt{1}\}$ given by $E^{\alpha}_{3s}(0) = M$ and $E^{\alpha}_{3s}(n) = \mathtt{D}_n$ (for $n \geq 1$).

A rational number with base-3s expansion of the form $M.\mathtt{D}_1\ \mathtt{D}_2 \dots \mathtt{D}_n\mathtt{111111}\dots$ has a *semi-finite* base-3s expansion (an analogue of a finite base-b expansion). A rational q has a semi-finite base-3s expansion iff q can be written in the form $q = \frac{2m+1}{2 \cdot 3^n}$ for some $m, n \in \mathbb{N}$; the least such n is the *length* of the semi-finite base-3s expansion of q. A rational that does not have a semi-finite base-3s expansion, has a periodic base-3s expansion. A period cannot consist entirely of $\mathtt{1}$s or entirely of $\widehat{\mathtt{1}}$s. If q has a periodic base-3s expansion, the length of the period can be chosen strictly less the denominator of q.

We can compare two distinct real numbers using the lexicographic ordering of their base-3s expansions exactly as for (standard) base-b expansions. In addition, for any irrational α and any rational q with semi-finite base-3s expansion of length n, we have $q < \alpha$ iff $q < \sum_{k=0}^{n} E^{\alpha}_{3s}(k)3^{-k}$. Moreover, given a rational q, we can elementarily compute the nth digit in the base-3s expansions of q; and if the expansion is periodic, we can elementarily compute the length of the period; and so on.

The following simple algorithm for converting base-3 expansions into base-3s expansions seems to be well known: First add $\frac{1}{2}$, and then subtract $\frac{1}{2}$. The addition which amounts to add $\mathtt{1}$ to each fractional digit, is carried out as a standard base-3 addition with carries. The subtraction which amounts to subtract $\mathtt{1}$ from each fractional digit, is done without carries: That will change the base-3 digits $\mathtt{0}, \mathtt{1}, \mathtt{2}$ into the base-3s digits $\widehat{\mathtt{1}}, \mathtt{0}, \mathtt{1}$, respectively.

Example. Given the base-3 expansion $\mathtt{0.20101201}\dots$, we add $\mathtt{1}$ to each digit and obtain $\mathtt{1.01220012}\dots$ (assuming no carry in the last displayed position). Thereafter we subtract $\mathtt{1}$ from each digit and obtain the base-3s expansion $\mathtt{1.}\widehat{\mathtt{1}}\mathtt{0111}\widehat{\mathtt{1}}\widehat{\mathtt{1}}\mathtt{01}\dots$.

Example. The base-3 expansion $\texttt{0.00(122)}^\omega$ is first converted to $\texttt{0.12(011)}^\omega$ by adding 1s; thereafter to the base-3s expansion $\texttt{0.01}(\widehat{\texttt{1}}\texttt{00})^\omega$ by subtracting 1s.

Conversely, we can convert a base-3s expansion into a base-3 expansion: first add 1 to each digit without carry, and then subtract 1 from each digit with carry.

Example. The base-3s expansion $\texttt{0.1}(\widehat{\texttt{1}}\texttt{0}\widehat{\texttt{1}})^\omega$ is first converted to $\texttt{0.2(010)}^\omega$ by adding 1s; thereafter to the base-3 expansion $\texttt{0.0(121)}^\omega$ by subtracting 1s.

It is easy to see that the indicated conversion algorithms are correct. Observe that they preserve the length of the period of periodic expansions.

Let $\mathcal{E}_{3s}$ denote the class of all base-3s expansions for irrational numbers in the interval $(0, 1)$. The considerations above show that oracle Turing machines can convert back and forth between base-3s and base-3 expansions of irrationals in the interval $(0, 1)$ (unbounded search will be needed to determine the carry). Hence, oracle Turing machines can also convert back and forth between base-3s expansions and Dedekind cuts of irrationals in the interval $(0, 1)$, and thus, $\mathcal{E}_{3s}$ is a representation according to Definition 1.

We round off this section by discussing a few representations based on trace functions. A function $T^{\uparrow} : \mathbb{Q} \cap [0, 1] \to \mathbb{Q} \cap [0, 1]$ is a *trace function from below* for an irrational $\alpha \in (0, 1)$ if $q < \alpha$ implies $q < T^{\uparrow}(q) < \alpha$; a function $T^{\downarrow} : \mathbb{Q} \cap [0, 1] \to \mathbb{Q} \cap [0, 1]$ is a *trace function from above* for an irrational $\alpha \in (0, 1)$ if $q > \alpha$ implies $q > T^{\downarrow}(q) > \alpha$. For any unary functions f, g, h, let $\langle f, g\rangle(x, y) = (f(x), g(y))$ and $\langle f, g, h\rangle(x, y, z) = (f(x), g(y), h(z))$.

Let $\mathcal{D}^{\uparrow}$ denote the set of all $\langle D, T^{\uparrow}\rangle$ such that: there exists irrational $\alpha \in (0, 1)$ such that D is the Dedekind cut of α and $T^{\uparrow}$ is a trace function from below for α. Let $\mathcal{D}^{\downarrow}$ denote the set of all $\langle D, T^{\uparrow}\rangle$ such that: there exists irrational $\alpha \in (0, 1)$ such that D is the Dedekind cut of α and $T^{\downarrow}$ is a trace function from above for α. Let $\mathcal{D}^{\uparrow\downarrow}$ denote the set of all $\langle D, T^{\uparrow}, T^{\downarrow}\rangle$ such that: there exists irrational $\alpha \in (0, 1)$ such that D is the Dedekind cut of α and $T^{\uparrow}, T^{\downarrow}$ are trace functions from below and above, respectively, for α.

It is easy to see that $\mathcal{D}^{\uparrow}$, $\mathcal{D}^{\downarrow}$, $\mathcal{D}^{\uparrow\downarrow}$ are representations, and it is known that $\mathcal{D}^{\uparrow}$ is S-equivalent to the representation by left best approximations; $\mathcal{D}^{\downarrow}$ is S-equivalent to the representation by right best approximations; $\mathcal{D}^{\uparrow\downarrow}$ is S-equivalent to the representation by continued fractions. Thus, $\mathcal{D}^{\uparrow}$ and $\mathcal{D}^{\downarrow}$ are S-incomparable and the degree of $\mathcal{D}^{\uparrow\downarrow}$ is the least upper bound of the degree of $\mathcal{D}^{\uparrow}$ and the degree of $\mathcal{D}^{\downarrow}$. For more details, see [6] and [2].

Note that we cannot compute the Dedekind cut of α in an arbitrary trace function from below for α (such a function will behave arbitrarily on rationals that lie above α). Hence, the class of trace functions from below for irrational numbers in the interval $(0, 1)$, will not be a representation according to Definition 1. The same goes for trace functions from above.

4 Difference Representations over Base-*b* Expansions

For any irrational $\alpha \in (0, 1)$ and any base b, we define the function $\mathrm{Diff}_b^\alpha : \mathbb{Q} \cap [0, 1) \to \mathbb{N}$ by

$$\mathrm{Diff}_b^\alpha(q) = \mu n[E_b^\alpha(n) \neq E_b^q(n)] \tag{3}$$

where $\mu n[E_b^\alpha(n) \neq E_b^q(n)]$ denotes the least n such that $E_b^\alpha(n) \neq E_b^q(n)$. Since base b-expansions are unique and α is irrational, the function Diff_b^α is well-defined, and we will refer to the function as the *base-b difference representation* of α. The base-b difference representation is the set $\mathcal{D}\mathrm{iff}_b$ of all Diff_b^α such that α is an irrational in the interval $(0, 1)$. It follows from (3) and algorithms given below that $\mathcal{D}\mathrm{iff}_b$ indeed is a representation according to Definition 1.

Proposition 4. $\mathcal{E}_b \preceq_S \mathcal{D}\mathrm{iff}_b$.

Proof. Assume $k \geq 1$ and that we have computed $E_b^\alpha(i) = \mathtt{D}_i$ for $i = 1, \ldots, k-1$. Then we can compute $E_b^\alpha(k)$ by the following subrecursive procedure:
For each $\mathtt{D} \in \{0, \ldots, b-1\}$ do

1. Compute $q = (0.\mathtt{D}_1\mathtt{D}_2 \ldots \mathtt{D}_{k-1}\mathtt{D})_b$.
2. If $\mathrm{Diff}_b^\alpha(q) > k$, return output $E_b^\alpha(k) = \mathtt{D}$.

This shows that the base-b expansion of α (the function E_b^α) can be computed subrecursively in the base-b difference representation of α (the function Diff_b^α). It is obvious that the indicated algorithm is correct and that no unbounded search is involved. Thus, we conclude that we have $\mathcal{E}_b \preceq_S \mathcal{D}\mathrm{iff}_b$. □

Let us comment on the proof of Proposition 4. The proof is based on a standard method that we will use over and over again: *We give a subrecursive algorithm which converts an R_2-representation into an R_1-representation. And then, with no further ado, we just conclude that $R_1 \preceq_S R_2$ holds.* Why can we do so? Well, by giving the conversion algorithm, we show that there exists an oracle Turing machine M such that Φ_M^f is an R_1-representation of α whenever the oracle f is an R_2-representation of α. Definition 2 states that $R_1 \preceq_S R_2$ holds if

$$\text{for any time-bound } t \text{ there exists time-bound } s \text{ s.t. } O(t)_{R_2} \subseteq O(s)_{R_1}. \quad (4)$$

Assume $\alpha \in O(t)_{R_2}$. Then, an R_2-representation g of α can be computed in time $O(t)$, and the oracle Turing machine M can be transformed into a standard Turing M' that computes an R_1-representation of α. In place of querying the oracle, M' computes g in time $O(t)$. Since the oracle Turing machine M does not undertake unbounded search, M' will run in time $O(s)$ for some time-bound s which will depend on the time-bound t, but not on the choice of α. Since M' runs in time s, we have $\alpha \in O(s)_{R_1}$. Since s does not depend on the choice of α, we have $O(t)_{R_2} \subseteq O(s)_{R_1}$.

These considerations show that (4) holds if we can show off a subrecursive algorithm that converts an R_2-representation into an R_1-representation; and if (4) holds, then $R_1 \preceq_S R_2$ holds by Definition 2.

Proposition 5. $\mathcal{D}^\uparrow \preceq_S \mathcal{D}\mathrm{iff}_b$.

Proof. The Dedekind cut D^α can be computed subrecursively in Diff_b^α by the following subrecursive procedure (input is $q \in \mathbb{Q}$, and Proposition 4 allows us to use E_b^α): (1) Compute $\mathrm{Diff}_b^\alpha(q) = n$. (2) If $E_b^q(n) < E_b^\alpha(n)$, then let $D^\alpha(q) = 0$. (3) If $E_b^q(n) > E_b^\alpha(n)$, then let $D^\alpha(q) = 1$. (It is obvious that procedure is correct and that unbounded search is not involved).

A trace function from below $T^\uparrow$ for α can be computed subrecursively in Diff_b^α by the following procedure (the input is $q \in \mathbb{Q}$, and procedures given above allow us to use E_b^α and D^α):

If $D^\alpha(q) = 1$, then return output 0.
If $D^\alpha(q) = 0$, then:

1. Compute $\mathrm{Diff}_b^\alpha(q) = n$. 2. Return output $q' = \sum_{k=1}^{n} E_b^\alpha(k)b^{-k}$.

In order to see that the procedure is correct and indeed computes a trace function from below, observe $q < \alpha$ implies $q < q' < \alpha$, since $E_b^q(n) < E_b^\alpha(n) = E_b^{q'}(n)$. It is obvious that unbounded search is not involved.

The two algorithms given above show that a function $\langle D^\alpha, T^\uparrow \rangle$, where $T^\uparrow$ is some trace function from below for α, can be computed subrecursively in Diff_b^α. Hence, we have $\mathcal{D}^\uparrow \preceq_S \mathcal{D}\mathrm{iff}_b$. □

Proposition 6. $\mathcal{D}^\downarrow \not\preceq_S \mathcal{D}\mathrm{iff}_b$.

Proof. Let f be an arbitrary honest function. By Theorem 3, it is sufficient to construct α such that a $\mathcal{D}\mathrm{iff}_b$-representation of α is elementary and no $\mathcal{D}^\downarrow$-representation of α is in $\mathcal{E}(f)$.

We define the sequence $\{d_i\}_{i\in\omega}$ by $d_0 = 1$ and $d_{i+1} = f'(d_i)$. The relation $d_x = y$ is elementary since f' has elementary graph, but the sequence $\{d_i\}_{i\in\omega}$ is not in $\mathcal{E}(f)$ since $f' \notin \mathcal{E}(f)$. The crucial α is given by the base-b expansion: $E_b^\alpha(x) = \mathtt{0}$ if $x = d_i$ for some i; and $E_b^\alpha(x) = \bar{\mathtt{0}}$ otherwise. Thus, the base-b expasion of α is of the form

$$\ldots\ldots\ldots\bar{\mathtt{0}}\,\bar{\mathtt{0}}\,\bar{\mathtt{0}}\underbrace{\;\mathtt{0}\;}_{d_i}\bar{\mathtt{0}}\,\bar{\mathtt{0}}\,\bar{\mathtt{0}}\ldots\ldots\ldots\bar{\mathtt{0}}\,\bar{\mathtt{0}}\,\bar{\mathtt{0}}\underbrace{\;\mathtt{0}\;}_{d_{i+1}}\bar{\mathtt{0}}\,\bar{\mathtt{0}}\,\bar{\mathtt{0}}\ldots\ldots\ldots\bar{\mathtt{0}}\,\bar{\mathtt{0}}\,\bar{\mathtt{0}}\underbrace{\;\mathtt{0}\;}_{d_{i+2}}\bar{\mathtt{0}}\,\bar{\mathtt{0}}\,\bar{\mathtt{0}}\ldots\ldots\ldots$$

and since the relation $d_x = y$ is elementary, the function E_b^α is also elementary.

We prove that α does not have a trace function from above in $\mathcal{E}(f)$. Assume otherwise, that is, we have a trace function from above $T^\downarrow$ for α in $\mathcal{E}(f)$. Then the following algorithm shows that we can compute d_{i+1} using d_i (and i): (1) Compute $q = b^{-d_i} + \sum_{k=1}^{d_i-1} E_b^\alpha(k)b^{-k}$. (2) Let $q' = T^\downarrow(q)$. (3) Search for the least $x < (q - q')^{-1} + 1$ such that $x = d_{i+1}$; give output x.

We argue that this algorithm works: Since $\alpha < q' < q$, the base-b expansions of α, q', q coincide in all positions less than d_i. But then $q' < q$ is only possible if $E_b^{q'}(d_i) = 0$. Moreover, $\alpha < q'$ implies that $E_b^{q'}(n) = \bar{\mathtt{0}}$ for all n with $d_i < n < d_{i+1}$ and $E_b^{q'}(m) > 0$ for some $m \geq d_{i+1}$. These considerations show that $q' + b^{-d_{i+1}+1} > q$ and $d_{i+1} < (q-q')^{-1} + 1$, and thus, the search for x is successful. It follows that the sequence $\{d_i\}_{i\in\omega}$ is in $\mathcal{E}(f)$, which is a contradiction. This proves that α does not have a trace function from above in $\mathcal{E}(f)$. It follows straightforwardly that α does not have a $\mathcal{D}^\downarrow$-representation in $\mathcal{E}(f)$.

Next we prove that α has an elementary $\mathcal{D}\mathrm{iff}_b$-representation. The following elementary algorithm computes Diff_b^α (input is $q = mn^{-1} \in \mathbb{Q}$):

1. Compute i and d_i such that $d_i \leq n < d_{i+1}$.
2. If $d_i + n < d_{i+1}$, let $p = d_i + n$.
3. Otherwise, compute d_{i+1} and let $p = d_{i+1} + n$.
4. Search for the least $x \leq p$ such that $E_b^q(x) \neq E_b^\alpha(x)$; give output x.

In Step 1, we can elementarily compute i and d_i from n using the graph of $\{d_i\}_{i\in\omega}$ (but d_{i+1} might not be accessible by elementary operations). In Step 2, we can elementarily check the inequality, again using the graph of $\{d_i\}_{i\in\omega}$. In Step 3, we have $d_{i+1} \leq d_i + n$ and in this case we can elementarily compute d_{i+1}. Since $d_{i+1} + n < 2d_{i+1} < d_{i+2}$, we will in both Step 2 and Step 3 have at least n consecutive positions, all of them less than or equal to p, where the base-b expansion of α has the digit $\overline{0}$. Assume that the search in Step 4 is not successful. Then q and α coincide in all positions $\leq p$, and thus, q has digit $\overline{0}$ in these positions. But each consecutive portion of n base-b digits of q must contain at least one period, therefore the period of q contains only $\overline{0}$, which by definition is not possible. This proves that the algorithm correctly computes Diff_b^α. Moreover, the algorithm is elementary. Thus, we conclude that Diff_b^α is an elementary function, and thus, α has an elementary $\mathcal{D}\mathrm{iff}_b$-representation. □

Proposition 7. $\mathcal{D}\mathrm{iff}_b \not\leq_S \mathcal{D}^\uparrow$.

Proof. We proceed as in the proof of Proposition 6. Now the crucial α is given by $E_b^\alpha(x) = 0$ if $x = d_i$ or $x = d_i - 1$; and $E_b^\alpha(x) = 1 \dot{-} E_b^\alpha(x-1)$ otherwise. Thus, the base-b expansion of α is of the form

$$\cdots\cdots\cdots \underbrace{\mathtt{0}}_{d_i}\ \mathtt{101}\cdots\cdots\cdots \mathtt{D}_i\,\mathtt{0}\ \underbrace{\mathtt{0}}_{d_{i+1}}\ \mathtt{101}\cdots\cdots\cdots \mathtt{D}_{i+1}\,\mathtt{0}\ \underbrace{\mathtt{0}}_{d_{i+2}}\ \cdots\cdots\cdots$$

where $\mathtt{D}_j = \mathtt{0}$ if $d_j + d_{j+1}$ is even; and $\mathtt{D}_j = \mathtt{1}$ if $d_j + d_{j+1}$ is odd. The function E_b^α is well-defined as $0 = d_0 - 1$, moreover, E_b^α is elementary as the relation $d_x = y$ is elementary.

First we give an elementary algorithm that computes $\langle D^\alpha, T^\uparrow\rangle$ where D^α is the Dedekind cut of α and $T^\uparrow$ is a trace function from below for α. This amounts to decide the inequality $q < \alpha$, and in the case $q < \alpha$, produce q' with $q < q' < \alpha$. On input $q = mn^{-1} \in \mathbb{Q}$ the algorithm carries out the following steps:

1. Compute i and d_i such that $d_i \leq n < d_{i+1}$.
2. If $d_i + n + 1 < d_{i+1}$, then take $p = d_i + n + 1$.
3. Otherwise, compute d_{i+1} and take $p = d_{i+1} + n + 1$.
4. Search for the least $x \leq p$ such that $E_b^q(x) \neq E_b^\alpha(x)$.
   - 4.1. If the search is not successful, return output $q > \alpha$.
   - 4.2. If the search is successful and $E_b^q(x) > E_b^\alpha(x)$, return output $q > \alpha$.
   - 4.3. If the search is successful and $E_b^q(x) < E_b^\alpha(x)$, return output $q < \alpha$ and $q' = \sum_{k=1}^{x} E_b^\alpha(k) b^{-k}$.

For the correctness, only the output in Step 4.1 requires explanation. In this case the base-b expansions of α and q coincide in all positions $\leq p$. Therefore, the base-b expansion of q contains the sequence 0101... of length $n+1$ (note that $d_{i+1}+n+1 < d_{i+2}$). The period of q is a part of this sequence, and that implies that the shortest period is 01 or 10. Indeed, the period cannot start and end with the same digit: the sequence in question has length $> n$ and does not contain successive equal digits. Now, after position p, the base-b expansion of q continues with alternating 0 and 1 digits, just as the expansion α might do, but the expansion of α will not continue like that forever, and at the first position of difference α has digit 0. We conclude that $q > \alpha$.

The algorithm executes nothing but elementary operations, and thus we conclude that α has an elementary $\mathcal{D}^{\uparrow}$-representation. Next we prove that α has no $\mathcal{D}\mathrm{iff}_b$-representation in $\mathcal{E}(f)$.

Assume for the sake of a contradiction that $\mathrm{Diff}_b^{\alpha} \in \mathcal{E}(f)$. Then the following algorithm shows that we can compute d_{i+1} given input d_i (and i):

1. Compute $q = \left(b^{d_i-1}(b^2-1)\right)^{-1} + \sum_{k=1}^{d_i-1} E_b^{\alpha}(k)b^{-k}$.
2. Compute $n = \mathrm{Diff}_b^{\alpha}(q)$.
3. If $n = d_{i+1}$, give output n. Otherwise, give output $n+1$.

Since the base-b expansion of $(b^2-1)^{-1}$ is $0.(01)^{\omega}$, the base-b expansions of q and α are identical up to position $d_{i+1}-2$. The first position where these two expansions differ is either $d_{i+1}-1$ or d_{i+1}. Hence, the output of the algorithm is correct.

The algorithm implies that the sequence $\{d_i\}_{i\in\omega}$ is in $\mathcal{E}(f)$, which is a contradiction (as $d_{i+1} = f'(d_i)$). Thus, $\mathrm{Diff}_b^{\alpha} \notin \mathcal{E}(f)$. Now we have proved that α has an elementary $\mathcal{D}^{\uparrow}$-representation, but no $\mathcal{D}\mathrm{iff}_b$-representation in $\mathcal{E}(f)$. Thus, by Theorem 3, we can conclude that $\mathcal{D}\mathrm{iff}_b \not\preceq_S \mathcal{D}^{\uparrow}$. □

Proposition 8. $\mathcal{D}\mathrm{iff}_b \preceq_S \mathcal{D}^{\uparrow\downarrow}$.

Proof. Let D^{α} be the Dedekind cut of α, and let $T^{\uparrow}$ and $T^{\downarrow}$ be trace functions from below and above, respectively, for α. The following algorithm computes Diff_b^{α} (on input $q \in \mathbb{Q}$) subrecursively in $\langle D^{\alpha}, T^{\uparrow}, T^{\downarrow}\rangle$:

1. If $D^{\alpha}(q) = 1$, let $q' = T^{\downarrow}(q)$; otherwise, let $q' = T^{\uparrow}(q)$.
2. Compute the least k s.t. $E_b^{q}(k) \neq E_b^{q'}(k)$.
3. Compute the least $n \leq k$ s.t. $E_b^{q}(n) \neq E_b^{\alpha}(n)$; return output $\mathrm{Diff}_b^{\alpha}(q) = n$.

Unbounded search is not required as k is bounded above by the product of the denominators of q and q'. For the correctness, note that the base-b expansions of α and q are identical in all positions less than n, and thus, the same goes for q and q'. Thus, in Step 3, $E_b^{q}(n) \neq E_b^{\alpha}(n)$ will hold for some n less than or equal to k. □

Theorem 9. *We have* $\mathcal{D}^{\uparrow} \prec_S \mathcal{D}\mathrm{iff}_b \prec_S \mathcal{D}^{\uparrow\downarrow}$*, and moreover,* $\mathcal{D}\mathrm{iff}_b$ *and* $\mathcal{D}^{\downarrow}$ *are S-incomparable, that is,* $\mathcal{D}\mathrm{iff}_b \not\preceq_S \mathcal{D}^{\downarrow}$ *and* $\mathcal{D}^{\downarrow} \not\preceq_S \mathcal{D}\mathrm{iff}_b$.

Proof. We have $\mathcal{D}^{\uparrow} \prec_S \mathcal{D}\text{iff}_b$ by Proposition 5 and Proposition 7. We have $\mathcal{D}\text{iff}_b \prec_S \mathcal{D}^{\uparrow\downarrow}$ by Proposition 8, Proposition 6 and the trivial fact $\mathcal{D}^{\downarrow} \preceq_S \mathcal{D}^{\uparrow\downarrow}$. We have $\mathcal{D}^{\downarrow} \not\preceq_S \mathcal{D}\text{iff}_b$ by Proposition 6.

We are left to prove that $\mathcal{D}\text{iff}_b \not\preceq_S \mathcal{D}^{\downarrow}$. Assume for the sake of a contradiction that $\mathcal{D}\text{iff}_b \preceq_S \mathcal{D}^{\downarrow}$. Thus, as $\mathcal{D}^{\uparrow} \preceq_S \mathcal{D}\text{iff}_b$, we also have $\mathcal{D}^{\uparrow} \preceq_S \mathcal{D}^{\downarrow}$. This contradicts that $\mathcal{D}^{\uparrow}$ is S-incomparable to $\mathcal{D}^{\downarrow}$, see Sect. 3. □

5 Base-b versus Base-$3s$

Proposition 10. *(1) $\mathcal{E}_{3s} \not\preceq_S \mathcal{E}_3$ and (2) $\mathcal{E}_3 \not\preceq_S \mathcal{E}_{3s}$.*

Proof. In order to prove (1), we use an arbitrary honest f to construct an irrational α with an elementary base-3 expansion, but no base-$3s$ expansion in $\mathcal{E}(f)$. As in preceding proofs let $d_0 = 1$ and $d_{i+1} = f'(d_i)$. In order to complete the construction we need $\Phi : \mathbb{N} \to \{1, 2\}$ such that $\Phi(i) = \phi(i, d_{i+1})$ for some $\phi \in \mathcal{E}$, but $\Phi(i) \neq \psi(i, d_i)$ for every $\psi \in \mathcal{E}(f)$. Such a Φ can be constructed by using diagonalization and the theory of honest functions.

Let $E_3^\alpha(x) = 0$ if $x = d_0$ or $x = d_i + 1$ for some i; $E_3^\alpha(x) = \phi(i, d_{i+1})$ if $x = d_{i+1}$ for some i; $E_3^\alpha(x) = 1$ otherwise. Thus, the base-3 expansion of α is of the form

$$0.\underbrace{0}_{d_0}\,011\ldots\ldots 1\underbrace{1\,\Phi(i-1)}_{d_i}\,011\ldots\ldots 1\underbrace{1\,\Phi(i)}_{d_{i+1}}\,011\ldots\ldots$$

and it follows straightforwardly from the definition of E_3^α that we have $E_3^\alpha \in \mathcal{E}$ (recall that the relation $d_x = y$ is elementary).

We obtain the base-$3s$ expansion of α with the algorithm described in Section 3: we add 1 to each digit with carry and then subtract 1 from each digit without carry. If $\Phi(i) = 1$, the portion of digits in positions $d_i + 1, d_i + 2, \ldots, d_{i+1}$ will remain unchanged. If $\Phi(i) = 2$, the same portion will be converted $20\ldots0$ during the addition. Note that there are no carries in positions d_i. Therefore, $\Phi(i) = 2$ iff $E_{3s}^\alpha(d_i + 1) = 1$. This entails that $E_{3s}^\alpha \notin \mathcal{E}(f)$ because $\Phi(i)$ is constructed such that we have $\Phi(i) \neq \psi(i, d_i)$ for every $\psi \in \mathcal{E}(f)$. We conclude that (1) holds.

In order to prove (2), we take a function $\Psi : \mathbb{N} \to \{1, \hat{1}\}$ with the same properties as Φ and construct α with base-$3s$ expansion of the form

$$0.\underbrace{0}_{d_0}\,100\ldots\ldots 0\underbrace{0\,\Psi(i-1)}_{d_i}\,100\ldots\ldots 0\underbrace{0\,\Psi(i)}_{d_{i+1}}\,100\ldots\ldots$$

It is obvious that $E_{3s}^\alpha \in \mathcal{E}$. To obtain the base-3 expansion of α we first add 1 to each digit, without carry, and then subtract 1 from each digit with carry. There are no carries in positions d_i, and thus, $\Psi(i) = \hat{1}$ iff $E_3^\alpha(d_i + 1) = 0$. It follows that $E_3^\alpha \notin \mathcal{E}(f)$. This proves (2). □

Proposition 11. *(1) $\mathcal{E}_{3s} \not\preceq_S \mathcal{E}_2$ and (2) $\mathcal{E}_2 \not\preceq_S \mathcal{E}_{3s}$.*

The proof of Proposition 11 is omitted due to the publisher's upper page limit.

Conjecture. The representation $\mathcal{E}_{3s}$ is S-incomparable to the representation $\mathcal{E}_b$ iff 6 does not divide b.

6 Difference Representations over Base-3s Expansions

For any irrational $\alpha \in (0,1)$, we define the function $\mathrm{Diff}_{3s} : \mathbb{Q} \cap [0,1) \to \mathbb{N}$ by

$$\mathrm{Diff}_{3s}^{\alpha}(q) \;=\; \mu n[E_{3s}^{\alpha}(n) \neq E_{3s}^{q}(n)]\,. \tag{5}$$

Since base $3s$-expansions are unique and α is irrational, the function $\mathrm{Diff}_{3s}^{\alpha}$ is well-defined, and we will refer to the function as the *base-*$3s$ *difference representation* of α. The base-$3s$ difference representation is the class $\mathcal{D}\mathrm{iff}_{3s}$ of all $\mathrm{Diff}_{3s}^{\alpha}$ such that α is an irrational in the interval $(0,1)$. It follows from (5) and algorithms given below that $\mathcal{D}\mathrm{iff}_{3s}$ is a representation according to Definition 1.

Proposition 12. *(1)* $\mathcal{E}_{3s} \preceq_S \mathcal{D}\mathrm{iff}_{3s}$ *and (2)* $\mathcal{D}^{\downarrow} \preceq_S \mathcal{D}\mathrm{iff}_{3s}$

Proof. The proof of (1) is an obvious modification of the proof of Proposition 4.

A trace function from above $T^{\downarrow}$ for α is computed subrecursively in $\mathrm{Diff}_{3s}^{\alpha}$ by the following algorithm (input $q \in \mathbb{Q}$):

If $D^{\alpha}(q) = 0$, then return output 0.

If $D^{\alpha}(q) = 1$, then:

Compute $n = \mathrm{Diff}_{3s}^{\alpha}(q)$ and return output $q' = \frac{1}{2}3^{-n} + \sum_{k=0}^{n} E_{3s}^{\alpha}(k)3^{-k}$.

It is easy to see that $q > \alpha$ implies $q > q' > \alpha$. Given this algorithm, the proof of (2) is an obvious modification of the proof of Proposition 5. □

Proposition 13. $\mathcal{D}^{\uparrow} \not\preceq_S \mathcal{D}\mathrm{iff}_{3s}$.

Proof. Let α be an irrational with a base-$3s$ expansion of the form

$$\ldots\ldots\hat{1}\,\hat{1}\,\hat{1}\underbrace{\;1\;}_{d_i}\hat{1}\,\hat{1}\,\hat{1}\ldots\ldots\hat{1}\,\hat{1}\,\hat{1}\underbrace{\;1\;}_{d_{i+1}}\hat{1}\,\hat{1}\,\hat{1}\ldots\ldots\hat{1}\,\hat{1}\,\hat{1}\underbrace{\;1\;}_{d_{i+2}}\hat{1}\,\hat{1}\,\hat{1}\ldots\ldots$$

where $d_0 = 1$ and $d_{i+1} = f'(d_i)$ for some arbitrary honest function f. Clearly, $E_{3s}^{\alpha} \in \mathcal{E}$ (recall that the relation $d_x = y$ is elementary). Assume for the sake of a contradiction that α has a trace function from below $T^{\uparrow} \in \mathcal{E}(f)$. Then the following algorithm computes d_{i+1} given input d_i (and i):

1. Compute $q = \sum_{k=0}^{d_i-1} E_{3s}^{\alpha}(k)3^{-k} + \frac{1}{2}3^{-d_i}$.
2. Let $q' = T^{\uparrow}(q)$.
3. Search for the least $x < (q'-q)^{-1} + 1$ such that $x = d_{i+1}$; give output x.

We have $q < q' < \alpha$, and the base-$3s$ expansions of α, q', q coincide in all positions less than d_i. This is only possible when $E_{3s}^{q'}(d_i) = 1$, and $E_{3s}^{q'}(n) = \hat{1}$ for all n with $d_i < n < d_{i+1}$, and $E_{3s}^{q'}(m) \neq 1$ for some $m \geq d_{i+1}$. Therefore, $q + 3^{-d_{i+1}+1} > q'$ and $d_{i+1} < (q'-q)^{-1} + 1$. It follows that the sequence $\{d_i\}_{i\in\omega}$ is in $\mathcal{E}(f)$, which is a contradiction (since $f' \notin \mathcal{E}(f)$). Hence α has no $\mathcal{D}^{\uparrow}$-representation in $\mathcal{E}(f)$.

The following algorithm shows that $\mathrm{Diff}_{3s}^{\alpha} \in \mathcal{E}$ (input is $q = mn^{-1} \in \mathbb{Q}$):

1. Compute i and d_i such that $d_i \leq n < d_{i+1}$.
2. If $d_i + n < d_{i+1}$, let $p = d_i + n$.
3. Otherwise, compute d_{i+1} and let $p = d_{i+1} + n$.
4. Search for the least $x \leq p$ s.t. $E^q_{3s}(x) \neq E^\alpha_{3s}(x)$; give output x.

The justification that the algorithm is elementary and correctly computes $\mathrm{Diff}^\alpha_{3s}$, is similar to the justification given for the algorithm computing Diff^α_b in the proof of Proposition 6. Thus, α has a $\mathcal{D}\mathrm{iff}_{3s}$-representation in $\mathcal{E}$, but no $\mathcal{D}^{\uparrow}$-representation in $\mathcal{E}(f)$. By Theorem 3, we conclude $\mathcal{D}^{\uparrow} \not\preceq_S \mathcal{D}\mathrm{iff}_{3s}$. □

Proposition 14. $\mathcal{D}\mathrm{iff}_{3s} \not\preceq_S \mathcal{D}^{\downarrow}$.

Proof. This proof is analogous to the proof of Proposition 7. Let α's base-$3s$ expansion be of the form

$$\ldots\ldots\ldots \underbrace{\ \widehat{1}\ }_{d_i} 0\,\widehat{1}\,0\,\widehat{1} \ldots\ldots\ldots \mathtt{D}_i\,\widehat{1} \underbrace{\ \widehat{1}\ }_{d_{i+1}} 0\,\widehat{1}\,0\,\widehat{1} \ldots\ldots\ldots \mathtt{D}_{i+1}\,\widehat{1} \underbrace{\ \widehat{1}\ }_{d_{i+2}} 0\,\widehat{1}\,0\,\widehat{1} \ldots\ldots\ldots$$

where $\mathtt{D}_j = \widehat{1}$ if $d_j + d_{j+1}$ is even; and $\mathtt{D}_j = 0$ if $d_j + d_{j+1}$ is odd. □

Proposition 15. $\mathcal{D}\mathrm{iff}_{3s} \preceq_S \mathcal{D}^{\uparrow\downarrow}$.

Proof. This proof is an obvious modification of the proof of Proposition 8. □

The next theorem follows from the propositions above (see the proof of Theorem 9).

Theorem 16. *We have $\mathcal{D}^{\downarrow} \prec_S \mathcal{D}\mathrm{iff}_{3s} \prec_S \mathcal{D}^{\uparrow\downarrow}$, and moreover, $\mathcal{D}\mathrm{iff}_{3s}$ and $\mathcal{D}^{\uparrow}$ are S-incomparable, that is, $\mathcal{D}\mathrm{iff}_{3s} \not\preceq_S \mathcal{D}^{\uparrow}$ and $\mathcal{D}^{\uparrow} \not\preceq_S \mathcal{D}\mathrm{iff}_{3s}$.*

Acknowledgements. The work of Ivan Georgiev is supported by the European Union-NextGenerationEU, through the National Recovery and Resilience Plan of the Republic of Bulgaria, project no. BG-RRP-2.004-0008-C01.

References

1. Austin, D.: Trees, teeth, and time: the mathematics of clock making. American Mathematical Society (2008). www.ams.org
2. Ben-Amram, A., Kristiansen, K., Simonsen, J.G.: On representations of real numbers and the computational complexity of converting between such representations. arXiv:2304.07227 [math.LO] (2023)
3. Ben-Amram, A., Kristiansen, L.: A degree structure on representations of irrational numbers. J. Log. Anal. **17** (2025)
4. Georgiev, I.: Subrecursive incomparability of the graphs of standard and dual baire sequences. Ann. Sofia Univ. Fac. Math. and Inf. 41–55 (2022). https://doi.org/10.60063/gsu.fmi.109.41-55
5. Georgiev, I.: Subrecursive graphs of representations of irrational numbers. In: Della Vedova, G., Dundua, B., Lempp, S., Manea, F. (eds.) CiE 2023: Unity of Logic and Computation. LNCS, vol. 13967, pp. 154–165. Springer, Cham (2023)

6. Georgiev, I., Kristiansen, L., Stephan, F.: Computable irrational numbers with representations of surprising complexity. Ann. Pure Appl. Log. **172**(2), 102893 (2021)
7. Hiroshima, K., Kawamura, A.: Elementarily traceable irrational numbers. In: Della Vedova, G., Dundua, B., Lempp, S., Manea, F. (eds.) CiE 2023: Unity of Logic and Computation. LNCS, vol. 13967, pp. 135–140. Springer, Cham (2023)
8. Kristiansen, L.: On subrecursive representability of irrational numbers. Computability **6**, 249–276 (2017)
9. Kristiansen, L.: On subrecursive representability of irrational numbers, part II. Computability **8**, 43–65 (2019)
10. Kristiansen, L.: On subrecursive representation of irrational numbers: contractors and baire sequences. In: CiE 2021: Connecting with Computability. LNCS, vol. 12813, pp. 308–317. Springer (2021)
11. Kristiansen, L., Schlage-Puchta, J.-C., Weiermann, A.: Streamlined subrecursive degree theory. Ann. Pure Appl. Logic **163**, 698–716 (2012)
12. Weihrauch, K.: Computable Analysis. Springer (2000)

Circuit Metaconstruction in Logspace for Rice-Like Complexity Lower Bounds in ANs and SGRs

Aliénor Goubault–Larrecq(✉) and Kévin Perrot

Aix Marseille Univ, CNRS, LIS, Marseille, France
{alienor.goubault-larrecq,kevin.perrot}@lis-lab.fr

Abstract. A new proof technique combining finite model theory and dynamical systems has recently been introduced to obtain general complexity lower bounds on *any* question one may formulate on the dynamics (seen as a graph) of a given automata network (AN). ANs are abstract finite dynamical systems of interacting entities whose evolution rules are encoded as circuits, hence the study also applies to succinct graph representations (SGRs). In this article, we detail the construction of circuits to obtain general complexity lower bounds (metareduction) and show that the reduction is feasible in logarithmic space.

Keywords: Automata networks · Complexity · Logspace reduction · Succinct graph representations

1 Introduction

Automata networks (AN) constitute a general framework for modeling interacting entities. They are widely used to model complex biological systems by representing interactions among components such as genes, proteins, and cells. They provide a powerful framework to study key dynamical properties happening in biological processes [6,12,13,16,22]. An automata network can be deterministic or non-deterministic, uniform (*e.g.* Boolean) or non-uniform, and it may follow various update policies. This allows automata networks to capture diverse dynamical behaviors across different contexts [12,14,17,18,23]. Beyond their biological applications, automata networks are also studied as computational models in theoretical computer science, their mathematical properties and algorithmic complexity being in the center of an active research field [1–3,10,21].

It is possible to express properties on the dynamics in the form of logical formulas. Given a formula ψ and an automata network F, knowing whether or not the dynamics of F (succinctly encoded in a circuit) verifies the property ψ is a theoretical problem meaningful for biological applications [1,3,8,9]. The question arises of how much time and space is necessary to compute its answer. For example, it has been proven that knowing if a dynamics contains a fixed point or not is NP-complete [1]. Other problems, still expressible as monadic second order formulas, have been studied. Questioning whether the update function of an automata

© The Author(s), under exclusive license to Springer Nature Switzerland AG 2025
A. Beckmann et al. (Eds.): CiE 2025, LNCS 15764, pp. 237–252, 2025.
https://doi.org/10.1007/978-3-031-95908-0_17

network is a bijection, the identity or a constant are coNP-complete [20]; and whether this function is nilpotent or not is PSPACE-complete [8].

Inspired by Rice's theorem, which highlights the undecidability of *any* non-trivial semantic property of programs, recent works have explored complexity-theoretic analogs for automata networks. Since these systems are finite, the focus shifts from computability to complexity. The aim of Rice-like metatheorems in this context is to show a clear dichotomy among the complexity of deciding properties. The first metatheorem, generalizing previous results, proved that in the deterministic case, AN properties expressible in first order logic are either trivial, NP-hard or coNP-hard to verify [8]. A second metatheorem proved a similar result in the non-deterministic case and with properties expressible in monadic second order logic, hence it includes more formulas [9]. A third metatheorem combined both techniques, for both deterministic and non-deterministic dynamics, at the level of monadic second order logic. A major improvement was to also fulfill a uniformity restriction, where all the automata have the same alphabet. The study focused on building models (that verify the property), and counter-models (that do not verify the property), using similar tools [11].

Based on these results, in this work we will detail the construction of circuits involved in these metareductions (reductions to obtain general complexity lower bounds for the metatheorems), while proving that it can be done in logspace (instead of the standard polytime considered so far). For hardness proofs, we reduce from the canonical problem **SAT**. We provide an explicit view of the circuits produced by the metareductions, including the new developments of [11]. It will also serve as a cornerstone to transfer the metatheorems to other models of computation. The main theorem that we will prove in this article is:

Theorem 1. *Let $q \geq 2$. For any q-non-trivial (resp. q-arborescent) MSO formula, the problem ψ-q-***AN-dynamics** *(resp. ψ-q-***NAN-dynamics***) is* NP*-hard or* coNP*-hard for logspace reductions.*

The notions are formally defined below. We present the proof of the deterministic case (AN) below, and the non-deterministic case (NAN) in analogous.

Outline In Sect. 2, we define the main concepts: automata networks, monadic second order logic, and the family of decision problems proven to be hard. In Sect. 3, we recall the models and counter-models constructed in [11], and provide all the necessary background and notations (the correctness of the metareduction is exposed in [11]). In Sect. 4, we detail how to compute the metareduction, *i.e.* how to construct the circuit of an automata network from a **SAT** formula. A fine study allows to conclude that logarithmic space is enough.

2 Definitions

For $n, q \in \mathbb{N}_+$, we denote $[n] = \{1, \ldots, n\}$ and $[\![q]\!] = \{0, \ldots, q-1\}$. For a graph G, we denote $V(G)$ is its set of vertices, $E(G)$ its set of arcs, and $|G| = |V(G)|$.

Circuits. A Boolean *circuit* is an acyclic graph with a type associated to every vertex: either input, output, $\wedge$, $\vee$ or $\neg$ (or usual syntactic sugars). The size of a circuit is its number of vertices also called *gates*. Its arcs are also called *wires*.

Fig. 1. Example deterministic AN of size $n = 2$ on alphabet $[\![q]\!] = \{0, 1, 2\}$ with $q = 3$. Circuit of the function $F : [\![q]\!]^n \to [\![q]\!]^n$ (left) and transition digraph dynamics $\mathcal{G}_F$ on configuration space $[\![q]\!]^n$ (right). A configuration is encoded on four bits, where inputs 0000 to 0001 correspond respectively to configurations 00 to 22. If we write a configuration of this AN $x_1x_2 \in [\![q]\!]^2$, a way to define the local functions is $F_1(x_1x_2) = x_1$, $F_2(x_1x_2) = (x_2 \mod 2) + 1$.

Inputs are source vertices, and outputs are sink vertices. A circuit computes a function from input words (one bit per input vertex) to output words (one bit per output vertex) with the standard semantics. It can simply be encoded by its adjacency matrix, together with a map associating a type to each gate vertex.

Automata Networks. A *deterministic automata network* (AN) F of size n contains n automata where the ith automaton (where $i \in [n]$) has a state from its finite alphabet $A_i = [\![q_i]\!]$ for some $q_i \in \mathbb{N}_+$. The AN is a function $F : X \to X$ where $X = \prod_{i \in [n]} A_i$ is the set of configurations of the AN. A *non-deterministic automata network* (NAN) F of size n is a relation $F \subseteq X \times X$. An AN F can be split into local functions $F_i : X \to A_i$ for each automaton $i \in [n]$, such that for any configuration $x \in X, F(x) = (F_1(x), F_2(x), \ldots, F_n(x))$. In the non-deterministic case (NAN), we do not suppose the existence of local relations.

An automata network (AN or NAN) F can be represented as a digraph called its *dynamics* and denoted $\mathcal{G}_F$, such that $V(\mathcal{G}_F) = X$ and $E(\mathcal{G}_F) = \{(x, F(x)) | x \in X\}$ in the deterministic case, or $E(\mathcal{G}_F) = F$ in the non-deterministic case. When there is $q \in \mathbb{N}_+$ such that $q_i = q$ for all $i \in [n]$, meaning that all automata have the same alphabet, we say that F is *q-uniform*. In this case we have $X = [\![q]\!]^n$. When $q = 2$, $X = \{0, 1\}^n$ and we say that F is a *Boolean* automata network.

The encoding of an automata network plays an important role here, given our focus on computational complexity. An AN F takes as input a configuration and outputs another configuration, hence it is encoded as a Boolean circuit with $\lceil \log_2 |X| \rceil$ input bits and $\lceil \log_2 |X| \rceil$ output bits. A NAN F takes as inputs two configurations c and c' and may or may not have a transition from c to c', hence it is encoded as a Boolean circuit with $2\lceil \log_2 |X| \rceil$ input bits and 1 output bit.

Succinct Graph Representations. The circuit encoding of a graph, namely $\mathcal{G}_F$, is also called a *succinct graph representation* (SGR) [7]. This forgets about the internal network structure made of interacting entities called automata. If G is a succinct graph representation of the dynamics of F, we write $G(c)$ in the deterministic case for the configuration such that $F(c) = G(c)$, and $G(c, c')$ in the non-deterministic case for the Boolean saying whether $(c, c') \in F$ or not. See an example on Fig. 1.

Tree-Decompositions and Treewidth. Tree-decompositions are widely used for the construction of graphs in [11]. Given $G = (V, E)$, its tree-decomposition $T = (V_T, E_T)$ is a tree such that every node $t \in V_T$ is associated to a bag $\mathcal{B}(t) \subseteq V$ of vertices of G verifying the following three properties: *(1)* For every $v \in V$, there exists a node $t \in V_T$ such that $v \in \mathcal{B}(t)$, *(2)* For every $(u, v) \in E$, there exists a node $t \in V_T$ such that $u \in \mathcal{B}(t)$ and $v \in \mathcal{B}(t)$, *(3)* For any $v \in V$, if T_v is the sub-tree of T induced by the nodes whose bag contains v then T_v is connected. The *width* of a tree-decomposition T is $w(T) = \max\{|\mathcal{B}(t)| \mid t \in V_T\} - 1$. A graph may admit several tree-decompositions of various widths but the most interesting decompositions are the one of minimal width. The *treewidth* of G is defined as $tw(G) = \min\{w(T) \mid T \text{ a tree-decomposition of } G\}$ [5].

Monadic Second Order Graph Logic. Our signature is $\{=, \to\}$ where $\to$ is the binary relation (arcs) of a graph, *i.e.* for $x, y \in V(G)$ the atom $x \to y$ is true if and only if $(x, y) \in E(G)$. Monadic second order (MSO) formulas are defined inductively by: $\psi = x \to y \mid x = y \mid x \in X \mid \exists x, \psi \mid \exists X, \psi \mid \psi \vee \psi \mid \neg\psi$, with all the usual syntactic sugars ($\forall, \wedge, \implies$), and where the quantification ($\exists$ or $\forall$) are on vertices or sets of vertices (usually we denote a vertex x and a set of vertices X). Given an MSO formula ψ, we say that a graph G is a *model* of ψ and write $G \vDash \psi$ when the graphical property described by ψ is true in G. If it is not, we say that G is a *counter-model* and write $G \nvDash \psi$ [15].

Example 1. For the AN F presented in Fig. 1 and the two formulas $\psi = \exists x, x \to x$ (existence of a fixed point) and $\varphi = \exists x, \exists y, x \to y \wedge y \to x$ (existence of a cycle of length 2), we have $\mathcal{G}_F \nvDash \psi$ and $\mathcal{G}_F \vDash \varphi$.

With our focus on computational complexity, we need to distinguish families of MSO formulas and especially define what we mean by triviality. A formula ψ is *non-trivial* when it has infinitely many models and infinitely many counter-models, among graphs of out-degree 1. A formula ψ is *arborescent* when there exists k such that ψ has infinitely many models of treewidth at most k and infinitely many counter-models of treewidth at most k. In the restriction to graphs of out-degree 1, ψ is either non-trivial or it has a finite number of models or a finite number of counter-models (then it is called *trivial*). However, without any graph restriction in the non-deterministic case, when ψ is not arborescent, it does not mean that ψ is trivial. We also restrict formulas to q-uniform automata networks, which is equivalent to considering only graphs $\mathcal{G}_F$ of size q^N for some $N \in \mathbb{N}_+$. We obtain the notions of *q-non-trivial* and *q-arborescent* formulas.

Problems. We will consider only the decision problems restricted to q-uniform networks, because the method identically applies to the weaker results without this restriction. We have two problems: one for the deterministic case (AN) and one for the non-deterministic case (NAN).

ψ-q-AN-dynamics	**ψ-q-NAN-dynamics**
Input: circuit of a q-AN f of size N.	*Input:* circuit of a q-NAN f of size N.
Question: does $\mathcal{G}_f \vDash \psi$?	*Question:* does $\mathcal{G}_f \vDash \psi$?

Boundaried Graphs. A *boundaried graph* G has two tuples of ports of the equal size: the primary ports $P_1(G) \subseteq V(G)$ and the secondary ports $P_2(G) \subseteq V(G)$. In general, we do not suppose that $P_1(G) \cap P_2(G) = \emptyset$, but we always have $|P_1(G)| = |P_2(G)|$. In the case where $k = |P_1(G)|$ we call G a *k-boundaried graph*, or *k-graph* for short. Given two k-graphs G and G', we call *gluing* the operation of taking their disjoint union, and merging the secondary ports of the first graph to the primary ports of the second graph. In is denoted $G \oplus G'$, and is formally the k-graph defined as:

$$\begin{aligned} V(G \oplus G') &= V(G) \cup (V(G') \backslash P_1(G')), \\ E(G \oplus G') &= E(G) \cup E(G'[P_2(G) \leftarrow P_1(G')]), \\ P_1(G \oplus G') &= P_1(G), \\ P_2(G \oplus G') &= P_2(G'[P_2(G) \leftarrow P_1(G')]), \end{aligned}$$

where if $P_2(G) = (p_1, \ldots, p_k)$ and $P_1(G') = (p'_1, \ldots, p'_k)$ then $G'[P_2(G) \leftarrow P_1(G')]$ denotes the graph G' where each vertex p'_i is renamed p_i for all $i \in [n]$, and the ports accordingly. Remark that $\oplus$ is associative but not commutative.

We define a notation for gluing several graphs. Let $\Gamma = \{G_i\}_{i \in I}$ be a finite family of k-graphs. For $\omega = \omega_1 \ldots \omega_\mu$ a nonempty word on alphabet I, the operation $\Delta^\Gamma(\omega)$ is defined by induction on the length $\mu \in \mathbb{N}_+$ of ω as follows:

$$\Delta^\Gamma(\omega_1) = G_{\omega_1} \quad \text{and} \quad \Delta^\Gamma(\omega_1 \ldots \omega_\mu) = \Delta^\Gamma(\omega_1 \ldots \omega_{\mu-1}) \oplus G_{\omega_\mu}.$$

In our main constructions, finite k-graphs will be glued to obtain a dynamics $\mathcal{G}_F$.

Remark 1. In the deterministic setting, we will always consider specific graphs such that, when glued together to produce the graph of a dynamics, we obtain a graph of out-degree one corresponding to a deterministic AN (Sect. 4 of [11]).

Logspace. Logspace is the complexity class of problems that can be solved with a deterministic Turing machine in space $\mathcal{O}(\log n)$ for inputs of size n. The space bound only applies to working tapes (the read-only input tape and write-only output tapes are not bounded, in particular to be able to read the input entirely) [19]. For example, copying a circuit only requires one bit of memory, and manipulating object sizes (such as the input size) is possible in logspace.

3 Description of the Metareduction

Rice-like complexity lower bounds for **ψ-dynamics** have been established in [8] (for deterministic SGRs and FO formulas) and in [9] (for non-deterministic SGRs and MSO formulas), and their results are strengthened in [11] from succinct graph representations to q-uniform automata networks for any $q \geq 2$. In the deterministic (resp. non-deterministic) setting it states that for any q-non-trivial (resp. q-arborescent) MSO formula ψ, the problem **ψ-q-AN-dynamics** (resp. **ψ-q-NAN-dynamics**) is either NP-hard or coNP-hard. The NP/coNP symmetry in this statement is necessary (unless $\mathsf{NP} = \mathsf{coNP}$), because some formulas are known to be NP-complete (*e.g.* deciding the existence of a fixed point: $\exists x, x \to x$ [1]) while others are known to be coNP-complete (*e.g.* deciding whether it is the identity: $\forall x, x \to x$ [20]).

All these hardness proofs consist in polynomial time many-one reductions from **SAT**, and the purpose of the present work is to explain them, meanwhile proving that they can be computed in logspace. Let $q \geq 2$ be a fixed alphabet size and ψ be a fixed q-non-trivial (resp. q-arborescent) MSO formula, hence defining a problem **ψ-q-AN-dynamics** (resp. **ψ-q-NAN-dynamics**). We consider the stronger general construction from [11] which, given an instance S of **SAT**, assembles the dynamics $\mathcal{G}_F$ of a q-uniform AN by gluing five fixed k-graphs denoted $\Gamma = \{G_0, G_1, G_2, G_3, G_4\}$. It is important to underline that the value of k and the k-graphs of Γ depend only on the formula ψ defining the problem, therefore they are of constant size for the purpose of the metareduction. We also have $|G_0| = |G_1|$. They are obtained using tools from finite model theory, and we refer to these articles for their definition (here their existence suffices).

For S a propositional formula with s variables, we denote $\overline{S}$ the word of length 2^s such that $\overline{S}_i$ is 1 if $S(i)$ is false, and 0 if $S(i)$ is true (viewing the binary expansion of i as a Boolean assignment for S). Then the metareduction consists in building the circuit encoding of a q-uniform AN (resp. NAN) whose dynamics is (see [11, Section 6.3]):

$$\mathcal{G}_F = \Delta^{\Gamma}(2 \cdot \overline{S} \cdot 4^{L(s)} \cdot 3) \tag{1}$$

where $L(s) = \frac{b}{a}(q^{(s+\log_q(\alpha)+1)\mu} - 1) - \alpha 2^s \geq 0$ and $a, b, \mu, \alpha, \log_q(\alpha)$ are constant integers depending only on ψ. This $\mathcal{G}_F$ is a model of ψ if and only if G_0 appears at least once, hence if and only if S is satisfiable. The purpose of adding $L(s)$ copies of G_4 is to obtain a dynamics on $|\mathcal{G}_F| = q^N$ vertices (configurations) for some $N \in \mathbb{N}$, hence consistent with the q-uniformity requirement. The main achievement of [11] is indeed to resolve these arithmetical constraints. From Eq. 1, summing graph sizes and subtracting merged port vertices, we obtain:

$$|\mathcal{G}_F| = q^N = |G_2| + 2^s\,|G_0| + L(s)\,|G_4| + |G_3| - k\,(2^s + L(s) + 1). \tag{2}$$

As we will deal with these quantities in the logspace reduction, it is important to note the following (although it is straightforward from [11]).

Proposition 1. *The binary values of 2^s, $L(s)$, N and q^N are computable in space $\mathcal{O}(\log s)$. Moreover, $\log_2(L(s))$ and $\log_2(q^N)$ are in $\mathcal{O}(s)$.*

In the next section, we will explain the circuit of the deterministic (resp. non-deterministic) automata network F constructed as an output of the reduction transformation, given S as input. This requires meticulous care and precise notations defined in the rest of this section. During the construction of the circuit encoding F, we will allocate each configuration label from 0 to $q^N - 1$ to some copy of a graph of Γ and explain how the circuit decodes a configuration $c \in \{0, \ldots, q^N - 1\}$ in order to identify which graph of Γ it belongs to, and the

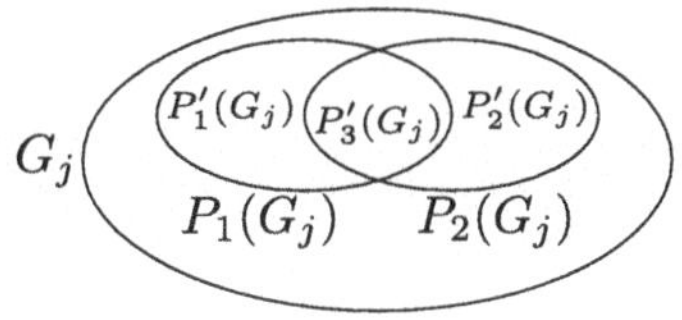

Fig. 2. Notation for the ports of G_j for $j \in \{0, 1, 4\}$.

relative position of that vertex in this graph (in particular, whether or not it is among the port vertices). With these two information, the circuit will be able to compute the image (resp. set of images) $F(c)$.

The construction of $\mathcal{G}_F$ consists in gluing a total of $2^s + L(s) + 2$ copies of the graphs G_0, G_1, G_2, G_3 and G_4. We denote $(H_i)_{i\in\{0,\ldots,2^s+L(s)+1\}}$ the graphs such that $\mathcal{G}_F = H_0 \oplus H_1 \oplus \ldots \oplus H_{2^s+L(s)+1}$:

- $H_0 = G_2$,
- $H_i \in \{G_0, G_1\}$ depending on $i \in \{1, \ldots, 2^s\}$, and
- $H_i = G_4$ for $i \in \{2^s + 1, \ldots, 2^s + L(s)\}$,
- $H_{2^s+L(s)+1} = G_3$.

For every G_j, its ports are written $P_1(G_j)$ and $P_2(G_j)$. From [11] the ports of G_0 and G_1 have the same labels as the ones of G_4 (precisely, we have $G_1 = \Delta^{\{G_4\}}(4^\alpha)$ and $G_0 = G_4 \sqcup \tilde{G_0}$ where the ports are taken in G_4). For convenience, we consider that the ports $P_1(G_3)$ have the same labels as the ports $P_1(G_4)$.

Recall that $P_1(G_j) \cap P_2(G_j)$ may be non-empty, and denote $P_3'(G_j) = P_1(G_j) \cap P_2(G_j)$, $P_1'(G_j) = P_1(G_j) \setminus P_3'(G_j)$ and $P_2'(G_j) = P_2(G_j) \setminus P_3'(G_j)$ for $j \in \{0, 1, 4\}$. See Fig. 2. We define without ambiguity $P_3' = P_3'(G_j)$ for any $j \in \{0, 1, 4\}$ because in $\mathcal{G}_F$ the vertices of $P_3'(G_j)$ for $j \in \{0, 1, 4\}$ will all be merged together. We have $P_1'(G_j) \cap P_2'(G_j) = \emptyset$ for $j \in \{0, 1, 4\}$. We extend the notation of P_1', P_2', P_3' to G_2 and G_3 as follows: $P_3'(G_2) \subseteq P_2(G_2)$ and $P_3'(G_3) \subseteq P_1(G_3)$ such that these are the vertices that will be merge into P_3' in $\mathcal{G}_F$. Also, set $P_2'(G_2) = P_2(G_2) \setminus P_3'(G_2)$ and $P_1'(G_3) = P_1(G_3) \setminus P_3'(G_2)$. We define neither $P_1'(G_2)$ nor $P_2'(G_3)$ because the primary ports of G_2 and secondary ports of G_3 are not relevant.

In the following we will use shortcuts for integer values relative to ports:

$$\begin{aligned}
g_j &= |G_j| \text{ for } j \in \{0,1,2,3,4\} \text{ hence with } g_0 = g_1 \\
k_2 &= |P_1'(G_j)| = |P_2'(G_j)| \text{ for any } j \in \{0,1,4\} \\
k_3 &= |P_3'| = k - k_2 \\
g_1' &= g_1 - 2k_2 - k_3 = |G_1 \setminus (P_1(G_1) \cup P_2(G_1))| \\
g_2' &= g_2 - k_2 - k_3 = |G_2 \setminus P_2(G_2)| \\
g_3' &= g_3 - k_2 - k_3 = |G_3 \setminus P_1(G_3)| \\
g_4' &= g_4 - 2k_2 - k_3 = |G_4 \setminus (P_1(G_4) \cup P_2(G_4))|
\end{aligned}$$

Without loss of generality (because our MSO graph logic on signature $\{=, \rightarrow\}$ is invariant by isomorphism), for any $j \in \{0, 1, 2, 3, 4\}$ we label the vertices

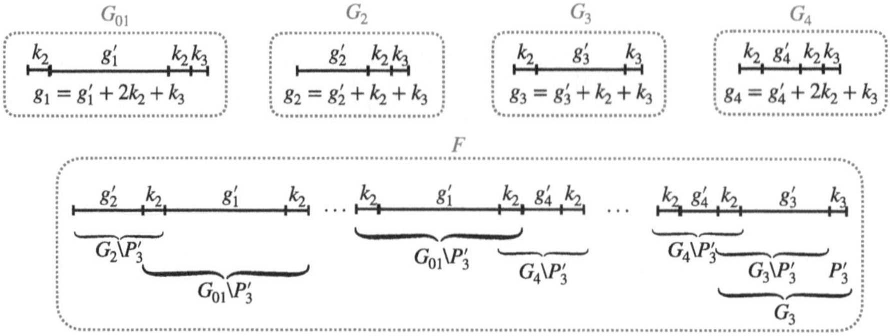

Fig. 3. Allocation of the configurations of F within the copies of G_0, G_1, G_2, G_3, G_4, for the purpose of constructing the dynamics $\mathcal{G}_F$ given by Eq. 1. Recall that G_{01} denotes G_0 or G_1 which have the same set of vertices (configurations). Observe that P_3', which are both primary and secondary ports that are merged in all copies, appear at the very end of the configuration space of F.

of G_j as $V(G_j) = \{0, \ldots, g_j - 1\}$, with $P_1'(G_j) = \{0, \ldots, k_2 - 1\}$, $P_2'(G_j) = \{g_j - k_2 - k_3, \ldots, g_j - k_3 - 1\}$ and $P_3' = \{g_j - k_3, \ldots, g_j - 1\}$, except for $P_1'(G_2)$ and $P_2'(G_3)$ which are not necessary. We also denote G_{01} for a graph which may be G_0 or G_1, depending on the evaluation of S in Eq. 1 (recall that $|G_0| = |G_1|$ and their configuration labels are compatible). See Fig. 3.

Remark 2. In the following we will consider that we construct a q-uniform AN or NAN because by taking $G_4 = G_0$ and replacing $L(s)$ by any integer we still construct a circuit suitable for the reduction without the q-uniform restriction.

4 Logspace Circuit Construction

We now prove that for any MSO formula ψ and any $q \geq 2$, the transformation from S to $\mathcal{G}_F$ described in Eq. 1 of Sect. 3 can be computed in logarithmic space. That is, the general NP or coNP hardness results on **ψ-q-AN-dynamics** and **ψ-q-NAN-dynamics** from [11] hold for logspace reductions. Recall that the five graphs of $\Gamma = \{G_0, G_1, G_2, G_3, G_4\}$ are constants depending only on the formula ψ defining the problem and that $L(s)$ and N can be computed in logspace from S (Proposition 1). Let S be a propositional formula on s variables. We define the circuit of F in the deterministic case, while proving some useful properties about the circuit. For the non-deterministic case it is analogous. Finally, we prove that these circuits can be computed in logspace. For each ψ, q the graphs of Γ are different but they always respect the conventions established in Sect. 3.

We now describe the circuit of the q-uniform AN F with N automata whose dynamics $\mathcal{G}_F$ is described by Eq. 1, *i.e.* the output of the reduction from S to prove that **ψ-q-AN-dynamics** is hard. In the deterministic case, each of the five graphs from Γ have out-degree at most one and the result of their gluing also has

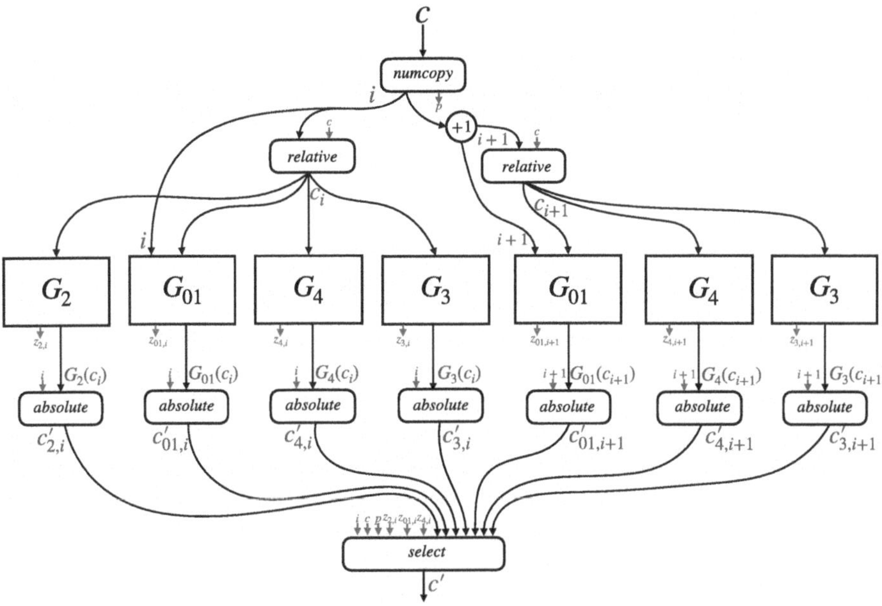

Fig. 4. Circuit of F, computing $F(c) = c'$. For sake of clarity, some wires are not drawn, but are replaced with references in blue (for the wires coming from c and i especially). We label in red the output of some subcircuits, these notations will be used in their formal description. (Color figure online)

out-degree exactly one (see [11, Section 4], some port vertex may have no out-neighbor but it appears after gluing). Since they are constants (do not depend on S), computing their adjacencies takes $\mathcal{O}(1)$ space. Figure 4 gives an overview of the construction, on $\lceil \log_2(q^N) \rceil$ input bits and $\lceil \log_2(q^N) \rceil$ output bits, with subcircuits represented as boxes. Let c be the input of the global circuit we are constructing.

- *numcopy* identifies the index i of the graph H_i to which c belongs.
- *relative* computes the label corresponding to c in $H_i \in \Gamma$.
- G_j computes the out-neighbor of a given vertex label in G_j.
- *absolute* is the inverse of *relative*.
- *select* selects among the subcircuits the correct out-neighbor of c.

The rest of this section provides a formal description of these subcircuits and an analysis of their complexity in order to eventually conclude that the whole circuit is computable in logspace from S. When the domain and range of a subcircuit are given, it means that other input words will have no effect on the final computation of the whole circuit and we do not even need to specify the output in this case.

Subcircuit G_j for $j \in \{01, 2, 3, 4\}$. Recall that the graphs from Γ have out-degree at most one. For $j \in \{2, 3, 4\}$, the subcircuit $G_j : V(G_j) \to V(G_j)$

simply computes $G_j(c)$ on input c. It has an extra output bit z equal to 1 when c has out-degree 0 in G_j (that is, when $G_j(c)$ is undefined). When $z = 1$, the other output bits supposed to encode $G_j(c)$ are set to 0. Subcircuit $G_{01} : \{1, \ldots, 2^s\} \times V(G_0) \to V(G_0)$ has the same logic but produces a graph that is a copy of either G_0 or G_1 (recall that $V(G_0) = V(G_1)$), depending on an evaluation of S. It has two inputs, i and c:

$$G_{01}(i, c) = \begin{cases} G_1(c) & \text{if } S(i-1) \text{ is true,} \\ G_0(c) & \text{otherwise.} \end{cases}$$

For $i \in \{1, \ldots, 2^s\}$, it means that H_i is a copy of G_1 if $S(i-1)$ is true and a copy of G_0 if $S(i-1)$ is false. The binary value of i is an input of the subcircuit G_{01} and its output will only be selected (by subcircuit *select*) on domain $i \in \{1, \ldots, 2^s\}$, where the binary value of $i - 1$ is a Boolean assignment for S.

Given that the graphs have constant size and that the formula S merely needs to be evaluated in order to branch on the case disjunction, it simply needs to be copied once inside the subcircuit G_{01}. Formally, the jth output bit of $G_{01}(i, c)$ is obtained as the following combination, using a copy of the circuit for G_1 evaluated on c with b_1 its jth output bit, a copy of the circuit for G_0 evaluated on c with b_2 its jth output bit, and a copy of S evaluated on assignment $i - 1$ (this bit is wired twice):

$$(S(i-1) \wedge b_1) \vee (\neg S(i-1) \wedge b_2).$$

This subcircuit can be assembled in logarithmic space as it only requires to copy S, and remember $|S|$ to place the decrement and constant circuits for G_0 and G_1 around, then add gates and wires for the final combination above.

Subcircuit *Numcopy*. The subcircuit $numcopy : \{0, \ldots, q^N - 1\} \to \{0, \ldots, 2^s + L(s) + 1\} \times \{0, 1\}$ computes, for each configuration c of $\mathcal{G}_F$, the smallest integer i such that c is in H_i, which is the $(i+1)^{\text{th}}$ copied graph in the definition of $\mathcal{G}_F$ in Eq. 1 (remark that it does not mean it is the only copied graph of $\mathcal{G}_F$ that contains c because gluing operations merge vertices). It has an extra output bit p equal to 1 when c is a secondary port in H_i. According to the nomenclature of Fig. 3, we have $numcopy(c) = (i, p)$ with i equal to:

$$\begin{cases} 0 & \text{if } c < g'_2 + k_2 \text{ or } c \geq q^N - k_3, \\ \left\lfloor \frac{c - g'_2 - k_2}{g'_1 + k_2} \right\rfloor & \text{if } g'_2 + k_2 \leq c < g'_2 + k_2 + 2^s(g'_1 + k_2), \\ \left\lfloor \frac{c - g'_2 - k_2 - 2^s(g'_1 + k_2)}{g'_4 + k_2} \right\rfloor + 2^s & \text{if } g'_2 + k_2 + 2^s(g'_1 + k_2) \leq c < q^N - (k_2 + g'_3 + k_3), \\ 2^s + L(s) + 1 & \text{if } q^N - (k_2 + g'_3 + k_3) \leq c < q^N - k_3, \end{cases}$$

and $p = 0$ if and only if c is part of an interval whose size is labeled by g'_2 or g'_1 or g'_4 or g'_3 on Fig. 3 (implemented as a straightforward case disjunction).

The subcircuit for *numcopy* can be constructed in logspace because the case disjunction relies on comparisons between the input c and integer values of linear size (this is important to underline as we are required to construct in logspace the circuits performing the comparisons) and computable in logspace by Proposition 1. Moreover, a circuit for the division (which is also computed by the circuit itself, because it depends on c, even though the denominator is a constant). of length-$\mathcal{O}(s)$ binary numbers (s is given in unary with S) can be constructed in logspace (see for example [24], and [4] for a recently improved logspace-uniform family of circuits for the division). Combining these elements as a unique circuit again requires to manipulate their number of gates in order to compute the adjacency matrix of disjoint unions (in terms of graph) and then wire them (*i.e.*, add arcs), but all these quantities are polynomial in s therefore encoded on $\mathcal{O}(\log(s))$ bits.

Subcircuit *Relative*. The subcircuit *relative* : $\{0,\ldots,2^s + L(s) + 1\} \times \{0,\ldots,2^N - 1\} \to V(G_j)$ computes, for each number of copy i and configuration c in $\mathcal{G}_F$, the label of c within the graph H_i, hence with range in the vertices of G_j such that $H_i = G_j$. So *relative*(i,c) equals:

$$\begin{cases} c & \text{if } i = 0 \text{ and } c < g'_2 + k_2, \\ c - g'_2 - k_2 - 2^s(g'_1 + k_2) - L(s)(g'_4 + k_2) - g'_3 & \text{if } i = 0 \text{ and } c \geq q^N - k_3, \\ c - g'_2 - k_2 - (i-1)(g'_1 + k_2) & \text{if } 1 \leq i \leq 2^s, \\ c - g'_2 - k_2 - 2^s(g'_1 + k_2) - (i - 2^s - 1)(g'_4 + k_2) & \text{if } 2^s < i \leq 2^s + L(s), \\ c - g'_2 - k_2 - 2^s(g'_1 + k_2) - L(s)(g'_4 + k_2) & \text{if } i = 2^s + L(s) + 1. \end{cases}$$

The two first cases correspond to $i = 0$, *i.e.* $H_i = G_2$, and they branch on the fact that c belongs to P'_3 or not, in accordance with our choice of placing P'_3 at the end of the configuration labels in $\mathcal{G}_F$. Observe that *relative* will work as expected, because in Fig. 4 it is the output i of *numcopy*(c), or $i+1$, that is given as an input to *relative*, hence this subcircuit will output *relative*(*numcopy*$(c), c)$, or *relative*(*numcopy*$(c) + 1, c)$. In particular, the case where $i = 0$ and $g'_2 + k_2 \leq c < q^N - k_3$ never occurs as an input to *relative*, therefore we do not need to define this case. We will use *relative* on i and $i+1$ to handle the fact that, when gluing two graphs H_i and H_{i+1}, a secondary port v from H_i may have, in $\mathcal{G}_F$, its out-neighbor defined in H_{i+1}. Indeed, when $p = 1$, that is when c is a secondary port vertex in H_i, the subcircuit *relative*(*numcopy*$(c) + 1, c)$ outputs the primary port vertex of H_{i+1} that will be merged with c.

By Proposition 1, the subcircuit *relative* is constructible in logspace.

Subcircuit *Absolute*. The subcircuit *absolute* : $\{0,\ldots,2^s + L(s) + 1\} \times V(G_j) \mapsto \{0,\ldots,q^N - 1\}$ computes, for each number of copy i and configuration c within $H_i = G_j$, the label of c in $\mathcal{G}_F$. The output of *absolute*(i,c) is defined as:

$$\begin{cases} c & \text{if } i = 0 \text{ and } c < g'_2 + k_2, \\ c + g'_2 + k_2 + 2^s(g'_1 + k_2) + L(s)(g'_4 + k_2) + g'_3 & \text{if } i = 0 \text{ and } c \geq g'_2 + k_2, \\ c + g'_2 + k_2 + (i-1)(g'_1 + k_2) & \text{if } 1 \leq i \leq 2^s, \\ c + g'_2 + k_2 + 2^s(g'_1 + k_2) + (i - 2^s - 1)(g'_4 + k_2) & \text{if } 2^s < i \leq 2^s + L(s), \\ c + g'_2 + k_2 + 2^s(g'_1 + k_2) + L(s)(g'_4 + k_2) & \text{if } i = 2^s + L(s) + 1. \end{cases}$$

Remark that for any given i the subcircuit *absolute* computes the inverse of *relative*, in the sense that we have $absolute(i, relative(i, c)) = c$. This is straightforward to check from their definition, except for $i = 0$ and $g'_2 + k_2 \leq c < q^N - k_3$ because *relative* is not defined, but as argued above it never occurs.

By Proposition 1, the subcircuit *absolute* is constructible in logspace.

Subcircuit *Select* and Assembly. Finally we present the function that will sort all the information computed in the assembly of previous subcircuits in order to output the correct out-neighbor of c in $\mathcal{G}_F$. Before giving a formal description of *select* and argue that the whole circuit indeed computes F with $\mathcal{G}_F$ given by Eq. 1, let us underline that we will consider the wirings given on Fig. 4, and the output labels it defines. The subcircuit:

$$select : \{0, \ldots, 2^s + L(s) + 1\} \times \{0, \ldots, q^N - 1\}^8 \times \{0, 1\}^4 \to \{0, \ldots, q^N - 1\}$$

takes as inputs:

- i and c such that c is in H_i (*i.e.*, i is the output of *numpcopy*(c)),
- the outputs from all the subcircuits for the graphs (after conversion to $\mathcal{G}_F$ through *absolute*), denoted $c'_{2,i}, c'_{01,i}, c'_{4,i}, c'_{3,i}, c'_{01,i+1}, c'_{4,i+1}, c'_{3,i+1}$,
- the bits p (whether c is a secondary port in H_i) and $z_{2,i}, z_{01,i}, z_{4,i}$ (whether c has out-degree zero as a primary port in G_2, G_{01}, G_4, respectively).

The purpose of *select* is to output only one configuration among the seven possible configurations $(c'_{2,i}, c'_{01,i}, c'_{4,i}, c'_{3,i}, c'_{01,i+1}, c'_{4,i+1}, c'_{3,i+1})$: the out-neighbor of c in $\mathcal{G}_F$.

Formally, the subcircuit *select* outputs:

$$c' = \begin{cases} c'_{2,i} & \text{if } i = 0 \text{ and } c < g'_2 + k_2 \text{ and } (p = 0 \text{ or } z_{2,i} = 0), & (1) \\ c'_{01,i+1} & \text{if } i = 0 \text{ and } c < g'_2 + k_2 \text{ and } p = 1 \text{ and } z_{2,i} = 1, & (2) \\ c'_{01,i} & \text{if } 1 \leq i \leq 2^s \text{ and } (p = 0 \text{ or } z_{01,i} = 0), & (3) \\ c'_{01,i+1} & \text{if } 1 \leq i < 2^s \text{ and } p = 1 \text{ and } z_{01,i} = 1, & (4) \\ c'_{4,i+1} & \text{if } i = 2^s \text{ and } p = 1 \text{ and } z_{01,i} = 1, & (5) \\ c'_{4,i} & \text{if } 2^s < i \leq 2^s + L(s) \text{ and } (p = 0 \text{ or } z_{4,i} = 0), & (6) \\ c'_{4,i+1} & \text{if } 2^s < i < 2^s + L(s) \text{ and } p = 1 \text{ and } z_{4,i} = 1, & (7) \\ c'_{3,i+1} & \text{if } i = 2^s + L(s) \text{ and } p = 1 \text{ and } z_{4,i} = 1, & (8) \\ c'_{3,i} & \text{if } i = 2^s + L(s) + 1, & (9) \\ c'_{2,i} & \text{if } i = 0 \text{ and } c \geq q^N - k_3 \text{ and } z_{2,i} = 0, & (10) \\ c'_{4,i+1} & \text{if } i = 0 \text{ and } c \geq q^N - k_3 \text{ and } z_{2,i} = 1 \text{ and } z_{4,i+1} = 0, & (11) \\ c'_{3,i+1} & \text{if } i = 0 \text{ and } c \geq q^N - k_3 \text{ and } z_{2,i} = 1 \text{ and } z_{4,i+1} = 1. & (12) \end{cases}$$

In this case disjunction, the cases on lines 1–2 handle H_0 excepted P'_3: if c is a secondary port ($p = 1$) and c has no out-neighbor in H_0 (which is a copy of G_2), then its out-neighbor is taken from H_1 (which is a copy of G_{01}). The cases on lines 3–5 handle H_i for $i \in \{1, \ldots, 2^s\}$: if c is a secondary port ($p = 1$) and c has out-neighbor in H_i (which is a copy of G_{01}), then its out-neighbor is taken from H_{i+1} (which is a copy of G_{01} when $i \neq 2^s$, and a copy of G_4 otherwise). The cases on lines 6–8 handle H_i for $i \in \{2^s+1, \ldots, 2^s+L(s)\}$: if c is a secondary port ($p = 1$) and c has out-neighbor in H_i (which is a copy of G_4), then its out-neighbor is taken from H_{i+1} (which is a copy of G_4 when $i \neq 2^s + L(s)$, and a copy of G_3 otherwise). The case on line 9 handles $H_{2^s+L(s)+1}$: configuration c is never a secondary port in this case (hence it is not necessary to precise that $p = 0$), and the out-neighbor is always in $H_{2^s+L(S)+1}$ (which is a copy of G_3). Finally, the cases lines 10–12 handle P'_3: configuration c is always a secondary port in this case (hence it is not necessary to precise that $p = 1$), and if its out-neighbor is not in H_0 (which is a copy of G_2), then it may be either be within P'_3 from the primary ports of H_1 (computed from a copy of G_4, which has the same port vertices as G_0 and G_1), or the out-neighbor is in $H_{2^s+L(S)+1}$ (which is a copy of G_3, whose port labels are identical to G_4, so that it is convenient here to also use c_{i+1} in the subcircuit for G_3).

From the explanations above, we conclude that the whole circuit computes F such that $\mathcal{G}_F$ is defined in Eq. 1. The argumentation relies on the fact that this $\mathcal{G}_F$ from [11] has, by hypothesis, out-degree exactly one, hence if a secondary port vertex has out-degree zero then its out-neighbor necessarily comes from the primary port vertex merged with it during the gluing operations (and if a secondary port vertex has an out-neighbor in two graphs H_i, H_{i+1}, then these two out-neighbors will be merged and our circuit considers the first one). The subcircuit for *select* is again a case disjunction with simple arithmetics on $\mathcal{O}(s)$-bits integers, constructible in logspace.

Remark 3. If $L(s) = 0$ then there is no copy of G_4, and *select* is simply patched by replacing $c'_{4,i+1}$ line 5 with $c'_{3,i+1}$, and deleting line 8 to avoid an ambiguity for $i = 2^s$ (lines 6–7 are vacuous, and line 11 is still correct).

Logspace Reduction. Each subcircuit has a size polynomial in s, hence encoded on $\mathcal{O}(\log(s))$ bits, therefore there is no difficulty in building a unique adjacency matrix (and gates vector) with additional wires for their assembly (between subcircuits inputs and outputs). For example, the $\lceil \log_2(q^N) \rceil$ output bits of each *absolute* can be grouped as consecutive gates, and remembered as a logspace pointer to the first of these output gate. There are finitely many such grouped wirings to implement. We conclude that, from an instance S of **SAT**, the circuit described on Fig. 4 can be constructed in logspace. Theorem 1 then follows from [11] (Sect. 3) regarding ψ-q-**AN-dynamics**.

5 Conclusion and Perspectives

While detailing the circuit metaconstruction of Rice-like complexity lower bounds introduced in [9] for succinct graph representations and refined in [11] for q-uniform automata networks, we have demonstrated that it is feasible in logspace. The main non-constant part of the metaconstruction given by Eq. 1

consists in evaluating S. Proving that this metareduction is computable in logspace may basically amount to exploiting the fact that **CVP** (Circuit Value Problem) is P-complete for many-one logspace reductions, but we have presented a much more direct construction of the circuit. Indeed, to formally exploit the fact that our circuits for $\mathcal{G}_F$ implement polytime algorithms, we would have to explain how to construct in logspace (from S) an algorithm for the circuit outputted by the metareduction (on input c in the deterministic case, or on input c, c' in the non-deterministic case), which requires an almost equal care, but is less explicit.

The final result of [11], whose reduction is improved in the current article, is that ψ-q-**AN-dynamics** and ψ-q-**NAN-dynamics** are NP-hard or coNP-hard if ψ is non-trivial in the former case, or arborescent in the latter case. In the deterministic case, if ψ is trivial, then ψ-q-**AN-dynamics** can be solved in constant time. However, in the non-deterministic case, if ψ is not arborescent, then the complexity of ψ-q-**NAN-dynamics** is still open. The difference comes form the fact that an arborescent formula can be seen as a non-trivial formula that has also infinitely many models and counter-models of bounded treewidth. Nevertheless, dynamics of unbounded treewidth may be pertinent, and the associated decision problem may be hard as well. A track of research consists in considering other graph parameters, such as cliquewidth or twinwidth, in order to push these results forward.

As a consequence of our results, the problems ψ-q-**AN-dynamics** and ψ-q-**NAN-dynamics** are P-hard (if ψ is respectively non-trivial or arborescent). More interestingly, it showcases how to obtain P-hardness with the technique of Rice-like metareduction for the study of the complexity of finite dynamical systems introduced in [9,11]. This is among current tracks of research for the development of Rice-like metatheorems, in particular for the introduction of new relations in the signature of monadic second order formulas. Indeed, introducing relations distinguishing configurations, for example a total or partial order $<$ on binary strings, results in breaking the isomorphism invariant of $\{=, \to\}$, that is the fact that $G \sim G'$ implies $G \models \psi \iff G' \models \psi$. It follows that "easy" questions can be asked, such as whether the minimum configuration 0^n is a fixed point: $\forall c : (\forall c' : c \neq c' \implies c < c') \implies (c \to c)$. This question is obviously P-complete, and we conjecture that a Rice-like result of the form "any non-trivial question on the signature $\{=, \to, <\}$ is P-hard" holds.

Finally, providing an explicit construction of the circuits succinctly encoding the ANs produced by metareductions is also a ground step to the transfer of Rice-like complexity lower bound metatheorems to other models of computation (*e.g.* reaction systems, membrane models, DNA folding, finite cellular automata...), which may require to accommodate additional constraints.

Acknowledgments. The authors thank projects ANR-24-CE48-7504 ALARICE, HORIZON-MSCA-2022-SE-01 101131549 ACANCOS, STIC AmSud CAMA 22-STIC-02.

References

1. Alon, N.: Asynchronous threshold networks. Graph. Combin. **1**, 305–310 (1985)
2. Aracena, J., Bridoux, F., Gómez, L., Salinas, L.: Complexity of limit cycles with block-sequential update schedules in conjunctive networks. Nat. Comput. **22**(3), 411–429 (2023)
3. Bridoux, F., Gaze-Maillot, C., Perrot, K., Sené, S.: Complexity of limit-cycle problems in Boolean networks. In: Bureš, T., et al. (eds.) SOFSEM 2021. LNCS, vol. 12607, pp. 135–146. Springer, Cham (2021). https://doi.org/10.1007/978-3-030-67731-2_10
4. Chiu, A., Davida, G., Litow, B.: Division in logspace-uniform NC^1. RAIRO - Theor. Inf. Appl. **35**(3), 259–275 (2001)
5. Downey, R.G., Fellows, M.R.: Fundamentals of Parameterized Complexity. Springer (2013)
6. Espinosa-Soto, C., Padilla-Longoria, P., Alvarez-Buylla, E.R.: A gene regulatory network model for cell-fate determination during Arabidopsis thaliana flower development that is robust and recovers experimental gene expression profiles. Plant Cell **16**, 2923–2939 (2004)
7. Galperin, H., Wigderson, A.: Succinct representations of graphs. Inf. Control **56**(3), 183–198 (1983)
8. Gamard, G., Guillon, P., Perrot, K., Theyssier, G.: Rice-like theorems for automata networks. In: Proceedings of STACS 2021. LIPIcs, vol. 187, pp. 32:1–32:17. Schloss Dagstuhl (2021)
9. Gamard, G., Guillon, P., Perrot, K., Theyssier, G.: Hardness of monadic second-order formulae over succinct graphs. arXiv preprint arXiv:2302.04522v2 (2023)
10. Goles, E., Montealegre, P., Ríos Wilson, M., Theyssier, G.: On the impact of treewidth in the computational complexity of freezing dynamics. In: De Mol, L., Weiermann, A., Manea, F., Fernández-Duque, D. (eds.) CiE 2021. LNCS, vol. 12813, pp. 260–272. Springer, Cham (2021). https://doi.org/10.1007/978-3-030-80049-9_24
11. Goubault-Larrecq, A., Perrot, K.: Rice-like complexity lower bounds for Boolean and uniform automata networks. arXiv preprint arXiv:2409.08762v1 (2024)
12. Karlebach, G., Shamir, R.: Modelling and analysis of gene regulatory networks. Nat. Rev. Mol. Cell Biol. **9**, 770–780 (2008)
13. Kauffman, S.A.: Metabolic stability and epigenesis in randomly constructed genetic nets. J. Theor. Biol. **22**, 437–467 (1969)
14. Lhoussaine, C., Remy, E.: Symbolic Approaches to Modeling and Analysis of Biological Systems. ISTE Consignment. Wiley (2023)
15. Libkin, L.: Elements of Finite Model Theory. Springer (2004)
16. Mendoza, L., Alvarez-Buylla, E.R.: Dynamics of the genetic regulatory network for Arabidopsis thaliana flower morphogenesis. J. Theor. Biol. **193**, 307–319 (1998)
17. Palma, E., Salinas, L., Aracena, J.: Enumeration and extension of non-equivalent deterministic update schedules in Boolean networks. Bioinformics **32**(5), 722–729 (2016)
18. Paulevé, L., Sené, S.: Boolean networks and their dynamics: the impact of updates. In: Systems Biology Modelling and Analysis: Formal Bioinformatics Methods and Tools, pp. 173–250. Wiley (2022)
19. Perifel, S.: Complexité algorithmique. Ellipses (2014)
20. Perrot, K.: Études de la complexité algorithmique des réseaux d'automates. Habilitation thesis, Aix-Marseille University (2022)

21. Salinas, L., Gómez, L., Aracena, J.: Existence and non existence of limit cycles in Boolean networks. In: Automata and Complexity, vol. 42, pp. 233–252. Springer (2022)
22. Thieffry, D., Thomas, R.: Dynamical behaviour of biological regulatory networks – II. Immunity control in bacteriophage lambda. Bull. Math. Biol. **57**, 277–297 (1995)
23. Thomas, R.: Boolean formalization of genetic control circuits. J. Theor. Biol. **42**, 563–585 (1973)
24. Wegener, I.: The Complexity of Boolean Functions. Wiley-Teubner (1987)

Binary Expansions of Regular Reals and Reordered Computable Numbers

Peter Hertling(✉) and Philip Janicki

Fakultät für Informatik, Institut für Theoretische Informatik,
Mathematik und Operations Research, Universität der Bundeswehr München,
85579 Neubiberg, Germany
{peter.hertling,philip.janicki}@unibw.de

Abstract. A real number is called *left-computable* if there exists a computable increasing sequence of rational numbers converging to it. In this article, we investigate the binary expansions of two kinds of left-computable numbers.

A real number is called *strongly left-computable* if its binary expansion is a computably enumerable set. A real number is called *regular* if it can be written as the sum of finitely many strongly left-computable numbers. Finally, a real number x is called *reordered computable* if there exist a computable function $f\colon \mathbb{N} \to \mathbb{N}$ with $\sum_{k=0}^{\infty} 2^{-f(k)} = x$ and a bijective function $\sigma\colon \mathbb{N} \to \mathbb{N}$ such that the rearranged series $\sum_{k=0}^{\infty} 2^{-f(\sigma(k))}$ converges computably. Every strongly left-computable number is regular and every regular number is reordered computable, but none of these implications can be reverted.

We clarify the relation between regular numbers, reordered computable numbers, and left-computable numbers that are immune or hyperimmune. In particular, we show that there exists a regular number which is immune and that every left-computable number which is not immune is reordered computable. On the other hand, we show that a regular number cannot be hyperimmune, but that there exists a reordered computable number which is hyperimmune. Finally, we show that there is a reordered computable and immune number that is neither regular nor hyperimmune.

Keywords: left-computable numbers · reordered computable numbers · regular numbers · immune sets · hyperimmune sets

1 Introduction and Statement of Results

For any set $A \subseteq \mathbb{N}$, we define $x_A := \sum_{n \in A} 2^{-(n+1)}$, which is a real number in the interval $[0, 1]$. It is well-known that A is a computable set if and only if x_A is a computable real number. We say that a real number $x \in [0, 1]$ is *strongly left-computable* if there exists a computably enumerable set $A \subseteq \mathbb{N}$ with $x_A = x$. Every strongly left-computable number is left-computable, that is, there exists a computable increasing sequence of rational numbers converging to it. However,

© The Author(s), under exclusive license to Springer Nature Switzerland AG 2025
A. Beckmann et al. (Eds.): CiE 2025, LNCS 15764, pp. 253–267, 2025.
https://doi.org/10.1007/978-3-031-95908-0_18

the converse is not true as pointed out by Jockusch (see Soare [5]). For the remainder of the article, whenever we speak about left-computable numbers we mean left-computable numbers in the unit interval.

Wu [9] has introduced the *regular* numbers.

Definition 1 (Wu [9]). *A real number is called* regular *if it can be written as the sum of finitely many strongly left-computable numbers.*

Wu [9] has also shown that the regular numbers form a proper hierarchy: for every $n \in \mathbb{N}$ there exists a number that can be written as the sum of $n+1$ strongly left-computable numbers but not as the sum of n strongly left-computable numbers.

Recall that an infinite set $A \subseteq \mathbb{N}$ is *immune* if it does not have a computably enumerable infinite subset; compare, e.g., [7]. This is equivalent to not having a computable infinite subset. We say that a real number $x \in [0,1]$ is *immune* if there exists an immune set $A \subseteq \mathbb{N}$ with $x_A = x$. First, we clarify the relation between regular numbers and immune numbers; compare Fig. 1. It is clear that the strongly left-computable numbers (which form the first level of the hierarchy of regular numbers) cannot be immune. Also in the next level of the hierarchy of regular numbers, there are numbers that are not immune. For example, the number x_A for the following set A is regular and not immune but not strongly left-computable: $A := \{2n \mid n \in K\} \cup \{1+2n \mid n \in \mathbb{N} \setminus K\}$, where K is an arbitrary c.e. set that is not computable (this number is regular because it is the sum of the strongly left-computable number x_{1+2K} and the computable number $x_{1+2\mathbb{N}} = 1/3$). But, in this level of the hierarchy of regular numbers there are immune numbers as well.

Theorem 1. *There exist a computable number and a strongly left-computable number such that their sum is immune.*

Corollary 1. *There exists a regular number which is immune.*

We already mentioned that Wu [9] has shown that the regular numbers form a proper hierarchy. In particular, there exist two strongly left-computable numbers whose sum is not strongly left-computable; actually, the example preceding Theorem 1 shows this as well. Theorem 1 strengthens this assertion considerably: there exist a computable and a strongly left-computable number such that their sum is even immune. Another strengthening has been obtained by Ambos-Spies and Zheng [1].

Of course, there are left-computable numbers that are immune and not regular. Examples will be given later; see Fig. 1. There also exists a left-computable number which is neither immune nor regular. This follows from two results by Hölzl, Janicki, Merkle and Stephan: [3, Theorem 8] together with [3, Theorem 15]. This observation can also be expressed as follows: for a left-computable number, being not immune does not imply regularity. But we will show that any left-computable number in $[0,1]$ that is not immune satisfies at least a somewhat weaker computability property with respect to convergence.

Definition 2 (Janicki [4]). *A real number x is called* reordered computable *if there exist a computable function $f\colon \mathbb{N} \to \mathbb{N}$ with $\sum_{k=0}^{\infty} 2^{-f(k)} = x$ and a bijective function $\sigma\colon \mathbb{N} \to \mathbb{N}$ such that the rearranged series $\sum_{k=0}^{\infty} 2^{-f(\sigma(k))}$ converges computably. In this case we call f a* name *for x.*

Recall that a sequence $(a_n)_n$ *converges computably* if it converges and if there exists a computable function $g : \mathbb{N} \to \mathbb{N}$ such that, for all m, n, the assumption $m \geq g(n)$ implies $|a_m - \lim_{k\to\infty} a_k| \leq 2^{-n}$. Every regular number is reordered computable but the converse is not true [4].

Theorem 2. *Every left-computable number that is not immune is reordered computable.*

The converse is not true according to Corollary 1 because every regular number is reordered computable.

Recall that an infinite set $A \subseteq \mathbb{N}$ is *hyperimmune* if there is no computable function $f\colon \mathbb{N} \to \mathbb{N}$ with $|A \upharpoonright f(n)| \geq n$ for all $n \in \mathbb{N}$. It is obvious that every hyperimmune set is immune and it is well known that the converse is not true; see, e.g., [7]. We say that a real number $x \in [0, 1]$ is *hyperimmune* if there exists a hyperimmune set $A \subseteq \mathbb{N}$ with $x_A = x$. Soare [6] has already shown that there exists a left-computable number that is hyperimmune. Stephan and Wu [8] have shown that every left-computable number that is nearly computable (a real number x is called *nearly computable* [2] if there exists a computable sequence $(a_n)_n$ of rational numbers converging to x such that for every increasing function $r : \mathbb{N} \to \mathbb{N}$ the sequence $(a_{r(n+1)} - a_{r(n)})_n$ converges computably to 0) but not computable is hyperimmune (we remark that in this result the assumption that the number is left-computable can be omitted [2]). In view of Corollary 1 the question arises whether there exists a regular number that is hyperimmune. The answer is no.

Theorem 3. *A regular number cannot be hyperimmune.*

But the answer is yes if we switch from regular numbers to the larger class of reordered computable numbers.

Theorem 4. *There is a reordered computable number which is hyperimmune.*

We wish to clarify completely the relation between regular numbers, reordered computable numbers and hyperimmune numbers. First we observe that the converse of Theorem 3 is not true. The left-computable Martin-Löf random numbers in $[0, 1]$, that is, the Ω numbers, serve as counterexamples because they are neither hyperimmune nor reordered computable [4], hence, not regular. This also complements Theorem 4 concerning the relationship between reordered computable numbers and hyperimmune numbers. The left-computable numbers in $[0, 1]$ that are nearly computable but not computable are hyperimmune [8] and not reordered computable [4].

The relations stated so far between left-computable numbers with the considered possible additional properties are all depicted in Fig. 1. In particular, the union of the set of regular numbers and the set of left-computable numbers that are not immune is a subset of the intersection of the set of reordered computable numbers and the set of left-computable numbers that are not hyperimmune. In fact, it is a proper subset.

Theorem 5. *There is a reordered computable and immune number that is neither regular nor hyperimmune.*

In the following section we introduce some more notation and formulate a simple but useful lemma and a useful second characterisation of reordered computable numbers. Then, in the next three sections Theorems 1, 2, and 3 are proved. In the final two sections we sketch the proof of Theorem 4 and give some hints concerning the proof of Theorem 5.

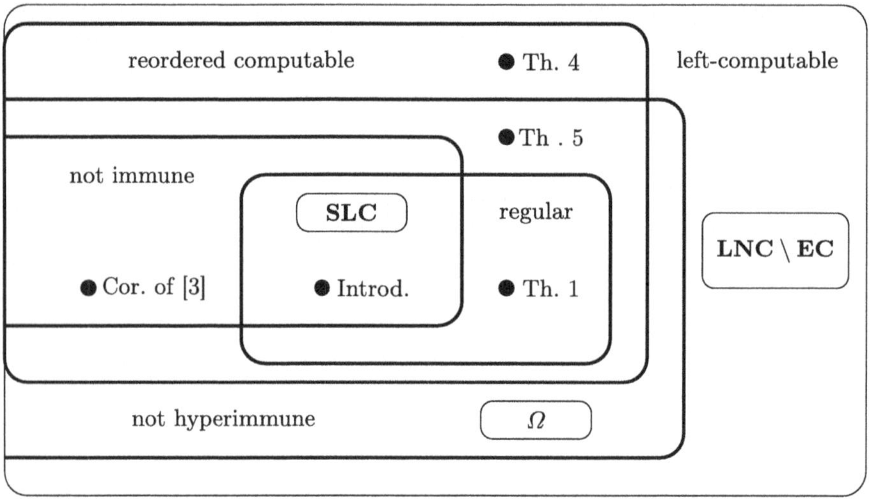

Fig. 1. The relationship between the sets of left-computable numbers that are (1) regular, (2) not immune, (3) reordered computable, or (4) not hyperimmune. The dots and the additional sets indicate examples of numbers in the respective subsets. Here **SLC** stands for the strongly left-computable numbers, Ω for the Ω numbers, and **LNC** $\setminus$ **EC** for the numbers that are left-computable and nearly computable but not computable.

2 Preliminaries

Let $\mathbb{N} = \{0, 1, 2, \ldots\}$ be the set of nonnegative integers. For $i, j \in \mathbb{N}$ we define $\langle i, j\rangle := \frac{(i+j)(i+j+1)}{2} + j$. Then $\langle \cdot, \cdot \rangle : \mathbb{N}^2 \to \mathbb{N}$ is a computable bijection with computable projection functions π_1, π_2 defined by $\pi_1(\langle i, j\rangle) := i$, $\pi_2(\langle i, j\rangle) := j$. For a set $A \subseteq \mathbb{N}$ and a number $i \in \mathbb{N}$ we write as usual $A \restriction i := A \cap \{0, \ldots, i-1\}$. Sometimes, in order to simplify the notation, we will identify subsets of the natural numbers with infinite binary sequences. For example, we call $A \upharpoonright i$ the *prefix of length* i *of* A. For a set $A \subseteq \mathbb{N}$ we write $1 + A := \{1 + n \mid n \in A\}$. Note that $i \in (1 + A)$ iff the sum $\sum_{j \in A} 2^{-(j+1)}$ contains the additive term 2^{-i} iff the ith digit after the binary point in the binary representation of x_A is equal to one. We shall make use of the following lemma. The easy verification is left to the reader.

Lemma 1. *Let* $A \subseteq \mathbb{N}$ *be an infinite set whose complement* $\mathbb{N} \setminus A$ *is infinite as well. Then for any sequence* $(B_n)_n$ *of sets* $B_n \subseteq \mathbb{N}$ *the following two conditions are equivalent.*

1. *The sequence of sets* $(B_n)_n$ *converges to the set* A.
2. *The sequence of real numbers* $(x_{B_n})_n$ *converges to the real number* x_A.

For a function $f :\subseteq \mathbb{N} \to \mathbb{N}$ that has either a finite domain of definition $\text{dom}(f)$ or tends to infinity, we can define a function $u_f : \mathbb{N} \to \mathbb{N}$ by $u_f(m) := |\{k \in \text{dom}(f) \mid f(k) = m\}|$. Note that if $f : \mathbb{N} \to \mathbb{N}$ is a function such that the series $\sum_k 2^{-f(k)}$ converges, then f tends to infinity.

Lemma 2 (Janicki [4]). *Let* $f : \mathbb{N} \to \mathbb{N}$ *be a function that tends to infinity. Then the following conditions are equivalent:*

1. *There exists a bijective function* $\sigma \colon \mathbb{N} \to \mathbb{N}$ *such that the series* $\sum_k 2^{-f(\sigma(k))}$ *converges computably.*
2. *The series* $\sum_k 2^{-f^*(k)}$ *converges computably, where* $f^* : \mathbb{N} \to \mathbb{N}$ *is the uniquely determined nondecreasing function with* $u_{f^*} = u_f$.
3. *The series* $\sum_m u_f(m) \cdot 2^{-m}$ *converges computably.*

Thus, a real number x is reordered computable iff there exists a computable function $f : \mathbb{N} \to \mathbb{N}$ such that $x = \sum_{k=0}^{\infty} 2^{-f(k)}$ and such that the series $\sum_m u_f(m) \cdot 2^{-m}$ converges computably.

Let $\varphi_0, \varphi_1, \varphi_2, \ldots$ be an effective list of all computable, possibly partial functions with domain and range in $\mathbb{N}$. Here "effective" means that there exists a computable injective function $h_u : \mathbb{N} \to \mathbb{N}^3$ with $h_u(\mathbb{N}) = \{(i, x, y) \mid \varphi_i(x) = y\}$. We write $\varphi_i(x)[t] \downarrow$ iff $(\exists s < t)(\exists y) h_u(s) = (i, x, y)$. We also define W_i as the domain of definition of φ_i. Then $W_0, W_1, W_2, \ldots$ is an effective list of all computably enumerable subsets of $\mathbb{N}$. Finally, we write $W_i[t] := \{x \in \mathbb{N} \mid \varphi_i(x)[t] \downarrow\}$. Then $(W_i[t])_t$ is a computable, nondecreasing sequence of finite sets converging to W_i.

3 Proof of Theorem 1

We shall construct a computable set $A \subseteq \mathbb{N}$ and a c.e. set $B \subseteq \mathbb{N}$ such that there is a unique set $C \subseteq \mathbb{N}$ with $x_A + x_B = x_C$, and this set C is immune. First, we explain the construction of C. In the end we will show how to construct suitable A and B satisfying $x_A + x_B = x_C$.

We have to make sure that C is infinite and that for every i such that W_i is infinite we have $W_i \not\subseteq C$. The set C will be constructed in stages, that is, we define a computable sequence $(C[t])_t$ of sets converging to C. Let the function $S : \mathbb{N} \to \mathbb{N}$ be defined by

$$S(i) := \sum_{k=1}^{i} k = \frac{i \cdot (i+1)}{2},$$

for $i \in \mathbb{N}$; note that $S(0) = 0$. For each $i \in \mathbb{N}$, let $I_i \subseteq \mathbb{N}$ be the interval defined by

$$I_i := \{S(i), \ldots, S(i) + i\} = \{S(i), \ldots, S(i+1) - 1\}.$$

For $i \in \mathbb{N}$, the set I_i has exactly $i+1$ elements. The set $\mathbb{N}$ is the disjoint union of all intervals I_i for $i \in \mathbb{N}$. We shall make sure that for every i and every t the set $C[t] \cap I_i$ contains exactly one element. We start with the largest element in each interval I_i and occasionally decrease it by 1, but at most i times. This ensures that $(C[t])_t$ converges to a set C such that, for every i, the set $C \cap I_i$ contains exactly one element as well. Thus, C is infinite. How do we ensure that for all i with W_i infinite we have $W_i \not\subseteq C$? Let $g :\subseteq \mathbb{N} \to \mathbb{N}$ be the partial computable function defined by

$$\mathrm{dom}(g) := \{i \in \mathbb{N} \mid W_i \not\subseteq \{0, \ldots, S(i+1) - 1\}\}$$

and, for $i \in \mathrm{dom}(g)$,

$$\begin{aligned} g(i) := \ & \text{the unique number } n \in W_i[t] \text{ with } n \geq S(i+1) \\ & \text{where } t = \min\{t' \mid W_i[t'] \not\subseteq \{0, \ldots, S(i+1) - 1\}\}. \end{aligned}$$

In other words,

- if $W_i \subseteq \{0, \ldots, S(i+1) - 1\}$, then $g(i)$ is undefined,
- if $W_i \not\subseteq \{0, \ldots, S(i+1) - 1\}$, then $g(i)$ is the first number greater than or equal to $S(i+1)$ enumerated by h into W_i.

In particular, if W_i is infinite, then $g(i)$ is defined. We shall make sure that for all i such that $g(i)$ is defined, $g(i) \notin C$. The idea is to decrease the current number in $C[t] \cap I_j$ by one if this number turns out to be equal to $g(i)$.

We define a function $r : \mathbb{N}^2 \to \mathbb{N}$ as follows in stages:

- In stage 0 we define $r(i)[0] := S(i+1) - 1$, for all $i \in \mathbb{N}$.
- In any stage $t > 0$, we define $e := \pi_1(t)$ and

$$r(i)[t] := \begin{cases} r(i)[t-1] - 1 & \text{if } W_e[t] \not\subseteq \{0, \ldots, S(e+1) - 1\} \\ & \quad \text{and } g(e) = r(i)[t-1], \\ r(i)[t-1] & \text{otherwise,} \end{cases}$$

  for all $i \in \mathbb{N}$. Finally, we define $C[t] := \{r(i)[t] \mid i \in \mathbb{N}\}$.

This ends the definition of r and $(C[t])_t$.

It is clear that r is computable. It is also clear that for every $i \in \mathbb{N}$ the sequence $(r(i)[t])_t$ is non-increasing. Let us fix a number i. Remember that I_i

contains exactly $i+1$ elements and that $(r(i)[t])_t$ starts with the largest value in I_i, that is, with the value $r(i)[0] = S(i+1)-1$. This value is decreased only when there is some e with $g(e)\downarrow$ and such that $g(e)$ is the current value of the sequence. But there can be at most i numbers e with $g(e)\downarrow$ and $g(e) \in I_i$, due to $g(e) \geq S(e+1)$ if $g(e)\downarrow$. Thus, $(r(i)[t])_t$ never leaves the interval I_i, in fact, $C[t] \cap I_i = \{r(i)[t]\}$. Furthermore, $(r(i)[t])_t$ is eventually constant with $R(i) := \lim_{t\to\infty} r(i)[t] \in I_i$. This implies that $C := \lim_{t\to\infty} C[t]$ exists and that $C \cap I_i = \{R(i)\}$. Hence, C is infinite. Let us consider a number e such that $g(e)$ is defined. This is in particular the case if W_e is infinite. Let i be the uniquely determined number with $g(e) \in I_i$. For any $i' \neq i$ and any t we have $r(i')[t] \in I_{i'}$, hence, $r(i')[t] \neq g(e)$, hence, $R(i') \neq g(e)$. We claim that also $R(i) \neq g(e)$. Otherwise, for some sufficiently large $t > 0$ we would have $e = \pi_1(t)$ and $W_e[t] \not\subseteq \{0,\ldots,S(e+1)-1\}$ and $r(i)[t-1] = g(e)$. But then by construction we would get $r(i)[t] = r(i)[t-1]-1 < g(e)$, which implies $R(i) < g(e)$, a contradiction. Thus, $R(i) \neq g(e)$ for all i. This implies $g(e) \notin C$, hence, $W_e \not\subseteq C$. We have shown that C is immune.

Finally, we define $A := \{S(i+1)-1 \mid i \in \mathbb{N}\}$ and

$$B := \{n \in \mathbb{N} \mid (\exists i,t)\,(r(i)[t] = n \text{ and } r(i)[t+1] \neq n)\}.$$

It is clear that A is computable, that B is computably enumerable, and that

$$\begin{aligned} x_A + x_B &= \sum_{i\in\mathbb{N}} 2^{-S(i+1)} + \sum_{i\in\mathbb{N}} \left(2^{-(2+R(i))} + \ldots + 2^{-S(i+1)}\right) \\ &= \sum_{i\in\mathbb{N}} 2^{-(1+R(i))} = x_C. \end{aligned}$$

4 Proof of Theorem 2

Let $A \subseteq \mathbb{N}$ be a set such that x_A is not immune. We can assume that neither A nor $\mathbb{N} \setminus A$ are finite, otherwise x_A is a dyadic number, which is reordered computable in particular. Since A is not immune, A has an infinite computable subset. Fix some computable increasing function $h\colon \mathbb{N} \to \mathbb{N}$ with $h(\mathbb{N}) \subseteq A$. Since x_A is left-computable, there exists a computable sequence $(B[n])_n$ of finite sets of natural numbers such that the sequence $(x_{B[n]})_n$ is increasing and converges to x_A. According to Lemma 1, the sequence $(B[n])_n$ converges to the set A. This allows us to define a function $s\colon \mathbb{N} \to \mathbb{N}$ recursively as follows:

$$\begin{aligned} s(0) &:= 0 \\ s(n+1) &:= \min\{m > s(n) \mid \{h(0),\ldots,h(n)\} \subseteq B[m]\} \end{aligned}$$

Obviously, s is well-defined, computable and increasing. Furthermore, we define the sequence $(C[n])_n$ as follows:

$$\begin{aligned} C[0] &:= \emptyset \\ C[n+1] &:= B[s(n+1)] \upharpoonright (h(n)+1) \end{aligned}$$

Clearly, $(C[n])_n$ is a computable sequence of finite sets of natural numbers, and the sequence $(x_{C[n]})_n$ is increasing—indeed, for any $n \in \mathbb{N}$ obviously $x_{B[s(n+1)]} > x_{B[s(n)]}$, and this implies

$$x_{B[s(n+1)]\restriction(h(n-1)+1)} \geq x_{B[s(n)]\restriction(h(n-1)+1)} = x_{C[n]},$$

furthermore $h(n) \in C[n+1] \setminus (B[s(n+1)] \restriction (h(n-1)+1))$—and converges to x_A.

We want to show that x_A is reordered computable. We define a computable function $f : \mathbb{N} \to \mathbb{N}$ witnessing this property in stages. In stage 0 we start with the function with empty domain of definition, and in any stage $t+1$ we append finitely many values as follows. Let $f[t]$ be the finite prefix of f that is defined at the end of stage t. For each $t \in \mathbb{N}$, let $D[t] \subseteq \mathbb{N}$ be the uniquely determined finite set of natural numbers with $x_{D[t]} = x_{C[t+1]} - x_{C[t]}$. Then $D[t] \subseteq \{0, \ldots, h(t)\}$ and $h(t) \in D[t]$. We fix the next $|D[t]| > 0$ values of f by demanding that these values should be exactly the values in the set $1 + D[t]$ in increasing order. In this way a computable function $f : \mathbb{N} \to \mathbb{N}$ is defined. Obviously, it satisfies $x_A = \sum_{k=0}^{\infty} 2^{-f(k)} = \sum_{m=0}^{\infty} u_f(m) \cdot 2^{-m}$. We claim that the series $\sum_{m=0}^{\infty} u_f(m) \cdot 2^{-m}$ converges computably. In fact, we claim that

$$\sum_{m=h(n)+2}^{\infty} u_f(m) \cdot 2^{-m} \leq 2^{-n-1}$$

for all $n \in \mathbb{N}$.

Fix some arbitrary $n \in \mathbb{N}$. First we will count how many carries occur at the position $h(n)$ during the approximation $(x_{C[t]})_t$.

Therefore, we define the sequence $(R[t])_t$ by

$$R[t] := \left\lfloor 2^{h(n)+1} \cdot \sum_{m=h(n)+2}^{\infty} u_{f[t]}(m) \cdot 2^{-m} \right\rfloor$$

for all $t \in \mathbb{N}$. We observe

$$\sum_{m=h(n)+2}^{\infty} u_{f[t]}(m) \cdot 2^{-m} < (R[t]+1) \cdot 2^{-(h(n)+1)}.$$

Note that the sum appearing in the definition of $R[t]$ is actually a finite sum because of $u_{f[t]}(m) = 0$ for almost all m. Obviously, $(R[t])_t$ is non-decreasing and bounded. Let $R := \lim_{t\to\infty} R[t]$. Then we obtain the inequality

$$\sum_{m=h(n)+2}^{\infty} u_f(m) \cdot 2^{-m} \leq (R+1) \cdot 2^{-(h(n)+1)}.$$

As for all $m, t \in \mathbb{N}$ we have $u_{f[t+1]}(m) \leq u_{f[t]}(m) + 1$, the difference between $\sum_{m=h(n)+2}^{\infty} u_{f[t+1]}(m) \cdot 2^{-m}$ and $\sum_{m=h(n)+2}^{\infty} u_{f[t]}(m) \cdot 2^{-m}$ is strictly less than $2^{-(h(n)+1)}$, which implies $R[t+1] \leq R[t] + 1$. Then the set

$$T := \{t \in \mathbb{N} \mid R[t+1] > R[t]\}$$

is finite and its cardinality is $|T| = R$. Note that the elements in T are exactly the stages when a carry occurs at the position $h(n)$ (that means: at the $(h(n)+1)$th position after the binary point). First, we claim that $t > n+1$ for all $t \in T$. In any stage $t \in T$ at least one number $m > h(n)+1$ must be enumerated by f. By the definition of $(D[t])_t$, we have $D[t] \subseteq \{0, \dots, h(t)\}$ for all $t \in \mathbb{N}$. Therefore, in any stage $t \leq n+1$ only numbers $m \leq h(n)+1$ can be enumerated by f. In other words, a number which is greater than $h(n)+1$ can only be enumerated by f after stage $n+1$. This shows that $t > n+1$ for all $t \in T$. Next, we observe that, whenever a carry at position $h(n)$ occurs, the binary expansion of $x_{C[t]}$ changes on the first $h(n)+1$ bits. But, for any stage $t > n+1$ the $n+1$ bits in the binary expansion of $x_{C[t]}$ at the positions $h(0), \dots, h(n)$ are fixed with value 1. There are exactly $2^{h(n)-n}$ binary strings of length $h(n)+1$ with this property. Therefore, there are at most $2^{h(n)-n} - 1$ stages after $n+1$ when the binary expansion of $(x_{C[t]})_t$ changes on the first $h(n)+1$ bits (remember that the sequence $(x_{C[t]})_t$ is increasing; this implies that the sequence $(x_{C[t]\restriction(h(n)+1)})_t$ is non-decreasing). So we have $|T| \leq 2^{h(n)-n} - 1$. Putting everything together, we obtain

$$\sum_{m=h(n)+2}^{\infty} u_f(m) \cdot 2^{-m} \leq (R+1) \cdot 2^{-(h(n)+1)}$$
$$= (|T|+1) \cdot 2^{-(h(n)+1)} \leq 2^{h(n)-n} \cdot 2^{-h(n)-1} = 2^{-n-1},$$

which shows that x_A is reordered computable.

5 Proof of Theorem 3

Let $x \in [0,1]$ be a regular number. We can assume that x is not strongly left-computable, since strongly left-computable numbers are not even immune. Let $A \subseteq \mathbb{N}$ be the unique set with $x_A = x$. Then both A and $\mathbb{N} \setminus A$ are infinite. As x is regular, there exists a computable name $f\colon \mathbb{N} \to \mathbb{N} \setminus \{0\}$ for x such that u_f is bounded. We define $c := \limsup_{n\to\infty} u_f(n)$. Note that we must have $c \geq 2$, since x is not strongly left-computable. For all $t \in \mathbb{N}$, let $A[t]$ be the unique finite set with $x_{A[t]} = \sum_{k=0}^{t-1} 2^{-f(k)} = \sum_{k\in\mathrm{dom}(f\restriction t)} 2^{-f(k)}$, where $f \restriction t : \{0, \dots, t-1\} \to \mathbb{N}$ is the restriction of f to the set $\{0, \dots, t-1\}$. According to Lemma 1 the sequence $(A[t])_t$ converges to A.

We will show that there exists a computable function $h\colon \mathbb{N} \to \mathbb{N}$ satisfying $|(1+A) \restriction h(n)| \geq n$ for all $n \in \mathbb{N}$. This suffices to show that A (and therefore x) is not hyperimmune.

Obviously, for any k the sequence $(u_{f\restriction t}(k))_t$ is nondecreasing and converges to $u_f(k)$. We choose a number $d \in \mathbb{N}$ with $d > \log_2(c) + 2$ and define $I_n := \{n \cdot d, \dots, (n+1) \cdot d - 1\}$ for all $n \in \mathbb{N}$. Then each I_n is an interval in $\mathbb{N}$ of length d and $\mathbb{N}$ is the disjoint union of all the intervals I_n. Furthermore, let $S := \{0, \dots, c\}$, and define on S^d the following partial order:

$$(x_1, \dots, x_d) \leq (y_1, \dots, y_d) :\Longleftrightarrow x_i \leq y_i \text{ for all } i \in \{1, \dots, d\}.$$

Fix some maximal $w \in S^d$ satisfying $(u_f(n \cdot d), \ldots, u_f((n+1) \cdot d - 1)) = w$ for infinitely many $n \in \mathbb{N}$. Then $w \neq 0^d$ because there are infinitely many n with $u_f(n) = c > 0$. The set

$$U := \{n \in \mathbb{N} \mid (u_f(n \cdot d), \ldots, u_f((n+1) \cdot d - 1)) = w\}$$

is infinite. We consider the following two disjoint cases.

Case 1: For almost all $n \in U$, we have $(1 + A) \cap I_n \neq \emptyset$.

In this case there is a number $D_1 \in \mathbb{N}$ satisfying $(1 + A) \cap I_n \neq \emptyset$ for all $n \in U$ with $n \geq D_1$. And there is a number $D_2 \in \mathbb{N}$ satisfying

$$(u_f(n \cdot d), \ldots, u_f((n+1) \cdot d - 1)) \neq u$$

for all $n \geq D_2$ and all $u > w$. Set $D := \max\{D_1, D_2\}$. We fix computable increasing functions $r, s \colon \mathbb{N} \to \mathbb{N}$ with $s(0) \geq D$ and

$$(u_{f \upharpoonright r(n)}(s(n) \cdot d), \ldots, u_{f \upharpoonright r(n)}((s(n)+1) \cdot d - 1)) = w$$

for all $n \in \mathbb{N}$. For all $n \in \mathbb{N}$ we have $s(n) \geq D$ and $s(n) \in U$, hence $(1+A) \cap I_{s(n)} \neq \emptyset$. Define the function $h \colon \mathbb{N} \to \mathbb{N}$ by $h(n) := (s(n)+1) \cdot d$, which is computable and increasing. We obtain

$$|(1 + A) \upharpoonright h(n)| \geq n + 1$$

for all $n \in \mathbb{N}$. Hence, A is not hyperimmune.

Case 2: For infinitely many $n \in U$, we have $(1 + A) \cap I_n = \emptyset$.

Remember that $(A[t])_t$ converges to A. Hence, in this case, for infinitely many $n \in U$ there is a number $s \in \mathbb{N}$ such that, for all $t \geq s$,

- $(u_{f \upharpoonright t}(n \cdot d), \ldots, u_{f \upharpoonright t}((n+1) \cdot d - 1)) = w$ and
- $(1 + A[t]) \cap I_n = \emptyset$.

Fix some number $D \in \mathbb{N}$ with

$$(u_f(n \cdot d), \ldots, u_f((n+1) \cdot d - 1)) \neq u$$

for all $n \geq D$ and all $u > w$, and such that $u_f(n) \leq c$ for all $n \geq D$. Then there exist computable increasing functions $r, s \colon \mathbb{N} \to \mathbb{N}$ satisfying the following four conditions:

1. $s(0) \geq D$,
2. $(u_{f \upharpoonright r(n-1)}(s(n) \cdot d), \ldots, u_{f \upharpoonright r(n-1)}((s(n)+1) \cdot d - 1)) = 0^d$, for $n > 0$,
3. $(u_{f \upharpoonright r(n)}(s(n) \cdot d), \ldots, u_{f \upharpoonright r(n)}((s(n)+1) \cdot d - 1)) = w$, for all n,
4. $(1 + A[r(n)]) \cap I_{s(n)} = \emptyset$, for all n.

Indeed, one observes that numbers $r(0)$ and $s(0)$ satisfying the conditions 1, 3, 4 must exist and can therefore be found. And, once for some $n > 0$ suitable numbers $r(n-1)$ and $s(n-1)$ have been defined, one observes that numbers $r(n) > r(n-1)$ and $s(n) > s(n-1)$ satisfying the conditions 2, 3, 4 must exist and can therefore be found.

Lemma 3. *For all n:* $(1 + A) \cap \{s(n) \cdot d, \ldots, s(n+1) \cdot d - 1\} \neq \emptyset$.

Before we prove this claim, let us conclude the proof. The function $h\colon \mathbb{N} \to \mathbb{N}$ defined by $h(n) := s(n) \cdot d$ is computable and increasing. We obtain:

$$|(1 + A) \upharpoonright h(n)| \geq n$$

for all $n \in \mathbb{N}$. Hence, A is not hyperimmune.

Proof (of Lemma 3). First, we explain the idea in terms of the binary representation of $x_{A[t]}$ for various stages t. When we speak about ones or zeros "in an interval J" we mean ones or zeros in the jth positions, for $j \in J$, of the binary representation of $x_{A[t]}$ at the currently considered stage t. For the following sketch the reader is advised to have a look at Fig. 2. For each n we can find a number $s(n)$ and a sufficiently large stage $r(n)$ so that, on the one hand, the additive terms of the form 2^{-i} for $i \in I_{s(n)}$ have already been enumerated (this is the third condition above) and, on the other hand, the binary representation of $x_{A[r(n)]}$ contains only zeros in the interval $I_{s(n)}$ (this is the fourth condition above). In addition, we choose $s(n)$ so that until the previously considered stage $r(n-1)$ no additive terms of the form 2^{-i} for $i \in I_{s(n)}$ had been enumerated (this is the second condition above). Then, due to the third condition applied to $n+1$, some ones will appear between the stages $r(n)$ and $r(n+1)$ in the interval $I_{s(n+1)}$. But, as at stage $r(n+1)$ the interval $I_{s(n+1)}$ does not contain any ones anymore (due to the fourth condition applied to $n+1$), the ones in this interval must have been moved to the left due to carries. We claim that some ones must remain in the interval $\{s(n) \cdot d, \ldots, s(n+1) \cdot d - 1\}$ (note that $s(n) \cdot d$ is the left endpoint of the interval $I_{s(n)}$ while $s(n+1) \cdot d - 1$ is one position to the left of the left endpoint of $I_{s(n+1)}$). Otherwise, these ones would have to have been moved by carries through the interval $I_{s(n)}$. But that is impossible because at stage $r(n)$ this interval contained only zeros and no terms of the form 2^{-i} for $i \in I_{s(n)}$ are added after stage $r(n)$ and the total sum of all smaller terms 2^{-j} is too small to force a carry through the whole interval $I_{s(n)}$.

Now we come to the formal proof. The assertion in Lemma 3 is equivalent to the assertion that for all n

$$x_{\{m \in A \mid 1+m \geq s(n) \cdot d\}} \geq 2^{-(s(n+1) \cdot d - 1)}; \tag{1}$$

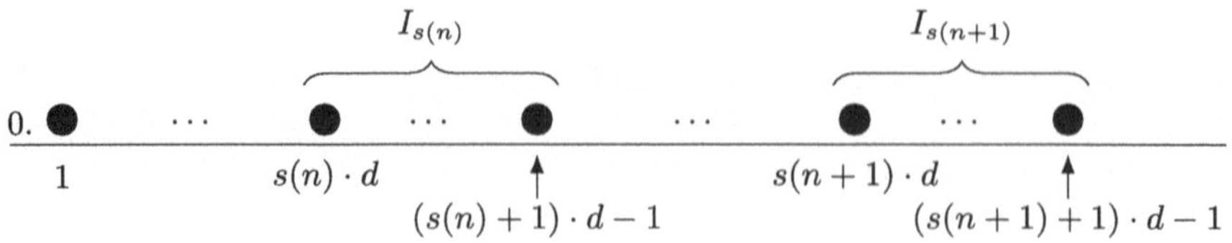

Fig. 2. The binary representation of x_A and the intervals $I_{s(n)}$ and $I_{s(n+1)}$.

compare Fig. 2.

Note that $x_{\{m \in A \mid 1+m \geq s(n) \cdot d\}}$ is the number that one obtains from x_A when in its binary expansion one sets the first $(s(n) \cdot d - 1)$ digits after the binary point to 0. Let us fix a number $n \in \mathbb{N}$ and let us look at the sequence

$$\left(x_{\{m \in A[t] \mid 1+m \geq s(n) \cdot d\}}\right)_{t \geq r[n]}.$$

It converges to $x_{\{m \in A \mid 1+m \geq s(n) \cdot d\}}$. Due to the fourth condition satisfied by r and s it starts with a small value:

$$x_{\{m \in A[r(n)] \mid 1+m \geq s(n) \cdot d\}} < 2^{-((s(n)+1) \cdot d - 1)}. \tag{2}$$

The number $x_{\{m \in A[t] \mid 1+m \geq s(n) \cdot d\}}$ for $t \geq r(n)$ can change only through additive terms of the form $2^{-f(i)}$ for $i \geq r(n)$ and $f(i) \geq s(n) \cdot d$. Can the value $x_{\{m \in A[t] \mid 1+m \geq s(n) \cdot d\}}$ ever decrease from step t to $t+1$? No, because for this there would have to be a carry to the $(s(n) \cdot d - 1)$th digit after the binary point, and this can happen only if $x_{\{m \in A[t] \mid 1+m \geq s(n) \cdot d\}}$ reaches or exceeds the value $2^{-(s(n) \cdot d - 1)}$. But this will never happen because of (2) and because for all $i \geq r(n)$ with $f(i) \geq s(n) \cdot d$ the third condition actually implies $f(i) \geq (s(n)+1) \cdot d$, hence, the sum of all terms $2^{-f(i)}$ for $i \geq r(n)$ with $f(i) \geq s(n) \cdot d$ is bounded by

$$\sum_{k=(s(n)+1) \cdot d}^{\infty} c \cdot 2^{-k} = c \cdot 2^{-((s(n)+1) \cdot d - 1)} < 2^{d-2} \cdot 2^{-((s(n)+1) \cdot d - 1)}$$

(remember that $u_f(k) \leq c$ for all $k \geq D$). We conclude that

$$\text{the sequence } \left(x_{\{m \in A[t] \mid 1+m \geq s(n) \cdot d\}}\right)_{t \geq r[n]} \text{ is nondecreasing.} \tag{3}$$

Now let us look at its value for $t = r(n+1)$. Due to the second and the third condition applied to $n+1$ the sum of the terms of the form $2^{-f(i)}$ for $i \geq r(n)$ and $f(i) \geq s(n) \cdot d$ that are added in stages t with $r(n) \leq t < r(n+1)$ is at least $\nu_{\mathbb{N}}(w) \cdot 2^{-((s(n+1)+1) \cdot d - 1)}$ (where here with $\nu_{\mathbb{N}}(w)$ we mean the natural number $\sum_{i=0}^{|w|-1} w(i) \cdot 2^{|w|-1-i}$). According to (3) adding all these terms does not lead to a carry to the $(s(n) \cdot d - 1)$th digit after the binary point. Hence, all of these additive terms contribute to $x_{\{m \in A[r(n+1)] \mid 1+m \geq s(n) \cdot d\}}$, and we obtain

$$x_{\{m \in A[r(n+1)] \mid 1+m \geq s(n) \cdot d\}} \geq \nu_{\mathbb{N}}(w) \cdot 2^{-((s(n+1)+1) \cdot d - 1)} \geq 2^{-((s(n+1)+1) \cdot d - 1)}.$$

But the fourth condition applied to $n+1$ says that $(1+A[r(n+1)])\cap I_{s(n+1)} = \emptyset$, and this means that

$$\begin{array}{ll} \text{if} & x_{m\in A[r(n+1)]|1+m\geq s(n)\cdot d\}} \geq 2^{-((s(n+1)+1)\cdot d-1)} \\ \text{then even} & x_{m\in A[r(n+1)]|1+m\geq s(n)\cdot d\}} \geq 2^{-(s(n+1)\cdot d-1)}. \end{array}$$

We conclude

$$x_{\{m\in A[r(n+1)]|1+m\geq s(n)\cdot d\}} \geq 2^{-(s(n+1)\cdot d-1)}.$$

Using (3) we conclude that (1) is true. Hence, Lemma 3 is proved.

6 Sketch of the Proof of Theorem 4

In the proof we use the following notation: for an infinite set $B \subseteq \mathbb{N}$ we denote by $p_B : \mathbb{N} \to \mathbb{N}$ the uniquely determined increasing function with $p_B(\mathbb{N}) = B$. In words: for every $i \in \mathbb{N}$ the number $p_B(i)$ is the $(i+1)$th number in B.

We will construct an infinite set $A \subseteq \mathbb{N}$ satisfying the following two conditions:

1. For all e, if $\varphi_e(e+2)\downarrow$ then $|A \upharpoonright \varphi_e(e+2)| < e+2$,
2. x_A is a reordered computable number.

It is clear that the first condition implies that A is hyperimmune.

The following algorithm for the construction of A performs a diagonalization and works in infinitely many stages. In every stage $t \in \mathbb{N}$, an infinite set $A[t] \subseteq \mathbb{N}$ is computed so that the sequence $(A[t])_t$ converges and its limit A is a set with the desired properties.

In stage 0, we define $A[0]$ to be the set described by the following binary sequence: 0100100010000100000 1..., that is,

$$A[0](s) := \begin{cases} 1 & \text{if } (\exists n) s = \frac{(n+2)(n+3)}{2} - 2, \\ 0 & \text{otherwise.} \end{cases}$$

In any stage $t+1$, we define $e := \pi_1(t+1)$ and check whether

- $\varphi_e(e+2)[t]\downarrow$
- $|A[t] \upharpoonright \varphi_e(e+2)| \geq e+2$.

If this is not the case then we set $A[t+1] := A[t]$ and continue with stage $t+2$. If this is the case then we say that e *is active in stage* $t+1$, and we set

$$A[t+1](s) := \begin{cases} 1 & \text{if } s = p_{A[t]}(e) - 1, \\ 0 & \text{if } p_{A[t]}(e) \leq s \leq \varphi_e(e+2), \\ A[t](s) & \text{otherwise,} \end{cases}$$

for $s \in \mathbb{N}$. In words: if the prefix of $A[t]$ of length $\varphi_e(e+2)$ contains at least $e+2$ numbers then we decrease the $(e+1)$th number by 1 (we will see that the resulting new number in $A[t+1]$ was not an element of $A[t]$) and remove the larger numbers in the prefix of length $\varphi_e(e+2)+1$ (note: not just of length $\varphi_e(e+2)$) of $A[t]$. This ends the description of the construction of the sequence of sets $(A[t])_t$.

Now one needs to show that the sequence $(A[t])_t$ converges and that its limit A has the desired properties. This is omitted due to space restrictions.

7 Some Hints on the Proof of Theorem 5

We shall construct a set $A \subseteq \mathbb{N}$ that is immune but not hyperimmune. In addition, we will ensure that the real number x_A is reordered computable but not regular. In order to ensure that A is immune we will ensure that $W_e \not\subseteq A$, for all $e \in \mathbb{N}$ such that W_e is infinite. We will proceed similarly as in the proof of Theorem 1. The set A will be constructed in stages: we define a computable sequence $(A[t])_t$ of sets converging to A. In the definition we make use of a function $g :\subseteq \mathbb{N} \to \mathbb{N}$, which is defined as in the proof of Theorem 1 with the exception that we do not use the function S considered in the proof of Theorem 1 but instead the following function $S : \mathbb{N} \to \mathbb{N}$ defined by $S(0) := 0$ and

$$S(k+1) := S(k) + 2 + (k+1)^2,$$

for $k \in \mathbb{N}$. For each $k \in \mathbb{N}$, let $I_k \subseteq \mathbb{N}$ be the interval defined by

$$I_k := \{S(k), \dots, S(k+1) - 1\}.$$

For $k \in \mathbb{N}$, the set I_k has exactly $S(k+1) - S(k) = 2 + (k+1)^2$ elements. The set $\mathbb{N}$ is the disjoint union of all intervals I_k for $k \in \mathbb{N}$.

While in the proof of Theorem 1 the set $C[t] \cap I_k$ (for the interval I_k considered in the proof of Theorem 1) contained exactly one element, for each k and t, here this single element is replaced by a small subinterval

$$J_k[t] := \{r(k)[t] - k, \dots, r(k)[t]\}$$

(containing exactly $k+1$ elements) of I_k such that $A[t] \cap I_k \subseteq J_k[t]$. We start with $r(k)[0] := S(k+1) - 2$ (the second largest number in I_k). The strategy for ensuring that A is immune in the proof of Theorem 1 was to move the single element in $C[t-1] \cap I_k$ one position to the left whenever it turned out to be equal to some $g(e)$. Here we proceed similarly: when necessary we delete all the elements in the current interval $J_k[t-1]$ and add instead the largest element in the interval $J_k[t]$ of the same length immediately to the left of $J_k[t-1]$; compare Fig. 3. We also make sure that $A[t] \cap I_k$ contains at least one element, for all $k \in \mathbb{N}$ and sufficiently large t. This ensures that the final set A will be infinite but not hyperimmune. The subintervals $J_k[t]$ of I_k are used in order to ensure that the number x_A is not regular. Finally, we shall show that x_A will be reordered computable by defining a computable name f of x_A that witnesses this. Actually, it is possible and makes sense to formulate the construction in terms of such a function f. A more detailed description of the idea of the construction and the precise formulation of the construction and the correctness proof are omitted due to space restrictions.

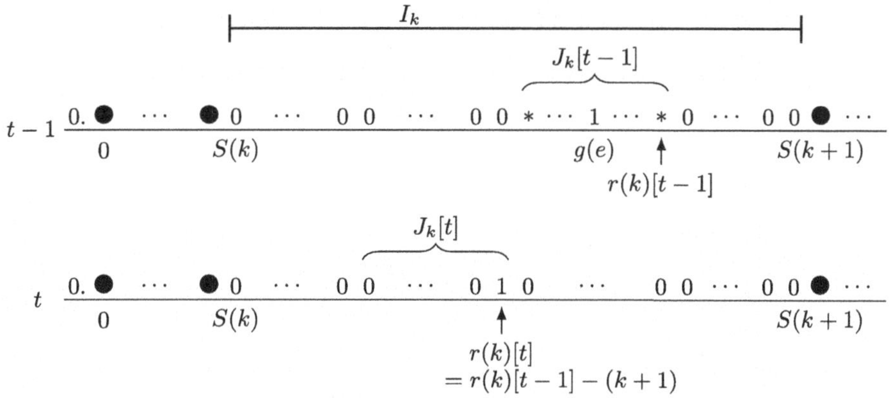

Fig. 3. A stage t in which e is active.

References

1. Ambos-Spies, K., Zheng, X.: On the differences and sums of strongly computably enumerable real numbers. In: Manea, F., Martin, B., Paulusma, D., Primiero, G. (eds.) CiE 2019. LNCS, vol. 11558, pp. 310–322. Springer, Cham (2019). https://doi.org/10.1007/978-3-030-22996-2_27
2. Hertling, P., Janicki, P.: Nearly computable real numbers. Computability **13**(1), 57–88 (2024). https://doi.org/10.3233/com-230445
3. Hölzl, R., Janicki, P., Merkle, W., Stephan, F.: Randomness versus superspeedability. In: Královič, R., Kučera, A. (eds.) 49th International Symposium on Mathematical Foundations of Computer Science (MFCS 2024). Leibniz International Proceedings in Informatics (LIPIcs), vol. 306, pp. 62:1–62:14. Schloss Dagstuhl – Leibniz-Zentrum für Informatik, Dagstuhl (2024). https://doi.org/10.4230/LIPIcs.MFCS.2024.62
4. Janicki, P.: Reordered computable numbers. Theory Comput. Syst. **68**(6), 1683–1708 (2024). https://doi.org/10.1007/s00224-024-10183-x
5. Soare, R.: Cohesive sets and recursively enumerable Dedekind cuts. Pacific J. Math. **31**, 215–231 (1969)
6. Soare, R.: Recursion theory and Dedekind cuts. Trans. Amer. Math. Soc. **140**, 271–294 (1969)
7. Soare, R.I.: Recursively enumerable sets and degrees. Perspectives in Mathematical Logic. Springer, Berlin (1987). https://doi.org/10.1007/978-3-662-02460-7
8. Stephan, F., Wu, G.: Presentations of K-trivial reals and Kolmogorov complexity. In: Cooper, S.B., Löwe, B., Torenvliet, L. (eds.) CiE 2005. LNCS, vol. 3526, pp. 461–469. Springer, Heidelberg (2005). https://doi.org/10.1007/11494645_57
9. Wu, G.: Regular reals. Math. Log. Q. **51**(2), 111–119 (2005)

Properties of Range Sets of Continuous Functions in Reverse Mathematics

Corrie Ingall(✉)

University of Connecticut, Storrs, CT 06269, USA
corrie.ingall@uconn.edu

Abstract. The range sets of continuous functions from $[0,1]$ into a complete separate metric space are closed, connected, compact, and bounded. The classification of these properties in reverse mathematics has not yet been fully explored prior to this work. Over RCA_0, the existence of closed set codes for the range sets of continuous functions is equivalent to WKL_0. The connectedness property is provable in RCA_0 and the compactness and boundedness properties are equivalent to WKL_0, as long as these properties are carefully defined to avoid the use of a closed code of $\mathsf{range(f)}$.

Keywords: Reverse Mathematics · Computability Theory · Computable Analysis

1 Introduction

1.1 Motivation

One of the first areas of interest in the early development of reverse mathematics was the classification of the most fundamental theorems of real analysis. A majority of these theorems are covered in great detail in Simpson's seminal text *Subsystems of Second Order Arithmetic* [6] and Brown's doctoral thesis *Functional Analysis in Weak Subsystems of Second Order Arithmetic* [1].

The Hahn-Mazurkiewicz Theorem (see [5]) is an early result in geometric analysis that has not yet been studied within the context of reverse mathematics.

Theorem 1 (Hahn-Mazurkiewicz). *A set $M \subseteq \mathbb{R}^n$ is the continuous image of $[0,1]$ if and only if it is compact, connected, and locally connected.*

The forward implication of the Hahn-Mazurkiewicz Theorem is, in standard mathematics, the simpler one. Proving that the image of $[0,1]$ is compact and connected is a fairly fundamental proof from a standard Real Analysis course. Local connectedness is generally not preserved by continuous functions, but if the domain is both locally connected and compact, the local connectedness is preserved. Regardless, proving a quality of the ranges of continuous functions (especially on a well-understood domain like $[0,1]$) is typically much simpler

© The Author(s), under exclusive license to Springer Nature Switzerland AG 2025
A. Beckmann et al. (Eds.): CiE 2025, LNCS 15764, pp. 268–282, 2025.
https://doi.org/10.1007/978-3-031-95908-0_19

than constructing a continuous function given a potential range set, which is required for the backwards implication.

This backwards implication has been explored within the context of computable analysis by Couch, Daniel, and McNicholl [3] and Daniel and McNicholl [4]. The forward implication, despite its relative simplicity in standard mathematics, still requires careful consideration within reverse mathematics. These considerations are the focus of this paper.

We will first provide some preliminaries concerning how we represent and work with common features of real analysis within reverse mathematics.

1.2 Preliminaries

In reverse mathematics, we work within subsystems of second order arithmetic (Z_2). The three subsystems of interest to this paper are RCA_0, WKL_0, and ACA_0.

Definition 1. RCA_0 *is the typical base system for reverse mathematics, the subsystem of* Z_2 *consisting of* PA^-*, the* Σ^0_1 *induction scheme, and the* Δ^0_1 *comprehension scheme. (See Definition I.7.4 in Simpson [6]).*

Note that RCA_0 is sufficient to prove bounded Σ^0_1 comprehension (Theorem II.3.9 in Simpson [6])

Definition 2. WKL_0 *is the subsystem of* Z_2 *consisting of* RCA_0 *and Weak König's Lemma, which states the existence of infinite paths of infinite binary trees* $T \subseteq 2^{<\mathbb{N}}$*. (See Definition I.10.1 in Simpson [6]).*

Definition 3. ACA_0 *is the subsystem of* Z_2 *consisting of* PA^-*, the arithmetical induction scheme, and the arithmetical comprehension scheme. (See Definition I.3.2 in Simpson [6]).*

Note that WKL_0 is an intermediary system, strictly stronger than RCA_0 but strictly weaker than ACA_0 [6]. Later, we will utilize the fact that WKL_0 is equivalent over RCA_0 to Bounded König's Lemma, which states that "If $T \subseteq \mathbb{N}^{<\mathbb{N}}$ is an infinite tree and $f : \mathbb{N} \to \mathbb{N}$ is a function such that for all $\sigma \in T$ and $n < |\sigma|$, $\sigma(n) \leq f(n)$, then T has an infinite path". (See Lemma IV.1.4 in Simpson [6]).

When working with real numbers in reverse mathematics, we work with them via a representation by rapidly Cauchy sequences of rationals.

Definition 4. (RCA_0) *A real number* r *is a sequence of rational numbers* $\langle q_k : k \in \mathbb{N}\rangle$ *such that for all* $k, i \in \mathbb{N}$*,* $|q_k - q_{k+i}| \leq 2^{-k}$*. (See Definition II.4.4 in Simpson [6]).*

We specify that these sequences representing real numbers are rapidly Cauchy as the explicit convergence bound of 2^{-k} makes these representations much easier to use in weak subsystems like RCA_0. For reals $x = \langle x_k : k \in \mathbb{N}\rangle$ and $y = \langle y_k : k \in \mathbb{N}\rangle$, the statement $x \leq y$ is expressed by the Π^0_1 formula $\forall k(x_k \leq y_k + 2^{-k+1})$ and the statement $x < y$ is expressed by the Σ^0_1 formula $\exists k(x_k + 2^{-k+1} < y_k)$.

As the statement $x \leq y$ is Π_1^0, it will be convenient to be able to reference statements of this form using a bounded quantifier. Given natural number m, we will state "$x \leq y$ as measured up to m" to mean $\forall k \leq m(x_k \leq y_k + 2^{-k+1})$.

We also use a similar coding for elements of any complete separable metric space. (See Definition II.5.1 in Simpson [6]).

Definition 5. (RCA$_0$) *A code for a complete separable metric space $\widehat{B}$ is given by a nonempty set $B \subseteq \mathbb{N}$ and a sequence $d : B \times B \to \mathbb{R}$ of reals satisfying the usual metric conditions.*

- $\forall a \in B(d(a, a) = 0)$
- $\forall a, b \in B(d(a, b) = d(b, a))$
- $\forall a, b, c \in B(d(a, c) \leq d(a, b) + d(b, c))$

A point $x \in \widehat{B}$ is a sequence $\langle b_k : k \in \mathbb{N}\rangle$ of elements of B such that for all $k, i \in \mathbb{N}$, $d(b_k, b_{k+i}) \leq 2^{-k}$.

Using this general definition, $\mathbb{R}$ is given by $\widehat{\mathbb{Q}}$.

Since reverse mathematics concerns second order arithmetic, we cannot work directly with open and closed subsets of $\mathbb{R}$ (or any other complete separable metric space $\widehat{B}$). Instead we consider codes of these subsets, relying on countable collections of basic open sets from which we can build our desired subsets.

Definition 6. (RCA$_0$) *A basic open subset of $\widehat{B}$ is an open metric ball with center $a \in B$ and positive rational radius r, represented as (a, r) or $B_r(a)$. An open set $U \subseteq \widehat{B}$ is given by a code $U \subseteq \mathbb{N} \times B \times \mathbb{Q}^+$ where for $x \in \widehat{B}$, $x \in U \iff \exists n, a, r$ such that $(n, a, r) \in U$ and $d(x, a) < r$. (See Definition II.5.6 in Simpson [6]).*

Closed sets are coded by the same type of sets, but the condition on membership is different.

Definition 7. (RCA$_0$) *A closed set $C \subseteq \widehat{B}$ is given by a code $C \subseteq \mathbb{N} \times B \times \mathbb{Q}^+$, where $x \in C \iff \forall (n, a, r) \in C$, $d(x, a) \geq r$. (See Definition II.5.12 in Simpson [6]).*

Note that the code of an open set provides basic open sets contained within the open set, while the code of a closed set is a code for its open complement. Even though a subset and its code are different kinds of objects, we will sometimes refer to both with the same variable as long as the context makes it clear which use is relevant.

There are two equivalent versions of Definition 7 that will be useful moving forward.

Lemma 1. (RCA$_0$) *Every closed set $C \subseteq \widehat{B}$ has a sequence code $C = \langle\langle a_n, r_n\rangle : n \in \mathbb{N}\rangle$, where each $a_n \in B$ and for $x \in \widehat{B}$, $x \in C \iff \forall n(d(x, a_n) \geq r_n)$.*

Lemma 2. (RCA$_0$) *Suppose $\varphi(x)$ is a Π^0_1 formula such that for all $x, y \in \widehat{B}$, if $x = y$ and $\varphi(x)$ then $\varphi(y)$. Then there is a closed $C \subseteq \widehat{B}$ such that for $x \in \widehat{B}$, $x \in C \iff \varphi(x)$. (See Lemma II.5.7 in Simpson [6]).*

Note that membership in a closed set is described by a Π^0_1 formula, while membership in an open set is described by a Σ^0_1 formula.

It is also necessary to work with codes for continuous functions on real numbers. We will code these functions as a set of quintuples following the approach in Simpson [6].

Definition 8. (RCA$_0$) *A code for a continuous function $f : [0,1] \to \widehat{B}$ is a set of quintuples $f \subseteq \mathbb{N} \times (\mathbb{Q} \cap [0,1]) \times \mathbb{Q}^+ \times B \times \mathbb{Q}^+$ which has the following properties. We use $(a,r)f(b,s)$ as short hand for $\exists n(n,a,r,b,s) \in f$.*

- *$(a,r)f(b,s) \wedge (a,r)f(b',s') \Longrightarrow d(b,b') \leq s + s'$*
- *$(a,r)f(b,s) \wedge (a',r') \subseteq (a,r) \Longrightarrow (a',r')f(b,s)$*
- *$(a,r)f(b,s) \wedge (b,s) \subseteq (b',s') \Longrightarrow (a,r)f(b',s')$*
- *For all $x \in [0,1]$ and for all $\epsilon > 0$, there exists $(a,r)f(b,s)$ such that $d(x,a) < r$ and $s < \epsilon$.*

(See Definition II.6.1 in Simpson [6]).

Intuitively, $(a,r)f(b,s)$ represents that the open ball $B_r(a)$ in $[0,1]$ is mapped into the closed ball $\overline{B}_s(b)$ in $\widehat{B}$. Like earlier, while a function and its code are different kinds of objects, we will often refer to both with the same variable as long as the context makes it clear which sort of use is relevant.

When working with compact sets in reverse mathematics, some caution is necessary. The Heine-Borel Theorem is equivalent to WKL$_0$ and the Bolzano-Weierstrass Theorem is equivalent to ACA$_0$ [6]. So if we want to work with compact sets within RCA$_0$, we need to avoid a definition of compactness that implies either Heine-Borel or Bolzano-Weierstrass within RCA$_0$. Therefore, the standard definition of compactness in reverse mathematics is a relatively weak one. It is usually stated for complete separable metric spaces (including in Definition III.2.3 in Simpson [6]), but here we state it for closed subsets of complete separable metric spaces.

Definition 9. (RCA$_0$) *A closed set C is a compact subset of $\widehat{B}$ if there exists an infinite sequence of finite sequences $\langle\langle x_{i,j} : i \leq n_j\rangle : j \in \mathbb{N}\rangle$ (with $x_{i,j} \in \widehat{B}$) such that for all $j \in \mathbb{N}$ and $x \in C$, there exists some $i \leq n_j$ such that $d(x_{i,j}, x) < 2^{-j}$.*

When it comes to the completeness of $\mathbb{R}$, a similar caution is needed. If we are working in RCA$_0$, we cannot think of $\mathbb{R}$ being complete in the sense of the Least Upper Bound Property as that is equivalent to ACA$_0$. Instead, we will use the Nested Interval Completeness of $\mathbb{R}$, which is provable in RCA$_0$.

Theorem 2 (Nested Interval Completeness of $\mathbb{R}$). (RCA$_0$) *Let $\langle a_n : n \in \mathbb{N}\rangle$ and $\langle b_n : n \in \mathbb{N}\rangle$ be sequences of real numbers such that for all n, $a_n \leq a_{n+1} \leq b_{n+1} \leq b_n$, and $\lim_n |a_n - b_n| = 0$. Then there exists a real number x such that $x = \lim_n a_n = \lim_n b_n$. (See Theorem II.4.8 in [6]).*

The nested intervals that are the name-sake of this theorem are the collection of $[a_n, b_n]$ for all n, and x is the unique intersection point of this collection.

Now we can move forward to the main properties of interest in this paper. In Sect. 2, we will show that the representation of $\mathsf{range(f)}$ as a closed set is equivalent to $\mathsf{WKL_0}$ over $\mathsf{RCA_0}$. This result will require us to adapt how we define properties of range sets to avoid the necessity of $\mathsf{WKL_0}$ to gain access to a code for $\mathsf{range(f)}$ as a closed set. In Sect. 3, we will show that a continuous function having connected range is provable in $\mathsf{RCA_0}$. In Sect. 4, we will show that a continuous function having compact or bounded range is equivalent to $\mathsf{WKL_0}$ over $\mathsf{RCA_0}$. In Sect. 5, we will discuss how we intend to use these results in our continued research.

2 range(f) as a Closed Set

One of the most familiar results in reverse mathematics about the range of functions is this common bridge theorem for $\mathsf{ACA_0}$.

Theorem 3. ($\mathsf{RCA_0}$) *The following are equivalent:*

- $\mathsf{ACA_0}$
- *For a one-to-one function $f : \mathbb{N} \to \mathbb{N}$, the range of f exists.*

(See Lemma III.1.3 in Simpson [6]).

Our main interest is in the range of continuous functions $f : [0, 1] \to \mathbb{R}^n$. Since $[0, 1]$ is compact, the range of f is closed. Therefore, the natural way to code $\mathsf{range(f)}$ in reverse mathematics is as a closed set.

Before we discuss our main result for this section, we will briefly discuss a special case when the code of the range set can exist within $\mathsf{RCA_0}$. The next lemma is an adaptation of the fact that $\mathsf{RCA_0}$ proves the existence of the range set of every increasing function $f : \mathbb{N} \to \mathbb{N}$.

Lemma 3. ($\mathsf{RCA_0}$) *Let $f : [0, 1] \to \mathbb{R}$ be increasing. Then there exists a code for* $\mathsf{range(f)}$ *as a closed set.*

Proof. Let f be as hypothesized. In $\mathsf{RCA_0}$ we can form the reals $f(0)$ and $f(1)$. Since f is increasing, $\mathsf{range(f)}$ is the closed set $[f(0), f(1)]$, which has a closed set code as defined by the following Π^0_1 formula.

$$y \in [f(0), f(1)] \iff f(0) \leq y \leq f(1) \tag{1}$$

□

In our proof that $\mathsf{WKL_0}$ suffices to form a code for $\mathsf{range(f)}$ as a closed set in the general case, we use $\mathsf{WKL_0}$ in two ways. First, we will use $\mathsf{WKL_0}$ to allow access to a modulus of uniform continuity of f.

Definition 10. *Let $f : [0,1] \to \widehat{B}$ be continuous. A modulus of uniform continuity of f is a function $h : \mathbb{N} \to \mathbb{N}$ where for all $n \in \mathbb{N}$ and all $x, y, \in [0,1]$, if $d(x,y) < 2^{-h(n)}$, then $d(f(x), f(y)) < 2^{-n}$. (See Definition IV.2.1 in [6]).*

Theorem 4. ($\mathsf{WKL_0}$) *For any continuous $f : [0,1] \to \widehat{B}$, there exists some modulus of uniform continuity h of f. (See Theorem IV.2.2 in [6]).*

This modulus of uniform continuity will be instrumental in the proof, but we do need the modulus to have some additional properties, which follow from Simpson's proof of Theorem 4.

Lemma 4. ($\mathsf{WKL_0}$) *Every continuous $f : [0,1] \to \widehat{B}$ has a modulus of uniform continuity h that satisfies all three of the following conditions:*

1. *$h(0) \geq 1$*
2. *h is strictly increasing*
3. *For every n and $i \leq 2^{h(n)}$ there is a point $b \in B$ such that $(\frac{i}{2^{h(n)}}, 2^{-h(n)+2}) f (b, 2^{-n})$*

There is one more necessary lemma, which clarifies some of the properties of an element $x \in [0,1]$ that we will construct as part of the proof of the main result of this section.

Lemma 5. ($\mathsf{RCA_0}$) *Let $g : \mathbb{N} \to \mathbb{N}$ be a strictly increasing function with $g(0) \geq 1$. If $x = \langle x_n : n \in \mathbb{N} \rangle$ is a sequence of rationals such that for all n, $|x_{n+1} - x_n| < 2^{-g(n)}$, then $x \in \mathbb{R}$ and for all $n \in \mathbb{N}$, $d(x, x_n) \leq 2^{-g(n)+1}$.*

Proof. First we will show that $x \in \mathbb{R}$. Let $n, i \in \mathbb{N}$ be arbitrary. Since $g(0) \geq 1$ and g is strictly increasing, we have $g(n) \geq n+1$ and $g(n+k) \geq g(n) + k$ for all n and k. From this, we have that

$$|x_{n+1} - x_n| < 2^{-g(n)} < 2^{-(n+1)} \tag{2}$$

Applying the triangle inequality results in the following:

$$|x_{n+i} - x_n| < \sum_{k=0}^{i-1} 2^{-(n+k+1)} = 2^{-n-1} \sum_{k=0}^{i-1} 2^{-k} \leq 2^{-n} \tag{3}$$

Therefore $x \in \mathbb{R}$. To complete the proof of the lemma, let $\epsilon > 0$, and let $i \in \mathbb{N}$ be sufficiently large such that $d(x, x_{n+i}) < \epsilon$. Applying the triangle inequality results in the following:

$$d(x, x_n) \leq d(x, x_{n+i}) + d(x_{n+i}, x_n) \leq \epsilon + 2^{-g(n)+1} \tag{4}$$

Since the above holds for all $\epsilon > 0$, $d(x, x_n) \leq 2^{-g(n)+1}$ as desired. □

Now we will prove the main theorem of interest in this section in two parts.

Theorem 5. ($\mathsf{WKL_0}$) *Every continuous $f : [0,1] \to \widehat{B}$ has a code for* $\mathsf{range(f)}$ *as a closed set.*

Proof. Assume WKL_0 and let f be given as hypothesized. By Theorem 4, f has a modulus of uniform continuity h. By Lemma 4, we can assume that h is strictly increasing, $h(0) \geq 1$, and for every n and $i \leq 2^{h(n)}$ there is a point $b \in B$ such that $(\frac{i}{2^{h(n)}}, 2^{-h(n)+2})f(b, 2^{-n})$. For each $n \in \mathbb{N}$ and $i \leq 2^{h(n)}$, define $b_{n,i} \in B$ to be a point such that $(\frac{i}{2^{h(n)}}, 2^{-h(n)+2})f(b_{n,i}, 2^{-n})$.

We define the closed set code C_f for $\mathsf{range(f)}$ with the following Π^0_1 formula. For $y \in \widehat{B}$, let $y \in C_f$ if and only if for all finite sequences $\langle\langle n_0, i_0\rangle, \dots \langle n_k, i_k\rangle\rangle$ such that $i_j \leq 2^{h(n_j)}$ and for all $j \leq k$ and the union of the balls $B_{2^{-h(n_j)}}(\frac{i_j}{2^{h(n_j)}})$ covers $[0, 1]$, there exists $j \leq k$ for which $d(y, b_{n_j,i_j}) \leq 2^{-n_j}$.

We need to verify that for $y \in \widehat{B}$, $y \in C_f$ if and only if there exists some $x \in [0, 1]$ such that $f(x) = y$.

First we will show that if $y \notin C_f$, then y is not in the range of f. Suppose $y \notin C_f$. Then there exists some sequence $\langle\langle n_0, i_0\rangle, \dots \langle n_k, i_k\rangle\rangle$ that witnesses this fact, where the union of the balls $B_{2^{-h(n_j)}}(\frac{i_j}{2^{h(n_j)}})$ covers $[0, 1]$ but for all $j \leq k$ we have that $d(y, b_{n_j,i_j}) > 2^{-n_j}$. For the sake of contradiction, suppose there exists some $x \in [0, 1]$ such that $f(x) = y$. Since the union of the balls $B_{2^{-h(n_j)}}(\frac{i_j}{2^{h(n_j)}})$ covers $[0, 1]$, there exists some $j \leq k$ such that $x \in B_{2^{-h(n_j)}}(\frac{i_j}{2^{h(n_j)}})$. As $(\frac{i}{2^{h(n_j)}}, 2^{-h(n)+2})f(b_{n_j,i_j}, 2^{-n_j})$, by the definition of image, $y \in \overline{B}_{2^{-n_j}}(b_{n_j,i_j})$. This is a contradiction to the fact that $d(y, b_{n_j,i_j}) > 2^{-n_j}$. Therefore y must not be in the range of f.

Lastly we will show that if $y \in C_f$, then y is in the range of f. Suppose $y \in C_f$. We will construct $x \in [0, 1]$ such that $f(x) = y$ using WKL_0, constructing x using an infinite path in an infinite bounded branching tree. First define a bounded branching tree $T \subseteq \mathbb{N}^{<\mathbb{N}}$ as follows. Let $\sigma \in T$ if and only if for all $n < |\sigma|$ the following three conditions hold:

1. $\sigma(n) \leq 2^{h(n)}$
2. $d(y, b_{n,\sigma(n)}) \leq 2^{-n}$ as measured up to $|\sigma|$
3. If $n + 1 < |\sigma|$, then $d(\sigma(n+1)/2^{h(n+1)}, \sigma(n)/2^{h(n)}) < 2^{-h(n)}$.

Note that, if T is infinite our desired result follows. Suppose T is infinite. By Condition 1, T has bounded branching by the function $2^{h(n)}$. So by WKL_0, T has some infinite path p. Define $x = \langle p(n)/2^{h(n)} : n \in \mathbb{N}\rangle$. Since $h(n)$ is strictly increasing, and $h(0) \geq 1$, it follows from Lemma 5 that $x \in \mathbb{R}$. Furthermore, $0 \leq p(n) \leq 2^{h(n)}$, so $x \in [0, 1]$. Finally, by Lemma 5, we have the following:

$$d(x, \frac{p(n)}{2^{h(n)}}) \leq 2^{-h(n)+1} < 2^{-h(n)+2} \tag{5}$$

Therefore $x \in B_{2^{-h(n)+2}}(\frac{p(n)}{2^{h(n)}})$ for all n. By Condition 2, $d(y, b_{n,\sigma(n)}) \leq 2^{-n}$ for all n. Since $(\frac{p(n)}{2^{h(n)}}, 2^{-h(n)+2})f(b_{n_j,p(n)}, 2^{-n})$ holds for all n, it follows that $f(x) = y$.

It remains to show that T is infinite. For the sake of contradiction, assume T is finite. Fix ℓ such that for all $\sigma \in T$, $|\sigma| < \ell$. Let $X = \{\langle n, i\rangle \ : \ n < \ell \wedge i \leq$

$2^{h(n)} \wedge d(b_{n,i}, y) > 2^{-n}\}$. Since n (and with it, i) are bounded and $d(b_{n,i}, y) > 2^{-n}$ is a Σ_1^0 formula, X exists via Bounded Σ_1^0 comprehension.

Claim. For every $j \leq 2^{h(\ell)}$, there is a $\langle n, i\rangle \in X$ such that $\frac{j}{2^{h(\ell)}} \in B_{2^{-h(n)}}(\frac{i}{2^{h(n)}})$.

Before proving this claim, we will first discuss how this claim leads to a desired contradiction. If this claim were true, then the finite collection of balls $B_{2^{-h(n)}}(\frac{i}{2^{h(n)}})$ for $\langle n, i\rangle \in X$ would cover $[0, 1]$. Since $y \in C_f$, it follows that $d(y, b_{n,i}) \leq 2^{-n}$ for some $\langle n, i\rangle \in X$. This inequality contradicts the definition of $\langle n, i\rangle \in X$, which necessitates that $d(b_{n,i}, y) > 2^{-n}$.

Finally, we will prove the claim. For the sake of contradiction, suppose there exists some $j \leq 2^{h(\ell)}$, such that if $\frac{j}{2^{h(\ell)}} \in B_{2^{-h(n)}}(\frac{i}{2^{h(n)}})$ then $\langle n, i\rangle \notin X$. We will construct a string $\sigma \in T$ of length ℓ, which will contradict the choice of ℓ.

Let i_k be the largest number such that $i_k \leq 2^{h(k)}$ and $i_k/2^{h(k)} \leq j/2^{h(\ell)}$. By the contradiction assumption, $\langle k, i_k\rangle \notin X$ for all $k < \ell$.

Let σ be defined as $\sigma(k) = i_k$ for all $k < \ell$. Thus $|\sigma| = \ell$. It is left to show that $\sigma \in T$, by verifying all three conditions. Condition 1 holds since $i_k \leq 2^{h(k)}$. Since $\langle k, i_k\rangle \notin X$, k and i_k cannot satisfy the conditions of the definition of X and we must have instead that $d(b_{k,i_k}, y) \leq 2^{-k}$. This satisfies Condition 2.

To verify Condition 3, let $k+1 < \ell$. Note that, if it is the case that $j/2^{h(\ell)} = i_{k+1}/2^{h(k+1)}$ then we have that

$$\frac{i_{k+1}}{2^{h(k+1)}} = \frac{j}{2^{h(\ell)}} \in B_{2^{-h(k)}}(\frac{i_k}{2^{h(k)}}) \tag{6}$$

Thus $d(i_{k+1}/2^{h(k+1)}, i_k/2^{h(k)}) < 2^{-h(k)}$ as desired.

So consider instead, the case that $j/2^{h(\ell)} \neq i_{k+1}/2^{h(k+1)}$. By definition of i_k, we know that both $i_k/2^{h(k)}$ and $i_{k+1}/2^{h(k+1)}$ are less than or equal to $j/2^{h(\ell)}$. Since i_{k+1} is chosen to be the largest possible value satisfying the relevant requirements, we have the following:

$$\frac{i_k}{2^{h(k)}} \leq \frac{i_{k+1}}{2^{h(k+1)}} \leq \frac{j}{2^{h(\ell)}} \tag{7}$$

Since $\frac{j}{2^{h(\ell)}} \in B_{2^{-h(k)}}(\frac{i_k}{2^{h(k)}})$, $i_{k+1}/2^{h(k+1)}$ is also contained in $B_{2^{-h(k)}}(\frac{i_k}{2^{h(k)}})$. Thus $d(i_{k+1}/2^{h(k+1)}, i_k/2^{h(k)}) < 2^{-h(k)}$ as desired. This completes the proof of the claim and with it, the proof of the theorem. □

Now we move on to the reversal of this result. This proof relies on the contrapositive approach from Simpson's proof that several properties on continuous functions on $[0, 1]$ reverse to WKL_0 as part of Theorem IV.2.3 in [6].

Theorem 6. (RCA_0) *The statement that for continuous $f : [0, 1] \to \widehat{B}$ there is a code for* range(f) *as a closed set implies* WKL_0.

Proof. We will demonstrate this implication by contrapositive in the case that $\widehat{B} = \mathbb{R}$. Assume $\neg WKL_0$. We will construct a continuous function $f : [0, 1] \to \mathbb{R}$ where range(f) cannot have a closed set code, following the method used by Simpson in his proof of Theorem IV.2.3 [6]. Since $\neg WKL_0$, we also have $\neg ACA_0$. Therefore by Theorem 3, there is some one-to-one function $g : \mathbb{N} \to \mathbb{N}$ such that the range of g does not exist. Define c_n as follows:

$$c_n = \begin{cases} 0 & n = 0 \\ \sum_{i=0}^{n} 2^{-g(i)} & n > 0 \end{cases} \quad (8)$$

Note that c_n is a bounded increasing sequence of rational numbers $c_0 < c_1 < \cdots < c_n < \cdots < 2$, and since the range of g does not exist, $\sup_{n \in \mathbb{N}} c_n$ also does not exist.

In the proof of Theorem IV.2.3, Simpson uses the sequence $\langle c_n \rangle$ and the existence of an infinite tree $T \subseteq 2^{<\mathbb{N}}$ with no infinite path to define a continuous function $\phi_4 : [0, 1] \to \mathbb{R}$ with three crucial properties. Letting $f = \phi_4$ for clarity of notation, these properties are:

1. $f(0) = 0$
2. $\forall n(0 \leq x \leq c_n \Rightarrow x \in \mathsf{range(f)})$
3. $\forall n(c_n < x) \Rightarrow x \notin \mathsf{range(f)}$

(For full details on the construction of f, reference the proof of Theorem IV.2.3 in [6]. However, we will only need the three properties above for the remainder of this proof).

Classically, the range set of this function f should be $[0, \sup_{n \in \mathbb{N}} c_n]$, however such a supremum does not exist by assumption. We claim that the range of f cannot have a closed set code.

For the sake of contradiction, let $C = \langle\langle a_n, r_n \rangle : n \in \mathbb{N}\rangle$ be a closed set code for $\mathsf{range(f)}$ (see Lemma 1). Recall that the balls $B_{r_n}(a_n)$ enumerated by this code cover the open complement of $\mathsf{range(f)}$. Define a sequence of real numbers $\langle \rho_n : n \in \mathbb{N}\rangle$ where

$$\rho_n = \begin{cases} a_n - r_n & a_n - r_n > 0 \\ 0 & a_n - r_n \leq 0 \end{cases} \quad (9)$$

Essentially, if $\overline{B}_{r_n}(a_n)$ contains any positive numbers, ρ_n is the left-most boundary point of that closed ball in $\mathbb{R}$. If not, we let ρ_n be 0. Since $f(0) = 0$, we know 0 cannot be contained in $B_{r_n}(a_n)$ for any n. So, if $\overline{B}_{r_n}(a_n)$ contains any positive numbers, it only contains positive numbers.

We now introduce and prove two claims that are necessary for the rest of the proof.

Claim. If q is an upper bound of $\langle c_n : n \in \mathbb{N}\rangle$, then there exists some k such that $0 < \rho_k < q$.

Let q be as hypothesized. Note that q must be positive. Since $c_n < q$ for all n, it must be that $q \notin \mathsf{range(f)}$ by Property 3. So there exists some k such that $q \in B_{r_k}(a_k)$. Thus $0 < a_k - r_k < q < a_k + r_k$. So $\rho_k = a_k - r_k$ is as desired.

Claim. For all m, k, if $\rho_k \neq 0$, then $c_m < \rho_k$.

For the sake of contradiction, suppose there exists m and k such that $\rho_k \neq 0$ but $\rho_k \leq c_m$. We will prove this in two cases regarding the position of c_m relative to the ball $B_{r_k}(a_k)$.

Case 1. $\rho_k < c_m$.

In this case, $(0, c_m) \cap B_{r_k}(a_k) \neq \emptyset$. However $(0, c_m) \subset \mathsf{range(f)}$, while $B_{r_k}(a_k)$ is disjoint from $\mathsf{range(f)}$. This is a contradiction.

Case 2. $\rho_k = c_m$.

In this case, since the sequence $\langle c_n \rangle$ is increasing, $\rho_k < c_{m+1}$. Thus we can proceed as in Case 1 relative to c_{m+1} as opposed to c_m to find a contradiction.

Note that by this claim, every $\rho_k \neq 0$ is an upper bound of $\langle c_n \rangle$. Thus, by the first claim, for all such k, there exists some ℓ such that $\rho_\ell < \rho_k$.

With these two claims proved, we now recursively define a function $g : \mathbb{N} \to \mathbb{N}$ that picks a decreasing subsequence of $\langle p_n \rangle$ in the following way:

$$g(k) = \begin{cases} \text{least n such that } \rho_n > 0 & k = 0 \\ \text{least } n > g(\ell) \text{ such that } 0 < \rho_n < \rho_{g(\ell)} & k = \ell + 1 \text{ for some } \ell \in \mathbb{N} \end{cases} \tag{10}$$

By the previous two claims, g is defined for all $k \in \mathbb{N}$. Thus $\langle \rho_{g(k)} \rangle$ is a decreasing sequence, and by the second claim, for all n, $c_n < c_{n+1} < \rho_{g(n+1)} < \rho_{g(n)}$.

The last claim follows immediately from the definition of g and the first claim.

Claim. If q is an upper bound of $\langle c_n \rangle$, then there exists some k such that $\rho_{g(k)} < q$.

In pursuit of using the Nested Interval Completeness of $\mathbb{R}$, we now claim that $\lim_n \rho_{g(n)} - c_n = 0$. For the sake of contradiction, suppose not. Then there exists some M such that $\rho_{g(n)} - c_n > M$ for all n. Let $q \in \mathbb{Q}$ be an upper bound of $\langle c_n \rangle$ for which there exists j such that $q - c_j < M$. By the last claim, there is a $k \geq j$ such that $\rho_{g(k)} < q$. However, then $\rho_{g(k)} - c_k \leq \rho_{g(k)} - c_j < q - c_j < M$, which contradicts the choice of M.

Thus, by Nested Interval Completeness, there exists some real number x such that $x = \lim c_n = \lim \rho_{g(n)}$. However, this x would be the least upper bound of $\langle c_n \rangle$, which is a contradiction. Therefore there must not exist any closed set code C for $\mathsf{range(f)}$. Thus this theorem holds by contrapositive. □

Combining the two main theorems in this section, we get the following result.

Theorem 7. ($\mathsf{RCA_0}$) *The following are equivalent:*

- $\mathsf{WKL_0}$
- *For continuous* $f : [0, 1] \to \widehat{B}$, *there is a code for* $\mathsf{range(f)}$ *as a closed set.*

The proof of Lemma 4 generalizes from $[0, 1]$ to any compact complete separable metric space. This leads to a more general result.

Theorem 8. ($\mathsf{RCA_0}$) *The following are equivalent:*

- $\mathsf{WKL_0}$
- *For continuous* $f : [0, 1] \to \widehat{B}$, *there is a code for* $\mathsf{range(f)}$ *as a closed set.*
- *For complete separable metric spaces* $\widehat{A}$ *and* $\widehat{B}$, *where* $\widehat{A}$ *is compact, if* $f : \widehat{A} \to \widehat{B}$ *is continuous, there is a code for* $\mathsf{range(f)}$ *as a closed set.*

3 The Connectedness of range(f)

Since using a closed set code for range(f) requires WKL_0, we will carefully define our sense of having a connected range such that it implies that the range set of a function is connected without relying on a code for the range set itself. This will allow us to work with the theorem of interest in this section within RCA_0 without the complications that we just discussed in Sect. 2.

Definition 11. *Let $\widehat{C}$ be a complete separable metric space. An open or closed set $U \subseteq \widehat{C}$ is connected in $\widehat{C}$ if and only if for all open sets $A, B \subseteq \widehat{C}$, if $U \subseteq A \cup B$ and $U \cap A \neq \emptyset$ and $U \cap B \neq \emptyset$, then $A \cap B \neq \emptyset$.*

Definition 12. *For continuous function $f : [0,1] \to \widehat{C}$, f has connected range if and only if for all open sets $A, B \subseteq \widehat{C}$, if*

1. *$\forall x \in [0,1](f(x) \in A \vee f(x) \in B)$,*
2. *$\exists x \in [0,1](f(x) \in A)$, and*
3. *$\exists x \in [0,1](f(x) \in B)$*

then there exists some $x \in [0,1]$ such that $f(x) \in A \cap B$.

We need two lemmas for the main result in this section. The first is Lemma 1.24 in Brown [1]. The second will allow us to use the connectedness of the domain of f in the main theorem of the section.

Lemma 6. (RCA_0) *If $f : [0,1] \to \widehat{C}$ is continuous, then for all nonempty open $U \subseteq \widehat{C}$, $f^{-1}(U) = \{x \in [0,1] \mid f(x) \in U\}$ is open in $[0,1]$.*

Lemma 7. (RCA_0) *$[0,1]$ is connected in $\mathbb{R}$*

Proof. For the sake of contradiction, assume $[0,1]$ is not connected. Then there exist open sets $A, B \subseteq \mathbb{R}$ such that $[0,1] \subseteq A \cup B$, $[0,1] \cap A \neq \emptyset$, and $[0,1] \cap B \neq \emptyset$, but $A \cap B = \emptyset$. Let A be coded by $\langle\langle a_n, r_n\rangle : n \in \mathbb{N}\rangle$ and B be coded by $\langle\langle b_n, s_n\rangle : n \in \mathbb{N}\rangle$. Let $x_0 \in A \cap [0,1]$ and $y_0 \in B \cap [0,1]$ where $x_0 \neq y_0$. Without loss of generality, assume $x_0 < y_0$.

By primitive recursion we define sequences $\langle x_n\rangle$ and $\langle y_n\rangle$ as follows, to satisfy the conditions in the Nested Interval Completeness Theorem.

$$x_{n+1} = \begin{cases} \frac{x_n+y_n}{2} & \text{if } \frac{x_n+y_n}{2} \in A \\ x_n & \text{otherwise} \end{cases} \tag{11}$$

$$y_{n+1} = \begin{cases} \frac{x_n+y_n}{2} & \text{if } \frac{x_n+y_n}{2} \in B \\ y_n & \text{otherwise} \end{cases} \tag{12}$$

From the definitions above, for all n we have that $x_n \in A$, $y_n \in B$, and $x_n \leq x_{n+1} \leq y_{n+1} \leq y_n$. In addition:

$$\lim_n |x_n - y_n| = \lim_n \frac{|x_0 - y_0|}{2^n} = 0 \tag{13}$$

With these properties of $\langle x_n \rangle$ and $\langle y_n \rangle$, in RCA_0, by the Nested Interval Completeness of $\mathbb{R}$ there exists a real number x such that $x = \lim_n x_n = \lim_n y_n$. Note that $x \in [0,1]$. Since A and B are disjoint either $x \in A$ or $x \in B$ but not both.

First, suppose $x \in A$. Since A is open, there exists $\langle a_n, r_n \rangle$ in the code of A such that $x \in B_{r_n}(a_n)$. However, since A and B are disjoint, $B_{r_n}(a_n)$ and B are disjoint. Since $\lim_n y_n \to x$, there exists some k such that $d(x, y_k) < r_n$. However, that implies that $y_k \in B_{r_n}(a_n)$. This is a contradiction.

If it is supposed instead that $x \in B$, we derive a contradiction in the same manner as above. As there is a contradiction in both cases, $[0,1]$ must be connected. □

With the lemmas above, we can now prove the main theorem of this section.

Theorem 9. (RCA_0) *For all continuous* $f : [0,1] \to \widehat{C}$, *$f$ has connected range.*

Proof. Fix open sets $A, B \subseteq \widehat{C}$ that satisfy Conditions 1 through 3 in Definition 12. By Lemma 6, $f^{-1}(A)$ and $f^{-1}(B)$ are open subsets of $[0,1]$. By Condition 1, $[0,1] \subseteq f^{-1}(A) \cup f^{-1}(B)$. By Conditions 2 and 3, $f^{-1}(A) \cap [0,1]$ and $f^{-1}(B) \cap [0,1]$ are nonempty. Thus, by Lemma 7, there is a real $x \in [0,1]$ such that $x \in f^{-1}(A) \cap f^{-1}(B)$. By applying f, we have that $f(x) \in A \cap B$. Therefore f has connected range. □

As in the previous section, this result extends to a more general case, as stated below:

Theorem 10. (RCA_0) *Let $\widehat{A}$ and $\widehat{B}$ be complete separable metric spaces and let $\widehat{A}$ be connected. For all continuous $f : \widehat{A} \to \widehat{B}$, f has connected range.*

Using the results of this section, we can now provide a simpler proof that WKL_0 is equivalent to the existence of a closed set code for $\mathsf{range(f)}$ in the special case when $f : [0,1] \to \mathbb{R}$.

Lemma 8. (RCA_0) *Let $f : [0,1] \to \mathbb{R}$ be continuous. If $a <_{\mathbb{R}} b$ and there exist $x_a, x_b \in [0,1]$ such that $f(x_a) = a$ and $f(x_b) = b$, then for all $y \in [a,b]$, there exists $x \in [0,1]$ such that $f(x) = y$.*

Proof. Let f, a, b, x_a, and x_b be as hypothesized. Let $y \in [a,b]$ be given. For the sake of contradiction suppose $f(x) \neq y$ for all $x \in [0,1]$. By Theorem 9, f has connected range. However, the open sets $A = (-\infty, y)$ and $B = (y, \infty)$ satisfy Conditions 1–3 in Definition 12 and are disjoint, providing the desired contradiction. □

Theorem 11. *(RCA_0) The following are equivalent.*

(1) WKL_0
(2) For continuous $f : [0,1] \to \mathbb{R}$, if f is non-constant, there exist $a <_{\mathbb{R}} b$ such that $\mathsf{range(f)}$ *is $[a,b]$.*
(3) For continuous $f : [0,1] \to \mathbb{R}$, there is a code for $\mathsf{range(f)}$ *as a closed set.*

Proof. Note that (3) $\Rightarrow$ (1) follows by Theorem 6. (2) $\Rightarrow$ (3) follows from the fact that RCA_0 suffices to prove there is a closed set code for $[a, b]$ whenever $a <_{\mathbb{R}} b$. In the case where f is constant, RCA_0 also suffices to show that singleton sets have closed set codes.

It remains to show that (1) $\Rightarrow$ (2). Let non-constant continuous $f : [0, 1] \to \mathbb{R}$ be given. Assume WKL_0. By Theorem IV.2.3 in Simpson [6], f has and attains its supremum, and by an equivalent argument, f also has and attains its infimum. Let a denote the infimum of f and let b denote the supremum of f. Fix $x_a, x_b \in [0, 1]$ such that $f(x_a) = a$ and $f(x_b) = b$. The range of f is contained in $[a, b]$, and by Lemma 8, $\mathsf{range(f)}$ must be equal to $[a, b]$. □

4 The Compactness and Boundedness of range(f)

Similarly to Sect. 3, we need to ensure our proof for the compactness of the range set does not rely on a code for the range set itself. So we will first state an adapted definition for compactness. While compactness is our main property of interest here, boundedness is essential for the proof and is a useful property to examine in its own right. So we will also provide the boundedness definition that is appropriate for our setting.

Definition 13. (RCA_0) *For a continuous function $f : [0, 1] \to \widehat{B}$, f has compact range if and only if there is a sequence of finite sequences $\langle\langle b_{i,j} : i \leq n_j\rangle : j \in \mathbb{N}\rangle$ (with $b_{i,j} \in \widehat{B}$) such that for all $j \in \mathbb{N}$ and all $x \in [0, 1]$, there exists $i \leq n_j$ such that $d(b_{i,j}, f(x)) < 2^{-j}$.*

Definition 14. (RCA_0) *For a continuous function $f : [0, 1] \to \widehat{B}$, f is bounded if and only if there exists $M \in \mathbb{Q}^+$ such that for all $x, y \in [0, 1]$, $d(f(x), f(y)) \leq M$.*

In the special case that $f : [0, 1] \to \mathbb{R}$, note that RCA_0 is sufficient to show that the closed interval $[a, b]$ is compact and bounded. (See Examples III.2.6 in [6]). Thus, by Theorem 11, WKL_0 suffices to show that for continuous functions $f : [0, 1] \to \mathbb{R}$, f has compact and bounded range.

We will prove the main result of this section by extending some standard results from Simpson, returning again to Theorem IV.2.3 [6].

Theorem 12. (RCA_0) *The following are equivalent:*

(1) WKL_0.
(2) If $f : [0, 1] \to \widehat{B}$ is continuous, then f is bounded.
(3) If $f : [0, 1] \to \widehat{B}$ is continuous, then f has compact range.

Proof. Note that Simpson proved (1) $\Leftrightarrow$ (2). (See Theorem IV.2.3 [6]). Thus, proving that (2) $\Leftrightarrow$ (3) is sufficient to prove the theorem.

First we will show that (2) $\Rightarrow$ (3). Since (1) $\Leftrightarrow$ (2), assume WKL_0. By Theorem 4, let h be a modulus of uniform continuity of f. By Lemma 4, for each $j \in \mathbb{N}$ and $i \leq 2^{h(j)}$, let $b_{i,j} \in B$ be a point such that $(\frac{i}{2^{h(j)}}, 2^{-h(j)})f(b_{i,j}, 2^{-j})$.

We claim that $\langle\langle b_{i,j+1} : i \leq 2^{h(j+1)}\rangle : j \in \mathbb{N}\rangle$ witnesses that f has compact range. Let $j \in \mathbb{N}$ and $x \in [0,1]$ be given. There exists some $i \leq 2^{h(j+1)}$ such that $x \in B_{2^{-h(j+1)}}(\frac{i}{2^{h(j+1)}})$. Therefore $d(b_{i,j}, f(x)) \leq 2^{-(j+1)} < 2^{-j}$. Thus f has compact range.

Next we will show that (3) $\Rightarrow$ (2), reasoning in RCA_0. Let $\langle\langle b_{i,j} : i \leq n_j\rangle : j \in \mathbb{N}\rangle$ be given demonstrating that f has compact range. Consider the first finite sequence $\langle b_{i,0} : i \leq n_0\rangle$ of this infinite sequence. We claim that any $M \in \mathbb{Q}$ such that $M \geq 2 + \max_{j,k \leq n_0} d(b_{j,0}, b_{k,0})$ is a bound for the range of f as in Definition 14. Let $x, y \in [0,1]$ be arbitrary and fix $j, k \leq n_0$ such that $d(b_{j,0}, f(x)) < 1$ and $d(b_{k,0}, f(y)) < 1$. By applying the triangle inequality we find the following:

$$d(f(x), f(y)) \leq d(f(x), b_{j,0}) + d(b_{j,0}, b_{k,0}) + d(b_{k,0}, f(y)) \leq 2 + \max_{j,k \leq n_0} d(b_{j,0}, b_{k,0}) \leq M \tag{14}$$

Thus f is bounded. □

Brown proved something similar to (1) $\Rightarrow$ (3) for functions between complete separable metric spaces. (See Theorem 3.20 in Brown [1]). However, Brown's proof invoked the code for the range set, so further analysis was needed to determine whether or not the implication held without the strength of the implicit existence of the code.

As in the previous sections, our result extends to the more general case with some adapted definitions.

Theorem 13. (RCA_0) *Let $\widehat{A}$ and $\widehat{B}$ be complete separable metric spaces, and let $\widehat{A}$ be compact. The following are equivalent:*

(1) WKL_0.
(2) If $f : \widehat{A} \to \widehat{B}$ is continuous, then f is bounded.
(3) If $f : \widehat{A} \to \widehat{B}$ is continuous, then f has compact range.

5 Future Directions

In this paper, we have classified several fundamental theorems about range sets of continuous functions. However, there are several questions still unanswered. First, there is an alternative coding scheme for closed sets in reverse mathematics. Sets with these codes are called separably closed sets to distinguish them from those coded using the definition we discussed in this paper. Since Brown [2] proved that all closed subsets of compact spaces being separably closed is equivalent to ACA_0 over RCA_0, the results in Sect. 2 and 4 imply that ACA_0 is sufficient for the range set of a continuous function on $[0,1]$ to have a separably closed code. It is currently unclear whether ACA_0 is necessary or if a weaker subsystem is sufficient to prove this.

However, the main interest of our current line of research is local connectedness: the remaining property of range sets in the forward direction of the Hahn-Mazurkiewicz Theorem. It is still unknown which subsystem is required to prove that the range of continuous functions on $[0,1]$ are locally connected. Complicating this analysis is the variety of equivalent definitions for local connectedness.

It is still unclear how best to adapt the definition of local connectedness to suit the context of reverse mathematics. Once these questions are answered, the next topic of focus is the backwards implication of the Hahn-Mazurkiewicz Theorem, and further still, the classification of additional theorems related to space-filling curves. Space-filling curves are an active area of research in geometric analysis with many cross-disciplinary applications in computer science.

Acknowledgments. The author was partially funded by the NSF DMS grant 2154047, and the author thanks principal investigator Matthew Badger for his interest and support of this project. The author would also like to thank Reed Solomon for his insights and advice throughout the development of these results and the preparation of this paper. Lastly, the author would like to thank the three anonymous referees for their valuable comments and suggestions.

Disclosure of Interests. The author has no competing interests to declare that are relevant to the content of this article.

References

1. Brown, D.: Functional analysis in weak subsystems of second order arithmetic. Doctoral Dissertation, The Pennsylvania State University (1987)
2. Brown, D.: Notions of closed sets of a complete separable metric space in weak subsystems of second order arithmetic. Contemp. Math. **106**, 39–50 (1990)
3. Couch, P.J., Daniel, B.D., McNicholl, T.: Computing space filling curves. Theory Comput. Syst. **50**, 370–386 (2012)
4. Daniel, D., McNicholl, T.: Effective versions of local connectivity properties. Theory Comput. Syst. **50**, 621–640 (2012)
5. Sagan, H.: Space-Filling Curves, 1st edn. Springer, New York (1994)
6. Simpson, S.: Subsystems of Second Order Arithmetic, 2nd edn. Cambridge University Press, Cambridge (2010)

The Theory of Reachability of Trace-Pushdown Systems

Dietrich Kuske(✉)

FG Automaten und Logik, Technische Universität Ilmenau, Ilmenau, Germany
dietrich.kuske@tu-ilmenau.de

Abstract. We consider pushdown systems that store, instead of a single word, a Mazurkiewicz trace on their stack. These systems are special cases of valence automata over graph monoids and subsume multi-stack systems. We identify a class of such systems that allow to decide the first-order theory of their configuration graph with reachability.

This result complements results by D'Osualdo, Meyer, and Zetzsche (namely the decidability for arbitrary pushdown systems under a severe restriction on the dependence alphabet).

Keywords: pushdown system · Mazurkiewicz trace · reachability

1 Introduction

Pushdown systems are a well-studied type of infinite state system with many variants and generalizations. Valence automata [5–8,11–13,19,21,22,25] allow to study many of these variants in a uniform setting: in a pushdown system, the possible pushdown contents come from a free monoid and can be accessed at the suffix, only. Differently, in a valence automaton, the free monoid is replaced by an arbitrary monoid. As an example, graph monoids allow to model pushdown systems, Petri nets, multi-stack automata, counter automata and many more.

For a pushdown system, the configuration graph with reachability can be interpreted in the full tree. Hence Rabin's tree theorem ensures that monadic second order properties of this graph are decidable. This decidability gets lost very soon once one considers valence automata over graph monoids. Such a valence automaton consists of a finite control and the graph monoid as generalization of the pushdown.

Therefore, D'Osualdo, Meyer, and Zetzsche [5] considered first-order logic. They were able to characterize those graph monoids that ensure the decidability, no matter what local control the monoid is attached to.

Differently here, we allow all loop-free graph monoids, but aim at properties of the finite control that ensure the decidability of the first-order theory of the configuration graph with reachability. These loop-free graph monoids are precisely Mazurkiewicz' trace monoids.

In [15,16], we considered a related problem: there, we were concerned with the question what properties of the finite control ensure the decidability of the

© The Author(s), under exclusive license to Springer Nature Switzerland AG 2025
A. Beckmann et al. (Eds.): CiE 2025, LNCS 15764, pp. 283–298, 2025.
https://doi.org/10.1007/978-3-031-95908-0_20

reachability relation. It turned out that this is the case for trace-pushdown systems [16]. To prove the decidability, we showed that the reachability relation in this case (i.e., a binary relation on the trace monoid) is "lc-rational", a non-trivial restriction of rational trace relations. In general terms, this proof proceeds as follows. One first handles two very special cases of systems where, either, the pushdown is shortend in every step (i.e., no transition writes anything to the pushdown) or, alternatively, the pushdown is never shortened, but all transitions read "similar" letters (in some precise meaning). The central argument in these two cases is that the class of lc-rational relations is closed under the componentwise product with direct products of a recognizable and a rational trace language (in general, the product of lc-rational relations need not be lc-rational). To handle the general case, one shows that the reachability relation can be obtained from the special cases by union and composition. Here, one uses that the class of lc-rational relations is closed under these two operations (in general, the composition of rational trace relations is not necessarily rational).

A first observation in this paper is that the said first-order theory is in general undecidable for trace-pushdown systems, the undecidability even holds for existential formulas (Example 2.5). To overcome this undecidability, we restrict attention to loop-connected trace-pushdown systems. The configuration graph of such systems is proved to be an automatic structure [1,10,14] implying the desired decidability.

The central argument in this proof is therefore the automaticity of the configuration graph of a loop-connected trace-pushdown system. To this aim, we use the Foata normal form of a trace as its word representation. Consequently, one has to verify that the reachability relation between such Foata normal forms can be accepted by a synchronous two-tape automaton, i.e., that the reachability relation is "fnf-automatic". This proof follows the line of argument of the lc-rationality for arbitrary trace-pushdown systems, but extends it at some central point. So one first has to handle the two special cases mentioned above. In both these cases, it is used that the class of fnf-automatic relations is closed under product with direct products of two recognizable trace languages (Lemma 3.3). Then the result for deleting systems follows immediately. For non-deleting systems, the new requirement to be loop-connected allows to show that the set of traces that can be written onto the pushdown is recognizable (without the requirement, it was known to be rational, only). Now the general case follows as before since the class of fnf-automatic relations is closed under composition and union.

2 Preliminaries

2.1 Mazurkiewicz Traces

Zero is a natural number and, for $n \in \mathbb{N}$, we set $[n] = \{1, 2, \ldots, n\}$.

Let Σ be some alphabet and $w \in \Sigma^*$ a word over Σ. The *alphabet of* w, denoted $\text{alph}(w) \subseteq \Sigma$, is the set of letters occuring in the word w. The number of occurrences of the letter a in the word w is denoted $|w|_a$.

A *dependence alphabet* is a pair (Σ, D) where Σ is a finite set of *letters* and $D \subseteq \Sigma \times \Sigma$ is a reflexive and symmetric relation, the *dependence relation.* For a letter $a \in \Sigma$, we write $D(a)$ for the set $\{b \in \Sigma \mid (a,b) \in D\}$ of letters dependent from a, $D(B) = \bigcup_{b \in B} D(b)$ is the set of letters dependent from some letter in $B \subseteq \Sigma$. A set of letters $A \subseteq \Sigma$ is *connected* if the graph $(A, D \cap A^2)$ is connected.

Two letters a and b are *twins* if $D(a) = D(b)$. The set of twins of a letter a is denoted $\mathrm{twns}(a) = \{b \in \Sigma \mid D(a) = D(b)\}$. The *twin index* of (Σ, D) is the number of sets $\mathrm{twns}(a)$ with $a \in \Sigma$.

The *independence relation* $I \subseteq \Sigma \times \Sigma$ is the set of pairs (a,b) of distinct letters with $(a,b) \notin D$. If A and B are two sets of letters with $A \times B \subseteq I$ (i.e., any letter from A is independent from any letter from B), then we write $A \parallel B$. For words $u, v \in \Sigma^*$, $u \parallel v$ abbreviates $\mathrm{alph}(u) \parallel \mathrm{alph}(v)$.

Let $\sim \subseteq \Sigma^* \times \Sigma^*$ denote the least monoid congruence with $ab \sim ba$ for all $(a,b) \in I$. In other words, $u \sim v$ holds for two words $u, v \in \Sigma^*$ iff u can be obtained from v by transposing consecutive independent letters. In particular, $u \sim v$ implies $|u| = |v|$, $|u|_a = |v|_a$ for all $a \in \Sigma$, and $\mathrm{alph}(u) = \mathrm{alph}(v)$. Furthermore, $u \parallel v$ implies $uv \sim vu$ (but the converse implication does not hold as the example $u = v = ab$ shows).

The *(Mazurkiewicz) trace monoid* induced by (Σ, D) is the quotient of the free monoid Σ^* wrt. the congruence $\sim$, i.e., $\mathbb{M} = \Sigma^*/{\sim}$. Its elements are equivalence classes of words denoted $[w]$; by $[w]$, we mean the equivalence class containing w, it is the *trace induced by* w. Note that $\mathbb{M}$ is indeed a monoid whose unit element is the trace $[\varepsilon]$; this trace is denoted 1.

Suppose $D = \Sigma \times \Sigma$, i.e., all letters are mutually dependent. Then $u \sim v \iff u = v$ holds for all words $u, v \in \Sigma^*$; hence $\mathbb{M} \cong \Sigma^*$ in this case.

The other extreme is $D = \{(a,a) \mid a \in \Sigma\}$ where any two distinct letters are independent. Then $u \sim v$ iff $|u|_a = |v|_a$ holds for any letter $a \in \Sigma$. Hence $\mathbb{M} \cong (\mathbb{N}^{|\Sigma|}, +)$ in this case.

To consider a case in between, let Σ be the disjoint union of A and B and suppose $D = A^2 \cup B^2$. Then $\mathbb{M}$ is isomorphic to the direct product of the free monoids A^* and B^*.

Automata and Word Languages. An *ε-NFA* or *nondeterministic finite automaton with ε-transitions* is a tuple $\mathcal{A} = (Q, \Sigma, I, T, F)$ where Q is a finite set of *states*, Σ is an alphabet, $I, F \subseteq Q$ are the sets of *initial* and *final* states, respectively, and

$$T \subseteq Q \times (\Sigma \cup \{\varepsilon\}) \times Q$$

is the set of *transitions*. The ε-NFA $\mathcal{A}$ is an *NFA* if $T \subseteq Q \times \Sigma \times Q$.

Let $\mathcal{A} = (Q, A, I, T, F)$ be an ε-NFA. A *path* is a sequence

$$(p_0, a_1, p_1)(p_1, a_2, p_2) \cdots (p_{n-1}, a_n, p_n)$$

of matching transitions (i.e., elements of T). Such a path is usually denoted

$$p_0 \xrightarrow{a_1} p_1 \xrightarrow{a_2} p_2 \cdots \xrightarrow{a_n} p_n$$

or, if the intermediate states are of no importance,

$$p_0 \xrightarrow{a_1 a_2 \cdots a_n} p_n\,.$$

This path is *accepting* if it connects an initial state with a final state, i.e., if $p_0 \in I$ and $p_n \in F$. It accepts the word $w = a_1 a_2 \cdots a_n$ (note that $a_i \in \Sigma \cup \{\varepsilon\}$ such that $|w| < n$ is possible). We denote by $L(\mathcal{A})$ the set of words w accepted by $\mathcal{A}$. A language $L \subseteq A^*$ is *regular* if it is accepted by some NFA, i.e., if there is some NFA $\mathcal{A}$ with $L = L(\mathcal{A})$.

A foundational result in the theory of finite automata states that ε-NFA and NFA are equally expressive. Even more, ε-NFA can be transformed into equivalent NFA in polynomial time.

A language $L \subseteq \Sigma^*$ is *rational* if it can be constructed from finite languages using the operations union, multiplication, and Kleene star. By Kleene's theorem, a language is regular if, and only if, it is rational.

Rational and Recognizable Trace Languages. We fix some dependence alphabet (Σ, D). For a word language $L \subseteq \Sigma^*$, we denote by $[L]$ the set of traces $[u]$ induced by words from L, i.e., $[L] = \{[u] \mid u \in L\} \subseteq \mathbb{M}$.

Now let $\mathcal{L} \subseteq \mathbb{M}$. The set $\mathcal{L}$ is *recognizable* if the word language $\{u \in A^* \mid [u] \in \mathcal{L}\}$ is regular; it is *rational* if there exists a regular word language $L \subseteq \Sigma^*$ with $\mathcal{L} = [L]$. It follows that every recognizable trace language is rational; the converse implication is known to fail (consider, e.g., the trace language $\{[ab]^n \mid n \in \mathbb{N}\}$ with $a \parallel b$ that is rational, but not recognizable).

Since rational trace languages are images of rational word languages, we obtain that $\mathcal{L} \subseteq \mathbb{M}$ is rational if, and only if, it can be constructed from finite trace languages using the operations union, multiplication, and Kleene star.

2.2 Trace-Pushdown Systems

In this paper, we consider pushdowns that hold a trace $[w] \in \mathbb{M}$ which can be accessed at the suffix, only.

A *pushdown system* is a tuple $\mathcal{P} = (Q, \Sigma, D, \Delta)$ where Q is a finite set of *states*, (Σ, D) a dependence alphabet, and $\Delta \subseteq Q \times \Sigma \times \Sigma^* \times Q$ a finite set of *transitions*. For $(p, a, w, q) \in \Delta$, we regularly write $p \xrightarrow{a|w} q$.

The set $Q \times \mathbb{M}$ of configurations of $\mathcal{P}$ is denoted $\mathrm{Conf}(\mathcal{P})$. For two configurations $(p, [u]), (q, [v]) \in \mathrm{Conf}(\mathcal{P})$, we set $(p, [u]) \vdash (q, [v])$ if there is a transition $(p, a, w, q) \in \Delta$ and a word $x \in \Sigma^*$ such that $[u] = [xa]$ and $[v] = [xw]$. Note that $[u] = [xa]$ is equivalent to saying $[u] = [x] \cdot [a]$ and similarly $[v] = [xw]$ is equivalent to $[v] = [x] \cdot [w]$. Hence $(p, s) \vdash (q, t)$ for $p, q \in Q$ and $s, t \in \mathbb{M}$ iff there is a transition $(p, a, w, q) \in \Delta$ such that the trace t results from the trace s by replacing the suffix $[a]$ by $[w]$. The reflexive and transitive closure of the one-step relation $\vdash$ is denoted $\vdash^*$. Furthermore, for $p, q \in Q$, we let $\mathrm{Reach}_{p,q} \subseteq \mathbb{M}^2$ denote the set of pairs (s, t) of traces such that $(p, s) \vdash^* (q, t)$.

Definition 2.1. *Let $\mathcal{P} = (Q, \Sigma, D, \Delta)$ be a pushdown system. Its* configuration graph with reachability *is the structure*

$$\mathbb{G}(\mathcal{P}) = (Q \times \mathbb{M}, \vdash, \vdash^*, (\mathrm{state}_p)_{p \in Q}, (c)_{c \in Q \times \mathbb{M}})$$

consisting of the set of configurations $\mathrm{Conf}(\mathcal{P}) = Q \times \mathbb{M}$*, the one-step relation* $\vdash$*, the reachability relation* $\vdash^*$*, the unary state predicate* $\mathrm{state}_p = \{p\} \times \mathbb{M}$*, and a constant for every configuration of* $\mathcal{P}$*.*

If $D = \Sigma \times \Sigma$, then this structure can be interpreted in the full tree. Hence, by Rabin's tree theorem, its monadic second order theory is decidable. This is not true in general as the following result shows.

Theorem 2.2 (cf. D'Osualdo, Meyer, Zetzsche [5]). *Let (Σ, D) be a dependence alphabet. The first-order theory of $\mathbb{G}(\mathcal{P})$ is decidable for every pushdown system $\mathcal{P} = (Q, \Sigma, D, \Delta)$ if, and only if, any letter from Σ is independent from at most one letter.*

In this paper, we take another point of view: we ask for properties of the transitions of a pushdown system $\mathcal{P}$ that guarantee the decidability of the theory of $\mathbb{G}(\mathcal{P})$. The decidability of the theory of $\mathbb{G}(\mathcal{P})$ implies in particular the decidablitity of the reachability relation. In [16], we presented a class of pushdown systems whose reachability relation is decidable. These *trace-pushdown systems* are defined as follows.

Definition 2.3. *A* trace-pushdown system *(or* tPDS, *for short) is a pushdown system $\mathcal{P} = (Q, \Sigma, D, \Delta)$ such that the following hold:*

(P1) whenever $p \xrightarrow{a|w} q$ is a transition from Δ, then $D(w) \subseteq D(a)$,
(P2) for any transitions $p \xrightarrow{a|v} q \xrightarrow{b|w} r$ from Δ with $a \parallel b$, there is a state $q' \in Q$ such that $p \xrightarrow{b|w} q' \xrightarrow{a|v} r$ are transitions from Δ.

The main result about these tPDS reads as follows:

Theorem 2.4 (Kuske and Köcher [16]). *The following problem is decidable in time polynomial in the size of $\mathcal{P}$ and exponential in the twin index of (Σ, D):*
input: trace-pushdown system $\mathcal{P} = (Q, \Sigma, D, \Delta)$
two configurations (p, s) and (q, t) of $\mathcal{P}$
question: Does $(p, s) \vdash^* (q, t)$ hold?

We also constructed pushdown systems that satisfy only one of the two conditions (P1) and (P2) and have an undecidable reachability relation.

The following example shows that the first-order theory of $\mathbb{G}(\mathcal{P})$ is not uniformly decidable for the class of all tPDS.

Example 2.5. We want to reduce Post's correspondence problem to the existential theory. To this aim, let $\Sigma = \{a, b, a', b', \top\}$ with

$$D = \{a, b\}^2 \cup \{a', b'\}^2 \cup (\Sigma \times \{\top\}) \cup (\{\top\} \times \Sigma).$$

Furthermore, let $f\colon \{a,b\}^* \to \{a',b'\}^*$ be the monoid homomorphism with $f(a) = a'$ and $f(b) = b'$.

An instance of Post's correspondence problem is a tuple $I = \big((u_i, v_i)\big)_{i\in[k]}$ with $k \in \mathbb{N}$ and $u_i, v_i \in \{a,b\}^*$ for all $i \in [k]$. A solution is a nonempty tuple $(i_1, \cdots, i_n)$ with entries in $[k]$ such that

$$u_{i_1}\, u_{i_2} \;\cdots\; u_{i_n} = v_{i_1}\, v_{i_2} \;\cdots\; v_{i_n}\,.$$

The existence of a solution is undecidable.

Now let $I = \big((u_i, v_i)\big)_{i\in[k]}$ be an instance of Post's correspondence problem. From this finite tuple of words, we construct the following tPDS $\mathcal{P}$. It has four states ι, p, q, r and the following transitions:

- $\iota \xrightarrow{\top|u_i\, f(v_i)\top} p \xrightarrow{\top|u_i\, f(v_i)\top} p \xrightarrow{\top|\top} r$ for all $i \in [k]$
- $\iota \xrightarrow{\top|x\, f(x)\top} q \xrightarrow{\top|x\, f(x)\top} q \xrightarrow{\top|\top} r$ for all $x \in \{a,b\}$.

We claim that the instance I has a solution if, and only if, there are configurations (p, t_1), (q, t_2), and (r, t_3) such that

$$(\iota, \top) \vdash^* (p, t_1) \vdash^* (r, t_3) \text{ and } (\iota, \top) \vdash^* (q, t_2) \vdash^* (r, t_3)\,.$$

In other words, iff $\mathbb{G}(\mathcal{P})$ satisfies the formula

$$\begin{aligned}\varphi = \exists c_1, c_2, c_3\colon\quad & \mathit{state}_p(c_1) \wedge \mathrm{state}_q(c_2) \wedge \mathrm{state}_r(c_3)\\ & \wedge (\iota, [\top]) \vdash^* c_1 \vdash^* c_3\\ & \wedge (\iota, [\top]) \vdash^* c_2 \vdash^* c_3\,.\end{aligned}$$

The formula φ expresses the existence of a solution since $(\iota, \top) \vdash^* (p, t_1) \vdash^* (r, t_3)$ is equivalent to the existence of $1 \le n \le N$ and $i_1, \dots, i_N \in [k]$ such that

$$\begin{aligned} t_1 &= [u_{i_1}\, f(v_{i_1})\, u_{i_2}\, f(v_{i_2}) \cdots u_{i_n}\, f(v_{i_n})\top] \text{ and}\\ t_3 &= [u_{i_1}\, f(v_{i_1})\, u_{i_2}\, f(v_{i_2}) \cdots u_{i_N}\, f(v_{i_N})\top]\,.\end{aligned}$$

Since $u_i \parallel f(v_j)$ for all $i, j \in [k]$, we have

$$\begin{aligned} t_3 &= [u_{i_1}\, f(v_{i_1})\, u_{i_2}\, f(v_{i_2}) \cdots u_{i_N}\, f(v_{i_N})\top]\\ &= [u_{i_1}\, u_{i_2} \cdots u_{i_N}\, f(v_{i_1})\, f(v_{i_2}) \cdots f(v_{i_N})\top]\\ &= [u_{i_1}\, u_{i_2} \cdots u_{i_N}\, f(v_{i_1}\, v_{i_2} \cdots v_{i_N})\top]\end{aligned}$$

Similarly, $(\iota, \top) \vdash^* (q, t_2) \vdash^* (r, t_3)$ is equivalent to the existence of $1 \le m \le M$ and $x_1, x_2, \dots, x_M \in \{a,b\}$ such that

$$\begin{aligned} t_2 &= [x_1\, f(x_1)\, x_2\, f(x_2) \cdots x_m\, f(x_m)\top] \text{ and}\\ t_3 &= [x_1\, f(x_1)\, x_2\, f(x_2) \cdots x_M\, f(x_M)\top]\\ &= [x_1 x_2 \cdots x_M\, f(x_1 x_2 \cdots x_M)\top]\,.\end{aligned}$$

Thus, the formula φ holds if, and only if, there exists a word $w \in \{a,b\}^*$ with

$$u_{i_1}\, u_{i_2} \cdots u_{i_N}\, f(v_{i_1}\, v_{i_2} \cdots v_{i_N}) = w\, f(w)\,.$$

Since f is bijective, this is equivalent to

$$u_{i_1}\, u_{i_2} \cdots u_{i_N} = v_{i_1}\, v_{i_2} \cdots v_{i_N}\,,$$

i.e., to the existence of a solution for I. □

In order to rule out this undecidability, we restrict the class of pushdown systems further.

Definition 2.6. *Let $\mathcal{P} = (Q, \Sigma, D, \Delta)$ be a tPDS.*

1. *The tPDS $\mathcal{P}$ is* saturated *if (for any $p, q, r \in Q$, $a, b \in \Sigma$, and $u, v \in \Sigma^*$)*

$$p \xrightarrow{a|ubv} q \xrightarrow{b|\varepsilon} r \text{ and } b \parallel v \text{ imply } p \xrightarrow{a|uv} r.$$

2. *The tPDS $\mathcal{P}$ is* loop-connected *if it is saturated and the following holds (for all $n \in \mathbb{N}$, $p_i \in Q$, $a_i \in \Sigma$, $u_i \in \Sigma^*$ for $i \in [n]$). If*

$$p_0 \xrightarrow{a_0|u_1 a_1} p_1 \xrightarrow{a_1|u_2 a_2} p_2 \xrightarrow{a_2|u_3 a_3} \cdots \xrightarrow{a_{n-1}|u_n a_n} p_n \tag{1}$$

with $(p_0, a_0) = (p_n, a_n)$, then the set $\mathrm{alph}(u_1 u_2 \cdots u_n)$ *is connected.*

Suppose we have $p \xrightarrow{a|ubv} q \xrightarrow{b|\varepsilon} r$ and $b \parallel v$. Then, for any trace s, one gets

$$(p, x \cdot [a]) \vdash (q, x \cdot [ubv]) = (q, x \cdot [uvb]) \vdash (r, x \cdot [uv]).$$

If $\mathcal{P}$ is saturated, we can take the "shortcut" $(p, x \cdot [a]) \vdash (r, x \cdot [uv])$ in a single step. From an arbitrary tPDS, one can compute a saturated tPDS by adding transitions that has the same reachability relation [16]. It therefore makes sense to only consider saturated systems.

The idea of loop-connectedness is as follows. If transitions as in the definition exist, then the tPDS can make the following computation

$$(p_0, [a_0]) \vdash (p_1, [u_1 a_1]) \vdash \cdots \vdash (p_n, [u_1 u_2 \cdots u_n a_n]) = (p_0, [u_1 u_2 \cdots u_n a_0]).$$

In other words, the tPDS can replace any trace $[u\, a_0]$ by the trace $[u\, u_1 \cdots u_n\, a_0]$ and therefore, by repeating this loop, by $[u\, (u_1 \cdots u_n)^k\, a_0]$ for any number k. The requirement is that such a trace shall be connected.

We can now formulate the main result of this paper.

Theorem 2.7. *The following problem is decidable.*
input: loop-connected tPDS $\mathcal{P}$ and first-order formula φ in the language of $\mathbb{G}(\mathcal{P})$
question: Does φ hold in $\mathbb{G}(\mathcal{P})$?

Before starting the proof, we demonstrate that this decidability does not hold for monadic second-order logic.

Example 2.8. Let $\mathcal{P} = (Q, \Sigma, D, \Delta)$ be the following loop-connected tPDS with seven transitions: $p \xrightarrow{\top|a\top} p \xrightarrow{\top|\top} q \xrightarrow{\top|b\top} q \xrightarrow{\top|b} r \xrightarrow{b|\varepsilon} r$, $q \xrightarrow{\top|\varepsilon} r$, and $r \xrightarrow{a|\varepsilon} r$. Furthermore, suppose $a \parallel b$ and $(a, \top), (b, \top) \in D$.

Then $(p, [\top]) \vdash^* (r, [a^m b^n])$ for all $m \geq 0, n \geq 1$ and $(r, [a^k b^\ell]) \vdash^* (r, [a^m b^n])$ if, and only if, $k \geq m$ and $\ell \geq n$. Hence $\mathbb{G}(\mathcal{P})$ contains the infinite grid $(\mathbb{N} \times \mathbb{N}, \geq)$ implying that the monadic second-order theory is undecidable (cf., e.g., [3, Thm. 5.6]). □

To prove Theorem 2.7, we work with the following structure.

Definition 2.9. *Let $\mathcal{P} = (Q, \Sigma, D, \Delta)$ be a pushdown system. For $p, q \in Q$, we let $\text{Reach}_{p,q} \subseteq \mathbb{M}^2$ denote the set of pairs (s, t) of traces such that $(p, s) \vdash^* (q, t)$. Furthermore, $\text{Step}_{p,q} \subseteq \mathbb{M}^2$ is the set of pairs (s, t) such that $(p, s) \vdash (q, t)$.*

Define the structure $\mathcal{S}(\mathcal{P}) = \big(\mathbb{M}, (\text{Step}_{p,q}, \text{Reach}_{p,q})_{p,q \in Q}, (t)_{t \in \mathbb{M}}\big)$ with universe the set of traces over (Σ, D), the one-step relations $\text{Step}_{p,q}$ and the reachability relations $\text{Reach}_{p,q}$ for all $p, q \in Q$, and all traces $t \in \mathbb{M}$ as constants.

To justify that we work with this structure, we first prove that the theory of $\mathbb{G}(\mathcal{P})$ can be reduced to that of this structure.

Lemma 2.10. *Let $\mathcal{P} = (Q, \Sigma, D, \Delta)$ be a trace pushdown system. From a monadic second-order formula $\varphi(x_1, \ldots, x_n)$ in the language of $\mathbb{G}(\mathcal{P})$ and a tuple of states $\overline{p} = (p_1, \ldots, p_n) \in Q^n$, one can compute a formula $\varphi_{\overline{p}}(x_1, \ldots, x_n)$ in the language of $\mathcal{S}(\mathcal{P})$ such that*

$$\mathbb{G}(\mathcal{P}) \models \varphi((p_i, t_i)_{i \in [n]}) \iff \mathcal{S}(\mathcal{P}) \models \varphi_{\overline{p}}(t_1, \ldots, t_n)$$

holds for any $t_1, \ldots, t_n \in \mathbb{M}$. If φ is first-order, then the same applies to $\varphi_{\overline{p}}$.

Therefore, if the theory of $\mathbb{G}(\mathcal{P})$ is undecidable, then, by the above lemma, the same applies to the theory of $\mathcal{S}(\mathcal{P})$. By Examples 2.5 and 2.8, it follows that the first-order theory (the monadic second-order theory, resp.) of $\mathcal{S}(\mathcal{P})$ is not uniformly decidable for all (all loop-connected, resp.) tPDS. Conversely, in order to prove Theorem 2.7, it suffices to demonstrate the following result.

Theorem 2.11. *The following problem is decidable.*
input: loop-connected tPDS $\mathcal{P}$ and first-order formula φ in the language of $\mathcal{S}(\mathcal{P})$
question: Does φ hold in $\mathcal{S}(\mathcal{P})$?

The proof of this theorem can be found in Sect. 5. It is based on a special class of trace relations (cf. Definition 3.2 below) that we introduce and study next. Later, it will turn out that the relations $\text{Step}_{p,q}$ and $\text{Reach}_{p,q}$ of a loop-connected tPDS belong to this class, which implies the decidability using the theory of automatic structures.

3 Fnf-Automatic Relations

Let $A = \{a_1, \ldots, a_n\} \subseteq \Sigma$. The set A is *independent* if $(a_i, a_j) \in I$ for all $1 \leq i < j \leq n$ (in particular $\emptyset$ is independent). Let $\mathcal{F} \subseteq 2^\Sigma$ denote the set of all independent sets. Let $[.]\colon \mathcal{F}^* \to \mathbb{M}$ be the unique monoid homomorphism with $[A] = [a_1 \cdots a_n]$ for all $A = \{a_1, \ldots, a_n\} \in \mathcal{F}$.

A (possibly empty) word $A_1\, A_2\, \cdots\, A_n \in \mathcal{F}^*$ is *in extended Foata normal form* or *in eFNF* if $D(A_i) \supseteq A_{i+1}$ for all $i \in [n-1]$.

Since $D(\emptyset) = \emptyset$, all words in eFNF belong to $(\mathcal{F} \setminus \{\emptyset\})^* \{\emptyset\}^*$. A word over $\mathcal{F}$ is *in Foata normal form* or *in FNF* if it is in eFNF and belongs to $(\mathcal{F} \setminus \{\emptyset\})^*$.

Let $x \in \mathbb{M}$. Then, by [2, Theorem 1.2], there is a unique word $\mathrm{fnf}(x) \in \mathcal{F}^*$ in FNF with $x = [\mathrm{fnf}(x)]$; this word is the *Foata normal form of* x.

Let $W = A_1\, A_2\, \cdots\, A_n$ be in FNF. Then any factorisation of W into U and V trivially satisfies $[W] = [U] \cdot [V]$. But there are factorisations of $[W]$ into two traces that do not correspond to any factorisation of the word W. The following lemma describes the factorisations of the trace $[W]$ in terms of the word W.

Lemma 3.1. *Let $W = A_1 \cdots A_n$ be a word in eFNF and $x, y \in \mathbb{M}$. Then the following are equivalent:*

(1) $x \cdot y = [W]$
(2) There are sets $B_i \subseteq A_i$ for $i \in [n]$ such that
  (a) the word $(A_1 \setminus B_1)(A_2 \setminus B_2) \cdots (A_n \setminus B_n)$ is in eFNF,
  (b) $B_i \parallel (A_j \setminus B_j)$ for $1 \le i < j \le n$, and
  (c) $x = \big[(A_1 \setminus B_1)(A_2 \setminus B_2) \cdots (A_n \setminus B_n)\big]$ and $y = [B_1\, B_2 \cdots B_n]$.

Proof. First, assume (2) holds. From (b), we obtain $[A_j \setminus B_j] \cdot [B_1] = [B_1] \cdot [A_j \setminus B_j]$ for all $j \in \{2, \ldots, n\}$. Hence

$$\begin{aligned} x \cdot [B_1] &= [A_1 \setminus B_1] \cdots [A_n \setminus B_n] \cdot [B_1] \\ &= [A_1 \setminus B_1] \cdot [B_1] \cdot [A_2 \setminus B_2] \cdots [A_n \setminus B_n] \\ &= [A_1] \cdot [A_2 \setminus B_2] \cdots [A_n \setminus B_n]\,. \end{aligned}$$

Now (1) follows by induction.

Conversely, suppose (1) holds. By Levi's Lemma (cf. [4, p. 74]), there are traces x_i and y_i for $i \in [n]$ such that

- $[A_i] = x_i \cdot y_i$ for all $i \in [n]$,
- $y_i \parallel x_j$ for all $1 \le i < j \le n$,
- $x = x_1 \cdot x_2 \cdots x_n$, and
- $y = y_1 \cdot y_2 \cdots y_n$.

Since, for $i \in [n]$, A_i is an independent set, there is $B_i \subseteq A_i$ with $x_i = [A_i \setminus B_i]$ and $y_i = [B_i]$.

Claim (b) holds since $B_i = \mathrm{alph}(y_i) \parallel \mathrm{alph}(x_j) = A_j \setminus B_j$ for all $1 \le i < j \le n$.

To show (a), let $i \in [n-1]$ and $b \in A_{i+1} \setminus B_{i+1} \subseteq A_{i+1}$. Since the word $A_1\, A_2\, \cdots\, A_n$ is in eFNF, there exists $a \in A_i$ with $(a, b) \in D$. Then $a \in B_i$ is not possible since this would contradict $B_i \parallel A_{i+1} \setminus B_{i+1} \ni b$.

Claim (c) is immediate by our choice of B_i. □

We next define what we mean by "fnf-automatic relations".

Let $\mathcal{R} \subseteq \mathbb{M}^k$ be a k-ary relation on the set of traces $\mathbb{M}$. From this relation, we first construct a language $L_{\mathcal{R}}$ over the alphabet $\mathcal{F}^k$. Note that the elements of $\mathcal{F}^k$ are k-tuples of independent sets that we write interchangeably by row or column vectors. Now consider a word

$$\begin{pmatrix} A_1^1 \\ A_2^1 \\ \vdots \\ A_k^1 \end{pmatrix} \begin{pmatrix} A_1^2 \\ A_2^2 \\ \vdots \\ A_k^2 \end{pmatrix} \cdots \begin{pmatrix} A_1^n < \\ A_2^n \\ \vdots \\ A_k^n > \end{pmatrix} \in (\mathcal{F}^k)^* . \tag{2}$$

It belongs to the language $L_{\mathcal{R}}$ if

- $W_i = A_i^1\, A_i^2 \cdots A_i^n$ is in eFNF for all $i \in [k]$ and
- the tuple $([W_1], [W_2], \ldots, [W_k])$ belongs to $\mathcal{R}$.

Intuitively, this means that all the "lines" in (2) are in eFNF and the tuple of values of these lines belongs to $\mathcal{R}$.

Definition 3.2. *A relation $\mathcal{R} \subseteq \mathbb{M}^k$ is* fnf-automatic *if the language $L_{\mathcal{R}} \subseteq (\mathcal{F}^k)^*$ is regular.*

By the very definition, any fnf-automatic relation $\mathcal{R}$ can be represented by an NFA $\mathcal{A}$ for the language $L_{\mathcal{R}}$; we will always assume that fnf-automatic relations are presented that way.

A very simple example of an fnf-automatic relation is the identity relation $\mathrm{Id}_{\mathbb{M}} = \{(x, x) \mid x \in \mathbb{M}\}$. It is not difficult to see that the class of fnf-automatic relations is closed under union, intersection, complementation, and inversion $\mathcal{R}^{-1} = \{(x_k, x_{k-1}, \ldots, x_1) \mid (x_1, x_2, \ldots, x_k) \in \mathcal{R}\}$; in addition, it is closed under composition of binary relations. Furthermore, any fnf-automatic relation is rational, but not vice versa.

In general, the product of two fnf-automatic relations is not fnf-automatic (consider, e.g., the unary relations $\mathcal{L}_1 = [a]^*$ and $\mathcal{L}_2 = [bc]^*$ with $(a, b) \in D$ and $\{a, b\} \parallel \{c\}$ since $[(ac)^m b^n] \in \mathcal{L}_1 \cdot \mathcal{L}_2$ iff $m = n$). The following lemma proves that certain products are fnf-automatic.

Lemma 3.3. *Let $\mathcal{R} \subseteq \mathbb{M}^2$ be fnf-automatic and $\mathcal{K}, \mathcal{L} \subseteq \mathbb{M}$ recognizable. Then the relation $\mathcal{R} \cdot (\mathcal{K} \times \mathcal{L}) = \{(x_1 y, x_2 z) \mid (x_1, x_2) \in \mathcal{R}, y \in \mathcal{K}, z \in \mathcal{L}\}$ is effectively fnf-automatic.*

Proof. It suffices to consider the case $\mathcal{L} = \{1\}$ since $\mathcal{R} \cdot (\mathcal{K} \times \mathcal{L})$ is the composition of $\mathcal{R} \cdot (\mathcal{K} \times \{1\})$ and $\left(\mathrm{Id}_{\mathbb{M}} \cdot (\mathcal{L} \times \{1\})\right)^{-1}$.

Since $\mathcal{R}$ is fnf-automatic and $\mathcal{K} \subseteq \mathbb{M}$ recognizable, there are NFAs $\mathcal{A}_1 = (Q_1, \mathcal{F}^2, I_1, \rightarrow_1, F_1)$ and $\mathcal{A}_2 = (Q_2, \Sigma, I_2, \rightarrow_2, F_2)$ that accept the languages $L_{\mathcal{R}} \subseteq (\mathcal{F}^2)^*$ and $\{w \in \Sigma^* \mid [w] \in \mathcal{K}\}$, respectively.

We now build a new NFA $\mathcal{A} = (Q, \mathcal{F}^2, I, \rightarrow, F)$ as follows:

- States are quadruples (p_1, p_2, X_1, X_2) with $p_i \in Q_i$ and $X_i \subseteq \Sigma$ for $i \in [2]$.

- Initial states are quadruples $(\iota_1, \iota_2, \emptyset, \Sigma)$ with $\iota_i \in I_i$ for $i \in [2]$.
- We have $(p_1, p_2, X_1, X_2) \xrightarrow{(A,C)} (q_1, q_2, Y_1, Y_2)$ for any $B \subseteq A$ with
  - $p_1 \xrightarrow{(A\setminus B,C)}_1 q_1$,
  - there is some word $w \in [B]$ with $p_2 \xrightarrow{w}_2 q_2$,
  - $Y_1 = X_1 \cup B$ and $X_1 \parallel A \setminus B$, and
  - $Y_2 = A \setminus B$ and $D(X_2) \supseteq A \setminus B$.
- Accepting states are quadruples (f_1, f_2, X_1, X_2) with $f_i \in F_i$ for $i \in [2]$.

Now let $U = A_1\, A_2\, \cdots\, A_n$ and $V = C_1\, C_2\, \cdots\, C_n$ be in eFNF and let

$$U \otimes V = \binom{A_1}{C_1} \binom{A_2}{C_2} \cdots \binom{A_n}{C_n} .$$

Then $U \otimes V \in L(\mathcal{A})$ if, and only if, there are $\iota_i \in I_i$, $g_i \in F_i$, and $Y_1, Y_2 \subseteq \Sigma$ such that

$$(\iota_1, \iota_2, \emptyset, \Sigma) \xrightarrow{U \otimes V} (q_1, q_2, Y_1, Y_2) . \tag{3}$$

But this is the case if, and only if, there are sets $B_i \subseteq A_i$ for $i \in [n]$ such that

(i) $\iota_1 \xrightarrow{\binom{A_1 \setminus B_1}{C_1} \binom{A_2 \setminus B_2}{C_2} \cdots \binom{A_n \setminus B_n}{C_n}}_1 q_1$,
(ii) for all $i \in [n]$, there are words $w_i \in [B_i]$ such that $\iota_2 \xrightarrow{w_1\, w_2\, \cdots\, w_n}_2 q_2$,
(iii) $Y_1 = \bigcup_{i \in [n]} B_i$ and $B_i \parallel (A_j \setminus B_j)$ for all $1 \le i < j \le n$, and
(iv) $Y_2 = A_n \setminus B_n$ and $(A_1 \setminus B_1)\,(A_2 \setminus B_2)\, \cdots\, (A_n \setminus B_n)$ is in eFNF.

Thus, we have to show that this holds if, and only if, $([U], [V]) \in \mathcal{R} \cdot (\mathcal{K} \times \{1\})$.

First assume (i-iv) hold. Then set $x = \big[(A_1 \setminus B_1)\,(A_2 \setminus B_2)\, \cdots\, (A_n \setminus B_n)\big]$, $y = [B_1\, B_2\, \cdots\, B_n]$, and $z = [C_1\, C_2\, \cdots\, C_n] = [V]$.

Since, by (iv), the word $(A_1 \setminus B_1) \cdots (A_n \setminus B_n)$ is in eFNF, (i) and our assumption on the NFA $\mathcal{A}_1$ imply $(x, z) \in \mathcal{R}$

Property (ii) implies $w_1 w_2 \cdots w_n \in L(\mathcal{A}_2)$ and therefore $y = [w_1 \cdots w_n] \in \mathcal{K}$.

Finally, Lemma 3.1 together with (iii) and (iv) and our definition of x and y yield $x \cdot y = [A_1] \cdot [A_2] \cdots [A_n] = [U]$.

Now $(x, z) \in \mathcal{R}$ and $y \in \mathcal{K}$ imply $([U], [V]) = (x \cdot y, z) \in \mathcal{R} \cdot (\mathcal{K} \times \{1\})$. This finishes the verification of the implication "$\Rightarrow$".

Conversely, suppose $([U], [V]) \in \mathcal{R} \cdot (\mathcal{K} \times \{1\})$. Set $z = [V]$ as above. There are traces x and y with $[U] = x \cdot y$, $(x, z) \in \mathcal{R}$, and $y \in \mathcal{K}$. Since U is in eFNF with $[U] = x \cdot y$, Lemma 3.1 implies the existence of sets $B_i \subseteq A_i$ for $i \in [n]$ such that

(a) the word $X := (A_1 \setminus B_1)\,(A_2 \setminus B_2)\, \cdots\, (A_n \setminus B_n)$ is in eFNF,
(b) $B_i \parallel A_j \setminus B_j$ for all $1 \le i < j \le n$, and
(c) $x = \big[(A_1 \setminus B_1)\,(A_2 \setminus B_2)\, \cdots\, (A_n \setminus B_n)\big]$ and $y = [B_1\, B_2\, \cdots\, B_n]$.

The words X and V both are in eFNF with $([X], [V]) = (x, z) \in \mathcal{R}$. Our assumption on the NFA $\mathcal{A}_1$ implies that the word from (i) belongs to $L(\mathcal{A}_1)$, i.e., there are $(\iota_1, g_1) \in I_1 \times F_1$ such that (i) holds.

Let $w_i \in [B_i]$ for $i \in [n]$. Then $[w_1 w_2 \cdots w_n] = y \in \mathcal{K}$. Hence the word $w_1 \cdots w_n$ is accepted by the NFA $\mathcal{A}_2$, i.e., there are $(\iota_2, g_2) \in I_2 \times F_2$ such that (ii) holds.

Setting $Y_1 = \bigcup_{i \in [n]} B_i$ and $Y_2 = A_n \setminus B_n$ ensures that also (iii) and (iv) hold. Thus, we have (3) and therefore $U \otimes V \in L(\mathcal{A})$. But this finishes the verification of the implication "$\Leftarrow$". □

As a consequence, we get the following relation between fnf-automatic and recognizable languages.

Corollary 3.4. *Any recognizable language in* $\mathbb{M}$ *is fnf-automatic, but there are fnf-automatic languages that are not recognizable.*

4 Reachability and Fnf-Automaticity

The aim of this section is the proof of the following result.

Theorem 4.1. *Let* $\mathcal{P} = (Q, \Sigma, D, \Delta)$ *be a loop-connected tPDS and* $p, q \in Q$ *two states. Then the one-step relation* $\mathrm{Step}_{p,q}$ *and the reachability relation* $\mathrm{Reach}_{p,q}$ *are effectively fnf-automatic.*

Note that the relation $\mathrm{Step}_{p,q}$ is the union of all relations $\mathrm{Id}_{\mathbb{M}} \cdot (\{[a]\} \times \{[w]\})$ with $(p, a, w, q) \in \Delta$. Hence $\mathrm{Step}_{p,q}$ is fnf-automatic by Lemma 3.3.

The proof of the claim regarding $\mathrm{Reach}_{p,q}$ is more involved. It uses constructions and results from [16] that were shown for all tPDS and it then refines them for the case of loop-connected systems. It starts by first demonstrating the theorem for two special cases.

The first special case assumes that all transitions of $\mathcal{P}$ shorten the pushdown, i.e., write the empty word. In that case, the proof of [16, Prop. 5.4] constructs a recognizable language $\mathcal{K}$ with $\mathrm{Reach}_{p,q} = \mathrm{Id}_{\mathbb{M}} \cdot (\mathcal{K} \times \{1\})$. Now Lemma 3.3 implies the following:

Proposition 4.2. *Let* $\mathcal{P} = (Q, \Sigma, D, \Delta)$ *be a tPDS with* $\Delta \subseteq Q \times \Sigma \times \{\varepsilon\} \times Q$*, and* $p, q \in Q$ *two states. Then the relation* $\mathrm{Reach}_{p,q}$ *is effectively fnf-automatic.*

While above we consider systems that never write anything onto the pushdown, we next want to study systems that always write something. In other words, these systems do not shorten the pushdown in any step. We make the additional assumption that all letters read from the pushdown are twins, i.e., if $p \xrightarrow{a|w} q$ and $p' \xrightarrow{a'|w'} q'$ are transitions from Δ, then $D(a) = D(a')$. In [16], it is shown that the reachability relation in such systems is a union of $\mathrm{Id}_{\mathbb{M}}$ and finitely many relations of the form $\mathrm{Id}_{\mathbb{M}} \cdot (\{a\} \times \mathcal{H}_a)$ for some rational set of traces $\mathcal{H}_a$. The additional assumption of loop-connectedness allows to show that $\mathcal{H}_a$ is even recognizable.

Lemma 4.3. *Let $\mathcal{P} = (Q, \Sigma, D, \Delta)$ be a loop-connected tPDS with $\Delta \subseteq Q \times \operatorname{twns}(a) \times \Sigma^+ \times Q$ for some $a \in \Sigma$, and $p, q \in Q$ be two states. There effectively exists a recognizable trace language $\mathcal{H}_a \subseteq \mathbb{M}$ such that, for any $x \in \mathbb{M}$,*

$$\{y \in \mathbb{M} \mid (x \cdot [a], y) \in \operatorname{Reach}_{p,q}\} = \{x\} \cdot \mathcal{H}_a\,. \tag{4}$$

Proof. [16, Lemma 5.5] constructs a *rational* set $\mathcal{H}_a$ with (4). Here, we show that this set is even *recognizable* provided $\mathcal{P}$ is loop-connected.

We next recall the construction of an ε-NFA $\mathcal{A} = (Q_{\mathcal{A}}, A, \{(p,a)\}, \delta, \{(q,\varepsilon)\})$ from the proof of [16, Lemma 5.5] (the set $\mathcal{H}_a$ equals $[L(\mathcal{A})]$).

The set $Q_{\mathcal{A}}$ of states of $\mathcal{A}$ contains all pairs (q,ε) and $(r,c) \in Q \times \Sigma$. It has c-labeled transitions from (q,c) to (q,ε) for any $c \in \Sigma$. In addition, for any $(r, c, udv, s) \in \Delta$ with $d \in \Sigma$ and $u, v \in \Sigma^*$ such that $v \parallel d$, we add a uv-labeled path from (r,c) to (s,d).

We set $\mathcal{H}_a = [L(\mathcal{A})]$. In the proof of [16, Lemma 5.5], it is then shown that Eq. (4) holds for all $x \in \mathbb{M}$, so it remains to be shown that $\mathcal{H}_a$ is recognizable.

To this aim, we first verify that the ε-NFA $\mathcal{A}$ satisfies the following: if q is a state of $\mathcal{A}$ and $w \in \Sigma^*$ such that $q \xrightarrow{w}_{\mathcal{A}} q$, then $\operatorname{alph}(w)$ is connected.

So suppose $(p_0, c_0) \xrightarrow{w}_{\mathcal{A}} (p_0, c_0)$. Then, for some $i \geq 1$, $p_i \in Q$, $c_i \in \Sigma$, and $v_i \in \Sigma^*$, we have

$$(p_0, c_0) \xrightarrow{u_1}_{\mathcal{A}} (p_1, c_1) \xrightarrow{u_2}_{\mathcal{A}} \cdots \xrightarrow{u_n}_{\mathcal{A}} (p_n, c_n) = (p_0, c_0)$$

such that no state from $Q \times \Sigma$ is visited between (p_{i-1}, c_{i-1}) and (p_i, c_i). Furthermore, $w = u_1\, u_2 \,\cdots\, u_n$.

The construction of $\mathcal{A}$ yields, for any $i \in [n]$, a factorisation $u_i = x_i\, y_i$ such that $(p_{i-1}, c_{i-1}, x_i c_i y_i, p_i) \in \Delta$ and $c_i \parallel y_i$.

The assumption on Δ implies $c_i \in \operatorname{twns}(a)$ for all $i \in [n]$. Since $\mathcal{P}$ is a tPDS, we have $\operatorname{alph}(y_i) \subseteq D(y_i) \subseteq D(x_i c_i y_i) \subseteq D(c_{i-1}) = D(a) = D(c_i)$. Now $c_i \parallel y_i$ and $u_i = x_i\, y_i$ imply $y_i = \varepsilon$ and $u_i = x_i$ and therefore $(p_{i-1}, c_{i-1}, u_i c_i, p_i) \in \Delta$.

These transitions form a loop as in (1) implying the connectedness of the set $\operatorname{alph}(u_1 u_2 \cdots u_n) = \operatorname{alph}(w)$.

Thus, indeed, any word labeling a loop of the ε-NFA $\mathcal{A}$ is connected.

It now follows from [20] that the set of traces $\mathcal{H}_a$ is effectively recognizable. □

Proposition 4.4. *Let $\mathcal{P} = (Q, \Sigma, D, \Delta)$ be a loop-connected tPDS with $\Delta \subseteq Q \times \operatorname{twns}(b) \times \Sigma^+ \times Q$ for some $b \in \Sigma$, and $p, q \in Q$ two states. Then the relation $\operatorname{Reach}_{p,q}$ is effectively fnf-automatic.*

Proof. From Lemma 4.3, we get

$$\operatorname{Reach}_{p,q} = \operatorname{Id}_{\mathbb{M}} \cup \bigcup_{a \in \operatorname{twns}(b)} \operatorname{Id}_{\mathbb{M}} \cdot (\{[a]\} \times \mathcal{H}_a)$$

for some effectively recognizable languages $\mathcal{H}_a \subseteq \mathbb{M}$. Recall that $\operatorname{Id}_{\mathbb{M}}$ is fnf-automatic and that the class of fnf-automatic relations is effectively closed under union. Hence Lemma 3.3 implies that $\operatorname{Reach}_{p,q}$ is effectively fnf-automatic. □

We now come to the proof of the central result of this section.

Proof (of Theorem 4.1). We first split the set of transitions of the loop-connected system $\mathcal{P} = (Q, \Delta)$ into finitely many sets:

$$\Delta_\varepsilon = \Delta \cap (Q \times \Sigma \times \{\varepsilon\} \times Q) \text{ and } \Delta_T = \Delta \cap (Q \times T \times \Sigma^+ \times Q)$$

where $T = \text{twns}(a)$ for some $a \in \Sigma$. For all possible X, $\mathcal{P}_X = (Q, \Delta_X)$ is a loop-connected tPDS. Let $\text{Reach}_{r,s}^X$ denote the reachability relation of the system $\mathcal{P}_X$. This relation is effectively fnf-automatic by Propositions 4.2 and 4.4, respectively.

In the proof of [16, Theorem 5.12], it is shown that the reachability relation $\text{Reach}_{p,q}$ of the tPDS $\mathcal{P}$ can effectively be obtained from these relations by composition and union. Hence, the theorem's claim follows. □

5 Proof of Theorem 2.11

We have to demonstrate that the first-order theory of the structure $\mathcal{S}(\mathcal{P})$ is uniformly decidable for all loop-connected tPDS. To this aim, we prove that the structure $\mathcal{S}(\mathcal{P})$ is automatic [1,10,14] for any loop-connected tPDS $\mathcal{P} = (Q, \Sigma, D, \Delta)$.

So let eFNF $\subseteq \mathcal{F}^*$ denote the regular language of all words in eFNF. By Theorem 4.1, also the languages $L_{\text{Step}_{p,q}}$ and $L_{\text{Reach}_{p,q}}$ for $p, q \in Q$ are regular. Since the mapping eFNF $\to \mathbb{M}\colon W \mapsto [W]$ is surjective, the structure $\mathcal{S}$ is automatic. Hence its first-order theory is decidable (uniformly in $\mathcal{P}$) [1,10,14]. □

Remark 5.1. Let $\mathcal{R} \subseteq \mathbb{M}^k$ be a k-ary fnf-automatic relation and $(p_1, \ldots, p_k)$ a tuple of states of the loop-connected tPDS $\mathcal{P}$. Extend the configuration graph $\mathbb{G}(\mathcal{P})$ with the relation $\{(p_i, t_i)_{1 \le i \le k} \mid (t_1, \ldots, t_k) \in \mathcal{R}\}$ and the structure $\mathcal{S}(\mathcal{P})$ with the relation $\mathcal{R}$. Then the above proof still gives the decidability of the first-order theory. Adding such relations seems pretty exotic, but it might be of interest in case $\mathcal{R}$ is a recognizable language (as then, by Corollary 3.4, it is a unary fnf-automatic relation).

Recall that in case $D = \Sigma^2$, even the monadic second-order theory of $\mathbb{G}(\mathcal{P})$ is decidable. By Example 2.8, this does not hold for arbitrary loop-connected tPDS. But another restriction of second-order logic yields a decidable theory of reachability. In this restriction, second-order quantification $\exists X$ infinite: φ is possible for relational variables X of arbitrary arity r, but with the condition that the atomic formula $(x_1, \ldots, x_r) \in X$ appears only negatively in φ. In [18], it is shown that the validity of such formulas is decidable for any automatic structure. Thus, this decidability also holds for the configuration graph with reachability of a loop-connected tPDS.

The work on automatic structures (see, e.g., [9] for a recent survey) revealed that additional types of quantifiers are possible: the infinity quantifier $\exists^\infty$ [1], modulo-counting quantifiers [23], Ramsey-quantifier [24], and a boundedness quantifier [17].

Disclosure of Interests. The author is not aware of any competing interests wrt. this article.

References

1. Blumensath, A., Grädel, E.: Finite presentations of infinite structures: automata and interpretations. Theory Comput. Syst. **37**(6), 641–674 (2004)
2. Cartier, P., Foata, D.: Problèmes combinatoires de commutation et réarrangements. Lecture Notes in Mathematics, vol. 85, Springer, Berlin - Heidelberg - New York (1969)
3. Courcelle, B., Engelfriet, J.: Graph Structure and Monadic Second-Order Logic. Cambridge University Press (2012)
4. Diekert, V., Rozenberg, G.: The Book of Traces. World Scientific Publ. Co. (1995)
5. D'Osualdo, E., Meyer, R., Zetzsche, G.: First-order logic with reachability for infinite-state systems. In: LICS 2016, pp. 457–466. ACM (2016)
6. Elder, M., Kambites, M., Ostheimer, G.: On groups and counter automata. Int. J. Algebra Comput. **18**(08), 1345–1364 (2008)
7. Elston, G., Ostheimer, G.: On groups whose word problem is solved by a counter automaton. Theoret. Comput. Sci. **320**(2–3), 175–185 (2004)
8. Gilman, R.: Formal Languages and Infinite Groups. DIMACS Series in Discrete Mathematics and Theoretical Computer Science, vol. 25. DIMACS (1996)
9. Grädel, E.: Automatic structures: twenty years later. In: LICS 2020, pp. 21–34. ACM (2020)
10. Hodgson, B.: On direct products of automaton decidable theories. Theoret. Comput. Sci. **19**, 331–335 (1982)
11. Ibarra, O., Sahni, S., Kim, C.: Finite automata with multiplication. Theor. Comput. Sci. **2**(3), 271–294 (1976)
12. Ito, M., Martín-Vide, C., Mitrana, V.: Group weighted finite transducers. Acta Inform. **38**, 117–129 (2001)
13. Kambites, M.: Formal languages and groups as memory. Commun. Algebra **37**, 193–208 (2009)
14. Khoussainov, B., Nerode, A.: Automatic presentations of structures. In: Logic and Computational Complexity. Lecture Notes in Computer Science, vol. 960, pp. 367–392.Springer (1995)
15. Köcher, C., Kuske, D.: Backwards-reachability for cooperating multi-pushdown systems. J. Comput. Syst. Sci. **148**, 103601 (2025). 14 pages
16. Köcher, C., Kuske, D.: Reachability in trace-pushdown systems (2025, submitted). http://eiche.theoinf.tu-ilmenau.de/kuske
17. Kuske, D.: Theories of automatic structures and their complexity. In: Bozapalidis, S., Rahonis, G. (eds.) CAI 2009. LNCS, vol. 5725, pp. 81–98. Springer, Heidelberg (2009). https://doi.org/10.1007/978-3-642-03564-7_5
18. Kuske, D., Lohrey, M.: Some natural decision problems in automatic graphs. J. Symb. Log. **75**(2), 678–710 (2010)
19. Mitrana, V., Stiebe, R.: Extended finite automata over groups. Discret. Appl. Math. **108**(3), 287–300 (2001)
20. Ochmański, E.: Regular behaviour of concurrent systems. Bull. Eur. Assoc. Theor. Comput. Sci. **27**, 56–67 (1985)
21. Render, E., Kambites, M.: Rational subsets of polycyclic monoids and valence automata. Inf. Comput. **207**(11), 1329–1339 (2009)

22. Render, E.: Rational monoid and semigroup automata. Ph.D. thesis, University of Manchester (2010)
23. Rubin, S.: Automatic structures. Ph.D. thesis, University of Auckland (2004)
24. Rubin, S.: Automata presenting structures: a survey of the finite string case. Bull. Symb. Log. **14**, 169–209 (2008)
25. Zetzsche, G.: The emptiness problem for valence automata over graph monoids. Inf. Comput. **277**, 104583 (2021)

A Point to Set Principle for Finite-State Dimension

Elvira Mayordomo(✉)

Departamento de Informática e Ingeniería de Sistemas, Instituto de Investigación en Ingeniería de Aragón, Universidad de Zaragoza, Zaragoza, Spain
elvira@unizar.es

Abstract. Effective dimension has proven very useful in geometric measure theory through the point-to-set principle [9] that characterizes Hausdorff dimension by relativized effective dimension. Finite-state dimension is the least demanding effectivization in this context [3] that among other results can be used to characterize Borel normality [2].

In this paper we prove a characterization of finite-state dimension in terms of information content of a real number at a certain precision. We then use this characterization to give a robust concept of relativized normality and prove a finite-state dimension point-to-set principle. We finish with an open question on the equidistribution properties of relativized normality.

Keywords: Algorithmic randomness · Algorithmic information theory · Effective dimension · Point to set principle · Borel normality

1 Introduction

Effective dimension was introduced in [7,8] as an effectivization of Hausdorff dimension. One of its generalizations is finite-state dimension [3] that is a robust notion that interacts with compression and characterizes Borel normality [2].

In [9] Lutz and Lutz proved a point-to-set principle that characterizes Hausdorff dimension in terms of relativized effective dimension. This principle has already produced a number of interesting results in geometric fractal theory through computability based proofs (see [6,10] and a number of more recent results such as [5]).

In this paper we provide a characterization of finite-state dimension on Euclidean space based on the finite-state information content of a real number at a certain precision, which also provides an alternative characterization of Borel normality. This new characterization gives rise to a natural and robust relativization of finite state dimension with the strong property of a finite-state dimension point-to-set principle. We finish with open questions on the equidistribution properties of the corresponding relativized normality.

Research supported in part by Spanish Ministry of Science and Innovation grant PID2019-104358RB-I00 and by the Science dept. of Aragon Government: Group Reference T64_20R (COSMOS research group).

© The Author(s), under exclusive license to Springer Nature Switzerland AG 2025
A. Beckmann et al. (Eds.): CiE 2025, LNCS 15764, pp. 299–304, 2025.
https://doi.org/10.1007/978-3-031-95908-0_21

2 Preliminaries

Let Σ be a finite alphabet. We write Σ^* for the set of all (finite) *strings* over Σ and Σ^∞ for the set of all (infinite) *sequences* over Σ. We write $|x|$ for the length of a string or sequence x, and we write λ for the *empty string*, the string of length 0. For $x \in \Sigma^* \cup \Sigma^\infty$ and $0 \le n < |x|$, we write $x \upharpoonright n = x[0..n-1]$. For $w \in \Sigma^*$ and $x \in \Sigma^* \cup \Sigma^\infty$, we say that w is a *prefix* of x, and we write $w \sqsubseteq x$, if $x \upharpoonright |w| = w$.

A Σ *finite-state transducer* (Σ*-FST*) is a 4-tuple $T = (Q, \delta, \nu, q_0)$, where

- Q is a nonempty, finite set of *states*,
- $\delta : Q \times \Sigma \to Q$ is the *transition function*,
- $\nu : Q \times \Sigma \to \Sigma^*$ is the *output function*, and
- $q_0 \in Q$ is the *initial state*.

For $q \in Q$ and $w \in \Sigma^*$, we define the *output* from state q on input w to be the string $\nu(q, w)$ defined by the recursion

$$\nu(q, \lambda) = \lambda,$$
$$\nu(q, wa) = \nu(q, w)\nu(\delta(q, w), a)$$

for all $w \in \Sigma^*$ and $a \in \Sigma$. We then define the *output* of T on input $w \in \Sigma^*$ to be the string $T(w) = \nu(q_0, w)$.

For each integer $b \ge 1$ we let $\Sigma_b = \{0, 1, \ldots, b-1\}$ be the alphabet of *base-b digits*. We use infinite sequences over Σ_b to represent real numbers in $[0, 1)$. Each $S \in \Sigma_b^\infty$ is associated the real number $\mathrm{real}_b(S) = \sum_{i=1}^{\infty} S[i-1]b^{-i}$ and for each $x \in [0, 1)$, $\mathrm{seq}_b(x)$ is the infinite sequence S that does not finish with infinitely many $b-1$ and such that $x = \mathrm{real}_b(S)$.

A set of real numbers $A \subseteq [0, 1)$ is represented by the set

$$\mathrm{seq}_b(A) = \{\mathrm{seq}_b(\alpha) \mid \alpha \in A\}$$

of sequences. If $X \subseteq \Sigma_b^\infty$ then

$$\mathrm{real}_b(X) = \{\mathrm{real}_b(x) \mid x \in X\}.$$

We will denote with $\mathcal{D}_b$ the set of rational numbers that have finite representation in base b, that is,

$$\mathcal{D}_b = \{q \,|\, \mathrm{seq}_b(q) = w0^\infty, w \in \Sigma_b^*\}.$$

We will write $\mathrm{real}_b(w)$ for $\mathrm{real}_b(w0^\infty)$ when $w \in \Sigma_b^*$.

For the sake of completeness we include the definition of Borel normality.

Definition 1 (Borel [1]). *A real number α is normal in base $b \in \mathbb{N}$ ($b \ge 2$) if, for every $m \ge 1$ and every length-m sequence w of base-b digits, the asymptotic, empirical frequency of w in the base-b expansion of α is b^{-m}.*

α is absolutely normal if it is normal in every base $b \ge 2$.

3 An Euclidean Characterization of Finite-State Dimension and Borel Normality

Finite-state dimension was introduced in [3] on the space of infinite sequences over a finite alphabet. The original definition in terms of gambling was proven robust by several characterizations in terms of information lossless compression [3] and several versions of entropy [2].

Here we present an alternative definition on the Euclidean space and then prove its equivalence with [3].

Definition 2. *Let T be a Σ-FST and let $w \in \Sigma^*$. The T-information content of w is*

$$\mathrm{K}^T(w) = \min\{|\pi| \mid T(\pi) = w\}.$$

Definition 3. *Let T be a Σ_b-FST, $\delta > 0$ and $x \in [0,1)$. The base-b T-information content of x at precision δ is*

$$\mathrm{K}^T_\delta(x) = \min\left\{\mathrm{K}^T(w) \mid |\,\mathrm{real}_b(w) - x| < \delta\right\}.$$

We next define the finite-state dimension of points and sets.

Definition 4. *Let $b \geq 1$. Let $x \in [0,1)$ and $A \subseteq [0,1)$. The base-b finite-state dimension of x is*

$$\dim^b_{\mathrm{FS}}(x) = \inf_{T\Sigma_b-\mathrm{FST}} \liminf_{\delta>0} \frac{\mathrm{K}^T_\delta(x)}{\log_b(1/\delta)},$$

the base-b finite-state dimension of A is

$$\dim^b_{\mathrm{FS}}(A) = \inf_{T\Sigma_b-\mathrm{FST}} \sup_{x\in A} \liminf_{\delta>0} \frac{\mathrm{K}^T_\delta(x)}{\log_b(1/\delta)}.$$

Observation 1. $\dim^b_{\mathrm{FS}}(x) = \inf_{T\Sigma_b-\mathrm{FST}} \liminf_n \frac{\mathrm{K}^T_{b^{-n}}(x)}{n}$.

The definition of finite-state dimension from [3] is usually done in a space of infinite sequences, while identifying $[0,1)$ and Σ_b^∞ through seq_b or base-b representation.

Doty and Moser [4] proved that finite dimension on sequences can be characterized in terms of finite-state transducers.

Theorem 2 ([4]). *Let $S \in \Sigma^\infty$,*

$$\dim_{\mathrm{FS}}(S) = \inf_{T\Sigma-\mathrm{FST}} \liminf_n \frac{\mathrm{K}^T(S \upharpoonright n)}{n}.$$

We next show that the notion of information content at a certain precision characterizes finite-state dimension.

Theorem 3. *For each* $b \geq 1$, $x \in [0,1)$, *and* $A \subseteq [0,1)$

$$\dim_{\mathrm{FS}}^b(x) = \dim_{\mathrm{FS}}(\mathrm{seq}_b(x)),$$

$$\dim_{\mathrm{FS}}^b(A) = \dim_{\mathrm{FS}}(\mathrm{seq}_b(A)).$$

Proof. Let $x \in [0,1)$, let $S = \mathrm{seq}_b(x)$. Then for every $n \in \mathbb{N}$ and T Σ_b-FST, $\mathrm{K}^T_{b^{-n}}(x) \leq \mathrm{K}^T(S \upharpoonright (n+1))$ and therefore $\dim_{\mathrm{FS}}^b(x) \leq \dim_{\mathrm{FS}}(S)$.

For each $w \in \Sigma_b^* \cup \Sigma_b^\infty$, let $comp(w)$ be the complement of w, that is, $comp(w)[i] = b - 1 - w[i]$ for $0 \leq i < |w|$.

Claim. $\dim_{\mathrm{FS}}(S) = \dim_{\mathrm{FS}}(comp(S))$. $\dim_{\mathrm{FS}}^b(x) = \dim_{\mathrm{FS}}^b(\mathrm{real}_b(comp(\mathrm{seq}_b(x))))$.

Claim. $\dim_{\mathrm{FS}}(S) \leq \dim_{\mathrm{FS}}^b(x)$.

To prove this claim, notice that $\dim_{\mathrm{FS}}^b(x)$ needs to be witnessed either by approximations from above or for approximations from below and that tighter approximations can be delayed.

That is, for every FST T there exist and infinitely many n_i such that

$$\lim_i \frac{\mathrm{K}^T_{b^{-n_i}}(x)}{n_i} = \liminf_n \frac{\mathrm{K}^T_{b^{-n}}(x)}{n}.$$

For each i let w_i be such that $\mathrm{K}^T(w_i) = \mathrm{K}^T_{b^{-n_i}}(x)$ and $|x - \mathrm{real}_b(w_i)| < b^{-n_i}$. Let m_i be such that $b^{-m_i-1} \leq |x - \mathrm{real}_b(w_i)| < b^{-m_i} \leq b^{-n_i}$. Then $\frac{\mathrm{K}^T_{b^{-m_i}}(x)}{m_i} \leq \frac{\mathrm{K}^T_{b^{-n_i}}(x)}{n_i}$, and for $T'(\pi) = comp(T(\pi))$, either

$$\liminf_n \frac{\mathrm{K}^T(S \upharpoonright n)}{n} \leq \lim_i \frac{\mathrm{K}^T_{b^{-m_i}}(x)}{m_i}$$

or

$$\liminf_n \frac{\mathrm{K}^{T'}(comp(S) \upharpoonright n)}{n} \leq \lim_i \frac{\mathrm{K}^T_{b^{-m_i}}(x)}{m_i}.$$

Therefore either $\dim_{\mathrm{FS}}(S) \leq \dim_{\mathrm{FS}}^b(x)$ or $\dim_{\mathrm{FS}}(comp(S)) \leq \dim_{\mathrm{FS}}^b(x)$ and the claim follows.

Since finite-state dimension in the space of sequences characterizes Borel normality [2], we have an alternative characterization of normality in terms of finite-state dimension in the Euclidean space.

Corollary 1. *Let* $b \geq 1$, $x \in [0,1)$. x *is* b*-normal if and only if* $\dim_{\mathrm{FS}}^b(x) = 1$, *that is,*

$$\inf_{T\Sigma_b-\mathrm{FST}} \liminf_{\delta>0} \frac{\mathrm{K}^T_\delta(x)}{\log_b(1/\delta)} = 1.$$

4 Point to Set Principle for Finite-State Dimension

We denote as *separator* a set $S \subseteq [0,1)$ such that S is countable and dense in [0,1].

Definition 5. *A separator enumerator (SE) is a function $f : \Sigma^* \to [0,1)$ such that $\mathrm{Im}(f)$ is a separator.*

For each separator enumerator f we can define information content in $[0,1)$ relative to f.

Definition 6. *Let $f : \Sigma^* \to [0,1)$ be a SE. Let T be a Σ-FST, $\delta > 0$ and $x \in [0,1)$. The f-T-information content of x at precision δ is*

$$\mathrm{K}_\delta^{T,f}(x) = \min\left\{\mathrm{K}^T(w) \,|\, |f(w) - x| < \delta\right\}.$$

Definition 7. *Let $f : \Sigma^* \to [0,1)$ be a SE. Let $x \in [0,1)$ and $A \subseteq [0,1)$. The f-enumerator finite-state dimension of x is*

$$\dim_{\mathrm{FS}}^f(x) = \inf_{T\Sigma-\mathrm{FST}} \liminf_{\delta>0} \frac{\mathrm{K}_\delta^{T,f}(x)}{\log_{|\Sigma|}(1/\delta)},$$

the f-enumerator finite-state dimension of A is

$$\dim_{\mathrm{FS}}^f(A) = \inf_{T\Sigma-\mathrm{FST}} \sup_{x\in A} \liminf_{\delta>0} \frac{\mathrm{K}_\delta^{T,f}(x)}{\log_{|\Sigma|}(1/\delta)}.$$

We can generalize Borel normality through the same relativization.

Definition 8. *Let $f : \Sigma^* \to [0,1)$ be a SE, let $x \in [0,1)$. x is f-normal if $\dim_{\mathrm{FS}}^f(x) = 1$.*

Given this natural relativization of finite-state dimension we next prove a point-to-set principle stating that for every set A there exists an SE f such that classical Hausdorff dimension of A is exactly f-finite-state dimension. This implies that classical geometrical measure theory results can be obtained using only finite-state dimension.

Theorem 4. *Let $A \subseteq [0,1)$.*

$$\dim_{\mathrm{H}}(A) = \min_{f\mathrm{SE}} \dim_{\mathrm{FS}}^f(A),$$

$$\dim_{\mathrm{H}}(A) = \min_{f:\{0,1\}^*\to\mathcal{D}_2} \dim_{\mathrm{FS}}^f(A).$$

Proof. Let C be such that $\dim_{\mathrm{H}}(A) = \dim^C(A)$ from the point-to-set principle in [9]. Le U be the universal oracle Turing Machine used in the definition of Kolmogorov Complexity for effective dimension dim. Let $h : \{0,1\}^* \to \{0,1\}^*$ be such that $h(w) = U^C(w)$ when $U^C(w)$ is defined, and $U^C(w) = 0$ otherwise. Then $f(w) = \mathrm{real}_2(h(w))$ is the required SE.

Notice that the previous theorem holds even when fixing a particular countable dense set. In terms of Borel normality, it shows that reordering the set $\mathcal{D}_b$ of base-b finite representation numbers is enough to obtain normality for any other base.

5 Conclusions and Open Questions

We expect that our main theorem will prove new lower bounds on Hausdorff dimension in different settings. Notice that the result can be directly translated into any separable metric space and any reasonable gauge family.

Our result helps clarify the oracle role in the point to set principles. The next step would be to classify the different enumerations of a countable dense set.

We believe that the notion of f-normal sequence can be of independent interest with robustness properties inherited from those of the original concept, for instance from the fact that x is b-normal exactly when the sequence $(b^n x)_n$ is equidistributed modulo 1.

Open Question. Let $f : \Sigma^* \to [0,1)$ be a SE, let $x \in [0,1)$. For each $n \in \mathbb{N}$, let $a_n(x) = f(w)$ for $|w| \leq n$ such that $f(w) \leq x$ and $x - f(w)$ is minimum. Can we characterize f-normality in terms of the equidistribution properties of $(|\Sigma|^n a_n(x))$?

References

1. Borel, E.: Sur les probabilités dénombrables et leurs applications arithmétiques. Rendiconti del Circolo Matematico di Palermo **27**(1), 247–271 (1909)
2. Bourke, C., Hitchcock, J.M., Vinodchandran, N.V.: Entropy rates and finite-state dimension. Theor. Comput. Sci. **349**(3), 392–406 (2005)
3. Dai, J.J., Lathrop, J.I., Lutz, J.H., Mayordomo, E.: Finite-state dimension. Theor. Comput. Sci. **310**, 1–33 (2004)
4. Doty, D., Moser, P.: Finite-state dimension and lossy decompressors. Technical Report CoRR abs/cs/0609096, Arxiv, 2006
5. Jacob, D.S., Fiedler, B.: Pinned distances of planar sets with low dimension. Technical Report arXiv:2408.00889 [math.CA], arXiv.org, 2024
6. Lutz, J., Lutz, N.: Who asked us? how the theory of computing answers questions about analysis. In: Complexity and Approximation: In Memory of Ker-I Ko, pp. 48–56. Springer, Ding-Zhu Du and Jie Wang (eds.) edition, 2020
7. Lutz, J.H.: Dimension in complexity classes. SIAM J. Comput. **32**(5), 1236–1259 (2003)
8. Lutz, J.H.: The dimensions of individual strings and sequences. Inf. Comput. **187**(1), 49–79 (2003)
9. Lutz, J.H., Lutz, N.: Algorithmic information, plane Kakeya sets, and conditional dimension. ACM Trans. Comput. Theory **10**, Article 7 (2018)
10. Lutz, J.H., Mayordomo, E.: Algorithmic fractal dimensions in geometric measure theory. In: Brattka, V., Hertling, P. (eds.), Handbook of Computability and Complexity in Analysis, pp. 271–302. Springer-Verlag, 2021

Some Structural Complexity Results for $\exists\mathbb{R}$

Klaus Meer and Adrian Wurm(✉)

BTU Cottbus-Senftenberg, Platz der Deutschen Einheit 1, 03046 Cottbus, Germany
{meer,wurm}@b-tu.de

Abstract. The complexity class $\exists\mathbb{R}$, standing for the complexity of deciding the existential first order theory of the reals as real closed field in the Turing model, has raised considerable interest in recent years. It is well known that NP $\subseteq \exists\mathbb{R} \subseteq$ PSPACE. In their compendium [23], Schaefer, Cardinal, and Miltzow give a comprehensive presentation of results together with a rich collection of open problems. Here, we answer some of them dealing with structural issues of $\exists\mathbb{R}$ as a complexity class. We show analogues of the classical results of Baker, Gill, and Solovay finding oracles which do and do not separate NP from $\exists\mathbb{R}$, of Ladner's theorem showing the existence of problems in $\exists\mathbb{R} \setminus$ NP not being complete for $\exists\mathbb{R}$ (in case the two classes are different), as well as a characterization of $\exists\mathbb{R}$ by means of descriptive complexity.

Keywords: Existential Theory of the Reals · structural complexity · descriptive complexity

1 Introduction

The existential theory of the reals collects all true sentences in existential first-order logic over $\mathbb{R}$, i.e., sentences of the form $\exists x_1 \dots \exists x_n \Phi(x_1, \dots, x_n)$, where Φ is a quantifier-free formula expressing a Boolean combination of polynomial equalities and inequalities. The complexity of deciding the truth of a given sentence plays a prominent role for several computational models and complexity classes. In the Blum-Shub-Smale model - henceforth BSS for short - over $\mathbb{R}$ basic arithmetic operations can be executed with unit cost and the size of Φ is basically the algebraic size of a dense encoding of all polynomials occuring in it. The above decision problem then turns out to be $\mathrm{NP}_{\mathbb{R}}$-complete, where the latter class is the real number analogue of classical NP [3]. However, the decision problem also makes perfect sense in the Turing model of computation if we restrict all coefficients for the polynomials in Φ to be rational. Then, its size is the usual bit-size and the problem of deciding truth is easily seen to be NP-hard. In the Turing model, the exact complexity of the existential theory for the reals is (wide) open. Using deep results on quantifier elimination it has been shown that it can be decided within PSPACE, see [4] and also [21]. The open placement of the problem somewhere between NP and PSPACE has led to the

© The Author(s), under exclusive license to Springer Nature Switzerland AG 2025
A. Beckmann et al. (Eds.): CiE 2025, LNCS 15764, pp. 305–320, 2025.
https://doi.org/10.1007/978-3-031-95908-0_22

definition of an own complexity class denoted $\exists\mathbb{R}$, see a precise definition in the next subsection.

Beside many other things, in [23] the authors ask whether several structural results well known to hold for the classical class NP in the Turing model also can be settled for $\exists\mathbb{R}$. The latter include an analogue of the Baker-Gill-Solovay result [1] on existence of oracles separating $\exists\mathbb{R}$ from NP and the question whether $\exists\mathbb{R}$ can be characterized by means of descriptive complexity as done in [7] for NP. In this paper we give positive answers to these questions and also show that Ladner's result [14] can be transformed showing that if we assume $\exists\mathbb{R} \neq \mathrm{NP}$ there exist problems in the set difference not being $\exists\mathbb{R}$-complete.

A few words on the proofs might be appropriate. All above mentioned results for NP turn out to hold in the corresponding variants as well for $\exists\mathbb{R}$. In some cases, though not completely straightforward, proofs follow from not too complicated modifications of known proofs of related results in the BSS model. One basic ingredient for the proofs to work is the characterization of $\exists\mathbb{R}$ via the BSS class $\mathrm{BP}(\mathrm{NP}^0_{\mathbb{R}})$ in Theorem 1 below together with the most important observation that the class of $\mathrm{NP}^0_{\mathbb{R}}$-machines in the BSS model is countable. Therefore, we see the main value of the present paper in working out the necessary proof details of expected results in order to enlarge the body of structural results known to hold for $\exists\mathbb{R}$.

1.1 Basic Definitions: $\exists\mathbb{R}$ and BSS-Machines Without Constants

We now define precisely the class $\exists\mathbb{R}$ together with BSS computations using only rational machine constants as well as related discrete complexity classes $\mathrm{BP}(\mathrm{P}^0_{\mathbb{R}})$ and $\mathrm{BP}(\mathrm{NP}^0_{\mathbb{R}})$.

The class $\exists\mathbb{R}$ was formally introduced in [22], but somehow was examined already earlier, see [23] for a historic account on questions related to this class being studied already in the 1990s. The class can most easily be defined starting from a complete problem and then building the downward closure under usual polynomial time many one reductions. In the meanwhile, several problems have been shown to be complete. For our purposes, the existence of real solutions of a polynomial system of degree-two equations is the most suitable one.

Definition 1. *a) The decision problem Quadratic Polynomial Systems* QPS^0 *is defined as follows: Given* $n, m \in \mathbb{N}$ *and polynomials* $f_i \in \mathbb{Q}[x_1, \ldots, x_n], 1 \leq i \leq m$ *of degree at most* 2, *is there an* $x^* \in \mathbb{R}^n$ *such that* $f_i(x^*) = 0$ *for all* i*? The problem is understood as a decision problem in the Turing model, i.e., instances are coded as binary strings over* $\{0,1\}^*$.[1] *b) The complexity class* $\exists\mathbb{R}$ *consists of all decision problems* $L \subseteq \{0,1\}^*$ *such that there exists a polynomial time many one reduction from* L *to* QPS^0.

[1] The superscript ^{0} in QPS^0 results from the close relation of the approach to Blum-Shub-Smale machines that are only using rational constants and is common in that context, see below. Here it indicates rational coefficients.

The completeness of QPS^0 for class $\exists\mathbb{R}$, i.e., for the existential theory of the reals easily follows from the proof in [3] that the corresponding problem for quadratic polynomial systems with arbitrary real coefficients is $\mathrm{NP}_{\mathbb{R}}$-complete in the BSS model over $\mathbb{R}$. In the corresponding proof, no real constants are introduced and starting from decision problems in $\{0,1\}^*$ the reduction runs in polynomial time also in the Turing model.

The class $\exists\mathbb{R}$ can easily be characterized via a restriction of the class $\mathrm{NP}_{\mathbb{R}}$ in the BSS model, see [3]. The full real number BSS model allows to compute with reals as entities, performing the basic arithmetic operations together with a test of the format 'is $x \geq 0$' at unit cost. Algorithms are allowed to use a fixed finite set of real numbers as so-called machine constants. The latter always include 0 and 1. The (algebraic) size of an instance then is the number of reals necessary to specify it (in a reasonable encoding), and the class $\mathrm{P}_{\mathbb{R}}$ of problems decidable in deterministic polynomial time is defined literally the same as P, i.e., there exists a polynomial time algorithm (measured with respect to the algebraic size measure for instances and the unit cost measure for algorithms) deciding the problem correctly. Similarly, $\mathrm{NP}_{\mathbb{R}}$ stands for problems verifiable in polynomial time when having access to a witness that is a real vector of polynomial (algebraic) size in the size of the given instance. If we want to handle discrete problems in this model, i.e., problems $L \in \{0,1\}^*$, it is appropriate to restrict inputs to binary strings and to disallow BSS algorithms to use non-rational machine constants. This leads to complexity classes highly important in connection with $\exists\mathbb{R}$.

Definition 2. *The class* $\mathrm{BP}(\mathrm{P}^0_{\mathbb{R}})$ *consists of all* $L \subseteq \{0,1\}^*$ *such that there exists a BSS algorithm running in (algebraic) polynomial time, using only rational machine constants and deciding* L *on inputs from* $\{0,1\}^*$*. Similarly, the class* $\mathrm{BP}(\mathrm{NP}^0_{\mathbb{R}})$ *consists of all* $L \subseteq \{0,1\}^*$ *such that there exists a BSS algorithm using only rational machine constants and verifying* L *on inputs from* $\{0,1\}^* \times \mathbb{R}^*$*. Moreover, for inputs* (y,z) *the (algebraic) running time is polynomially bounded in the size of* y*.*

Above, BP stands for 'Boolean Part' and $\mathbb{R}^*$ denotes the real number analogue of $\{0,1\}^*$, i.e., finite sequences of real numbers. Note that for discrete inputs $y \in \{0,1\}^*$ the bit-size and the algebraic size coincide. However, the cost measure for algorithms working with discrete data is the algebraic one. It is also important to stress again that for verification algorithms the certificate $z \in \mathbb{R}^*$ is allowed to be a vector of reals. This is the reason for the tight relation of $\mathrm{BP}(\mathrm{NP}^0_{\mathbb{R}})$ to $\exists\mathbb{R}$. It follows immediately from the original completeness proof in [3]:

Theorem 1. $\mathrm{BP}(\mathrm{NP}^0_{\mathbb{R}}) = \exists\mathbb{R}$.

A decisive aspect of this theorem is that $\exists\mathbb{R}$ can be characterized through a complexity class defined by a model which has a countable number of machines.

2 Relativization of NP Versus $\exists\mathbb{R}$

The first question from the compendium [23] we want to deal with is whether there exist oracles which do and do not separate $\exists\mathbb{R}$ from NP and PSPACE, respectively. Comparing P and NP in the Turing model, these questions were answered affirmatively long ago [1], in the BSS model an analogue result holds [6], see also [8]. Usually, the construction of an oracle separating the representative classes is done via a diagonalization argument. As consequence, the separating oracles in both settings are not very natural (with an exception in [8], where a Knapsack problem is used to separate other classes). As we shall see, for NP and $\exists\mathbb{R}$ a quite natural oracle work. The definition of oracle classes like P^A and NP^A naturally relies on a basic model for algorithms defining a complexity class, like deterministic or non-deterministic polynomial time Turing machines. Then, one equips them with the additional feature to use intermediate results as queries to an oracle. Following the same nearby ongoing for $\exists\mathbb{R}$ we should work with the following definition.

Definition 3. *Let $A \subseteq \mathbb{R}^*$. An $\mathrm{NP}^0_{\mathbb{R}}$-oracle algorithm with oracle A is an $\mathrm{NP}^0_{\mathbb{R}}$-algorithm M which in addition has an oracle state. If M enters this state, it can ask the oracle whether an element $y \in \mathbb{R}^*$ previously computed belongs to A, receives the correct answer in one step, and continues its computation. A problem $L \subseteq \{0,1\}^*$ belongs to the oracle class $\exists\mathbb{R}^A$, if there exists such an M such that for $x \in L$ there exists a computation of M which accepts x and for $x \in \{0,1\}^* \setminus L$ no computation accepts.*

Note that though this is the straightforward definition of oracle classes $\exists\mathbb{R}^A$, some subtleties are hidden. Since the basic machine is an $\mathrm{NP}^0_{\mathbb{R}}$-algorithm, it is allowed to produce real number oracle questions, something a Turing machine cannot. Vice versa, since a Turing machine only computes with discrete data, we know in advance that an NP-oracle machine will never produce arbitrary non-rational queries, an information which is undecidable for a BSS machine to know in general. Our results below establishing the integers as separating oracle implicitly make use of such effects. We further comment on this at the end of this section.

Theorem 2. *Given the above definition of oracle classes based on $\exists\mathbb{R}$, the following hold:*

a) There exists an oracle A such that $\mathrm{P}^A = \mathrm{NP}^A = \exists\mathbb{R}^A = \mathrm{PSPACE}^A$.
b) The oracle $\mathbb{Z}$, seen as subset of $\mathbb{R}$, both satisfies $\mathrm{NP} = \mathrm{NP}^{\mathbb{Z}} \subsetneq \exists\mathbb{R}^{\mathbb{Z}}$ and $\exists\mathbb{R}^{\mathbb{Z}} \neq \mathrm{PSPACE}^{\mathbb{Z}}$.
c) In the full BSS model, $\mathrm{P}^{\mathbb{Z}}_{\mathbb{R}} \subsetneq \mathrm{NP}^{\mathbb{Z}}_{\mathbb{R}}$.

Proof. For a) we can choose A as the PSPACE-complete problem Quantified Boolean Formulas QBF. Since PSPACE = NPSPACE = co-NPSPACE it is well known that $\mathrm{P}^{\mathrm{QBF}} = \mathrm{NP}^{\mathrm{QBF}} = \mathrm{PSPACE}^{\mathrm{QBF}} = \mathrm{PSPACE}$. Arguing about $\exists\mathbb{R}^{\mathrm{QBF}}$ needs some care since the underlying machine model in the definition changes.

Suppose M to be a basic $NP^0_{\mathbb{R}}$-oracle machine using oracle QBF. Let $p(n)$ be the polynomial time running bound of M for inputs in $\{0,1\}^n$. W.l.o.g. suppose M for such inputs asks $m \le p(n)$ oracle queries, all being of polynomial size at most $p(n)$. A PSPACE-algorithm deciding M^{QBF} on input $x \in \{0,1\}^n$ works as follows: First, consider a computable enumeration of bit-vectors $b \in \{0,1\}^m$ and of m strings $(q_1,\ldots,q_m)$, all q_i strings over alphabet $\{0,1,*\}$ and of size at most $p(n)$. Below, we interpret the q_i as oracle queries M poses during its computation and b_i as the answers. A component $*$ in a query indicates that it does not belong to $\{0,1\}$. In case that M computes such a query, the answer from the QBF-oracle is 0 since the query does not code a correct QBF-instance. Now, given a fixed tuple $(b_1,\ldots,b_m,q_1,\ldots,q_m)$ in the enumeration, consider an existential first-order sentence over $\mathbb{R}$ without non-rational constants claiming that there exists a computation of M on x which produces the queries q_i under the assumption that $b_1,\ldots,b_{i-1}$ are the correct answers to the previous queries, and then finally accepts. Since the q_i are vectors over $\{0,1,*\}$, the requirement that a query computed by M equals q_i is easily expressible with such a first-order sentence. This holds for queries being bit-vectors as well as for a query containing a component $*$, in which case the corresponding part of the first-order sentence has to express that the component does not belong to $\{0,1\}$. This is easily doable. The resulting entire sentence is an instance of $\exists\mathbb{R}$ and thus can be decided in PSPACE by [4]. It remains to guarantee in addition that the b_i are the correct answers to QBF-queries q_i. This of course can be done in PSPACE. Note that if a q_i contains a non-binary component, the corresponding b_i must be 0. Now the algorithm does the above for all elements in the enumeration by re-using its space and accepts, if for at least one tuple $(b_1,\ldots,b_m,q_1,\ldots,q_m)$ all conditions are satisfied. It follows $\exists\mathbb{R}^{QBF} = $ PSPACE as well.

For part b) note that both $NP^{\mathbb{Z}}$ and $PSPACE^{\mathbb{Z}}$ only contain problems being decidable in the Turing model. Moreover, if we consider a Turing machine over $\{0,1\}^*$ coding computations over $\mathbb{Q}$ in any usual way, then it is easy to decide for an intermediate result whether it encodes an integer or not. Since for an NP-machine the size of a query is polynomial in the input, we can also check in polynomial time whether a query is integral, thus NP $= NP^{\mathbb{Z}}$. But $\exists\mathbb{R}^{\mathbb{Z}}$ contains problems being undecidable in the Turing model such as Hilbert's 10th problem: Given a polynomial $f \in \mathbb{Z}[x_1,\ldots,x_n]$ in n variables with integer coefficients, an $NP^{0,\mathbb{Z}}_{\mathbb{R}}$-algorithm can guess in a non-deterministic computation n real numbers $x_1^*,\ldots,x_n^*$ and then ask the oracle whether all x_i^* are integers. If the answer is positive the algorithm evaluates $f(x_1^*,\ldots,x_n^*)$ in polynomial time in the algebraic model. It accepts iff the result is 0. Thus, the problem to decide whether there is an integer zero of f belongs to $\exists\mathbb{R}^{\mathbb{Z}}$, but is well known to be undecidable [16].

For c) we can simplify Emerson's construction by using a result from [17]. Consider the set $A := \{t \in [0,2\pi] | \exists k \in \mathbb{N} \text{ s.t. } \frac{k\cdot t}{2\pi} \in \mathbb{N}\}$. It is easy to see that the decision problem: Given $x \in \mathbb{R}$, is $x \in A$? belongs to $NP^{\mathbb{Z}}_{\mathbb{R}}$: For input x guess a $k \in \mathbb{R}$ and check whether $k \ge 0$ and $k \in \mathbb{Z}$ by asking the oracle. Then compute $\frac{kx}{2\pi}$ and ask the oracle again whether it belongs to $\mathbb{Z}$. All this can be done in constantly many algebraic steps. Note that in the BSS model $x \in \mathbb{R}$

has input size 1. However, in order to belong to $P_{\mathbb{R}}^{\mathbb{Z}}$ the question should be decided in constant time using the oracle. In [17] it is shown by using a 'typical path argument' that this problem can not be decided by a polynomial time BSS machine which is allowed in addition to evaluate the sine-function in one step. Since in this 'sine-model' one can decide in constant time whether a number is integral, it follows can $P_{\mathbb{R}}^{\mathbb{Z}}$ can be decided by a sine-machine, but A can not. We conclude that $A \notin P_{\mathbb{R}}^{\mathbb{Z}}$. □

Note that part c) alternatively to the proof in [6] gives a more natural problem yielding the separation. As also for part b), this seems to be an effect observable at least for some results in the BSS setting, compare a similar statement for a real number version of Post's problem [19]. A strange effect why the separation for NP and $\exists\mathbb{R}$ works is the fact that even though the former is a subset of the latter, this does not any longer hold for arbitrary relativized versions. Clearly, the reason for this is the use of different machine models to define the corresponding oracle classes. Such an effect concerning relativized classes is also known from classical complexity theory for less prominent classes and studied in [25]. It might be interesting to both find other oracles yielding the separation and to study the above effect in more detail in our framework. We close the section pointing to recent work by Hamm et al. [12], in which oracle separations for a polynomial hierarchy based on $\exists\mathbb{R}$ have been studied as well. They restrict the way a real BSS machine can produce oracle queries resulting in discrete oracles to be considered.

3 Descriptive Complexity for $\exists\mathbb{R}$

The next question from the compendium [23] is whether $\exists\mathbb{R}$ can be characterized by purely logical means in the sense of descriptive complexity theory. Corresponding results were given in the early years of finite model theory for classical NP [7] and later on for the real number version $NP_{\mathbb{R}}$ in the BSS model [10], based on the development of meta-finite model theory in [9]; see [11] for a more intensive introduction into such results. One special feature, given the above mentioned characterization of $\exists\mathbb{R}$ as $BP(NP_{\mathbb{R}}^0)$, is the mixture of discrete inputs coded as in classical complexity theory over $\{0,1\}^*$, and an algebraic cost measure for the arithmetic computations performed by the underlying machine. It turns out that this split can be modelled by a restriction of so-called $\mathbb{R}$-structures used in [10] to characterize (full) $NP_{\mathbb{R}}$ and respective logics on them. Thus, the focus below will be on defining the corresponding restriction which we call *discrete* $\mathbb{R}$*-structures*, and the suitable logics on them used to capture both $\exists\mathbb{R}$ and $BP(P_{\mathbb{R}}^0)$. Corresponding proofs then are quite similar to those given in [10] and due to space limitations not worked out again. Though below we work with restricted $\mathbb{R}$-structures, we have to recall their definition in full generality, except for disallowing arbitrary real constants in the so-called secondary part due to the fact that the BSS algorithms studied here do not use such constants.

Definition 4. *Let L_s, L_f be finite vocabularies, where L_s may contain relation and function symbols, and L_f contains function symbols only. An* ℝ-structure of signature $\sigma = (L_s, L_f)$ *is a triple* $\mathfrak{D} = (\mathcal{A}, \mathcal{R}, \mathcal{F})$ *consisting of*

(i) *a finite structure $\mathcal{A}$ of vocabulary L_s, called the* primary part *or* skeleton *of $\mathfrak{D}$; its universe A is also said to be the* universe *of $\mathfrak{D}$;*
ii) *the infinite structure* $\mathcal{R} = (\mathbb{R}, 0, 1, +, -, \cdot, /, \text{sign}, <)$ *called the* secondary part. *Here,* $\text{sign} : \mathbb{R} \mapsto \{0, 1\}$ *denotes the sign-function being 0 for negative values and 1 otherwise;*
(iii) *a finite set $\mathcal{F}$ of functions $X : A^k \to \mathbb{R}$ interpreting the function symbols in L_f, where k depends on the respective symbol only.*

We denote the set of all ℝ-structures of signature σ by $\text{Struct}_{\mathbb{R}}(\sigma)$.
For $\mathfrak{D} \in \text{Struct}_{\mathbb{R}}(\sigma)$ *the* size *of $\mathfrak{D}$ is* $|\mathfrak{D}| := |A|$.

We now restrict ℝ-structures to so-called *discrete* ℝ*-structures* suitable for modelling computations in $\text{P}^0_{\mathbb{R}}$ and $\text{NP}^0_{\mathbb{R}}$ on Boolean languages. In order to model bit-strings as structures the universe always will have the form $A = \{0, 1, \ldots, n-1\}$ and L_s contains a unary relation $X \subseteq A$ interpreted as a bit-vector in $\{0,1\}^n$ via $X(i) = 1 \leftrightarrow i \in X, X(i) = 0 \leftrightarrow i \notin X$. In addition, for well known reasons we include a linear order $< \in L_s$ in order to capture polynomial time computations below. Finally, as the only element in L_f we need a function $real : A \mapsto \mathbb{R}$ which is able to change the type of an element in A to become a real (like the cost operator in a programming language like C_{++}) : We define $real(i) := \begin{cases} 1 \in \mathbb{R} \text{ for } i \neq 0, i \in A \\ 0 \in \mathbb{R} \text{ for } i = 0 \in A \end{cases}$

Definition 5. *A* discrete ℝ-structure $\mathfrak{D}$ *is an ℝ-structure over vocabulary $\sigma = (L_s, L_f)$, where $A = \{0, 1, \ldots, n-1\}$ for some $n \in \mathbb{N}$, L_s contains at least a unary relation X and a binary relation $<$ interpreted as bit string and linear order, respectively, and $L_f = \{real\}$, where $real : A \mapsto \mathbb{R}$ is interpreted as above.*

We denote by $\text{Struct}^0_{\mathbb{R}}(\sigma)$ *all discrete ℝ-structures with vocabulary σ.*

Note that below when we use existential second-order (so) logic, both $<$ and *real* can be avoided by claiming their existence as functions from A to ℝ using an existential quantifier together with a fo-formula.

Discrete ℝ-structures and logics on them provide the right framework to model BSS-computations on Boolean languages that do not use constants other than rational ones. Inputs being discrete and measured by their usual bit size, the used logics will transfer those inputs into the real number part of a discrete ℝ-structure so that computations are modelled therein, including the use of the algebraic unit cost measure. Let $V = \{v_0, v_1, \ldots\}$ denote a countable set of variables. The v_i are supposed to be fo-variables ranging over the discrete universe of a discrete ℝ-structure.

Definition 6. *Fix* $\mathcal{R} = (\mathbb{R}, 0, 1, +, -, \cdot, /, \text{sign}, <)$. *The language* $\text{FO}^0_{\mathbb{R}}$ *contains, for each signature $\sigma = (L_s, L_f)$, a set of formulas and terms. Each term t takes,* *when interpreted in some discrete* $\mathbb{R}$*-structure, values in either the skeleton, in which case we call it an* index term, *or in* $\mathbb{R}$*, in which case we call it a* number term. *Terms are defined inductively as follows*

(i) *The set of index terms is the closure of the set* V *of variables under applications of function symbols of* L_s.
(ii) *Any rational number is a number term.*
(iii) *If* $h_1, \ldots, h_k$ *are index terms and* X *is a* k*-ary function symbol of* L_f*, then* $X(h_1, \ldots, h_k)$ *is a number term. In particular, for* h *an index term,* $real(h)$ *is a number term with value in* $\{0,1\} \subset \mathbb{R}$.
(iv) *If* t, t' *are number terms, then so are* $t + t'$, $t - t'$, $t \cdot t'$, *and* $\text{sign}(t)$. *Here, the sign function is defined as* $\text{sign}(t) := \begin{cases} 1 & \text{if } t \geq 0 \\ 0 & \text{if } t < 0 \end{cases}$.

Atomic formulas are equalities $h_1 = h_2$ *of index terms, equalities* $t_1 = t_2$ *and inequalities* $t_1 < t_2, t_1 \leq t_2$ *of number terms, and expressions* $P(h_1, \ldots, h_k)$*, where* P *is a* k*-ary predicate symbol in* L_s *and* $h_1, \ldots, h_k$ *are index terms.*

The set of formulas of $\text{FO}^0_{\mathbb{R}}$ *is the smallest set containing all atomic formulas and which is closed under Boolean connectives and quantification* $(\exists v)\psi$ *and* $(\forall v)\psi$.

Remark 1. Having a linear order available on A it is folklore in finite model theory to define the first and the last element (denoted by 0 and $n-1$, respectively) of A with respect to the order, as well as a linear order on every $A^k, k \in \mathbb{N}$ by means of fo-formulas. This is tacitly used below.

In order to capture complexity class $\exists\mathbb{R} = \text{BP}(\text{NP}^0_{\mathbb{R}})$ as well as $\text{BP}(\text{P}^0_{\mathbb{R}})$ we have to extend fo-logic to existential second-order logic and first order logic, respectively. This is done similarly as in [10] and the proofs showing that the resulting logics capture the intended classes is a nearby variation of the corresponding ones in the full BSS model. The only technical aspect to respect is that for discrete $\mathbb{R}$-structures and the logics used we cannot introduce arbitrary real constants. This is rather a restriction for fixed point logic than for existential so-logic since for the latter we can existentially quantify such real objects and then express their desired properties via a fo-formula. Since fixed point logic also needs to work with so-objects, we first define existential so logic.

Definition 7. *Second-order logic* $\text{SO}^0_{\mathbb{R}}$ *on discrete* $\mathbb{R}$*-structures is obtained starting from* $\text{FO}^0_{\mathbb{R}}$ *logic by adding the possibility to quantify over function symbols. More precisely, given a vocabulary* $\sigma = (L_s, L_f)$*, where* L_f *contains a function symbol* Y *interpreted as a function from some* $A^k \to \mathbb{R}$*, together with a first-order formula* ϕ *over* σ*, both* $\exists Y \phi(Y)$ *and* $\forall Y \phi(Y)$ *are second-order formulas. If all quantified function symbols are existentially quantified we get* existential second-order logic $\exists\text{SO}^0_{\mathbb{R}}$.

Example 1. We express the core problem of $\exists\mathbb{R}$, namely real solvability of an instance of QPS^0 (Definition 1), as an existential so-property on suitable discrete $\mathbb{R}$-structures. Some (easy) technical aspects are only sketched. Consider as input instance a system of $m \in \mathbb{N}$ polynomials $p_1, \ldots, p_m \in \mathbb{Z}[x_1, \ldots, x_n]$ in some $n \in \mathbb{N}$ real variables having integer coefficients. All p_i are supposed to have degree at most 2 and the question is to decide whether there exists a common real zero. Recall that this is an $\exists\mathbb{R}$-complete problem. Rational coefficients for sake of easiness in the further description can be removed by multiplying all p_i by the least common multiple of the denominators of their respective coefficients. We represent the system as follows as a discrete $\mathbb{R}$-structure $\mathfrak{D} = (\mathcal{A}, \mathcal{R}, \{real\})$, where $\mathcal{A}$ has the discrete set $A = \{1, \ldots, n\} \times \{1, \ldots, m\} \times \{0, \ldots, L\}$ as universe. Here, $L \in \mathbb{N}$ is an upper bound for the bit-size of the coefficients of all p_i. The vocabulary L_s contains a linear ordering and a coefficient function $C : A^4 \mapsto \{0, 1\} \subset A$, where for $i, j \in \{1, \ldots, n\}, k \in \{1, \ldots, m\}$ and $r \in \{0, \ldots, L\}$ we interpret $C(i, j, k, 0)$ as sign of the coefficient of monomial $x_i x_j$ of p_k; here $C(i, j, k, 0) = 0 \in A$ codes sign -1 and $C(i, j, k, 0) = 1 \in A$ codes sign 1. Furthermore, $C(i, j, k, r)$ is the bit (as element in A) of 2^{r-1} in the binary representation of the coefficient of $x_i x_j$ in p_k. Some technical comments are necessary. Formally, the above discrete part looks different from the definition of a discrete $\mathbb{R}$-structure. However, it is an easy task to change the above structure to one with a universe $\tilde{A} = \{0, 1, \ldots, K-1\}$ for some $K \in \mathbb{N}$ and an $\tilde{X} \subset \tilde{A}$ interpreted as bit-string coding the instance. Towards this aim, one has to use an additional relation coding the values n, m and L in $\tilde{A}$. Using the linear order on $\tilde{A}$ it is easy to express via fo-logic components representing variables, polynomials, and coefficients, respectively. Similarly, for a function $\tilde{C} : \tilde{A}^4 \mapsto \{0, 1\} \subset \tilde{A}$ it is easy to describe the correct arguments and their meaning using fo-logic to define $\tilde{C}$ arbitrarily on arguments having no interpretation. The interested reader might try to elaborate this coding; for sake of readability we prefer to use the above coding instead.

In the existential second order part we now ask for the existence of three functions mapping some A^k to $\mathbb{R}$. First, $C_{\mathbb{R}} : A^4 \mapsto \mathbb{R}$ should represent the coefficients as real numbers in order to use them in number terms for expressing the evaluation of the p_k in real arguments. This can be done using the following fo-formulas. Note that properties like $i, j \in \{1, \ldots, n\}, k \in \{1, \ldots, m\}, r \in \{0, \ldots, L\}$ can be expressed using the linear ordering; similarly, expressions like $r+1$ can be defined to have the natural meaning.

First, the correct sign of a coefficient is expressed as

$$\begin{aligned} \forall i, j \in \{1, \ldots, n\}, k \in \{1, \ldots, m\}\ C(i, j, k, 0) = 0 \Leftrightarrow C_{\mathbb{R}}(i, j, k, 0) &= -1 \in \mathbb{R} \\ &\text{(as real number term)} \\ \wedge\ C(i, j, k, 0) = 1 \Leftrightarrow C_{\mathbb{R}}(i, j, k, 0) &= 1 \in \mathbb{R} \end{aligned}$$

Next, $\forall i, j \in \{1, \ldots, n\}, k \in \{1, \ldots, m\}, r \in \{1, \ldots, L-1\}$ fixing the highest bit as real and describing the real represented by bit vector $C(i, j, k, *)$ is done via

$$\begin{aligned} &C_{\mathbb{R}}(i, j, k, L) = real(C(i, j, k, L)) \text{ and} \\ &C_{\mathbb{R}}(i, j, k, r) = C_{\mathbb{R}}(i, j, k, r+1) \cdot 2 + real(C(i, j, k, r)), \end{aligned}$$

where of course 2 is a real number term for 1+1. This way $C_{\mathbb{R}}(i, j, k, 1)$ represents (as real) the absolute value of the coefficient of monomial $x_i \cdot x_j$ in p_k. Finally, we require for all $i, j, k : C_{\mathbb{R}}(i, j, k, L + 1) = C_{\mathbb{R}}(i, j, k, 0) \cdot C_{\mathbb{R}}(i, j, k, 1)$ to get the correct sign.

Next, use $\exists Y : A \mapsto \mathbb{R}$ in order to existentially quantify a potential common real zero $(Y(1), \ldots, Y(n))$ of all p_k. Finally, we have to express the evaluation of all p_k in Y. However, this is easy and can be done similarly to the foregoing construction. Existentially quantify a function $Z : A^3 \mapsto \mathbb{R}$ which represents the partial sums when evaluating all polynomials in Y. One component in the arguments of Z again addresses the respective polynomial to evaluate, the two others code $x_i \cdot x_j$. Since all p_k have degree at most 2, we need for every $k \in \{1, \ldots, m\}$ to cycle through $\{1, \ldots, n\}^2$, and this can be expressed similarly to the computation of a coefficient from a binary representation above. The proceeding is the same as that in [10] for evaluating a single degree-4 polynomial.

A logical characterization of $\exists\mathbb{R}$ now is possible in almost the same way as for $\mathrm{NP}_{\mathbb{R}}$ in [10].

Theorem 3. *Let (F, F^+) be a decision problem of discrete $\mathbb{R}$-structures, i.e., F is the set of such structures over a given vocabulary, instance structures are coded as strings in $\{0,1\}^*$ as explained above, and $F^+ \subset F$. Then $(F, F^+) \in \exists\mathbb{R}$ if and only if there is an existential second-order sentence ψ such that $F^+ = \{\mathfrak{D} \in F | \mathfrak{D} \models \psi\}$.*

Dealing with a logical characterization of $\mathrm{BP}(\mathrm{P}^0_{\mathbb{R}})$ a bit more care for technical details is necessary. For example, we cannot define immediately an ordering on A. Nevertheless, repairing those technical problems is at most tedious. The extension of $\mathrm{FO}^0_{\mathbb{R}}$ to fixed point logic necessary in order to capture polynomial time in our setting once again was basically introduced already in [10], here we outline the technical differences due to the lack of arbitrary real constants. The only new technical demand in order to describe $\mathrm{P}^0_{\mathbb{R}}$-computations using a logic on discrete $\mathbb{R}$-structures is to change at certain places discrete data (i.e., objects formalized in the discrete part of an $\mathbb{R}$-structure) to real numbers. Suppose, for example, we use the given order on some A^k and express by $\mathrm{FO}_{\mathbb{R}}$ logic that $\underline{t} \in A^k$ is the successor of $\underline{s}$ in this order, i.e., we have a formula saying $\underline{t} = \underline{s}+1$. We then often want to use the characteristic value $\chi[\underline{t} = \underline{s} + 1]$ as real number. Formally, this can be expressed in our logics as follows: $\underline{t} = \underline{s} + 1$ is an abbreviation of a formula on the discrete part of an $\mathbb{R}$-structure. We can introduce a new discrete variable σ together with the fo-formula $\sigma \Leftrightarrow \underline{t} = \underline{s}+1$. Now, $real(\sigma)$ is a number term expressing the truth value of $\underline{t} = \underline{s}+1$ as real (!) number 0 or 1. We express this by writing $real([\chi[\underline{t} = \underline{s} + 1])$. Similar constructions are used below and should be clear from the context.

We need the following definitions, compare [10]. In order to deal with partially defined functions from some $A^k \to \mathbb{R}$, enlarge $\mathbb{R}$ by an element *undef* and extend the arithmetic operations via $r + undef := r - undef := undef, r \cdot undef := r/undef := \begin{cases} 0 & r = 0 \\ undef & r \neq 0 \end{cases}$ and $sgn(undef) := undef$. For a number term $F(\underline{s}, \underline{t})$ with free first order variables $\underline{s}, \underline{t}$, the max operator $\max\limits_{\underline{s}} F(\underline{s}, \underline{t})$ is a number term with free variables $\underline{t}$ and obvious semantics. Below, we use max in order to cycle through some A^k. Typically, we want $\underline{s}$ to be the direct predecessor of $\underline{t}$, then express the characteristic value $\chi[\underline{t} = \underline{s} + 1]$ via $\mathrm{FO}^0_{\mathbb{R}}$-logic, take the corresponding number term $real(\chi[\underline{t} = \underline{s} + 1])$ as explained above, and then identify $\underline{s}$ via $\max_{\underline{s}} real(\chi[\underline{t} = \underline{s} + 1])$. That way, when a computation is in step $\underline{s}$ we can formalize the next step $\underline{t}$. In order to define fixed point logic on discrete $\mathbb{R}$-structures $\mathfrak{D} = (L_s, \mathcal{R}, \{real\})$ let Z be a function symbol of arity r interpreted on $\mathfrak{D}$ as function from $A^r \to \mathbb{R}$ and let $F(Z, \underline{x})$ be a number term in $\mathrm{FO}^0_{\mathbb{R}}$ + max logic on signature $(L_s, \{real, Z\})$ (i.e., the closure of $\mathrm{FO}^0_{\mathbb{R}}$ logic under use of the max operator) with free first order variables $\underline{x} = (x_1, \ldots, x_r) \in V^r$.

Definition 8. *The fixed point Z^∞ with respect to $F(Z, \underline{x})$ is a function symbol of arity r. Its interpretation on a discrete $\mathbb{R}$-structure $\mathfrak{D}$ is defined iteratively as follows: Set $Z^{(0)}(\underline{x}) := undef\ \forall \underline{x} \in A^r$ and for $i \in \mathbb{N}$: $Z^{(i)}(\underline{x}) := \begin{cases} Z^{(i-1)}(\underline{x}) \text{ if it is defined already} \\ F(Z^{(i-1)}, \underline{x}) \text{ if } Z^{(i-1)}(\underline{x}) = undef \end{cases}$.*

After at most $j \le |A|^r$ iterations this process becomes saturated, i.e., $Z^{(j)} = Z^{(j+1)}$. This fixed point then is Z^∞.

Definition 9. *Functional fixed-point logic* $\mathrm{FFP}^0_{\mathbb{R}}$ *is obtained as closure of the set of first-order number terms under the maximization rule and the fixed-point rule.* $\mathrm{FP}^0_{\mathbb{R}}$ *denotes the class of characteristic functions definable in functional fixed-point logic.*

Theorem 4. *On ranked discrete $\mathbb{R}$-structures the functions in* $\mathrm{FFP}^0_{\mathbb{R}}$ *are exactly those computable in polynomial time in the constant-free BSS model. In particular, the characteristic functions in* $\mathrm{FFP}^0_{\mathbb{R}}$ *are those of polynomial time solvable decision problems in $\{0,1\}^*$, i.e.* $\mathrm{FP}^0_{\mathbb{R}} = \mathrm{BP}(\mathrm{P}^0_{\mathbb{R}})$.

We finally note that another approach to characterize a certain fragment of ∃ℝ by means of a so-called probabilistic independence logic was given in [13].

4 A Ladner Like Theorem for NP and ∃ℝ

Ladner's theorem [14] guarantees the existence of intermediate problems between P and the NP-complete ones. The range of the classical proof technique has been analyzed in [24]. In relation with the BSS model of computation over uncountable structures the theorem has been shown in [15] to hold as well over the complex numbers. Over ℝ it is open, but [2,5,18] give partial results. Especially, [2] addresses the importance of quantifier elimination QE algorithms when constructing intermediate problems. In the case of separating the class of ∃ℝ-complete problems from NP under the assumption ∃ℝ ≠ NP a combination of the arguments from [2,15] works as well.

Theorem 5. *If* ∃ℝ ≠ NP, *there exists a problem in* ∃ℝ \ NP *not being* ∃ℝ-*complete.*

For the rest of this section we outline the proof idea of how to construct such a problem. Technical details then easily can be filled following the presentations in [2,15]. Our construction starts from the $\exists\mathbb{R}$-complete problem QPS^0 of Definition 1. As mentioned before, the $\text{NP}_{\mathbb{R}}$-completeness proof of the corresponding problem for the class $\text{NP}_{\mathbb{R}}$ in [3] shows as well completeness of QPS^0 in $\exists\mathbb{R}$ since given an $\text{NP}^0_{\mathbb{R}}$-machine the reduction to the full QPS problem, i.e., where polynomials are allowed to have real coefficients as well, does not introduce real constants and runs in polynomial time also in the Turing model; intermediate results can be subsumed under real existential quantification.

The main reasons why a proof of Theorem 5 can be done similarly to showing related results are the decidability of $\exists\mathbb{R}$ within a computable time bound in the Turing model due to real quantifier elimination, and effective countability of NP-machines together with polynomial time bounds. Starting from QPS^0 we build certain restricted languages $L(a)$ depending on strictly increasing sequences $(a) := (a_i)_{i\in\mathbb{N}}$ of natural numbers. For input dimensions $n \in \{a_{2i-1}, \ldots, a_{2i} - 1\}$ we let $L(a)$ be the QPS^0 problem, for the remaining input dimensions $L(a)$ equals the empty set. The strategy now is to define two such sequences $(a), (b) \in \mathbb{N}^{\mathbb{N}}$ so that both have the desired properties.

Definition 10 (cf. [15]). *Let $(a), (b) \in \mathbb{N}^{\mathbb{N}}$ be two sequences of natural numbers.*

a) The sequences have an exponential gap *iff both are strictly increasing and for all $i \in \mathbb{N}$ satisfy $a_{2i+1} \geq 2^{b_{2i}}$ as well as $b_{2i+1} \geq 2^{a_{2i+2}}$, i.e., $a_1 < a_2 < 2^{a_2} \leq b_1 < b_2 < 2^{b_2} \leq a_3 \ldots$.*

b) The decision problem $L(a)$ (and $L(b)$ similarly) is defined dimension-wise for inputs of size n as follows:

$$L(a) \cap \{0,1\}^n := \begin{cases} \text{QPS}^0 \cap \{0,1\}^n & \text{if } \exists i \in \mathbb{N} \text{ s.t. } a_{2i-1} \leq n < a_{2i} \\ \emptyset & \text{otherwise} \end{cases},$$

Since $\exists\mathbb{R}$-completeness is defined downward from QPS^0 by usual polynomial time reductions, it is easy to see that assuming $\exists\mathbb{R} \neq \text{NP}$ neither $L(a)$ nor $L(b)$ can be $\exists\mathbb{R}$-complete if (a) and (b) have an exponential gap.

Lemma 1. *Suppose $\exists\mathbb{R} \neq$ NP and let sequences $(a), (b)$ have an exponential gap. Suppose furthermore that $L(a), L(b)$ belong to $\exists\mathbb{R} \setminus$ NP. Then both are not $\exists\mathbb{R}$-complete.*

Proof. Suppose one problem, say $L(a)$, is complete. Then there exists a polynomial time reduction f from $L(b)$ to $L(a)$. Given an input $x \in \{0,1\}^n$ for $L(b)$, there are two cases. If x is a positive QPS^0-instance, i.e., n belongs to some $\{b_{2i-1}, \ldots, b_{2i} - 1\}$, then $f(x)$ must be a positive instance for $L(a)$. This is only possible if its size $|f(x)|$ is at most $a_{2i} \leq \log n$. In this case we can solve the QPS^0-instance $f(x)$ by an algorithm like [4,21], which are PSPACE-algorithms and thus run in exponential time in $\log n$, i.e., in polynomial time in n. If the size of $f(x)$ is larger than $\log n$, then $a_{2i} \leq n < b_{2i-1}$ and thus $x \notin L(b)$. This gives a polynomial time decision algorithm for $L(b)$ and thus $L(b) \in \text{P} \subseteq \text{NP}$, contradicting our assumption $L(b) \in \exists\mathbb{R} \setminus \text{NP}$. □

The main task is to construct (a) and (b) so that $L(a)$ and $L(b)$ belong to $\exists\mathbb{R} \setminus$ NP. Below, we define both sequences stepwise in the order $a_1, a_2, b_1, b_2, a_3, a_4, b_3, \ldots$ The construction will guarantee the following properties:

i) $b_{2i-1} \geq 2^{a_{2i}}, a_{2i+1} \geq 2^{b_{2i}} \; \forall i \in \mathbb{N}$;
ii) the following task is efficiently solvable: Given $n \in \mathbb{N}$ in unary notation, compute the maximal j, ℓ such that $a_j \leq n$ and $b_\ell \leq n$;
iii) let $(M_i, p_i)_i$ be an effective enumeration of all NP-machines running in polynomial time (M_i is restricted to run in time p_i for a polynomial $p_i \in \mathbb{Q}[x]$). Then for all $i \in \mathbb{N}$ there exists a QPS^0-instance $x \in \{0,1\}^n$ with $n \in \{a_{2i+1}, \ldots, a_{2i+2} - 1\}$ such that M accepts x iff x is unsolvable, i.e., x is an instance for which M_i does not work correctly in time $p_i(|x|)$.

Lemma 2. *If ii) above can be accomplished, then $L(a)$ and $L(b)$ belong to $\exists\mathbb{R}$.*

Proof. A polynomial time reduction from $L(a)$ to QPS^0 works as follows: Given an $x \in \{0,1\}^n$, compute the maximal j and corresponding a_j such that $a_j \leq n$. If $j = 2j' - 1$ is odd, then $n \in \{a_{2j'-1}, \ldots, a_{2j'} - 1\}$ and thus $L(a) \cap \{0,1\}^n$ equals $\mathrm{QPS}^0 \cap \{0,1\}^n$. The reduction then outputs x. If $j = 2j'$ is even, then $L(a) \cap \{0,1\}^n = \emptyset$ and the reduction outputs a fixed unsolvable QPS^0 instance. Similarly for $L(b)$. □

Now let $\{(M_i, p_i) | i \in \mathbb{N}\}$ be an effective enumeration of all NP-Turing machines with running time bounded by a polynomial. W.l.o.g. we can assume for M_i that it is a non-deterministic Turing machine which is forced to stop its computation on x after at most $p_i(|x|)$ steps and rejects in case the computation has not been completed. We describe how to define sequences (a) and (b) such that i)-iii) above are satisfied. The construction relies on known results that allow to describe computations of both NP-algorithms and of $\mathrm{NP}^0_{\mathbb{R}}$-algorithms via existential first-order sentences over the real numbers.

Proposition 1. *a) Let* QPS^0 *be as above and $n \in \mathbb{N}$. Then there exists an effectively computable existential fo-formula $\varphi^{(n)}$ in the theory of $\mathbb{R}$ as a real closed field with coefficients from $\mathbb{Q}$ only such that $\forall x \in \{0,1\}^n : \varphi^{(n)}(x) \Leftrightarrow$ x is solvable as* QPS^0*-instance.*

b) For every NP*-machine M_i (in the Turing model) with running time bounded by a polynomial p_i and for every $n \in \mathbb{N}$ there exists an effectively computable fo-formula $\phi_i^{(n)}$ in the theory of $\mathbb{R}$ as a real closed field with coefficients from $\mathbb{Q}$ only such that $\forall x \in \{0,1\}^n : \phi_i^{(n)}(x) \Leftrightarrow$ there is an accepting computation of M_i on x in at most $p_i(|x|)$ steps.*

In both statements, effectively computable *means computable in the Turing model given the machines and n using a time bound depending on the input size only.*

Proof. a) Let M be the $\mathrm{NP}^0_{\mathbb{R}}$ BSS-machine for QPS^0 existing according to Theorem 1. Existence and effective computability of $\varphi^{(n)}$ follows from the analogue statement for general BSS-machines given in [20], see also [2]. Even the formula obtained for the latter result in the more general setting does not involve non-rational constants, so the same holds for $\varphi^{(n)}$ and its construction can be performed by a Turing machine. Note that the variables existentially quantified in $\varphi^{(n)}$ describe both the potential real solution of an instance and the intermediate results of a computation of M.

b) Follows easily from a) by considering M_i as an $\mathrm{NP}^0_{\mathbb{R}}$-algorithm. A (discrete) guess y of M_i can be expressed via existential quantification ranging over $\mathbb{R}$ by requiring the equation $y(y-1)=0$ to be fulfilled. □

For given $i, n \in \mathbb{N}$ define another fo-formula $\Theta_i^{(n)}$ over $\mathbb{R}$ as

$$\Theta_i^{(n)} \equiv \exists x \in \{0,1\}^n \ \neg(\varphi^{(n)}(x) \Leftrightarrow \phi_i^{(n)}(x))$$

It states that there exists an instance $x \in \{0,1\}^n$ for which NP-machine M_i within $p_i(n)$ steps gives the false result w.r.t. x as instance for QPS^0, i.e., (M_i, p_i) makes an error concerning verifying QPS^0 on inputs of dimension n. Thus, if we define (a) in such a way that there is an $n = n(i)$ in $\{a_{2i-1}, \ldots a_{2i}-1\}$ such that $\Theta_i^{(n(i))}$ holds, $L(a)$ will not be correctly decided by M_i within time bound p_i; and similarly for $L(b)$.

We now describe an algorithm which at the same time defines and computes (a) and (b), thereby proving condition ii) above to hold. Its structure is such that given $n \in \mathbb{N}$ or later an input $x \in \{0,1\}^n$, respectively, the computation of all $a_i, b_j < n$ is done anew.

Algorithm defining and computing $(a), (b)$:

(1) Define $a_1 := 0$.
(2) For given $n > a_1$ do the following for every $k = a_1 + 1, \ldots, n:$
(2.1) Decide validity of $\Theta_k^{(1)}$ by using any of the quantifier elimination algorithms mentioned above. If this algorithm needs more than n steps, then a_2 is not yet defined, but is guaranteed to satisfy $a_2 > n$.
(2.2) If a $k \leq n$ is found within at most n steps such that $\Theta_k^{(1)}$ holds, then define $a_2 := n$ as well as $b_1 := 2^{a_2}$.
(3) Compute similarly b_2 w.r.t. finding a k with $b_1 \leq k < b_2$.

For a general n suppose within at most n steps numbers $a_1, \ldots, a_{2i+1}, b_1, \ldots, b_{2i}$ have been computed. Then a_{2i+2} is computed as follows for $n \geq a_{2i+1}:$

(4) Do the following for every $k = a_{2i+1}, \ldots, n:$
(4.1) Decide validity of $\Theta_k^{(i)}$ by quantifier elimination. If the execution of both the computation of $a_1, \ldots, b_{2i}$ and the quantifier elimination exceeds n steps, then a_{2i+2} is not yet defined but guaranteed to satisfy $n < a_{2i+2}$.
(4.2) If $k \leq n$ is found within n steps such that $\Theta_k^{(i)}$ holds, then define $a_{2i+2} := n$ as well as $b_{2i+1} := 2^{a_{2i+2}}$.
(5) Compute b_{2i+2} similarly.

Theorem 6. *Assume* $\exists\mathbb{R} \neq \mathrm{NP}$ *and define sequences* (a) *and* (b) *as above. Then* $L(a)$ *and* $L(b)$ *belong to* $\exists\mathbb{R} \setminus \mathrm{NP}$*, but none is* $\exists\mathbb{R}$*-complete.*

Proof. Given the way the algorithm defines $(a), (b)$, condition ii) above is satisfied, i.e., the assumption of Lemma 2 holds. Thus both problems belong to $\exists\mathbb{R}$. Assuming $\exists\mathbb{R} \neq \mathrm{NP}$, the construction of both sequences implies that no NP-machine M with a polynomial time bound p will verify $L(a)$ and neither $L(b)$ correctly according to the requirements of being in NP since any such (M, p) occurs in the effective enumeration as some (M_i, p_i). By construction (a) and (b) have an exponential gap, so by Lemma 1 none of them is $\exists\mathbb{R}$-complete. □

References

1. Baker, T., Gill, J., Solovay, R.: Relativizations of the $P =?NP$ question. SIAM J. Comput. **4**, 431–442 (1975)
2. Ben-David, S., Meer, K., Michaux, C.: A Note on Non-complete Problems in $NP_{\mathbb{R}}$. J. Complex. **16**(1), 324–332 (2000)
3. Blum, L., Cucker, F., Shub, M., Smale, S.: Complexity and Real Computation. Springer-Verlag, Berlin (1998)
4. Canny, J.: Some algebraic and geometric computations in PSPACE. In: Proceedings of the 27th ACM Symposium on Theory of Computing STOC, pp. 460–469 (1988)
5. Chapuis, O., Koiran, P.: Saturation and stability in the theory of computation over the reals. Ann. Pure Appl. Log. **99**(1–3), 1–49 (1999)
6. Emerson, T.: Relativizations of the P=?NP question over the reals (and other ordered rings). Theor. Comput. Sci. **133**(1), 15–22 (1994)
7. Fagin, R.: Generalized first-order spectra and polynomial-time recognizable sets. Proc. SIAM-AMS **7**, 43–73 (1974)
8. Gassner, C.: The separation of relativized versions of P and DNP for the ring of the reals. J. Univers. Comput. Sci. **16**(18), 2563–2568 (2010)
9. Grädel, E., Gurevich, Y.: Metafinite model theory. In: D. Leivant (ed.), Logic and Computational Complexity, pp. 313–366, Springer (1996)
10. Grädel, E., Meer, K.: Descriptive complexity theory over the real numbers. In: Renegar, J., Shub, M., Smale, S. (eds.): The Mathematics of Numerical Analysis, Lectures in Applied Mathematics, vol. 32, AMS, pp. 381–404 (1996)
11. Grädel, E., et al.: Finite Model Theory and Its Applications. Springer (2007)
12. Hamm, T., Meijer, L., Miltzow, T., Patro, S.: Oracle Separations for RPH, Preprint, arXiv:2502.09279 (2025)
13. Hannula, M., Kontinen, J., Van den Busche, J., Virtema, J.: Descriptive complexity of real computation and probabilistic independence logic. In: Proceedings of the 35th ACM/IEEE Symposium LICS, pp. 550–563 (2020)
14. Ladner, R.: On the structure of polynomial time reducibility. J. ACM **22**, 155–171 (1975)
15. Malajovich, G., Meer, K.: On the structure of $\mathrm{NP}_{\mathbb{C}}$. SIAM J. Comput. **28**(1), 27–35 (1999)
16. Matiyasevich, Y.: Enumerable sets are Diophantine. Soviet Math. Doklady **11**, 354–358 (1970)

17. Meer, K.: Real number models under various sets of operations. J. Complex. **9**(3), 366–372 (1993)
18. Meer, K.: On Ladner's result for a class of real machines with restricted use of constants. Inf. Comput. **210**, 13–20 (2012)
19. Meer, K., Ziegler, M.: An explicit solution to Post's problem over the reals. J. Complex. **24**(1), 3–15 (2008)
20. Michaux, C.: $P \neq NP$ over the nonstandard reals implies $P \neq NP$ over $\mathbb{R}$. Theor. Comput. Sci. **133**, 95–104 (1994)
21. Renegar, J.: On the computational complexity and geometry of the first-order theory of the reals I–III. J. Symb. Comput. **13**, 255–352 (1992)
22. Schaefer, M.: Complexity of some geometric and topological problems. In: Eppstein, D., Gansner, E.R. (eds) Graph Drawing. GD 2009. Lecture Notes in Computer Science, vol. 5849, pp. 334–344, Springer, Berlin, Heidelberg, (2010)
23. Schaefer, M., Cardinal, J., Miltzow, T.: The Existential Theory of the Reals as a Complexity Class: A Compendium, Preprint, arXiv:2407.18006 (2024)
24. Schöning, U.: A uniform approach to obtain diagonal sets in complexity classes. Theor. Comput. Sci. **18**, 95–103 (1982)
25. Vereshchagin, N.K.: Relativizable and nonrelativizable theorems in the polynomial theory of algorithms. Russian Acad. Sci. Izv. Math. **42**(2), 261–298 (1994)

On the Computational Power of C-Random Strings

Alexey Milovanov(✉)

LASIGE, Faculdade de Ciências, Universidade de Lisboa, Lisbon, Portugal
almas239@gmail.com

Abstract. Denote by H the halting problem. Let $R_U := \{x | \mathrm{C}_U(x) \geq |x|\}$, where $\mathrm{C}_U(x)$ represents the plain Kolmogorov complexity of x under a universal decompressor U. We demonstrate the existence of a universal U such that H is solvable in polynomial time with access to the oracle R_U. This result resolves a problem posed by Eric Allender in [1] regarding the computational power of Kolmogorov complexity-based oracles.

Keywords: Kolmogorov complexity · Computability · Computation complexity

1 Introduction

At an informal level, the Kolmogorov complexity of string x is defined as the minimal length of a program that outputs x on the empty input. This definition requires clarification: there are various types of complexity (plain, prefix, and others), but even when fixed, there remains a dependency on the choice of programming language or decompressor. We will give all the definitions needed for this paper in the next section; we also refer the reader to [8,10] for formal definitions and the main properties of Kolmogorov complexity.

The concept of Kolmogorov complexity allows us to define the notion of an individual random string. This can be done as follows: string x is called random if its Kolmogorov complexity $\gtrsim |x|$, where $|x|$ is the length of x. Of course, the formal definition again requires clarifications: in addition to the aforementioned nuances with the definition of complexity, it is necessary to formally define what $\gtrsim$ means. Let us assume that we have clarified all the details. Define by R the set of all random strings. The question arises: How powerful is the oracle R? For example, what languages belong to $\mathbf{P}^R$?

This and other similar questions were investigated in [2,3,6].

In these papers some lower bounds for $\mathbf{P}^R, \mathbf{P/poly}^R$ and other complexity classes were proven. These results are robust in the following sense: the results are valid for all reasonable definitions of R, in particular, it does not matter what type of Kolmogorov complexity we consider.

The situation is different for upper bounds. The results in [4,5,9] show some limits on the computational power of R only for *prefix complexity*.

© The Author(s), under exclusive license to Springer Nature Switzerland AG 2025
A. Beckmann et al. (Eds.): CiE 2025, LNCS 15764, pp. 321–332, 2025.
https://doi.org/10.1007/978-3-031-95908-0_23

For example, the results in [4,5] together with the lower bound obtained in [6] show that

$$\mathrm{EXP}^{\mathrm{NP}} \subseteq \bigcap_U \mathrm{P}^{R_{\mathrm{K}_U}} \subseteq \mathrm{EXPSPACE}.$$

Here, the intersection is for all universal prefix-free decompressors U.

The lower and upper bounds are significantly different for the prefix complexity but at least we understand that the intersection does not contain undecidable languages. For the plain complexity even this is unknown. Let us introduce some notations.

Denote by H the halting problem:

$$H := \{x | \text{ program } x \text{ halts on the empty input}\}.$$

Denote $R_U := \{x | \mathrm{C}_U(x) \geq |x|\}$, where $\mathrm{C}_U(x)$ is the plain Kolmogorov complexity of x under a universal decompressor U.

Open question [1]
Is $H \notin \mathrm{P}^{R_U}$ for some universal U?

There is some evidence that the answer to this question should be affirmative. Kummer [7] showed that for each universal machine U, there is a time-bounded disjunctive truth-table reduction from H to R_U: That is, there is a computable function that takes x as input, and produces a list of strings, with the property that $x \in H$ if and only if at least one of the strings is in R_U. However, it was shown in [2] that, no matter what computable time bound t one picks, there is some U such that the disjunctive truth-table reduction from H to R_U requires more time than t. It should be noted that the analogue of Kummer's theorem does not hold for prefix complexity [11].

Let us consider another question. How powerful can R_U be for *some* universal U? It was shown in [2] that there exists a universal U such that H can be reduced to R_U by disjunctive truth-table in double exponential time.

Our main result is the following

Theorem 1. *There exists a universal decompressor U such that $H \in \mathrm{P}^{R_U}$.*

At the simplest level, the idea of the proof can be explained as follows.

The universal decompressor is optimal up to an additive constant, which allows us to construct certain exotic decompressors with specific properties. For example, there exists a decompressor U_{even} such that for every x the Kolmogorov complexity $\mathrm{C}_{U_{\text{even}}}(x)$ is always an even number (Lemma 3). We will also construct some exotic decompressor that allows to solve the halting problem.

To explain the idea, we will first prove a simpler result: there exists a universal decompressor U such that H can be solved by the *oracle function* F_U; that is, on input x the function F_U outputs its Kolmogorov complexity $\mathrm{C}_U(x)$.

Let U_{even} be the decompressor mentioned above, where every minimal description has even length.

The decompressor U is defined as follows for arbitrary $d \in \{0,1\}^*$:

- if $d = 00d'$ then $U(d) := U_{\text{even}}(d')$.
- if $d = 1d'$ then $U(d) := U_{\text{even}}(d') = x$ if $x \in H$.

Note that U is computable because H is enumerable. We claim that U is universal and $x \in H$ iff $C_U(x)$ is odd. Indeed, since U_{even} is universal then U is also universal (we need only descriptions that starts with 00). If $x \in H$, then the minimal description of x under U starts with 1 and hence has odd length. Conversely, if $C_U(x)$ is odd, then $x \in H$.

We, however, need to solve H using a weaker oracle that only allows to distinguish random strings from not-random. The challenge arises when the string x is non-random, as the oracle R_U distinguishes only between random and non-random strings. Now we describe some ideas that allow to get around the problem.

One useful tool is the result from [3] stating that, for any universal decompressor U, the complexity classes BPP^{R_U} and P^{R_U} are equal.

Another tool is Lemma 4, which we state informally here: Let x be a string of length n with complexity k, and let A be a random binary matrix of dimensions $k \times n$. Then, with high probability, the string $A \circ Ax$ (where "$\circ$" denotes concatenation) is algorithmically random, i.e., its length is close to its Kolmogorov complexity.

This observation allows to construct a universal decompressor U such that $H \in \text{P}^{R_U}$. We use a similar approach as for oracle function but instead of string x we directly look at $A \circ Ax$. This string can be considered as a "finger print" of x (note that matrices are the family of 2-universal hash-functions) and also this string is close to random by the previously-mentioned properties of Lemma 4. We construct U in such a way that $A \circ Ax$ is random iff $x \in H$.

The remainder of this paper is organized as follows. In the next section we recall basic definitions and properties of Kolmogorov complexity.

Then in Sect. 3 we prove some technical lemmas that are needed to define the universal decompressor U for Theorem 1 (Sect. 4). Then we describe a polynomial-time algorithm that solves H by using oracle R_U (Sect. 5) and prove its correctness (Sect. 6).

2 Definitions and Some Properties of Kolmogorov Complexity

Let us recall the definition of plain conditional and unconditional Kolmogorov complexity.

Let U be an algorithm whose inputs and outputs are binary strings. We will call such U *decompressor* and define the (unconditional) complexity $C_U(x)$ of a binary string x with respect to U as follows:

$$C_U(x) := \min\{|y| : U(y) = x\}$$

Any y such that $U(y) = x$ we call a description (or U-description) of x.

Therefore, the complexity of x is defined as the length of the shortest (or minimal) description of x.

A decompressor U is called *universal* if for every decompressor U' there exists a constant M such that for every string x it holds that

$$C_U(x) \leq C_{U'}(x) + M.$$

Now we recall the definition of the *conditional* Kolmogorov complexity. Let $D(p, x)$ be a computable partial function (that we will also call decompressor) of two string arguments; its values are also binary strings. We may think of U as an interpreter of some programming language. The first argument p is considered as a program and the second argument is an input for this program. Then we define the complexity function

$$C_D(x|y) := \min\{|p| : D(p, y) = x\};$$

here $|p|$ stands for the length of a binary string p, so the right hand side is the minimal length of a program that produces output x given input y.

Theorem 2 (Kolmogorov-Solomonoff). *There exists a universal decompressor D such that for every other decompressor D' there exists a constant M such that*

$$C_D(x|y) \leq C_{D'}(x|y) + M$$

for all strings x and y.

The proofs for all statements of this section can be found in [10].

Every decompressor U for the conditional complexity defines unconditional complexity can be naturally derived from the conditioned version by considering the condition as the empty string:

$$C_U(x) := C_U(x|\text{ the empty string}).$$

It is easy to see that if U is a universal decompressor for the conditional Kolmogorov complexity then the corresponding decompressor for the unconditional Kolmogorov complexity is also universal.

We will use the following basic properties of Kolmogorov complexity, which hold for any universal decompressor U:

- **Bounded by length:** There exists a constant M' such that:

$$C_U(x) \leq |x| + M'.$$

- **Computable transformations:** For every total computable function f, there exists a constant M' such that:

$$C_U(f(x)) \leq C_U(x) + M'.$$

- **Counting descriptions:** For every string y and every k, the number of strings x satisfying $C_U(x|y) \leq k$ is at most 2^{k+1}.

The Kolmogorov complexity $C_U(x, y)$ of a pair of strings x and y is defined as follows. Let $x, y \to [x, y]$ be an injective computable function that maps a pair of strings into a single string. Then:

$$C_U(x, y) := C_U([x, y]).$$

This definition depends on the choice of the pairing function $[,]$, but only up to an additive $O(1)$ term. Similarly, the complexity of triples of strings, integer numbers or any finite object can be defined.

For every natural number M its Kolmogorov complexity $C_U(M)$ is not greater than $\log M + O(1)$ (as the length of its binary representation).

We will use the following fact about the complexity of pairs.

Theorem 3.

$$C_U(x, y) \leq C_U(x) + C_U(y|x) + 2 \log C_U(y) + O(1).$$

Moreover, it is known that this inequality is actually an equality up to $O(\log(C_U(x) + C_U(y)))$ term which is known as the *Symmetry of Information.*

The next two statements are known but not so common. We will provide proofs for these statements for the reader's convenience.

Lemma 1. *For every string x it holds that*

$$C_U(x, C_U(x)) = C_U(x) + O(1).$$

Proof. The inequality

$$C_U(x, C_U(x)) \geq C_U(x) + O(1)$$

obviously holds: every description for the pair $(x, C_U(x))$ provides a description for x that has the same length up to an additive constant.

On the other hand, the shortest description of x (of length $C_U(x)$) determines both x and $C_U(x)$.

Lemma 2. *Let $C_U(y|n) = s$. Then*

$$C_U(y|n, s) = s - O(1).$$

Proof. Assume that there is a description q for y given n, s of length $s - M$ for some M. We claim that there is a description for y given n of length $s - M + O(\log M)$. Indeed, knowing q and M it is possible to find s (as, $s = |q| + M$). So:

$$C_U(y|n) \leq C_U(q, M) + O(1).$$

Now we estimate

$$C_U(q, M) \leq |q| + O(\log M) = s - M + O(\log M).$$

Since $s - M + O(\log M)$ must be not smaller than s we conclude that M is constant.

3 Preliminary Lemmas

In this section we prove some technical lemmas that will be used in the next sections. We start with the following folklore fact.

Lemma 3. *There exists a universal decompressor V for the* conditional *Kolmogorov complexity such that for every string x and y, the value $C_V(x|y)$ is even.*

Proof. Let V_{opt} be an arbitrary universal decompressor for the conditional Kolmogorov complexity. Define $V(d, y)$ for arbitrary d and y in the following way:

- if $d = 0 \circ d'$ for odd $|d'|$ then $V(d, y) := V_{\text{opt}}(d', y)$;
- if $d = 11 \circ d'$ for even $|d'|$ then $V(d, y) := V_{\text{opt}}(d', y)$;
- Otherwise $V(d, y)$ is undefined.

From this definition it follows that for arbitrary x and y the value $C_V(x|y)$ is even and it holds that:

$$C_V(x|y) \le C_{V_{\text{opt}}}(x|y) + 2.$$

Therefore, V is a universal decompressor.

As in the previous section we define $C_V(x)$ for arbitrary x as $C_V(x|$ empty word).

This decompressor V will be instrumental in constructing a universal decompressor U that enables solving the halting problem.

We now prove two lemmas and define two constants (D and G), which will be used in the next section to define U.

We will use the following simple fact about matrices.

Fact 1. *Let $b_1 \neq b_2$ be two binary strings of length n. Let A be a random under the uniform distribution binary $k \times n$ matrix. Then the probability of the event $[Ab_1 = Ab_2]$ is equal to 2^{-k}.*

Lemma 4. *Let y be a string of length n. Denote $k = C_V(y|n) + 5$. Assume that $n + 1$ is a prime number and $k < n$.*

Let A be a random binary $k \times n$ matrix (under the uniform distribution).

Let f be the string Ay. Then for some constant D the following holds with probability at least $\frac{9}{10}$:

$$C_V(A \circ f) \ge nk + k - D.$$

Here A is viewed as a string of length nk.

Proof. **First** we **claim** that with probability at least $\frac{15}{16} = 1 - \frac{1}{2^4}$ there is no $t \neq y$ such that $At = f$ and $C_V(t|n) \le k - 5$. Indeed, the number of strings t such that $C_V(t|n) \le k - 5$ is at most 2^{k-4}. Now we just use Fact 1.

The **second claim** is that for some constant C_1 with probability at least $\frac{99}{100}$ it holds that

$$C_V(A \circ y) \ge nk + k - C_1.$$

Let $C_V(A \circ y) \leq nk + k - R$ for some R and for 1 percent of all possible A. We want to show that $R = O(1)$. Note that

$$C_V(y|n, k, R) \leq k - R + O(1).$$

Indeed, the number of strings h of length n such that there are at least 1 percent of all matrices A such that $C_V(A \circ y) \leq nk + k - R$ is at most $2^{k-R+O(1)}$. We can enumerate all such strings given n, k, R and describe y as the corresponding number in this enumeration.

By Lemma 2 $C_V(y|n, k) = k - O(1)$. By using the same technique as in the proof of Lemma 2 we conclude that R is constant.

Now we prove the statement of the lemma. Let

$$C_V(A \circ f) = nk + k - D \text{ for some } D.$$

We need to prove that $D = O(1)$ with high probability.

For this, we estimate the complexity of tuple $(A \circ f, n, k)$.

First, note that by Lemma 1 we have

$$C_V(A \circ f, nk + k - D) = nk + k - D + O(1).$$

By Theorem 3 we can estimate the complexity of the following triple:

$$C_V(A \circ f, nk + k - D, D) \leq nk + k - D + O(\log D).$$

Knowing $(n + 1)k - D$ and D it is simple to find n and k since $n + 1$ is a large prime. Hence,

$$C_V(A \circ f, n, k) < nk + k - D + O(\log D).$$

With high probability (at least $1 - \frac{1}{16} - \frac{1}{100} > \frac{9}{10}$) both claims above hold, i.e. y does not have collisions among strings t such that $C_V(t|n) \leq k - 5$ and

$$C_V(A \circ y) \geq nk + k - O(1).$$

Therefore, it is possible to find y by $(A \circ f, n, k)$ as the unique preimage of A among strings with $C_V(y|n) \leq k - 5$.

Hence, D is constant.

Now, we state a simple lemma that is an easy consequence of the fact that strings x and $x \circ 0^9$ (just appending 9 zeros to x) have the complexity up to some additive constant.

We need this lemma to define constant G that we will use in the next section.

Lemma 5. *For every D there exists G such that the following holds.*
For every x if $C_V(x) \geq |x| - D$ then $C_V(x \circ 0^9) \geq (|x| + 9) - G$.

4 Definition of Universal Decompressor U

Here we describe universal decompressor U that allows to solve the halting problem in polynomial time. We will use decompressor V from the previous section.

This decompressor U will have the following properties:

- every string $A \circ f$ of length $nk + k$ as in Lemma 4 such that $C_V(A \circ f) \geq nk + k - D$ is U-random (i.e. $C_U(A \circ f) \geq |A \circ f| = nk + k$).
- Lemma 7 (stated below) holds.

Odd numbers m such that $m = p \cdot u$, where $u < p$ and p is a prime number we call *specific*. For specific number m the number p is called its *large prime factor.*

We define $U(d)$ for an arbitrary $d \in \{0,1\}^*$ as follows (using constants D and G defined in the previous section):

1. If $d = 1^{D-1}0d'$ for some d' then $U(d)$ runs $V(d')$.
   Assume that $V(d')$ outputs some string y (if $V(d')$ is undefined, then $U(d)$ is also undefined).
   If y does not have a form $y = y'0^9$ where $|y'|$ is specific then $U(d) := V(d')$.
   Otherwise, (if y has this form) we compare $|y|$ with $|d'|$.
   If $|y| > |d'| - G$ then $U(d) := V(d')$. Otherwise $U(d)$ is undefined.
2. If $d = 1^D0d'$ for some d' then $U(d) := d'$.
3. If $d = 0d'$ for some d' and $|d'| + 5$ is specific then U works as follows. Let p_l— the lth prime number—be the large prime factor of $|d'|+5$. Let $|d'| = p_l \cdot k - 5$. Denote $n = p_l - 1$. Consider d' as the concatenation $d' = A \circ d''$, where A is the $k \times n$ matrix and d'' is a string of length $k - 5$.
   Run $V(d''|n)$. If it halts consider the output y. If $|y| \neq n$ then U is undefined in this case. Otherwise we consider x — the first l bits of y. If $x \in H$ then $U(d') := A \circ Ay \circ 0^9$.

Note that there are no strings with specific lengths that have U-description of the third type: all strings that have descriptions of the third type have even length:

$$|A \circ Ay \circ 0^9| = nk + k + 9 = (n+1)k + 9 = p_l \cdot k + 9.$$

Since $k - 5$ is the length of some V-description we conclude that k is odd. So, $p_l \cdot k + 9$ is even.

Lemma 6. *Decompressor U is universal.*

Proof. To prove this lemma we need only the descriptions of the first and second type in the definition of U.

Indeed, consider some string y. Let d' be a minimal V-description of y.

If $|y| > |d'| - G$ then string $1^{D-1}0d'$ is an U-description of y and therefore

$$C_U(y) \leq C_V(y) + D.$$

If $|y| \leq |d'| - G$ then $1^D 0y$ is an U-description and therefore

$$C_U(y) \leq |y| + D + 1 \leq |d'| - G + D + 1 = C_U(y) - G + D + 1.$$

So, in both cases

$$C_U(y) \leq C_V(y) + D + 1.$$

Therefore, U is universal since V is universal.

Lemma 7. *Let b be some string such that $|b|$ is specific and $|b| = p_l \cdot k$, where p_l is a large prime. Consider b as the concatenation $b = A \circ c$, where $|c| = k$.*

Assume that $C_U(b) \geq |b|$ but $C_U(b \circ 0^9) < |b| + 9$.

Then there exists a string y of length $n := p_l - 1$ such that the first l bits of y describe a program x that belongs to H.

Moreover, $C_V(y|n) \leq k - 5$ and $Ay = c$.

Proof. We claim that the inequality $C_U(b) \geq |b|$ implies that $C_V(b) \geq |b| - D$. Indeed, otherwise b has an U-description of length less then b of the first type.

(Note that since $|b|$ is specific, then there is no b' such that $b = b' \circ 0^9$ and $|b'|$ is specific because the lengths of b and b' can not be both odd. Hence, the exception about G does not apply to b.)

Then by Lemma 5 we have that $C_V(b \circ 0^9) \geq |b| + 9 - G$. Hence, there are no V-descriptions of the first type for $b \circ 0^9$.

Also, the minimal V-description of $b \circ 0^9$ is not of the second type because this string has a rather short description . Hence, $b \circ 0^9$ has a description of the third type that yields the conclusion of the lemma.

5 Algorithm

We claim that the following probabilistic polynomial-time algorithm with access to R_U solves the halting problem:

Let x be an input instance for the halting problem. Denote by l the length of x.

Denote by p_l the lth prime number and set $n := p_l - 1$.

Finally, denote by y the following string of length n: $y := x \circ 0^{n-l}$.

Then, for every $k = 1, \ldots, n-1$ and for $m = n^{50}$ we consider m random matrices $A_1, \ldots, A_m$ of size $k \times n$.

If for some k and for at least half of $i \in \{1, \ldots, m\}$ it holds that

- $A_i \circ A_i y \in R_U$ but
- $A_i \circ A_i y \circ 0^9 \notin R_U$;

then the algorithm outputs that the program x halts; otherwise, it outputs that x does not halt.

6 Correctness

Proof (Proof of Theorem 1). By Lemma 6 the decompressor U is universal and as it was mentioned in the Introduction $\mathrm{P}^{R_U} = \mathrm{BPP}^{R_U}$. So, it is enough to show that the algorithm in the previous section outputs the correct answer with high probability.

If x Halts. We need to prove that for some k and for random A the event

$$A \circ Ay \in R_U \text{ and } A \circ Ay \circ 0^9 \notin R_U. \tag{1}$$

happens with high probability.

Set $k = \mathrm{C}_V(y|n)+5$. Then, by Lemma 4, with probability at least $\frac{9}{10}$ it holds that

$$\mathrm{C}_V(A \circ f) \geq nk + k - D.$$

Thus, by the definition of U the event $A \circ Ay \in R_U$ holds.

We claim that $A \circ Ay \circ 0^9 \notin R_U$ with probability 1.

Note that the first l bits of y is a string x that belongs to H. Consider the following U-description of type 3: the string $0 \circ A \circ d''$, where d'' is the minimal V-description of y conditional to n. By the definition of U this is a description of $A \circ Ay \circ 0^9$. The length of the description is

$$1 + k \cdot n + k - 5$$

that is less than

$$|A \circ Ay \circ 0^9| = k \cdot n + k + 9.$$

Thus, we conclude that $A \circ Ay \circ 0^9 \notin R_U$, proving correctness in this case.

If x does not Halt. We claim that, in this case, for every k and random A the probability of the event (1) is small.

More precisely, we show that this probability is not greater than $\frac{1}{16}$.

This implies that the algorithm works correctly with high probability.

Indeed, the algorithm fails only if event (1) holds for at least half of the matrices $A_1, \ldots, A_m$. Since matrices are chosen independently and m is a sufficiently large polynomial Hoeffding's inequality implies that this probability is exponentially small in n.

(In more detail, define independent random variables $I_1, I_2, \ldots, I_m$, where

$$I_j = \begin{cases} 1, & \text{if event (1) occurs for matrix } A_j, \\ 0, & \text{otherwise.} \end{cases}$$

Since the expectation of each I_j is at most $\frac{1}{16}$, it follows that

$$\mathbb{E}\left[\sum_{i=1}^{m} I_i\right] \leq \frac{m}{16}.$$

By Hoeffding's inequality, the probability that

$$\sum_{i=1}^{m} I_i \geq \frac{m}{2}$$

is exponentially small.)

Thus, the algorithm works correctly for all k with high probability.

It remains to prove the claim.

To analyze when event (1) can happen, assume that for some k and A, it does occur. By Lemma 7, this implies the existence of some $y' \neq y$ of length n such that

- $C(y'|n) \leq k - 5$, and
- $Ay' = Ay$.

(Since y must also satisfy additional conditions, we only focus on these two properties.)

For a fixed k, there are at most 2^{k-4} strings y' such that $C(y'|n) \leq k - 5$. Thus, by Fact 1, for a random matrix A, the probability of $Ay' = Ay$ for some such y' is at most $\frac{1}{16}$.

Acknowledgments. I would like to thank Eric Allender for the formulation of the question, and Bruno Bauwens for useful discussions.

This work was funded by the European Union (ERC, HOFGA, 101041696). Views and opinions expressed are however those of the author(s) only and do not necessarily reflect those of the European Union or the European Research Council. Neither the European Union nor the granting authority can be held responsible for them.

This work was also supported by FCT through the LASIGE Research Unit (ref. UIDB/00408/2025 and UIDP/00408/2025), and by CMAFcIO, FCT Project UIDB/04561/2020 (https://doi.org/10.54499/UIDB/04561/2020).

Disclosure of Interests. The author has no competing interests to declare that are relevant to the content of this article.

References

1. Allender, E.: Parting Thoughts and Parting Shots (Read On for Details on How to Win Valuable Prizes!), Guest Piece for the Complexity Theory Column, edited by Lane Hemaspaandra, SIGACT NEWS 54, March 2023, pp. 63–81. https://people.cs.rutgers.edu/~allender/papers/sigact.news.draft.pdf
2. Allender, E., Buhrman, H., Koucký, M.: What can be efficiently reduced to the Kolmogorov-random strings? Ann. Pure Appl. Log. **138**, 2–19 (2006)
3. Allender, E., Buhrman, H., Koucký, M., van Melkebeek, D., Ronneburger, D.: Power from random strings. SIAM J. Comput. **35**, 1467–1493 (2006)
4. Allender, E., Friedman, L., Gasarch, W.: Limits on the computational power of random strings. Inf. Comput. **222**, 80–92 (2013)
5. Cai, M., Downey, R.G., Epstein, R., Lempp, S., Miller, J.: Random strings and tt-degrees of Turing complete c.e. sets. Log. Methods Comput. Sci. **10**(3) (2014). lmcs:1126; https://doi.org/10.2168/LMCS-10(3:15)2014

6. Hirahara, S.: Unexpected hardness results for Kolmogorov complexity under uniform reductions. In: Proceedings of the 52nd Annual ACM SIGACT Symposium on Theory of Computing (STOC 2020), pp. 1038–1051
7. Kummer, M.: On the complexity of random strings. In: Symposium on Theoretical Aspects of Computer Science (STACS), volume 1046 of Lecture Notes in Computer Science, pp. 25–36. Springer, 1996
8. Li, M., Vitányi, P.: An Introduction to Kolmogorov complexity and its applications, 3rd ed., Springer, 2008 (1 ed., 1993; 2 ed., 1997), xxiii+790 pp. ISBN 978-0-387-49820-1
9. Milovanov, A.: Some Games on Turing Machines and Power from Random Strings, Lecture Notes in Computer Science,Unity of Logic and Computation, pp. 105–119, Springer Nature, Switzerland, 2023
10. Shen, A., Uspensky, V., Vereshchagin, N.: Kolmogorov Complexity and Algorithmic Randomness, ACM (2017)
11. Muchnik, A.A., Positselsky, S.Y.: Kolmogorov entropy in the context of computability theory. Theor. Comput. Sci. **271**, 15–35 (2002)

Generalized Learnability of Stochastic Principles

Morenikeji Neri[1], Nicholas Pischke[1], and Thomas Powell[2(✉)]

[1] Department of Mathematics, Technische Universität Darmstadt, Darmstadt, Germany
{neri,pischke}@mathematik.tu-darmstadt.de

[2] Department of Computer Science, University of Bath, Bath, UK
trjp20@bath.ac.uk

Abstract. Motivated by recent applications of proof theory in probability, we introduce a novel computational interpretation of probabilistic $\exists\forall$-formulas, called dependent learnability. This encompasses several important notions of quantitative stochastic convergence, where it represents a generalized version of the property – widely studied in probability and ergodic theory – that a sequence of random variables has bounded fluctuations. We study both deterministic and stochastic variants of this notion and relate these to other computational interpretations of $\exists\forall$-formulas from the literature. In particular, we prove dependent learnability to be primitive recursively equivalent to the influential notion of metastability, which in conjunction with results from applied proof theory highlights that dependently learnable rates can be extracted from large classes of nonconstructive proofs of $\exists\forall$-formulas. Furthermore, we present a primitive recursive algorithm for joining two (and thus finitely many) dependently learnable rates, which in particular proves to be considerably more mathematically intuitive than the corresponding functional for joining rates of metastability. Finally, we discuss our results in the light of game semantics.

Keywords: Quantitative convergence · Proof theory · Computability in probability theory

1 Introduction

Results on controlling the oscillation behaviour of a sequence of random variables are central to probability theory. One of the most well-known examples of such results is Doob's upcrossing inequality for supermartingales, asserting that the number of times an L_1-bounded supermartingale upcrosses some fixed interval

The first and third authors were partially funded by the EPSRC (grant numbers EP/L016540/1 and EP/W035847/1 respectively). For the purpose of open access, the authors have applied a Creative Commons Attribution (CC-BY) licence to any Author Accepted Manuscript version arising.

© The Author(s), under exclusive license to Springer Nature Switzerland AG 2025
A. Beckmann et al. (Eds.): CiE 2025, LNCS 15764, pp. 333–348, 2025.
https://doi.org/10.1007/978-3-031-95908-0_24

$[\alpha, \beta]$ for $\alpha < \beta$ is bounded in mean, which in turn is used to establish the fundamental result that L_1-bounded supermartingales converge almost surely.

Doob's upcrossing inequalities, and similar results in general, further gain significance because they provide explicit quantitative convergence information in a situation where *direct, computable* rates of convergence generally do not exist (for the simple reason that supermartingales immediately inherit the Specker phenomena that apply to monotone bounded sequences of reals [23]). Indeed, such quantitative results on the oscillatory behaviour of stochastic processes abound in the probability literature. In particular, there are further inequalities that express oscillatory behavior in martingale theory [5,7], as well as in the convergence of ergodic averages [4,10,11], along with results from information theory [8], just to name a few.

The computability theory of convergence has been long studied by logicians. Confronted with a convergence theorem for which no computable rates exist, there are several options available. One is to follow the approach of computable analysis and rebuild the underlying notions and the theory in a computable way (as done for measure theory e.g. in [9]). Another is to work in a general mathematical setting but to adjust what is meant by computable convergence, with the benefit of remaining compatible with the usual mathematical practice.

It is the second approach that is adopted in the proof mining program, a general methodology developed by Kohlenbach and his collaborators (see [13] for a comprehensive monograph on the subject and [14] for a recent survey) which applies ideas and techniques from proof theory to extract quantitative information from nonconstructive proofs in mathematics, substantiated and (re-)enforced by underlying proof-theoretic results known as general logical metatheorems.

In the context of this second approach, it has been shown that in the *non-stochastic* setting, there is in fact a hierarchy of quantitative effective data that one can associate to a convergent sequence, classified by the amount of classical logic permissible in proofs from which these data can be potentially extracted [15]. Concretely, this hierarchy is populated with computable rates of convergence at the top (which in general can only be extracted from semi-constructive proofs) and so-called rates of metastability at the bottom (which can be extracted in very general situations, in particular in the presence of full classical logic). Intermediate to these lies the property of a sequence having a computable bound on the number of *fluctuations* it experiences (which is extractable in contexts with a restricted use of the law of excluded middle).

It has been recently shown by the first and third author [20] that a natural lifting of this property to the stochastic setting – called uniform learnability – is enjoyed by large family of stochastic processes whose oscillatory behaviour is bounded in mean, including L_1-bounded supermartingales. To be more precise, for such processes $\{X_n\}$ it is possible to construct a computable function $\phi : (\mathbb{Q}^+)^2 \to \mathbb{N}$ such that for any $\varepsilon, \lambda \in \mathbb{Q}^+$ and sequences $a_0 < b_0 \leq a_1 < b_1 \leq \dots$ of natural numbers, there exists some $n \leq \phi(\varepsilon, \lambda)$ such that

$$\mathbb{P}(\exists i, j \in [a_n; b_n] \ (|X_i - X_j| \geq \varepsilon)) < \lambda,$$

or in other words, the probability of an ε-fluctuation occurring in the interval $[a_n; b_n]$ is very low. However, there are many convergent stochastic processes and algorithms that do not posses natural upcrossing inequalities in mean, and the question then arises: What kind of computational information can we hope to provide in those cases, i.e. what mathematically natural notion populates the bottom of an analogous hierarchy of quantitative effective data that one can associate to stochastic processes?

In this paper, we answer that question by introducing a new concept of computation for probabilistic formulas, which we call *dependent learnability.* This is defined abstractly in terms of measurable sets, but applies in particular to the convergence of stochastic processes, where we obtain a strict generalization of uniform learnability. We prove that dependent learnability is in fact equivalent to the natural lifting(s) of the metastability property in the probabilistic setting, thereby demonstrating that rates of dependent learnability can be extracted from rather general proofs (that make use of full classical logic and comprehension principles), and thus in principle can be found for any stochastic process whose convergence is provable in a theory amenable to proof mining.

We then provide a "joining theorem", which sets out a primitive recursive algorithm for combining a finite number of rates of dependent learnability. Informally speaking, this allows us to combine individual quantitative bounds for properties $A_1, \ldots, A_n$ into bounds for their conjunction $A_1 \wedge \ldots \wedge A_n$, which in the concrete setting of convergence enables us, for example, to combine quantitative fluctuation information about two individual stochastic processes into fluctuation information for their sum or product. In the simple case of uniform learnability this recently proved crucial for obtaining rates for almost supermartingales [19], and such combinations naturally feature in many other proof mining case studies. Our construction is extremely general and is intended to form a useful tool for future case studies of proof mining in probability, in addition to being of interest in its own right.

Finally, we discuss the logical and semantic aspects of our work. We show that dependent learnability has an elegant reading in terms of game semantics, as a winning strategy in a two-player probabilistic game. This interpretation extends to our joining constructions, highlighting that dependent learnability is a mathematically natural and intuitive concept, and providing insight into programs extracted from proofs in probability.

All of our main definitions and results can be formulated within the logical framework for probability developed by the first two authors [18]. In particular, we only make use of finite unions and intersections of events, giving a logical explanation of the uniformities underlying our constructions.

2 Fluctuations and Generalized Learnability

In this section, we now survey the well-known notion of fluctuations for sequences of real numbers as well as the related notion of a learnable rate of convergence, which we take as motivation to introduce our new generalized notion of learnability, both in a deterministic and a probabilistic variant, which are then compared to other such computational interpretations of (probabilistic) principles.

2.1 The Deterministic Setting

A sequence of real numbers $\{x_n\}$ is said to be *(Cauchy) convergent* if

$$\forall \varepsilon > 0\, \exists n \in \mathbb{N}\, \forall i, j \geq n\, (|x_i - x_j| < \varepsilon).$$

Results establishing the convergence of sequences of real numbers are fundamental to analysis, and obtaining computational information on the asymptotic behavior of such sequences and their convergence from such theorems is of great interest and importance. The most direct computational interpretation one can give here is a *rate of convergence*, which is a function $r : \mathbb{Q}^+ \to \mathbb{N}$ such that:

$$\forall \varepsilon \in \mathbb{Q}^+\, \forall i, j \geq r(\varepsilon)\, (|x_i - x_j| < \varepsilon).$$

However, it is well known that there are many convergence results for which computable rates do not exist. Thus, one is tasked to seek other computational interpretations for the convergence of sequences.

A natural interpretation which has found favour amongst proof-theorists [2, 15,20] as well as probability theorists [5,11,12] is that of a bound on the number of fluctuations, namely a function $\phi : \mathbb{Q}^+ \to \mathbb{N}$ such that for any $\varepsilon \in \mathbb{Q}^+$ and any $i_0 < j_0 \leq i_1 < j_1 \leq \ldots \leq i_{k-1} < j_{k-1}$ with $|x_{i_n} - x_{j_n}| \geq \varepsilon$ for all $n = 0, \ldots, k-1$, it holds that $k \leq \phi(\varepsilon)$.

Fluctuation bounds offer an intuitive computational interpretation for many theorems for which one cannot obtain a direct computable rate of convergence. The prime example here is the case of a monotone sequence $\{x_n\} \subseteq [0, 1]$, where one can easily show that a fluctuation bound is given by $\phi(\varepsilon) := \lceil 1/\varepsilon \rceil$, but following the fundamental results of Specker [23], one can construct a computable monotone sequence of rational numbers in $[0, 1]$ with no computable rate of convergence. In the stochastic setting, it is supermartingales and related classes of process that represent the canonical examples of sequences with controlled fluctuation behaviour, as we will discuss further in Sect. 2.2 below.

We now present fluctuations formally, lifting them from the specific setting of convergent sequences and defining them on arbitrary $\exists\forall$-formulas. We then show that the closely related concept of *learnability* represents an alternative way of capturing bounded fluctuations.

Definition 1 (Fluctuations for formulas). *For an arbitrary formula $A(i, j)$*[1] *on pairs of natural numbers, and $N \in \mathbb{N}$, we define $J_{N,A}$ to be the maximal $k \in \mathbb{N}$ for which there exist*

$$a_0 < b_0 \leq a_1 < b_1 \leq \ldots \leq a_{k-1} < b_{k-1} < N$$

[1] These formulas could represent formal statements in an axiomatic theory, but for the purpose of this paper we just consider them as arbitrary properties of pairs of natural numbers, i.e. as subsets of $\mathbb{N} \times \mathbb{N}$.

such that for all $n = 0, \ldots, k-1$, *there exist*[2] $i, j \in [a_n; b_n]$ *such that* $A(i,j)$ *holds. We write*

$$J_{\infty,A} := \lim_{N\to\infty} J_{N,A},$$

where this quantity could be infinite. We say that the fluctuations are bounded if $J_{\infty,A} \leq e$ *for some* $e \in \mathbb{N}$, *which we call a fluctuation bound.*

Here we imagine $A(i,j)$ as representing a "fluctuation" from i to j. While we suppress it in this paper, the formula A could naturally depend on outside parameters. In that vein, we recover the previous notion of a sequence of real numbers $\{x_n\}$ having bounded fluctuations: Setting $A_\varepsilon(i,j) :\equiv |x_i - x_j| \geq \varepsilon$ for $\varepsilon \in \mathbb{Q}^+$, a function ϕ is a fluctuation bound for $\{x_n\}$ if, and only if, we have

$$J_{\infty,A_\varepsilon} \leq \phi(\varepsilon)$$

for all $\varepsilon \in \mathbb{Q}^+$. An alternative and particularly intuitive way of presenting fluctuation bounds (and so in particular bounds on the number of ε-fluctuations of the sequence $\{x_n\}$), is the following, which was first isolated in this context in [19]:

Definition 2 (Learnability of formulas). *A formula* $\exists n \forall i, j \geq n\, A(i,j)$ *is learnable if there exists a number* $e \in \mathbb{N}$ *such that for all sequences of natural numbers* a, b *with* $a_0 < b_0 \leq a_1 < b_1 \leq \ldots$, *we have:*

$$\exists n \leq e\ \forall i, j \in [a_n; b_n]\ A(i,j).$$

The number e *is called a* learnability bound *for* A.

The intuition behind the term "learnability" is that $a_0 < a_1 < \ldots$ represent guesses of points for which $\forall i, j \geq a_n\, A(i,j)$, with $b_0 < b_1 < \ldots$ counterexamples to those guesses. Whenever a guess fails, we learn from that guess and move on to the next. A learnability bound represents the maximum number of mind-changes required before a guess works. This semantic meaning is discussed in greater depth in Sect. 4, but for now we simply note that learnability represents a simple and intuitive alternative to the property of the (negated) formula having finite fluctuations, as it can be easily shown that

$$e \text{ is a learnability bound for } A \iff J_{\infty,\neg A} \leq e.$$

While fluctuations bounds allow one to give a computational interpretation to results where one cannot obtain direct rates, there are still cases of computable converging sequences for which one cannot obtain an effective fluctuation bound (see e.g. [15]). In such cases, inspired by the no counterexample interpretation of Kreisel [16,17], one arrives at the computationally weaker concept of a *rate of metastability*, also discussed in more detail later on, the existence of which can be broadly substantiated through the perspective of proof mining, already for

[2] Here, and in the following, we write $[k; l] := [k, l] \cap \mathbb{N}$.

large classes of non-effective proofs in analysis (see also the discussion in Sect. 2.4 later on).

The following definition now offers a new computational interpretation for the convergence of sequences which do not possess effective learnable rates of convergence, which is mathematically different to the notion of metastability (albeit, as we will later show, being computationally and proof-theoretically equivalent) and in a way more closely represents this paradigm of learnability.

Definition 3 (Dependent learnability of formulas). *A formula* $\exists n \forall i, j \geq n\, A(i,j)$ *is dependently learnable if for all sequences of natural numbers* a, b *with* $a_0 < b_0 \leq a_1 < b_1 \leq \dots$, *we have:*

$$\exists n \in \mathbb{N}\ \forall i, j \in [a_n; b_n]\ A(i,j).$$

A function $e : \mathbb{N}^{\mathbb{N}} \times \mathbb{N}^{\mathbb{N}} \to \mathbb{N}$ *which bounds* $n \leq e(a,b)$ *in terms of* a, b *is called a* dependently learnable rate *for* A.

In the special case that e is a constant function, it is a learnability bound for A or equivalently a bound for $J_{\infty,\neg A}$ as before, and therefore we can view dependent learnability as a generalization of the property of having bounded fluctuations. Rephrased in terms of fluctuations, a formula is dependently learnable if, and only if, for any fixed sequences $a_0 < b_0 \leq a_1 < b_1 \leq \dots$, the length of fluctuation sequences with the property that each fluctuation is contained within a unique interval $[a_n; b_n]$ is bounded.

2.2 The Stochastic Setting

We now arrive at the first main contribution of the paper: A concept of generalized learnability in the stochastic setting, which in particular represents a broader notion of "good fluctuation behaviour" for stochastic processes.

Everything that follows now takes place over an arbitrary probability space $(\Omega, \mathcal{F}, \mathbb{P})$. While not explicitly highlighted any further in this paper, all of the main definitions and theorems that follow will only involve finite unions and intersections of events. This means that technically we can restrict $\mathbb{P}$ to being a probability *content* (also called a charge) [3], that is a finitely additive function $\mathbb{P} : \mathcal{F} \to [0,1]$ over some algebra $\mathcal{F}$. This then explains our results in light of the recent logical metatheorem for probability [18].

In any case, we say that a logical formula $\varphi(\omega, x_1, \dots, x_n)$, with parameters $x_1, \dots, x_n$ and ω a variable taking values in Ω, is *measurable* if for all parameters $x_1, \dots, x_n$, we have $\varphi(x_1, \dots, x_n) := \{\omega \in \Omega : \varphi(\omega, x_1, \dots, x_n)\} \in \mathcal{F}$. If $\varphi(n)$ is a measurable formula, with $n \in \mathbb{N}$, and $p, q \in \mathbb{N}$ we define $\neg\varphi := \varphi^c$ as well as

$$\exists n \in [p;q]\ \varphi(n) := \bigcup_{n \in [p;q]} \varphi(n) \quad \text{and} \quad \forall n \in [p;q]\ \varphi(n) := \bigcap_{n \in [p;q]} \varphi(n).$$

We now come straight to our definition of generalized stochastic learnability. As with many concepts from probability, a single definition in the nonstochastic setting (in this case Definition 3) splits into multiple stochastic notions, in this case *uniform* and *pointwise*, which we justify and explain in more detail below.

Definition 4 (Dependent learnability of measurable formulas). *Let $A(i,j)$ be a measurable formula. A function $e : \mathbb{Q}^+ \times \mathbb{N}^\mathbb{N} \times \mathbb{N}^\mathbb{N} \to \mathbb{N}$ is*

(a) a uniform dependent learnable rate for $\exists n \forall i,j \geq n\, A(i,j)$ if

$$\exists n \leq e(\lambda, a, b)\, (\mathbb{P}(\exists i,j \in [a_n; b_n]\, \neg A(i,j)) < \lambda)$$

for any $\lambda \in \mathbb{Q}^+$ and $a_0 < b_0 \leq a_1 < b_1 \leq \ldots$ (and if e is independent of the sequences a, b then it is a uniform learnable rate for A),

(b) a pointwise dependent learnable rate for $\exists n \forall i,j \geq n\, A(i,j)$ if

$$\mathbb{P}(\forall n \leq e(\lambda, a, b)\ \exists i,j \in [a_n; b_n]\, \neg A(i,j)) < \lambda$$

for any $\lambda \in \mathbb{Q}^+$ and $a_0 < b_0 \leq a_1 < b_1 \leq \ldots$ (and if e is independent of the sequences a, b then it is a pointwise learnable rate for A).

The simpler notions of pointwise and uniform learnability for arbitrary measurable formulas, which represent stochastic versions of Definition 2, were introduced in [20], and these were in turn inspired by the quantitative formulations of pointwise and uniform convergence of measurable functions in [1].

Analogously to the nonstochastic case, also pointwise and uniform learnability are intimately connected to the previously introduced notion of fluctuations. More precisely, for a measurable formula A, define $J_{\infty,A}$ just as in Definition 1, where it now becomes a random variable. In [20, Theorem 3.6 (i)], it is shown that if $\mathbb{E}[J_{\infty,\neg A}] < K$, then $e(\lambda) := \lceil K/\lambda \rceil$ is a uniform learnable rate for $\exists n \forall i,j \geq n\, A(i,j)$, and in the case of stochastic processes, uniform learnable rates exist generally when those processes enjoy L_1-bounds on fluctuations or upcrossings (thus L_1-bounded sup- or supermartingales can be given uniform learnable rates cf. [20, Theorem 7.4]). Similarly, it is shown in [20, Theorem 3.6 (ii)] that if e is a rate of convergence for

$$\lim_{n \to \infty} \mathbb{P}\left(J_{\infty,\neg A} \geq n\right) = 0,$$

then e is also a pointwise learnable rate for $\exists n \forall i,j \geq n\, A(i,j)$.

However, not all convergent stochastic processes enjoy the kind of strong quantitative fluctuation behaviour from which we can expect to obtain simple pointwise or uniform rates, in the same way that not all deterministic sequences posses computable learnability rates. We now show that dependent learnability is a much broader notion.

2.3 The Computational Strength of Dependent Learnability

We prove that having a uniform/pointwise dependent learnable rate for A is *computationally* equivalent to having a uniform/pointwise *rate of metastability* for the statement that $\exists n \forall i,j \geq n\, A(i,j)$ holds almost surely. We thereby demonstrate that dependent learnable rates can be in principle extracted from general convergence proofs, and thus represent a kind of quantitative fluctuation behaviour that is enjoyed by *any* stochastic processes that converge almost surely, provably so within a system amenable to proof mining.

Metastability for stochastic processes was introduced (in the context of logic) by Avigad et al. in [1], and adapted to arbitrary $\exists\forall$-formulas in [20]. We briefly recall the key definitions as they relate to the notions presented here. We first observe that $\exists n \forall i, j \geq n\, A(i,j)$ holding with probability one can be equivalently formulated in terms of finite unions and intersections as

$$\forall \lambda \in \mathbb{Q}^+ \; \exists n \in \mathbb{N} \; \forall m \in \mathbb{N} \, (\mathbb{P}\,(\exists i, j \in [n; n+m] \; \neg A(i,j)) < \lambda) \,. \qquad (*)$$

As recently confirmed by the results of [18], in many situations we can reasonably expect that even when a proof of the above statement is nonconstructive (and so the above n cannot be computably witnessed), we can still obtain a witness for the monotone Dialectica interpretation of

$$\forall \lambda \in \mathbb{Q}^+ \; \neg\neg\exists n \in \mathbb{N} \; \forall m \in \mathbb{N} \, (\mathbb{P}\,(\exists i, j \in [n; n+m] \; \neg A(i,j)) < \lambda) \qquad (**)$$

which is a functional $\varphi : \mathbb{Q}^+ \times (\mathbb{N} \to \mathbb{N}) \to \mathbb{N}$ satisfying

$$\forall \lambda \in \mathbb{Q}^+, g : \mathbb{N} \to \mathbb{N} \, \exists n \leq \varphi(\lambda, g) \, (\mathbb{P}\,(\exists i, j \in [n; n+g(n)] \neg A(i,j)) < \lambda) \,.$$

We call such a functional a *rate of uniform metastability* for $\exists n \forall i, j \geq n\, A(i,j)$. In many cases, notably when analysing convergence proofs that work in a pointwise way on elements ω of the underlying probability space, it is considerably easier to instead produce a functional φ satisfying the following apparently weaker condition

$$\forall \lambda \in \mathbb{Q}^+ \; \forall g : \mathbb{N} \to \mathbb{N} \, (\mathbb{P}\,(\forall n \leq \varphi(\lambda, g) \; \exists i, j \in [n; n+g(n)] \; \neg A(i,j)) < \lambda)$$

and we label such a functional a *rate of pointwise metastablity* for $\exists n \forall i, j \geq n\, A(i,j)$. It is immediate that a rate of uniform metastability is also a pointwise rate, and in fact, the two are computationally equivalent, where (as made explicit in [20, Theorem 3.2]) the route from the latter to the former uses a computational version of Egorov's theorem given in [1] and inspired by a construction of Tao [24, Theorem A.2], though with the apparent cost of a significant blowup in complexity (where the construction in [1] uses bar recursion).

In the main result of this subsection, we demonstrate that one can easily translate between uniform (resp. pointwise) dependent learnability and uniform (resp. pointwise) metastability, using the same construction in both cases. In particular, this means that the complexity difference between pointwise and uniform dependent learnability matches that between the two corresponding forms of metastability. This generalises a similar correspondence between nondependent learnability and metastability proven recently as [20, Lemma 3.5].

Theorem 1. *Suppose $A(i,j)$ is a measurable formula and e is a pointwise (resp. uniform) dependent learnable rate for $\exists n \forall i, j \geq n\, A(i,j)$. Then*

$$\varphi(\lambda, g) := a^g_{e(\lambda, a^g, b^g)} = \tilde{g}^{(e(\lambda, a^g, b^g))}(0)$$

is a rate of pointwise (resp. uniform) metastability for $\exists n \forall i, j \geq n\, A(i,j)$*, with* a^g, b^g *defined by* $a^g_i := \tilde{g}^{(i)}(0)$ *and* $b^g_i := \tilde{g}^{(i+1)}(0)$*, where* $\tilde{g}(n) := n + g(n) + 1$*. Conversely, if* φ *is a rate of pointwise (resp. uniform) metastability for* $\exists n \forall i, j \geq n\, A(i,j)$*. Then*

$$e(\lambda, a, b) := k_{a,b}(\varphi(\lambda, g_{a,b}))$$

is a rate of pointwise (resp. uniform) dependent learnability for $\exists n \forall i, j \geq n\, A(i,j)$*, with* $k_{a,b}, g_{a,b}$ *defined by* $g_{a,b}(n) := b_{k_{a,b}(n)} - n$ *and* $k_{a,b}(n) := \min\{i : n \leq a_i\}$*.*

Proof. We only consider the case of pointwise rates, the case of uniform rates follows analogously. To see the first part, assume that e is a pointwise dependent learnable rate for $\exists n \forall i, j \geq n\, A(i,j)$ and let $\lambda > 0$ and $g : \mathbb{N} \to \mathbb{N}$ be fixed. It is easy to see that a^g and b^g satisfy $a^g_0 < b^g_0 \leq a^g_1 < b^g_1 \leq \dots$. We therefore have

$$\mathbb{P}(\forall i \leq e(\lambda, a^g, b^g)\, \exists k, l \in [a^g_i; b^g_i]\, \neg A(k,l)) < \lambda.$$

Let $\omega \in \Omega$ be such that

$$\exists i \leq e(\lambda, a^g, b^g)\, \forall k, l \in [a^g_i; b^g_i]\, A(\omega, k, l).$$

We then have $[a^g_i; b^g_i] = [a^g_i; \tilde{g}(a^g_i)] = [a^g_i; a^g_i + g(a^g_i) + 1] \supset [a^g_i; a^g_i + g(a^g_i)]$ and as a^g is monotone increasing, we have $a^g_i \leq a^g_{e(\lambda, a^g, b^g)} = \varphi(\lambda, g)$ so that we have shown

$$\exists n \leq \varphi(\lambda, g)\, \forall k, l \in [n; n + g(n)]\, A(\omega, k, l).$$

As ω was arbitrary, we in particular have

$$\begin{aligned} &\mathbb{P}(\forall n \leq \varphi(\lambda, g)\, \exists k, l \in [n; n + g(n)]\, \neg A(k,l)) \\ &\quad \leq \mathbb{P}(\forall i \leq e(\lambda, a^g, b^g)\, \exists k, l \in [a^g_i; b^g_i]\, \neg A(k,l)) < \lambda. \end{aligned}$$

To see the second part, assume that φ is a rate of pointwise metastability for $\exists n \forall i, j \geq n\, A(i,j)$ and let $\lambda > 0$ as well as a, b with $a_0 < b_0 \leq a_1 < b_1 \leq \dots$ be fixed. Note first that $k_{a,b}$ and $g_{a,b}$ are well-defined since a is strictly increasing and $n \leq a_{k_{a,b}(n)} < b_{k_{a,b}(n)}$, and thus in particular we have

$$\mathbb{P}(\forall n \leq \varphi(\lambda, g_{a,b})\, \exists k, l \in [n; n + g_{a,b}(n)]\, \neg A(k,l)) < \lambda.$$

Let $\omega \in \Omega$ be such that

$$\exists n \leq \varphi(\lambda, g_{a,b})\, \forall k, l \in [n; n + g_{a,b}(n)]\, A(\omega, k, l).$$

By definition we have $n \leq a_{k_{a,b}(n)}$ and $b_{k_{a,b}(n)} = n + g_{a,b}(n)$ and therefore $A(\omega, k, l)$ holds for all $k, l \in [a_{k_{a,b}(n)}; b_{k_{a,b}(n)}]$. As $k_{a,b}$ is monotone, we have $k_{a,b}(n) \leq k_{a,b}(\varphi(\lambda, g_{a,b})) = e(\lambda, a, b)$ so that we have shown

$$\exists i \leq e(\lambda, a, b)\, \forall k, l \in [a_i; b_i]\, A(\omega, k, l).$$

As ω was arbitrary, we again have

$$\begin{aligned} &\mathbb{P}(\forall i \leq e(\lambda, a, b)\, \exists k, l \in [a_i; b_i]\, \neg A(k,l)) \\ &\quad \leq \mathbb{P}(\forall n \leq \varphi(\lambda, g_{a,b})\, \exists k, l \in [n; n + g_{a,b}(n)]\, \neg A(k,l)) < \lambda. \end{aligned}$$

This completes the proof.

Considering the one-point space immediately allows us to derive an analogous result for the deterministic case, showing the equivalence for dependent learnability as discussed in Definition 3 and rates of metastability for formulas $\exists n \forall i, j \geq n\, A(i,j)$ in the form of functionals $\varphi : (\mathbb{N} \to \mathbb{N}) \to \mathbb{N}$ satisfying

$$\exists n \leq \varphi(g)\ \forall i, j \in [n; n + g(n)]\ A(i,j)$$

for all $g : \mathbb{N} \to \mathbb{N}$. More specifically, in one direction, a rate of metastability for $\exists n \forall i, j \geq n\, A(i,j)$ is given by $\varphi(g) := a^g_{e(a^g, b^g)}$, and conversely a rate of dependent learnability by $k_{a,b}(\varphi(g_{a,b}))$. Remaining this time in the stochastic setting but restricting our attention to *nondependent* learnability, a similar and simpler variant of the proof of Theorem 1 was given in [20] to show that $e : \mathbb{Q}^+ \to \mathbb{N}$ is a pointwise (resp. uniform) learnable rate for $\exists n \forall i, j \geq n\, A(i,j)$ if and only if $\varphi(\lambda, g) := \tilde{g}^{(e(\lambda))}(0)$ is a rate of pointwise (resp. uniform) metastability. By considering the one-point space as before, one regains a deterministic correspondence result from this which was previously obtained in [15].

2.4 Dependent Learnability and Program Extraction

We conclude by making more precise the significance of dependent learnability from the perspective of program extraction, particularly in light of the recent logical approach to probability in [18]. Almost sure convergence of a stochastic process $\{X_n\}$ is equivalent to the following property (analogous to $(*)$ above):

$$\forall \varepsilon, \lambda > 0\ \exists n \in \mathbb{N}\ \forall m \in \mathbb{N}\, (\mathbb{P}\, (\exists i, j \in [n; n + m]\ \ (|X_i - X_j| \geq \varepsilon)) < \lambda)\,.$$

As shown in [18, Section 9], one can formally represent this property in a logical system for reasoning about probability contents, such that the usual program extraction mechanism of proof mining (arising typically by a combination of a negative translation with a monotone variant of Gödel's functional interpretation, see [13] for more details) guarantees the existence of a computable rate of uniform metastability for the convergence property, whenever the above is provable in that system. Theorem 1 then guarantees the existence of a computable uniform dependent learnable rate. A similar argument holds in the pointwise setting.

As such, unlike for ordinary learnable rates, which require strong assumptions on the fluctuation behaviour of stochastic processes, the existence of computable rates of dependent learnability relies only on a stochastic process being provably convergent in a system amenable to proof mining. The existence of a core formal system for probability theory [18] along with empirical evidence from recent case studies [19,20] suggest that dependent learnable rates should therefore be broadly extractible from convergence proofs in core probability theory.

3 Joining Learnable Rates

We now move on to constructions on rates of dependent learnability, demonstrating how two dependent rates for formulas $\exists n \forall i, j \geq n\, A_1(i,j)$ and $\exists n \forall i, j \geq$

$n\, A_2(i,j)$ respectively can be combined to obtain a dependent rate for $\exists n \forall i,j \geq n\ (A_1(i,j) \wedge A_2(i,j))$ (and hence iteratively combined to produce a dependent rate for finite conjunctions).

Constructions of this kind are crucial in proof mining. In the deterministic setting, the joining of rates of metastability are discussed in abstract terms in [21], where it is connected to the finite double negation shift, with a concrete example of the deterministic joining presented in detail in e.g. [22], among many other examples. In the stochastic setting, the joining of uniform learnable rates was central in the recent analysis of the Robbins-Siegmund theorem [19], where multiple instances were required to arrive at a uniform rate of learnability for almost-supermartingales.

We now present a series of joining results covering all notions of learnability introduced in this paper. For a sequence of natural numbers a and $n \in \mathbb{N}$, let us write $a^{(n)} := (a_i^{(n)})$ for the sequence defined by $a_i^{(n)} := a_{n+i}$ for all $i \in \mathbb{N}$.

Theorem 2. *Suppose $A_1(i,j), A_2(i,j)$ are measurable formulas and let e_1 and e_2, respectively, be pointwise dependent learnable rates for $\exists n \forall i,j \geq n\, A_1(i,j)$ and $\exists n \forall i,j \geq n\, A_2(i,j)$. Then $\exists n \forall i,j \geq n\ (A_1(i,j) \wedge A_2(i,j))$ has a pointwise dependent learnable rate given by*

$$e(\lambda, a, b) := p_{2e_1(\lambda/2, a', b')+1}$$

where a', b' are defined by $a'_i := a_{p_{2i}}$, $b'_i := b_{p_{2i+1}}$ for all $i \in \mathbb{N}$, and with (p_i) defined inductively by

$$p_0 := 0 \text{ and } p_{i+1} := p_i + e_2(\lambda/2^{i+1}, a^{(p_i)}, b^{(p_i)}) + 1.$$

Proof. Suppose a, b are sequences of natural numbers satisfying $a_0 < b_0 \leq a_1 < b_1 \leq \dots$ and let $\lambda \in \mathbb{Q}^+$ be given. Set

$$Q := \{\omega \in \Omega : \exists k \leq e_1(\lambda/2, a', b')\ \forall i,j \in [a'_k; b'_k]\ A_1(\omega, i, j)\}.$$

From the fact that e_1 is a pointwise dependent learnable rate for $\exists n \forall i,j \geq n\, A_1(i,j)$, we have $\mathbb{P}(\neg Q) < \lambda/2$. For this, note in particular that a', b' satisfy

$$a'_i = a_{p_{2i}} < b_{p_{2i}} < b_{p_{2i+1}} = b'_i \leq a_{p_{2i+2}} = a'_{i+1},$$

for any $i \in \mathbb{N}$ as we have $p_i < p_{i+1}$ for any $i \in \mathbb{N}$. Further, for each $k \in \mathbb{N}$, set

$$R_k := \{\omega \in \Omega : \exists l \leq e_2(\lambda/2^{k+1}, a^{(p_k)}, b^{(p_k)})\ \forall i,j \in [a_l^{(p_k)}; b_l^{(p_k)}]\ A_2(\omega, i, j)\}.$$

From the fact that e_2 is a pointwise dependent learnable rate for $\exists n \forall i,j \geq n\, A_2(i,j)$, we get $\mathbb{P}(\neg R_k) < \lambda/2^{k+1}$ where, for this, we note that $a_0^{(p_k)} < b_0^{(p_k)} \leq a_1^{(p_k)} < b_1^{(p_k)} \leq \dots$ holds for any $k \in \mathbb{N}$.

Now, let

$$\omega \in P := Q \cap \bigcap_{k \leq e_1(\lambda/2, a', b')} R_{2k}.$$

As $\omega \in Q$, we can take $k \leq e_1(\lambda/2, a', b')$ such that

$$\forall i, j \in [a'_k; b'_k]\ A_1(\omega, i, j) \equiv \forall i, j \in [a_{p_{2k}}; b_{p_{2k+1}}]\ A_1(\omega, i, j).$$

As also $\omega \in R_{2k}$, we can take $l \leq e_2(\lambda/2^{2k+1}, a^{(p_{2k})}, b^{(p_{2k})})$ such that

$$\forall i, j \in [a_l^{(p_{2k})}; b_l^{(p_{2k})}]\ A_2(\omega, i, j) \equiv \forall i, j \in [a_{p_{2k}+l}; b_{p_{2k}+l}]\ A_2(\omega, i, j).$$

Now we have $a_{p_{2k}} \leq a_{p_{2k}+l}$ and $b_{p_{2k}+l} \leq b_{p_{2k}+e_2(\lambda/2^{2k+1}, a^{(p_{2k})}, b^{(p_{2k})})+1} = b_{p_{2k+1}}$. Thus, we also have $A_1(\omega, i, j)$ for all $i, j \in [a_{p_{2k}+l}, b_{p_{2k}+l}]$ and so combined with the above we have

$$\forall i, j \in [a_{p_{2k}+l}; b_{p_{2k}+l}]\ (A_1(\omega, i, j) \wedge A_2(\omega, i, j)).$$

As furthermore

$$p_{2k}+l \leq p_{2k}+e_2(\lambda/2^{2k+1}, a^{(p_{2k})}, b^{(p_{2k})})+1 = p_{2k+1} \leq p_{2e_1(\lambda/2, a', b')+1} = e(\lambda, a, b),$$

we have thereby shown that

$$P \subseteq \{\omega \in \Omega : \exists n \leq e(\lambda, a, b)\ \forall i, j \in [a_n; b_n]\ (A_1(\omega, i, j) \wedge A_2(\omega, i, j))\}.$$

Thus we conclude that

$$\begin{aligned} &\mathbb{P}(\forall n \leq e(\lambda, a, b)\ \exists i, j \in [a_n; b_n]\ (\neg A_1(i, j) \vee \neg A_2(i, j))) \\ &\quad \leq \mathbb{P}(\neg Q) + \mathbb{P}\left(\bigcup_{i \leq e_1(\lambda/2, a', b')} \neg R_{2i}\right) < \frac{\lambda}{2} + \sum_{i=0}^{\infty} \mathbb{P}(\neg R_{2i}) < \lambda. \end{aligned}$$

By considering the one-point probability content space, an immediate consequence of the above is the following result in the deterministic case:

Theorem 3. *Suppose $A_1(i, j), A_2(i, j)$ are formulas and let e_1 and e_2, respectively, be dependent learnable rates for $\exists n \forall i, j \geq n\, A_1(i, j)$ and $\exists n \forall i, j \geq n\, A_2(i, j)$. Then $\exists n \forall i, j \geq n\ (A_1(i, j) \wedge A_2(i, j))$ has a dependent learnable rate given by*

$$e(a, b) := p_{2e_1(a', b')+1}$$

where a', b' are defined by $a'_i := a_{p_{2i}}$, $b'_i := b_{p_{2i+1}}$ for all $i \in \mathbb{N}$, and with (p_i) defined inductively by

$$p_0 := 0 \text{ and } p_{i+1} := p_i + e_2(a^{(p_i)}, b^{(p_i)}) + 1.$$

The case for uniform dependent learnable rates can be obtained in a similar manner to the pointwise case, in which case however some simplifications apply. We collect this in the following theorem, the proof of which is a straightforward adaptation of that of Theorem 2.

Theorem 4. *Suppose $A_1(i,j), A_2(i,j)$ are measurable formulas and let e_1 and e_2, respectively, be uniform dependent learnable rates for $\exists n \forall i,j \geq n\, A_1(i,j)$ and $\exists n \forall i,j \geq n\, A_2(i,j)$. Then $\exists n \forall i,j \geq n\, (A_1(i,j) \wedge A_2(i,j))$ has a uniform dependent learnable rate given by*

$$e(\lambda, a, b) := p_{2e_1(\lambda/2, a', b')+1}$$

where a', b' are defined by $a'_i := a_{p_{2i}}$, $b'_i := b_{p_{2i+1}}$ for all $i \in \mathbb{N}$, and with (p_i) defined inductively by

$$p_0 := 0 \text{ and } p_{i+1} := p_i + e_2(\lambda/2, a^{(p_i)}, b^{(p_i)}) + 1.$$

In the case of pointwise learnable rates in the ordinary sense, one can also drastically simplify the above bounds. Again, the proof is similar to that of Theorem 2.

Theorem 5. *Suppose $A_1(i,j), A_2(i,j)$ are measurable formulas and let e_1 and e_2, respectively, be pointwise learnable rates for $\exists n \forall i,j \geq n\, A_1(i,j)$ and $\exists n \forall i,j \geq n\, A_2(i,j)$. Then $\exists n \forall i,j \geq n\, (A_1(i,j) \wedge A_2(i,j))$ has a pointwise learnable rate given by*

$$e(\lambda) := \left(e_2\left(\frac{\lambda}{2(e_1(\lambda/2)+1)}\right) + 1\right)(2e_1(\lambda/2) + 1).$$

The last of the further simplifications that we want to highlight is the case of uniform learnable rates in the ordinary sense where the above bound just collapses to a sum. As this follows rather immediately using arguments similar to [19, Lemma 2.10] and [20, Lemma 7.3], we omit the proof.

Theorem 6. *Suppose $A_1(i,j), A_2(i,j)$ are measurable formulas and let e_1 and e_2, respectively, be uniform learnable rates for $\exists n \forall i,j \geq n\, A_1(i,j)$ and $\exists n \forall i,j \geq n\, A_2(i,j)$. Then $\exists n \forall i,j \geq n\, (A_1(i,j) \wedge A_2(i,j))$ has a uniform learnable rate given by*

$$e(\lambda, a, b) := e_1(\lambda/2) + e_2(\lambda/2).$$

4 Our Results in the Light of Game Semantics

Another benefit of our various notions of learnability is that they have an intuitive reading in terms of game semantics. This was already alluded to in Sect. 2.1 above, and this in short section, we extend this intuition to our new notion of stochastic learnability and our joining constructions.

It has long been known that one can view the computational content of classical reasoning in terms of winning strategies in backtracking games, an idea that is already implicit Hilbert's epsilon calculus, and has been explored in numerous different settings since. A particularly readable account is given by Coquand in [6], where an explicit connection is made between games and double negative translations (as represented in the present paper by (∗∗) above), also in the context of a simple "joining" operation (cf. [6, Section 3]). While we do not go into details of the general situation here, we discuss how both our notions of learning and the stochastic joining constructions can be connected back to game semantics in the sense of [6].

For all of our main definitions of learnability, we now interpret $a_0 < b_0 \leq a_1 < b_1 \leq \dots$ as a dialogue between two players (E) and (A), where $a_0 < a_1 < \dots$ represent attempts by (E) to find a number a such that $\forall i, j \geq a\, A(i, j)$, while $b_0 < b_1 < \dots$ represent counter-moves by (A) to refute these attempts, i.e. for each a_n find some b_n such that $\exists i, j \in [a_n; b_n] \neg A(i, j)$. In this sense, a_n represents a winning move whenever $\forall i, j \in [a_n; b_n]\, A(i, j)$. With these intuitions in mind, all of our learnability bounds represents the maximum number of mind changes that (E) might need to make before finding a winning move, with slight differences in meaning as set out below:

- **Pointwise dependent:** For any λ and dialogue a, b, the probability that (E) wins with at most $e(\lambda, a, b)$ mind changes is $> 1 - \lambda$.
- **Uniform dependent:** For any λ and dialogue a, b, with at most $e(\lambda, a, b)$ mind changes (E) finds a move that wins with probability $> 1 - \lambda$.

In the nonstochastic case both variants collapse to the same notion, namely that (E) wins with at most $e(a, b)$ mind changes, and whenever $e(\lambda, a, b)$ is independent of the dialogue, we obtain the previously studied notions of pointwise/uniform learnability.

More interestingly, the main constructions of Sect. 3 can be understood as constructions on games. For example, informally speaking, in Theorem 2, we decompose the dialogue into a sequence of consecutive finite subdialogues marked by the points (p_i), where within each subdialogue the probability that (E) finds a winning move for A_2 is $> 1 - \lambda/2^{i+1}$. The endpoints (p_i) are then used to define a sparser dialogue, for which we can provide a bound such that the probability that (E) finds a winning move for A_1 on this sparse dialogue is $> 1 - \lambda/2$. The proof of Theorem 2 makes formal the simple idea that if (E) fails to find a winning move for $A_1 \wedge A_2$ within the stated number of mind changes, then this represents a failure either in the sparse dialogue for A_1, or one of the subdialogues for A_2, the total probability of which is bounded by $\lambda/2 + \sum_{i=0}^{\infty} \lambda/2^{i+1} = \lambda$.

The other constructions in Sect. 3 can be interpreted in a similar way, where the situation for uniform rates (Theorem 6) is notable in that the construction is *symmetric* in A_1 and A_2, which as discussed in [6, Section 3.2] is not generally the case for strategies arising from double negative translations.

5 Concluding Remarks

The main achievements of this paper have been a new notion of learnability in the stochastic setting, which is general enough that we can expect computable bounds to be extractable for any convergent stochastic process for which there exists a convergence proof that lies within the scope of techniques from proof mining, along with a series of joining results which we expect to be crucial for future applications of proof theory in probability.

Several interesting questions directly leading off from the work in this paper present themselves: Are the constructions given in Sect. 3 optimal? This is particularly relevant given that, in practice, the joining constructions would likely be iterated to form a finite joining operation. Can we flesh out our hierarchy of quantitative notions for stochastic convergence with further levels? For example, the hierarchy for deterministic convergence in [15] has a more complex notion of learnability that, computationally, lies strictly between learnability and dependent learnability in our sense. Finally, can we give a more rigorous definition of a learning semantics for classical probability theory, along the lines of [6] but applicable to a logical system for probability as in [18]?

References

1. Avigad, J., Dean, E., Rute, J.: A metastable dominated convergence theorem. J. Log. Anal. **4**(3), 19 (2012)
2. Avigad, J., Gerhardy, P., Towsner, H.: Local stability of ergodic averages. Trans. Am. Math. Soc. **362**(1), 261–288 (2010)
3. Bhaskara Rao, K., Bhaskara Rao, M.: Theory of Charges, Pure and Applied Mathematics, vol. 109. Elsevier, Amsterdam (1983)
4. Bishop, E.: Foundations of Constructive Analysis. Mcgraw-Hill, New York, NY, USA (1967)
5. Chashka, A.: Fluctuations in martingales. Uspekhi Matematicheskikh Nauk **49**(2), 179–180 (1994). english translation in Russian Mathematical Surveys 49:2 (1994)
6. Coquand, T.: A semantics of evidence for classical arithmetic. J. Symb. Log. **60**(1), 325–337 (1995)
7. Dubins, L.E.: Some upcrossing inequalities for uniformly bounded martingales. In: Symposia Mathematica, vol. 9, pp. 169–177. Academic Press (1972)
8. Hochman, M.: Upcrossing inequalities for stationary sequences and applications. Ann. Probab. **37**(6), 2135–2149 (2009)
9. Hoyrup, M., Rute, J.: Handbook of Computability and Complexity in Analysis. Computable Measure Theory and Algorithmic Randomness, pp. 227–270. Springer (2021)
10. Ivanov, V.: Oscillations of averages in the ergodic theorem. Dokl. Akad. Nauk SSSR **347**(6), 736–738 (1996)
11. Jones, R., Kaufman, R., Rosenblatt, J., Wierdl, M.: Oscillation in ergodic theory. Ergod. Theory Dynam. Syst. **18**(4), 889–935 (1998)
12. Kachurovskii, A.: The rate of convergence in ergodic theorems. Russ. Math. Surv. **51**(4), 653–703 (1996)
13. Kohlenbach, U.: Applied Proof Theory: Proof Interpretations and their Use in Mathematics. Springer Monographs in Mathematics, Springer (2008)
14. Kohlenbach, U.: Proof-theoretic methods in nonlinear analysis. In: Proceedings ICM 2018. vol. 2, pp. 61–82 (2019)
15. Kohlenbach, U., Safarik, P.: Fluctuations, effective learnability and metastability in analysis. Ann. Pure Appl. Log. **165**, 266–304 (2014)
16. Kreisel, G.: On the interpretation of non-finitist proofs, Part I. J. Symb. Log. **16**, 241–267 (1951)
17. Kreisel, G.: On the Interpretation of Non-Finitist Proofs: Part II. Interpretation of Number Theory. Applications J. Symb. Log. **17**, 43–58 (1952)

18. Neri, M., Pischke, N.: Proof mining and probability theory (2024). preprint, available at https://arxiv.org/abs/2403.00659
19. Neri, M., Powell, T.: A quantitative Robbins-Siegmund theorem (2024). preprint, available at https://arxiv.org/abs/2410.15986
20. Neri, M., Powell, T.: On quantitative convergence for stochastic processes: crossings, fluctuations and martingales. Trans. Am. Math. Soc. (2025). to appear, preprint available at https://arxiv.org/abs/2406.19979
21. Oliva, P.: Understanding and using Spector's bar recursive interpretation of classical analysis. In: Logical Approaches to Computational Barriers, pp. 423–434. Lecture Notes in Computer Science, Springer (2006)
22. Powell, T.: A note on the finitization of Abelian and Tauberian theorems. Math. Log. Q. **66**(3), 300–310 (2020)
23. Specker, E.: Nicht konstruktiv beweisbare Sätze der Analysis. J. Symb. Log. **14**, 145–158 (1949)
24. Tao, T.: Norm convergence of multiple ergodic averages for commuting transformations. Ergod. Theory Dynam. Syst. **28**, 657–688 (2008)

Computably Discrete Represented Spaces

Eike Neumann, Arno Pauly, Cécilia Pradic(✉), and Manlio Valenti

Swansea University, Swansea, UK
c.pradic@swansea.ac.uk

Abstract. In computable topology, a represented space is called computably discrete if its equality predicate is semidecidable. While any such space is classically isomorphic to an initial segment of the natural numbers, the computable-isomorphism types of computably discrete represented spaces exhibit a rich structure. We show that the widely studied class of computably enumerable equivalence relations (ceers) corresponds precisely to the computably Quasi-Polish computably discrete spaces. We employ computably discrete spaces to exhibit several separating examples in computable topology. We construct a computably discrete computably Quasi-Polish space admitting no decidable properties, a computably discrete and computably Hausdorff precomputably Quasi-Polish space admitting no computable injection into the natural numbers, a two-point space which is computably Hausdorff but not computably discrete, and a two-point space which is computably discrete but not computably Hausdorff. We further expand an example due to Weihrauch that separates computably regular spaces from computably normal spaces.

Keywords: Computable analysis · Topology

1 Introduction

A represented space is computably discrete if equality is semidecidable. Up to homeomorphism, this notion is not particularly interesting: since every computably discrete represented space is countable and classically discrete, computably discrete spaces are classically isomorphic to the natural numbers or a finite initial segment thereof. However, in the realm of computable topology, we can provide a much more fine-grained analysis showcasing how, in fact, computably discrete spaces exhibit a rich structure and can be used as a source of counterexamples to separate the computable counterparts of classical topological notions.

After a short introduction on the notation and the main relevant notions (Sect. 2), in Sect. 3, we focus on computably discrete computably Quasi-Polish spaces. In particular, we show that such spaces are isomorphic to quotients of $\mathbb{N}$ by a computably enumerable equivalence relation (ceer). The theory of ceers has received a lot of attention over the past few years. In particular, there is an extensive literature on the structure of ceers under so-called *computable reducibility*,

© The Author(s), under exclusive license to Springer Nature Switzerland AG 2025
A. Beckmann et al. (Eds.): CiE 2025, LNCS 15764, pp. 349–364, 2025.
https://doi.org/10.1007/978-3-031-95908-0_25

a notion of reducibility between ceers that can be seen as a computable counterpart of Borel reducibility on equivalence relations, widely studied in descriptive set theory. In particular, given two ceers R and S, the computable reducibility of R to S can be restated as the existence of a computable injection between the quotients $\mathbb{N}/R$ and $\mathbb{N}/S$. For a more thorough overview of the theory of ceers, we refer the reader to [1,5]. It is easy to see that computably discrete spaces are not necessarily computably Hausdorff, *i.e.*, computable discreteness does not imply that equality is decidable. We significantly improve this result by constructing an infinite computably discrete computably Quasi-Polish space admitting no non-trivial decidable properties at all, answering a question by Emmanuel Rauzy.

In Sect. 4, we turn our attention to computably discrete precomputably (*i.e.*, not necessarily computably overt) Quasi-Polish spaces. We show that there is a computably discrete and computably Hausdorff precomputably Quasi-Polish space admitting no computable injection into the natural numbers. We also explore some computational properties of an example employed by Weihrauch to separate computably regular spaces from computably normal spaces.

In Sect. 5, we focus on finite spaces. In particular, we show that computably Hausdorff and computably discrete are incomparable notions by building a two-point space that is computably Hausdorff but not computably discrete and a two-point space that is computably discrete but not computably Hausdorff.

Finally, in Sect. 6, we conclude with some observations on the computable-isomorphism types of $\mathbb{N}$.

A slightly extended version of this paper with full proofs may be found on arXiv [7].

2 Background

We assume some familiarity with the theory of represented spaces as presented e.g. in [8]. In the sequel, we will frequently refer to represented spaces simply as *spaces*. We refer to [4] for details on (pre)computably Quasi-Polish spaces, a special kind of represented spaces.

2.1 Effectively Countably-Based Spaces

Definition 1. *A space* $\mathbf{X}$ *is called effectively countably-based, if there is a computable map* $B_{\cdot} : \mathbb{N} \to \mathcal{O}(\mathbf{X})$ *such that the induced computable map*

$$A \mapsto \bigcup_{n \in A} B_n : \mathcal{V}(\mathbb{N}) \to \mathcal{O}(\mathbf{X})$$

has a computable multi-valued right-inverse. In this case, we call $(B_n)_n$ *an* effective basis *of* $\mathbf{X}$.

In general, effectively countably based spaces are not closed under subspaces. However, if $\mathbf{X}$ is effectively countably based and (computably) admissible, then so is any subspace of $\mathbf{X}$.

Definition 2. *A space* $\mathbf{X}$ *admits an effectively fibre-overt representation if there is a computable surjection* $\delta :\subseteq \mathbb{N}^{\mathbb{N}} \to \mathbf{X}$ *such that* $\delta^{-1} : \mathbf{X} \rightrightarrows \mathbb{N}^{\mathbb{N}}$ *is computable and* $x \mapsto \overline{\delta^{-1}(x)} : \mathbf{X} \to \mathcal{V}(\mathbb{N}^{\mathbb{N}})$ *is computable.*

Proposition 1. *The following are equivalent for a computably admissible space* $\mathbf{X}$*:*

1. $\mathbf{X}$ *is effectively countably-based.*
2. $\mathbf{X}$ *computably embeds into* $\mathcal{O}(\mathbb{N})$*, i.e., there is a computable map* $i \colon \mathbf{X} \to \mathcal{O}(\mathbb{N})$ *with a computable inverse* $i^{-1} \colon i(\mathbf{X}) \to \mathbf{X}$.
3. $\mathbf{X}$ *admits an effectively fibre-overt representation.*

Subspaces of countably based admissible represented spaces carry the subspace topology. The next proposition effectivises this fact:

Proposition 2. *If* $\mathbf{Y}$ *is computably admissible and effectively countably-based and* $\mathbf{X} \subseteq \mathbf{Y}$ *is a subspace, then the restriction map* $U \mapsto (X \cap U) : \mathcal{O}(\mathbf{Y}) \to \mathcal{O}(\mathbf{X})$ *has a computable multi-valued right-inverse.*

Proof. We may assume without loss of generality that the representation $\delta_{\mathbf{Y}}$ of $\mathbf{Y}$ is effectively fibre-overt. Assume we are given a name of an open set $V \in \mathcal{O}(\mathbf{X})$. Then as in the proof of ($3 \Rightarrow 1$) in Proposition 1, we can compute a total function $G \colon \mathbb{N}^{\mathbb{N}} \to \mathbb{N}^{\mathbb{N}}$ such that for all $q \in \delta_{\mathbf{Y}}^{-1}(\mathbf{X})$ we have $\delta_{\mathbf{X}}(q) \in V$ if and only if there exists some k such that $G(q)(k) \neq 0$.

Using computability of $\overline{\delta_{\mathbf{Y}}^{-1}(y)}$ as an overt set and the definition of overtness, we can compute the open set

$$U = \left\{ y \in \mathbf{Y} \mid \exists q \in \overline{\delta_{\mathbf{Y}}^{-1}(y)}.\exists k \in \mathbb{N}.G(q)(k) \neq 0 \right\}.$$

This set is computable as an open set since $\overline{\delta_{\mathbf{Y}}^{-1}(y)}$ is computable as an overt set. Observe that by continuity we have

$$U = \left\{ y \in \mathbf{Y} \mid \exists q \in \delta_{\mathbf{Y}}^{-1}(y).\exists k \in \mathbb{N}.G(q)(k) \neq 0 \right\}.$$

If $x \in \mathbf{X}$ and $q \in \mathbb{N}^{\mathbb{N}}$ with $\delta_{\mathbf{Y}}(q) = x$ then we have by definition of G that $x \in V$ if and only if $G(q)(k) \neq 0$ for some k. This shows that $U \cap \mathbf{X} = V$. □

2.2 Computably Quasi-Polish Spaces

Quasi-Polish spaces were introduced by de Brecht [3] as a unifying framework for descriptive set theory over Polish spaces and continuous domains. A topological space is Quasi-Polish if and only if it is countably based and admits a total admissible representation (see [3, Theorem 49]). de Brecht, Pauly, and Schröder [4] have proposed an effectivisation of Quasi-Polish spaces, closely resembling the ideal presentation of effective domains, based on the following characterisation. Let $\ll$ be a transitive relation over the natural numbers. An *ideal* with respect to $\ll$ is a non-empty subset I of $\mathbb{N}$ which is *downwards closed*, i.e., for all $x \in \mathbb{N}$, if there exists $y \in I$ with $x \ll y$, then $x \in I$, and *upwards directed*, *i.e.*, for all $x, y \in I$ there exists $z \in I$ with $x \ll z$ and $y \ll z$.

The *space of ideals* $\mathbf{I}(\ll)$ is the represented space whose underlying set is the set of ideals of $\ll$. A point $p \in \mathbb{N}^{\mathbb{N}}$ represents an ideal I if and only if $I = \{p(k) \mid k \in \mathbb{N}\}$. A represented space is Quasi-Polish if and only if it is isomorphic to $\mathbf{I}(\ll)$ for some transitive relation $\ll$. Further, a topological space is Quasi-Polish if and only if it is isomorphic to $\mathbf{I}(\ll)$ for some transitive relation $\ll$. Now, a space $\mathbf{X}$ which is computably isomorphic to $\mathbf{I}(\ll)$ for some *computably enumerable* transitive relation $\ll$ on $\mathbb{N}$ is called a *precomputably Quasi-Polish.* If $\mathbf{X}$ is precomputably Quasi-Polish and computably overt, then $\mathbf{X}$ is called *computably Quasi-Polish.* All precomputably Quasi-Polish spaces are computably admissible and effectively countably based.

2.3 Computable Metric Spaces

A *(complete) computable metric space* is specified by a computable pseudometric $d\colon \mathbb{N} \times \mathbb{N} \to \mathbb{R}$ on the natural numbers. The metric induces a represented space $\mathbf{X}_d$ as follows: The underlying set of $\mathbf{X}_d$ is the Cauchy completion of $(\mathbb{N}, d)$. A sequence $p \in \mathbb{N}^{\mathbb{N}}$ is a name of a point $x \in \mathbf{X}_d$ if and only if $d(p(n), x) < 2^{-n}$ for all n. The space $\mathbf{X}_d$ is a complete metric space. The distance function $d\colon \mathbf{X}_d \times \mathbf{X}_d \to \mathbb{R}$ is computable. The space $\mathbf{X}_d$ is a computably Hausdorff computably Quasi-Polish space. More generally, we call a represented space $\mathbf{X}$ a computable metric space if it is isomorphic to $\mathbf{X}_d$ for some computable pseudometric d on $\mathbb{N}$. A represented space is a complete computable metric space if and only if it is isomorphic to a computably overt computably Π^0_2-subset of the Hilbert cube $[0,1]^\omega$.

2.4 Computably Normal Spaces

In Sect. 4 below we will present a discrete precomputably Quasi-Polish space which was (essentially) exhibited by Weihrauch as an example to separate "computably regular" from "computably metrizable". We now introduce computably regular and computably normal spaces.

Definition 3. *We call* $\mathbf{X}$ *computably regular, if the multi-valued map* $\mathrm{Reg} :\subseteq \mathbf{X} \times \mathcal{A}(\mathbf{X}) \rightrightarrows \mathcal{O}(\mathbf{X}) \times \mathcal{O}(\mathbf{X})$ *with* $(x, A) \in \mathrm{dom}(\mathrm{Reg})$ *iff* $x \notin A$ *and* $(U, V) \in \mathrm{Reg}(x, A)$ *iff* $x \in U$, $A \subseteq V$ *and* $U \cap V = \emptyset$ *is computable.*

Proposition 3. *If* $\mathbf{X}$ *is computably Hausdorff and computably discrete, then it is computably regular.*

Proof. Given the input (x, A), we ignore A and return the pair $\{x\}$ and $\mathbf{X} \setminus \{x\}$. The former we obtain by computable discreteness of $\mathbf{X}$, the latter by computable Hausdorffness. □

Definition 4. *We call* $\mathbf{X}$ *computably normal iff the map* $\mathrm{Norm} :\subseteq \mathcal{A}(\mathbf{X})^2 \rightrightarrows \mathcal{O}(\mathbf{X})^2$ *with* $(A, B) \in \mathrm{dom}(\mathrm{Norm})$ *iff* $A \cap B = \emptyset$ *and* $(U, V) \in \mathrm{Norm}(A, B)$ *iff* $U \cap V = \emptyset$ *and* $A \subseteq U$ *and* $B \subseteq V$ *is computable.*

Definition 5. *We call* $\mathbf{X}$ *computably hereditarily normal iff the map* $\mathrm{HeNorm}: \mathcal{A}(\mathbf{X})^2 \rightrightarrows \mathcal{O}(\mathbf{X})^2$ *with* $(U,V) \in \mathrm{HeNorm}(A,B)$ *iff* $U \cap V = \emptyset$ *and* $A \setminus B \subseteq U$ *and* $B \setminus A \subseteq V$ *is computable.*

A space is computably hereditarily normal if and only if all of its open subspaces are computably normal in a uniform way [7, Proposition 10].

Proposition 4. *If* $\mathbf{Y}$ *is effectively countably based, computably admissible, and computably hereditarily normal, then every subspace of* $\mathbf{Y}$ *is computably hereditarily normal.*

Proof. Let $\mathbf{X}$ be a subspace of $\mathbf{Y}$. It follows immediately from Proposition 2 that the restriction map $\mathcal{A}(\mathbf{Y}) \to \mathcal{A}(\mathbf{X}), A \mapsto A \cap X$ has a computable multi-valued right-inverse $s\colon \mathcal{A}(\mathbf{X}) \rightrightarrows \mathcal{A}(\mathbf{Y})$. Letting $r\colon \mathcal{O}(\mathbf{Y}) \to \mathcal{O}(\mathbf{X}), r(U) = U \cap X$ denote the restriction map on open sets, it is straight-forward to check that the multi-valued map

$$(r \times r) \circ \mathrm{HeNorm} \circ (s \times s) \quad : \quad \mathcal{A}(\mathbf{X})^2 \rightrightarrows \mathcal{O}(\mathbf{X})^2$$

witnesses that $\mathbf{X}$ is effectively hereditarily normal. □

Proposition 5. $[0,1]^\omega$ *is effectively hereditarily normal.*

Proof. Given $A, B \in \mathcal{A}([0,1]^\omega)$ we can (non-deterministically) compute an enumeration $(\widehat{V}_n)_n$ of all closed balls with rational radius and centre that are contained in $[0,1]^\omega \setminus A$, and an enumeration $(\widehat{U}_n)_n$ of all such closed balls that are contained in $[0,1]^\omega \setminus B$.

For a closed ball $\widehat{B}$, let B denote the corresponding open ball. Let

$$U = \bigcup_{n\in\mathbb{N}} \left(U_n \setminus \bigcup_{k<n} \widehat{V}_k \right) \qquad \text{and} \qquad V = \bigcup_{m\in\mathbb{N}} \left(V_m \setminus \bigcup_{k\leq m} \widehat{U}_k \right).$$

Clearly, U and V are uniformly computable in A and B.

We claim that U and V are disjoint. Consider a set of the form $U_n \setminus \bigcup_{k<n} \widehat{V}_k$. By construction, this set is disjoint from $V_m \setminus \bigcup_{k\leq m} \widehat{U}_k$ for $m \geq n$. Again by construction, it is disjoint from V_m for $m < n$, so it is a fortiori disjoint from $V_m \setminus \bigcup_{k\leq m} \widehat{U}_k$. This shows that U and V are disjoint.

Now, let $x \in A \setminus B$. Since $x \notin B$ and $(\widehat{U}_n)_n$ enumerates *all* closed rational balls that are contained in $[0,1]^\omega \setminus B$, we have $x \in U_n$ for some n. Since $x \in A$ we have $x \notin \widehat{V}_k$ for all k. So $x \in U_n \setminus \bigcup_{k<n} \widehat{V}_k$ and hence $x \in U$. Thus, $U \supseteq A \setminus B$. An analogous argument establishes $V \supseteq B \setminus A$. □

3 Computably Discrete Computably Quasi-Polish Spaces

To avoid drowning the text in occurrences of the word *computably*, we adopt the convention that from this point onwards, "discrete", "Hausdorff", "admissible", "overt", "fibre-overt", and "isomorphic" all refer to the computable version, and use the modifier *classical* to identify the rare cases where we do not mean the computable version.

We start our investigation by looking at discrete computably Quasi-Polish spaces. Computably Quasi-Polish spaces tend to be a setting where everything in computable topology works out very nicely. They are the computable version of Quasi-Polish spaces [3] proposed in [4,6]. Indeed, we can obtain several characterizations. We first start with the following proposition.

Proposition 6. *A quotient of* $\mathbb{N}$ *by an equivalence relation* R *is admissible iff it is discrete.*

Proof. For the left-to-right implication, we first observe that, given $n, m \in \mathbb{N}$, we can compute a name for

$$\mathcal{F} = \{U \in \mathcal{O}(\mathbb{N}/R) \mid [n]_R \in U \wedge [m]_R \in U\} \in \mathcal{O}(\mathcal{O}(\mathbb{N}/R)).$$

Let Φ be a computable functional witnessing the admissibility of $\mathbb{N}/R$. It is not hard to check that $\Phi(\mathcal{F})$ produces an output iff $[n]_R = [m]_R$. Indeed, if $x := [n]_R = [m]_R$, then $\mathcal{F}$ is the neighbourhood filter of x, hence $\Phi(\mathcal{F}) = x$. Conversely, assume that $\Phi(\mathcal{F})$ commits to some y. Since any prefix of a name for $\mathcal{F}$ can be extended to a name for a neighbourhood filter of $[n]_R$ or $[m]_R$, the admissibility of $\mathbb{N}/R$ implies that $y = [n]_R = [m]_R$. This proves that $\mathbb{N}/R$ is discrete.

For the converse direction, assume that $\mathbb{N}/R$ is discrete. In particular, for every $m \in \mathbb{N}$, we can compute $U_m := \{[m]_R\} \in \mathcal{O}(\mathbb{N}/R)$. Given $\mathcal{U} := \{U \in \mathcal{O}(\mathbb{N}/R) \mid [n]_R \in U\} \in \mathcal{O}(\mathcal{O}(\mathbb{N}/R))$, we can computably search for some $m \in \mathbb{N}$ such that $U_m \in \mathcal{U}$. Clearly, any such m is such that $[m]_R = [n]_R$. Since we will eventually find one, this witnesses the admissibility of $\mathbb{N}/R$. □

Theorem 1. *For discrete* $\mathbf{X}$*, the following are equivalent:*

1. $\mathbf{X}$ *admits a computably equivalent representation with domain* $\mathbb{N}^{\mathbb{N}}$.
2. *There exists a computable surjection* $s : \mathbb{N} \to \mathbf{X}$.
3. $\mathbf{X}$ *is computably Quasi-Polish.*
4. $\mathbf{X}$ *is isomorphic to a discrete quotient of* $\mathbb{N}$.
5. $\mathbf{X}$ *is isomorphic to an admissible quotient of* $\mathbb{N}$.

Proof. (1 ⇒ 2): Let $\delta : \mathbb{N}^{\mathbb{N}} \to \mathbf{X}$ be a total representation for $\mathbf{X}$ and let isEqual : $\mathbf{X} \times \mathbf{X} \to \mathbb{S}$ be a computable map witnessing the discreteness of $(\mathbf{X}, \delta)$. Using a realizer $E : \mathbb{N}^{\mathbb{N}} \times \mathbb{N}^{\mathbb{N}} \to \{0,1\}^{\mathbb{N}}$ for isEqual, we can compute a list $(w_i)_{i \in \mathbb{N}}$ of all $w \in \mathbb{N}^{<\mathbb{N}}$ such that E writes the first 1 upon reading the pair (w, w). Since δ is total, for every $p \in \mathbb{N}^{\mathbb{N}}$ we know that $E(p,p)$ contains a 1. In particular, the sequence $(w_i)_{i\in\mathbb{N}}$ identifies an open cover of $\mathbb{N}^{\mathbb{N}}$. Moreover, the discreteness assumption implies that δ is constant on each cylinder $w_i\mathbb{N}^{\mathbb{N}}$. As such, we can define a computable surjection $s : \mathbb{N} \to \mathbf{X}$ as the map realized by $i \mapsto w_i 0^\omega$.

($2 \Rightarrow 1$): Let $\delta_{\mathbb{N}}$ be a total representation of $\mathbb{N}$. We claim that $\delta'_{\mathbf{X}} := s \circ \delta_{\mathbb{N}}$ is a total representation of $\mathbf{X}$. The translation from $\delta'_{\mathbf{X}}$ to $\delta_{\mathbf{X}}$ is realized by a realizer of s. For the converse, given a $\delta_{\mathbf{X}}$-name p we exhaustively generate all $s(n) \in \mathbf{X}$, and use isEqual to identify some n with $s(n) = \delta_{\mathbf{X}}(p)$, and pick some $q \in \delta_{\mathbb{N}}^{-1}(\{n\})$. Then q is a $\delta'_{\mathbf{X}}$-name for the same point as q, given the translation in the other direction.

($1 \wedge 2 \Rightarrow 3$): We first show that 2. implies $\mathbf{X}$ being admissible. Assume we are given $\{U \in \mathcal{O}(\mathbf{X}) \mid x \in U\} \in \mathcal{O}(\mathcal{O}(\mathbf{X}))$. Using s and discreteness of $\mathbf{X}$, we can generate all $\{s(n)\} \in \mathcal{O}(\mathbf{X})$ for $n \in \mathbb{N}$, and search for some n with $\{s(n)\} \in \{U \in \mathcal{O}(\mathbf{X}) \mid x \in U\}$. Since s is surjective, such an n needs to exist, and it obviously satisfies that $s(n) = x$. As such, we can compute x, thus witnessing the admissibility of $\mathbf{X}$.

Next, we observe that the representation $\delta : \mathbb{N}^{\mathbb{N}} \to \mathbf{X}$ from point 1 is fibre-overt. Given a prefix w and a point $x \in \mathbf{X}$ we can search for extensions p of w such that the equality test on $\mathbf{X}$ confirms that $\delta(p) = x$. As $\mathbb{N}^{\mathbb{N}}$ itself is computably overt, this is an effective process and yields the claim. By [4, Theorem 14] a total admissible fibre-overt representation characterizes computably Quasi-Polish spaces.

($3 \Rightarrow 1$): By [4, Theorem 14] a computably Quasi-Polish spaces admits a total representation.

($2 \Rightarrow 4$): Let $n \cong m$ iff $s(n) = s(m)$. We claim that $\mathbf{X}$ is isomorphic to $\mathbb{N}/\cong$. Clearly, the computable map $\varphi : \mathbb{N}/\cong \to \mathbf{X}$ induced by s is a bijection. Conversely, given $x \in X$ we can use discreteness of $\mathbf{X}$ to (non-deterministically) identify some $n \in \mathbb{N}$ with $s(n) = x$. Hence, φ^{-1} is computable, too. The discreteness of $\mathbb{N}/\cong$ follows immediately from the discreteness of $\mathbf{X}$.

($4 \Rightarrow 2$): Straight-forward.

($4 \Leftrightarrow 5$): This follows by Proposition 6. □

We can observe that there is a close connection between discrete Quasi-Polish spaces and the represented spaces of equivalence classes of computably enumerable equivalence relations on $\mathbb{N}$ (ceers). In particular, if R is a ceer, we write $\mathbb{N}/R$ for the represented space of R-equivalence classes, where each class is represented via any of its representatives. As mentioned in the introduction, ceers exhibit a rich structure and they have received significant attention in recent years.

The equivalence of the points 3, 4, and 5 in the previous theorem can be restated as follows:

Corollary 1. *Let* $\mathbf{X}$ *be a represented space. The following are equivalent:*

1. $\mathbf{X}$ *is a discrete Quasi-Polish space.*

2. *$\mathbf{X}$ there is an equivalence relation R such that $\mathbb{N}/R$ is admissible and $\mathbf{X}$ is isomorphic to $\mathbb{N}/R$.*
3. *$\mathbf{X}$ there is a ceer R such that $\mathbf{X}$ is isomorphic to $\mathbb{N}/R$.*

Proposition 7. *For an infinite discrete computably Quasi-Polish space $\mathbf{X}$ the following are equivalent:*

1. *$\mathbf{X} \cong \mathbb{N}$.*
2. *There is a computable injection $\iota : \mathbf{X} \to \mathbb{N}$.*
3. *$\mathbf{X}$ is a computable metric space.*
4. *$\mathbf{X}$ is Hausdorff.*

Proof. It is immediate that 1 implies 2 and that 2 implies 4. It is also immediate that 1 implies 3 and that 3 implies 4. We only need to show that 4 implies 1. By Theorem 1 we have a computable surjection $s : \mathbb{N} \to \mathbf{X}$. Since $\mathbf{X}$ is discrete and Hausdorff, we find that $s(n) = s(m)$ is decidable for $n, m \in \mathbb{N}$. Consequently, the set $S = \{n \in \mathbb{N} \mid \forall i < n \ s(i) \neq s(n)\}$ is a decidable infinite subset of $\mathbb{N}$, which means there is a (computable) isomorphism $\sigma : \mathbb{N} \to S$, which we can lift to yield an isomorphism between $\mathbf{X}$ and $\mathbb{N}$. □

In computable topology, being discrete does not imply being Hausdorff. But there is even more, and we can exhibit a discrete computably Quasi-Polish space having no decidable non-trivial properties at all. This answers a question posed to us by Emmanuel Rauzy.

Example 1. There is an infinite discrete computably Quasi-Polish space $\mathbf{X}$ such that every computable $f : \mathbf{X} \to \mathbb{N}$ is constant.

Proof. We build the space $\mathbf{X}$ from a directed graph G with vertex set $\mathbb{N}$. Each vertex will have out-degree at most 1. For the vertex n, we first wait for confirmation that the n-th Turing machine halts on n and outputs some number a_n. If this never happens, n will have out-degree 0. If it does happen, we then search for some $m \neq n$ such that the n-th TM halts on m and outputs some $a_m \neq a_n$. If we do find such an m, we add an edge $n \mapsto m$ for the first candidate found.

Let $n \equiv m$ if there is an *undirected* path between n and m in G. This is a computably enumerable relation. Let $\mathbf{X} = \mathbb{N}/\equiv$. By Theorem 1, this yields a discrete computably Quasi-Polish space.

To see that $\mathbf{X}$ is infinite, we observe that there are infinitely many n such that the n-th TM does not halt on n. This means that there are infinitely many vertices in G with out-degree 0, and – since all vertices in G have out-degree at most 1 – no two such vertices can be connected by an undirected path. This means there are infinitely many connected components in the graph that is obtained from G by forgetting edge directions, and thus infinitely many points in $\mathbf{X}$.

For the final claim, assume for the sake of a contradiction that the n-th Turing machine computes a non-constant function $f : \mathbf{X} \to \mathbb{N}$. Because f is total, the n-th TM must halt on n and output some a_n. Because f is not constant, there is some m on which the n-th TM outputs some different value a_m, which means that there will be an edge from n to one such m in G. But that means that n and m denote the same point in $\mathbf{X}$, and thus the n-th TM doesn't actually compute a function due to the failure of extensionality. □

Instead of directly constructing a space to witness the claim in Example 1, we could instead have used the connection to ceers we established and import known results from the literature via the observation that the following are equivalent for a ceer R:

1. Any distinct equivalence classes of R are recursively inseparable.
2. Every computable multi-valued function $F : (\mathbb{N}/R) \rightrightarrows \mathbf{2}$ has a constant choice function.

Ceers whose equivalence classes are not only recursively inseparable, but are so in an effective or uniformly effective way have been constructed and studied previously, see e.g. [2] and Proposition 3.13 therein (the examples built there are far more complicated as they are meant to satisfy much more specific properties).

4 Computably Discrete Precomputably Quasi-Polish Spaces

The difference between a computably Quasi-Polish space and a precomputably Quasi-Polish is that the former is required to be overt, but the latter might not be. It is straight-forward to come up with separating example:

Example 2 (Discrete, Hausdorff, admissible, not overt). The complement of the Halting problem $\mathbb{H}^c = \{n \in \mathbb{N} \mid \Phi_n \uparrow\}$ is an infinite Hausdorff discrete precomputably Quasi-Polish space which is not overt.

With some more work we can see that the equivalence of being Hausdorff and admitting a computable injection into $\mathbb{N}$ we established for discrete computably Quasi-Polish spaces does not extend to precomputably Quasi-Polish spaces:

Example 3. There is a discrete Hausdorff precomputably Quasi-Polish space $\mathbf{X}$ such that there is no computable injection $\iota : \mathbf{X} \to \mathbb{N}$.

Proof. Given $n \in \mathbb{N}$, we can uniformly build a discrete Hausdorff precomputably Quasi-Polish space $\mathbf{X}_n$ such that the n-th computable function does not realize an injection $\iota_n : \mathbf{X}_n \to \mathbb{N}$. Then $\sum_{n \in \mathbb{N}} \mathbf{X}_n$ is the desired example. The space $\mathbf{X}_n$ could be the singleton $\{\mathbb{N}\} \subseteq \mathcal{O}(\mathbb{N})$, or it could be $\{I, \mathbb{N} \setminus I\} \subseteq \mathcal{O}(\mathbb{N})$ where I is a finite set given via a 2-c.e. code[1]. We also built a witness for computable Hausdorffness. The construction proceeds in phases:

Phase 1 We search for some finite set J such that Φ_n outputs some $m \in \mathbb{N}$ upon seeing some enumeration of J. As long as we have not yet found one, $\mathbf{X}_n$ is $\{\mathbb{N}\}$ and the witness of computable Hausdorffness is trivial. If we do find such a J, we proceed to Phase 2.

[1] Recall that a 2-c.e. set is the difference of two computably enumerable (c.e.)sets, see e.g. [10, Sec. 3.8.4]. A 2-c.e. set can be represented with a pair of names for c.e. sets.

Phase 2 We state that $I := \{0, 1, \ldots, \max J + 1\}$, and add to the witness of computable Hausdorffness that enumerations containing all of $\{0, 1, \ldots, \max J + 1\}$ refer to different points than enumerations containing $\max J + 2$. We search for some finite set $K \subseteq \mathbb{N} \setminus I$ such that Φ_n outputs some $\ell \neq m$ upon seeing some enumeration of K. If we do find such a K, we proceed to Phase 3.
Phase 3 We change our mind regarding I and assert instead that $I = J \cup K$. The witness of Hausdorffness will now also declare that enumerations containing all of $J \cup K$ refer to different points than those containing $\max K + 1$.

This does yield a valid witness of computable Hausdorffness, as the condition added in Phase 2 is not actually met by any point if we do reach Phase 3. If we remain in Phase 1 forever, then Φ_n is undefined everywhere. If we remain in Phase 2 forever, then Φ_n is either undefined on some names in $\mathbf{X}_n$ or is constant m. If we reach Phase 3, Φ_n does not act extensionally on names for $I \in \mathbf{X}_n$. □

For the remainder of this section, we investigate a family of discrete spaces S_A parameterized by some $A \subseteq \mathbb{N}$. This generalizes an example of Weihrauch which is computably normal but not computably regular [11, Example 5.4].

Definition 6. *Given* $A \subseteq \mathbb{N}$, *let* $S_A := \{(n, \top) \mid n \in A\} \cup \{(n, \bot) \mid n \notin A\} \subseteq \mathbb{N} \times \mathbb{S}$.

Regardless of the choice of A, the space S_A is effectively countably-based and admissible as it inherits these properties as a subspace of $\mathbb{N} \times \mathbb{S}$. The space S_A is Hausdorff and discrete, as the projection $\pi_1 : S_A \to \mathbb{N}$ is a computable injection into a Hausdorff and discrete space. It follows that S_A is computably regular. Beyond that, we observe the following:

Proposition 8. *The following are equivalent:*

1. $S_A \cong \mathbb{N}$.
2. S_A *is a complete computable metric space.*
3. S_A *is computably Quasi-Polish.*
4. S_A *is computably separable.*
5. S_A *is overt.*
6. A *is computably enumerable.*

Proof. The implications $1 \Rightarrow 2 \Rightarrow 3 \Rightarrow 4 \Rightarrow 5$ are all trivial. If S_A is overt we can recognize $n \in A$ by asking whether the open set accepting only $(n, \top)$ is non-empty. If A is computably enumerable, the map $n \mapsto (n, [n \in A]) : \mathbb{N} \to \mathbb{N} \times \mathbb{S}$ is computable, and together with the projection $\pi_1 : \mathbb{N} \times \mathbb{S} \to \mathbb{N}$, it witnesses that $\mathbb{N} \cong S_A$. □

Proposition 9. *The following are equivalent:*

1. S_A *computably embeds into* $[0, 1]^\omega$.
2. S_A *is computably normal.*
3. A *is in the first level of the difference hierarchy over* Σ_1.

Proof. 1 ⇒ 2 By Propositions 4 and 5.

2 ⇒ 3 We are given some $n \in \mathbb{N}$, and initially proclaim that $n \notin A$. We run the algorithm for normality of S_A on the closed sets $(\{n\} \times \mathbb{S}) \cap S_A$ and $\{(n, \bot)\} \cap S_A$. If it is actually the case that $n \in A$, then the algorithm must react by making the first open set it returns accept $(n, \top)$. If this ever happens, we change our proclamation to be that $n \in A$, and we adjust the closed sets we feed to the algorithm to be $\emptyset$ and $\{(n, \bot)\} \cap S_A$. Then let us show that the algorithm reacts by letting the second open set it returns accept $(n, \bot)$ if and only if $n \notin A$, which is enough to prove that A is a computable difference of opens. The right-to-left direction is obvious. For the left-to-right, assume that $n \in A$ and that the code for the second open set returned by Norm accepts $(n, \bot)$; by continuity of the realizer, it means it must also accept $(n, \top)$. But then it means that Norm returns open two sets overlapping on $(n, \top)$ for a valid input, which is a contradiction.

3 ⇒ 1 We describe an embedding $\iota : S_A \to \mathbb{N} \times [0, 1]$ instead. The point (n, b) gets mapped somewhere into $\{n\} \times [0, 1]$. We run our 2–c.e. procedure for A. In the first phase, we believe that $n \notin A$. We provide approximations to ι that map (n, b) to $(n, 0)$ and approximations to ι^{-1} that map (n, x) to $(n, \bot)$ for all $x \in [0, 1]$. If the 2–c.e. procedure for A ever declares that $n \in A$, we advance to Phase 2. There is some $\varepsilon > 0$ such that declaring $\iota(n, \top) = (n, \varepsilon)$ is still compatible with the information provided so far. We do this to one name of $(n, \top)$ at a time, and also adjust ι^{-1} such that $\iota^{-1}(n, x) = (n, \top)$ for $x > \frac{\varepsilon}{2}$. We do not provide any additional information on what $\iota(n, \bot)$ would be – in Phase 2, we believe this value does not need to be defined. If we do learn that after all, $n \notin A$, we enter the third and final phase. We stop setting $\iota(n, \top) = \varepsilon$, and instead make $\iota(n, \bot) = (n, 0)$ and $\iota^{-1}(n, 0) = (n, \bot)$ true. □

Proposition 10. *S_A is precomputably Quasi-Polish iff A is Δ^0_2.*

Proof. As $\mathbb{N} \times \mathbb{S}$ is computably Quasi-Polish, S_A is precomputably Quasi-Polish iff it is a Π^0_2-subspace of $\mathbb{N} \times \mathbb{S}$. The latter implies that A is Δ^0_2, since $n \in A \Leftrightarrow (n, \top) \in S_A$ and $n \notin A \Leftrightarrow (n, \bot) \in S_A$. Conversely, if A is Δ^0_2, then so is S_A. □

5 Finite Spaces

Proposition 11. *When $\mathbf{X}$ has cardinality n, is classically discrete and has a fibre-overt representation, there is a computable injection $\iota : \mathbf{X} \to \mathbf{n}$.*

Proof. Classically, we can select names $p_1, \ldots, p_n$ for the n points in $\mathbf{X}$. The classical discreteness ensures that these have prefixes $w_1, \ldots, w_n$ such that no w_i extends to a prefix of some p_j for $i \neq j$. Since the $(w_i)_{i \leq n}$ contain only finite information, we can use these as parameters for ι.

Given some $x \in \mathbf{X}$, we can use the fibre-overtness of the representation to ask for each i whether x has a name starting with w_i. Since this holds for exactly one i, we can determine this effectively – this process computes the desired ι. □

Corollary 2. *If* $\mathbf{X}$ *is finite, classically discrete and effectively countably-based, then it is already discrete and Hausdorff.*

Proposition 12. *If* $s : \mathbf{n} \to \mathbf{X}$ *is a computable surjection, then* $\mathbf{X}$ *is discrete iff it is Hausdorff.*

5.1 Computably Discrete but Not Computably Hausdorff Two-Point Space

Definition 7. *For an infinite and co-infinite set* $A \subset \mathbb{N}$*, we define the space* D_A *with underlying set* $\{a, b\}$ *where* p *is a name for* a *if* p *is obtained from the characteristic function of* A *by replacing finitely many* 1*s with* 0*s, and* p *is a name for* b *if it is obtained from the characteristic function of* $\mathbb{N} \setminus A$ *by replacing finitely many* 1*s with* 0*s.*

Observation 1. *The space* D_A *is discrete.*

Proof. Whenever two valid names have a 1 in the same position, they denote the same point in D_A. Conversely, any two names for the same point in D_A share a 1 somewhere. □

Proposition 13. *There exists* A *such that* D_A *is not Hausdorff.*

Proof. A realizer of the Hausdorffness of D_A essentially consists of an effective enumeration of pairs (w_i, u_i) such that seeing names with such prefixes leads to the claim that they refer to distinct points in D_A. We will diagonalize against all of them by determining $A \cap \{0, 1, \ldots, s_n\}$ for some $s_n \in \mathbb{N}$ in stage n. If the n-th candidate witness does not include a pair (w_i, u_i) such that w_i and u_i start with $0^{s_{n-1}}$, the witness fails to provide a required answer for some names. We assert $s_{n-1} + 1 \in A$, $s_{n-1} + 2 \notin A$ and set $s_n = s_{n-1} + 2$. Otherwise, we pick such a pair (w_i, u_i) and set $k \in A$ for every k such that $w_i(k) = 1$ or $u_i(k) = 1$. For every other k between s_{n-1} and $s_n := \max\{|u_i|, |w_i|\} + 2$ (exclusive of the bounds) we assert $k \notin A$. We also set $s_n \in A$. This ensures that the n-th witness incorrectly asserts that some names both denoting a actually refer to distinct points. Since at any stage we place some numbers into A and some into its complement, we do build an infinite and co-infinite set. □

We can adjust the construction of D_A by partitioning $\mathbb{N} \setminus A$ into finitely or infinitely many sets $B_0, B_1, \ldots$ and correspondingly the point b into $b_0, b_1, \ldots$ where p is a name for b_i if is obtained from the characteristic function of B_i by replacing finitely many 1s by 0s. This retains discreteness. If we construct A as in Proposition 13, this yields a discrete countably infinite space without any non-trivial decidable properties.

5.2 Computably Hausdorff but Not Computably Discrete Two-Point Space

The notion of Hausdorffness for represented spaces does not line up precisely with the topological definition. If a represented space is Hausdorff, the corresponding topological space is only sequentially Hausdorff (this difference does not appear for countably-based spaces). A stronger notion of effective Hausdorffness that also implies topological Hausdorffness was proposed by Schröder in [9, Definition 1]:

Definition 8. *We say* $\mathbf{X}$ *admits a computable witness of Hausdorffness if there are computable sequences* $(U_i)_{i\in\mathbb{N}}, (V_i)_{i\in\mathbb{N}} \in \mathcal{O}(\mathbf{X})^{\mathbb{N}}$ *of opens such that* $\bigcup_{i\in\mathbb{N}} U_i \times V_i = \mathbf{X}^2 \setminus \Delta_{\mathbf{X}}$.

At the Oberwolfach Meeting 2117 "Computability Theory" in 2021, Brattka raised the question whether a space which is (computably) Hausdorff and topologically Hausdorff will admit a computable witness of Hausdorffness. We provide a counterexample in Corollary 3.

For this subsection, by enumerations of sets $A \subseteq \mathbb{N}$, we mean sequences $s : \mathbb{N} \to \mathbb{N} \uplus \{\bot\}$ such that $A = \{s_n \mid n \in \mathbb{N}, s_n \neq \bot\}$.

Definition 9. *Fix some non-empty* $A \subsetneq \mathbb{N}$*. We define the two-point space* H_A *with underlying set* $\{a, b\}$ *as follows: A name for* a *starts with some* $n \notin A$ *followed by an enumeration of* A*. A name for* b *starts with some* $n \in A$ *followed by an enumeration of some* $B \subseteq \mathbb{N} \setminus A$*, possibly empty.*

Proposition 14. H_A *is (computably) Hausdorff.*

Proof. Assume we are given two names $\langle n, U\rangle$ and $\langle m, V\rangle$. If we find that $n \in V$ or $m \in V$ holds, then these have to be names for distinct points in H_A. Conversely, if they are names for distinct elements, then one of them is a name for a, and its corresponding enumerated set will contain the index of the other. As $n \in V$ or $m \in V$ is recognizable, the claim follows. □

Here H_A is also topologically Haudorff as it is easily seen to be classically homeomorphic to $\mathbf{2}$.

Proposition 15. *The sets of names for* $b \in H_A$ *and enumerations of* A *are Medvedev-equivalent.*

Proof. Clearly, if we have an enumeration of A we build a name for $\{b\} \subseteq H_A$ by accepting (n, V) whenever we notice $n \in A$. Conversely, any realizer for $\{b\} \in \mathcal{O}(H_A)$ must accept $(n, \emptyset)$ if and only if $n \in A$; this is because a prefix of the enumeration of $\emptyset$ can always be extended to an enumeration of A, and thus a valid name of a in case that $n \notin A$. □

Corollary 3. *For non-c.e. $A \subseteq \mathbb{N}$, the space H_A is topologically Hausdorff and (computably) Hausdorff but admits no witness of computable Hausdorffness.*

Proof. From the definition, it is clear that a finite space admits a computable witness of Hausdorffness if and only if every singleton is computably open. But $\{b\} \in \mathcal{O}(H_A)$ is not computably open if $A \subseteq \mathbb{N}$ is not computably enumerable. □

Corollary 4. *For non-c.e. $A \subseteq \mathbb{N}$, the space H_A is finite, Hausdorff but not discrete.*

Proof. In a discrete space $\mathbf{X}$, the singleton $\{x\}$ needs to be computably open for every computable point $x \in \mathbf{X}$. The point $b \in H_A$ is computable, but here $\{b\}$ is not computably open. Thus, H_A cannot be discrete. □

Corollary 5. *If A is not c.e., then H_A cannot be both overt and admissible.*

Proof. Note that a Hausdorff finite space can only fail to be discrete if some of its points are non-computable (in this case, this is because computing $\{a\} \in H_A$ is as hard as enumerating A). In an admissible and overt space, every computably open singleton contains a computable point. □

We do not know whether H_A is overt, admissible or neither in general.

6 Characterizing $\mathbb{N}$ up to Computable Isomorphism

It would be very pleasing to have a simple characterization of $\mathbb{N}$ up to isomorphism in a general computable topology setting. Proposition 7 tells us that $\mathbb{N}$ is, up to isomorphism, the only discrete Hausdorff non-compact computably Quasi-Polish space. We raise the question whether this result can be improved.

Question 1. Is $\mathbb{N}$ the only represented space (up to isomorphism) which is discrete, Hausdorff, overt and admissible but not compact?

We have already presented examples showing that none of the criteria in our question can be dropped, with the exception of admissibility. This is covered by the following:

Example 4 (Discrete, overt, Hausdorff, no onto surjection, not admissible). Pick a non-computable $p \in \mathbb{N}^{\mathbb{N}}$. Let $\delta_{p\mathbb{N}}(0^n1q) = n$ if $p \equiv_T q$, and $q \notin \mathrm{dom}(\delta_{p\mathbb{N}})$ for $q \not\equiv_T p$. The resulting space $p\mathbb{N}$ is discrete, Hausdorff and overt. It is not admissible, and there is no computable surjection $s : \mathbb{N} \to p\mathbb{N}$ (in fact, any function from $\mathbb{N}$ to $p\mathbb{N}$ needs to compute p).

To round off the discussion, let us point out that there are even more ways to construct spaces which are classically homeomorphic to $\mathbb{N}$, yet behave very differently in computable topology. A last example, which is overt but not Hausdorff nor discrete is the following.

Example 5. There exists a space $\mathbb{N}'$ with underlying set $\mathbb{N}$ such that $\mathrm{id} : \mathbb{N} \to \mathbb{N}'$ is computable, and $\mathrm{id} : \mathbb{N}' \to \mathbb{N}$ is computable relative to $\emptyset'$ such that $\emptyset$ and $\mathbb{N}$ are the only computable elements in $\mathcal{O}(\mathbb{N}')$.

Proof. Let $\mathrm{BB} : \mathbb{N} \to \mathbb{N}$ be the busy beaver function. We define $\delta : \mathbb{N}^{\mathbb{N}} \to \mathbb{N}$ by $\delta(p) = p(\mathrm{BB}(p(0)))$, and let $\mathbb{N}' = (\mathbb{N}, \delta)$. Any constant sequence n^ω is name for n, and having access to $\emptyset'$ lets us compute BB and thereby decode δ-names. Now let $U \in \mathcal{O}(\mathbb{N}')$ be a non-empty open set, and let $n \in U$. A computable realizer for U needs to accept any sequence mn^ω. If it accepts any such sequence before having read a prefix of length $\mathrm{BB}(m)$, then it will accept names for all numbers, *i.e.*, $U = \mathbb{N}$. But that means if $U \neq \mathbb{N}$, then by counting when a realizer accepts mn^ω, we obtain an upper bound for $\mathrm{BB}(m)$. This means that any realizer for non-empty $U \neq \mathbb{N}$ computes $\emptyset'$. □

Acknowledgements. The authors thank Emmanuel Rauzy and Zhifeng Ye for fruitful discussions. The authors grant to Swansea University a non-exclusive, irrevocable, sub-licensable, worldwide license to make the accepted manuscript available on its institutional repository.

References

1. Andrews, U., Badaev, S., Sorbi, A.: A survey on universal computably enumerable equivalence relations. In: Day, A., Fellows, M., Greenberg, N., Khoussainov, B., Melnikov, A., Rosamond, F. (eds.) Computability and Complexity. LNCS, vol. 10010, pp. 418–451 Springer, Cham (2017). https://doi.org/10.1007/978-3-319-50062-1_25. MR 3629733
2. Andrews, U., et al.: Universal computably enumerable equivalence relations. J. Symb. Log. **79**(1), 60–88 (2014)
3. de Brecht, M.: Quasi-Polish spaces. Ann. Pure Appl. Log. **164**(3), 356–381 (2013). https://doi.org/10.1016/j.apal.2012.11.001. MR 3001551
4. de Brecht, M., Pauly, A., Schröder, M.: Overt choice. Computability **9**(3-4), 169–191 (2020). https://doi.org/10.3233/com-190253. MR 4133712
5. Gao, S., Gerdes, P.: Computably enumerable equivalence relations. Stud. Log. **67**(1), 27–59 (2001). https://doi.org/10.1023/A:1010521410739. MR 1833652
6. Hoyrup, M., Rojas, C., Selivanov, V., Stull, D.M.: Computability on quasi-polish Spaces. In: Hospodár, M., Jirásková, G., Konstantinidis, S. (eds.) Descriptional Complexity of Formal Systems. DCFS 2019. LNCS, vol. 11612, pp. 171–183 Springer, Cham (2019). https://doi.org/10.1007/978-3-030-23247-4_13. MR 3989092
7. Neumann, E., Pauly, A., Pradic, C., Valenti, M.: Computably discrete represented spaces (2025)
8. Pauly, A.: On the topological aspects of the theory of represented spaces. Computability **5**(2), 159–180 (2016). https://doi.org/10.3233/COM-150049. MR 3602282
9. Schröder, M.: Computable Multifunctions on Effectively Hausdorff Spaces, 2021. available at http://cca-net.de/cca2021/abstracts/Schroeder.pdf

10. Soare, R.I.: Turing Computability: Theory and Applications, 1 ed., Theory and Applications of Computability. Springer-Verlag, Berlin (2016). https://doi.org/10.1007/978-3-642-31933-4. MR 3496974
11. Weihrauch, K.: Computably regular topological spaces. Log. Methods Comput. Sci. **9**(3), 3:5, 24 (2013). https://doi.org/10.2168/LMCS-9(3:5)2013. MR 3103837

Multidimensional Tilings and MSO Logic

Rémi Pallen[1(✉)] and Ilkka Törmä[2]

[1] ENS Paris-Saclay, Gif-sur-Yvette, France
remi.pallen@loria.fr
[2] Department of Mathematics and Statistics, University of Turku, Turku, Finland
iatorm@utu.fi

Abstract. We define sets of coulourings of the infinite discrete plane using monadic second order (MSO) formulas. We determine the complexity of deciding whether such a formula defines a subshift, parametrized on the quantifier alternation complexity of the formula. We also study the complexities of languages of MSO-definable sets, giving either an exact classification or upper and lower bounds for each quantifier alternation class.

Keywords: MSO logic · subshifts · symbolic dynamics · tilings

1 Introduction

A *tiling* or *configuration* is a colouring of the two-dimensional plane $\mathbb{Z}^2$ with finitely many colours. A *subshift* is the set of tilings such that no pattern from a set of *forbidden patterns* appears. These notions were first introduced by Wang [10] to study first order logic, and to find an algorithm which computes whether a formula is a tautology or not. Berger proved in 1966 [1] the undecidability of the domino problem, and many other problems were proved undecidable too thanks to this result in the following decades.

This article continues the line of research in [3,9]. The idea is to define sets of configurations using *monadic second order* (*MSO* for short) logic. MSO definability is a classical and broad area of research; see [8] for an overview in the context of finite and infinite languages. MSO formulas on finite words correspond to regular languages, and those on infinite words correspond to ω-regular languages. Subshifts can be seen as languages of two-dimensional infinite words, so it makes sense to ask which of them can be defined by MSO formulas. In the formalism of [3], an MSO formula always defines a translation invariant set, but if it has a first order existential quantifier, it may not define a topologically closed set, i.e. a subshift. This is why [3,9] mostly restrict their study to the case where the MSO formulas do not have first order existential quantifiers. In this article, we study the case where they are allowed.

Our first result is to determine the exact complexity of deciding whether an MSO formula defines a subshift or not, depending on the complexity (number of second-order quantifier alternations) of the formula. We also study the complexity of languages of subshifts and sets these formulas can define.

© The Author(s), under exclusive license to Springer Nature Switzerland AG 2025
A. Beckmann et al. (Eds.): CiE 2025, LNCS 15764, pp. 365–379, 2025.
https://doi.org/10.1007/978-3-031-95908-0_26

One motivation for this research is *Griddy* (formerly called Diddy), a Python library for discrete dynamical systems research developed by Ville Salo and the second author [6,7]. Griddy allows the definition and manipulation of multidimensional subshifts defined by first order logical formulas. Even though most properties of multidimensional subshifts are undecidable, partial algorithms can often give good solutions or approximations in "naturally occurring" cases. This raises the question whether the same holds for subshifts defined by MSO formulas, or whether the complexity is too high to obtain even partial algorithms.

2 Preliminaries

2.1 Subshifts

We review some basic facts from multidimensional symbolic dynamics. See [5], especially Section A.6, for a more comprehensive overview.

Let $\mathcal{A}$ be a finite set of colours called *alphabet*. A *pattern* is a map $P : D \to \mathcal{A}$ where $D \subset \mathbb{Z}^2$. D is called *support* of the pattern and is denoted $D(P)$. A *configuration* is a pattern x such that $D(x) = \mathbb{Z}^2$. We say that a pattern P *appears* in a configuration $x \in \mathcal{A}^{\mathbb{Z}^2}$ if there exists $\boldsymbol{v} \in \mathbb{Z}^2$ such that $P_{\boldsymbol{u}} = x_{\boldsymbol{u}+\boldsymbol{v}}$ holds for all $\boldsymbol{u} \in \text{supp}(P)$. The number of occurrences of P in x (which may be infinite) is denoted $\#_x(P)$. A *cell* or *position* is an element $\boldsymbol{v} \in \mathbb{Z}^2$.

For a set of patterns $\mathcal{F}$, we define $X_{\mathcal{F}} \subseteq \mathcal{A}^{\mathbb{Z}^2}$ as the set of those $x \in \mathcal{A}^{\mathbb{Z}^d}$ in which no pattern from $\mathcal{F}$ appears. Such a set is called a *subshift*. When $\mathcal{F}$ is finite, $X_{\mathcal{F}}$ is a *subshift of finite type (SFT)*.

Property 1. For each Turing machine M, there exists a SFT (see Fig. 1 for the colours, a pattern of size 1×2 or 2×1 is forbidden if the tiles do not match), such that the *initial tile* may appear in a tiling if and only if M does not halt on the empty input. There exists a slightly different version for which an input is written on the computation tape. In this case, if the initial tile appears on a configuration with a finite input x written on it, M does not halt on x.

For $\boldsymbol{v} \in \mathbb{Z}^2$ and $x \in \mathcal{A}^{\mathbb{Z}^2}$, denote by $\sigma_{\boldsymbol{v}}(x)$ the configuration $\sigma_{\boldsymbol{v}}(x)_{\boldsymbol{u}} = x_{\boldsymbol{u}+\boldsymbol{v}}$ for all $\boldsymbol{u} \in \mathbb{Z}^2$. The function $\sigma_{\boldsymbol{v}} : \mathcal{A}^{\mathbb{Z}^2} \to \mathcal{A}^{\mathbb{Z}^2}$ is called the *shift by* $\boldsymbol{v}$. A set of configurations E is *shift-invariant* if $\sigma_{\boldsymbol{v}}(x) \in E$ for all $x \in E$ and $\boldsymbol{v} \in \mathbb{Z}^2$.

We give a distance function d on the space $\mathcal{A}^{\mathbb{Z}^2}$. For two configurations $x \neq y$, let $d(x, y) = 2^{-\min\{|a|+|b| | x_{(a,b)} \neq y_{(a,b)}\}}$ and $d(x, x) = 0$ for all x. The set $\mathcal{A}^{\mathbb{Z}^2}$ is therefore a metric space and has a corresponding topology, which is generated by the clopen *cylinder sets* $[P] = \{x \in \mathcal{A}^{\mathbb{Z}^2} \mid \forall \boldsymbol{v} \in \text{supp}(P)\ x_{\boldsymbol{v}} = P_{\boldsymbol{v}}\}$ for finite patterns P.

Property 2. A set of configurations X is a subshift if and only if it is shift-invariant and topologically closed.

The *language* $\mathcal{L}(X)$ of a set of configurations X is the set of finite patterns P such that P appears in at least one $x \in X$.

Property 3. If X and Y are subshifts such that $\mathcal{L}(X) = \mathcal{L}(Y)$, then $X = Y$.

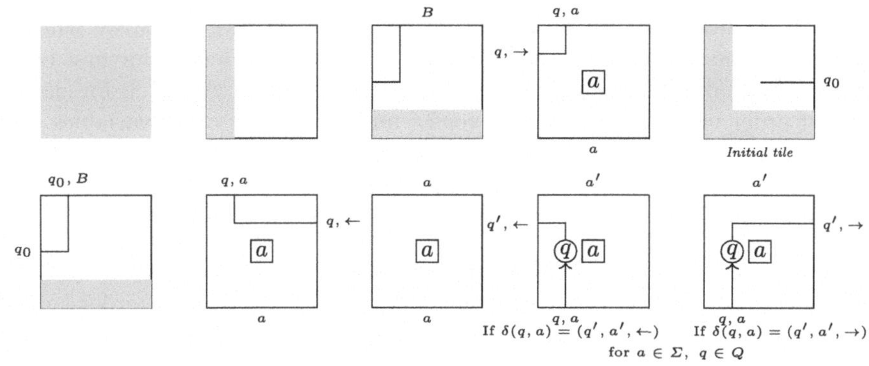

Fig. 1. Computation tiles

2.2 Computability

A decision problem is a predicate with some free variables (first or second order). The *arithmetic hierarchy* for decision problems is defined inductively as follows:

- Σ_0^0 and Π_0^0 are the set of decidable problems.
- R is in Π_{n+1}^0 if there exists $S \in \Sigma_n^0$ such that $R \Leftrightarrow \forall x\ S(x)$.
- R is in Σ_{n+1}^0 if there exists $S \in \Pi_n^0$ such that $R \Leftrightarrow \exists x\ S(x)$.

Here, quantifiers range over natural numbers. We define in a simililar way the analytical hierarchy:

- Σ_0^1 and Π_0^1 are the set of problems in the arithmetical hierarchy.
  R is in Π_{n+1}^1 if there exists $S \in \Sigma_n^1$ such that $R \Leftrightarrow \forall_1 X\ S(X)$.
- R is in Σ_{n+1}^1 if there exists $S \in \Pi_n^1$ such that $R(x) \Leftrightarrow \exists_1 X\ S(X)$.

Here, $\forall_1$ and $\exists_1$ quantify over sets of natural numbers. We often write simply $\exists$ and $\forall$ with a capital letter variable when the context is clear.

We use *many-one reductions* to compare the computational complexity of decision problems. We denote $Q \leq P$ is there exists a computable function $f : \mathbb{N} \to \mathbb{N}$ such that $Q(x) \Leftrightarrow P(f(x))$ for all inputs x. For a class C and a problem P, if $Q \leq P$ holds for all $Q \in C$, then P is *C-hard.* If P is in C and C-hard, we say that it is *C-complete.*

2.3 MSO Logic

We now review the formalism used in [3,9] for defining subshifts (and, more generally, shift-invariant sets of configurations) using logical formulas. Let $\mathcal{A}$ be a finite set of colours. A *term* is either a first order variable or one of East(t), West(t), North(t) and South(t), where t is a term. An *atomic formula* is $t = t'$ or $P_c(t)$, where t and t' are terms and $c \in \mathcal{A}$ is a colour. A *monadic second order* (*MSO* for short) formula is constructed from terms using the logical connectives

$\neg$, $\wedge$ and $\vee$, plus first and second order quantification. In general, we will use uppercase letters for second order variables and lowercase letters for first order variables. Second order quantification is sometimes denoted $\forall_1$ and $\exists_1$ for clarity. The first order variables will range over $\mathbb{Z}^2$ and the second order variables over subsets of $\mathbb{Z}^2$. The *radius* of a formula is the maximal depth of terms t in it. A formula is *closed* if all variables inside the formula are bound by a quantifier. It is *first order* (*FO* for short) if it contains no second order quantifier.

Remark 1. We concentrate on MSO formulas of the form $Q_1X_1Q_2X_2\cdots Q_nX_n\psi$, where the Q_i are second order quantifiers and ψ is first order, in order to discuss its number of second order quantifier alternations. A formula with mixed quantifier order can always be put into this form by replacing each out-of-order first order variable with a second order variable constrained to be a singleton. For example, $\forall x\ \exists Y\ \psi$ with ψ first order becomes $\forall X\ \exists Y\ \neg(\exists n\ \forall m\ m \in X \Leftrightarrow m = n) \vee (\exists x\ x \in X \wedge \psi)$.

We define a hierarchy called *MSO hierarchy* of the complexity of formulas:

- $\bar{\Sigma}_0$ and $\bar{\Pi}_0$ are the set of first order formulas.
- $\phi \in \bar{\Sigma}_{n+1}$ if it is of the form $\exists X_1\ \ldots \exists X_n\ \psi$ with $\psi \in \bar{\Pi}_n$.
- $\phi \in \bar{\Pi}_{n+1}$ if it is of the form $\forall X_1\ \ldots \forall X_n\ \psi$ with $\psi \in \bar{\Sigma}_n$.

For any configuration $x \in \mathcal{A}^{\mathbb{Z}^2}$, we consider the model $\mathcal{M}_x = (\mathbb{Z}^2, \tau)$, whose domain is $\mathbb{Z}^2$ and whose signature τ contains

- four unary functions East, West, North and South, whose interpretations are $\text{East}^{\mathcal{M}_x}(a,b) = (a+1,b)$,$\text{West}^{\mathcal{M}_x}(a,b) = (a-1,b)$, $\text{North}^{\mathcal{M}_x}(a,b) = (a,b+1)$ and $\text{South}^{\mathcal{M}_x}(a,b) = (a,b-1)$, and
- a unary predicate symbol P_c for each $c \in \mathcal{A}$, with the interpretation $P_c^{\mathcal{M}_x}(\boldsymbol{v}) = \top$ if and only if $x_{\boldsymbol{v}} = c$.

For a formula ϕ, we write $x \models \phi$ if $\mathcal{M}_x \models \phi$, that is, ϕ holds in the model $\mathcal{M}_x$. For each MSO formula ϕ, denote $X_\phi = \{x \in \mathcal{A}^{\mathbb{Z}^2} \mid x \models \phi\}$. The set X_ϕ is always shift-invariant but not always closed, and so not always a subshift (see Example 3). For a set of configurations X, if $X = X_\phi$ for some ϕ, we say that X is *MSO-definable*, and C*-definable* for a class of formulas C if $\phi \in C$.

Every SFT is definable by an MSO formula of the form $\forall v\ \psi$, where ψ is quantifier-free. This is essentially the format used by Griddy [7], with syntactic support for easily defining complex quantifier-free formulas.

If $X_\phi = X_\psi$, we say ϕ and ψ are *equivalent*. If Q is a quantifier, then $\bar{Q} = \forall$ if $Q = \exists$, $\bar{Q} = \exists$ otherwise. We say that a quantifier Q in a formula ϕ is *irrelevant* if ϕ is equivalent to the formula where that quantifier has been replaced by $\bar{Q}$.

Our second order variables range over subsets of $\mathbb{Z}^2$. It is equivalent to quantify over configurations in the following sense. Consider a formula $\exists X \in X_\phi\ \psi$, where X ranges over configurations of X_ϕ and ψ may contain subformulas $P_c(X(t))$ with the interpretation that the term t is coloured with the colour c in the configuration X. This formula be rewritten as an equivalent $\bar{\Sigma}_n$ formula if $\phi, \psi \in \bar{\Sigma}_n$, and analogously for $\bar{\Pi}_n$, by replacing X with $|\mathcal{A}|$ subsets of $\mathbb{Z}^2$, one for each color, with the requirement that the sets form a partition of $\mathbb{Z}^2$. Thus, we allow quantification over configurations since it is often more convenient. Furthermore, we may restrict any such quantification to be over a set of configurations defined by a first order formula ψ without changing the second-order quantifier complexity: $\forall_1 X \in X_\psi\ \phi$ can be implemented as $\forall_1 X(\neg\psi(X) \vee \phi)$, and $\exists_1 X \in X_\psi\ \phi$ as $\exists_1 X(\psi(X) \wedge \phi)$, after which the quantifiers of ϕ can be moved over $\psi(X)$ to bring it to prenex normal form.

Example 1. We define $\text{fin}(E)$, a $\bar{\Sigma}_1$ formula with a free second order variable E which is true if and only if the set $E \subset \mathbb{Z}^2$ is finite. The formula is $\text{fin}(E) = \exists N \exists S\ \phi_1 \wedge \phi_2 \wedge \phi_3 \wedge \phi_4 \wedge \phi_5$, where

- $\phi_1 = \forall v\ (v \in N \Leftrightarrow \text{West}(v) \in N \wedge \text{South}(v) \in N)$
- $\phi_2 = \forall v\ (v \in S \Leftrightarrow \text{East}(v) \in S \wedge \text{North}(v) \in S)$
- $\phi_3 = \exists v\ v \in N \wedge \neg\,\text{North}(v) \in N \wedge \neg\,\text{East}(v) \in N$
- $\phi_4 = \exists v\ v \in S \wedge \neg\,\text{South}(v) \in S \wedge \neg\,\text{West}(v) \in S$
- $\phi_5 = \forall v\ v \in E \Rightarrow v \in N \wedge v \in S$

The idea is that $N \subset \mathbb{Z}^2$ is a nondegenerate southwest quarter-plane, $S \subset \mathbb{Z}^2$ is a nondegenerate northeast quarter-plane, and E is contained in their intersection.

Example 2. We define $\text{Stable}(X) = \forall n\ n \in X \Leftrightarrow \text{East}(n) \in X$ where X is a second order variable. The predicate $\text{Stable}(X)$ is valuated to true if, whenever an element is in X, then all the elements of the same line are in X too. We define also $\text{Line}(X, Y)$ as the formula

$$\text{Stable}(X) \wedge ((\text{Stable}(Y) \wedge (\exists n\ n \in X \wedge n \in Y)) \Rightarrow (\forall n\ n \in X \Rightarrow n \in Y)).$$

Then $\forall Y\ \text{Line}(X, Y)$ is true if and only if X is exactly the set of elements of a line, or is empty. Finally, define the formula $\phi = \exists X\ \forall Y\ \text{Line}(X, Y) \wedge (\forall n\ P_{\blacksquare}(n) \Rightarrow n \in X)$. The subshift X_ϕ consists of exactly those configurations where every symbol $\blacksquare$ appears in the same line.

Example 3. Define the formula $\phi = \exists x\ \forall y\ P_{\blacksquare}(y) \Leftrightarrow y = x$. Then X_ϕ is the set of configurations where there is exactly one position colored $\blacksquare$. It is not a subshift since it is not topologically closed. Indeed, there is a sequence of configurations in X_ϕ which converges to the all-$\square$ configuration, which is not in X_ϕ.

3 First Order Formulas

In this section, we analyze the structure of FO-definable sets. The main result is that they are exactly the sets that can be defined by restricting the number of occurrences of a finite number of finite patterns up to some finite upper bound. Our techniques come from [2], where similar results were proved in the context of *picture languages*.

Definition 1. *Let $k \geq 0$. For $0 \leq a \leq k$ and a finite pattern P on an alphabet $\mathcal{A}$, define $S_a^k(P) \subset \mathcal{A}^{\mathbb{Z}^2}$ as the set of configurations where P appears exactly a times if $a < k$, or at least k times if $a = k$.*

Let also $n \geq 0$. For a function $f : \mathcal{A}^{n \times n} \to [0, k]$, define $C_f = \bigcap_P S_{f(P)}^k(P)$. Then $\{C_f \mid f : \mathcal{A}^{n \times n} \to [0, k]\}$ is a partition of $\mathcal{A}^{\mathbb{Z}^2}$ into equivalence classes, and if two configurations $x, y \in \mathcal{A}^{\mathbb{Z}^2}$ are in the same class, we denote $x \sim_{(n,k)} y$.

In other words, $x \sim_{(n,k)} y$ means that each $n \times n$ pattern either occurs exactly $t < k$ times in both x and y, or occurs at least k times in both.

Lemma 1. *Let ϕ be a first order formula. There exists a couple $(n,k) \in \mathbb{N}^2$ such that X_ϕ is a union of $\sim_{(n,k)}$-equivalence classes. Moreover, this union is computable from ϕ.*

We stress that apart from the computability of the union, this result essentially appeared already in [2,3].

Corollary 1. *Let ϕ be a first order formula. We can compute two formulas of the form $\exists x_1 \ldots \exists x_n \forall y_1 \ldots \forall y_m\ \psi_1$ and $\forall x_1 \ldots \forall x_n \exists y_1 \ldots \exists y_m\ \psi_2$ with ψ_1 and ψ_2 quantifier-free which are equivalent with ϕ.*

Proof. For all $n, k \in \mathbb{N}$, $0 \geq a \geq k$ and P pattern of size $n \times n$, $S_a(P)$ is first-order definable by formulas of the form $\exists x_1 \ldots \exists x_n \forall y_1 \ldots \forall y_m\ \psi_1$ and of the form $\forall x_1 \ldots \forall x_n \exists y_1 \ldots \exists y_m\ \psi_2$, with ψ_1 and ψ_2 quantifier-free, and these formulas are computable from $S_a(P)$. It follows that it is also the case for every first-order formula using Lemma 1.

4 MSOSUB

In this section, we are interested in the following problems:

Definition 2 (MSOSUB). *Let C be a class of MSO formulas. We define the following problem, called C-SUB: given a formula $\phi \in C$, is X_ϕ a subshift?*

If C is $\bar{\Sigma}_0$ or $\bar{\Pi}_0$, we prefer to call this problem FOSUB.

4.1 FOSUB

Theorem 1. FOSUB *is Π^0_4-complete.*

We split the proof into Lemmas 2 and 3.

Lemma 2. FOSUB *is Π^0_4.*

Proof (sketch). Let ϕ be a first order formula. Using Lemma 1 we can compute $n, k \in \mathbb{N}$ and disjoint $\sim_{(n,k)}$-equivalence classes $C_1, \ldots, C_r$ such that $X_\phi = \bigcup_{i=1}^r C_i$. We want to prove that knowing whether $\bigcup_{i=1}^r C_i$ is closed is Π^0_4.

For a $\sim_{(n,k)}$-equivalence class $C = \bigcap_{P \in \mathcal{A}^{n^2}} S_{a_P}(P)$, define the function $f_C : \mathcal{A}^{n^2} \to \{0, \ldots, k\}$ by $f_C(P) = a_P$. A *map* of C is a function $M : \mathcal{A}^{n^2} \to \mathcal{P}_{\text{fin}}(\mathbb{Z}^2)$. For $\ell \in \mathbb{N}$ and two maps M_1, M_2 of C, we say that (C, ℓ, M_1, M_2) *tiles the plane* if there exists $x \in \mathcal{A}^{\mathbb{Z}^2}$ such that for each $P \in \mathcal{A}^{n^2}$:

- if $f_C(P) < k$, then the pattern P appears at positions $M_1(P) \cup M_2(P)$ and no other positions,

- if $f_C(P) = k$, then the pattern P appears on positions $M_1(P) \cup M_2(P)$ (with $|M_1(P) \cup M_2(P)| \geq k$), and if $|M_1(P)| < k$, the only positions of $[-\ell, \ell - 1]^2$ at which P appears are in $M_1(P)$.

To decide whether a given quadruple tiles the plane is Π_1^0.

Now let us consider the following arithmetical formula F: there exist $m \in \mathbb{N}$, an index $i \in \{1, \ldots, r\}$ and two functions $I_1, I_2 : \mathcal{A}^{n^2} \to \{0, \ldots, k\}$ such that $I_1 + I_2 = f_{C_i}$, and for all $\ell > m$ there exist two maps M_1, M_2 of C_i such that

- $|M_j(P)| = I_j(P)$ for each $j = 1, 2$ and $P \in \mathcal{A}^{n^2}$,
- $M_1(P) \subseteq [-m, m]^2$ and $M_2(P) \cap [-\ell, \ell - 1]^2 = \emptyset$ for each $P \in \mathcal{A}^{n^2}$,
- (C_i, ℓ, M_1, M_2) tiles the plane, and
- there is no $j \in \{1, \ldots, r\}$ such that $(C_j, 0, M_1, P \mapsto \emptyset)$ tiles the plane.

By its form, F is Σ_4^0. One can prove that F holds if and only if $\bigcup_{i=1}^r C_i$ is not closed. The idea is that if F holds, then one obtains a sequence x_ℓ of configurations in which the maps M_1 and M_2 control the positions of each pattern $P \in \mathcal{A}^{n^2}$, and the tiling conditions make sure that at least one pattern "escapes to infinity" in such a way that no limit point of the sequence is in $\bigcup_{i=1}^r C_i$.

Lemma 3. FOSUB *is Π_4^0-hard.*

Proof (sketch). We reduce the Π_4^0-complete problem $\forall COF$: given a Turing Machine M which takes two inputs, does $M(v, \_)$ have a cofinite language for all v? For this, we define an SFT X and add on it a first order formula to get a formula ϕ. Then, we will prove that $M \in \forall COF$ if and only if $\phi \in$ FOSUB.

Let M be a turing machine. The SFT X is the one described by Fig. 2 (the only allowed patterns are those of size 2×2 in this figure, but other inputs v and w are possible, even the infinite one). The input v is copied to the southeast until it hits the gray diagonal signal emitted by the $*$-symbol at the end of the second input w, where it is erased. The diagonal turns black after the point of erasure. The machine M is not allowed to halt, so in all configurations containing two finite inputs v and w, $M(v, w)$ does not halt.

Let ϕ' be the formula which defines this SFT, and let $\phi = \phi' \wedge ((\exists x\ \exists y\ P_1(x) \wedge P_2(y)) \Rightarrow \exists z\ P_3(z))$, where tiles 1, 2 and 3 are showed in Fig. 2). Intuitively, ϕ requires the configuration to satisfy ϕ', and if there is a diagonal with a finite input v written on it, then there also exists a finite input w.

Now $M \in \forall COF$ if and only if $\phi \in MFOSUB$, since the only reason for X_ϕ not to be closed is that for some input v, there are infinitely many inputs w such that $M(v, w)$ does not halt, and a limit point of the correspondng configurations violates the second condition of ϕ.

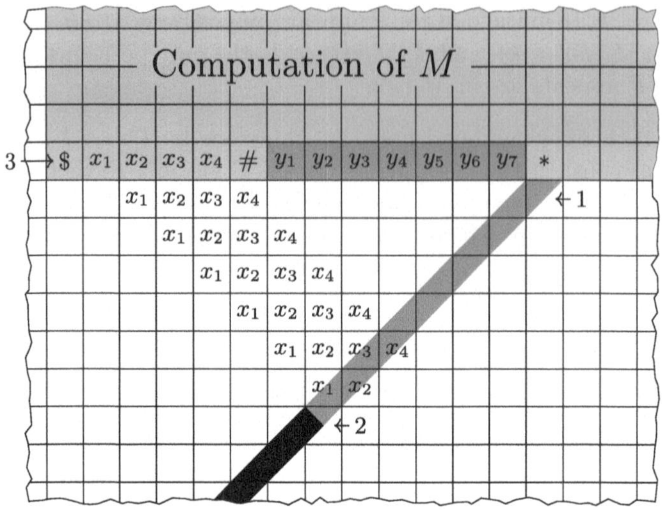

Fig. 2. A configuration of the SFT X in the proof of Lemma 3.

4.2 Higher Order Cases

Theorem 2. *For all $n \geq 1$, $\bar{\Pi}_n$-SUB and $\bar{\Sigma}_n$-SUB are Π^1_n-complete.*

We split the proof into Lemmas 4, 5 and 6.

Remark 2. All problems of the form $\exists_1 X_1 \exists x_1 \forall x_2\ R$ with R a decidable predicate, are in Σ^0_2. Dually, all problems of the form $\forall_1 X_1 \forall x_1 \exists x_2\ R$ with R a decidable predicate, are in Π^0_2.

Lemma 4. *For all $n \geq 1$, $\bar{\Pi}_n$-SUB and $\bar{\Sigma}_n$-SUB are in Π^1_n.*

Proof. Let $\phi \in \bar{\Pi}_n$ with m second order quantifiers (the $\bar{\Sigma}_n$ case is similar). Denote by ϕ^* the first order formula on $\mathcal{A} \times \{0,1\}^m$ obtained from ϕ by removing second order quantifiers and replacing each $v \in X_i$ with $\bigvee_{c \in \pi_i^{-1}(1)} P_c(v)$, where $\pi_i(c_0, \ldots, c_m) = c_i$. For a configuration $x \in \mathcal{A}^{\mathbb{Z}^2}$, we have $x \vDash \phi$ if and only if

$$\forall X_1\ \ldots\ QX_m\ x \times X_1 \times \ldots \times X_m \vDash \phi^* \tag{1}$$

for $Q = \exists$ if n is even, $Q = \forall$ otherwise. From Corollary 1 we deduce that $x \times X_1 \times \ldots \times X_m \vDash \phi^*$ is a Δ^0_2 condition, when ϕ^* is given as input and x and the X_i are given as oracles. With Remark 2 it follows that (1), and hence $x \vDash \phi$, is Π^0_2 for $n = 1$, and Π^1_{n-1} for $n \geq 2$, with ϕ as input and x as an oracle.

The set X_ϕ is closed if and only if for all sequences x_n of configurations such that $x_n \vDash \phi$ for all n and $x_n|_{[-n,n]^2} = x_m|_{[-n,n]^2}$ for all $m \geq n$, we have $\lim_n x_n \vDash \phi$. By the above, this condition is Π^1_n.

Lemma 5. *For all $n \geq 1$, $\bar{\Pi}_n$-SUB is Π^1_n-hard.*

Proof. We reduce an arbitrary problem in Π_n^1. Let R be a computable predicate. We construct a formula $\phi \in \bar{\Pi}_n$ such that $\forall A_1 \dots Q A_n\ \bar{Q}k\ Qm\ R$ holds if and only if X_ϕ is a subshift, with $Q = \exists$ if n is even and $Q = \forall$ otherwise.

Let ϕ' be the formula which describes a tiling where a northeast quarter-plane is coloured with 0 and 1 such that each row is coloured in a same way, and the other tiles are blank. Since the quarter-plane must appear, $X_{\phi'}$ is not a subshift. The infinite binary sequence repeated on each line is interpreted as the oracle A_1 in the following.

Let $\phi = \forall X_2 \in X_{\phi'} \dots\ \bar{Q}X_n \in X_{\phi'}\ \bar{Q}Z\ \phi' \wedge \psi$, where ψ is the formula stating the following:

- each of $X_2, \dots, X_n$ has a northeast quarter-plane at the same position as the main configuration $x \in X_{\phi'}$,
- the configuration Z contains, in the same quarter-plane, a simulated computation of a Turing machine M that can use the binary sequences of x and the X_i as oracles,
- if $\bar{Q} = \exists$, then M visits a special state q_s an infinite number of times (a $\bar{\Pi}_1$ condition by Example 1), and
- if $\bar{Q} = \forall$, then M visits the state q_s a finite number of times.

Notice that the configuration Z is determined by what was quantified before, so its quantifier is irrelevant. Hence ϕ is $\bar{\Pi}_n$.

We specify the machine M. Starting with $k = 0$, it enumerates $m = 0, 1, 2, \dots$ until it finds one such that $R(A_1, \dots, A_n, k, m)$ does not hold. Whan this happens, it visits the state q_s, increments k, and starts the enumeration again.

Notice that $x \models \phi$ if and only if $\forall A_2 \dots\ \bar{Q}A_n\ \ Qk\ \bar{Q}m\ \neg R$ holds. Hence, if $\forall A_1 \exists A_2 \dots Q A_n\ \bar{Q}k\ Qm\ R$ holds, then X_ϕ is empty, and so a subshift. If the formula does not hold, then at least one configuration with a quarter-plane is in X_ϕ. But the blank configuration is not in X_ϕ, so X_ϕ is not a subshift.

Lemma 6. *For all $n \geq 1$, $\bar{\Sigma}_n$-SUB is Π_n^1-hard.*

The proof is similar to that of Lemma 5, but instead of emptiness, we use a separate binary layer with at most one 1-symbol.

5 Complexity of the MSO Languages

In this section, we consider the complexity of the C-definable sets and subshifts for C classes of the MSO hierarchy.

5.1 Maximal Complexity

Here we determine the maximal complexity of languages of formulas in the different classes of the MSO hierarchy.

Theorem 3. *Languages of FO and $\bar{\Sigma}_1$ sets are Σ_2^0. Languages of $\bar{\Sigma}_n$ sets for $n \geq 2$ are Σ_{n-1}^1. Languages of $\bar{\Pi}_n$ sets for $n \geq 1$ are Σ_n^1.*

Proof. Let $\phi \in \bar{\Sigma}_n$. For a pattern P, we have $P \in \mathcal{L}(X_\phi)$ if there exists $x \in \mathcal{A}^{\mathbb{Z}^2}$ such that $x \models \phi$ and P appears in x. Given x as an oracle, P appearing in x is a Σ_1^0 predicate, and $x \models \phi$ is Σ_{n-1}^1 if $n \geq 2$, and Σ_2^0 if $n \leq 1$ thanks to Corollary 1 and Remark 2. We conclude that $\mathcal{L}(X_\phi)$ is Σ_{n-1}^1 if $n \geq 2$ and Σ_2^0 if $n \leq 1$ (we need once more Remark 2 for the case $n = 1$).

Let $\phi \in \bar{\Pi}_n$ for $n \geq 1$. Given $x \in \mathcal{A}^{\mathbb{Z}^2}$ as an oracle, $x \models \phi$ is Π_{n-1}^1 if $n \geq 2$, and Π_2^0 if $n = 1$ thanks to Corollary 1 and Remark 2. We conclude that $\mathcal{L}(X_\phi)$ is Σ_n^1 for all $n \geq 1$.

Theorem 4. *There exists an FO-definable subshift with a Σ_2^0-hard language.*

Proof. Let ϕ' be the first order formula which defines an SFT X containing, in a northeast quarter-plane, a simulated computation of a turing machine M'. Figure 3 contains asample configuration. The machine M' takes the code $[M]$ of another Turing machine M, and starting with $k = 0$, simulates M with input k. If M halts on k, then M' enters a special state q_H, increments k and starts again. The tile containing q_H sends a signal (dark gray in Fig. 3) to the west, which turns south when it encounters the left side of the quarter-plane, merging with another southward signal if one is present. We forbid to have such a signal under the diagonal of the quarter-plane.

Now let $\phi = \phi' \wedge ((\exists u\ \exists v\ P_1(u) \wedge P_2(v) \wedge P_3(\mathrm{South}(v))) \Rightarrow \exists w\ \neg P_4(w) \wedge P_3(\mathrm{West}(w)))$ (tiles 1, 2, 3 and 4 are shown on Fig. 3). Intuitively, ϕ requires that ϕ' is true and if the computation starts and the input is finite, then there exists a tile on the west border of the quarter-plane with no signal on it.

We prove first that X_ϕ is a subshift. Suppose for a contradiction that there is a sequence $(x_n)_{n\in\mathbb{N}}$ such that $x_n \models \phi$ for all n and $x_n \overset{n\to\infty}{\longrightarrow} x \not\models \phi$. Since $X_{\phi'}$ is a subshift, $x \models \phi'$. Hence x shows a computation with a Turing machine M on which the west border of the quarter-plane is covered by the signal. As x_n converges to x, from some n onward, the code $[M]$ is entirely written on the south border of the quarter-planes of the x_n and the computation of M' is determined by this. Since $x_n \models \phi$, the west borders are not covered by the signal. Then the the sequence x_n must be eventually constant. With $x \not\models \phi$ we reach a contradiction, so X_ϕ is a subshift.

Notice that $[M]$ is in $\mathcal{L}(X_\phi)$ if and only if $M \in$ coTOTAL, i.e. if M does not halt on at least one input. This problem is known to be Σ_2^0-complete.

Theorem 5. *For all $n \geq 1$, there exists a $\bar{\Pi}_n$-definable set with a Σ_n^1-hard language.*

Proof. Let ϕ' be the first order formula which defines the set of configurations that are blank except for an infinite binary word w_1, and the code $[M]$ of a total Turing machine M written just above this word, at the left end. As no other configurations are allowed, $X_{\phi'}$ is not a subshift.

Let $\phi = \phi' \wedge \forall X_2\ \exists X_3 \ldots\ Q\ X_n\ Q\ Z\ \psi$, where $Q = \forall$ if n is even and $Q = \exists$ otherwise, and ψ specifies that each X_i contains an infinite binary word w_i at the same position as the main configuration x, and Z contains a simulated computation which verifies $\exists k\ \forall m\ M(w_1, \ldots, w_n, k, m)$ if n is even, and

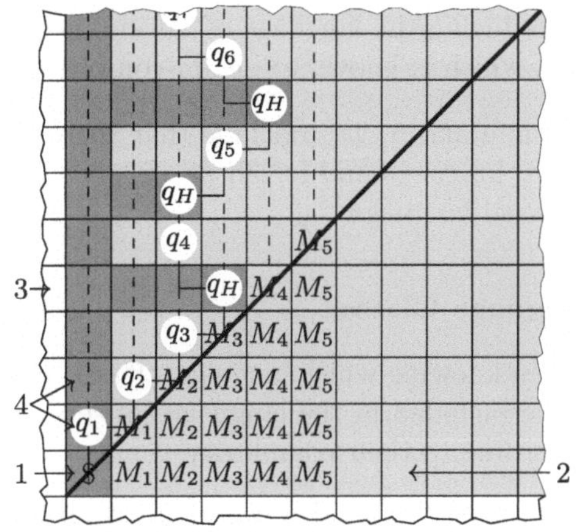

Fig. 3. Computation of M'.

$\forall k\ \exists m\ M(w_1, \ldots, w_n, k, m)$ if n is odd. We use the same technique as in Lemma 5 to design ϕ so that it is $\bar{\Pi}_n$.

Now $[M]$ is in $\mathcal{L}(X_\phi)$ if and only if

$$\exists w_1\ \forall w_2\ \ldots\ Q\ w_n\ \bar{Q}\ k\ Q\ m\ M(w_1, \ldots, w_n, k, m).$$

This condition is Σ_n^1-hard, so $\mathcal{L}(X_\phi)$ is Σ_n^1-hard too.

Theorem 6. *There exists a $\bar{\Pi}_1$-definable subshift with a Π_3^0-hard language.*

Proof (sketch). Let X be the subshift in which all the configurations are blank except at most one infinite word, which contains the code $[M]$ of a Turing machine M followed by an infinite sequence of tiles coloured with $\{0, 1, |\}$ (plus all limit configurations of these). Let ψ be the first order formula such that $X = X_\psi$.

Let ϕ_1 be a $\bar{\Pi}_1$ formula which requires that if $[M]$ is finite, then there are infinitely many $|$-tiles. We interpret this as $[M]$ being followed by an infinite list of natural numbers in binary.

Let ϕ_2 be a $\bar{\Pi}_1$ formula which requires that if $[M]$ is finite, then for each natural number n in the list, M never halts on n. This can be done by universally quantifying on a configuration containing a simulated computation of $M(n)$.

Let ϕ_3 be a $\bar{\Pi}_1$ formula which requires that if $[M]$ is finite, then for each natural number n, either n occurs in the list, or $M(n)\downarrow$. This can be done by simulating $M(n)$ and simultaneously searching for n in the list, and requiring that this computation halts at some point.

Let ϕ_4 be a $\bar{\Pi}_1$ formula which requires that if $[M]$ is finite, the list of natural numbers written after M is strictly increasing. Once again, we can do this using a single universal quantifier.

Then let ϕ be a $\bar{\Pi}_1$ formula equivalent with $\psi \wedge \phi_1 \wedge \phi_2 \wedge \phi_3 \wedge \phi_4$ (just rewriting this formula in the prenex form). One can prove that X_ϕ is a subshift with a

Π^0_3-hard language by reducing the language of Turing machines that diverge on infinitely many inputs, which is known to be Π^0_3-complete.

Thanks to Theorems 4 and 5, we conclude that there exists a $\bar{\Sigma}_1$-definable subshift with a Σ^0_2-hard language and for all $n \geq 2$, there exists a $\bar{\Sigma}_n$-definable subshift with a Σ^1_{n-1}-hard language.

5.2 Definable Language Classes

Here we are interested in knowing which language classes of the arithmetical and analytical hierarchy are included in the languages of sets or subshifts definable by MSO formulas according to their complexity.

Theorem 7. *Some decidable languages are not $\bar{\Sigma}_1$-definable by sets. Some decidable languages are not $\bar{\Pi}_1$-definable by sets.*

Proof. We prove here the $\bar{\Sigma}_1$ case; the $\bar{\Pi}_1$ case is similar. The proof is based on the technique of [4].

Let X be the subshift on $\mathcal{A} = \{0, 1, \#\}$ which is the closure of the set of configurations with two square $\{0, 1\}$-patterns of the same size aligned and at distance one where everything around these patterns is colored by # and where each pattern is the mirror image of the other one. We denote the mirror image of a pattern P by P^R. It is clear that $\mathcal{L}(X)$ is decidable.

Suppose for a contradiction that there exists a $\bar{\Sigma}_1$ formula $\phi = \exists Y \in \Sigma^{\mathbb{Z}^2}\ \psi \in \bar{\Sigma}_1$, where ψ is FO, such that $\mathcal{L}(X) = \mathcal{L}(X_\phi)$. For $N \geq 1$ and $P \in \{0, 1\}^{N^2}$, let $x_P \in X$ be the configuration where $x|_{[-N,-1]\times[0,N-1]} = P$ and $x|_{[0,N-1]^2} = P^R$. The pattern $\rho_P = x|_{[-N-1,N]\times[-1,N]}$, which contains P and P^R surrounded by #-symbols, is in $\mathcal{L}(X) = \mathcal{L}(X_\phi)$, so it appears in some configuration of X_ϕ. We see that the only configuration of X_ϕ in which ρ_P appears in the same position as in x_P must be x_P itself.

As ψ is FO, by Lemma 1 there exist $n, k \in \mathbb{N}$ such that X_ψ is a union of $\sim_{(n,k)}$-equivalence classes. For $P \in \{0, 1\}^{N^2}$, let $y_P \in \Sigma^{\mathbb{Z}^2}$ be such that $x_P \times y_P \models \psi$. For N large enough there exist $P \neq Q \in \{0, 1\}^{N^2}$ such that the pattern $(x_P \times y_P)|_P$ has the same border of size n as $(x_Q \times y_Q)|_Q$ and the same number of each $n \times n$ pattern counted up to k (since $(2|\Sigma|)^{4Nn-4n^2}(k+1)^{(2|\Sigma|)^{n^2}} < 2^{N^2}$ for large N).

Now consider the configuration $z = (z_1, z_2)$ defined by $z_{\boldsymbol{v}} = (x_P \times y_P)_{\boldsymbol{v}}$ for $\boldsymbol{v} \notin [0, N-1]^2$ and $z_{\boldsymbol{v}} = (x_Q \times y_Q)_{\boldsymbol{v}}$ for $\boldsymbol{v} \in [0, N-1]^2$. In other words, we have replaced the P^R-patch of x_P with Q^R, and made the corresponding replacement on the y-layer as well. Let us prove that $z \sim_{(n,k)} x_P \times y_P$. They are equal outside $C_{\text{in}} = [n, N-n-1]^2$ (see Fig. 4) and they have the same number of patterns of size $n \times n$ counted up to k inside $C_{\text{out}} = \mathbb{Z}^2 \setminus [0, N-1]^2$. So if z and $x_P \times y_P$ are not equivalent, this is due to a pattern of size $n \times n$ which overlaps C_{in} and C_{out}, which is impossible.

Thus $z \in X_\psi$ and $z_1 \in X_\phi \subseteq X$. But z_1 consists of the patterns P and Q^R surrounded by #-symbols, so it is not in X, and we have a contradiction.

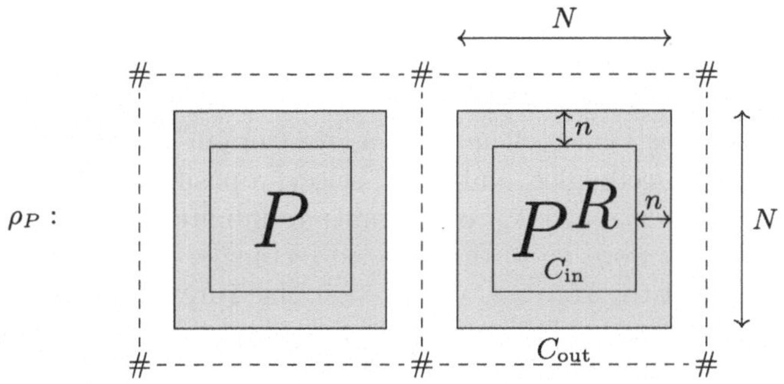

Fig. 4. Non $\bar{\Sigma}_1$-definability of a decidable language

Theorem 8. *For all $n \geq 2$, all Π^1_{n-1} subshifts are $\bar{\Pi}_n$-definable, and all Σ^1_{n-1} subshifts are $\bar{\Sigma}_n$-definable.*

Proof (sketch). We sketch the proof for $n = 3$. In the first case, we have a subshift $X \subseteq \mathcal{A}^{\mathbb{Z}^2}$ whose language is defined by a Π^1_2 formula $\forall_1 A_1 \exists_1 A_2 \forall k \exists m \xi$, where ξ is computable. We construct an MSO formula $\phi = \forall X_1 \exists X_2 \forall X_3 \psi$, where the X_i are constrained by FO conditions, that defines X.

We follow the technique of [9, Theorem 3]. The configuration X_1 contains a finite rectangle R, an encoding w of a finite rectangular pattern over $\mathcal{A}$, and an infinite binary word w_1. The configuration X_2 contains either a gadget witnessing that w is not the encoding of the pattern $x|_R$ in the main configuration $x \in \mathcal{A}^{\mathbb{Z}^2}$ (either due to having the wrong dimensions, or differing at a single coordinate), or an infinite binary word w_2. The configuration X_3 contains an infinite simulated computation of a Turing machine that can access the infinite words w_1 and w_2 as oracle tapes. The machine checks sequentially for $k = 0, 1, 2, \ldots$ that there exists $m \in \mathbb{N}$ such that ξ holds. Whenever such an m is found, the machine enters a special state q_s and moves to the next value of k. The configuration also contains an upper half-plane H. The formula F checks that if X_2 contains the word w_2, then X_3 contains a q_s within the half-plane H.

Suppose $x \models \phi$. Given a rectangle R, take an X_1 that correctly encodes $x|_R$ into a word w and contains some w_1. There exists an X_2 that necessarily contains a w_2 (because a witness for w not encoding the pattern does not exist), and wherever we place the upper half-plane in X_3, it contains q_s. Hence the machine in X_3 enters q_s an infinite number of times, and $x|_R \in \mathcal{L}(X)$.

Suppose then $x \in X$. Every valid configuration X_1 contains some rectangle R and some words w and w_1 above it. If w does not encode $x|_R$, we can find a witness and choose it as X_2. If it does, we can choose X_2 to contain the word w_2 that corresponds to a suitable A_2. Then every X_3 satisfies ψ, since either it does not have the correct form, or the Turing machine enters q_s an infinite number of times. Hence $x \models \phi$.

In the second case, we have a subshift $X \subseteq \mathcal{A}^{\mathbb{Z}^2}$ whose language is defined by a Σ^1_2 formula $\exists_1 A_1 \forall_1 A_2 \exists k \forall m \xi$, where ξ is computable. We construct an

MSO formula $\phi = \exists X_1 \forall X_2 \exists X_3 \psi$ that defines X. Here, X_1 contains the infinite computation of a Turing machine that simply lists all finite rectangles R and nondeterministically guesses a pattern $P \in \mathcal{A}^R$ and an infinite binary word w_1 for each. It also selects one cell $\boldsymbol{w} \in \mathbb{Z}^2$ as the "origin". The configuration X_2 selects one of these rectangles, and either selects a position $\boldsymbol{v} \in R$ and contains a witness gadget for $x_{\boldsymbol{w}+\boldsymbol{v}} = P_{\boldsymbol{v}}$, or contains an infinite binary word w_2 at the same position as the word w_1 of the encoded rectangle. The configuration X_3 is similar to the X_3 of the Π^1_2 case, except that the Turing machine looks for an m such that ξ does not hold. The formula F is almost as before: if X_2 contains a w_2, then X_3 does not contain a q_s within H.

Suppose $x \models \phi$, and choose an X_1 from ϕ with origin $\boldsymbol{w}$. Given a rectangle R, we can find it encoded in X_1 together with a pattern P and a word w_1. For each X_2 that selects the encoding of R and P and a position $\boldsymbol{v} \in R$, we get a witness for $x_{\boldsymbol{w}+\boldsymbol{v}} = R_{\boldsymbol{v}}$. For those X_2 that contain a w_2 instead, we get an X_3 whose simulated Turing machine enters q_s a finite number of times. Hence $x|_R \in \mathcal{L}(X)$.

Suppose then $x \in X$, and choose a X_1 with origin $\mathbf{0}$ that correctly encodes every rectangular pattern P in x, and for each, contains a word w_1 corresponding to the set A_1 given by the Σ^1_2 formula of $\mathcal{L}(X)$. Every X_2 either contains a witness for the correctness of a single coordinate in some encoded pattern (which are all correct by definition) or a word w_2. Then the machine simulated in X_3 enters q_s only a finite number of times, so we can place the half-plane H above the last occurrence to satisfy ψ. Hence $x \models \phi$.

Table 1. Summary of the complexity of MSO languages.

	Sets		Subshifts	
	Max complexity	Includes all in	Max complexity	Includes all in
FO	Σ^0_2	$\Delta^0_1 \not\subseteq$	Σ^0_2	$\Delta^0_1 \not\subseteq$
$\bar{\Sigma}_1$	Σ^0_2	$\Delta^0_1 \not\subseteq$	Σ^0_2	$\Delta^0_1 \not\subseteq$
$\bar{\Pi}_1$	Σ^1_1	$\Delta^0_1 \not\subseteq$	$[\Pi^0_3, \Sigma^1_1]$	$\Delta^0_1 \not\subseteq$
$\bar{\Sigma}_n, n \geq 2$	Σ^1_{n-1}	Σ^1_{n-1}	Σ^1_{n-1}	Σ^1_{n-1}
$\bar{\Pi}_n, n \geq 2$	Σ^1_n	Π^1_{n-1}	$[\Pi^1_{n-1}, \Sigma^1_n]$	Π^1_{n-1}

6 Conclusion

In Sect. 4 we determined the complexity of C-SUB for various classes C. FOSUB turned out to be Π^0_4-complete, while $\bar{\Pi}_n$-SUB and $\bar{\Sigma}_n$-SUB for $n \geq 1$ are both Π^1_n-complete.

The results of Sect. 5 are summarized in Table 1. Note that we have completely characterized the $\bar{\Sigma}_n$ subshifts with $n \geq 2$, as well as the languages of $\bar{\Sigma}_n$ sets. We have also pinpointed the maximal complexity of the languages of first order and $\bar{\Sigma}_1$ subshifts, which are both Σ^0_2, higher than the well known Σ^0_1 bound of SFTs and sofic shifts. The main remaining question concerns the maximal complexity of $\bar{\Pi}_1$ subshifts: is it Σ^1_n, the same as for general $\bar{\Pi}_1$ sets, or is it strictly lower?

Acknowledgement. Ilkka Törmä was supported by the Academy of Finland grant 359921.

Disclosure of Interests. The authors have no competing interests.

References

1. Berger, R.: The Undecidability of the Domino Problem. Memoirs of the American Mathematical Society. American Mathematical Society (1966). https://books.google.fi/books?id=mLfTCQAAQBAJ
2. Giammarresi, D., Restivo, A., Seibert, S., Thomas, W.: Monadic second-order logic over rectangular pictures and recognizability by tiling systems. Inf. Comput. **125**(1), 32–45 (1996). https://doi.org/10.1006/inco.1996.0018
3. Jeandel, E., Theyssier, G.: Subshifts as models for MSO logic. Inf. Comput. **225**, 1–15 (2013)
4. Kass, S., Madden, K.: A sufficient condition for non-soficness of higher-dimensional subshifts. Proc. Am. Math. Soc. **141**(11), 3803–3816 (2013). https://doi.org/10.1090/S0002-9939-2013-11646-1
5. Lind, D., Marcus, B.: An Introduction to Symbolic Dynamics and Coding, 2 edn. Cambridge Mathematical Library. Cambridge University Press, Cambridge (2021). https://doi.org/10.1017/9781108899727
6. Salo, V., Törmä, I.: Diddy: a python toolbox for infinite discrete dynamical systems. In: Cellular Automata and Discrete Complex Systems, vol. 14152 of LNCS, pp. 33–47. Springer, Cham (2023). https://doi.org/10.1007/978-3-031-42250-8_3
7. Salo, V., Törmä, I.: Griddy (2023). GitHub repository. https://github.com/ilkka-torma/griddy
8. Thomas, W.: Languages, automata, and logic. In: Handbook of Formal Languages, vol. 3, pp. 389–455. Springer, Berlin (1997)
9. Törmä, I.: Subshifts, MSO logic, and collapsing hierarchies. In: Diaz, J., Lanese, I., Sangiorgi, D. (eds.) TCS 2014. LNCS, vol. 8705, pp. 111–122. Springer, Heidelberg (2014). https://doi.org/10.1007/978-3-662-44602-7_10
10. Wang, H.: Proving theorems by pattern recognition - II. Bell Syst. Techn. J. **40**(1), 1–41 (1961). https://doi.org/10.1002/j.1538-7305.1961.tb03975.x

Injectivity of Polynomials over Finite Discrete Dynamical Systems

Antonio E. Porreca and Marius Rolland(✉)

Aix-Marseille Université, CNRS, LIS, Marseille, France
marius.rolland@lis-lab.fr

Abstract. The analysis of observable phenomena (for instance, in biology or physics) allows the detection of dynamical behaviors and, conversely, starting from a desired behavior allows the design of objects exhibiting that behavior in engineering. The decomposition of dynamics into simpler subsystems allows us to simplify this analysis (or design). Here we focus on an algebraic approach to decomposition, based on alternative and synchronous execution as the sum and product operations; this gives rise to polynomial equations (with a constant side). In this article we focus on univariate, injective polynomials, giving a characterization in terms of the form of their coefficients and a polynomial-time algorithm for solving the associated equations.

Keywords: Finite dynamical systems · Functional digraphs · Graph direct product

1 Introduction

Finite discrete-time dynamical systems (FDDS) are pairs (X, f) where X is a finite set of states and $f : X \to X$ is a transition function (when f is implied, we denote (X, f) simply as X). We find these systems in the analysis of concrete models such as Boolean networks [1,2] and we can apply them to biology [3–5] to represent, for example, genetic regulatory networks or epidemic models. They also appear in chemistry [6], or information theory [7]. It is often useful to decompose large and complex systems into smaller subsystems, easier and computationally less expensive to analyze; this would allow us to determine global dynamical properties (*i.e.*, the existence and number of fixed points and the time of convergence to limit behaviors) in terms of the corresponding properties of the subsystems.

In the following, we focus on deterministic FDDS, and often identify them with their transition digraphs, which have uniform out-degree one. These graphs are also known as functional digraphs. Their general shape is a collection of cycles, where each node of a cycle is the root of a finite in-tree (a directed tree with the edges directed toward the root). The nodes of the cycles are periodic states, while the others are transient states. Figure 1 shows an example of FDDS.

The set $(\mathbb{D}, +, \times)$ of FDDS up to isomorphism, with the alternative execution of two systems as *addition* and the synchronous parallel execution of two FDDS

© The Author(s), under exclusive license to Springer Nature Switzerland AG 2025
A. Beckmann et al. (Eds.): CiE 2025, LNCS 15764, pp. 380–394, 2025.
https://doi.org/10.1007/978-3-031-95908-0_27

as *multiplication* is a commutative semiring [8]. As all semirings, we can define a semiring of polynomials over $\mathbb{D}$ and thus polynomial equations. Hence, in this paper we decompose FDDS from this algebraic perspective. For instance, the number of fixed points of a product of two FDDS is simply the product of the numbers of fixed points of the two factors.

Although it has already been proven that general polynomial equations over $\mathbb{D}$ (with variables on both sides of the equation) are undecidable, it is easily proved that, if one member of the equation is constant, then the problem is decidable (there is just a finite number of possible solutions) [8]. This variant of the problem is actually in NP because, since sums and products can be computed in polynomial time, we can just guess the values of the variables (their size is bounded by the constant side of the equation), evaluate the polynomial, and compare it with the constant side (in our case, isomorphism can be checked in linear time [9]).

However, more restricted equations are not yet classified in terms of complexity. For example, we do not know if monomial equations of the form $AX = B$ are NP-hard, in P, or possible candidates for an intermediate difficulty. However, it has been shown that when A and B are connected and with a cycle of length one (i.e., they are trees with a loop, or fixed point, on the root), the equation $AX = B$ can be solved in polynomial time [10]. It has also been proved that, if X is connected, we can solve $AX^k = B$ in polynomial time with respect to the size of the inputs (A and B) for all positive integer k [11].

These results share a common feature: all these equations have *at most one solution*. It is natural, then, to investigate whether *all* polynomial equations with at most one solution can be solved in polynomial time. However, we still lack a characterization for this type of equation; indeed, for the same polynomial on one side of the equation, if we change the constant side, the number of solutions can change. For this reason, considering the case where the polynomial is injective seems to be a relevant starting point, requiring a characterization of this class of polynomials, at least for the univariate case.

In this paper, as a starting point towards the solution of larger and larger classes of equations, we focus on univariate injective polynomials. After giving the necessary preliminaries (Sect. 2), we first focus on the semiring of unrolls (equivalence classes of FDDS sharing the same transient behavior, introduced in [10]) and prove that all univariate polynomials over unrolls are injective and that every equation $\sum_{i=0}^{m} \mathcal{U}(A_i)\mathcal{U}(X)^i = \mathcal{U}(B)$ can be solved in polynomial time (Sect. 3). Then, we prove that all univariate polynomial $P = \sum_{i=0}^{m} A_i X^i$ with at least one cancelable A_i (for some $i > 0$) is injective, and this will also give us a polynomial-time algorithm for solving the associated equations (Sect. 4). Finally, we show that this is also a necessary condition for injectivity (Sect. 5).

2 Preliminaries

In this paper we compose and decompose FDDS in terms of two algebraic operations. Thee *sum* of two FDDS consists in the mutually exclusive alternative

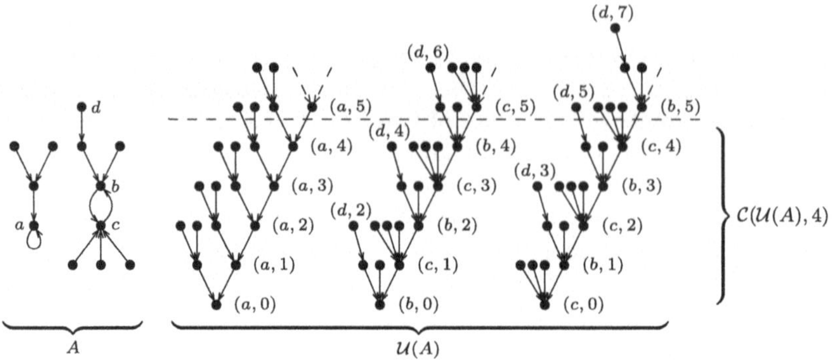

Fig. 1. An FDDS A with two connected components (on the left), a finite portion of its unroll $\mathcal{U}(A)$ (on the right) and the cut $\mathcal{C}(\mathcal{U}(A), 4)$ (below the dashed line). A few vertex names are shown in order to highlight their contribution to the unroll.

between their behaviors, while the *product* is their synchronous parallel execution. The set of FDDS taken up to isomorphism, with these two operations, forms a semiring [8]. This semiring is isomorphic to the semiring of functional digraphs with the operations of disjoint union and direct product. We recall [12] that the direct product of two graphs A and B is the graph C having vertices $V(C) = V(A) \times V(B)$ and edges

$$E(C) = \{((u, u'), (v, v')) \mid (u, v) \in E(A), (u', v') \in E(B)\}.$$

Within the class of *connected* FDDS, we further distinguish those having a (necessarily unique) cycle of size 1 (*i.e.*, a fixed point in terms of dynamics) and refer to them as *dendrons.*

In this semiring, the structure of the product is particularly rich. For example, the semiring is not factorial, *i.e.*, there exist irreducible FDDS A, B, C, D such that $A \neq C$ and $A \neq D$ but $AB = CD$. This richness might originate from the interactions between the periodic and the transient behaviors of the two FDDS being multiplied.

For the periodic behavior of FDDS, the product of a connected A, with a cycle of size p, and a connected B, with a cycle of size q, generates $\gcd(p, q)$ connected components, each with a cycle of size $\text{lcm}(p, q)$ [12]. For the analysis of transient behaviors, we use the notion of unroll introduced in [10].

Definition 1 (Unroll). *Let $A = (X, f)$ be an FDDS. For each state $u \in X$ and $k \in \mathbb{N}$, we denote by $f^{-k}(u) = \{v \in X \mid f^k(v) = u\}$ the set of k-th preimages of u. For each u in a cycle of A, we call the* unroll tree of A in u *the infinite tree $\mathbf{t}_u = (V, E)$ having vertices $V = \{(s, k) \mid s \in f^{-k}(u), k \in \mathbb{N}\}$ and edges $E = \{((v, k), (f(v), k-1))\} \subseteq V^2$. We call* unroll of A, *denoted $\mathcal{U}(A)$, the set of its unroll trees (see Fig. 1).*

In the rest of this paper, *unrolls will always be taken up to isomorphism, i.e.*, as multisets or sums (disjoint unions) of unlabeled trees; the equality sign will thus represent the isomorphism relation.

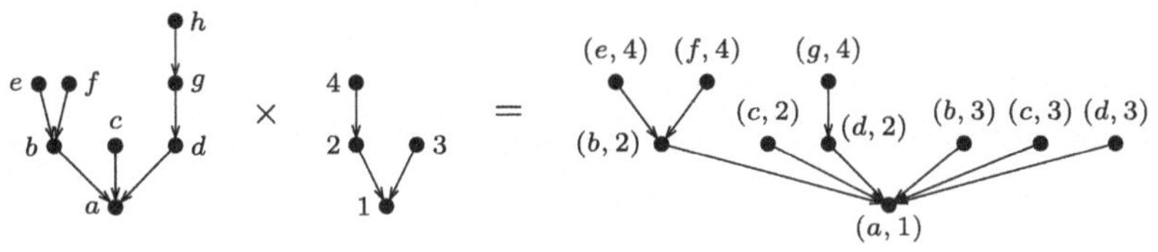

Fig. 2. The levelwise product of two finite trees. Remark how the depth of the result is given by the minimum depth of the two factors.

An unroll tree contains exactly one infinite branch, onto which are periodically rooted the trees representing the transient behaviors of the corresponding connected component. We also exploit a notion of "levelwise" product on trees. For readability, we will denote trees and forests using bold letters (in lower and upper case respectively) to distinguish them from FDDS.

Definition 2 (Product of trees). *Let* $\mathbf{t}_1 = (V_1, E_1)$ *and* $\mathbf{t}_2 = (V_2, E_2)$ *be two trees with roots* r_1 *and* r_2*, respectively. Their* product *is the tree* $\mathbf{t}_1 \times \mathbf{t}_2 = (V, E)$ *with vertices* $V = \{(u, v) \in V_1 \times V_2 \mid \mathrm{depth}_u = \mathrm{depth}_v\}$*, where* depth_u *is the length of the shortest path between* u *and the root of its tree, and edges* $E = \{((u, u'), (v, v')) \mid (u, v) \in E_1, (u', v') \in E_2\} \subseteq V^2$ *(see Fig. 2).*

We deduce from [10] that the set of unrolls up to isomorphism, with disjoint union for addition and product of trees for multiplication, is a semiring with the infinite path as the multiplicative identity. In addition, the unroll operation is a homomorphism between the semiring $\mathbb{D}$ of FDDS and the semiring of unrolls.

Unrolls have one major algorithmic issue: they are infinite. To overcome this obstacle, we will consider only a finite portion of the unrolls. For this purpose, we extend the notion of *depth* to forests of finite trees by taking the maximum depth of its trees. We also define the *depth of an FDDS* as the maximum depth among the trees rooted in one of its periodic states.

We remark that the set of forests of bounded depth d up to isomorphism is a also semiring (with the restrictions of the same tree sum and product operations) for all depth d, where the multiplicative identity is the path of depth d.

We can now define the *cut* of $\mathbf{t}$ at depth $k \geq 0$, denoted $\mathcal{C}(\mathbf{t}, k)$, as the induced sub-tree of $\mathbf{t}$ restricted to the vertices of depth lesser or equal to k (see Fig. 1). This operation generalizes (homomorphically) to forests $\mathbf{F} = \mathbf{t}_1 + \ldots + \mathbf{t}_n$ as $\mathcal{C}(\mathbf{F}, k) = \mathcal{C}(\mathbf{t}_1, k) + \ldots + \mathcal{C}(\mathbf{t}_n, k)$. The cut of an unroll at a certain depth has already been used to exhibit properties of unrolls [10,11]. This approach notably gave a characterization of *cancelable* FDDS, *i.e.* FDDS A such that $AB = AC$ implies $B = C$ for all FDDS B, C. Indeed, a FDDS A is cancelable if and only if at least one of its connected components is a dendron [10, Theorem 30].

One of the tools for the analysis of trees and forests is the total order $\leq$ on finite and infinite trees introduced in [10]. Indeed, this order is compatible with the product for infinite trees[1], *i.e.*, if $\mathbf{t}_1, \mathbf{t}_2$ are two infinite trees then $\mathbf{t}_1 \leq \mathbf{t}_2$ if and only if $\mathbf{t}_1\mathbf{t} \leq \mathbf{t}_2\mathbf{t}$ for all tree $\mathbf{t}$ [10, Lemma 24]. A similar property has been proved for the case of finite trees:

Lemma 1 ([10]). *Let* $\mathbf{t}_1, \mathbf{t}_2, \mathbf{t}$ *then* $\mathcal{C}(\mathbf{t}_1, \text{depth}_\mathbf{t}) \leq \mathcal{C}(\mathbf{t}_2, \text{depth}_\mathbf{t})$ *if and only if* $\mathbf{t}_1\mathbf{t} \leq \mathbf{t}_2\mathbf{t}$.

In the rest of the paper all polynomials will be implicitly univariate and, by convention, the symbol $\mathbf{A}^0$ (i.e., the 0-th power of a forest of finite trees $\mathbf{A}$) will denote the path of length $\text{depth}_\mathbf{A}$, which behaves as an identity for the product with forests of equal or less depth.

3 Characterization of Injective Polynomials over Unrolls

The search for a solution to polynomial equations over FDDS can be seen as the search for a compatible solution between solutions of polynomial equations of transient states and solutions of polynomial equations of periodic states. This is why we can start by looking for solutions for transient states. In order to do this, we will use the notion of unroll and determine the number of solutions to polynomial equations over unrolls.

More precisely, our final goal is to prove the following theorem:

Theorem 1. *All univariate polynomials over unrolls are injective.*

For the proof of this theorem, we begin by remarking that for all polynomial P over unrolls, and for all FDDS X, Y the equality $P(\mathcal{U}(X)) = P(\mathcal{U}(Y))$ implies $\mathcal{C}(P(\mathcal{U}(X)), n) = \mathcal{C}(P(\mathcal{U}(Y)), n)$ for all n. In order to study polynomial equations over forests of finite trees, we start by proving one technical lemma on the behavior of the order on powers of trees with the same depth.

Lemma 2. *Let* $\mathbf{x}, \mathbf{y}$ *be two finite trees with the same depth and* k *be a positive integer. Then* $\mathbf{y}^k \geq \mathbf{x}^k$ *if and only if* $\mathbf{y} \geq \mathbf{x}$.

Proof. If $\mathbf{y} \geq \mathbf{x}$ and $\text{depth}_\mathbf{x} = \text{depth}_\mathbf{y}$, by [10, Corollary 20], we have $\mathbf{y}^2 \geq \mathbf{x}^2$ and $\text{depth}_{\mathbf{y}^2} = \text{depth}_{\mathbf{x}^2}$. Then, by induction, we obtain $\mathbf{y}^k \geq \mathbf{x}^k$.

For the other direction, assume that $\mathbf{y}^k \geq \mathbf{x}^k$ and, by contradiction, that $\mathbf{x} > \mathbf{y}$. Since $\text{depth}_\mathbf{x} = \text{depth}_\mathbf{y}$, by [10, Corollary 20], we have $\mathbf{y}^k\mathbf{x} \geq \mathbf{y}^k\mathbf{y}$. Thus, by [10, Lemma 21], we have $\mathbf{y}^{k-1} \geq \mathbf{x}^{k-1}$. By induction, we obtain $\mathbf{y} \geq \mathbf{x}$, a contradiction. □

Let $\mathbf{A}$ be a forest and $i \geq 0$ be an integer. We denote by $\Gamma(\mathbf{A}, i)$ the sub-multiset of trees of $\mathbf{A}$ where each tree has depth *at least* i. We focus our attention on the trees in $P(\mathbf{X})$ with a certain depth and form.

[1] This compatibility with the product of $\leq$ is sufficient for most applications in this paper; we refer the reader to the original paper for the actual definition.

Lemma 3. *Let $P = \sum_{i=1}^{m} \mathbf{A}_i\mathbf{X}^i$ be a polynomial without constant term over forests and $\mathbf{X} = \mathbf{x}_1 + \ldots + \mathbf{x}_n$ be a forest sorted in nondecreasing order. Let $\mathbf{A}_i = \sum_{j=1}^{m_i} \mathbf{a}_{i,j}$ be in nondecreasing order, hence $\mathbf{a}_{i,1} = \min(\mathbf{A}_i)$, for all i. Then, there exists $\alpha \in \{1, \ldots, m\}$ such that $\min_i(\mathbf{A}_i\mathbf{X}^{i-1}\mathbf{x}_k)$ is isomorphic to $\mathbf{a}_{\alpha,1}\mathbf{x}_1^{\alpha-1}\mathbf{x}_k$.*

Proof. Let $\mathbf{t}_{min}$ be the smallest tree in $P(\mathbf{X})$. Hence, there exists $\alpha \in \{1, \ldots, m\}$ such that $\mathbf{t}_{min} \in \mathbf{A}_\alpha\mathbf{X}^\alpha$. In addition, by [10, Lemma 24], we have $\mathbf{x}_1^2 \leq \mathbf{x}_1\mathbf{x}_j$ for all j. By induction, we deduce that $\mathbf{t}_{min} \in \mathbf{A}_\alpha\mathbf{x}_1^\alpha$. Analogously, we have $\mathbf{a}_{\alpha,1}\mathbf{x}_1^\alpha \leq \mathbf{a}_{\alpha,j}\mathbf{x}_1^\alpha$ for all $\mathbf{a}_{\alpha,j} \in \mathbf{A}_\alpha$, and we conclude that $\mathbf{t}_{min} = \mathbf{a}_{\alpha,1}\mathbf{x}_1^\alpha$.

We recall that $\mathbf{t}_1\mathbf{t}_3 < \mathbf{t}_2\mathbf{t}_3$ if and only if $\mathcal{C}(\mathbf{t}_1, \mathrm{depth}_{\mathbf{t}_3}) < \mathcal{C}(\mathbf{t}_2, \mathrm{depth}_{\mathbf{t}_3})$ (Lemma 1). Therefore, for all j and $\mathbf{a} \in \mathbf{A}_j$, we have:

$$\mathbf{a}_{\alpha,1}\mathbf{x}_1^\alpha \leq \mathbf{a}\mathbf{x}_1^j \Leftrightarrow \mathcal{C}(\mathbf{a}_{\alpha,1}\mathbf{x}_1^{\alpha-1}, \mathrm{depth}_{\mathbf{x}_1}) \leq \mathcal{C}(\mathbf{a}\mathbf{x}_1^{j-1}, \mathrm{depth}_{\mathbf{x}_1})$$

Three cases are possible. First, if $\alpha = j$ then $\mathbf{a}_{\alpha,1} \leq \mathbf{a}$ by minimality of $\mathbf{a}_{\alpha,1}$. This implies that $\mathbf{a}_{\alpha,1}\mathbf{x}_1^{\alpha-1}\mathbf{x}_k \leq \mathbf{a}\mathbf{x}_1^{\alpha-1}\mathbf{x}_k$.

Second, if $\alpha \neq j$ and $\alpha \neq 1$, then $\mathcal{C}(\mathbf{a}_{\alpha,1}\mathbf{x}_1^{\alpha-1}, \mathrm{depth}_{\mathbf{x}_1}) = \mathbf{a}_{\alpha,1}\mathbf{x}_1^{\alpha-1}$. Since $\mathcal{C}(\mathbf{a}\mathbf{x}_1^{j-1}, \mathrm{depth}_{\mathbf{x}_1}) \leq \mathbf{a}\mathbf{x}_1^{j-1}$, we have $\mathbf{a}_{\alpha,1}\mathbf{x}_1^{\alpha-1}\mathbf{x}_k \leq \mathbf{a}\mathbf{x}_1^{j-1}\mathbf{x}_k$.

Thirdly, if $\alpha \neq j$ but $\alpha = 1$, then either

$$\mathcal{C}(\mathbf{a}_{\alpha,1}\mathbf{x}_1^{\alpha-1}, \mathrm{depth}_{\mathbf{x}_1}) = \mathcal{C}(\mathbf{a}\mathbf{x}_1^{j-1}, \mathrm{depth}_{\mathbf{x}_1}) = \mathbf{a}\mathbf{x}_1^{j-1}$$

which implies $\mathbf{a}_{\alpha,1}\mathbf{x}_1^\alpha = \mathbf{a}_{j,1}\mathbf{x}_1^j$, and we can just exchange α and j, which is already covered by the second case. Otherwise $\mathcal{C}(\mathbf{a}_{\alpha,1}\mathbf{x}_1^{\alpha-1}, \mathrm{depth}_{\mathbf{x}_1}) < \mathbf{a}\mathbf{x}_1^{j-1}$. In this case, we need to recall the definition of the order from [10] and the code upon which it is defined. Indeed, by [10, Lemma 17], we know that the leftmost difference between the codes of $\mathcal{C}(\mathbf{a}_{\alpha,1}\mathbf{x}_1^{\alpha-1}, \mathrm{depth}_{\mathbf{x}_1})$ and $\mathcal{C}(\mathbf{a}\mathbf{x}_1^{j-1}, \mathrm{depth}_{\mathbf{x}_1})$ corresponds to a node of depth at most $\mathrm{depth}_{\mathbf{x}_1} - 1$. Therefore, if $\mathrm{depth}_{\mathbf{x}_k} \geq \mathrm{depth}_{\mathbf{x}_1} - 1$ this difference is propagated into $\mathcal{C}(\mathbf{a}_{\alpha,1}\mathbf{x}_1^{\alpha-1}, \mathrm{depth}_{\mathbf{x}_1})\mathbf{x}_k$ and $\mathbf{a}\mathbf{x}_1^{j-1}\mathbf{x}_k$, and $\mathbf{a}_{\alpha,1}\mathbf{x}_1^{\alpha-1}\mathbf{x}_k < \mathbf{a}\mathbf{x}_1^{j-1}\mathbf{x}_k$; otherwise, we have $\mathbf{a}_{\alpha,1}\mathbf{x}_1^{\alpha-1}\mathbf{x}_k = \mathbf{a}\mathbf{x}_1^{j-1}\mathbf{x}_k$.

Hence, in all cases we have $\mathbf{a}_{\alpha,1}\mathbf{x}_1^{\alpha-1}\mathbf{x}_k \leq \mathbf{a}\mathbf{x}_1^{j-1}\mathbf{x}_k$. Since for all p we have $\mathbf{x}_1^{p-1} \leq \mathbf{t}$ with $\mathbf{t} \in \mathbf{X}^{p-1}$, the lemma follows. □

Thanks to Lemma 3 we can prove the first important result of this section. Remark that, due to closure under sums and products, the set of forests of finite trees of depth at most d_{max} is a semiring for every natural number d_{max}.

Proposition 1. *Let $P = \sum_{i=0}^{m} \mathbf{A}_i\mathbf{X}^i$ with $d_{max} = \max_{i>0}(\mathrm{depth}_{\mathbf{A}_i})$ and $\mathrm{depth}_{\mathbf{A}_0} \leq d_{max}$. Then P is injective over the semiring of forests of finite trees of depth at most d_{max}.*

Proof. Let us first suppose that P has no constant term. Let $\mathbf{X} = \mathbf{x}_1 + \ldots + \mathbf{x}_{m_x}$ and $\mathbf{Y} = \mathbf{y}_1 + \ldots + \mathbf{y}_{m_y}$ be two forest (sorted in nondecreasing order), and assume $P(\mathbf{X}) = P(\mathbf{Y})$. Thus, by Lemma 3, there exist $i, j \in \{1, \ldots, m\}$, with $i \leq j$ such that, for each k, the smallest tree $\mathbf{t}_s$ with factor $\mathbf{x}_k$ (resp., $\mathbf{y}_k$) in $\sum_{i=1}^{m} \mathbf{A}_i\mathbf{X}^i$ is isomorphic to $\mathbf{a}_{i,1}\mathbf{x}_1^{i-1}\mathbf{x}_k$ (resp., $\mathbf{a}_{j,1}\mathbf{y}_1^{j-1}\mathbf{y}_k$), with $\mathbf{a}_{i,1} = \min(\mathbf{A}_i)$ and $\mathbf{a}_{j,1} =$

$\min(\mathbf{A}_j)$. We deduce that $\mathbf{a}_{i,1}\mathbf{x}_1 = \min(P(\mathbf{X})) = \min(P(\mathbf{Y})) = \mathbf{a}_{j,1}\mathbf{y}_1$ from the hypothesis.

First, we show that $\mathbf{x}_1 = \mathbf{y}_1$. By contradiction, and without loss of generality, assume $\mathbf{x}_1 < \mathbf{y}_1$. Then, by Lemma 2, it follows that $\mathbf{x}_1^j < \mathbf{y}_1^j$. By [10, Lemma 24], we deduce that $\mathbf{a}_{j,1}\mathbf{x}_1^j < \mathbf{a}_{j,1}\mathbf{y}_1^j = \mathbf{a}_{i,1}\mathbf{x}_1^i$. Nevertheless, $\mathbf{a}_{j,1}\mathbf{x}_1^j \in P(\mathbf{X})$, contradicting the minimality of $\mathbf{a}_{i,1}\mathbf{x}_1^i$.

Finally, since $\mathbf{x}_1 = \mathbf{y}_1$, we deduce that $P(\mathbf{X}) - P(\mathbf{x}_1) = P(\mathbf{Y}) - P(\mathbf{y}_1)$. In addition, the smallest tree in $P(\mathbf{X}) - P(\mathbf{x}_1)$ is $\mathbf{a}_{i,1}\mathbf{x}_1^{i-1}\mathbf{x}_2$ and the smallest tree in $P(\mathbf{Y}) - P(\mathbf{y}_1)$ is $\mathbf{a}_{j,1}\mathbf{y}_1^{j-1}\mathbf{y}_2$. Since $\mathbf{x}_1 = \mathbf{y}_1$, we have $\mathbf{a}_{i,1}\mathbf{x}_1^i = \mathbf{a}_{j,1}\mathbf{y}_1^j = \mathbf{a}_{j,1}\mathbf{x}_1^j$. This implies that we can define that $i = j$ and therefore $\mathbf{a}_{i,1}\mathbf{x}_1^{i-1} = \mathbf{a}_{j,1}\mathbf{y}_1^{j-1}$. Thus, by Lemma 2, we deduce that $\mathbf{x}_2 = \mathbf{y}_2$. By induction, we conclude that $\mathbf{X} = \mathbf{Y}$.

This proof extends to polynomials with a constant term $\mathbf{A}_0$. Indeed, let $Q = P + \mathbf{A}_0$ and suppose $Q(\mathbf{X}) = Q(\mathbf{Y})$. Then $Q(\mathbf{X}) - \mathbf{A}_0 = Q(\mathbf{Y}) - \mathbf{A}_0$, that is $P(\mathbf{X}) = P(\mathbf{Y})$, which implies $\mathbf{X} = \mathbf{Y}$. Thus, Q is also injective. □

Our main theorem follows directly from Proposition 1.

Proof of Theorem 1. Suppose $P(\mathcal{U}(X)) = P(\mathcal{U}(Y))$. Then, for all n we have $\mathcal{C}(P(\mathcal{U}(X)), n) = \mathcal{C}(P(\mathcal{U}(Y)), n)$. If $\mathcal{U}(X)$ and $\mathcal{U}(Y)$ were different, there would exist an n such that $\mathcal{C}(\mathcal{U}(X), n) \neq \mathcal{C}(\mathcal{U}(Y), n)$. However, by Proposition 1 and since $\mathcal{C}(\cdot, n)$ is a homomorphism, we have $\mathcal{C}(\mathcal{U}(X), n) = \mathcal{C}(\mathcal{U}(Y), n)$ for all n. □

The proof of Proposition 1 also suggests an algorithm (inspired by [11, Algorithm 1]) for solving polynomial equations over forests of finite trees. Indeed, starting with the maximum depth d_{max} (Fig. 3 ①), we can solve the equation $\Gamma(P(\mathbf{X}), d_{max}) = \Gamma(\mathbf{B}, d_{max})$ ②. In fact, we construct the solution $\Gamma(\mathbf{X}, d_{max})$ by first solving the equation $\min(\Gamma(\mathbf{A}_i, d_{max}))\mathbf{x}_1^i = \min(\Gamma(\mathbf{B}, d_{max}))$ ③④ for an $i \in \{1, \dots, m\}$. The tree $\mathbf{x}_1$ is minimal among the trees of maximum depth appearing in the solution $\mathbf{X}$. Thus, by Lemma 3, either we can construct the solution *Sol* inductively ④ or $P(Sol) \not\subseteq \mathbf{B}$ ③, and we can deduce that this i is not the one appearing in the statement of Lemma 3, and we try the next value of i, if any. Indeed, the j-th tree $\mathbf{x}_i$ of the solution is equal to

$$\frac{\min(\Gamma(\mathbf{B}, d_{max}) - \Gamma(P(\mathbf{x}_1 + \ldots + \mathbf{x}_{j-1}), d_{max}))}{\min(\Gamma(\mathbf{A}_i, d_{max}))\mathbf{x}_1^{i-1}}.$$

Then, we can inductively construct $\mathbf{x} = \Gamma(\mathbf{X}, d)$ from $\Gamma(\mathbf{X}, d')$ where $d < d_{max}$ is the depth of a tree in $\mathbf{B}$ and d' the smallest depth of a tree in $\mathbf{B}$ greater than d ⑤. In order to do so, we need to find $\min(\Gamma(\mathbf{X}, d))$. Let $\mathbf{a} = \min(\Gamma(\mathbf{A}_j, d))$ for a $j \in \{1, \dots, m\}$ and $\mathbf{b} = \min(\Gamma(\mathbf{B}, d))$.

- If $\mathbf{b}$ has depth at least d' ⑥, or $\mathbf{a}$ has depth d and $\mathbf{a} \times \min(\Gamma(\mathbf{X}, d'))^j \leq \mathbf{b}$, then $\mathbf{x}$ has depth at least d' and thus is already in $\Gamma(\mathbf{X}, d')$.
- Otherwise $\mathbf{b}$ necessarily has depth d. In that case, either $\mathbf{a}$ has depth at least d', or $\mathbf{a} \times \min(\Gamma(\mathbf{X}, d'))^j > \mathbf{b}$, and in both cases we have $\mathbf{x} = \sqrt[j]{\mathbf{b}/\mathbf{a}}$.

Then, similarly to the step for depth d_{max}, we can construct a portion of the solution $\mathbf{X}$ ⑧, or find a mistake ⑦ and try another value of j, if any. If during the iteration we find mistakes for all coefficients, this implies that the equation does not have a solution.

Moreover, the number of iterations is bounded by the number of coefficients m times the depth d_{max}, which are both polynomial with respect to the size of the inputs, and the operations carried out in each iteration can be performed in polynomial time [10,11].

We cannot directly apply this algorithm to polynomial equations over unrolls, since they contain infinite trees, but we can show that, in fact, we only need to consider a finite cut at depth polynomial with respect to the sizes of the FDDS in order to decide if a solution exists:

Proposition 2. *Let* $P = \sum_{i=0}^{m} A_i X^i$*, let* B *be an FDDS and* α *the number of unroll trees of* $\mathcal{U}(B)$*. Let* $n \geq 2 \times \alpha^2 + \text{depth}_{\mathcal{U}(B)}$ *be an integer. Then, there exists a forest* $\mathbf{Y}$ *with* $\text{depth}_{\mathbf{Y}} \leq n$ *such that* $\sum_{i=0}^{m} \mathcal{C}(\mathcal{U}(A_i), n)\mathbf{Y}^i = \mathcal{C}(\mathcal{U}(B), n)$ *if and only if there exists a FDDS* X *such that* $\mathcal{U}(P(X)) = \mathcal{U}(B)$*. Furthermore, if such an* X *exists, then necessarily* $\mathcal{C}(\mathcal{U}(X), n) = \mathbf{Y}$*.*

Proof. Assume that $\sum_{i=0}^{m} \mathcal{C}(\mathcal{U}(A_i), n)\mathbf{Y}^i = \mathcal{C}(\mathcal{U}(B), n)$. First, remark that we can apply a similar reasoning to [11, Lemma 11] in order prove that if there exists $\mathbf{X}$ such that $\sum_{i=0}^{m} \mathcal{U}(A_i)\mathbf{X} = \mathcal{U}(B)$ then there exists an FDDS Y such that $\mathcal{U}(Y) = \mathbf{X}$. Since α bounds the greatest cycle length in $A_0, \ldots, A_m, B$ and since, from the hypothesis, $\text{depth}_{\mathcal{U}(B)}$ bounds $\text{depth}_{\mathcal{U}(A_i)}$ for each i, we have all the tools needed in order to prove (similarly to [11, Theorem 5]) that there exists a FDDS X such that $\mathcal{U}(P(X)) = \mathcal{U}(B)$. The inverse implication is trivial. □

This proposition has two consequences. First of all, the solution (if it exists) to an equation $\sum_{i=0}^{m} \mathcal{U}(A_i)\mathbf{X}^i = \mathcal{U}(B)$ is always the unroll of an FDDS. Furthermore, by employing the same method for rebuilding $\mathcal{U}(X)$ from the solution $\mathbf{Y}$ with $\text{depth}_{\mathbf{Y}} < n$ of $\sum_{i=0}^{m} \mathcal{C}(\mathcal{U}(A_i), n)\mathbf{Y}^i = \mathcal{C}(\mathcal{U}(B), n)$ exploited in the proof of [11, Theorem 6], we deduce an algorithm for solving equations over unrolls. In order to encode unrolls (which are infinite graphs), we represent them finitely by FDDS having those unrolls.

Theorem 2. *Let* P *be a polynomial over unrolls (encoded as a polynomial over FDDS) and* B *an FDDS. Then, we can solve the equation* $P(\mathcal{U}(X)) = \mathcal{U}(B)$ *in polynomial time.*

4 Sufficient Condition for the Injectivity of Polynomials

Even if all polynomials over unrolls are injective, we can easily remark that is not the case for the FDDS. For example, let A be the cycle of length 2; then, the polynomial AX is not injective since $A^2 = 2A$. Hence, since the existence of injective polynomials is trivial (just consider the polynomial X), we want to obtain a characterization of this class of polynomials.

Proposition 40 of [10] gives a sufficient condition for the injectivity of polynomials without constant term, namely if the coefficient A_1 is cancelable, *i.e.*, at

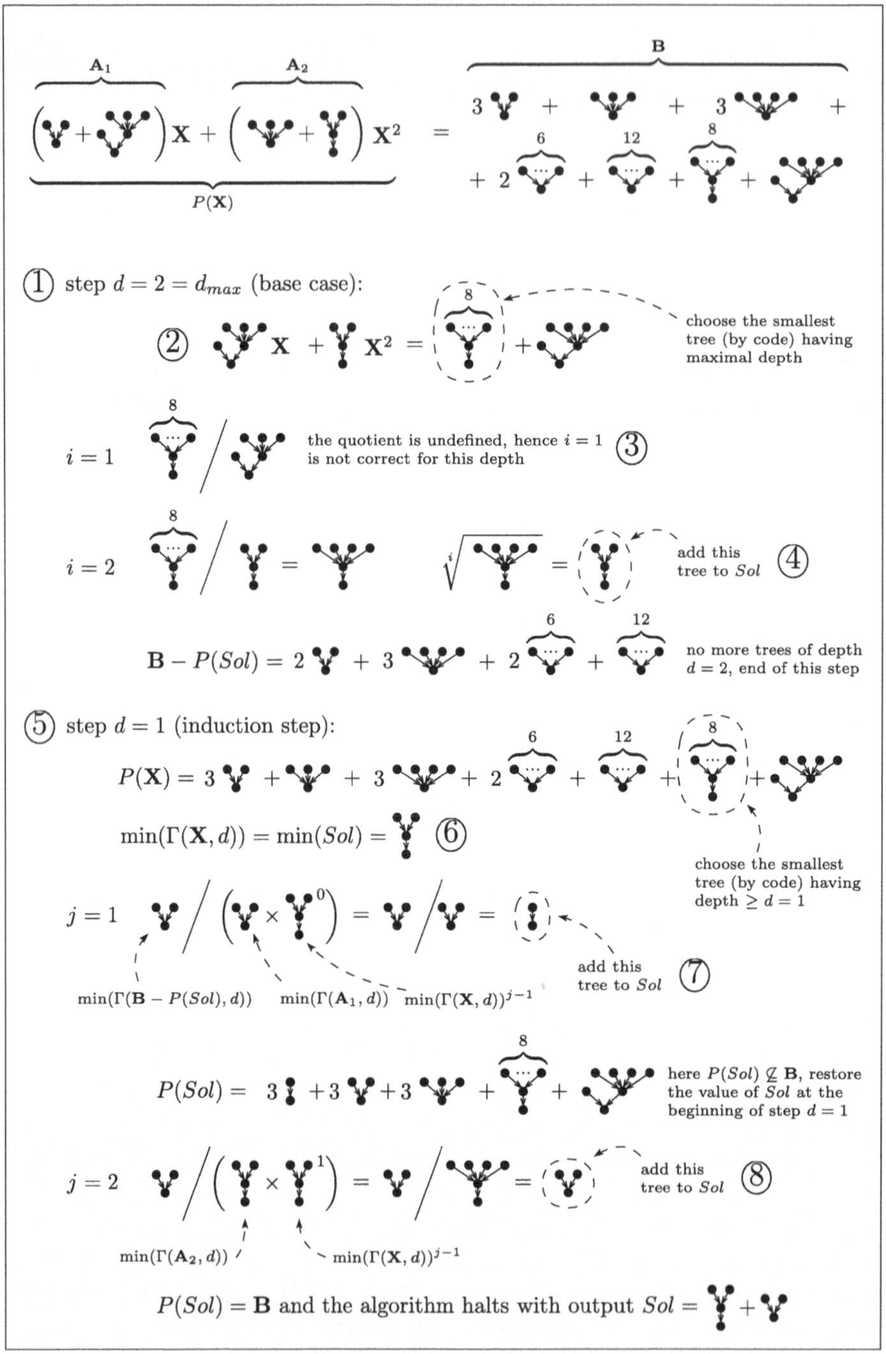

Fig. 3. A run of the algorithm for polynomial equations over forests of finite trees.

least one of its connected component is a dendron. However, since X^k is injective for all $k > 0$ (by uniqueness of k-th roots [10]), this condition is not necessary. Hence, we start by giving a sufficient condition which covers these cases.

Proposition 3. *Let $P = \sum_{i=0}^{m} A_i X^i$ be a polynomial. If at least one coefficient of P is cancelable then P is injective.*

For the proof of this proposition, we start by showing the case of polynomials $P = \sum_{i=1}^{m} A_i X^i$ without a constant term. For that, we need to analyze the structure of the possible evaluations of P, and more precisely their multisets of cycles. We denote by $\ell(A)$ the length of the cycle of a connected FDDS A. Let us introduce two total orders over connected FDDS:

Definition 3. *Let A, B be connected FDDS, $a = \ell(A)$, $b = \ell(B)$, $\mathbf{t} = \min(\mathcal{U}(A))$, and $\mathbf{t}' = \min(\mathcal{U}(B))$. We define:*

- *$A \geq_c B$ if and only if $a > b$ or $a = b$ and $\mathbf{t} \geq \mathbf{t}'$.*
- *$A \geq_t B$ if and only if $\mathbf{t} > \mathbf{t}'$ or $\mathbf{t} = \mathbf{t}'$ and $a \geq b$.*

For an FDDS X and a positive integer p, let us consider the function $(X)(p)$, which computes the multiset of connected components of X with cycles of length p, and the function $(X)\{p\}$, computing the multiset of connected components of X with cycles of length *dividing* p.

Lemma 4. *Let $p > 0$ be an integer; then $(\cdot)\{p\}$ is an endomorphism over FDDS.*

Proof. First, if we denote by $\mathbf{1}$ the fixed point (the multiplicative identity in $\mathbb{D}$), we obviously have $(\mathbf{1})\{p\} = \mathbf{1}$. Now, let A, B be two FDDS. It is clear that $(A + B)\{p\} = (A)\{p\} + (B)\{p\}$ since the sum is the disjoint union. All that remains to prove is that $(AB)\{p\} = (A)\{p\}(B)\{p\}$. By the distributivity of the sum over the product, it is sufficient to show the property in the case of connected A and B. Recall that, by the definition of product of FDDS, the product of two connected FDDS U, V has $\gcd(\ell(U), \ell(V))$ connected component with cycle length $\text{lcm}(\ell(U), \ell(V))$. We deduce that all elements in $(AB)\{p\}$ are generated by the product of elements in $(A)\{p\}$ and $(B)\{p\}$. Hence $(AB)\{p\} \subseteq (A)\{p\}(B)\{p\}$. Furthermore, by the definition of $(A)\{p\}$, each element of $(A)\{p\}$ has a cycle length dividing p. Thus, the cycle length of the product of elements of $(A)\{p\}$ and $(B)\{p\}$ is bounded by p. Since the lcm of integers dividing p is also a divisor of p, we have $(A)\{p\}(B)\{p\} \subseteq (AB)\{p\}$. □

Our goal is to prove, by a simple induction, the core of the proof of Proposition 3. Lemma 4 is a first step to show the induction case. Indeed, it allows us to treat sequentially the different cycle lengths. We use the order $\leq_c$ to process sequentially connected components having the same cycle length, as shown by the following lemma.

Lemma 5. *Let $X = X_1 + \cdots + X_k$ be an FDDS with k connected components sorted by $\leq_c$ and $P = \sum_{i=1}^{m} A_i X^i$ a polynomial over FDDS without constant term and with at least one cancelable coefficient. Then $B = \min_{\geq_c}(P(X) - P(\sum_{i=1}^{n-1} X_i))$ implies $\ell(X_n) = \ell(B)$ for all $1 \leq n \leq k$.*

Proof. We have $\ell(X_n) \leq \text{lcm}(a, \ell(X_i))$ for all integers i with $n \leq i \leq k$ and $a > 0$. It follows that $\ell(X_n) \leq \ell(B)$. In addition, since at least one coefficient of P, say A_c, is cancelable, it contains one connected component whose cycle length is one. This implies that $A_c X_n^c$ contains a connected component C such that $\ell(C) = \ell(X_n)$. If B is minimal, then $\ell(X_n) \geq \ell(B)$. □

We can now identify recursively and unambiguously each connected component in X. Therefore, we can use Lemma 5, as part of the induction step for the following lemma.

Lemma 6. *Let* $P = \sum_{i=1}^{m} A_i X^i$ *be a polynomial without constant term. If at least one coefficient of* P *is cancelable then* P *is injective.*

Proof. Let $X = X_1 + \cdots + X_{n_1}$, resp., $Y = Y_1 + \cdots + Y_{n_2}$ be FDDS consisting of n_1 (resp., n_2) connected components sorted by $\leq_c$, and let B be an FDDS. Suppose $P(X) = B = P(Y)$. We prove by induction on the number n_1 of connected components in X that $X = Y$.

If X has 0 connected components, since P does not have constant term, we have $P(X) = 0$. In addition, since P has at least a cancelable coefficient, it trivially contains at least one nonzero coefficient. Thus, since $P(Y) = P(X) = 0$, we deduce that $Y = 0$, hence $X = Y$. The property is true in the base case.

Let n be a positive integer. Suppose that the properties is true for n, *i.e.* $X_1 + \cdots + X_n = Y_1 + \cdots + Y_n$. Then, $P(X_1 + \cdots + X_n) = P(Y_1 + \cdots + Y_n)$. This implies that $P(X) - P(X_1 + \cdots + X_n) = B - P(X_1 + \cdots + X_n) = P(Y) - P(Y_1 + \cdots + Y_n)$. Let p be the length of the cycle of the smallest connected component of $B - P(X_1 + \cdots + X_n)$ according to $\geq_c$. By Lemma 5, we have $\ell(X_{n+1}) = \ell(Y_{n+1}) = p$.

Besides, by Lemma 4, we deduce that $(P(X))\{p\} = \sum_{i=1}^{m} (A_i)\{p\}(X^i)\{p\}$ and $(P(Y))\{p\} = \sum_{i=1}^{m} (A_i)\{p\}(Y^i)\{p\}$. Since $(P(X))\{p\} = (B)\{p\} = (P(Y))\{p\}$, we deduce that $\sum_{i=1}^{m} (A_i)\{p\}(X^i)\{p\} = \sum_{i=1}^{m} (A_i)\{p\}(Y^i)\{p\}$. By taking unrolls, we obtain $\sum_{i=1}^{m} \mathcal{U}((A_i)\{p\})\mathcal{U}((X^i)\{p\}) = \sum_{i=1}^{m} (A_i)\{p\}(Y^i)\{p\}$. By injectivity of polynomials over unrolls (Theorem 1), we have $\mathcal{U}((X)\{p\}) = \mathcal{U}((Y)\{p\})$. Thus, the smallest tree of $\mathcal{U}((X)\{p\}) - \mathcal{U}((X_1 + \cdots + X_n)\{p\})$ is equal to the smallest of $\mathcal{U}((Y)\{p\}) - \mathcal{U}((Y_1 + \cdots + Y_n)\{p\})$. However, the smallest tree of $\mathcal{U}((X)\{p\}) - \mathcal{U}((X_1 + \cdots + X_n)\{p\})$ is the smallest tree of $\mathcal{U}(X_{n+1})$ and of $\mathcal{U}(Y_{n+1})$. This implies that X_{n+1} and Y_{n+1} are two connected components whose cycle length is p and have the same minimal unroll tree, so $X_{n+1} = Y_{n+1}$. The induction step and the statement follow. □

What remains in order to conclude the proof of Proposition 3 is to extend Lemma 6 to polynomials with a constant term.

Proof of Proposition. 3 Let $P = \sum_{i=0}^{m} A_i X^i$ be a polynomial with at least one non-constant cancelable coefficient. Let $P' = \sum_{i=1}^{m} A_i X^i$ be the polynomial obtained from P by removing the constant term A_0. Hence $P = P' + A_0$. If there exist two FDDS X, Y such that $P(X) = P(Y)$, then $P'(X) + A_0 = P'(Y) + A_0$. This implies that $P'(X) = P'(Y)$. Thus, by Lemma 6, we conclude $X = Y$. □

From the proof of Lemma 5 and Theorem 2, we deduce that we can solve in polynomial time all equations of the form $P(X) = B$ with P having a cancelable non-constant coefficient. Indeed, we can just select the connected components with the correct cycle lengths, and cut their unrolls to the correct depth. Then, we solve the equation over the forest thus obtained, select the smallest tree of the result, and re-roll it to the correct period. Finally, we remove the corresponding multiset of connected components and reiterate until either all connected components have been removed, or we find a connected component that we cannot remove (which implies that the equation has no solution).

5 Necessary Condition for the Injectivity of Polynomials

We will now show that the sufficient condition for injectivity of Proposition 3 is actually necessary. As previously, we start by considering polynomials without constant terms. Our starting point for this proof is given by [10, Theorem 34], which characterizes the injectivity of linear monomials. Furthermore [10, Lemma 33] proves that this condition is necessary for linear monomials. We begin by extending the proof of [10, Lemma 33] to show that this is necessary for *all* monomials.

Let δ_J be recursively defined by $\delta_\emptyset = 1$ and $\delta_{J\cup\{a\}} = \gcd(a, \operatorname{lcm}(J))\delta_J$ for all $a \in \mathbb{N}$. Now let $\mathcal{A}$ be a set of integers strictly larger than 1 and for $I \subseteq \mathcal{A}$ we define $\alpha_I = \delta_{\mathcal{A}} \prod_{a\in\mathcal{A}} a$ and $\beta_I = \alpha_I + (-1)^{|I|}\delta_I \prod_{a\in\mathcal{A}-I} a$. We will use α_I and β_I to construct two different FDDS X and Y such that $AX = AY$ if A is not cancelable. We will start with the case where A is a cycle, denoted by C_b where $b = \ell(C_b)$.

Lemma 7. *Let $k \geq 1$ be an integer, $b \in \mathcal{A}$, $I \subseteq \mathcal{A} - \{b\}$ and $J = I \cup \{b\}$. Then $C_b(\beta_I C_{\operatorname{lcm}(I)} + \beta_J C_{\operatorname{lcm}(J)})^k = C_b(\alpha_I C_{\operatorname{lcm}(I)} + \alpha_J C_{\operatorname{lcm}(J)})^k$.*

Proof. We have $C_b(\beta_I C_{\operatorname{lcm}(I)} + \beta_J C_{\operatorname{lcm}(J)}) = C_b(\alpha_I C_{\operatorname{lcm}(I)} + \alpha_J C_{\operatorname{lcm}(J)})$ by the proof of [10, Lemma 33]. Thus, we can replace one instance of the term $C_b(\beta_I C_{\operatorname{lcm}(I)} + \beta_J C_{\operatorname{lcm}(J)})$ in the equation of the statement above and we obtain:

$$C_b(\beta_I C_{\operatorname{lcm}(I)} + \beta_J C_{\operatorname{lcm}(J)})^k = (\alpha_I C_{\operatorname{lcm}(I)} + \alpha_J C_{\operatorname{lcm}(J)}) C_b(\beta_I C_{\operatorname{lcm}(I)} + \beta_J C_{\operatorname{lcm}(J)})^{k-1}.$$

By repeating the same substitution, the thesis follows. □

Now we can prove that monomials of the form AX^k, with $k \geq 1$ and where A is a sum of cycles (*i.e.*, a *permutation*) with no cycle of length 1, are never injective.

Lemma 8. *Let $A = A_1 + \cdots + A_m$ be a sum of cycles of length greater than 1 and $k \geq 1$ be an integer. Then, there exist two different FDDS X, Y such that $A_j X^k = A_j Y^k$ for all $1 \leq j \leq m$ and, by consequence, $AX^k = AY^k$.*

Proof. Let $\mathcal{A}$ be the set of cycle lengths of A. Let $X = \sum_{I \subseteq \mathcal{A}} \alpha_I C_{\mathrm{lcm}(I)}$ and $Y = \sum_{I \subseteq \mathcal{A}} \beta_I C_{\mathrm{lcm}(I)}$. Remark that $\alpha_I \neq \beta_I$ for all $I \subseteq \mathcal{A}$, since β_I is α_I plus a nonzero term; in particular, $\alpha_\emptyset \neq \beta_\emptyset$ and, since these two integers are the number of fixed points of X and Y respectively, we have $X \neq Y$.

Let us consider a generic term A_j, which is a cycle C_b for some $b > 1$ for all j. Let $I_1, \dots, I_n$ be an enumeration of the subsets of $\mathcal{A} - \{b\}$ and let $J_i = I_i \cup \{b\}$ for all $1 \leq i \leq n$. Then $Y = \sum_{i=1}^{n} (\beta_{I_i} C_{\mathrm{lcm}(I_i)} + \beta_{J_i} C_{\mathrm{lcm}(J_i)})$

In order to compute $A_j Y^k = C_b Y^k$ we can distribute the power over the sum, then the product over the sum, and we obtain:

$$C_b Y^k = \sum_{k_1+\cdots+k_n=k} \binom{k}{k_1,\dots,k_n} C_b(\beta_{I_1} C_{\mathrm{lcm}(I_1)} + \beta_{J_1} C_{\mathrm{lcm}(J_1)})^{k_1} \prod_{i=2}^{n} (\beta_{I_i} C_{\mathrm{lcm}(I_i)} + \beta_{J_i} C_{\mathrm{lcm}(J_i)})^{k_i}.$$

We have $C_b(\beta_{I_1} C_{\mathrm{lcm}(I_1)} + \beta_{J_1} C_{\mathrm{lcm}(J_1)})^{k_1} = C_b(\alpha_{I_1} C_{\mathrm{lcm}(I_1)} + \alpha_{J_1} C_{\mathrm{lcm}(J_1)})^{k_1}$ by Lemma 7. Thus, by replacing this in the previous equation, we have

$$C_b Y^k = \sum_{k_1+\cdots+k_n=k} \binom{k}{k_1,\dots,k_n} C_b(\alpha_{I_1} C_{\mathrm{lcm}(I_1)} + \alpha_{J_1} C_{\mathrm{lcm}(J_1)})^{k_1} \prod_{i=2}^{n} (\beta_{I_i} C_{\mathrm{lcm}(I_i)} + \beta_{J_i} C_{\mathrm{lcm}(J_i)})^{k_i}.$$

If we repeat this substitution for all $2 \leq i \leq n$ and factor C_b from each term of the sum we obtain

$$C_b Y^k = C_b \sum_{k_1+\cdots+k_n=k} \binom{k}{k_1,\dots,k_n} \prod_{i=1}^{n} (\alpha_{I_i} C_{\mathrm{lcm}(I_i)} + \alpha_{J_i} C_{\mathrm{lcm}(J_i)})^{k_i}.$$

Since $\sum_{k_1+\cdots+k_n=k} \binom{k}{k_1,\dots,k_n} \prod_{i=1}^{n} (\alpha_{I_i} C_{\mathrm{lcm}(I_i)} + \alpha_{J_i} C_{\mathrm{lcm}(J_i)})^{k_i}$ is just X^k, we conclude that $C_b Y^k = C_b X^k$, *i.e.*, $A_j Y^k = A_j X^k$.

This holds separately for each connected component A_j of A; by adding all terms and factoring X^k and Y^k we obtain $AX^k = AY^k$. □

We remark that the X and Y which we have built are just permutations. An interesting property of permutations is that they are closed under sums and products (and thus nonnegative integer powers). Thus X^k and Y^k are also permutations. Another useful property of permutations is, intuitively, that if we multiply a sum of connected components $A = A_1 + \cdots + A_n$ with a sum of cycles $X = X_1 + \cdots + X_m$, the trees rooted in each connected component of $A_i X_j$ are just a periodic repetition of the trees rooted in the cycle of A_i for all i, j [10, Corollary 14]. Thanks to these properties, we can characterize the injectivity of monomials.

Proposition 4. *A monomial over FDDS is injective if and only if its coefficient is cancelable.*

Proof. Assume that the monomial AX^k, where $A = A_1 + \cdots + A_n$ as a sum of connected components and $k > 0$, is not cancelable. Hence, by [10, Theorem 34], A does not have a cycle of length 1 (a fixed point). Let $B = B_1 + \cdots + B_n$ the permutation such that $\ell(B_i) = \ell(A_i)$ for all i. By Lemma 8, there exist two distinct permutations X, Y such that $B_i X^k = B_i Y^k$ for all i. Thus, by [10, Corollary 14], we have $A_i X^k = A_i Y^k$ and, by summing, $AX^k = AY^k$.

The inverse implication was already proved in Proposition 3. □

In the rest of this section, we will prove that a polynomial is injective if and only if a certain monomial is injective. Remark that the sum of any FDDS with a cancelable FDDS is also cancelable. Hence, the sum of the coefficients of a polynomial is cancelable if and only if at least one of them is cancelable.

From the characterization of injective monomials (Proposition 4) we can deduce that there exists an positive integer k such that AX^k is injective if and only if AX^n is injective for *all* positive integer n.

Proposition 5. *Let* $P = \sum_{i=1}^{m} A_i X^i$ *be a polynomial without constant term and consider the monomial* $M = (\sum_{i=1}^{m} A_i)X$. *Then,* P *is injective if and only if* M *is injective.*

Proof. Assume that M is injective. Thus, by the Proposition 4, its coefficient is cancelable. Hence, at least one coefficient of P is cancelable. By Proposition 3, we conclude that P is injective.

For the other direction, assume that M is not injective. Let X and Y the FDDS constructed in the proof of Proposition 4. As previously, $X \neq Y$ but $A_i X^i = A_i Y^i$ for all $i \in \{1, \dots, m\}$. So, by sum, we have $P(X) = P(Y)$. □

Thanks to this characterization, we conclude that the condition given in Proposition 3 is also necessary for polynomials without a constant term. All it remains to prove is that this is also necessary in the presence of a constant term.

Theorem 3. *Let* $P = \sum_{i=0}^{m} A_i X^i$ *a polynomial. Then* P *is injective if and only if at least one of* A_i *is cancelable for some* $i \geq 1$.

Proof. Assume that no non-constant coefficient of P is cancelable, and let $P' = \sum_{i=1}^{m} A_i X^i$ be the polynomial P without the term A_0. By the Proposition 5, there exist two different FDDS X, Y such that $P'(X) = P'(Y)$. Then $P'(X) + A_0 = P'(Y) + A_0$, which implies $P(X) = P(Y)$. Thus, P is not injective. The inverse implication is due to Proposition 3. □

6 Conclusion

In this article we proved that all univariate polynomials over unrolls are injective and we characterized the injectivity of univariate polynomials over FDDS. Furthermore, thanks to the proof of Lemma 5, we can use a technique similar to [11] to compute a solution (if it exists) to injective polynomial equations over unrolls and FDDS. This result suggests interesting directions for future work, namely, characterizing the injectivity of multivariate polynomials, and finding efficient algorithms for other classes of equations involving non-injective polynomials (or the reason why they do not exist, if this is the case).

Furthermore, in the literature, there exists a procedure for solving and enumerating the solutions of an equation of the form $P(X_1, \dots, X_n) = B$ where polynomial P is a sum of univariate monomials [13]. Thus, we can hopefully find an efficient algorithm for the special case where P is a sum of injective univariate monomials. Since polynomial equations with an unbounded number of variables are NP-hard [14], one could try to improve these results for equations with a bounded number of variables, as well as finding exactly how many variables are necessary for the problem to become NP-hard.

Acknowledgment. This work has been partially supported by the HORIZON-MSCA-2022-SE-01 project 101131549 "Application-driven Challenges for Automata Networks and Complex Systems (ACANCOS)" and by the project ANR-24-CE48-7504 "ALARICE".

References

1. Gershenson, C.: Introduction to random Boolean networks. arXiv e-prints (2004)
2. Goles, E., Martìnez, S.:Neural and Automata Networks: Dynamical Behavior and Applications. Kluwer Academic Publishers (1990)
3. Thomas, R., D'Ari, R.: Biological Feedback. CRC Press, Boca Raton (1990)
4. Thomas, R.: Boolean formalization of genetic control circuits. J. Theor. Biol. **42**(3), 563–585 (1973)
5. Bernot, G., et al.: Modeling and analysis of gene regulatory networks: a multidisciplinary endeavor. In: Modeling in Computational Biology and Biomedicine, pp. 47–80. Springer, Heidelberg (2012)
6. Ehrenfeucht, A., Rozenberg, G.: Reaction systems. Fund. Inf. **75**, 263–280 (2007)
7. Gadouleau, M., Riis, S.: Graph-theoretical constructions for graph entropy and network coding based communications. IEEE Trans. Inf. Theory **57**(10), 6703–6717 (2011)
8. Dennunzio, A., Dorigatti, V., Formenti, E., Manzoni, L., Porreca, A.E.: Polynomial equations over finite, discrete-time dynamical systems. In: Mauri, G., El Yacoubi, S., Dennunzio, A., Nishinari, K., Manzoni, L. (eds.) ACRI 2018. LNCS, vol. 11115, pp. 298–306. Springer, Cham (2018). https://doi.org/10.1007/978-3-319-99813-8_27
9. Hopcroft, J.E., Wong, J.K.: Linear time algorithm for isomorphism of planar graphs (preliminary report). In: Proceedings of the Sixth Annual ACM Symposium on Theory of Computing, pp. 172–184 (1974)
10. Naquin, É., Gadouleau, M.: Factorisation in the semiring of finite dynamical systems. Theor. Comput. Sci. **998**, 114509 (2024)
11. Doré, F., Perrot, K., Porreca, A.E., Riva, S., Rolland, M.: Roots in the semiring of finite deterministic dynamical systems. arXiv e-prints (2025). https://arxiv.org/abs/2405.09236
12. Hammack, R., Imrich, W., Klavžar, S.: Handbook of Product Graphs. Discrete Mathematics and Its Applications, 2nd edn. CRC Press, Boca Raton (2011)
13. Dennunzio, A., Formenti, E., Margara, L., Riva, S.: An algorithmic pipeline for solving equations over discrete dynamical systems modelling hypothesis on real phenomena. J. Comput. Sci. **66**, 101932 (2023)
14. Porreca, A.E.: Composing behaviours in the semiring of dynamical systems. In: International Workshop on Boolean Networks (IWBN 2020) (2020). https://doi.org/10.5281/zenodo.3934396

Weihrauch Problems as Containers

Cécilia Pradic[(✉)] and Ian Price

Swansea University, Swansea, UK
c.pradic@swansea.ac.uk

Abstract. We note that Weihrauch problems can be regarded as containers over the category of projective represented spaces and that Weihrauch reductions correspond exactly to container morphisms. We also show that Bauer's extended Weihrauch degrees and the posetal reflection of containers over partition assemblies are equivalent. Using this characterization, we show how a number of operators over Weihrauch degrees, such as the composition product, also arise naturally from the abstract theory of polynomial functors.

Keywords: Weihrauch reducibility · Containers · Polynomial functors

A slightly extended version of this paper may be found on arXiv [23].

Weihrauch reducibility allows to compare the computational strength of partial multi-valued functions over Baire space, which are thus called *Weihrauch problems*. Much like Reverse Mathematics, this can be used as a framework to gauge the relative power of certain Π^1_2 statements such as "for every sequence in $\{0,1\}^{\mathbb{N}}$, we can produce a cluster point" [4].

Concretely, a partial multi-valued function $f :\subseteq X \rightrightarrows Y$ is a map $f : X \to \mathcal{P}(Y)$. Its domain $\mathrm{dom}(f) \subseteq X$ is the set of elements such that $f(x) \neq \emptyset$. When $X = Y = \mathbb{N}^{\mathbb{N}}$, we call such f *Weihrauch problems* and define reduction between those as follows.

Definition 1 ([15]). *If f and g are two Weihrauch problems, f is said to be* Weihrauch reducible to g *if there exists a pair of partial type 2 computable maps (φ, ψ) such that φ is a map $\mathrm{dom}(f) \to \mathrm{dom}(g)$ and for every $i \in \mathrm{dom}(f)$ and $j \in g(\varphi(i))$, $\psi(i,j)$ is defined and belongs to $f(i)$.*

We write $f \leq_{\mathrm{W}} g$ when a reduction from f to g exists. Since Weihrauch reductions compose as depicted in Fig. 1, Weihrauch problems and reductions form a preorder. Equivalence classes of problems are called Weihrauch degrees.

The Weihrauch degrees actually form a distributive lattice, where meets and joins can be regarded as natural operators on problems: the join $P \sqcup Q$ allows to ask a question either to P or Q and get the relevant answer, while $P \sqcap Q$ requires to ask one question to each and get only one answer. Many other natural operators have been introduced on the Weihrauch lattice, including a parallel product $P \times Q$ (ask questions to both P and Q, get both answers), a composition

© The Author(s), under exclusive license to Springer Nature Switzerland AG 2025
A. Beckmann et al. (Eds.): CiE 2025, LNCS 15764, pp. 395–409, 2025.
https://doi.org/10.1007/978-3-031-95908-0_28

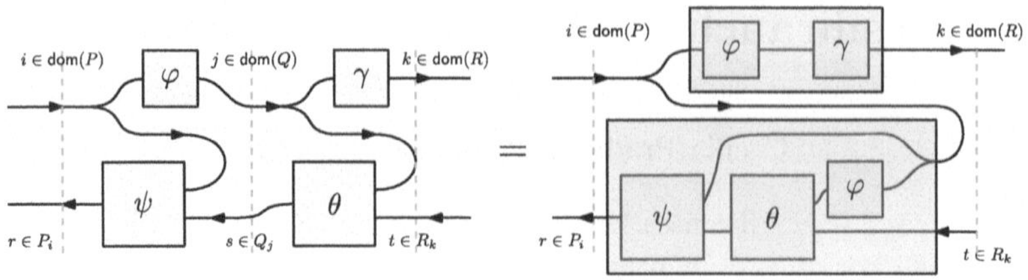

Fig. 1. String diagram representing the composition of Weihrauch reductions from P to Q and Q to R.

operator $P \star Q$ (ask a question to Q and then, depending on the answer you got, a question to P) and many others, including residuals and fixpoints of other operators [5, Definition 1.2].

The definition of a number these operators often does not seem to depend much on the specifics of type 2 computability. They also tend to satisfy nice equational theories whose soundness can solely be established through abstract nonsense. It has also been remarked that Weihrauch reducibility shares a likeness with Gödel's Dialectica interpretation [5,26, resp. §1.9 and 6]. This hints that "the category of Weihrauch reductions" should be constructible by applying a functorial construction $\mathsf{Cat} \to \mathsf{Cat}$ (where Cat is the category of all locally small categories) to a somewhat canonical category of spaces and type 2 computable functions between them.

In this paper, we observe that this construction has been known as the category of *containers* [1][1] over projective represented spaces (that is, the full subcategory of represented spaces containing all subspaces of $\mathbb{N}^{\mathbb{N}}$).

Theorem 1. *The Weihrauch degrees are isomorphic to the posetal reflection of the category of answerable*[2] *containers over the category of projective represented spaces and (type 2) computable functions.*

The proof consists mostly of unpacking definitions; let us try to give a rough idea of its content. Containers over a category $\mathcal{C}$ with pullbacks are bundles, that is, morphisms in $\mathcal{C}$. A morphism from the container $P : X \to U$ to $Q : Y \to V$ consists of a forward map from the base U of P to the base V of Q and a backward map $Y \times_V U \to X$ involving the total spaces X and Y. One can turn multi-valued functions f into the bundle $F : \{(x, y) \mid y \in f(x)\} \xrightarrow{\pi_1} \mathrm{dom}(f)$ to get half of the correspondence we want and check that this extends to a functor: Weihrauch reductions *are* morphisms between the containers built this way.

[1] While the name container comes from research on generic programming, they sometimes also are known under different names in other communities. They may be called *polynomials* due to their correspondence with polynomial functors over locally cartesian closed categories [9].

[2] Meaning those that can be represented by surjections between carriers, or equivalently, those pullback-stable epimorphisms; see Definition 7.

Example 1. One can check that the container corresponding to LPO is

$$\{(0^\omega, 0^\omega)\} \cup \{(p, 1^\omega) \mid p \in \{0,1\}^{\mathbb{N}} \backslash \{0^\omega\}\} \longrightarrow \{0,1\}^{\mathbb{N}}$$

It is equivalent (in the sense that there are morphisms both ways) to the surjective bundle $l : \mathbb{N} \twoheadrightarrow \mathbb{N}_\infty \subseteq 2^{\mathbb{N}}$ defined by $l(0) = 0^\omega$ and $l(n+1) = 0^n 1 0^\omega$.

This correspondence applies not only to Weihrauch reducibility, but also to some closely connected variants. One is continuous Weihrauch reducibility, the notion obtained by replacing "computable" by "continuous" in Definition 1, for which the theorem can be adapted straightforwardly as follows.

Theorem 2. *The continuous Weihrauch degrees are isomorphic to the posetal reflection of the category of answerable containers over projective represented spaces and continuous functions.*

Another closely related notion are the extended Weihrauch degrees introduced in [3] by means of a more general notion of reduction between predicates in a topos. Again a correspondence can be made with containers over a relatively natural category of projectives, although this requires a bit of massaging.

Theorem 3. *Assuming the axiom of choice, the extended Weihrauch degrees over a filtered PCA* $(\mathbb{A}', \mathbb{A})$ *are isomorphic to the posetal reflection of the category of containers over partitioned assemblies* $\mathsf{pAsm}(\mathbb{A}', \mathbb{A})$.

In making those folklore correspondences explicit, we hope among other things to link the literature on Weihrauch reducibility and containers. After providing the necessary background definitions (Sect. 1) and proving the theorems stated thus far (Sect. 2), we discuss some of the algebraic operators of the Weihrauch lattice in the context of the theory of containers (Sect. 3). We will only focus on the two lattice operations, the parallel product and the composition product. While we will obtain clean correspondences to known functorial constructions for the first three, we will hit a snag when it comes to the composition product. This is due to the fact that the categories of spaces we are taking containers over are only *weakly* locally cartesian closed. This means that the composition product will only be a quasi-functor, which complexifies its systematic study; we leave that for further work.

Related Work. The idea of regarding a bundle as a problem to be solved by finding one of its section is old one that predates the "container" terminology. For instance, Hirsch [10, Definition 3.4] defines an equivalent category to study the topological complexity of problems and reductions between those. This perspective also already appeared in the literature on Weihrauch problems (see for instance [13, Remark 2.8]), although most of the recent efforts we are aware of to "categorify" Weihrauch reducibility and operators on problems tend to use other tools instead.

One such natural categorical construction that captured Weihrauch degrees is the restriction of Bauer's extended Weihrauch problems [3] to objects that are actually Weihrauch problems, which are characterized as the $\neg\neg$-dense predicates over modest sets. Interestingly, Ahman and Bauer also linked generalized Weihrauch reducibility to containers [2], but by way of the more general notion of instance reducibility that works over families of truth values. They get back to generalized Weihrauch predicated by specializing their constructions to assemblies which do have a natural truth value object which is not partitioned while here we work directly with bundles of partitioned assemblies.

Trotta et al. also formally linked (extended) Weihrauch reducibility with the Dialectica interpretation [26], which can be regarded bicompletions of fibrations by simple products and sums [11]. Aside from the fact that they work in a posetal setting throughout, it is interesting to note that the category of containers over $\mathcal{C}$ can be recovered by completing the terminal fibration over $\mathcal{C}$ by arbitrary products and then sums and taking the fiber over 1. They also manage to capture *strong* Weihrauch reducibility, which seems hard to achieve directly with our approach.

Pauly also studied a generic notion of reducibility that encompass Weihrauch reducibility, starting from categories of multivalued functions [21], in which he derived the lattice operators as well as finite parallelizations in a generic way. In contrast, we try to stay as far away as possible from multi-valued functions as a notion of morphism in this paper.

1 Categories for Computable Analysis

We review briefly some standard terminology from category theory and realizability that we will use in this paper. In the rest of this paper we will favor terminology from realizability rather than computable analysis for convenience as it allows us to state Theorem 3. Aside from the use of partial combinatory algebras, the main terminological differences are summarized in Fig. 2. We assume that the reader is familiar with the notion of what are categories, functors, natural transformations, finite (co)limits and what is type-2 computability.

1.1 Categories for Type-2 Computability

Since we are going to be borrowing notions from realizability, we are going to use definition based on the notion of partial combinatory algebras (PCAs). Without going into details (which the interested reader can find in e.g. [28]), a PCA is given by a set $\mathbb{A}$ (its *carrier*) and a partial binary operation $\cdot : \mathbb{A}^2 \rightharpoonup \mathbb{A}$. The intuition is that elements of $\mathbb{A}$ are to be regarded as codes for programs and that $\cdot$ denotes function application; as such we simply call it application in the sequel. In addition to having this operation, a PCA $\mathbb{A}$ is required to have certain distinguished elements s and k satisfying certain equations, essentially so that the untyped λ-calculus is interpretable in $\mathbb{A}$.

One of the PCAs we will be concerned with will be $\mathcal{K}_2$ (Kleene's second algebra), whose carrier is $\mathbb{N}^{\mathbb{N}}$. $\mathcal{K}_2$ is meant to capture all partial continuous functions $\mathbb{N}^{\mathbb{N}} \rightharpoonup \mathbb{N}^{\mathbb{N}}$. See [28] for the details of the definition.

For the purpose of computable analysis, we need to use computable continuous functions in a setting where all elements of $\mathbb{N}^{\mathbb{N}}$ are in the background. For this we need the notion of a *filtered PCA* $(\mathbb{A}', \mathbb{A})$ [20, Definition 1.5], which is a pair where $\mathbb{A}$ is a PCA and $\mathbb{A}'$ is a subPCA of $\mathbb{A}$ (i.e. it is stable by $\cdot$ and contains the distinguished k and s, i.e., the interpretation of untyped λ-calculus with parameters in $\mathbb{A}'$ using the structure of $\mathbb{A}$ lands in $\mathbb{A}'$).

The one filtered PCA we shall use in this paper is $(\mathcal{K}_2^{\text{rec}}, \mathcal{K}_2)$ where $\mathcal{K}_2^{\text{rec}}$ consists of the set of computable elements of $\mathbb{N}^{\mathbb{N}}$. It can be checked that they correspond, through applications, to functions computable by type-2 Turing machines. When defining a category $\mathcal{C}(\mathbb{A}', \mathbb{A})$ parameterized by a filtered PCA $(\mathbb{A}', \mathbb{A})$, we may write $\mathcal{C}(\mathbb{A})$ for $\mathcal{C}(\mathbb{A}, \mathbb{A})$ for brevity's sake.

Realizability	modest sets (Mod)	assemblies (Asm)	partitioned modest sets (pMod)
Computable analysis	represented spaces	multi-represented spaces	represented subspaces of $\mathbb{N}^{\mathbb{N}}$ (projective represented spaces)

Fig. 2. Categories used in realizability and their counterpart, once they are specialized to the filtered PCA $(\mathcal{K}_2^{\text{rec}}, \mathcal{K}_2)$, in computable analysis.

With this machinery in place, we can start to define categories of sets represented by codes in PCAs. We begin with the more general notion we will encounter in this paper and decline the useful subcategories we will discuss.

Definition 2. *An assembly over a PCA $\mathbb{A}$ is a pair $(X, \Vdash_X)$ where X is a set called the* carrier *and $\Vdash_X \subseteq \mathbb{A} \times X$ is a relation whose image is X.*

A map f between the carriers of the assemblies $(X, \Vdash_X)$ and $(Y, \Vdash_Y)$ is tracked *by a code $\lceil f \rceil \in \mathbb{A}$ if, whenever $e \Vdash_X x$, we have $\lceil f \rceil \cdot e \Vdash_Y f(x)$.*

Assuming $(\mathbb{A}', \mathbb{A})$ is a filtered PCA, the category $\mathsf{Asm}(\mathbb{A}', \mathbb{A})$ has assemblies over $\mathbb{A}$ as objects and maps tracked by codes in $\mathbb{A}'$ as morphisms.

Given an assembly A, write $|A|$ for its carrier.

Remark 1. An object of $\mathsf{Asm}(\mathcal{K}_2^{\text{rec}}, \mathcal{K}_2)$ is essentially the same thing as a multi-represented space [12, Remark 5.2] and the morphisms of $\mathsf{Asm}(\mathcal{K}_2^{\text{rec}}, \mathcal{K}_2)$ (respectively $\mathsf{Asm}(\mathcal{K}_2)$) are the computable maps between them.

Definition 3. *Given a filtered PCA $(\mathbb{A}', \mathbb{A})$, the category of modest sets $\mathsf{Mod}(\mathbb{A}', \mathbb{A})$ is the full subcategory of assemblies $(X, \Vdash_X)$ such that any element of $\mathbb{A}$ realizes at most one element (i.e., $e \Vdash_X x$ and $e \Vdash_X x'$ imply $x = x'$).*

Remark 2. The objects of $\mathsf{Mod}(\mathcal{K}_2)$ are essentially *represented spaces*.

Definition 4. *Call an assembly* partitioned *when every of its element have at most one realizer. We write $\mathsf{pMod}(\mathbb{A}', \mathbb{A})$ and $\mathsf{pAsm}(\mathbb{A}', \mathbb{A})$ for the respective full subcategories of partitioned modest sets $\mathsf{Mod}(\mathbb{A}', \mathbb{A})$ and assemblies $\mathsf{Asm}(\mathbb{A}', \mathbb{A})$.*

Remark 3. The objects of $\mathsf{pMod}(\mathcal{K}_2)$ are essentially *represented subspaces* of $\mathbb{N}^{\mathbb{N}}$, which can also be characterized as the regular projective objects of $\mathsf{Mod}(\mathcal{K}_2)$.

1.2 Notations for Category-Theoretical Notions

We assume familiarity with the notion of cartesian products and coproducts. We write $\langle f_1, f_2\rangle : Z \to A_1 \times A_2$ for the pairing of $f_i : Z \to A_i$ $(i \in \{1,2\})$ and $g : Z \to A_2$ and $\pi_i : A_1 \times A_2 \to A_i$ $(i \in \{1,2\}$. Dually we write $[f_1, f_2] : A_1 + A_2 \to Z$ and $\mathsf{in}_i : A_i \to A_1 + A_2$ for the copairing and coprojections.

We will heavily use the notion of pullbacks throughout this paper as well as standard notation that involves them. A pullback square is a commuting square as found in the diagram below satisfying the following universal property: if there are morphisms α, β as depicted such that $f \circ \beta = g \circ \alpha$, then there is a unique γ such that $\alpha = k \circ \gamma$ and $\beta = h \circ \gamma$.

$$\begin{array}{ccc} Z & & \\ & A \times_C B & \xrightarrow{k} B \\ & \downarrow h & \downarrow g \\ & A & \xrightarrow{f} C \end{array}$$

with $\alpha : Z \to B$, $\beta : Z \to A$ and $\exists! \gamma : Z \dashrightarrow A \times_C B$.

In this diagram $A \times_C B$ (together with the projections h and k) is a pullback of f and g. We use the notation around the top-left corner of the commuting square to indicate we intend this to be a pullback in a diagram. Pullbacks are only determined up to unique isomorphisms. Concretely, pullbacks in all categories we are interested in here can be built by taking the cartesian products of the domains of the maps f and g and restricting to the sets of pairs (a, b) with $f(a) = g(b)$. When we fix a choice of pullbacks for each pair of morphisms, we write $f^*(g)$ for the morphism corresponding to h (read "the pullback of g along f"); f^* actually extends to a functor $\mathcal{C}/_C \to \mathcal{C}/_A$ where $\mathcal{C}/_A$ ("the slice over A") is the category whose objects are morphisms of $\mathcal{C}$ with codomain A and morphisms are maps between domains of objects that make the obvious triangle diagram commute.

We recall that an *epimorphism* is a morphism $f : A \to B$ such that, whenever we have $h, k : B \to Z$ with $h \circ f = k \circ f$, we actually know that $h = k$.

Definition 5. *An epimorphism $e : X \twoheadrightarrow I$ is called* pullback-stable *if for any $f : J \to I$, $f^*(e)$ is also an epimorphism.*

In topological spaces (and a fortiori in represented spaces), the pullback-stable epimorphisms can be characterized as the maps that are surjective in Set. This does not cover all epimorphisms.

Example 2. $\mathbb{Q} \subseteq \mathbb{R}$ gives rise to an epimorphism due to the fact that all maps are continuous, but that epimorphism is not pullback-stable.

Finally, let us recall that preorders (and a fortiori posets) can be regarded as categories whose objects are elements of the preorder and with a unique morphism between two objects x and y if and only if $x \leq y$. Posets form a reflective subcategory of the category of small categories Cat, which means that the inclusion of posets into categories has a left adjoint functor $\mathcal{C} \mapsto \mathcal{C}/_{\leftrightarrow}$. $\mathcal{C}/_{\leftrightarrow}$ is the quotient of $\mathcal{C}$ where objects X and Y are equivalent when there exists morphisms $X \to Y$ and $Y \to X$.

2 Reducibility as Container Morphisms

2.1 A Container Is an Internal Family

At the intuitive level, a container in a category $\mathcal{C}$ is a family of objects of $\mathcal{C}$ indexed by an object of $\mathcal{C}$. Since objects of $\mathcal{C}$ are not necessarily sets, to make this formal, one needs to define what is a family *internal* to a category. This is done by regarding objects of $\mathcal{C}/_I$ as I-indexed families. The basic idea is that in Set, given a family $(A_i)_{i\in I}$, we can form the disjoint union $\sum_{i\in I} A_i$ and encode the family in this formalism by taking the projection $\sum_{i\in I} A_i \to I$. Conversely, to show that the categorical point of view captures something equivalent to the usual notion in Set, we can perform an inverse operation turning any set-theoretic map $f : X \to I$ into the family $(f^{-1}(i))_{i\in I}$.

A fundamental operation one may perform on families is *reindexing*: given a map of index sets $f : I \to J$ and a family over J, one may produce a family over I. Then the fundamental operation of reindexing families along a map $f : I \to J$ is given by taking a pullback.

Families of sets	Families internal to $\mathcal{C}$
Families over J $\to$ Families over I $(B_j)_{j\in J} \quad \mapsto \quad (B_{f(i)})_{i\in I}$	$\mathcal{C}/_J \to \mathcal{C}/_I$ $\begin{array}{c} B \\ \downarrow{\scriptstyle\alpha} \\ J \end{array} \mapsto \begin{array}{ccc} B' & \longrightarrow & B \\ {\scriptstyle f^*\alpha}\downarrow & \lrcorner & \downarrow{\scriptstyle\alpha} \\ I & \longrightarrow & J \end{array}$

For any choice[3] of pullbacks, this yield functors $f^* : \mathcal{C}/_J \to \mathcal{C}/_I$ which we call *reindexing along* f. Those functor always have a left adjoint $\Sigma_f : \mathcal{C}/_J \to \mathcal{C}/_I$ given by precomposition that corresponds to *sums along* f. In Set, this corresponds to being able to take set-indexed disjoint unions: given a family $(A_i)_{i\in I}$, f allows us to consider it as a family of families over J we may define the family of families, $((A_i)_{i\in f^{-1}(j)})_{j\in J}$ and then perform the set-theoretic disjoint union component-wise.

[3] Only unique up to unique isomorphism in $\mathcal{C}/_J$.

Families of sets	Families internal to $\mathcal{C}$
Families over $I \to$ Families over J	$\mathcal{C}/_I \to \mathcal{C}/_J$
$(A_i)_{i \in I} \mapsto \left(\sum_{i \in f^{-1}(j)} A_i \right)_{j \in J}$	$\begin{array}{c} A \\ \downarrow \alpha \\ I \end{array} \mapsto \begin{array}{c} A \\ \downarrow \alpha \\ I \\ \downarrow f \\ J \end{array}$

The existence of right adjoints Π_f to reindexing functors f^* is however not always guaranteed for every category with pullbacks. It is the case if and only if all slice categories of $\mathcal{C}$ are cartesian closed, in which case we call $\mathcal{C}$ *locally cartesian closed*. Having access to the functors $\Sigma_f \dashv f^* \dashv \Pi_f$ means that we can interpret closed type theoretic-expressions that involve equalities, sums and products over specified objects of $\mathcal{C}$ as objects of $\mathcal{C}$; open expressions containing variables are interpreted as families whose indexing objects correspond to the variables.

2.2 The Category of Containers: Morphisms

As said previously, the objects of the category $\mathsf{Cont}(\mathcal{C})$ of containers over $\mathcal{C}$ will be morphisms in $\mathcal{C}$. Following terminology from [19], for such a container $P : X \to U$, we will call U the object of *positions* and X the object of *directions*. Now the important thing with containers is the notion of morphism that comes with them. In sufficiently structured categories, they correspond to strong natural transformations between functors induced by the containers [9, Theorem 2.12], but they also admit a more general low-level characterization that is easily seen to correspond to notions of reducibility[4]. With sets, assuming we have $Q : Y \to V$, $P_u = P^{-1}(u)$ and $Q_v = Q^{-1}(v)$, a morphism of container from P to Q is given by a pair of maps

$$\varphi : U \to V \qquad \text{and} \qquad \psi \in \prod_{u \in U} (Y_{\varphi(u)} \to X_u)$$

Note that the backward map ψ is here typed using dependent products, which only exist internally in locally cartesian closed categories. We adopt the more elementary equivalent definition which works in all categories with pullbacks.

Definition 6 ([9, Theorem 2.12]). *A morphism representative from a container $P : X \to U$ to $Q : Y \to V$ is a pair (φ, ψ) making the following diagram commute, the rightmost square being a pullback*[5]*:*

[4] And was also independently defined for such purposes, see e.g. [10, Definition 3.4].

[5] While we give a suggestive type-theoretic name for the relevant object, it is determined up to unique isomorphism. This is why we need to quotient morphism representatives; with chosen pullbacks, we could have dispensed with this step.

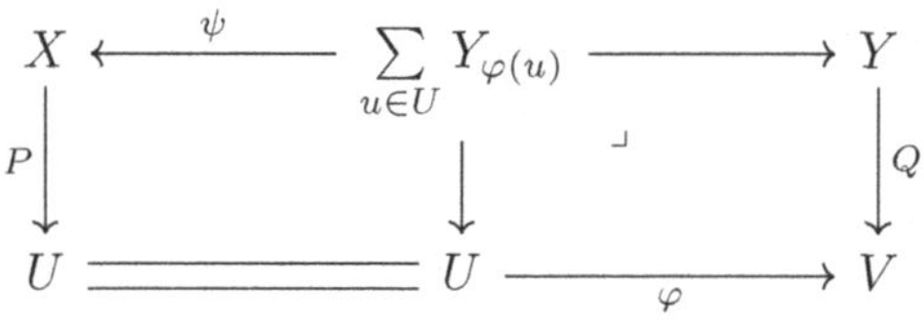

Two morphisms representatives (φ, ψ) and (φ', ψ') from P to Q are defined to be equivalent if and only if $\varphi = \varphi'$ and $\psi \circ \theta = \psi'$ when θ is the canonical isomorphism between the domain of ψ to the domain of ψ' that witnesses that they both are pullbacks of Q and φ. A morphism of container $P \to Q$ is an equivalence class of morphisms representatives.

Given a morphism (representative) (φ, ψ), we will call φ the forward map and ψ the backward map. Composition is defined essentially as suggested in Fig. 1; more formally, it is determined as per the categorical diagram in Fig. 3. We will call (φ, ψ) *horizontal* when ψ is the identity and *vertical* when φ is the identity.

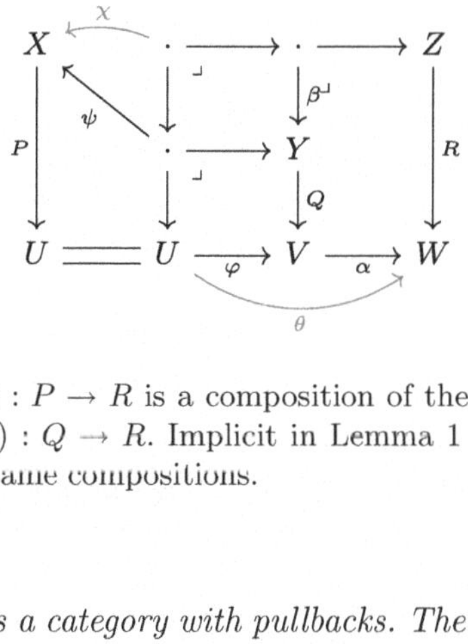

Fig. 3. We say that $(\theta, \chi) : P \to R$ is a composition of the morphism representatives $(\varphi, \psi) : P \to Q$ and $(\alpha, \beta) : Q \to R$. Implicit in Lemma 1 is that pairs of equivalent representative admit the same compositions.

Lemma 1. *Assume $\mathcal{C}$ is a category with pullbacks. Then we can form a category $\mathcal{C}$ where the objects are maps in $\mathcal{C}$, morphisms are given as in Definition 6, identities over $P : X \to U$ are represented by (id_U, id_X) and the composition is given as Fig. 3.*

Proof. It can be checked by unpacking the definitions that $\mathsf{Cont}(\mathcal{C})$ is exactly the total category of the fibration $\mathrm{cod}^{\mathrm{op}}$ over $\mathcal{C}$, which is obtained by taking opposite of the codomain fibration over $\mathcal{C}$ (which exists because of pullbacks); see [25, §1 & §5] for details. □

2.3 Weihrauch Problems are Containers Over Partitioned Modest Sets

We are now almost ready to detail how to prove the main theorem, except for one thing: some of the containers in $\mathsf{Cont}(\mathsf{pMod}(\mathcal{K}_2^{\mathrm{rec}}, \mathcal{K}_2))$ may represent families of subsets of $\mathbb{N}^{\mathbb{N}}$ that may have the empty set as a member. Weihrauch problems do not allow that, so we need to restrict the class of objects to rule out those containers.

Definition 7. *We call a container* answerable *if the underlying map is a pullback-stable epimorphism.*

As previously mentioned, in Asm and a fortiori pMod, the pullback-stable epimorphisms are surjections over the relevant carriers; this ensures each fiber is non-empty. It will be sometimes interesting in the sequel to also consider containers of $\mathsf{Cont}(\mathcal{K}_2^{\mathrm{rec}}, \mathcal{K}_2)$ which are not answerable - so when needed we will refer to objects of (the posetal reflection of) $\mathsf{Cont}(\mathcal{K}_2^{\mathrm{rec}}, \mathcal{K}_2)$ as *slightly extended Weihrauch problems (degrees).*

Theorem 1. *The Weihrauch degrees are isomorphic to the posetal reflection of the category of answerable*[6] *containers over the category of projective represented spaces and (type 2) computable functions.*

Proof. Recall that the category of projective represented spaces is isomorphic to $\mathsf{pMod}(\mathcal{K}_2^{\mathrm{rec}}, \mathcal{K}_2)$. To each answerable container $P : X \to U$ over $\mathsf{pMod}(\mathcal{K}_2^{\mathrm{rec}}, \mathcal{K}_2)$, we can map the P to the Weihrauch problem $\mathsf{w}(P) :\subseteq \mathbb{N}^{\mathbb{N}} \rightrightarrows \mathbb{N}^{\mathbb{N}}$ with $\mathsf{w}(P)(p) = \{q \in \mathbb{N}^{\mathbb{N}} \mid \exists x \in |X|.\ P(x) \Vdash_U p \wedge x \Vdash_X q\}$. For any other $Q : Y \to V$ and a reduction representative $(\varphi, \psi) : P \to Q$, we can find an equivalent one (φ, ψ') where the domain of ψ' is actually the set of pairs (u, y) with $Q(y) = \varphi(u)$ as it is a pullback. Then (φ, ψ') is literally a Weihrauch reduction from $\mathsf{w}(P)$ to $\mathsf{w}(Q)$.

In the other direction, given a Weihrauch problem $f :\subseteq \mathbb{N}^{\mathbb{N}} \rightrightarrows \mathbb{N}^{\mathbb{N}}$, we have a map $\mathsf{c}(f) : \sum_{u \in \mathrm{dom}(f)} f(u) \xrightarrow{\pi_1} \mathrm{dom}(f)$; we can regard its domain and codomain as carriers of partitioned modest sets by taking the trivial realizability relations $\langle e, e' \rangle \Vdash_{\mathrm{dom}(\mathsf{c}(f))} (e, e')$ and $e \Vdash_{\mathrm{dom}(f)} e$. By definition, if a reduction $f \leq_{\mathrm{W}} g$ is represented by a pair of computable maps (φ, ψ), we have $\varphi : \mathrm{dom}(f) \to \mathrm{dom}(g)$ and $\psi : \sum_{u \in \mathrm{dom}(f)} g(\varphi(u)) \to \bigcup_{u \in \mathrm{dom}(f)} f(u)$. Since $\psi(u, y)$ is meant to give an answer to u, by combining ψ with the projection $\sum_{u \in \mathrm{dom}(f)} g(\varphi(u)) \to \mathrm{dom}(f)$, we can form a map $\widetilde{\psi} : \sum_{u \in \mathrm{dom}(f)} g(\varphi(u)) \to \sum_{u \in \mathrm{dom}(f)} f(u)$ so that (φ, ψ) is a morphism representative $\mathsf{c}(f) \to \mathsf{c}(g)$.

Now, to conclude is suffices to show that $\mathsf{c}(\mathsf{w}(P)) \cong P$ and $\mathsf{w}(\mathsf{c}(f)) =_{\mathrm{W}} f$ for all P and f.

For the first part, for $P : X \to U$, because P is answerable, we do also have $\mathrm{dom}(\mathsf{w}(P)) = \{p \in \mathbb{N}^{\mathbb{N}} \mid \exists u \in |U|.\ p \Vdash_U u\}$; this is enough to derive an isomorphism $\theta : \mathrm{dom}(\mathsf{w}(P)) \cong U$ in $\mathsf{pMod}(\mathcal{K}_2^{\mathrm{rec}}, \mathcal{K}_2)$ which is tracked by the code for the identity. Then we can easily check that $\mathsf{c}(\mathsf{w}(P))$ is a pullback of P along θ, hence (θ, id) and $(\theta^{-1}, \mathrm{id})$ represent an isomorphism $\mathsf{c}(\mathsf{w}(P)) \cong P$ as expected.

[6] Meaning those that can be represented by surjections between carriers, or equivalently, those pullback-stable epimorphisms; see Definition 7.

For the second part we have that $\mathsf{w}(\mathsf{c}(f))(p) = \{\langle p, x\rangle \mid x \in f(p)\}$ for every $p \in \mathbb{N}^{\mathbb{N}}$. So the forward component of both reductions we want can be taken to be the identity, the backward component of the reduction $f \leq_{\mathrm{W}} \mathsf{w}(\mathsf{c}(f))$ can project away the first component of the output of $\mathsf{w}(\mathsf{c}(f))$ while the backward part of the converse can be the identity. □

We can note that in the proof above did not use computability aside from the fact type-2 computable maps correspond to morphisms in the category in question. As partial continuous maps are all representable in $\mathcal{K}_2$, Theorem 2 can be proven in the same way.

3 Operators on Containers and Weihrauch Problems

In this section, we list a few classical algebraic operators that appear naturally under some form both in the literature on the Weihrauch lattice [6] and containers [19]. Those operators will typically correspond to (quasi-)functors over containers, implying in particular that they are monotonous over degrees. Those functors can also be generically defined without reference to the specifics of type-2 computability; thus we will define them over $\mathsf{Cont}(\mathcal{C})$, making minimal assumptions about $\mathcal{C}$ while staying consistent with $\mathsf{pMod}(\mathbb{A}', \mathbb{A})$ and $\mathsf{pAsm}(\mathbb{A}', \mathbb{A})$.

3.1 The Commutative Tensor Products

One basic fact about Weihrauch degrees is that they form a distributive lattice. This is reflects the fact that under some mild assumptions, $\mathsf{Cont}(\mathcal{C})$ has products and coproducts that distribute over one another.

Definition 8 ([7]). *A category with finite sums and products is called* extensive *when the canonical functor* $+ : \mathcal{C}/A \times \mathcal{C}/B \to \mathcal{C}/(A+B)$ *is an isomorphism. We say a category is* lextensive *if it has all finite limits, all finite coproducts and is extensive.*

Proposition 1 (See also [24, (3), (6) and (7)] for a particular case). *If $\mathcal{C}$ is lextensive,* $\mathsf{Cont}(\mathcal{C})$ *has finite products, coproducts and they distribute.*

Proposition 2. *Assuming $\mathcal{C}$ is lextensive, answerable containers of* $\mathsf{Cont}(\mathcal{C})$ *are stable under binary products and coproducts. The initial object is an answerable container but not the terminal object.*

Containers also admit another natural monoidal product, which is sometimes called the *parallel product*. The parallel product of $P_1 : X_1 \to U_1$ and $P_2 : X_2 \to U_2$, which we write $P_1 \otimes P_2$, is represented by the morphism $P_1 \times P_2 : X_1 \times X_2 \to U_1 \times U_2$ obtained by the functoriality of $\times$ in $\mathcal{C}$ (assuming $\mathcal{C}$ has cartesian products)[19, Definition 3.65]. Regarded as a Weihrauch problem, this means we may ask a question to both P_1 and P_2 and get back answers for both.

It is straightforward to check that $\otimes$ is a symmetric monoidal product in $\mathcal{C}$ (see [16, Chapter VII]) and that answerable containers are stable under $\otimes$. If we further assume that $\mathcal{C}$ is lextensive, we can show that there are further distributive laws.

Proposition 3 (See also [24, (6) and (28)] for a particular case). *If $\mathcal{C}$ is lextensive, then we have natural transformations $P \otimes Q + P \otimes R \cong P \otimes (Q + R)$ and $(P \otimes Q) \times R \to P \otimes (Q \times R)$ in* $\mathsf{Cont}(\mathcal{C})$.

Note that with this and the functoriality of $\times$, when moving to $\mathsf{Cont}(\mathcal{C})/_{\leftrightarrow}$ we recover all of the axioms listed in [18] pertaining to binary joins, meets and parallel products.

3.2 The Composition Product in Nicer Categories of Containers

Another monoidal product which appears in the literature on containers is the *composition product.* The most natural way to introduce this is maybe by making the link between containers and polynomial functors [9, Theorem 2.12]. In short, assuming that $\mathcal{C}$ is locally cartesian closed, we map any container $P : X \to U$ to the endofunctor over $\mathcal{C}$ defined by $\Sigma_{!_U} \circ \Pi_P \circ !_X^* : \mathcal{C}/_1 \to \mathcal{C}/_1$ and the isomorphism $\mathrm{dom} : \mathcal{C}/_1 \cong \mathcal{C}$. Regarding P as an internal family $(X_u)_{u \in U}$ to $\mathcal{C}$, on objects this functor maps A to $\sum_{u \in U} A^{X_u}$. Those endofunctors are called *polynomial endofunctors* over $\mathcal{C}$, and one important fact is that strong natural transformations between those correspond exactly to container morphisms. One important theorem is that polynomial endofunctors are stable under composition, which induces a monoidal product on polynomials sometimes called the *substitution* or *sequential product* [9, (9) Proposition 1.12].

In the spirit of Weihrauch reducibility, the sequential product $P \star Q$ can be regarded as the problem whose inputs are composed of an input v for Q together with a function f turning a solution for said input into an input for P. A solution for $P \star Q$ then consists of a solution for v and a solution for $f(v)$. In sets, assuming we have $Q : Y \to V$, $X_u = P^{-1}(u)$ and $Y_v = Q^{-1}(v)$ for every u, v, $P \star Q$ is the family $\left(\sum_{y \in Y_v} X_{f(y)}\right)_{(v,f)}$ whose indexing set is $\sum_{v \in V} (Y_v \to X)$. This can be internalized in any locally cartesian closed category.

In short, reducing to $P \star Q$ is the ability to make an oracle call to Q and then P. The sequential product of Weihrauch degrees is designed to capture this intuition and is thus defined this way [29, Definition 3][7], but for a small wrinkle: $\mathsf{pMod}(\mathcal{K}_2^{\mathrm{rec}}, \mathcal{K}_2)$ is not locally cartesian closed!

Proposition 4. $\mathsf{pMod}(\mathcal{K}_2)$, $\mathsf{pAsm}(\mathcal{K}_2)$, $\mathsf{pMod}(\mathcal{K}_2^{\mathrm{rec}}, \mathcal{K}_2)$ *and* $\mathsf{pAsm}(\mathcal{K}_2^{\mathrm{rec}}, \mathcal{K}_2)$ *are not cartesian closed.*

The failure of cartesian closure is essentially due to a lack of quotients rather than a lack of power: it is not that we cannot build modest sets of codes for function spaces, it is that we cannot guarantee that a function has a unique code. It means in particular that we still have a weak version of local cartesian closure for pMod, which is enough to talk about spaces of codes for functions; this is why the same intuition can be used to build the sequential product of Weihrauch problems as a *quasi-functor* $\star_p$ rather than a functor. But a disadvantage of

[7] The original definition in [6] actually has a slightly different flavour, but they correspond to the same Weihrauch degree.

this approach is that we then need to reconstruct from scratch certain proofs, for instance that the sequential product on the left distributes over coproducts. Another strategy that one could take would be to observe that pMod is the full subcategory of *projective* objects of Mod and that every object of Mod is *covered* by a partitioned modest set. This allows to define $\star_p$ over pMod directly from the definition of the composition product $\star$ in Mod and transfer some properties of $\star$ to $\star_p$; we invite the interested reader to consult the appendices for some preliminary steps in this direction.

4 Conclusion

We hope that making the connection explicit between Weihrauch reducibility and containers will allow to connect more tightly the literature on the two topics. For instance, it would be nice to relate formally universal equational theories of the Weihrauch lattice with operators, like the ones studied in [18,22], to similar theories interpreted in categories of containers over toposes with $\mathcal{W}/\mathcal{M}$-types. Also, since we have shown that Weihrauch problems are a definable subclasses of containers in assemblies over $(\mathcal{K}_2^{\mathrm{rec}}, \mathcal{K}_2)$, it might be a way to formalize the theory of Weihrauch reducibility in in proof assistants in a synthetic style (see e.g. [8] for similar work for Turing-reducibility in Rocq and [14] for a development in a synthetic style in $\mathsf{Asm}(\mathcal{K}_2^{\mathrm{rec}}, \mathcal{K}_2)$).

Another area which might be worth investigating is the connections between Weihrauch reducibility and proof theory, as one can naturally interpret linear type theories in fibrations that are closely linked to categories of containers [17]. Could this be exploited to characterize Weihrauch reducibility in a type theory, a bit in the spirit of [27,30]? One difficulty with a naive attempt is that interpreting types directly in $\mathfrak{Fam}(\mathfrak{Fam}^{\mathrm{op}}(\mathrm{id}_{\mathsf{pAsm}(\mathcal{K}_2^{\mathrm{rec}}, \mathcal{K}_2)}))$ (see [11] for a definition) seems somewhat counter-intuitive, since $\mathbf{I} \leq_{\mathrm{W}} P \multimap Q$ holds iff P reduces to ("is less powerful than") Q.

Acknowledgments. Many thanks to Eike Neumann, Arno Pauly and Manlio Valenti for discussions (especially on Proposition 4) and encouragements. Many thanks as well to Takayuki Kihara and an anonymous reviewer for pointers to the literature and comments that helped improve the presentation of the paper.

The authors acknowledge support by Swansea University and grant to Swansea University a non-exclusive, irrevocable, sub-licensable, worldwide license to make the accepted manuscript available on its institutional repository.

References

1. Abbott, M., Altenkirch, T., Ghani, N.: Categories of containers. In: Gordon, A.D. (ed.) FoSSaCS 2003. LNCS, vol. 2620, pp. 23–38. Springer, Heidelberg (2003). https://doi.org/10.1007/3-540-36576-1_2
2. Ahman, D., Bauer, A.: Comodule representations of second-order functionals (2024). https://arxiv.org/abs/2409.17664

3. Bauer, A.: Instance reducibility and Weihrauch degrees. Log. Methods Comput. Sci. **18**(3) (2022). https://doi.org/10.46298/LMCS-18(3:20)2022
4. Brattka, V., Gherardi, G., Marcone, A.: The Bolzano-Weierstrass theorem is the jump of Weak Kőnig's Lemma. Ann. Pure Appl. Logic **163**(6), 623–655 (2012). https://doi.org/10.1016/j.apal.2011.10.006
5. Brattka, V., Gherardi, G., Pauly, A.: Weihrauch complexity in computable analysis. CoRR **abs/1707.03202** (2017)
6. Brattka, V., Pauly, A.: On the algebraic structure of Weihrauch degrees. Logical Methods Comput. Sci. **14**(4) (2018). https://doi.org/10.23638/LMCS-14(4:4)2018
7. Carboni, A., Lack, S., Walters, R.: Introduction to extensive and distributive categories. J. Pure Appl. Algebra **84**(2), 145–158 (1993). https://doi.org/10.1016/0022-4049(93)90035-R
8. Forster, Y., Kirst, D., Mück, N.: Oracle computability and turing reducibility in the calculus of inductive constructions. In: Hur, C. (ed.) Programming Languages and Systems - 21st Asian Symposium, APLAS 2023, Taipei, Taiwan, November 26-29, 2023, Proceedings. LNCS, vol. 14405, pp. 155–181. Springer, Cham (2023)
9. Gambino, N., Kock, J.: Polynomial functors and polynomial monads. Math. Proc. Cambridge Philos. Soc. **154**(1), 153–192 (2012). https://doi.org/10.1017/s0305004112000394
10. Hirsch, M.D.: Applications of topology to lower bound estimates in computer science. Ph.D. thesis, University of California, Berkeley (1990)
11. Hofstra, P.: The Dialectica monad and its cousins. In: Makkai, M., Hart, B. (eds.) Models, Logics, and Higher-dimensional Categories: A Tribute to the Work of Mihály Makkai. American Mathematical Society, CRM Proceedings & Lecture Notes (2011)
12. Iljazović, Z., Kihara, T.: Computability of subsets of metric spaces. In: Handbook of Computability and Complexity in Analysis, pp. 29–69. Springer (2021)
13. Kihara, T.: Borel-piecewise continuous reducibility for uniformization problems. Logical Methods Comput. Sci. **12**(4) (2017). https://doi.org/10.2168/lmcs-12(4:4)2016
14. Konečný, M., Park, S., Thies, H.: Certified computation of nondeterministic limits. In: NASA Formal Methods, pp. 771–789. Springer (2022). https://doi.org/10.1007/978-3-031-06773-0_41
15. Lempp, S., Miller, J., Pauly, A., Soskova, M., Valenti, M.: Minimal covers in the Weihrauch degrees. Proc. Am. Math. Soc. **152**(11), 4893–4901 (2024)
16. Mac Lane, S.: Categories for the Working Mathematician, vol. 5. Springer, Cham (2013)
17. Moss, S.K., von Glehn, T.: Dialectica models of type theory. In: Proceedings of the 33rd Annual ACM/IEEE Symposium on Logic in Computer Science, LICS 2018, Oxford, UK, July 09-12, 2018, pp. 739–748 (2018)
18. Neumann, E., Pauly, A., Pradic, C.: The equational theory of the Weihrauch lattice with multiplication. CoRR **abs/2403.13975** (2024). https://doi.org/10.48550/ARXIV.2403.13975
19. Niu, N., Spivak, D.I.: Polynomial functors: a mathematical theory of interaction (2024). https://arxiv.org/abs/2312.00990
20. van Oosten, J., Zou, T.: Classical and relative realizability (2016). https://arxiv.org/abs/1603.03621
21. Pauly, A.: Many-one reductions between search problems. CoRR **abs/1102.3151** (2011). http://arxiv.org/abs/1102.3151
22. Pradic, C.: The equational theory of the Weihrauch lattice with (iterated) composition (2025)

23. Pradic, C., Price, I.: Weihrauch problems as containers (2025). https://arxiv.org/abs/2501.17250
24. Spivak, D.I.: A reference for categorical structures on **Poly** (2024)
25. Streicher, T.: Fibered categories à la Jean Bénabou (2023)
26. Trotta, D., Valenti, M., de Paiva, V.: Categorifying computable reducibilities. arXiv preprint arXiv:2208.08656 (2022)
27. Uftring, P.: The characterization of Weihrauch reducibility in systems containing $\mathrm{E} - \mathrm{PA}^{\omega} + \mathrm{QF} - \mathrm{AC}^{0,0}$. J. Symbolic Logic **86**(1), 224–261 (2020). https://doi.org/10.1017/jsl.2020.53
28. Van Oosten, J.: Realizability: An Introduction To its Categorical Side, vol. 152. Elsevier (2008)
29. Westrick, L.: A note on the diamond operator. Computability **10**(2), 107–110 (2021). https://doi.org/10.3233/COM-200295
30. Yoshimura, K.: General treatment of non-standard realizabilites (2016). unpublished

On the Reverse Mathematics of Cut-Elimination and Determinacy

Philipp Provenzano(✉)

Ghent University, 9000 Gent, Belgium
Philipp.Provenzano@UGent.be

Abstract. We introduce a non-wellfounded proof system $\mathrm{G}_\omega^{\searrow\mathrm{P}<\omega}$ based on a non-wellfounded language $\mathcal{L}\left(\mathrm{G}_\omega^{\searrow\mathrm{P}<\omega}\right)$, having roughly the expressive strength of the language of finitely iterated inductive definitions $\mathrm{ID}_{<\omega}$. As a measure of its strength, we analyze the proof theoretic properties of consistency, cut-elimination and soundness of $\mathrm{G}_\omega^{\searrow\mathrm{P}<\omega}$ from a reverse mathematical point of view. Establishing a natural semantic for the non-wellfounded language through games of infinite length, we show that all three properties are equivalent to the determinacy assertion $\forall n.(\Pi^0_1)_n-\mathrm{Det}$ over Baire space and to the system $\Pi^1_3-\mathrm{RFN}\left(\Pi^1_1-\mathrm{CA}_0\right)$ of second order arithmetic.

For a restricted system $\mathrm{G}_2^{\searrow\mathrm{P}<\omega}$ based on a non-wellfounded language $\mathcal{L}\left(\mathrm{G}_2^{\searrow\mathrm{P}<\omega}\right)$ with binary connectives, we show that syntactic cut-elimination versus semantic cut-admissibility have vastly different reverse mathematical strength. The former still corresponds to $\Pi^1_3-\mathrm{RFN}\left(\Pi^1_1-\mathrm{CA}_0\right)$. The latter, together with consistency and soundness for the system $\mathrm{G}_2^{\searrow\mathrm{P}<\omega}$, correspond to the determinacy assertion $\forall n.(\Pi^0_1)_n-\mathrm{Det}^\star$ over Cantor space and to the second order theory $\Pi^1_2-\mathrm{RFN}(\mathrm{ACA}_0) \equiv \mathrm{ACA}_0'$.

Keywords: Reverse mathematics · Proof theory · Cut-elimination · Determinacy

1 Introduction

The discipline of non-wellfounded proof theory has recently gained attention in studies of arithmetic [5,8], linear logic [3,4,9,10] and the modal μ-calculus [1,15]. In this paper, we introduce a non-wellfounded system $\mathrm{G}_\omega^{\searrow\mathrm{P}<\omega}$. It is roughly an ω-branching propositional version of LKID^ω, a non-wellfounded system introduced by Brotherston and Simpson in [5] capable of interpreting $\mathrm{ID}_{<\omega}$. A novelty in the present paper is that in addition to proofs, also the language $\mathcal{L}\left(\mathrm{G}_\omega^{\searrow\mathrm{P}<\omega}\right)$ is non-wellfounded, leading to natural semantics through some games of infinite length.

Consider a least or greatest fixpoint formula $\varphi(n) \equiv n \in \mu X.\mathcal{O}(X)$ or $\varphi(n) \equiv n \in \nu X.\mathcal{O}(X)$, respectively, in the language of ID_1, where $\mathcal{O}$ is a positive arithmetical operator in negation normal form. That is, the formula $n \in \mathcal{O}(X)$ is built from $\exists, \forall, \vee, \wedge$ and possibly negated atoms, except those of the form

© The Author(s), under exclusive license to Springer Nature Switzerland AG 2025
A. Beckmann et al. (Eds.): CiE 2025, LNCS 15764, pp. 410–424, 2025.
https://doi.org/10.1007/978-3-031-95908-0_29

$t \in X$ which only appear positively. Intuitively, we want to define its infitinary unraveling $\varphi^\infty(n) :\equiv n \in \mathcal{O}(\mathcal{O}(\mathcal{O}(\cdots)))$ with bound variables in the different copies of $\mathcal{O}$ suitably renamed. This is to be understood as a coinductive definition on some class of non-wellfounded formulas, e.g., coded by possibly illfounded binary trees where nodes are labelled with propositional connectives and quantifiers, and leaves are labelled by atomic formulas. In particular, $\varphi^\infty(n)$ usually contains infinite quantifier alternations. Note that, syntactically, we use the same infinitary expansion for both least and greatest fixpoints. It will be up to our proposed semantics to distinguish them.

As customary in ω-logic, we may translate $\varphi^\infty(\underline{n})$ for a fixed $n \in \omega$ into a purely propositional non-wellfounded formula $\varphi^\omega(\underline{n})$. This is achieved by replacing existential quantifiers $\exists m.\, \psi(m)$ with an infinite disjunction over all its instances $\bigvee_{\underline{m} \in \omega} \psi(\underline{m})$ and similar for universals. Atomic formulas then have definite truth values and are replaced by an empty conjunction ($\top$) or an empty disjunction ($\bot$), respectively. The resulting syntactic objects may be coded by possibly illfounded ω-branching trees where nodes are labelled by propositional connectives.

For a closed propositional formula φ of finite depth but possibly infinite width, i.e., ω-branching connectives, its Tarski truth definition gives rise to a game between player or prover P and opponent or falsifier O. Starting with φ, the two players subsequently choose subformulas of it. If the current formula is disjunctive, P chooses a direct subformula to continue the play on. Otherwise, if it is conjunctive, O chooses the next subformula. Since φ has finite depth, the game eventually reaches a leaf. If it is conjunctive, P wins, as an empty conjunction corresponds to $\top$, otherwise O wins. P has a winning strategy in this game precisely when φ is true.

The dynamics of this game, i.e., the two players subsequently choosing subformulas, readily generalizes to the non-wellfounded formula $\varphi^\omega(\underline{n})$. However, the game no longer necessarily terminates, and we have to specify the winner in an infinite play. This is where we can distinguish between least and greatest fixpoints. In the μ-evaluation game for $\varphi^\omega(\underline{n})$, P wins a play if it reaches an empty conjunction and O wins if it reaches an empty disjunction or if it continues forever. Dually, P wins a play in the ν-evaluation game precisely if it does not end in an empty disjunction. One checks that $\mathbb{N} \models \varphi(\underline{n}) \equiv \underline{n} \in \mu X.\mathcal{O}(X)$ iff P has a winning strategy in the μ-evaluation game for $\varphi^\omega(\underline{n})$, and analogously for ν-formulas. In Definition 1, we introduce the language $\mathcal{L}\left(\mathrm{G}_\omega^{\searrow \mathrm{P}<\omega}\right)$ of non-wellfounded formulas that handles both cases uniformly and also allows for nesting, but not interleaving, of fixpoints.

Based on this language, we define the non-wellfounded system $\mathrm{G}_\omega^{\searrow \mathrm{P}<\omega}$. As a measure of its strength, we analyze its proof-theoretic properties of consistency, cut-elimination and soundness from the point of view of reverse mathematics. That is to say, we find a system of second-order arithmetic that is necessary and sufficient to prove these properties. Based on the semantics outlined above, we show that consistency, cut-elimination and soundness of $\mathrm{G}_\omega^{\searrow \mathrm{P}<\omega}$ are all equivalent to determinacy of evaluation games for $\mathcal{L}\left(\mathrm{G}_\omega^{\searrow \mathrm{P}<\omega}\right)$-formulas. Determinacy of a game asserts that either P or O has a winning strategy. While readily proved for finite games, determinacy for infinite games has a rich study of reverse mathematical results [12,13,16,19]. We rely on these to connect our results to the more well known systems of second order arithmetic.

A general connection between questions of determinacy and cut-elimination was previously observed by Takeuti [18, §23]. This was extended by Aguilera in [2]. Comparable to our language $\mathcal{L}\left(G_\omega^{\searrow P<\omega}\right)$, their system of determinate logic DL also uses formulas with quantifier-chains of length ω. However, the rules of DL operate on the whole chain of quantifiers at once, essentially quantifying over strategies in some game. This makes the connection between determinacy and cut-elimination for DL feel somewhat artificial. Moreover, while proofs in DL are wellfounded - by which it does not actually make use of the non-wellfoundedness of its language - sequents are possibly infinite, even uncountable. A main goal of the present paper is to provide a more natural instance of this connection between determinacy and proof theory through the study of $G_\omega^{\searrow P<\omega}$.

Finally, we introduce a natural variant $G_2^{\searrow P<\omega}$ based on the non-wellfounded language $\mathcal{L}\left(G_2^{\searrow P<\omega}\right)$ with binary connectives only. It exhibits the proof-theoretically uncommon phenomenon that semantic cut-admissibility and syntactic cut-elimination have vastly different strength. The former states that every provably sequent is also cut-free provable, and we show that it is equivalent to consistency and soundness of $G_2^{\searrow P<\omega}$. Syntactic cut-elimination, on the other hand, asserts that a concrete Gentzen/Tait style algorithm produces a valid cut-free proof. We show that it is equivalent to cut-elimination for $G_\omega^{\searrow P<\omega}$. This situation may be seen as an analog to the vastly different strength between consistency and strong normalization of type systems like System F. The former states that there is some uninhabited type and just follows from soundness of classical propositional logic. In the case of System F, strong normalization, on the other hand, implies the consistency of second order arithmetic, as shown by Girard [11].

2 Definitions

A common base system for the study of reverse mathematical results is recursive comprehension RCA_0 which besides basic arithmetical axioms can form computable sets and has induction for Σ_1^0-formulas. Its first order part corresponds to primitive recursive arithmetic PRA. Other systems of interest are arithmetical comprehension ACA_0 and Π_1^1-comprehension $\Pi_1^1-\mathrm{CA}_0$ which can additionally form sets defined by arithmetical and Π_1^1-formulas, respectively. The Π_n^1-reflection principle for a theory T, $\Pi_n^1-\mathrm{RFN}\,(T)$, asserts that every Π_n^1-formula φ which is provable in T is true. This depends on an internalized notion of T-provability. Reflection principles commonly arise from internalizing universal quantification over some axiom scheme. For example, $\Pi_2^1-\mathrm{RFN}\,(\mathrm{ACA}_0)$ is more commonly known as ACA_0' which adds to ACA_0 the axiom

$$\forall n. \text{“for every set } X\text{, the } n\text{'th Turing jump } X^{(n)} \text{ of } X \text{ exists”}.$$

A general overview for these systems can be found in Simpson [17].

Definition 1. *A formula, or ω-formula, $\varphi \in \mathcal{L}\left(\mathrm{G}_\omega^{\searrow\mathrm{P}<\mathrm{n}}\right)$ is an (at most) ω-branching tree T_φ together with a labelling of nodes by*

- *a propositional connective, $c : T_\varphi \to \{\vee, \wedge\}$*
- *a natural number, called priority, $p : T_\varphi \to \omega$ such that*
  - *it is non-increasing along branches, $s \sqsubseteq t \in T_\varphi \Rightarrow p(s) \geq p(t)$*
  - *the root priority $p(\langle\rangle)$ is less than n*

A binary, or 2-formula, $\varphi \in \mathcal{L}\left(\mathrm{G}_2^{\searrow\mathrm{P}<\mathrm{n}}\right)$ is an ω-formula $\varphi \in \mathcal{L}\left(\mathrm{G}_\omega^{\searrow\mathrm{P}<\mathrm{n}}\right)$ such that T_φ is (at most) binary branching. We define $\mathcal{L}\left(\mathrm{G}_\omega^{\searrow\mathrm{P}<\omega}\right) := \bigcup_{n\in\omega} \mathcal{L}\left(\mathrm{G}_\omega^{\searrow\mathrm{P}<\mathrm{n}}\right)$ and $\mathcal{L}\left(\mathrm{G}_2^{\searrow\mathrm{P}<\omega}\right) := \bigcup_{n\in\omega} \mathcal{L}\left(\mathrm{G}_2^{\searrow\mathrm{P}<\mathrm{n}}\right)$.

The priority labels are meant to model the nesting of fixpoints, and they will also allow us to distinguish the μ- and ν-evaluation games from the introduction uniformly by parity. Non-increasingness of priorities models that $\mathrm{ID}_{<\omega}$ allows for nested but not interleaved fixpoints, and we indicate it in the name of the proof system by $\searrow$P.

Definition 2. *The evaluation game for a formula $\varphi \in \mathcal{L}\left(\mathrm{G}_\omega^{\searrow\mathrm{P}<\omega}\right)$ is the following game between two players, called Player P and Opponent O. Every position in the game is a node of φ, starting at the root. If the current position σ is labelled by $\vee$, it is P's turn to move, and otherwise it is O's turn. The respective player then selects a direct child of σ. If the play reaches a leaf labelled by $\wedge$ or $\vee$, respectively, then the game ends and P, respectively O, wins. For an infinite play, the priority label eventually becomes constant by non-increasingness. The play is won by P if the eventual priority is even, and won by O if it is odd.*

The winning condition for P or O in such an evaluation game can be expressed within the Π^0_1 difference hierarchy $\left(\Pi^0_1\right)_{<\omega} := \bigcup_{n\in\omega} \left(\Pi^0_1\right)_n$ where $\left(\Pi^0_1\right)_1 :\equiv \Pi^0_1$ and $\left(\Pi^0_1\right)_{n+1} :\equiv \Pi^0_1 \backslash \left(\Pi^0_1\right)_n$. $\left(\Pi^0_1\right)_{<\omega}$ is properly contained in Π^0_2. Given Σ^0_1-formulas $\varphi_1(x), \ldots, \varphi_n(x)$, the difference hierarchy can essentially check the parity of the smallest i such that $\varphi_i(x)$, i.e.,

$$\varphi(x) :\equiv \min\{i \leq n \mid \varphi_i(x)\} \equiv_2 0$$

is in $\left(\Pi^0_1\right)_{<\omega}$. For fixed n, this is precisely what we need to express the winning condition for evaluation games of $\mathcal{L}\left(\mathrm{G}_\omega^{\searrow\mathrm{P}<\mathrm{n}}\right)$-formulas.

For $\varphi, \psi \in \mathcal{L}\left(\mathrm{G}_\omega^{\searrow\mathrm{P}<\omega}\right)$, we define propositional connectives like $\varphi \wedge \psi$ in the obvious way, as a tree with direct subtrees φ and ψ, the root labelled by $\wedge$ and a priority that is at least as large as the root priority of φ and ψ. By abuse of notation, we omit the ambiguous root priority. Negation is defined by flipping all propositional labels and the parity of all priorities, e.g., incrementing them by one. Quantifiers are treated as infinitary propositional connectives. Given a formula φ, its subformulas correspond to the labelled subtrees rooted at nodes of φ.

Definition 3. *A sequent in $\mathcal{L}\left(\mathrm{G}_{\omega}^{\searrow \mathrm{P}<\mathrm{n}}\right)$ is a finite multiset Γ of $\mathcal{L}\left(\mathrm{G}_{\omega}^{\searrow \mathrm{P}<\mathrm{n}}\right)$-formulas. It is said to be true if $\bigvee \Gamma$ is, in the sense of evaluation games.*

We can now define a non-wellfounded proof system operating on these sequents. To this end, we first ignore the priority labels and use an inference system commonly known from ω-logic to deal with the propositional part. We choose to work with a monotone system, in the sense that the conclusion of an inference also appears in the premises, which is a way to automatically admit contraction. Concretely, the class of pre-proofs in $\mathrm{G}_{\omega}^{\searrow \mathrm{P}<\mathrm{n}}$ is defined over the language $\mathcal{L}\left(\mathrm{G}_{\omega}^{\searrow \mathrm{P}<\mathrm{n}}\right)$ by the following three rule schemas

$$\frac{\Gamma, \bigwedge_{i \in I} \varphi_i, \varphi_i \quad (\text{all } i \in I)}{\Gamma, \bigwedge_{i \in I} \varphi_i} (\wedge)$$

$$\frac{\Gamma, \bigvee_{i \in I} \varphi_i, \varphi_i \quad (\text{some } i \in I)}{\Gamma, \bigvee_{i \in I} \varphi_i} (\vee)$$

$$\frac{\Gamma, \varphi \quad \Gamma, \neg\varphi}{\Gamma} (\mathrm{Cut})$$

The priority labels now allow us to single out the correct proofs from the above pre-proofs via the following trace condition. For a proof P, we identify its inferences with the subproofs rooted at them. A branch in P is then a sequence of consecutive inferences or subproofs, starting at P.

Definition 4. *A pre-proof P in $\mathrm{G}_{\omega}^{\searrow \mathrm{P}<\mathrm{n}}$ is a correct proof if for every branch $(P_n)_{n \in \mathbb{N}}$, there exists a sequence $(\varphi_n)_{n \geq N}$ for some N, called a trace, such that*

- *φ_N is a formula in the conclusion of P_N*
- *all φ_n have the same even priority*
- *for all $n \geq N$*
  - *i) either $\varphi_{n+1} \equiv \varphi_n$ is a side formula of P_n*
  - *ii) or φ_{n+1} and φ_n are the main premise and conclusion of the inference P_n, respectively*
- *ii) happens infinitely often*

Provability of a sequent Γ is denoted $\mathrm{G}_{\omega}^{\searrow \mathrm{P}<\mathrm{n}} \vdash \Gamma$ and cut-free provability by $\mathrm{G}_{\omega}^{\searrow \mathrm{P}<\mathrm{n}} \vdash^0 \Gamma$. We define the limit system $\mathrm{G}_{\omega}^{\searrow \mathrm{P}<\omega} := \bigcup_{n \in \omega} \mathrm{G}_{\omega}^{\searrow \mathrm{P}<\mathrm{n}}$, which is more restrictive than defining an analogous system over the language $\mathcal{L}\left(\mathrm{G}_{\omega}^{\searrow \mathrm{P}<\omega}\right)$.

The systems $\mathrm{G}_{2}^{\searrow \mathrm{P}<\mathrm{n}}$ are analogously defined over the language $\mathcal{L}\left(\mathrm{G}_{2}^{\searrow \mathrm{P}<\mathrm{n}}\right)$, and $\mathrm{G}_{2}^{\searrow \mathrm{P}<\omega} := \bigcup_{n \in \omega} \mathrm{G}_{2}^{\searrow \mathrm{P}<\mathrm{n}}$.

3 Determinacy and Soundness

The following result connects this provability notion to the concept of determinacy.

Proposition 1 (RCA$_0$). *Given a formula $\varphi \in \mathcal{L}\left(\mathrm{G}_\omega^{\searrow \mathrm{P}<\omega}\right)$, it is cut-free provable in $\mathrm{G}_\omega^{\searrow \mathrm{P}<\omega}$ iff Opponent does not have a winning strategy in its evaluation game.*

Proof. ($\Rightarrow$) : Given an Opponent strategy σ in the evaluation game, it selects a branch through the cut-free proof of φ. By the correctness condition, there exists a trace through a subformula of φ that can be reached by Player against σ. Following it then gives a counter strategy against σ.

($\Leftarrow$) : Suppose φ is not cut-free provable. There exists an exhaustive pre-proof P for φ, exploring every disjunct exactly once on every branch of P. By assumption, this cannot be a correct proof and thus has a branch without a trace. This branch then defines a winning strategy for Opponent in the evaluation game.

Corollary 1 (RCA$_0$). *Given a formula $\varphi \in \mathcal{L}\left(\mathrm{G}_\omega^{\searrow \mathrm{P}<\omega}\right)$, it is determined, i.e., one player has a winning strategy in the evaluation game, iff there are not both cut-free proofs for φ and $\neg\varphi$.*

Note that the latter condition is a consistency assertion about our proof system.

In the following, Cantor space is $2^{\mathbb{N}}$, so that games on it only involve binary choices. In contrast, games on Baire space $\mathbb{N}^{\mathbb{N}}$ involve countable choices. We use Det* to refer to Cantor space and Det for Baire space.

Theorem 1 (RCA$_0$). *The following are equivalent for $\mathrm{G}_\omega^{\searrow \mathrm{P}<\omega}$*

i) soundness, i.e., $\forall n.\, \mathrm{G}_\omega^{\searrow \mathrm{P}<\mathrm{n}} \vdash \varphi \Rightarrow$ "P has a winning strategy for φ"
ii) semantic cut-admissibility, i.e., $\forall n.\, \mathrm{G}_\omega^{\searrow \mathrm{P}<\mathrm{n}} \vdash \Gamma \Rightarrow \mathrm{G}_\omega^{\searrow \mathrm{P}<\mathrm{n}} \vdash^0 \Gamma$
iii) consistency, i.e., $\forall n.\, \mathrm{G}_\omega^{\searrow \mathrm{P}<\mathrm{n}} \nvdash \emptyset$

as well as

iv) $\forall n.\, \left(\Pi_1^0\right)_n$ −Det over Baire space
v) Π_3^1−RFN$\left(\Pi_1^1 - \mathrm{CA}_0\right)$

Proof. *i)* $\Rightarrow$ *ii)* follows from Proposition 1 and *ii)* $\Rightarrow$ *iii)* is clear. *iii)* $\Rightarrow$ *iv)* uses a normal form for $\left(\Pi_1^0\right)_{<\omega}$-games, representing them as evaluation games for formulas in $\mathcal{L}\left(\mathrm{G}_\omega^{\searrow \mathrm{P}<\omega}\right)$, and then concludes by Proposition 1. The equivalence between *iv)* and *v)* was established by Pacheco and Yokohama in [16].

For the remaining *iv)* $\Rightarrow$ *i)*, we first establish some uniform determinacy. Given a countable family $(\varphi_i)_{i\in\omega}$ of $\mathcal{L}\left(\mathrm{G}_\omega^{\searrow \mathrm{P}<\mathrm{n}}\right)$-formulas, there is a set determining for each φ_i the winner in the corresponding evaluation game, together with a winning strategy. Indeed, we look at the ω-formula $\bigwedge_{i\in\omega} \varphi_i \vee \neg\varphi_i \in \mathcal{L}\left(\mathrm{G}_\omega^{\searrow \mathrm{P}<\mathrm{n}}\right)$. Opponent cannot have a winning strategy, as this would be a pair of winning strategies for φ_i and $\neg\varphi_i$ for some i. Thus, by *iv*), Player has a winning strategy which is the desired set.

Now, given an unsound proof P in $\mathrm{G}_\omega^{\searrow\mathrm{P}<\omega}$, we have uniform determinacy for all formulas appearing in it. From this, we can construct a branch where the end sequent as well as all the cut-formulas appearing on it are falsified, thus falsifying the trace condition for P.

Remark 1. If in *i*) – *iii*) we had used schemas, taking the union $\mathrm{G}_\omega^{\searrow\mathrm{P}<\omega} = \bigcup_{n\in\omega}\mathrm{G}_\omega^{\searrow\mathrm{P}<\mathrm{n}}$ externally instead of internally, then in *iv*) we would have gotten $\bigcup_{n\in\omega}\left(\Pi^0_1\right)_n$ –Det. By Tanaka [19], in place of *v*) we would then get $\Pi^1_1 - \mathrm{CA}_0$.

Theorem 2 (RCA_0). *The following are equivalent for* $\mathrm{G}_2^{\searrow\mathrm{P}<\omega}$

i) soundness*
ii) semantic cut-admissibility*
iii) consistency*

as well as

iv)* $\forall n.\left(\Pi^0_1\right)_n$ $-\mathrm{Det}^\star$ *over Cantor space*
v)* Π^1_2–RFN (ACA_0)

Proof. The equivalence between *iv**) and *v**) is again established in [16]. Items *i**) – *iv**) work as in Proposition 1.

Note that in the proof of uniform determinacy in *iv*) ⇒ *i*), we may regard the formula $\bigwedge_{i\in\omega}\varphi_i\vee\neg\varphi_i$ as a formula in $\mathcal{L}\left(\mathrm{G}_2^{\searrow\mathrm{P}<\mathrm{n}+1}\right)$. Indeed, it is equivalent (in the sense of evaluation games) to the binary formula Φ_0 which we define coinductively as

$$\Phi_n :\equiv (\varphi_n \vee \neg\varphi_n) \wedge \Phi_{n+1}$$

where the conjunctions are assigned priority $n-1$ or n, whichever of the two is even.

Remark 2. As in Remark 1, if in *i**) – *iii**) we had used schemas instead of internal quantification, then in *iv**) we would have gotten $\bigcup_{n\in\omega}\left(\Pi^0_1\right)_n$ $-\mathrm{Det}^\star$. As established by Nemoto et al. [14], we would then get ACA_0 in place of *v**).

4 Cut-elimination

What is missing from these equivalences is syntactic cut-elimination. Note that the cut-free proof in *ii*) was established through a generic exhaustive proof search, independent of the original proof.

We employ the following Gentzen or Tait-style reduction steps on proofs. Given a proof containing a cut

$$\mathrm{C}\;\dfrac{\mathrm{P}\;\dfrac{\begin{matrix}\vdots\ \mathrm{P}_i\\ \Gamma,\bigvee_{i\in I}\varphi_i,\varphi_i\ \ (\text{some } i\in I)\end{matrix}}{\Gamma,\bigvee_{i\in I}\varphi_i}(\vee)\qquad \mathrm{Q}\;\dfrac{\begin{matrix}\vdots\ \mathrm{Q}_i\\ \Gamma,\bigwedge_{i\in I}\neg\varphi_i,\neg\varphi_i\ \ (\text{all } i\in I)\end{matrix}}{\Gamma,\bigwedge_{i\in I}\neg\varphi_i}(\wedge)}{\Gamma}(\mathrm{Cut})$$

we obtain a reduct by replacing this subproof with

$$
\dfrac{
\mathrm{C}_l\ \dfrac{\dfrac{\vdots\ \mathrm{P}_i}{\Gamma, \bigvee_{i\in I} \varphi_i, \varphi_i} \quad \dfrac{\vdots\ \mathrm{Q}^{\varphi_i}}{\Gamma, \bigwedge_{i\in I} \neg\varphi_i, \varphi_i}}{\Gamma, \varphi_i}\,(\mathrm{Cut})
\qquad
\mathrm{C}_r\ \dfrac{\dfrac{\vdots\ \mathrm{P}^{\neg\varphi_i}}{\Gamma, \bigvee_{i\in I} \varphi_i, \neg\varphi_i} \quad \dfrac{\vdots\ \mathrm{Q}_i}{\Gamma, \bigwedge_{i\in I} \neg\varphi_i, \neg\varphi_i}}{\Gamma, \neg\varphi_i}\,(\mathrm{Cut})
}{\Gamma}\,(\mathrm{Cut})
$$

where Q^{φ_i} is obtained from Q by adding φ_i to every sequent and analogously for $\mathrm{P}^{\neg\varphi_i}$. We chose here a variant which symmetrically duplicates the original cut on both premises. We call subproofs of P_i and $\mathrm{P}^{\neg\varphi_i}$ in the reduct copies of their corresponding subproofs in P_i and P within the original proof, and analogously for Q. Further, we call the cuts C_l and C_r in the reduct descendants of the original cut C.

Customarily, one avoids the additional cut on the right premise by employing an inversion lemma, stating that the rule

$$
\dfrac{\Gamma, \bigwedge_{i\in I} \varphi_i}{\Gamma, \varphi_i}
$$

is admissible without increasing the proof complexity. For our purposes, the symmetric variant is slightly better, as it essentially forces both premises to be looked at during the cut-elimination process. The more common variant could be recovered by preprocessing the proofs to delay introductions of conjunctions as long as possible by the use of inversion lemmas.

Non-cut inferences with major formula different from the cut can be commuted as usual, for example from

$$
\mathrm{C}\ \dfrac{\dfrac{\dfrac{\vdots\ \mathrm{P}_i}{\Gamma, \psi, \bigwedge_{i\in I} \varphi_i, \varphi_i \quad (\text{all } i \in I)}}{\Gamma, \psi, \bigwedge_{i\in I} \varphi_i}\,(\wedge) \qquad \dfrac{\vdots\ \mathrm{Q}}{\Gamma, \neg\psi, \bigwedge_{i\in I} \varphi_i}}{\Gamma, \bigwedge_{i\in I} \varphi_i}\,(\mathrm{Cut})
$$

we obtain the following as reduct

$$
\dfrac{\mathrm{C}'\ \dfrac{\dfrac{\vdots\ \mathrm{P}_i}{\Gamma, \psi, \bigwedge_{i\in I} \varphi_i, \varphi_i} \quad \dfrac{\vdots\ \mathrm{Q}^{\varphi_i}}{\Gamma, \neg\psi, \bigwedge_{i\in I} \varphi_i, \varphi_i}}{\Gamma, \bigwedge_{i\in I} \varphi_i, \varphi_i}\,(\mathrm{Cut}) \qquad (\text{all } i \in I)}{\Gamma, \bigwedge_{i\in I} \varphi_i}\,(\wedge) \quad .
$$

The notion of copies and descendant cut is defined as before. We further define all copies to be descendants and close both relations under reflexivity and transitivity. The ancestor relation is defined dually to descendants.

We do not allow here to commute two cuts. By abuse of notation, we identify a cut-reduction sequence with the corresponding sequence of proofs.

Definition 5. *Given a proof P and a possibly infinite sequence $(P_n)_{n\in\omega}$ of cut-reductions, we define its result P_ω as the partial pre-proof consisting of those inferences which are eventually stable, i.e., which are eventually not in the sub-proof changed by any reduction.*

Note that P_ω is generally neither recursive nor cut-free or even a pre-proof; it could a priori happen that some sequent in it is not the result of an inference.

For the purpose of this work, correctness of syntactic cut-elimination shall mean the following.

Definition 6. *For a proof P, syntactic cut-elimination is admissible if there exists a cut-reduction sequence whose result is a correct cut-free proof.*

While syntactic cut-elimination for $\mathrm{G}_\omega^{\searrow\mathrm{P}<\omega}$ fits well within the other equivalences, for $\mathrm{G}_2^{\searrow\mathrm{P}<\omega}$ it massively blows up in strength.

Theorem 3 (RCA_0). *The following are equivalent*

- *any of the statements in Theorem 1*
- *vi) syntactic cut-elimination for $\mathrm{G}_\omega^{\searrow\mathrm{P}<\omega}$*
- *vi*) syntactic cut-elimination for $\mathrm{G}_2^{\searrow\mathrm{P}<\omega}$*

In vi) and vi), the unions $\mathrm{G}_\omega^{\searrow\mathrm{P}<\omega} = \bigcup_{n\in\omega}\mathrm{G}_\omega^{\searrow\mathrm{P}<\mathrm{n}}$, and analogously for the binary system, are taken by internal quantification.*

The proof is covered in the remainder of this paper.

4.1 Reduction to Soundness

Given the resemblance of $\mathrm{G}_\omega^{\searrow\mathrm{P}<\omega}$ with $\mathrm{ID}_{<\omega}$, it should be possible to employ methods from its ordinal analysis, e.g. [6], to derive *vi*) from *v*). Nevertheless, we want to sketch a different idea here, with the aim to explain the blowup of *vi**). To this end, we want to reduce syntactic cut-elimination *vi*) to soundness *i*). The technique sketched here seems to be similar to the cut elimination proof of $\mu\mathrm{MALL}^\infty$, a non-wellfounded system for multiplicative-additive linear logic with fixpoints, given by Baelde, Doumane and Saurin in [4]. I had not been aware of this work when first writing up the present argument.

We first define a particular cut-reduction sequence. It is roughly inspired by the notion of debates in game semantics, e.g., [7]. There, one considers two-player games with backtracking. Cut-free proofs correspond to strategies and cut-elimination corresponds to a debate conducted between strategies. Control is handed over from one strategy to the other whenever progress is made on the cut formula.

We argue informally inside RCA_0. For the remainder of this section, fix a proof P in $\mathrm{G}_\omega^{\searrow \mathrm{P}<\omega}$. We inductively define a cut-reduction sequence $(P_n)_{n\in\omega}$ of P. Suppose we have already defined P_n. Given some cut

$$\dfrac{\begin{array}{cc} \vdots\, \mathrm{P}_\varphi & \vdots\, \mathrm{P}_{\neg\varphi} \\ \Gamma, \varphi & \Gamma, \neg\varphi \end{array}}{\Gamma}\,(\mathrm{Cut})$$

in P_n on a disjunctive formula φ, we define $\mathrm{P}_{\neg\varphi}$ to have local control in P_n, unless its last inference is the $(\wedge)$-rule introducing $\neg\varphi$. In the latter case, P_φ has local control. A subproof Q of P_n has global control in P_n if

- it is rooted in the cut-free part of P_n, or
- it has local control and is a premise of a cut-inference with global control.

Note that P_n still remains correct, provably over RCA_0, because it is obtained from P by finitely many cut-reduction steps. In order to define P_{n+1}, select a cut inference C rooted in the cut-free part of P_n. This selection must be exhaustive, in the sense that every such cut-inference C will get selected infinitely often. In the subproof C, the inferences of global control form a finite branch. Indeed, an infinite tail of cuts is forbidden by the trace condition. Thus, there is a unique maximal cut-inference C' of global control in the subproof C. Then, C' admits a cut-reduction or commutation step. We define P_{n+1} as the resulting proof. Observe that the subproof resulting from C' after the reduction step retains global control.

Suppose now for contradiction that the final result P_ω of this cut-reduction sequence is incorrect. I.e., there exists some branch B in the cut-free part of P_ω which is either infinite without trace, or it ends in a sequent which is not the result of any non-cut inference in P_ω. We fix such a branch B for the remainder of this section. Note that for any P_n appearing in the cut-reduction sequence, its cut-free part is an initial segment of B. We now further drop all cut-reduction steps where a cut-inference C outside B was selected, resulting in a partial pre-proof $\hat{P}$. Note that the B-parts of P_ω and $\hat{P}$ coincide. Moreover, the resulting cut-reduction sequence is infinite. Otherwise, $\hat{P}$ would be a correct proof and B cannot end in a cut-inference C as it would be selected infinitely often by the exhaustive procedure. Thus, B could not invalidate $\hat{P}$, hence not P_ω.

Suppose that the endsequent of P is Γ. The rough idea in [4] is to truncate Γ by removing all conjuncts and disjuncts not appearing in B such that the resulting $[\Gamma]$ would be falsified by B. Moreover, the proof P is truncated into a proof $[P]$ of $[\Gamma]$ by additionally removing all parts of cut-formulas that do not contribute to the cut-reduction sequence, and then restrict $(\wedge)$-inferences to the remaining conjuncts. This would then finally contradict soundness.

An additional complication we encounter for $\mathrm{G}_\omega^{\searrow \mathrm{P}<\omega}$ is that we work with classical instead of linear logic. The same formula occurrence might participate in multiple cut-reduction steps, whereas in (multiplicative-additive) linear logic it can participate in at most one. As a consequence, we not only need to drop unused parts of cut-formulas, but also multiply the remaining parts. This essentially allows employing distinct copies of cut-formulas for every step in the cut-reduction sequence, and thus to account for the generally different behavior of each such interaction. Creating these copies is essentially responsible for the blowup of vi^*).

In order to keep track of these interactions, we define a notion of *handoff*, which essentially tells when a cut-inference switches local control from one premise to the other.

Definition 7. *Let C be a cut-inference in P_n on the disjunctive cut-formula φ. We say that C is a handoff for φ if the φ-subproof of C has global control in P_n and this is not the case for any proper ancestor of C, c.f. the beginning of Sect. 4 for the ancestor relation.*

We can now attach this information to formulas via the following translation.

Definition 8. *Let C be a cut-inference in P_n with disjunctive cut-formula φ. Define H^{φ}_{C} as the set of all descendants of C which are handoffs for φ. Further, define R^{φ}_{C} as the set of descendants C' of C which have global control and for which the φ-subproof of C' $(\vee)$-introduces φ as its last inference.*

Finally,

$$[\neg\varphi]_C :\equiv \bigvee_{C' \in H^{\varphi}_{C}} \bigwedge_{C'' \in R^{\varphi}_{C'}} [\neg\varphi_{C''}]^{+}_{\hat{C}}$$

where $\varphi_{C''}$ is the disjunct introduced as the last inference on the φ subproof of C'' and $\hat{C}$ is the new cut produced in the reduction of C''. Priorities are inherited from the original formulas. $[\varphi]_C$ is defined dually.

We further need a transformation of formulas in the endsequent Γ of P.

Definition 9. *Let φ be a formula occurrence in Γ or in the cut-free part of some P_n.*

- *If φ is disjunctive, define $[\varphi] :\equiv \bigvee_{t \in T^{\varphi}}[\varphi_t]$ where T^{φ} is the set of cut-reduction steps/times t where a $(\vee)$-inference is introduced in the cut-free part of P_t with this occurrence of φ as its conclusion and φ_t as its main formula.*
- *If φ is conjunctive, define $[\varphi] :\equiv \bigvee_{t \in T^{\varphi}}[\varphi_t]$ where T^{φ} is defined analogously. φ_t is the conjunct chosen by the path B for the $(\wedge)$-inference introduced at time t.*

We can now state the main technical result as follows.

Lemma 1 (RCA_0)**.** *The sequent $[\Gamma]$ is provable but falsifiable.*

Proof (Sketch). By construction, formulas in $[\Gamma]$ have no conjunctions and their leaves are empty disjunctions. A winning path through one such formula would give rise to a trace in B, thus cannot exist.

The difficult part is to modify the original proof P of Γ into a proof $[P]$ of $[\Gamma]$. Due to space constraints, we do not attempt to provide a proper account of this technical result. We merely want to outline the transformation for the first few steps.

The initial cut-free part of P is readily translated by choosing $t = 0$ as time witness in the T^{φ}'s. Consider the first cut C, i.e., nearest to the root, say on a disjunctive formula φ. We copy it as a cut on $[\varphi]_C$.

In our cut-reduction sequence, the subproof $\mathrm{P}_{\neg\varphi}$ starts with global control. Thus, all inferences until the first $(\wedge)$-introduction of $\neg\varphi$ are witnessed by the handoff at C. Once a copy of the $(\wedge)$-introduction of $\neg\varphi$ obtains global control, we get a handoff for φ, say a descendant C' of C. We can then introduce the leading disjunction in $[\neg\varphi]_C$, as well as the following conjunction, thus introducing $[\neg\varphi]_C$.

For the translation on the P_{φ} branch, we start by asking for a handoff for φ. I.e., we start by $(\wedge)$-introducing $\bigwedge_{C' \in H^{\varphi}_{C}} \bigvee_{C'' \in R^{\varphi}_{C'}} [\varphi_{C''}]^{+}_{C'}$, as in the definition of $[\varphi]_C$. For the remainder of this subproof, we thus obtain a handoff C', say in P_t, for φ. We then know that the steps after t in our cut-reduction sequence will process the subproof P_{φ}, up to the point where φ is introduced, say at step t'. Hence, we find time witnesses for the inferences up to this point, which will be commuted into the cut-free part of $P_{t'}$. The $(\vee)$-inference C'' introducing φ then gives an inhabitant of $R^{\varphi}_{C'}$. Thus, we can use it to finally introduce $[\varphi]_C$.

Generally, we want to define $[Q]$ coinductively in such a way that at the beginning of $[Q]$ we know some copy of Q having global control in some P_n. This way, we locally know the behavior of the cut-reduction sequence. The trace condition for $[P]$ will be satisfied if every branch in $[P]$ essentially contains a branch in P.

4.2 Blowup to ω-Formulas

Showing that vi^{*}) blows up to the same strength as vi) relies on encoding proofs for infinitary formulas into proofs of binary ones in a way that is compatible with cut-elimination. This technique is similar to Girard's consistency proof of second-order arithmetic through strong normalization of system F [11].

Concretely, assuming vi^{*}), we want to show consistency iii). Given some proof of contradiction P, we want to transform it into a proof of $\top$ - which is obviously sound - in such a way that the result of cut-elimination is the empty (partial pre-)proof. This contradicts correctness of syntactic cut-elimination. Essentially, we want to move some type checking from formulas to proofs, allowing to switch from $\mathrm{G}_{\omega}^{\searrow \mathrm{P}<\omega}$ to the less expressive system $\mathrm{G}_{2}^{\searrow \mathrm{P}<\omega}$.

In a first step, we want to embed concrete ω-formulas φ into some *generic* formula G. We then translate proofs of φ into proofs of $G, \top$ by checking whether the moves in G originate from valid proofs in φ and otherwise abort, using the additional formula $\top$.

Definition 10. *For $n < \omega$, we define by simultaneous coinduction*

$$G_n^\omega :\equiv G_{n-1}^\omega \wedge \bigvee_\omega \bigwedge_\omega G_n^\omega.$$

The connectives are assigned priority n. For G_0^ω, the G_{-1}^ω clause is omitted.

Arguing by contradiction, assume there is a proof of contradiction $Q \in \mathrm{G}_\omega^{\searrow\mathrm{P}<\omega}$. By adding some superfluous nodes to its formulas, we can transform Q into a proof of contradiction $P \in \mathrm{G}_\omega^{\searrow\mathrm{P}<\omega}$ where all formulas are subformulas of initial segments of G_{n-1}^ω. We define a map from such P to proofs $\tilde{P} \in \mathrm{G}_\omega^{\searrow\mathrm{P}<\omega}$ of $\top$ where all formulas, except $\top$, are now subformulas of the full G_{n-1}^ω. Indeed, on the additional conjuncts not corresponding to nodes of the formulas in P, we may simply abort by introducing $\top$. Observe that every cut-reduction sequence $((\tilde{P})_n)_{n\in\omega}$ of $\tilde{P}$ originates from a cut-reduction sequence $(P_n)_{n\in\omega}$ of P such that $(\tilde{P})_n \equiv \widetilde{(P_n)}$. Thus, no $(\tilde{P})_n$ introduces $\top$ in its cut-free part, as claimed.

It remains to collapse the infinitary formula G_n^ω into a binary version G_n.

Definition 11. *Let* Nat $:\equiv \top \vee$ Nat *with assigned priority 1. Note that* $\neg$Nat *may be defined with priority 0. Define*

$$G_n :\equiv G_{n-1} \wedge (\mathrm{Nat} \rightarrow (\mathrm{Nat} \wedge G_n))$$

with $G_0 :\equiv \top$. The connectives are assigned priority n.

Nat is essentially the standard encoding of natural numbers when dealing with propositional fixpoint logics. We transform subformulas φ of G_n^ω into subformulas φ^2 of G_{n+2} by raising all priorities by 2 to account for the priority of Nat, and then replacing ω-disjunctions $\bigwedge_\omega \psi$ with Nat$\vee\psi^2$, and $\bigvee_\omega \psi$ with Nat $\rightarrow \psi^2$. We then transform the previous proof $\tilde{P}$ into a proof $P^2 \in \mathrm{G}_2^{\searrow\mathrm{P}<\omega}$ of $\top$. An inference $(\bigwedge_\omega \varphi)$ can be simulated by an inference $(\vee)$ of Nat $\rightarrow \varphi \equiv \neg$Nat $\vee \varphi$ by first reading the input from Nat, say n, and then continuing on the subproof for φ_n. Similarly, $(\bigvee_\omega \varphi)$ can be simulated by a $(\wedge)$-inference of Nat $\wedge \varphi$, indicating the disjunctive choice through Nat.

For the sake of presentation, we look at the cut-reduction sequence $(P_n^2)_{n\in\omega}$ of P^2 which is defined as in the previous section. Then, there is a subsequence n_k and a reduction-sequence $\left(\tilde{P}_{n_k}\right)_{k\in\omega}$ of $\tilde{P}$ such that $\left(\tilde{P}_{n_k}\right)^2 \equiv (P^2)_{n_k}$. The subsequence is needed because cuts on Nat in $\mathrm{G}_2^{\searrow\mathrm{P}<\omega}$ take finitely many steps to reduce, which has no counterpart in $\mathrm{G}_\omega^{\searrow\mathrm{P}<\omega}$. It follows that no $(P^2)_n$ can introduce $\top$, thus falsifying *vi**).

When dealing with arbitrary cut-reduction sequences rather than the concrete procedure from the previous section, the precise technical statement is more complicated. That is because a cut on Nat may no longer be fully reduced before processing other cuts. We do not wish to spell out the precise technical statement here.

Acknowledgements. This project has been funded by the FWO (Research Foundation - Flanders) under file number 11K7425N.

I also want to thank Anupam Das and Giovanni Soldà for discussions related to this work, as well as the reviewers for their detailed feedback.

References

1. Afshari, B., Leigh, G.E.: Cut-free completeness for modal mu-calculus. In: Proceedings of the 32nd Annual ACM/IEEE Symposium on Logic in Computer Science. LICS '17. IEEE Press (2017)
2. Aguilera, J.P.: Determinate logic and the axiom of choice. Ann. Pure Appl. Logic **171**(2), 102745 (2020). https://doi.org/10.1016/j.apal.2019.102745
3. Baelde, D., Doumane, A., Kuperberg, D., Saurin, A.: Bouncing threads for circular and non-wellfounded proofs: towards compositionality with circular proofs. In: Proceedings of the 37th Annual ACM/IEEE Symposium on Logic in Computer Science. LICS '22. Association for Computing Machinery, New York (2022). https://doi.org/10.1145/3531130.3533375
4. Baelde, D., Doumane, A., Saurin, A.: Infinitary proof theory: the multiplicative additive case. In: Talbot, J.M., Regnier, L. (eds.) 25th EACSL Annual Conference on Computer Science Logic (CSL 2016). Leibniz International Proceedings in Informatics (LIPIcs), vol. 62, pp. 42:1–42:17. Schloss Dagstuhl – Leibniz-Zentrum für Informatik, Dagstuhl, Germany (2016). https://doi.org/10.4230/LIPIcs.CSL.2016.42. https://drops.dagstuhl.de/entities/document/10.4230/LIPIcs.CSL.2016.42
5. Brotherston, J., Simpson, A.: Sequent calculi for induction and infinite descent. J. Logic Comput. **21**(6), 1177–1216 (2010). https://doi.org/10.1093/logcom/exq052
6. Buchholz, W., Feferman, S., Pohlers, W., Sieg, W.: Iterated Inductive Definitions and Subsystems of Analysis: Recent Proof-Theoretical Studies. LNM, vol. 897. Springer, Heidelberg (1981). https://doi.org/10.1007/BFb0091894
7. Coquand, T.: A semantics of evidence for classical arithmetic. J. Symb. Log. **60**(1), 325–337 (1995). https://doi.org/10.2307/2275524
8. Das, A.: On the logical complexity of cyclic arithmetic. Log. Methods Comput. Sci. **16**(1) (2020). https://doi.org/10.23638/lmcs-16(1:1)2020
9. Das, A., De, A., Saurin, A.: Comparing Infinitary Systems for Linear Logic with Fixed Points. In: Bouyer, P., Srinivasan, S. (eds.) 43rd IARCS Annual Conference on Foundations of Software Technology and Theoretical Computer Science (FSTTCS 2023). Leibniz International Proceedings in Informatics (LIPIcs), vol. 284, pp. 40:1–40:17. Schloss Dagstuhl – Leibniz-Zentrum für Informatik, Dagstuhl (2023). https://doi.org/10.4230/LIPIcs.FSTTCS.2023.40. https://drops.dagstuhl.de/entities/document/10.4230/LIPIcs.FSTTCS.2023.40
10. Fortier, J., Santocanale, L.: Cuts for circular proofs: semantics and cut-elimination. In: Ronchi Della Rocca, S. (ed.) Computer Science Logic 2013 (CSL 2013). Leibniz International Proceedings in Informatics (LIPIcs), vol. 23, pp. 248–262. Schloss Dagstuhl – Leibniz-Zentrum für Informatik, Dagstuhl (2013). https://doi.org/10.4230/LIPIcs.CSL.2013.248. https://drops.dagstuhl.de/entities/document/10.4230/LIPIcs.CSL.2013.248
11. Girard, J.Y., Lafont, Y., Taylor, P.: Proofs and Types. No. 7 in Cambridge Tracts in Theoretical Computer Science. Cambridge University Press, Cambridge (1989)
12. Medsalem, M.O., Tanaka, K.: Δ^0_3-determinacy, comprehension and induction. J. Symb. Log. **72**(2), 452–462 (2007). https://doi.org/10.2178/jsl/1185803618

13. Montalbán, A., Shore, R.A.: The limits of determinacy in second order arithmetic: consistency and complexity strength. Israel J. Math. **204**(1), 477–508 (2014). https://doi.org/10.1007/s11856-014-1117-9
14. Nemoto, T., Ould MedSalem, M., Tanaka, K.: Infinite games in the cantor space and subsystems of second order arithmetic. Math. Log. Q. **53**(3), 226–236 (2007). https://doi.org/10.1002/malq.200610041
15. Niwiński, D., Walukiewicz, I.: Games for the μ-calculus. Theor. Comput. Sci. **163**(1), 99–116 (1996)
16. Pacheco, L., Yokoyama, K.: Determinacy and reflection principles in second-order arithmetic (2023). https://doi.org/10.48550/arXiv.2209.04082
17. Simpson, S.G.: Subsystems of Second Order Arithmetic, Perspectives in Logic, 2nd edn. Cambridge University Press, Cambridge (2009)
18. Takeuti, G.: Proof Theory. Dover Books on Mathematics, 2nd edn. Dover Publications, Mineola (2013)
19. Tanaka, K.: Weak axioms of determinacy and subsystems of analysis i: Δ_2^0 games. Math. Log. Q. **36**(6), 481–491 (1990). https://doi.org/10.1002/malq.19900360602

$L_{\omega\omega}$, $L_{\omega_1\omega}$, and the Wadge Hierarchy

Victor Selivanov[1,2(✉)]

[1] Department of Mathematics and Computer Science, St.Petersburg State University, Saint Petersburg, Russia
vseliv@yandex.ru
[2] A.P. Ershov Institute of Informatics Systems, Novosibirsk, Russia

Abstract. Continuing earlier works of several authors, we investigate various relationships between finitary sentences, infinitary sentences, and the sets of countable models of sentences. In particular, we partially extend a result of Keisler [5] and some results of Vaught [22] and Miller [9] from the quantifier-alternation and difference hierarchies to finer hierarchies, and confirm two conjectures formulated by Wadge [23].

Keywords: First-order logic · infinitary logic · countable model · Wadge hierarchy · fine hierarchy

1 Introduction

Continuing earlier works of several authors, we investigate various relationships between hierarchies of finitary sentences, hierarchies of infinitary sentences, and hierarchies of sets of countable models of sentences. The most popular such hierarchies are induced by the quantifier-alternation classifications of formulas, but we also discuss refined classifications like the difference hierarchy and Wadge hierarchy.

Let L be a finite or countable language without functional symbols of positive arity and T a first-order theory of L. Let $\{\Sigma_{1+n}(T)\}$ (resp. $\{\Sigma^{inf}_{1+n}(T)\}$) be the quantifier-alternation hierarchy of sentences of $L_{\omega\omega}$ (resp. of $L_{\omega_1\omega}$) modulo the equivalence in T (i.e., in all models of T). Let $\{S_\alpha(T)\}_{\alpha<\varepsilon_0}$ (resp. $\{S^{inf}_\alpha(T)\}_{\alpha<\varepsilon_0}$) be the fine hierarchy (FH) over $\{\Sigma_{1+n}(T)\}$ (resp. $\{\Sigma^{inf}_{1+n}(T)\}$), see Sect. 2 for the definition and properties of the FHs. Intuitively, the FHs are iterated versions of the difference hierarchies over any level $\Sigma_{1+n}(T)$. In particular, all levels of the quantifier-alternation and difference hierarchies are among the levels of FHs.

Keisler [5] established a surprising fact that if a finitary sentence is in $\Sigma^{inf}_{1+n}(T)$ then it is already in $\Sigma_{1+n}(T)$. Recently, alternative proofs of Keisler's theorem appeared in [4] and [2][1]. Our first result (the proof of which uses the proof of Theorem 1.1 in [4] rather than Keisler's original proof) is the following partial extension of Keisler's theorem from levels $\Sigma_{1+n}(T)$ to all levels of the FH over $\{\Sigma_{1+n}(T)\}$.

This research was supported by the Russian Science Foundation, project 19-71-30002.

[1] I thank an anonymous reviewer for the hint to papers [2,5].

© The Author(s), under exclusive license to Springer Nature Switzerland AG 2025
A. Beckmann et al. (Eds.): CiE 2025, LNCS 15764, pp. 425–439, 2025.
https://doi.org/10.1007/978-3-031-95908-0_30

Theorem 1. *For any* $\alpha < \varepsilon_0$ *we have:* $S_\alpha(T) = L^0_{\omega\omega} \cap S^{inf}_\alpha(T)$, *where* $L^0_{\omega\omega}$ *is the set of finitary L-sentences.*

Our second result (partially, but see also the remarks at the end of this section) extends classical facts from [9,22] about the (invariant) Borel and difference hierarchies to the FHs. For any Polish group action $G \times X \to X$, $(g, x) \mapsto g \cdot x$, the result relates the FH $\{\mathbf{S}_\alpha(X)\}$ over the finite Borel hierarchy $\{\mathbf{\Sigma}_{1+n}(X)\}_{n<\omega}$ in a Polish space X to the G-invariant version $\{\mathbf{S}^G_\alpha(X)\}$ of this FH defined as follows. Let Inv^G be the set of G-invariant subsets of X (a subset A is *G-invariant*, if it is closed under the equivalence relation $x \sim_G y \leftrightarrow \exists g \in G(y = g \cdot x)$), and let $\{\mathbf{S}^G_\alpha(X)\}$ be the FH over the base $\{Inv^G \cap \mathbf{\Sigma}_{1+n}(X)\}$.

Theorem 2. *For any Polish group action* $G \times X \to X$ *we have:* $\mathbf{S}^G_\alpha(X) = Inv^G \cap \mathbf{S}_\alpha(X)$ *for every* $\alpha < \varepsilon_0$.

Our third result (in which $S_\alpha = S_\alpha(T)$ for $T = \emptyset$) extends Theorem I.F.4 in [23], the conjecture about the difference hierarchies formulated after that theorem, and (partially) extends the classical theorems of Vaught [22] about $\Sigma^{inf}_{1+\alpha}$-sentences and of Miller [9] about the difference hierarchy over $\Sigma^{inf}_{1+\alpha}$. It relates the FH $\{S_\alpha\}_{\alpha<\varepsilon_0}$ of finitary L-sentences to the FH $\{\mathbf{S}_\alpha(2^\omega)\}_{\alpha<\varepsilon_0}$ in the Cantor space. The relation is provided by the mapping that associates with any L-sentence ϕ the set $Mod(\phi)$ of models of ϕ with universe ω (the L-structures with universe ω are identified with the elements of 2^ω via a natural coding, see e.g. [12]).

Theorem 3. *For any* $\phi \in L^0_{\omega\omega}$ *we have:* $\phi \in S_\alpha \iff Mod(\phi) \in \mathbf{S}_\alpha(2^\omega)$, *for every* $\alpha < \varepsilon_0$.

Our fourth result states that some of the above-mentioned FHs do not collapse, continuing a series of non-collapse results for the FHs in [11,19]. The non-collapse is certainly a fundamental property of any natural hierarchy.

Theorem 4. *There are a finite language L and a finitary* Π_2*-sentence* θ *of L such that the corresponding FH of sentences* $\{S_\alpha(T)\}$, *where* $T = \{\theta\}$, *does not collapse, i.e., for every* $\alpha < \varepsilon_0$ *there is* $\phi \in S_\alpha(T)$ *such that* $\neg\phi \notin S_\alpha(T)$.

Our fifth result confirms another conjecture of Wadge formulated after Theorem I.F.4 in [23]. Informally, this result characterizes the topological complexity of the sets of countable models of L-sentences in terms of the Wadge reducibility $\leq_W$ on subsets of 2^ω defined by: $A \leq_W B$, if $A = f^{-1}(B)$ for some continuous function f on 2^ω. It is known [23] that the class of Borel subsets of 2^ω under $\leq_W$ is well founded and has no antichains of size 3.

Theorem 5. *If L is a finite language that contains a predicate symbol of arity* ≥ 2, *then the rank of the quotient-poset of* $(\{Mod(\phi) \mid \phi \in L^0_{\omega\omega}\}; \leq_W)$ *coincides with* ε_0.

The results above were formulated for "finitary" versions of the FHs but Theorems 2–5 are also true for the "infinitary" FHs (IFHs) introduced and studied in [20]. They are constructed over ω_1-*bases* that are sequences of lattices $\{L_\alpha\}_{\alpha<\omega_1}$ (with natural additional properties) in a Boolean σ-algebra $\mathbb{B}$. The typical examples of ω_1-bases are Borel hierarchies $\{\mathbf{\Sigma}_{1+\alpha}(X)\}_{\alpha<\omega_1}$ in quasi-Polish spaces X and the full quantifier-alternation hierarchies $\{\Sigma^{inf}_{1+\alpha}(T)\}_{\alpha<\omega_1}$ of infinitary L-sentences. The IFH over the first ω_1-base coincides with the (extended to any quasi-Polish space X) Wadge hierarchy $\{\mathcal{S}_\alpha(X)\}_{\alpha<\upsilon}$ [20], where υ is the rank of the Wadge hierarchy of Borel sets in 2^ω. The IFH over the second ω_1-base looks as $\{\mathcal{S}_\alpha(T)\}_{\alpha<\upsilon}$.

The infinitary versions of Theorems 2—5 are obtained by replacing in the formulations of these theorems FHs by the corresponding IFHs, $L_{\omega\omega}$ by $L_{\omega_1\omega}$, and ε_0 by υ. In this paper, we give proofs only for the finitary versions. The ideas of proofs of the infinitary versions remain the same but technical details become too cumbersome to be included in this conference paper. The infinitary versions of Theorems 2 and 3 provide full extensions of the corresponding results in [9,22] to the IFHs. The infinitary version of Theorem 5 characterizes the topological complexity of the infinitary logic via the Wadge ordinal υ. Theorem 5 and its infinitary version may be informally formulated as follows: the topological complexity of $L_{\omega\omega}$ (resp. of $L_{\omega_1\omega}$) is ε_0 (resp. υ). Theorems 4 and 5 provide a partial answer to the question 4.20 in [2].

After necessary preliminaries in the next section, we prove Theorems 1–5 in Sections 3–7, respectively. Note that some results of this paper were announced in Section of [16] but only now we provide their proofs in sufficient (modulo some earlier results and proofs) details. At the end of any Section 4–7 we give additional comments concerning infinitary versions of Theorems 2–5, respectively.

2 Preliminaries

We assume the reader to be familiar with basic notions of finitary and infinitary logic (see e.g. [3,6,12,21]), and of the Wadge hierarchy [23]. Here we recall some less known definitions and facts about FHs that are important for precise formulations and serve as major technical tools.

Let $\mathbb{B} = (B; \cup, \cap, \bar{\cdot}, 0, 1)$ be a Boolean algebra. We often use abbreviation $ab = a \cap b$. For any sublattice L of the bounded lattice $(B; \cup, \cap, 0, 1)$ and any k, let $D_k(L)$ be the set of elements of the form $\bigcup_i (a_{2i} \cap \bar{a}_{2i+1})$, where $a_0 \supseteq a_1 \supseteq \cdots$ and $a_k = 0$. The sequence $\{D_k(L)\}_{k\in\omega}$ is called the *difference hierarchy over* L.

By a *base in* $\mathbb{B}$ we mean a sequence $L = \{L_n\}_{n<\omega}$ of sublattices of $(B; \cup, \cap, 0, 1)$ such that $L_n \cup \check{L}_n \subseteq L_{n+1}$ for each $n < \omega$, where $\check{L}_n = \{\bar{a} \mid a \in L_n\}$. The base L is *reducible*, if any L_n has the reduction property (i.e., for all $x_0, x_1 \in L_n$ there are disjoint $y_0, y_1 \in L_n$ (*i.e.*, $y_0 y_1 = 0$) such that $y_i \subseteq x_i$ and $x_0 \cup x_1 = y_0 \cup y_1$. The base L is *interpolable* if for every $n < \omega$ any two disjoint elements in $\check{L}_{n+1}$ are separable by a Boolean combination of elements in L_n.

The FH is defined in terms of the operators *bisep* and *Bisep* on subsets of B which are finitary versions of the corresponding infinitary operators popular in the Wadge theory:

$$bisep(X, Y_0, Y_1, Y_2) = \{x_0y_0 \cup x_1y_1 \cup \bar{x}_0\bar{x}_1y_2 \mid x_i \in X, y_j \in Y_j, x_0x_1 = 0\},$$

$$Bisep(X, Y_0, Y_1, Y_2) = \{x_0y_0 \cup x_1y_1 \cup \bar{x}_0\bar{x}_1y_2 \mid x_i \in X, y_j \in Y_j, x_0x_1y_0 = x_0x_1y_1\}.$$

Operation *Bisep* is more general and applies to FH over arbitrary base, operation *bisep* leads to an equivalent easier definition of FH over reducible bases (see Lemma 1(4) below).

Definition 1. *The FH over a base L is the sequence $\{S_\alpha\}_{\alpha<\epsilon_0}$, where $S_\alpha = S_\alpha^0$ and the classes $S_\alpha^n(n < \omega)$ are defined by induction on α:*
$S_0^n = \{0\}$; $S_{\omega^\gamma}^n = S_\gamma^{n+1}$ for $\gamma > 0$;
$S_{\beta+1}^n = Bisep(L_n, S_\beta^n, \check{S}_\beta^n, S_0^n)$ for each $\beta < \epsilon_0$;
$S_{\beta+\omega^\gamma}^n = Bisep(L_n, S_\beta^n, \check{S}_\beta^n, S_{\omega^\gamma}^n)$ for $\gamma > 0$ and β of the form $\beta = \omega^\gamma \cdot \beta_1 > 0$.

The FH may be considered as the collection of iterated difference hierarchies over the lattices L_n. We recall from [13–16] some known properties of the FH.

Lemma 1. *1. If $\alpha < \beta < \varepsilon_0$, then $S_\alpha \cup \check{S}_\alpha \subseteq S_\beta$.*
2. For any $n < \omega$, the sequence $\{S_k^n\}_{k<\omega}$ is the difference hierarchy over L_n.
3. If $\{L_n\}$ and $\{L_n^{'}\}$ are bases in Boolean algebras B and B' respectively and $g : B \to B^{'}$ is a homomorphism satisfying $g(L_n) \subseteq L_n^{'}$ for all $n < \omega$, then $g(S_\alpha) \subseteq S_\alpha^{'}$ for all $\alpha < \varepsilon_0$.
4. Classes of the fine hierarchy over a reducible base L coincide with the corresponding classes obtained by using the operator bisep in place of Bisep in Definition 1.
5. If L is interpolable then $S_\lambda \cap \check{S}_\lambda = \bigcup_{\alpha<\lambda} S_\alpha$ for every limit ordinal $\lambda < \varepsilon_0$.
6. Let $\alpha < \varepsilon_0$ be a limit ordinal, $a \in S_\alpha^n$ (resp. $a \in \check{S}_\alpha^n$), and $x \in L_{n+1} \cap \check{L}_{n+1}$. Then $ax \in S_\alpha^n$ (resp. $ax \in \check{S}_\alpha^n$).
7. Let $\alpha < \varepsilon_0$ be a limit ordinal and $n \in \omega$. Then $S_{\alpha+n} \cdot \check{S}_1 = S_{\alpha+n}$ and $\check{S}_{\alpha+n} \cdot \check{S}_1 = \check{S}_{\alpha+n}$, where $X \cdot Y = \{xy \mid x \in X, y \in Y\}$.

Next we recall an alternative description of the FH from [13,15]. Let $\omega^{<\omega}$ (resp. $2^{<\omega}$) be the set of all finite strings of numbers (resp. of bits). By $\sigma \subseteq \tau$ we denote that a string σ is an initial segment of a string τ. By $|\sigma|$ we denote the length of a string σ, by $\sigma\tau$ (σk)—the concatenation of strings σ, τ (respectively of a string σ and a number k). For $k < 2$ let $\bar{k} = 1 - k$, and let $\bar{\sigma}$ be defined by $\bar{\emptyset} = \emptyset$ and $\overline{\sigma k} = \bar{\sigma}\bar{k}$.

By *a tree* (in $\mathbb{B}$) we mean a family $\{a_\sigma\}_{\sigma\in 2^{<\omega}}$ of elements of B almost all of which are zeros; the elements $\tilde{a}_\tau = a_\tau \setminus (a_{\tau 0} \cup a_{\tau 1})$ are called *components* of the tree. Such a tree is called: *normal*, if $a_\emptyset = 1$; *monotone*, if it is normal and

$a_\sigma \supseteq a_{\sigma k}$; *reduced*, if it is monotone and $a_{\sigma 0}a_{\sigma 1} = 0$; a *$\mu$-tree* for a base L (where $\mu \in \omega^{<\omega}$), if $a_\sigma = 0$ for $|\sigma| > |\mu|$ and $a_{\sigma k} \in L_{\mu(|\sigma|)}$ for $|\sigma| < |\mu|$. We say that an element $a \in B$ *is determined by a tree* $\{a_\sigma\}$, if $a \subseteq a_0 \cup a_1$, $aa_{\sigma 0} \subseteq a_{\sigma 00} \cup a_{\sigma 01}$ and $\bar{a}a_{\sigma 1} \subseteq a_{\sigma 10} \cup a_{\sigma 11}$.

Note that a tree $\{a_\sigma\}$ determines an element a iff the tree $\{b_\sigma\}$, defined by $b_\emptyset = 1$ and $b_{\sigma k} = a_{\sigma k}$, determines a (indeed, a does not depend on $a_\emptyset$). Below we for simplicity sometimes identify $\{a_\sigma\}$ with the normal tree $\{b_\sigma\}$. Let us recall some properties of the introduced notions.

Let M_μ (resp. R_μ) be the set of all elements defined by the monotone μ-trees (resp. by the reduced μ-trees) over a base L. Define strings $\mu^n_\alpha (n < \omega)$ by induction on α as follows:

$\mu^n_0 = \emptyset$, $\mu^n_{\alpha+1} = n\mu^n_\alpha$, $\mu^n_{\omega^\gamma} = \mu^{n+1}_\gamma$ for $\gamma > 0$, $\mu^n_{\delta+\omega^\gamma} = \mu^n_{\omega^\gamma} n \mu^n_\delta$ for $\delta = \omega^\gamma \cdot \delta' > 0$, $\gamma > 0$.

Let $\mu_\alpha = \mu^0_\alpha$. The following characterization of levels S_α of the FH over L was established Theorem 4.3 in [15]; we cite it for further references.

Lemma 2. *For any base L and for any $\alpha < \varepsilon_0$ we have $S_\alpha = M_{\mu_\alpha}$; if L is reducible then $S_\alpha = R_{\mu_\alpha}$.*

Note that by the Stone Representation Theorem, we may (up to isomorphism) think of the levels L_n of a base as of lattices of sets (i.e., sublattices of the power-set Boolean algebra $P(X)$ for some set X) which suffices for many examples of FHs. Nevertheless, considering of "abstract" FHs as above is sometimes useful, e.g. when we deal with hierarchies of formulas, as we do in this paper.

With any base L in $\mathbb{B}$ we associate some other bases as follows. For any $m < \omega$, let $L^m(S) = \{L_{m+n}\}_n$; we call this base in $\mathbb{B}$ the *m-shift of $\mathcal{L}(S)$*. By Definition 1, the FH over the shifted base L^m coincides with $\{S^m_\alpha\}_{\alpha<\varepsilon_0}$. For any $u \in L_0$, let $L(u) = \{L_n(u)\}_{n<\omega}$ where $L_n(u) = \{u \cap v \mid v \in L_n\}$; we call this base in $\mathbb{B}_u = \{u \cap b \mid b \in B\}$ the *u-restriction of L*.

For any subalgebra $\mathbb{B}'$ of $\mathbb{B}$, let $L' = \{B' \cap L_n\}_n$; we call this base in $\mathbb{B}'$ the *$\mathbb{B}'$-fragment of L*. It is easy to see that for the FH $\{S'_\alpha\}$ over L' we have $S'_\alpha \subseteq B' \cap S_\alpha$ for every $\alpha < \varepsilon_0$. The converse inclusions do not hold in general but if they do, then we call the Boolean algebra $\mathbb{B}'$ *L-precise*. Theorems 1 and 2 are examples of this, i.e. $L^0_{\omega\omega}$ is $\{\Sigma^{inf}_{1+n}\}$-precise and Inv^G is $\{\mathbf{\Sigma}^0_{1+n}\}$-precise.

Note that the trees μ_α in Lemma 2 are symmetric. More liberal non-symmetric trees are also useful in the study of FHs, and especially of their extensions from hierarchies of sets to hierarchies of k-partitions for any $k \geq 2$, or from finitary to the infinitary FHs in [20]). We briefly recall the characterization of FHs through via (in general) non-symmetric trees only for the FHs in $P(X)$.

By a *tree* (that is is now non-symmetric in general) we mean a nonempty initial segment of $(\omega^{<\omega}; \subseteq)$. For any preorder Q, let $\mathcal{T}(Q)$ be the set of finite Q-labeled trees (T, t), where T is a finite tree and $t : T \to Q$ a labeling. Let $\leq_h$ be the h-preorder on $\mathcal{T}(Q)$ defined by: $(T, t) \leq_h (U, u)$, if there is a monotone function $\varphi : T \to U$ such that $t(x) \leq_Q u(\varphi(x))$ for each $x \in T$. Interesting preorders are obtained by iterating the construction $Q \mapsto \mathcal{T}(Q)$ starting with $Q = \bar{k} = (\{0, \ldots, k-1\}; =)$. Namely, we define the sequence $\{\mathcal{T}_m(\bar{k})\}_{m<\omega}$ of h-preorders as follows: $\mathcal{T}_0(\bar{k}) = \bar{k}$ and $\mathcal{T}_{m+1}(\bar{k}) = \mathcal{T}(\mathcal{T}_m(\bar{k}))$. The ω-th iterate $\mathcal{T}_\omega(\bar{k})$ and the h-preorder on it may also be naturally defined.

With any base $\mathcal{L} = \{\mathcal{L}_n\}_{n<\omega}$ in $P(X)$ we now associate the *FH of k-partitions over* $\mathcal{L}$ which is a family $\{\mathcal{L}(X,T)\}_{T \in \mathcal{T}_\omega(\bar{k})}$ of subsets of k^X such that $T \leq_h S$ implies $\mathcal{L}(X,T) \subseteq \mathcal{L}(X,S)$. We recall the precise definitions from [18,19] only for bases in $P(X)$ but they can be modified for bases in arbitrary Boolean algebras in the obvious way.

For any finite tree $T \subseteq \omega^{<\omega}$ and any T-family $\{U_\tau\}_{\tau \in T}$ of subsets of X, we define the T-family $\{\tilde{U}_\tau\}$ of subsets of X by $\tilde{U}_\tau = U_\tau \setminus \bigcup\{U_{\tau'} \mid \tau \subset \tau' \in T\}$. The T-family $\{U_\tau\}$ is *monotone* if $U_\tau \supseteq U_{\tau'}$ for all $\tau \subseteq \tau' \in T$. We associate with any T-family $\{U_\tau\}$ the monotone T-family $\{U'_\tau\}$ by $U'_\tau = \bigcup_{\tau' \supseteq \tau} U_{\tau'}$. A T-family $\{V_\tau\}$ is *reduced* if it is monotone and satisfies $V_{\tau i} \cap V_{\tau j} = \emptyset$ for all $\tau i, \tau j \in T$. Obviously, for any reduced T-family $\{V_\tau\}$ the components $\tilde{V}_\tau$ are pairwise disjoint. Let $A \in k^X$, $T \in \mathcal{T}_k(1)$, and let $\{U_\tau\}$ be a monotone T-family of subsets of X. We say that $\{U_\tau\}$ determines A, if $A(x) = t(\tau)$, for any $\tau \in T$ with $x \in \tilde{U}_\tau$. Note that $t : T \to \bar{k}$ and that x may belong to different components $\tilde{U}_\tau$, $\tilde{U}_\sigma$ with incomparable τ, σ.

We define the notion "F is a T-family in $\mathcal{L}(X)$" by induction as follows: If $T \in \mathcal{T}_0(\bar{k})$ then $F = \{X\}$; If $(T,t) \in \mathcal{T}_{n+1}(\bar{k})$ then $F = (\{U_\tau\}, \{F_\tau\})$ where $\{U_\tau\}$ is a monotone T-family of $\mathcal{L}_0$-sets with $U_\varepsilon = X$ and, for each $\tau \in T$, F_τ is a $t(\tau)$-family in the 1-shifted base $\mathcal{L}^1(\tilde{U}_\tau)$. The notion of a reduced T-family in $\mathcal{L}(X)$ is obtained from this definition by requiring $\{U_\tau\}$ and F_τ in item (2) to be reduced. We also define the notion "a T-family F in $\mathcal{L}(X)$ determines a partition $A : X \to \bar{k}$" as follows: If $T \in \mathcal{T}_0(\bar{k})$, $T = i < k$ (so $F = \{X\}$), then T determines the constant partition $A = \lambda x.i$; If $(T,t) \in \mathcal{T}_{n+1}(\bar{k})$ (so F is of the form $(\{U_\tau\}, \{F_\tau\})$) then T determines the k-partition A such that $A|_{\tilde{U}_\tau} = B_\tau$ for every $\tau \in T$, where $B_\tau : \tilde{U}_\tau \to \bar{k}$ is the k-partition of $\tilde{U}_\tau$ determined by F_τ. Finally, we define *the FH of k-partitions over* $\mathcal{L}(X)$ as the family $\{\mathcal{L}(X,T)\}_{T \in \mathcal{T}_\omega)(\bar{k})}$ of subsets of k^X where $\mathcal{L}(X,T)$ is the set of $A : X \to \bar{k}$ determined by some T-family in $\mathcal{L}(X)$.

Since inductive proofs according to the given definitions sometimes hide the ideas behind them, let us give examples of explicit descriptions of the introduced notions. For $T = i \in \mathcal{T}_0(\bar{k})$, there is only one T-family $\{X\}$ in $\mathcal{L}(X)$ which determines the constant partition $\lambda x.i$.

For $T \in \mathcal{T}_1(\bar{k})$, a T-family F in $\mathcal{L}(X)$ is essentially a monotone family $\{U_\tau\}$ of $\mathcal{L}_0(X)$-sets whose components $\tilde{U}_\tau$ cover X. Such a family determines A if $A(x) = t(\tau)$, for any $\tau \in T$ with $x \in \tilde{U}_\tau$. Note that $t : T \to \bar{k}$ and that x may belong to different components $\tilde{U}_\tau$, $\tilde{U}_\sigma$ with incomparable τ, σ.

For $T \in \mathcal{T}_2(\bar{k})$, a T-family F in $\mathcal{L}(X)$ consists of a family $\{U_\tau\}$ as above, and, for each $\tau_0 \in T$, a family $\{U_{\tau_0\tau_1}\}_{\tau_1 \in t_0(\tau_0)}$ of $\mathcal{L}_1(X)$-sets whose components (which we call second-level components) $\tilde{U}_{\tau_0\tau_1}$ cover $\tilde{U}_{\tau_0}$ (called first-level components). Such an F determines A if $A(x) = t_1(\tau_1)$, for all $\tau_0 \in T, \tau_1 \in t_0(\tau_0)$ with $x \in \tilde{U}_{\tau_0\tau_1}$. Note that $t_0 : T \to \mathcal{T}_k(1), t_1 : t_0(\tau_0) \to \bar{k}$.

Intuitively, the T-family F that determines A provides a mind-change "algorithm" for computing $A(x)$ for a given $x \in X$ as follows. First, we search for a component $\tilde{U}_{\tau_0}$ containing x; this is the usual mind-change procedure working with differences of $\mathcal{L}_0$-sets. While x sits in $\tilde{U}_{\tau_0}$, we search for a component $\tilde{U}_{\tau_0\tau_1}$ containing x; this is a harder mind-change procedure working with differences of $\mathcal{L}_1$-sets, and so on. Note that if F is reduced then the computation is "linear" since the components of each level are pairwise disjoint, otherwise the algorithm is "parallel" since already at the first level x may belong to several components $\tilde{U}_{\tau_0}$. With this interpretation in mind, we consider the FH as an "iterated difference hierarchy", one should only make precise how to "iterate" them.

The FH of sets is obtained from this construction for $k = 2$, in accordance with the fact that the quotient-poset of $(\mathcal{T}_\omega(\bar{2}); \leq_h)$ has order type $\bar{2} \cdot \varepsilon_0$. More precisely, there exist sequences of trees $\{T_\alpha\}, \{T'_\alpha\}_{\alpha<\varepsilon_0}$ in $\mathcal{T}_\omega(\bar{2})$ such that: $T_\alpha, T'_\alpha <_h T_\beta, T'_\beta$ for $\alpha < \beta < \varepsilon_0$; any element of $\mathcal{T}_\omega(\bar{2})$ is h-equivalent to precisely one of T_α, T'_α; the sets $\mathcal{S}_\alpha = \mathcal{L}(X, T_\alpha)$, $\alpha < \varepsilon_0$, are precisely the levels of the FH over $\mathcal{L}$. For details see Definition 8.27 and Proposition 8.28 in [18].

We conclude this section by giving some intuition about the IFHs. Let $\mathbb{B} = (B; \cup, \cap, \bar{\ }, 0, 1)$ be a Boolean σ-algebra (i.e., countable unions and intersections are allowed). By an *ω_1-base in $\mathbb{B}$* we mean a sequence $L = \{L_\alpha\}_{\alpha<\omega_1}$ of sublattices of $(B; \cup, \cap, 0, 1)$ such that every L_α is closed under countable unions and $L_\alpha \cup \check{L}_\alpha \subseteq L_\alpha$ for all $\alpha < \beta < \omega_1$. Natural examples of ω_1-bases were given in the Introduction. In Sect. 3 of [20] the machinery of iterated trees (building on [7]) and the definition of IFHs of k-partitions were given in much the same way as above, only now, instead of finite trees, we work with finite or countable well-founded trees. For $k = 2$ the IFH degenerates to the semi-well-order of height υ which resembles the Wadge hierarchy. In particular, the IFH over the Borel hierarchy in 2^ω coincides with the Wadge hierarchy.

3 Proof of Theorem 1

Proof of Theorem 1 is easily deduced from some related results in [4,14], hence we start with recalling them. In Sect. 2 of [14], the relations $\subseteq_n$ on L-structures were defined as follows: $\mathbb{A} \subseteq_n \mathbb{B}$, if $\mathbb{A} \subseteq \mathbb{B}$, and $\varphi^{\mathbb{A}}(\bar{a}) = \varphi^{\mathbb{B}}(\bar{a})$ for all finitary Δ_{n+1}-formulas $\varphi(\bar{x})$ and all $\bar{a} \in A$. In Sect. 2.2 of [4], the relations $\preceq_n$ on L-structures were defined as follows: $\mathbb{A} \preceq_n \mathbb{B}$, if $\mathbb{A} \subseteq \mathbb{B}$, and $\varphi^{\mathbb{A}}(\bar{a}) = \varphi^{\mathbb{B}}(\bar{a})$ for all finitary formulas $\varphi(\bar{x}) \in \Sigma_n \cup \Pi_n$ and all $\bar{a} \in A$. It is easy to check the following.

Lemma 3. *The relations $\subseteq_n$ and $\preceq_n$ coincide for every $n < \omega$. Furthermore, $\mathbb{A} \preceq_n \mathbb{B}$ iff $\mathbb{A} \subseteq \mathbb{B}$, and $\mathbb{A} \models \varphi(\bar{a}) \implies \mathbb{B} \models \varphi(\bar{a})$ for all finitary Σ_{n+1}-formulas $\varphi(\bar{x})$ and all $\bar{a} \in A$.*

As noticed in [4], the relations $\preceq_n$ (and the relation of being an elementary substructure) are not respected by satisfiability of infinitary formulas. To overcome this obstacle, some forcing relations were proposed in [4]. For our purposes here, the so called weak forcing relation $\mathbb{A} \Vdash^* \phi(\bar{a})$ (where ϕ is an infinitary formula in general) turns out useful. We do not reproduce the definition of weak forcing but cite some of its properties from [4].

Lemma 4. *1. Let $\mathbb{A} \preceq_n \mathbb{B}$ and $\phi \in \Sigma_n^{inf}$. Then $\mathbb{A} \Vdash^* \phi \iff \mathbb{B} \Vdash^* \phi$. (Follows from Lemma 4.1 in [4]).*

2. For any finitary sentence ϕ, $\mathbb{A} \models \phi \iff \mathbb{A} \Vdash^* \phi$. (Follows from Lemma 3.2 in [4]).

3. If infinitary sentences ϕ and ψ are equivalent in $\mathbb{A}$, then $\mathbb{A} \Vdash^* \phi \iff \mathbb{A} \Vdash^* \psi$. (Follows from Lemma 3.15 in [4]).

4. For any infinitary sentence ϕ, $\mathbb{A} \Vdash^* \neg\phi \iff \mathbb{A} \nVdash^* \phi$, and similarly for $\wedge, \vee$. (Follows from theorem 3.13 in [4]).

Next we recall a model-theoretic characterization of the classes $S_\alpha(T)$ in terms of the relations $\subseteq_n$ (Theorem 2 from [14]) and formulate a partial infinitary analogue of this characterization.

Let $\mu \in \omega^{<\omega}$ and ϕ be a finitary L-sentence. By a *μ-alternating tree for ϕ in T* we mean a family $\{\mathbb{A}_\rho \mid \rho \in 2^{<\omega}, |\rho| \leq |\mu|\}$ of T-models such that $\mathbb{A}_\rho \subseteq_{\mu(|\rho|)} \mathbb{A}_{\rho k}$ for all $|\rho| < |\tau|$ and $k < 2$, $\mathbb{A}_\emptyset \models \phi$, $\mathbb{A}_{\rho 0} \models \phi$, and $\mathbb{A}_{\rho 1} \models \neg\phi$ for all $|\rho| < |\mu|$.

We say that the family $\{\mathbb{A}_\rho \mid \rho \in 2^{<\omega}, |\rho| \leq |\mu|\}$ of T-models is a *μ-alternating tree for ϕ in T with respect to* $\Vdash^*$ if it has the properties from the previous paragraph with $\models$ replaced by $\Vdash^*$.

Lemma 5. *1. For any first order theory T and any $\alpha < \varepsilon_0$, the set $S_\alpha(T)$ coincides with the set of L-sentences that do not have μ_α-alternating trees in T, where μ_α is the string defined before Lemma 2.*

2. If $\{\mathbb{A}_\rho \mid \rho \in 2^{<\omega}, |\rho| \leq |\mu|\}$ is a μ-alternating tree for ϕ in T, then it is also is a μ-alternating tree for ϕ in T with respect to $\Vdash^*$.

3. If $\{\mathbb{A}_\rho \mid \rho \in 2^{<\omega}, |\rho| \leq |\mu_\alpha|\}$ is a μ_α-alternating tree for ϕ in T with respect to $\Vdash^*$, then $\phi \notin S_\alpha^{inf}(T)$.

Proof. (1) This item is Theorem 2 from [14].

(2) Let $\{\mathbb{A}_\rho \mid \rho \in 2^{<\omega}, |\rho| \leq |\mu|\}$ be a μ-alternating tree for ϕ in T. By Lemma 4(2), $\mathbb{A}_\emptyset \Vdash^* \phi$, $\mathbb{A}_{\rho 0} \Vdash^* \phi$, and $\mathbb{A}_{\rho 1} \Vdash^* \neg\phi$. By Lemma 4(3), $\mathbb{A}_\emptyset \Vdash^* \psi$, $\mathbb{A}_{\rho 0} \Vdash^* \psi$, and $\mathbb{A}_{\rho 1} \Vdash^* \neg\psi$. Thus, $\{\mathbb{A}_\rho \mid \rho \in 2^{<\omega}, |\rho| \leq |\mu|\}$ is a μ-alternating trees in T for ψ w.r.t. $\Vdash^*$.

(3) Let $\{\mathbb{A}_\rho \mid \rho \in 2^{<\omega}, |\rho| \leq |\mu_\alpha|\}$ be a μ_α-alternating tree for $\phi \in L^0_{\omega_1\omega}$ in T with respect to $\Vdash^*$. Repeating the corresponding direction of the proof of Theorem 2 in [14] with $\models$ replaced by $\Vdash^*$, we obtain $\phi \notin S_\alpha^{inf}(T)$. □

For instance, a simplified version of item (1) of the previous lemma for the difference hierarchies $D_m(\Sigma_n(T))$ looks as follows. By an alternating n-chain of length m for ϕ we mean a sequence $\mathbb{A}_0 \subseteq_n \cdots \subseteq_n \mathbb{A}_m$ of T-models such that $\phi \models \mathbb{A}_{2i}$ and $\phi \not\models \mathbb{A}_{2i+1}$. Then for all $n, m > 0$, the set $D_m(\Sigma_n(T))$ coincides with the set of L-sentences that do not have alternating n-chains of T-models of length m. This follows from Lemma 1(2).

Proof of Theorem 1. Let ϕ be a finitary sentence which is equivalent to $\psi \in S_\alpha^{inf}(T)$ in every model of T; we have to show that $\phi \in S_\alpha(T)$. By Lemma 5(1), it suffices to show that ϕ does not have μ_α-alternating trees in T. Suppose the contrary: a family $\{\mathbb{A}_\rho \mid \rho \in 2^{<\omega}, |\rho| \leq |\mu_\alpha|\}$ is such a tree. By Lemma 5(2), this family is a μ_α-alternating trees in T for $\psi \equiv \phi$ w.r.t. $\Vdash^*$. By Lemma 5(3), $\psi \notin S_\alpha^{inf}(T)$. Contradiction. □

Remark 1. Note that, by the downward Löwenheim-Skolem theorem for $L_{\omega\omega}$ and its version for $L_{\omega_1\omega}$, all results of this section remain true if restricted to countable structures.

4 Proof of Theorem 2

Let $G \times X \to X$, $x \mapsto g \cdot x$, be a Polish group action. In [22], Vaught defined two important functions $^{*,\Delta} : P(X) \to Inv^G$, that became known as *Vaught's transforms*, as follows: $A^* = \{x \mid \{g \mid g \cdot x \in A\}$ is comeager$\}$ and $A^\Delta = \{x \mid \{g \mid g \cdot x \in A\}$ is non-meager$\}$. We state some properties of the Vaught transforms, with the emphasis to the second transform, which is crucial for the proof of Theorem 2. We call a k-partition $A \in k^X$ G-invariant, if $A(x) = A(y)$ whenever $x \sim_G y$.

Lemma 6. *1. A^* and A^Δ are G-invariant for each $A \subseteq X$.*
2. $X \setminus A^\Delta = (X \setminus A)^*$ and $X \setminus A^* = (X \setminus A)^\Delta$.
3. If $A \subseteq B \subseteq X$ then $A^\Delta \subseteq B^\Delta$.
4. If $A = \bigcup_{n<\omega} A_n \subseteq X$ then $A^\Delta = \bigcup_{n<\omega} A_n^\Delta$.
5. If $A \in \mathbf{\Sigma}_\alpha(X)$ then $A^\Delta \in \mathbf{\Sigma}_\alpha(X)$.
6. Let $A \in k^X$ be a G-invariant k-partition determined by a monotone T-family $\{U_\tau\}$ of subsets of X, where $T \in \mathcal{T}_k(1)$. Then A is also determined by the monotone T-family $\{U_\tau^\Delta\}$ of G-invariant sets.

Proof. Proof of items (1)–(5) may be found (among other interesting assertions) in Section 1 of [22] or in Section 3.2 of [12].

(6) It is easy to check that A is G-invariant iff all components $A_0, \ldots, A_{k-1}$, $A_i = A^{-1}(i)$, are G-invariant. We have $A_i = \bigcup\{\tilde{U}_\tau \mid \tau \in T, t(\tau) = i\}$ for each $i < k$ or, equivalently, $\tilde{U}_\tau \subseteq A_{t(\tau)}$ for each $\tau \in T$.

From (1), (3) and (4) it follows that $\{U_\tau^\Delta\}$ is a monotone T-family of G-invariant sets, hence it remains to check that $\widetilde{U_\tau^\Delta} \subseteq A_{t(\tau)}$ for each $\tau \in T$. Let $x \in \widetilde{U_\tau^\Delta}$, i.e. $x \in U_\tau^\Delta$ and $x \notin U_{\tau \cdot i}^\Delta$ for all $\tau \cdot i \in T$. Then $g \cdot x \in U_\tau$ for nonmeager many $g \in G$ and, by (2), $g \cdot x \notin U_{\tau \cdot i}$ for comeager many $g \in G$ (for each $\tau \cdot i \in T$). Then $g \cdot x \in U_\tau$ for nonmeager many $g \in G$ and $g \cdot x \in \bigcup\{U_{\tau \cdot i} \mid \tau \cdot i \in T\}$ for meager many $g \in G$. Therefore, $g \cdot x \in \tilde{U}_\tau \subseteq A_{t(\tau)}$ for nonmeager many $g \in G$, hence $g \cdot x \in A_{t(\tau)}$ for some $g \in G$. Since $A_{t(\tau)}$ is G-invariant, $x \in A_{t(\tau)}$. □

Proof of Theorem 2. The inlcusion $\mathbf{S}_\alpha^G(X) \subseteq Inv^G \cap \mathbf{S}_\alpha(X)$ follows from Lemma 1(3). For the converse inclusion, it suffices to show that any G-invariant set $A \in S_\alpha(X)$ is in $S_\alpha^G(X)$. According to the characterization of FH in terms of non-symmetric trees in the end of Sect. 2, it suffices to show that any G-invariant set A (identified with the 2-partition $A : X \to \bar{2}$ via the characteristic function) from $\mathcal{L}(X, T)$ is in $\mathcal{L}^G(X, T)$, for every $T \in \mathcal{T}_m(\bar{2})$ and $m < \omega$ (where $\mathcal{L}(X) = \{\mathbf{\Sigma}_{1+n}(X)\}$ and $\mathcal{L}^G(X) = \{Inv^G \cap \mathbf{\Sigma}_{1+n}(X)\}$). We prove the more general assertion for k-partitions $A : X \to \bar{k}$, for every $k \geq 2$ and $T \in \mathcal{T}_m(\bar{k})$. Simplifying notation, we stick to the particular cases $m = 1, 2$ from Sect. 2; the proof easily extends to arbitrary m.

Let first $m = 1$ and A be a G-invariant k-partition from $\mathcal{L}(X, T)$, where $T \in \mathcal{T}_1(\bar{k})$; we have to show that A is in $\mathcal{L}^G(X, T)$, where $T \in \mathcal{T}_1(\bar{k})$. The k-partition A is determined by a monotone T-family $\{U_\tau\}$ of $\mathcal{L}_0(X)$-sets. By the proof of Lemma 6, A is determined by the monotone T-family $\{U^\Delta_\tau\}$ of G-invariant sets. By Lemma 6(5), all sets U^Δ_τ are in $\mathcal{L}_0(X)$. Therefore, $A \in \mathcal{L}^G(X, T)$.

Let now $m = 2$ and A be a G-invariant k-partition from $\mathcal{L}(X, T)$, where $T \in \mathcal{T}_2(\bar{k})$; we have to show that A is in $\mathcal{L}^G(X, T)$. The k-partition A is determined by a T-family $(\{U_{\tau_0}\}, \{U_{\tau_0\tau_1}\})$ in $\mathcal{L}(X)$, see Sect. 2. Consider the T-family $(\{U^\Delta_{\tau_0}\}, \{U^\Delta_{\tau_0\tau_1}\})$. By Lemma 6(5), the sets $U^\Delta_{\tau_0}$ are in $\mathcal{L}_0(X)$, while the sets $U^\Delta_{\tau_0\tau_1}$ are in $\mathcal{L}_1(X)$. Repeating the proof of Lemma 6, we obtain that $\widetilde{U^\Delta_{\tau_0\tau_1}} \subseteq A_{t_1(\tau_1)}$, for all $\tau_0 \in T, \tau_1 \in t_0(\tau_0)$. This means that A is determined by the T-family $(\{U^\Delta_{\tau_0}\}, \{U^\Delta_{\tau_0\tau_1}\})$ in $\mathcal{L}(X)$ with G-invariant components. Therefore, $A \in \mathcal{L}^G(X, T)$. □

From the proof we immediately obtain the following.

Corollary 1. *Theorem 2 holds not only for hierarchies of sets but also for hierarchies of k-partitions for any $k \geq 3$.*

Remark 2. The following infinitary version of Theorem 2 holds: Let $\{\mathcal{S}_\alpha(X)\}_{\alpha<\upsilon}$ (resp. $\{\mathcal{S}^{inv}_\alpha(X)\}_{\alpha<\upsilon}$) be the IFH over the ω_1-base $\{\mathbf{\Sigma}_{1+\alpha}(X)\}_{\alpha<\omega_1}$ (resp. $\{Inv^G \cap \mathbf{\Sigma}_{1+\alpha}(X)\}_{\alpha<\omega_1}$) (see Sect. 3 of [20] for the detailed definitions), where $G \times X \to X$ is a Polish group action. Then $\mathcal{S}^G_\alpha(X) = Inv^G \cap \mathcal{S}_\alpha(X)$ for every $\alpha < \upsilon$.

The proof of Theorem 2 is adjusted to the infinitary case in a straightforward way because item (4) of Lemma 6 holds for countable unions and intersections which is crucial for the infinitary version of item (6) of Lemma 6 to hold for countable well-founded trees used for the descriptions of levels of the IFHs in Sect. 3 of [20].

5 Proof of Theorem 3

The proof of Theorem 3 is obtained by a combination of some known properties of the map $\phi \mapsto Mod(\phi)$ and the particular case of Theorem 2 when $X = 2^\omega$ (identified with the set of L-structures with universe ω) and $G = S_\infty$ is the group of permutations of ω. The group S_∞ acts naturally on $Mod(\phi)$ by permuting the universes of structures so that the orbit equivalence relation $\sim_{S_\infty}$ corresponds with the isomorphism relation (see e.g. Section 11.3 of [12] for additional details).

Proof of Theorem 3. For the implication $\Longrightarrow$ note that, by the properties of $\phi \mapsto Mod(\phi)$ in [12], this map is a homomorphism from $L^0_{\omega\omega}$ into $P(2^\omega)$ such that $\phi \in \Sigma_{1+n} \implies Mod(\phi) \in \mathbf{\Sigma}_{1+n}(2^\omega)$ for each $n < \omega$. By Lemma 1(3), $\phi \in S_\alpha \implies Mod(\phi) \in \mathbf{S}_\alpha(2^\omega)$ for each $\alpha < \varepsilon_0$.

In the other direction, we use the characterization of FHs in terms of non-symmetric trees from the end Sect. 2. For a better correspondence with notation of Sect. 2 we denote the bases $\{\Sigma_{1+n}\}$, $\{\Sigma^{inf}_{1+n}\}$, and $\{\mathbf{\Sigma}_{1+n}(2^\omega)\}$ respectively by $\mathcal{L}(L^0_{\omega\omega})$, $\mathcal{L}^{inf}(L^0_{\omega_1\omega})$, and $\mathcal{L}(2^\omega)$.

Let $Mod(\phi) \in \mathcal{L}(2^\omega, T)$ for some $T \in \mathcal{T}_m(\bar{2})$; we have to show that $\phi \in \mathcal{L}(L^0_{\omega\omega}, T)$. Sticking for simplicity again to the particular case $m = 2$, let $(\{U_{\tau_0}\}, \{U_{\tau_0\tau_1}\})$ be a T-family in $\mathcal{L}(2^\omega)$ that determines $Mod(\phi)$. By the proof of Theorem 2, $(\{U^\Delta_{\tau_0}\}, \{U^\Delta_{\tau_0\tau_1}\})$ is a T-family in $\mathcal{L}^{S_\infty}(2^\omega)$ that determines $Mod(\phi)$.

By the basic case of the Lopez-Escobar theorem (Vaught's version proved in [22]) there are families $\{\phi_{\tau_0}\}_{\tau_0 \in T}$ of infinitary Σ_1-sentences and $\{\phi_{\tau_0\tau_1}\}_{\tau_1 \in t(\tau_0)}$ of infinitary Σ_2-sentences such that ϕ is determined by the T-family $(\{\phi_{\tau_0}\}, \{\phi_{\tau_0\tau_1}\})$ in $\mathcal{L}^{inf}(L^0_{\omega_1\omega})$. Thus, $\phi \in \mathcal{L}^{inf}(L^0_{\omega_1\omega}, T)$. Since ϕ is finitary, by Theorem 1 we obtain that $\phi \in \mathcal{L}(L^0_{\omega\omega}, T)$. □

Remark 3. The following infinitary version of Theorem 3 holds: Let $\{\mathcal{S}_\alpha\}_{\alpha<\upsilon}$ (resp. $\{\mathcal{S}_\alpha(2^\omega)\}_{\alpha<\upsilon}$) be the IFH over the ω_1-base $\{\Sigma^{inf}_{1+\alpha}\}_{\alpha<\omega_1}$ (resp. $\{\mathbf{\Sigma}_{1+\alpha}(2^\omega)\}_{\alpha<\omega_1}$) (see Sect. 3 of [20] for the detailed definitions). Then $\phi \in S^{inf}_\alpha \iff Mod(\phi) \in \mathcal{S}_\alpha(2^\omega)$ for every $\alpha < \upsilon$.

The proof of Theorem 3 is adjusted to the infinitary case in a straightforward way.

6 Proof of Theorem 4

Let $L = \{\prec^{(2)}, S^{(2)}, P^{(1)}\}$ (in particular, P is a unary predicate symbol) and let θ be the conjunction of L-sentences saying the following:

1) $\prec$ is a (strict) linear ordering such that $\forall x \exists y (x \prec y)$;

2) S is a binary predicate such that $S(x, y) \iff x \prec y \wedge \neg\exists z (x \prec z \prec y)$.

Clearly, θ is equivalent to a Π_2-sentence, hence for proving Theorem 4 it suffices to construct, for any $\alpha < \varepsilon_0$, an L-sentence $\phi_\alpha \in S_\alpha(T) \setminus \check{S}_\alpha(T)$. The construction and proof are tedious but rather straightforward and resemble similar constructions in automata theory (see e.g. [11,17]), so we give only a sketch. Note that T-models are essentially infinite binary words on some linear orderings (i.e., the 2-labeled partial orderings satisfying 1)).

In fact, the assertion about ϕ_α follows from its weaker version about words over alphabets $\bar{k}$, $k \geq 2$, i.e. structures of $L_k = \{\prec^{(2)}, Q^{(1)}_0, \ldots, Q^{(1)}_{k-1}\}$ with the unary predicates interpreted on a k-labeled linear ordering $((A; \prec), l)$, $l : A \to \bar{k}$, satisfying 1), in such a way that $l(a) = i$ iff $Q_i(a)$ is true; for the obvious axioms on Q_i see Sect. 2 of [17]). The weaker version states that for any $\alpha < \varepsilon_0$ and

$n < \omega$ there exist $k = k(n, \alpha) \geq 2$ and $\phi^n_\alpha \in S^n_\alpha(T_k) \setminus \check{S}^n_\alpha(T_k)$ where T_k is the L_k-theory with axiom 1) and the axioms about Q_i's from Sect. 2 of [17]. Using the predicate S, it is easy to encode countable k-words (i.e., models of T_k) by countable binary words (i.e., models of T) in such a way that the assertion of the previous paragraph follows from the weaker alphabet-dependent version (for details see e.g. [11] where this trick is used for the case of ω-words).

The weaker version is proved by induction on α, by cases $\alpha \in [1, \omega), \alpha \in [\omega, \omega^\omega), \alpha \in [\omega^\omega, \omega^{\omega^\omega})$, and so on (obviously, we have to define ϕ^n_0 as a false sentence). We first define $k(n, 1) = n + 2$ and L_{n+2}-sentences ϕ^n_1 by induction on n:

$$\phi^0_1 = \exists x Q_1(x) \text{ and } \phi^{n+1}_1 = \exists x \exists y (x \prec y \wedge Q_{n+2}(x) \wedge Q_{n+2}(y) \wedge \neg \phi^{n,(x,y)}_1),$$

where $\phi^{n,(x,y)}_1$ is the relativization of ϕ^n_1 to the interval (x, y). Easy modifications of the argument of Theorem 3.3 in [17], based on the Ehrenfucht-Fraisse games, show that $\phi^n_1 \in \Sigma_{n+1}(T_k) \setminus \Pi_{n+1}(T_k)$ (the modifications are needed because that theorem was about finite words while our words here are more complex). The sentence $\phi^n_\alpha \in D_\alpha(\Sigma_{n+1}) \setminus \check{D}_\alpha(\Sigma_{n+1})$ for $\alpha \in [2, \omega)$ is again essentially constructed in [17] (see Sect. 5). This concludes the case $\alpha \in [1, \omega)$, according to Lemma 1(2).

Let now $\alpha \in [\omega, \omega^\omega)$, then α is uniquely representable as $\omega^{\alpha_1} + \cdots + \omega^{\alpha_m}$ for some $m \in [1, \omega)$ and $\alpha_m \leq \cdots \leq \alpha_1 < \omega$. We argue by induction on m. The case $m = 1$ is trivial since we can take $\phi^n_{\omega^{\alpha_1}} = \phi^{n+1}_{\alpha_1}$, according to Definition 1.

For $m > 1$, we have to apply $m - 1$ times the operation *Bisep* (the simpler version *bisep* does not work since the base $\{\Sigma_{1+n}\}$ is not reducible). This leads to some technical problems (caused by the condition $x_0 x_1 y_0 = x_0 x_1 y_1$ from the definition of *Bisep*) which may be overcome by further alphabet extensions. For an L_k-sentence ϕ and $i < k \leq j$, let $\phi[i \mapsto j]$ be obtained from ϕ by replacing Q_i by Q_j. Note that ϕ and $\phi[i \mapsto j]$ are in the same level of the corresponding FHs.

We illustrate the idea by defining $\phi = \phi^0_{\omega+1}$ to be the L_6-sentence $(\chi_0 \wedge \psi_0) \vee (\chi_1 \wedge \psi_1)$ (thus, $k(0, \omega + 1) = 6$), where

$$\chi_0 = \phi^0_1[1 \mapsto 4]), \chi_1 = \phi^0_1[1 \mapsto 5], \psi_0 = \phi^1_1 \wedge \neg\chi_1, \psi_1 = \neg\phi^1_1[2 \mapsto 3] \wedge \neg\chi_0.$$

Note that the minimal alphabets for which these formulas make sense are $\{0, 4\}$, $\{0, 5\}$, $\{0, 1, 2, 5\}$, $\{0, 1, 3, 4\}$, respectively. Since

$\chi_0, \chi_1 \in S^0_1(T_6)$, $\psi_0 \in S^1_1(T_6) = S^0_\omega(T_6)$, $\psi_1 \in \check{S}^0_\omega(T_6)$, and $\chi_0 \wedge \chi_1 \wedge \psi_0 \equiv \chi_0 \wedge \chi_1 \wedge \psi_0 \equiv$ false,

we get $\phi \in S^0_{\omega+1}(T_6)$ by Definition 1 and Lemma 1(6,7).

It remains to show that $\phi \notin \check{S}_{\omega+1}$. By Theorem 2 from [14] (see Sect. 3), it suffices to find a $\tau_{\omega+1}$-alternating tree $\{\mathbb{A}_\rho \mid \rho \in 2^{<\omega}, |\rho| \leq 2\}$ of T_6-models for ϕ; note that $\tau_{\omega+1} = 01$. Since $\phi^1_1 \notin \check{S}_\omega(T_3)$, there are words $\mathbb{B}_0 \subseteq_1 \mathbb{B}_1$ over alphabet $\bar{3}$ such that $\mathbb{B}_0 \not\models \phi^1_1$ and $\mathbb{B}_1 \models \phi^1_1$. Clearly, we may assume $0^\omega \subseteq \mathbb{B}_0$, where 0^ω is the ω-word consisting only of 0's. Also, we may expand the words

$\mathbb{B}_0, \mathbb{B}_1$ to words over alphabet $\bar{6}$ (denoted also by $\mathbb{B}_0, \mathbb{B}_1$) in such a way that they both contain letter 4. Similarly, since $\phi_1^1[2 \mapsto 3] \notin \check{S}_\omega(T_{\{0,1,3\}})$, there are words $\mathbb{C}_0 \sqsubseteq_1 \mathbb{C}_1$ over alphabet $\bar{6}$ such that $\mathbb{C}_0 \not\models \phi_1^1[2 \mapsto 3]$, $\mathbb{C}_1 \models \phi_1^1[2 \mapsto 3]$, $0^\omega \sqsubseteq \mathbb{C}_0$, and they both contain letter 5. Therefore, we can take

$\mathbb{A}_\emptyset = 0^\omega$, $\mathbb{A}_0 = \mathbb{A}_{00} = \mathbb{B}_0$, $\mathbb{A}_{01} = \mathbb{B}_1$, $\mathbb{A}_1 = \mathbb{A}_{11} = \mathbb{C}_0$, $\mathbb{A}_{10} = \mathbb{C}_1$.

The ideas from the example of the previous paragraph work for arbitrary $\alpha \in [\omega, \omega^\omega)$ and $n < \omega$ though it requires additional unary predicates and causes notational difficulties (cf. [11]). Moreover, these ideas work in a similar way for $\alpha \in [\omega^\omega, \omega^{\omega^\omega})$ (and the remaining cases). Such an α is uniquely representable as $\omega^{\alpha_1} + \cdots + \omega^{\alpha_m}$ for some $m \in [1, \omega)$ and $\alpha_m \leq \cdots \leq \alpha_1 < \omega^\omega$, $\alpha_1 \geq \omega$. We again argue by induction on m, the case $m = 1$ being trivial by induction. For $m > 1$, we argue as in the previous case, only many more (but finitely many for each n, α) additional unary predicate symbols are required. □

Remark 4. I guess that the following infinitary version of Theorem 4 holds: Let $\{\mathcal{S}_\alpha(T)\}_{\alpha<\upsilon}$ be the IFH over the ω_1-base $\{\Sigma_{1+\alpha}^{inf}(T)\}_{\alpha<\omega_1}$, where T is the same theory as above. Then the IFH does not collapse, i.e. $\mathcal{S}_\alpha(T) \not\subseteq \check{\mathcal{S}}_\alpha(T)$ for every $\alpha < \upsilon$.

Although the detailed proof of the infinitary version of the proof of Theorem 4 is probably rather cumbersome, I do not see a principal obstacle in extending the proof sketch above.

7 Proof of Theorem 5

We need Theorem 1 in [14] formulated as follows: If T is axiomatizable by Π_2-sentences then $S_\lambda(T) \cap \check{S}_\lambda(T) = \bigcup_{\alpha<\lambda} S_\alpha(T)$ for every limit ordinal $\lambda < \varepsilon_0$.

First we prove Theorem 5 for the language $L = \{\prec^{(2)}, S^{(2)}, P^{(1)}\}$ from the proof of Theorem 4. By Theorem 1 in [14] (for $T = \emptyset$), any $\phi \in L^0_{\omega\omega}$ is in precisely one of

$$S_\alpha \setminus \check{S}_\alpha,\ \check{S}_\alpha \setminus S_\alpha,\ (S_{\alpha+1} \cap \check{S}_{\alpha+1}) \setminus (S_\alpha \cup \check{S}_\alpha),$$

for some $\alpha < \varepsilon_0$. By Theorem 3, any $Mod(\phi)$ is in precisely one of

$$\mathbf{S}_\alpha \setminus \check{\mathbf{S}}_\alpha,\ \check{\mathbf{S}}_\alpha \setminus \mathbf{S}_\alpha,\ (\mathbf{S}_{\alpha+1} \cap \check{\mathbf{S}}_{\alpha+1}) \setminus (\mathbf{S}_\alpha \cup \check{\mathbf{S}}_\alpha).$$

Since $\mathbf{S}_\alpha$ and $\mathbf{S}_{\alpha+1}$ are subsequent levels of the Wadge hierarchy by Definition 1, properties of this hierarchy imply that $rk \leq \varepsilon_0$, where rk is the rank of $(\{Mod(\phi) \mid \phi \in L^0_{\omega\omega}\}; \leq_W)$. For the estimate $rk \geq \varepsilon_0$, by Definition 1 and Theorem 3 it suffices to show that, in the notation of Theorem 4, $\theta \wedge \phi^3_\alpha \in S^3_\alpha \setminus \check{S}^3_\alpha$ for every $\alpha < \varepsilon_0$. By Theorem 4 and Definition 1, $\phi^3_\alpha \in S^3_\alpha(T) \setminus \check{S}^3_\alpha(T)$, hence $\phi^3_\alpha \in S^3_\alpha$. Since $\theta \in \Pi_2$, $\theta \wedge \phi^3_\alpha \in S^3_\alpha$ by Lemma 1(6). Since $\phi^3_\alpha \notin \check{S}^3_\alpha(T)$, by Theorem 2 from [14] (see Sect. 3) there is a τ^3_α-tree of T-models $\{\mathbb{A}_\rho \mid \rho \in 2^{<\omega}, |\rho| \leq |\tau^3_\alpha|\}$ for ϕ^3_α. Since $\mathbb{A}_\rho \models \theta$ for each ρ, this family is also a τ^3_α-tree of L-structures for $\theta \wedge \phi^3_\alpha$. By Theorem 2 from [14] we obtain $\theta \wedge \phi^3_\alpha \notin \check{S}^3_\alpha$.

Next we prove Theorem 5 for $L = \{\prec^{(2)}\}$ (equivalently, for $L = \{R^{(2)}\}$, where $R^{(2)}$ is an arbitrary binary relation symbol). There is a standard interpretation of $\{\prec^{(2)}, S^{(2)}, P^{(1)}\}$-structures in L-structures, using finitely many L-formulas (see e.g. Section 6.3 of [10] or Section 1.3 in [8]); let all those formulas be Δ_m-formulas, $m \geq 3$. It is not hard to see that this interpretation induces an effective transformation of the $\{\prec^{(2)}, S^{(2)}, P^{(1)}\}$-sentences into L-sentences that preserves all levels S^m_α of the corresponding FHs. Thus, $S^m_\alpha \not\subseteq \check{S}^m_\alpha$ for every $\alpha < \varepsilon_0$ also for the L-sentences, which suffices for the lower bound. The upper bound is obtained again by Theorem 1 in [14].

Finally, if L is a finite language that contains a predicate symbol of arity ≥ 2 then the language $\{R^{(2)}\}$ has a trivial interpretation in L that preserves the levels of the FHs of sentences. Thus, Theorem 5 holds for L.

Remark 5. Once we have a proof of the infinitary version of Theorem 4, the proof of Theorem 5 will imply the following infinitary version of Theorem 5: If L is a finite language that contains a predicate symbol of arity ≥ 2, then the rank of the quotient-poset of $(\{Mod(\phi) \mid \phi \in L^0_{\omega_1\omega}\}; \leq_W)$ coincides with v.

Acknowledgement. I thank the anonymous referees for their careful reading and valuable suggestions that certainly improved the text.

References

1. Addison, J.W.: The method of alternating chains. In: The Theory of Models, Amsterdam, North Holland, pp. 1–16 (1965)
2. Andrews, U., Gonzalez D., Lempp, S., Rossegger, D., Zhu, H.: The Borel Complexity of the Class of Models of First-Order Theories (2024). https://doi.org/10.48550/arXiv.2402.10029
3. Chang, C.C., Keisler, H.J.: Model Theory, third ed., Studies in Logic and the Foundations of Mathematics, vol. 73, North-Holland Publishing Co., Amsterdam (1990)
4. Harrison-Trainor, M., Kretschmer, M.: Infinitary logic has no expressive efficiency over finitary logic (2023)
5. Keisler, H.J.: Finite approximations of infinitely long formulas. In: Addison, J.W., Henkin, L., Tarski, A. (eds.) The Theory of Models. Proceedings of the 1963 International Symposium at Berkeley, pp. 158–169. North-Holland Publishing Co., Amsterdam (1965)
6. Keisler, H.J.: Model theory for infinitary logic. Logic with countable conjunctions and finite quantifiers. In: Studies in Logic and the Foundations of Mathematics, vol. 62. North-Holland Publishing Co., Amsterdam-London (1971)
7. Kihara, T., Montalbán, A.: On the structure of the Wadge degrees of bqo-valued Borel functions. Trans. Amer. Math. Soc. **371**(11), 7885–7923 (2019)
8. Marker, D.: Model Theory. Springer, New York (2002)
9. Miller, D.E.: The invariant Π^0_α separation principle. Trans. Am. Math. Soc. **242**, 185–204 (1978)
10. Montalbán, A.: Computable Structure Theory. Perspectives in Logic, Cambridge (2021)

11. Orekhovskii, V., Selivanov, V.: Logic vs topology on regular ω-languages. G. Della Vedova et al. (eds.) CiE 2023, LNCS, vol. 13967, pp. 141–153 (2023)
12. Gao, S.: Invariant Descriptive Set Theory. Pure and Applied Mathematics, A Series of Monographs and Textbooks, vol. 293. Taylor & Francis Group (2009)
13. Selivanov, V.L.: Fine hierarchies of arithmetical sets and definable index sets. Trudi Mat. Inst. SO AN USSR **12**, 165–185 (1989). (Russian)
14. Selivanov, V.L.: Refined hierarchy of formulas. Algebra Logic **30**, 368–378 (1991)
15. Selivanov, V.L.: Fine hierarchies and Boolean terms. J. Symb. Logic **60**(1), 289–317 (1995)
16. Selivanov, V.L.: Fine hierarchies and m-reducibilities in theoretical computer science. Theor. Comput. Sci. **405**, 116–163 (2008)
17. Selivanov, V.L.: Hierarchies and reducibilities on regular languages related to modulo counting. RAIRO Theor. Inf. Appl. **41**, 95–132 (2009)
18. Selivanov, V.L.: Fine hierarchies via Priestley duality. Ann. Pure Appl. Logic **163**, 1075–1107 (2012)
19. Selivanov, V.: Non-collapse of the effective Wadge hierarchy. Computability **11**, 335–358 (2022)
20. Selivanov, V.: A Q-Wadge hierarchy in quasi-Polish spaces. J. Symb. Logic (2), 732–757 (2022)
21. Shoenfield, J.R.: Mathematical Lodic. Addison-Wesley, Boston (1967)
22. Vaught, R.L.: Invariant sets in topology and logic. Fund. Math. **82**, 269–294 (1974)
23. Wadge, W.: Reducibility and Determinateness in the Baire Space. PhD thesis, University of California, Berkely (1984)

Ordinal Invariants of the h-Preorder on k-Labeled Forests

Victor Selivanov(✉) and Ilya Smirnov

Department of Mathematics and Computer Science, St. Petersburg University, St. Petersburg 199178, Russia
v.selivanov@spbu.ru

Abstract. We compute (using the Veblen ordinal functions) the maximal order type and the width of the quotient-posets of the h-preorder on finite k-labeled forests, for every integer $k \geqslant 3$.

Keywords: k-labeled forest · h-preorder · well partial order · height · width · maximal order type

1 Introduction

The h-preorder $(\mathcal{F}_k; \leqslant_h)$ on finite k-labeled forests (and its quotient-poset $(\mathbb{F}_k; \leqslant_h)$) was studied in several papers (see e.g. [5,10,11] and references therein), motivated mainly by its close connection with the Wadge reducibility and hierarchies of k-partitions. Recall that the relation $F \leqslant_h G$ between two finite k-labeled forests means that there is a monotone function from F to G that preserves the labels. So, in our main results we deal with the simplest case when the labels are pairwise incomparable.

The picture below shows initial segments of $(\mathbb{F}_2; \leqslant_h)$ and of $(\mathbb{F}_3; \leqslant_h)$. At this picture, triangles and circles mean equivalence classes, and the forests inside (which have minimal possible size among the elements in a class), show the canonical representatives. The canonical elements in triangles (resp. circles) are trees (resp. forests not h-equivalent to trees). Simplifying notation, we often write $\mathbb{F}_k$ instead of $(\mathbb{F}_k; \leqslant_h)$. Along with these posets, we consider in this paper their substructures $(\mathbb{T}_k; \leqslant_k)$ formed by the k-labeled trees; the latter poset is often denoted by $\mathbb{T}_k$.

Apparently, $\mathbb{F}_3$ is much more complex than $\mathbb{F}_2$, $\mathbb{F}_4$ is more complex than $\mathbb{F}_3$ and so on, but how can one capture this distinction in precise terms? There are different approaches to this question.

In [6] it was shown that the natural encoding of any of the posets $(\mathbb{F}_k; \leqslant_h)$ by words in a finite alphabet yields a presentation computable in cubic time (and also some important predicates on the structures are computable in cubic time), so all these structures are feasible and the "presentation complexity" of any of these posets is essentially the same (Fig. 1).

The work of both authors was supported by the Russian Science Foundation, project 23-11-00133.

© The Author(s), under exclusive license to Springer Nature Switzerland AG 2025
A. Beckmann et al. (Eds.): CiE 2025, LNCS 15764, pp. 440–451, 2025.
https://doi.org/10.1007/978-3-031-95908-0_31

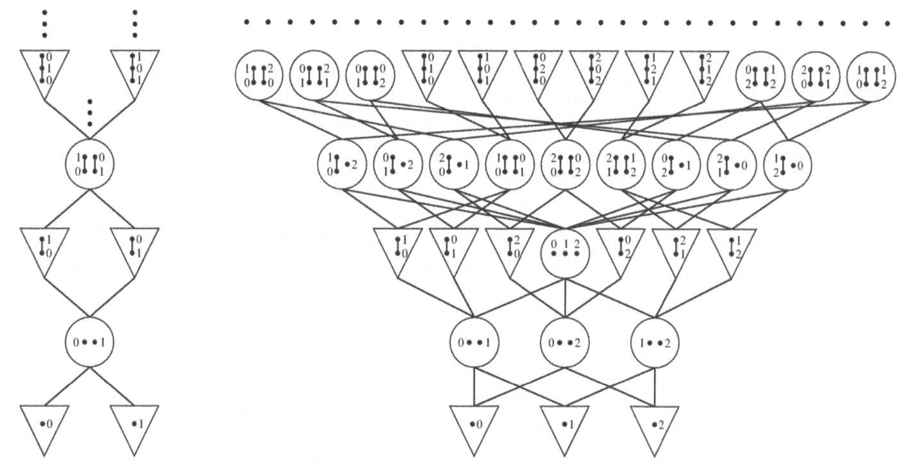

Fig. 1. $(\mathbb{F}_2; \leqslant_h)$ and $(\mathbb{F}_3; \leqslant_h)$ (The pictures were made by Anton Zhukov, we thank him for the permission to reproduce them)

The distinction in "logical complexity" was described in [8] by showing, in particular, that the first-order theory of $(\mathbb{F}_k; \leqslant_h)$ for $k \geqslant 3$ is computably isomorphic to the first-order arithmetic (while the first-order theory of $(\mathbb{F}_2; \leqslant_h)$ is easily seen to be decidable). The results in [8] show that $(\mathbb{F}_k; \leqslant_h)$ for $k \geqslant 3$ is logically much more complex than $(\mathbb{F}_2; \leqslant_h)$ but all $(\mathbb{F}_k; \leqslant_h)$ for $k \geqslant 3$ have the same logical complexity.

In this paper, we describe the distinction in "combinatorial complexity", based on the fact that any $\mathbb{F}_k$ is a well partial order (WPO), i.e., it has neither infinite descending chains nor infinite antichains. Such partial orders are studied in detail within WPO-theory (better known as the WQO-theory), a popular branch of infinite combinatorics with applications to several fields of mathematics and theoretical computer science [15]. With any WPO $(A, \leqslant)$, one can associate some ordinals (known as the ordinal invariants of A), including the *height* $h(A)$, *width* $w(A)$, and *maximal ordinal type* $o(A)$ of A. We define some of them in the next section; for more information and references see e.g. [3].

It is well known and easy to see that $h(\mathbb{F}_k) = \omega$ for every $k \geqslant 2$ [5,10], $w(\mathbb{F}_2) = 2$, and $o(\mathbb{F}_2) = \omega$. The following main result of this paper computes the remaining ordinal invariants of $\mathbb{F}_k$ in terms of Veblen's unary functions $\{\varphi_n\}_{n \geqslant 0}$ on ordinals defined by induction on n as follows: $\varphi_0(\alpha) = \omega^\alpha$, and $\varphi_{n+1}(\alpha)$ is the α-th fixed point of φ_n.

Theorem 1. *For any integer $k \geqslant 3$ we have: $o(\mathbb{F}_k) = w(\mathbb{F}_k) = \varphi_{k-1}(0)$. The same estimates hold for the posets $\mathbb{T}_k$.*

This result shows that the ordinal invariants distinguish all structures $\mathbb{F}_k$ for different $k \geqslant 3$ by their combinatorial complexity, in contrast with the representation complexity and logical complexity. It fits well to the popular programme of WQO-theory which aims to compute the ordinal invariants of naturally arising WPOs. This task is often non-trivial and requires technically involved proofs and notions (see e.g. [1–4,9,16]).

At the beginning of WQO-theory, the main goal was to show that many natural preorders are WQOs. Among the most prominent examples are the following (for other interesting examples see papers in [15]):

1) If Q is a WQO then $(Q^*; \leqslant^*)$ is also a WQO where Q^* is the set of finite sequences in Q and $(x_1, \ldots, x_m) \leqslant^* (y_1, \ldots, y_n)$ means that for some strictly increasing $\varphi : \{1, \ldots, m\} \to \{1, \ldots, n\}$ we have $x_i \leqslant y_{\varphi(i)}$ (Higman).
2) If Q is a WQO then the finite structured Q-labeled trees ordered similarly to the words in 1) form a WQO (Kruskal).
3) The finite graphs ordered by the graph-minor relation form a WQO (Robertson-Seymour).
4) The countable linear orders with the embeddability relation form a WQO (Laver).

At a later stage of WQO-theory, several complexity measures for WPOs appeared. Along with those mentioned above, the "proof complexity" of WQOs (which aims to classify WQO using the powerful tools of reverse mathematics) appeared. The complexity of many WQOs were determined, including those mentioned in 1)—4).

The h-preorders discussed in our paper (which may be defined for the Q-labeled forests where Q is an arbitrary WQO, not just a finite antichain as above) should be compared with the Kruskal preorders in 2), the ordinal invariants for which were investigated in details [1,3,9]. Though the h-preorder looks at first glance similar to the Kruskal preorder, their ordinal invariants are in fact very different.

In Sect. 2 we recall relevant notions and facts but also make some new technical observations important for the sequel. In Sect. 3 we try to informally explain ideas used in the proof of Theorem 1. In Sect. 4 we describe the main technical notion and facts that are crucial for our estimates. In Sect. 5 we present proofs of our main results, and we conclude in Sect. 6 with a discussion of the remaining open questions.

2 Preliminaries and Auxiliary Facts

We use standard set-theoretical notation, in particular $P(X)$ denotes the set of subsets of a set X. We also use the following less standard notation. For $f : X \to Y$ and $x \in X$, let $f - x$ denote the restriction of f to $X \setminus \{x\}$.

We will use some well-known facts about ordinal arithmetic. As usual, $\alpha + \beta$, $\alpha \cdot \beta$ and α^β denote the ordinal addition, multiplication and exponentiation of ordinals α and β, respectively. The first infinite ordinal is denoted ω and coincides with $\{0, 1, \ldots\}$.

Any positive ordinal α is uniquely representable in the Cantor normal form $\alpha = \omega^{\alpha_0} + \cdots + \omega^{\alpha_n}$ where $n < \omega$ and $\alpha \geqslant \alpha_0 \geqslant \ldots \geqslant \alpha_n$. The natural sum

$\alpha \oplus \beta$ of positive ordinals is defined from the Cantor normal forms for α and β by treating them formally as polynomials on ω; we also define $0 \oplus \alpha = \alpha \oplus 0 = \alpha$.

In our estimates below, it is often convenient to denote by $(< \alpha)$ an arbitrary ordinal less than α. For instance, we have the estimate $(< \omega^{\beta}) \oplus (< \omega^{\beta}) < \omega^{\beta}$.

In our notation below, we identify a WQO $(A; \leqslant)$ with its set of points A, and identify a set $X \subseteq A$ with the WQO $(X; \leqslant_X)$ where $\leqslant_X$ is the induced preorder on X.

For a WPO $(A; \leqslant)$, its *height* $h(A)$ is the supremum of ordinals embeddable into A, while its *maximal order type* $o(A)$ is the supremum of order types of linearisations of $\leqslant$ (i.e., of linear orders on A which extend $\leqslant$); the latter definition is correct because WPOs are known to coincide with the partial orders $(A; \leqslant)$ such that all linearisations of $\leqslant$ are well orders. We skip the definition of the width $w(A)$ of A (which may be found e.g. in [3]) because our estimations of this ordinal invariant below depend only on its properties established in [3].

The definitions of ordinal invariants of WPOs apply to WQOs in the usual way, by taking the standard quotient-poset of a WQO.

Definition 1. *For a WQO $(A; \leqslant)$, and $x \in A$, let $L_A(x) = \{y \mid y \in A, x \nleqslant y\}$, and $l_A(x) = o(L_A(x))$.*

We cite from [2,3] some important properties of the introduced notions (below $A + B$ is the lexicographic sum of WQOs $(A, \leqslant_A)$ and $(B, \leqslant_B)$, which is the order $\leqslant$ on the disjoint union of A and B such that $x \leqslant y$ iff $x, y \in A$ and $x \leqslant_A y$, or $x, y \in B$ and $x \leqslant_B y$, or $x \in A$ and $y \in B$).

Lemma 1. *1. If $(A, \leqslant)$ is a WPO then there is a linearisation of $\leqslant$ of order type $o(A)$.*
2. $o(A) \leqslant \alpha \iff \forall x \in A(l_A(x) < \alpha)$.
3. If A and B are WPOs then $o(A + B) = o(A) + o(B)$.
4. If A is a WPO and $X, Y \subseteq A$, then $o(X \cup Y) \leqslant o(X) \oplus o(Y)$.

For a WQO A, let $\mathcal{S}(A)$ be the WQO of finite subsets of A, with the *domination preorder* $X \leqslant Y \iff \forall a \in X \exists b \in Y(a \leqslant b)$. Then for any ordinal α we have the following lemma, which is essential for proving the upper bounds in Theorem 1.

Lemma 2. *1. If $o(A) \leqslant \alpha$, then $o(\mathcal{S}(A)) \leqslant \omega^{\alpha}$.*
2. If $X \in \mathcal{S}(A)$ and $\forall x \in X(l_A(x) < \alpha)$, then $l_{\mathcal{S}(A)}(X) < \omega^{\alpha}$.

Proof. Clearly, (2) and Lemma 1(2) imply (1) for any fixed α, hence it suffices to prove (2) by induction on α.

For $X = \varnothing$, we have $l_{\mathcal{S}(A)}(X) = 0 < \omega^{\alpha}$.

For $X = \{x\}$, we have $L_{\mathcal{S}(A)}(X) = \mathcal{S}(L_A(x))$. Since $o(L_A(x)) = l_A(x) < \alpha$, we get by induction $o(\mathcal{S}(L_A(x))) \leqslant \omega^{l_A(x)} = \omega^{(<\alpha)} < \omega^{\alpha}$, hence $l_{\mathcal{S}(A)}(X) \leqslant o(\mathcal{S}(L_A(x))) < \omega^{\alpha}$.

Let now $X = \{x_0, \dots, x_n\}$, $n > 1$. For any $Y \in L_{\mathcal{S}(A)}(X)$, we have $X \not\leqslant Y$, so there exists at least one x_i such that $\{x_i\} \not\leqslant Y$. Hence $L_{\mathcal{S}(A)}(X) = L_{\mathcal{S}(A)}(\{x_0\}) \cup \dots \cup L_{\mathcal{S}(A)}(\{x_n\})$, and by Lemma 1(4)

$$l_{\mathcal{S}(A)}(X) \leqslant l_{\mathcal{S}(A)}(\{x_0\}) \oplus \dots \oplus l_{\mathcal{S}(A)}(\{x_n\}) = (< \omega^\alpha) \oplus \dots \oplus (< \omega^\alpha) < \omega^\alpha.$$

□

We need one more lemma on the domination preorder, which is essential to prove the lower bounds in Theorem 1.

Lemma 3. *Let A be a WPO of height $\leqslant \omega$ and let $o(A) = \omega \cdot \alpha + n$ for some $n < \omega$. Then $o(\mathcal{S}(A)) \geqslant \omega^\alpha \cdot (n+1)$.*

Proof. We argue by induction on $o(A)$. If $o(A) = 0$, then $o(\mathcal{S}(A)) \geqslant 1 = \omega^0$ because $\varnothing \in \mathcal{S}(A)$.

If $o(A) > 0$ and $n = 0$, then for all $\beta < \alpha$ and $m < \omega$ according to Lemma 1(2), there exists $x \in A$ such that $\omega \cdot \beta + m \leqslant l_A(x) < o(A)$. Let $l_A(x) = \omega \cdot \gamma + k$ for $k < \omega$. Then by induction $o(\mathcal{S}(L_A(x)) \geqslant \omega^\gamma \cdot (k+1) \geqslant \omega^\beta \cdot (m+1)$. Since $S(L_A(x)) \subset S(A)$, we get $o(\mathcal{S}(A)) \geqslant o(\mathcal{S}(L_A))$, so

$$o(\mathcal{S}(A)) \geqslant \sup\{\omega^\beta \cdot (n+1) \mid \beta < \alpha, n < \omega\} = \omega^\alpha.$$

Now, let $o(A) > 0$ and $n > 0$. Then A has an element x that is not less than any other element. As the set $M = \{y \in A \mid y \leqslant x\}$ is finite, $o(A \setminus M) \geqslant \omega \cdot \alpha$. Let $\leqslant'$ be the following extension of $\leqslant$ order in $\mathcal{S}(A)$: $X \leqslant' Y$ iff $X \leqslant Y$ or $x \in Y \setminus X$. Then $(\mathcal{S}(A), \leqslant')$ is isomorphic to the lexicographic sum of $(\mathcal{S}(A \setminus \{x\}), \leqslant)$ and $(\{X \in \mathcal{S}(A) \mid x \in X\}, \leqslant) \cong (\mathcal{S}(A \setminus M), \leqslant)$, so by Lemma 1(3) and induction we get

$$o(\mathcal{S}(A)) \geqslant o((\mathcal{S}(A), \leqslant')) = o(\mathcal{S}(A \setminus \{x\})) + o(\mathcal{S}(A \setminus M)) \geqslant \omega^\alpha \cdot n + \omega^\alpha = \omega^\alpha \cdot (n+1).$$

□

We conclude this section with some information on the h-preorder (for more details, see e.g. [10,13]). Let $\omega^{<\omega}$ be the set of finite strings of natural numbers that include the empty string ε. By a *tree* we mean a nonempty initial segment of $(\omega^{<\omega}; \sqsubseteq)$ where $\sqsubseteq$ is the prefix relation. The *rank* of a finite tree is the length of a chain that is the longest in the tree. It is often useful to consider the "abstract" trees (a special kind of posets) together with the "concrete" trees defined above (initial segments of $\omega^{<\omega}$). Obviously, any abstract tree is isomorphic to a concrete one. In some definitions below, we implicitly use this obvious relation between "abstract" and "concrete" trees.

For a finite tree T and $\tau \in T$, define the tree $T(\tau) = \{\sigma \in \omega^{<\omega} \mid \tau \cdot \sigma \in T\}$ where $\cdot$ is the concatenation of strings. Then any non-singleton finite tree T is determined by the singleton tree $\{\varepsilon\}$ and the finite trees $T(i)$, $i \in T$ of lesser tree ranks than T, so $T = \{\varepsilon\} \cup \bigcup_{i \in T} T(i)$. By a *forest* we mean an initial segment of $(\omega^{<\omega} \setminus \{\varepsilon\}; \sqsubseteq)$. Note that there is a unique empty forest, and for any forest F

there is a unique tree T with $F = T \setminus \{\varepsilon\}$. Any non-empty forest F is isomorphic to the disjoint union of trees $F(i)$, $i \in \omega \cap F$ (here we identify i with the string (i)).

Definition 2. *For any preorder Q, let $\mathcal{T}(Q)$ be the set of finite Q-labeled trees (T, t), where T is a finite tree and $t : T \to Q$ a labeling. Let $\leqslant_h$ be the h-preorder on $\mathcal{T}(Q)$ defined by: $(T, t) \leqslant_h (U, u)$, if there is a monotone function $\varphi : T \to U$ such that $t(x) \leqslant_Q u(\varphi(x))$ for each $x \in T$.*

The construction $Q \mapsto \mathcal{T}(Q)$ was introduced by the first-named author of this paper in [12] and studied in several subsequent publications. Kruskal's tree theorem (see example 2) in Sect. 1) implies that if Q is a WQO then $(\mathcal{T}(Q); \leqslant_h)$ is a WQO too. Similarly for the h-preorder of Q-labeled forests.

In this paper, we deal with the simplest particular case (first considered in [5]) $Q = \bar{k}$, where $\bar{k} = (\{i \mid i < k\}; =)$ is the antichain with k elements, $2 \leqslant k < \omega$. We denote $\mathcal{T}(\bar{k})$ and $\mathcal{F}(\bar{k})$ resp. as $\mathcal{T}_k$ and $\mathcal{F}_k$, and the corresponding quotient-posets as $\mathbb{T}_k$ and $\mathbb{F}_k$. More complex structures $\mathcal{T}(Q)$ are discussed in Sect. 6.

Below we often identify elements $b = \{t_0, \ldots, t_{n-1}\} \in \mathcal{S}(\mathcal{T}_k)$ with the k-labeled forests $t_0 \sqcup \cdots \sqcup t_{n-1} \in \mathcal{F}_k$. For $a \in \bar{k}$ and $b \in \mathcal{S}(\mathcal{T}_k)$, let ab be a tree with root label a, whose successor trees are the members of b, in some order (all such trees are h-equivalent). Note that any concrete k-forest may be displayed using only labels, operation $\sqcup$, and concatenation, e.g.: $1, 120, 01 \sqcup 12, 0(12 \sqcup 01 \sqcup 210)$, and so on. Note that, for instance, $0011 \equiv_h 01$, $0(1 \sqcup 0) \equiv_h 01$, and $0(10 \sqcup 12) <_h 01(0 \sqcup 2)$.

We may stick to repetition-free k-forests (those that have distinct labels on all neighbor vertices) instead of arbitrary ones. In fact, the "minimal forests", which were characterized in [10], are repetition-free and provide the canonical elements for the h-equivalence. The correctness of the mentioned informal identifications and notational simplifications follow from the properties observed in [10, 13]. We cite some of these properties (which are, in fact, easy exercises).

Lemma 4. *1. Any k-labeled tree (forest) is h-equivalent to some repetition-free k-labeled tree (forest).*
2. For any $t, t_0, \ldots, t_{n-1} \in \mathcal{T}_k$ we have: $t \leqslant_h t_0 \sqcup \cdots \sqcup t_{n-1}$ iff $t \leqslant_h t_i$ for some $i < n$.
3. For all $a, a_1 \in \bar{k}$ and $b, b_1 \in \mathcal{S}(\mathcal{T}_k)$ we have: $ab \leqslant_h a_1 b_1$ iff $(a \neq a_1 \wedge ab \leqslant_h t$ for some $t \in b_1)$ or $(a = a_1 \wedge t \leqslant_h a_1 b_1$ for all $t \in b)$.
4. The poset $(\mathbb{F}_k; \leqslant_h)$ is isomorphic to the quotient-poset of $(\mathcal{S}(\mathcal{T}_k); \leqslant)$.

3 Proof Overview

In this section, we provide a brief informal overview of the proof of Theorem 1.

- The goal is to find $o(\mathbb{F}_k)$.
- Using Lemmas 4(4) and 2(1), the goal could be reduced to finding $o(\mathbb{T}_k)$.

- The upper bound for $o(\mathcal{T}_k)$ (obtained in Proposition 1) is obtained using Lemma 1(2) and the upper bound for the order types of some subsets of trees.
  - Such subsets of trees, named *trees with constraints* and denoted $\mathcal{T}^i(f)$, are defined in Sect. 4. They are parametrized by functions f, which map some of the labels to constraints.
  - The upper bounds for the trees with constraints are given in Lemma 5. They have the form $o(\mathcal{T}^i(f)) \leqslant \varphi_{u-1}(r)$, where u is the number of unconstrained labels and r is some measure of the "weakness of constraints" (the tighter the constraints—the smaller r is; the precise statement is given in the lemma).
  - We prove Lemma 5 by induction on $\varphi_{u-1}(r)$ (formally, we prove the statement "for all i and f, if $\alpha = \varphi_{u-1}(r)$ then $o(\mathcal{T}^i(f)) \leqslant \alpha$" by induction on α).
  - Veblen's ordinals appear because, when u decreases by 1 in the induction step, r may increase by any ordinal less than $o(\mathcal{T}^i(f))$, so $o(\mathcal{T}^i(f))$ should be a fixed point of φ_{u-2}.
- The lower bound for $o(\mathcal{T}_k)$ (established in Proposition 2) is based on the lower bounds for $o(\mathcal{T}^i(f))$.
  - For a suitable f, the lower bound $o(\mathcal{T}^i(f)) \geqslant \varphi_{k-2}(\dots)$ is given in Lemma 6. The proof uses the induction on k and on a measure of "weakness of constraints".
  - The induction step on k proves that $o(\mathcal{T}^i(f))$ should be a fixed point of φ_{k-3}.
  - In the case $k = 3$ we cannot use induction hypothesis. However, by Lemmas 3 and 4(3) we obtain (informally) that $o(\mathcal{T}^i(f)) \geqslant \omega^{\frac{o(\mathcal{T}^i(f))}{\omega}}$, which is enough to prove that $o(\mathcal{T}^i(f))$ is a fixed point of φ_0.

4 Main Technical Facts

We define sets of *trees with constraints* $\mathcal{T}^i(f)$ which serve as main technical tools in proving our estimates below. Informally, we want to restrict possible subtrees of a tree. For some vertices we impose the restriction that the successors of that vertex form a forest from some set.

Definition 3. *For $k \leqslant m$, $f \colon \overline{k} \to P(\mathcal{S}(\mathcal{T}_m)) \cup \{\bot\}$, (where P denotes the powerset operator) and $f(i) = \bot$, let $\mathcal{T}^i(f)$ be the set of m-labeled trees with the following constraints:*

1. *The root label is distinct from i.*
2. *If all predecessors of some vertex have labels in $f^{-1}(\bot)$, then the label x of this vertex is in $\overline{k}$.*
3. *If also $f(x) \neq \bot$, then the successors of this vertex form a forest from $f(x)$.*

The first condition is just a technical restriction that makes induction easier and does not matter much.

For example, if $\forall x(f(x) = \bot)$, then the second constraint means that all labels are in $\overline{k}$, and the third constraint is trivial, so $\bigcup_{i=0}^{k-1} \mathcal{T}^i(f) = \mathcal{T}_k$. For another example, if $f(x) = \varnothing$ for some x, then the tree could contain label x only as the successor of other constrained vertices.

The next lemma is crucial for getting the upper bounds in Theorem 1.

Lemma 5. *Let* $f\colon \overline{k} \rightarrow P(\mathcal{S}(\mathcal{T}_k)) \cup \{\bot\}$. *If* $f(i) = \bot$ *and* $\alpha = \varphi_{|f^{-1}(\bot)|-1}\left(\bigoplus_{f(x)\neq\bot} o(f(x)) + 1\right)$, *then* $o\left(\mathcal{T}^i(f)\right) \leqslant \alpha$.

Proof. We argue by induction on α. By Lemma 1(2), it suffices to show that $l_{\mathcal{T}^i(f)}(a) < \alpha$ for each $i \in \overline{k}, a \in \mathcal{T}^i(f)$. This we check by induction on the number of vertices in a repetition-free tree a. We distinguish two cases.

Let first $a = mb, f(m) = \bot$. Then $|f^{-1}(\bot)| > 1$, since otherwise $m = i$. Observe that $L_{\mathcal{T}^i(f)}(a) \subseteq \mathcal{T}^i\left(f + \left(m \mapsto L_{\mathcal{S}(\mathcal{T}^m(f))}(b)\right)\right)$. Indeed, if we have a repetition-free tree t from $L_{\mathcal{T}^i(f)}(a)$, then its root label is distinct from i and constraints on vertices with label $x \notin f^{-1}(\bot)$ remain, because it is still a tree from $\mathcal{T}^i(f)$ by definition of $L_{\mathcal{T}^i(f)}(a)$. Additional constraints on vertices with label m arise from the fact that $mb \nleqslant t$, so the successors of vertex with label m form a forest from $\mathcal{S}(\mathcal{T}^m(f))$, which is not greater than or equal to b, so it is a member of $L_{\mathcal{S}(\mathcal{T}^m(f))}(b)$, as stated.

By induction, we have $\forall t \in b(l_{\mathcal{T}^m(f)}(t) < \alpha)$, hence by Lemma 2(2) $l_{\mathcal{S}(\mathcal{T}^m(f))}(b) < \omega^\alpha$. Since α is an ε-ordinal (i.e., a fixed point of φ_0), we get $l_{\mathcal{S}(\mathcal{T}^m(f))}(b) \oplus \bigoplus_{f(x)\neq\bot} o(f(x)) = (< \alpha) \oplus \bigoplus_{f(x)\neq\bot} (< \alpha) < \alpha$. Then by induction we get

$$
\begin{aligned}
l_{\mathcal{T}^i(f)}(a) &\leqslant o(\mathcal{T}^i\left(f + \left(m \mapsto L_{\mathcal{S}(\mathcal{T}^m(f))}(b)\right)\right)) < \\
&\leqslant \varphi_{|f^{-1}(\bot)|>-2}\left(l_{\mathcal{S}(\mathcal{T}^m(f))}(b) \oplus \bigoplus_{f(x)\neq\bot} o(f(x))\right) \\
&= \varphi_{|f^{-1}(\bot)|-2}(< \alpha) < \alpha.
\end{aligned}
$$

Let now $a = mb, f(m) \neq \bot$. Then $L_{\mathcal{T}^i(f)}(a) \subseteq \mathcal{T}^i\left(f + \left(m \mapsto L_{f(m)}(b)\right)\right)$, for the same reason as in the previous case.

Indeed, the only difference from the previous case is that we already have a constraint on sub-trees of vertices with label m. So by additional constraint from the fact that $mb \nleqslant t$, we have that successors of a vertex with label m, whose all predecessors have labels from $f^{-1}(\bot)$, form a tree from $f(m)$, which is not greater than or equal to b, so it is a member of $L_{f(m)}(b)$.

Then by induction we get

$$l_{\mathcal{T}^i(f)}(a) \leqslant o(\mathcal{T}^i(f + (m \mapsto L_{f(m)}(b))))$$
$$\leqslant \varphi_{|f^{-1}(\bot)|>-1}\left(l_{f(m)}(b) \oplus \bigoplus_{\substack{x \neq m \\ f(x)\neq\bot}} o(f(x)) + 1\right)$$
$$= \varphi_{|f^{-1}(\bot)|-1}\left(< \bigoplus_{f(x)\neq\bot} o(f(x)) + 1\right) < \alpha.$$

□

For any $A \subseteq \mathcal{S}(\mathcal{T}_m)$ and $j < k$, we define $f = f_{A,j} \colon \overline{k} \to P(\mathcal{S}(\mathcal{T}_m)) \cup \{\bot\}$ as follows: $f(j) = A$ and $f(x) = \bot$ for $x \in \overline{k} \setminus \{j\}$. The next lemma is crucial for getting the lower bounds in Theorem 1.

Lemma 6. *Let $m \geqslant k \geqslant 3$, $A \subseteq \mathcal{S}(\mathcal{T}_m)$, $i \neq j \in \overline{k}$, and $f = f_{A,j}$. Then $o(\mathcal{T}^i(f)) \geqslant \omega \cdot 2$ for $o(A) = 1$, and $o(\mathcal{T}^i(f)) \geqslant \varphi_{k-2}(\alpha)$ for $o(A) \geqslant 2 + \alpha$.*

Proof. We argue by induction on k, and on $o(A)$. It suffices to prove the assertion for $i = 0$, $j = k - 1$, $o(\mathcal{T}^i(f_{A,j})) = \min_{x\neq j} o(\mathcal{T}^x(f_{A,j}))$.

Let first $o(A) = 1$, then $A = \{a\}$ for some forest a. The order $1 < 10 < 101 < \cdots < 1ja < 10ja < 101ja < \cdots$ on the displayed subset of $\mathcal{T}^0(f)$ extends the h-preorder on this subset, hence $o(\mathcal{T}^0(f)) \geqslant \omega \cdot 2$.

Suppose now that $o(A) \geqslant 2+\alpha$ and consider a tree $a = jb$. Then $L_{\mathcal{T}^0(f)}(a) \supseteq \mathcal{T}^0(f_{L_A(b),j})$ (moreover, they are equal, but we would not use that fact). Indeed, any member of $\mathcal{T}^0(f_{L_A(b),j})$ is obviously also a member of $\mathcal{T}^0(f)$, and it is not greater than or equal to a, since the successors of the lowest vertex with label j form a forest that is not greater than or equal to b.

Then by Lemma 1(2), for any $\beta < \alpha$ there exists some $b \in A$, such that $l_A(b) \geqslant 2 + \beta$. Hence, by induction $o(\mathcal{T}^0(f)) > l_{\mathcal{T}^0(f)}(a) \geqslant o(\mathcal{T}^0(f_{L_A(b),j})) \geqslant \varphi_{k-2}(\beta)$, hence $o(\mathcal{T}^0(f)) \geqslant \max(\sup\{\varphi_{k-2}(\beta)+1 \mid \beta < \alpha\}, \omega\cdot 2)$ by the induction hypothesis.

We distinguish cases $k = 3$ and $k > 3$. Let $k = 3$ and let $\omega^\gamma \cdot n$ (where $n < \omega$) be the main monomial of the Cantor normal form for $o(\mathcal{T}^0(f))$. If $\omega \cdot \gamma + n \geqslant o(\mathcal{T}^0(f))$, then $\omega \cdot \gamma \geqslant \omega^\gamma$, then $\gamma = 1$ or $\gamma = \omega^\gamma$. The first case in impossible since $o(\mathcal{T}^0(f)) \geqslant \omega \cdot 2 > \omega + n$. In the second case, we get $\gamma + n \geqslant o(\mathcal{T}^0(f)) \geqslant \gamma$. Since $o(\mathcal{T}^0(f))$ is a limit ordinal $(\forall a \in \mathcal{T}^0(f)(a < 10a))$, we see that $o(\mathcal{T}^0(f)) = \gamma$ is an ε-ordinal above all ε-ordinals less than ε_α, i.e. $o(\mathcal{T}^0(f)) \geqslant \varepsilon_\alpha$.

Assume now that $\omega \cdot \gamma + n < o(\mathcal{T}^0(f)) \leqslant o(\mathcal{T}^1(f))$. Then by Lemma 1(2), for some $b \in \mathcal{T}^1(f)$ we have $l_{\mathcal{T}^1(f)}(b) \geqslant \omega \cdot \gamma + n$. Consider the tree $a = 1b$. By Lemma 4, the trees from $L_{\mathcal{T}^0(f)}(a)$ with the root label 1 coincide with the trees $1c$, where $c \in \mathcal{S}(L_{\mathcal{T}^1(f)}(b))$, and the order between them is isomorphic to the order between c-s. Since $l_{\mathcal{T}^1(f)}(b) \geqslant \omega \cdot \gamma + n$, we get by Lemma 3

$$o(\mathcal{T}^0(f)) > l_{\mathcal{T}^0(f)}(a) \geqslant o(\mathcal{S}(L_{\mathcal{T}^1(f)}(b))) \geqslant \omega^\gamma \cdot (n+1) > o(\mathcal{T}^0(f)).$$

A contradiction.

Let now $k > 3$. If $o(\mathcal{T}^0(f))$ is a φ_{k-2}-ordinal, then it is larger than all ordinals $\varphi_{k-2}(\beta)$ for $\beta < \alpha$, hence $o(\mathcal{T}^0(f)) \geqslant \varphi_{k-2}(\alpha)$. Otherwise, for some $\beta < o(\mathcal{T}^0(f))$ we have $\varphi_{k-3}(\beta) \geqslant o(\mathcal{T}^0(f))$. Since $o(\mathcal{T}^1(f)) \geqslant o(\mathcal{T}^0(f)) > \beta$, by Lemma 1(2), for some $b \in \mathcal{T}^1(f)$ we have $l_{\mathcal{T}^1(f)}(b) \geqslant \beta$. Consider the tree $a = 1b$. Observe that $L_{\mathcal{T}^0(f)}(a) \supseteq \mathcal{T}^0\left(f_{\mathcal{S}(L_{\mathcal{T}^1(f)}(b)),1} - j\right)$ which is exactly $\mathcal{T}^0\left(f_{\mathcal{S}(L_{\mathcal{T}^1(f)}(b)),1}\right)$ for $k-1$. Indeed, a tree from $\mathcal{T}^0\left(f_{\mathcal{S}(L_{\mathcal{T}^1(f)}(b)),1} - j\right)$ may contain a vertex with label j only as a successor of vertex with label 1. But the successors of a vertex with label 1 are members of $L_{\mathcal{T}^1(f)}(b)$, so they satisfy the restriction on j-labeled vertices and also are not greater than or equal to b.

So by induction on k, we obtain

$$\begin{aligned} o(\mathcal{T}^0(f)) > l_{\mathcal{T}^0(f)}(a) &\geqslant \varphi_{k-3}(o(\mathcal{S}(L_{\mathcal{T}^1(f)}(b)))) \\ &\geqslant \varphi_{k-3}(l_{\mathcal{T}^1(f)}(b)) \geqslant \varphi_{k-3}(\beta) \geqslant o(\mathcal{T}^0(f)). \end{aligned}$$

A contradiction. □

5 Proof of Theorem 1

Here we prove assertions which imply Theorem 1.

Proposition 1. *For any $k \geqslant 2$ we have $o(\mathcal{T}_k) \leqslant \varphi_{k-1}(0)$.*

Proof. By Lemma 1(2) it suffices to show that $l_{\mathcal{T}_k}(a) < \varphi_{k-1}(0)$ for each $a \in \mathcal{T}_k$. This is checked by induction on the number of vertices in a repetition-free tree a.

Let $a = mb$. Then $L_{\mathcal{T}_k}(a) = \bigcup_{i \in \overline{k} \setminus \{m\}} \mathcal{T}^i\left(f_{L_{\mathcal{S}(\mathcal{T}^k)}(b),m}\right)$, hence

$$\begin{aligned} l_{\mathcal{T}_k}(a) &\leqslant \bigoplus_{i \in \overline{k} \setminus \{m\}} o\left(\mathcal{T}^i\left(f_{L_{\mathcal{S}(\mathcal{T}_k)}(b),m}\right)\right) && \text{(Lemma 1(4))} \\ &\leqslant \bigoplus_{i \in \overline{k} \setminus \{m\}} \varphi_{k-2}\left(l_{\mathcal{S}(\mathcal{T}_k)}(b) + 1\right) && \text{(Lemma 5)} \\ &= \varphi_{k-2}\left(\left(< \omega^{\varphi_{k-1}(0)}\right) + 1\right) \cdot (k-1) && \text{(Lemma 2(2) and induction)} \\ &= \varphi_{k-2}(< \varphi_{k-1}(0)) \cdot (k-1) < \varphi_{k-1}(0). \end{aligned}$$

□

From Proposition 1 and Lemma 2 we immediately obtain the following.

Corollary 1. *For any $k \geqslant 2$, we have $o(\mathcal{S}(\mathcal{T}_k)) \leqslant \varphi_{k-1}(0)$.*

Proposition 2. *For any $k \geqslant 3$ we have $o(\mathcal{T}_k) \geqslant \varphi_{k-1}(0)$.*

Proof. For some ordinal γ we have $\omega^\gamma \leqslant o(\mathcal{T}_k) < \omega^{\gamma+1}$. Let $X_j \subset \mathcal{T}_k$ be the set of trees whose root label is distinct from j. Since $\bigcup_j X_j = \mathcal{T}_k$, by Lemma 1(4) we get that $\bigoplus_j o(X_j) \geqslant o(\mathcal{T}_k) \geqslant \omega^\gamma$, hence $o(X_j) \geqslant \omega^\gamma$ for some $j \in \overline{k}$. Let $j \neq i \in \overline{k}$, then by Lemma 6

$$\omega^{\gamma+1} > o(\mathcal{T}_k) \geqslant o(\mathcal{T}^i(f_{X_j,j})) \geqslant \varphi_{k-2}(o(X_j)) \geqslant \varphi_{k-2}(\omega^\gamma).$$

Then $\omega^\gamma \geqslant \varphi_{k-2}(\omega^\gamma)$, hence $o(\mathcal{T}_k) \geqslant \omega^\gamma \geqslant \varphi_{k-1}(0)$. □

From Proposition 2 we immediately obtain the following.

Corollary 2. *For any $k \geqslant 3$, we have: $o(\mathcal{S}(\mathcal{T}_k)) \geqslant \varphi_{k-1}(0)$.*

Corollaries 1 and 2 together with Lemma 4(4) imply the estimates of Theorem 1 concerning the maximal order types of $\mathbb{F}_k$ and $\mathbb{T}_k$. The next proposition establishes the remaining estimates of Theorem 1.

Proposition 3. *For any integer $k \geqslant 3$ we have: $w(\mathbb{F}_k) = \varphi_{k-1}(0)$.*

Proof. According to Corollary 3.9 in [3], if a WPO A is such that $h(A) < o(A)$ and $o(A)$ is closed under ordinal multiplication, then $o(A) = w(A)$. Since the WPOs $\mathbb{F}_k$ for $k \geqslant 3$ satisfy these conditions, by Propositions 1 and 2 we are done. □

6 Future Work

There are several natural and useful variations of h-preorders on the labeled trees (and forests) for which the ordinal invariants remain unknown.

In particular, this is the case for the preorder $(\mathcal{T}(Q); \leqslant_h)$, where Q is a finite WPO which is not an antichain. We do not know the answer even for the particular case $Q = (P(X); \subseteq)$ where X is a finite set with at least 3 elements.

Another interesting series of WQOs is obtained by iterating the construction $Q \mapsto \mathcal{T}(Q)$ starting with $Q = \overline{k}$ (see e.g. [13] for more information and references). Namely, we define the sequence $\{\mathcal{T}_m(\overline{k})\}_{m<\omega}$ of h-preorders as follows: $\mathcal{T}_0(\overline{k}) = \overline{k}$ and $\mathcal{T}_{m+1}(\overline{k}) = \mathcal{T}(\mathcal{T}_m(\overline{k}))$. The ω-th iterate $\mathcal{T}_\omega(\overline{k})$ and the h-preorder on it may also be naturally defined. For such iterates (with the exception of the zero and first iterates), and also for the forest variants of them, all three ordinal invariants are currently not known.

Finally, all the mentioned variations of h-preorders have important extensions to the infinite well founded trees and forests which are intimately related to the Wadge hierarchy and its extensions (see e.g. [7,11,14] and references therein).

Among these variations, the simplest are the structures of countable k-labeled forests introduced in [11]. We hope that the ideas and methods of this paper could help to compute the ordinal invariants of these structures (so far, only the height was computed in [11]). For the other variations, the problem is widely open and looks challenging.

Acknowledgment. We thank the anonymous referees for their careful reading and valuable suggestions that certainly improved the text.

References

1. Abriola, S., Halfon, S., Lopez, A., Schmitz, S., Schnoebelen, P., Vialard, I.: Measuring well-quasi-ordered finitary powersets, arXiv preprint (2023)
2. de Jongh, D.H.J., Parikh, R.: Well-partial orderings and hierarchies. Nederl. Akad. Wetensch. Proc. Ser. A 80 = Indag. Math. **39**(3), 195–207 (1977)
3. Dzamonja, M., Schmitz, S., Schnoebelen, P.: On ordinal invariants in well quasi orders and finite antichain orders. In [15], pp. 29–54
4. Friedman, H.M., Weiermann, A.: Some independence results related to finite trees. Phil. Trans. R. Soc. A **381**, 20220017 (2023)
5. Hertling, P.: Topologische Komplexitätsgrade von Funktionen mit endlichem Bild. Informatik-Berichte **152**, 34 p, Fernuniversität Hagen (1993)
6. Hertling, P., Selivanov, V.: Complexity issues for preorders on finite labeled forests. In: Brattka, V., Diener, H., Spreen, D. (eds.) Logic, Computation, Hierarchies, pp. 165–190. Ontos Publishing, de Gruiter, Boston-Berlin (2014)
7. Kihara, T., Montalbán, A.: On the structure of the Wadge degrees of bqo-valued Borel functions. Trans. Amer. Math. Soc. **371**(11), 7885–7923 (2019)
8. Kudinov, O.V., Selivanov, V.L.: A gandy theorem for abstract structures and applications to first-order definability. In: Ambos-Spies, K., Löwe, B., Merkle, W. (eds.) CiE 2009. LNCS, vol. 5635, pp. 290–299. Springer, Heidelberg (2009). https://doi.org/10.1007/978-3-642-03073-4_30
9. Schmidt, D.: Well-partial orderings and their maximal order types. Habilitationsschrift, University of Heidelberg (1979). See also a later release in [15], 351–391
10. Selivanov, V.L.: Boolean hierarchies of partitions over reducible bases. Algebra Logic **43**(1), 44–61 (2004)
11. Selivanov, V.L.: The quotient algebra of labeled forests modulo h-equivalence. Algebra Logic **46**(2), 120–133 (2007)
12. Selivanov, V.: A fine hierarchy of ω-regular k-partitions. In: Löwe, B., Normann, D., Soskov, I., Soskova, A. (eds.) CiE 2011. LNCS, vol. 6735, pp. 260–269. Springer, Heidelberg (2011). https://doi.org/10.1007/978-3-642-21875-0_28
13. Selivanov, V.: Q-Wadge degrees as free structures. Computability **9**, 327–341 (2020)
14. Selivanov, V.L.: A Q-Wadge hierarchy in quasi-Polish spaces. J. Symb. Log. **87**(2), 732–757 (2022)
15. Schuster, P.M., Seisenberger, M., Weiermann, A. (eds.): Well-Quasi Orders in Computation, Logic, Language and Reasoning. Springer (2020)
16. Weiermann, A.: A computation of the maximal order type of the term ordering on finite multisets. In: Ambos-Spies, K., Löwe, B., Merkle, W. (eds.) CiE 2009. LNCS, vol. 5635, pp. 488–498. Springer, Heidelberg (2009). https://doi.org/10.1007/978-3-642-03073-4_50

Completeness Theorems for Modal Logic in Second-Order Arithmetic

Sho Shimomichi, Yuto Takeda(✉), and Keita Yokoyama

Tohoku University, Sendai, Miyagi, Japan
{sho.shimomichi.s8,yuto.takeda.t8,keita.yokoyama.c2}@dc.tohoku.ac.jp

Abstract. This paper investigates the logical strength of completeness theorems for modal propositional logic within second-order arithmetic. We demonstrate that the weak completeness theorem for modal propositional logic is provable in RCA_0, and that, over RCA_0, ACA_0 is equivalent to the strong completeness theorem for modal propositional logic using canonical models. We also consider a simpler version of the strong completeness theorem without referring to canonical models and show that it is equivalent to WKL_0 over RCA_0.

Keywords: second-order arithmetic · reverse mathematics · completeness theorems · modal logic

1 Introduction

In this paper, we investigate the strength of completeness theorems for modal propositional logic with respect to Kripke semantics in second-order arithmetic. There have been several attempts to analyze logic in second-order arithmetic. In [3], Simpson studied the relations of several well-known theorems of classical logic in second-order arithmetic and showed that, over RCA_0, the Gödel's completeness theorem is equivalent to WKL_0. Similar studies have been conducted on intuitionistic logic, and in [4], Yamazaki showed, over RCA_0, the strong completeness theorem for intuitionistic predicate logic is equivalent to ACA_0. In this paper, we extend these studies to modal propositional logic. We start with the weak completeness theorem for modal propositional logic which is provable in RCA_0. Then we see that the strong completeness theorem by means of canonical models requires ACA_0, while the strong completeness theorem itself is equivalent to WKL_0 over RCA_0.

In Sect. 2, we formalize modal logic in second-order arithmetic. In Sect. 3, we show the weak completeness theorems for all modal logics built from the axiom K using modal formulas T, B, 4, and D in RCA_0. We formalize the proof of weak completeness by Moss [2] within RCA_0. In Sect. 4, we prove that ACA_0 is equivalent to the strong completeness theorem using canonical models over RCA_0. The idea of the equivalence proof depends on Yamazaki [4]. In Sect. 5, we consider a simpler version of the strong completeness theorem.

© The Author(s), under exclusive license to Springer Nature Switzerland AG 2025
A. Beckmann et al. (Eds.): CiE 2025, LNCS 15764, pp. 452–466, 2025.
https://doi.org/10.1007/978-3-031-95908-0_32

2 Modal Logic in Second-Order Arithmetic

In this section, we set up the basic notions of modal logic within RCA_0. For basic knowledge of modal logic, see [1], and for the formalization of mathematical logic in RCA_0, see [3]. As usual, we identify finite sequences of symbols of language for modal logic with natural numbers. Note that Seq is a set of all finite sequences of natural numbers.

Let $\mathrm{Prop} = \{p, q, r, \dots\}$ be an infinite set of propositional variables, and let $\mathrm{C} = \{\alpha_0, \alpha_1, \dots\}$ be an infinite set of propositional constants.[1] Thus in RCA_0, the set of all atomic formulas, $\mathrm{Atm} = \mathrm{Prop} \cup \mathrm{C} \cup \{\bot\}$, and the set of all modal formulas, Fml, exist as the sets of natural numbers.

Fix $p, q, r \in \mathrm{Prop}$. Let Axm, the axioms of normal modal logic, be the set of the following formulas:

$$p \to (q \to p), \qquad (p \to (q \to r)) \to ((p \to q) \to (p \to r)),$$
$$(\neg p \to \neg q) \to (q \to p), \qquad \Box(p \to q) \to (\Box p \to \Box q).$$

As in standard studies of modal logic, we also deal with the following formulas.

$$T \equiv \Box p \to p, \quad B \equiv p \to \Box\Diamond p, \quad D \equiv \Box p \to \Diamond p, \quad 4 \equiv \Box p \to \Box\Box p,$$
$$5 \equiv \Diamond p \to \Box\Diamond p, \quad .2 \equiv \Diamond\Box p \to \Box\Diamond p, \quad L \equiv \Box(\Box p \to p) \to \Box p.$$

In dealing with modal logic, we introduce the uniform substitution rule. Let $\mathcal{F}$ be the set of all functions from Prop to Fml with finite domains.

Proposition 2.1 (RCA_0). *There exists a function $f : \mathcal{F} \times \mathrm{Fml} \to \mathrm{Fml}$ such that for all finite function $\sigma :\subseteq \mathrm{Prop} \to \mathrm{Fml}$, $f(\sigma, \cdot)$ is the* uniform substitution *for σ, i.e.,*

- $f(\sigma, p) = \sigma(p)$ *if* $p \in \mathrm{Prop}$ *and* $p \in \mathrm{dom}(\sigma)$,
- $f(\sigma, \alpha) = \alpha$ *if* $\alpha \in \mathrm{C}$,
- $f(\sigma, \bot) = \bot$,
- $f(\sigma, \varphi \to \psi) = f(\sigma, \varphi) \to f(\sigma, \psi)$,
- $f(\sigma, \Box\varphi) = \Box f(\sigma, \varphi)$.

Henceforth, for all $\varphi \in \mathrm{Fml}$ and $\sigma \in \mathcal{F}$, we denote $f(\sigma, \varphi)$ as $\overline{\sigma}(\varphi)$. Note that for all $\sigma \in \mathrm{Seq}$, we denote the length of σ as $lh(\sigma)$. We now formalize the notion of provability.

[1] We distinguish propositional variables (may be seen as place holders) which allow uniform substitution and atomic constants. This helps finite axiomatization under the uniform substitution rule, but it does not change provability.

Definition 2.2 (RCA$_0$). *Let* $\Gamma \subset \mathrm{Fml}$ *and* $\Sigma \subseteq \{T, B, 4, 5, D, .2, L\}$. *We will define the following predicates:*

$$\begin{aligned}\mathrm{Prf}_{\mathbf{K}\Sigma}(\Gamma, p) \equiv\ & p \in \mathrm{Seq} \wedge \forall k(k < lh(p) \to p(k) \in \mathrm{Fml}) \\ & \wedge \forall k\Big(k < lh(p) \to \Big(p(k) \in \Gamma \vee p(k) \in \Sigma \vee p(k) \in \mathrm{Axm} \\ & \vee (\exists i < k \exists j < k)(p(i) = p(j) \to p(k)) \vee (\exists i < k)(p(k) = \Box p(i)) \\ & \vee (\exists i < k)(\exists \sigma : \{q < p(i) \mid q \in \mathrm{Prop}\} \to \{\varphi < p(k) \mid \varphi \in \mathrm{Fml}\}) \\ & (p(k) = \overline{\sigma}(p(i)))\Big)\Big),\end{aligned}$$

$$\begin{aligned}\mathrm{Pbl}_{\mathbf{K}\Sigma}(\Gamma, \varphi) \equiv\ & \exists \psi_1, \dots, \exists \psi_n \in \Gamma \exists p \\ & \big(\mathrm{Prf}_{\mathbf{K}\Sigma}(\Gamma, p) \wedge (\exists i < lh(p))(p(i) = (\psi_1 \wedge \cdots \wedge \psi_n \to \varphi))\big).\end{aligned}$$

Let $\Gamma \subset \mathrm{Fml}$ and $\Sigma \subseteq \{T, B, 4, 5, D, .2, L\}$. Then,

- Γ is $\mathbf{K}\Sigma$*-consistent* if $\neg\mathrm{Pbl}_{\mathbf{K}\Sigma}(\Gamma, \bot)$ holds,
- Γ is *closed under deduction* if $\forall \varphi \in \mathrm{Fml}(\mathrm{Pbl}_{\mathbf{K}\Sigma}(\Gamma, \varphi) \to \varphi \in \Gamma)$ holds,
- Γ is *complete* if $\forall \varphi \in \mathrm{Fml}(\varphi \in \Gamma \vee \neg\varphi \in \Gamma)$ holds.
- Γ is *maximally* $\mathbf{K}\Sigma$*-consistent* if Γ is $\mathbf{K}\Sigma$-consistent, closed under deduction, and complete.

We next formalize the Kripke semantics.

Definition 2.3 (RCA$_0$). *A* Kripke model *is a tuple* $M = (W, R, V)$ *satisfying the following conditions:*

1. $W \subseteq \mathbb{N}$ *is a non-empty set,*
2. R *is a binary relation on* W*, i.e.,* $R \subseteq W \times W$*,*
3. $V : W \times \mathrm{Fml} \to \{0, 1\}$*,*
4. $V(w, \bot) = 0$ *for any* $w \in W$*,*
5. $V(w, \varphi \to \psi) = 1 - V(w, \varphi)(1 - V(w, \psi))$ *for any* $w \in W$ *and any* $\varphi, \psi \in \mathrm{Fml}$*,*
6. $V(w, \Box\varphi) = 1 \iff \forall v \in W(wRv \to V(v, \varphi) = 1)$ *for any* $w \in W$ *and any* $\varphi \in \mathrm{Fml}$.

The pair (W, R) *is called a* frame *and the function* V *is called a* valuation *with* (W, R)*.*

We also use the following notations. Let F be a frame, M be a model, $w \in W$, and $\varphi \in \mathrm{Fml}$, we define

$$M, w \Vdash \varphi \equiv V(w, \varphi) = 1, \qquad M \Vdash \varphi \equiv (\forall w \in W)(V(w, \varphi) = 1),$$
$$F \Vdash \varphi \equiv (\forall V\text{: valuation with} F)(\forall w \in W)(V(w, \varphi) = 1).$$

Remark 1. In the above definition, the valuation V should cover the truth values of all formulas. Within RCA$_0$, one cannot extend the valuation for atomic formulas to the full valuation.

Definition 2.4 (RCA_0)**.** *Let $F = (W, R)$ be a frame.*

- *F is* appropriate to *T if R is reflexive,*
- *F is* appropriate to *B if R is symmetric,*
- *F is* appropriate to *4 if R is transitive,*
- *F is* appropriate to *5 if R is Euclidean,*
- *F is* appropriate to *D if R is serial,*
- *F is* appropriate to *$.2$ if R is directed,*
- *F is* appropriate to *L if R is transitive and W has no infinite ascending sequences by R.*

In general, F is appropriate to *$\Sigma \subseteq \{T, B, 4, 5, D, .2, L\}$ if for all $\varphi \in \Sigma$, F is appropriate to φ.*

Given a frame F, a set of formulas Γ, $\Sigma \subseteq \{T, B, 4, 5, D, .2, L\}$, and $\varphi \in \mathrm{Fml}$, we let

$$F \Vdash \Gamma \equiv (\forall \psi \in \Gamma)(F \Vdash \psi),$$
$$\Gamma \Vdash_{\mathbb{F}_\Sigma} \varphi \equiv (\forall F\text{: frame which is appropriate to } \Sigma)(F \Vdash \Gamma \to F \Vdash \varphi).$$

Now we are ready to state the soundness theorem.

Theorem 2.5 (Soundness Theorem, RCA_0**).** *Let $\Sigma \subseteq \{T, B, 4, 5, D, .2, L\}$ and $\varphi \in \mathrm{Fml}$. If $\mathrm{Pbl}_{\mathbf{K}\Sigma}(\emptyset, \varphi)$ holds, then $\Vdash_{\mathbb{F}_\Sigma} \varphi$ holds.*

To prove the soundness theorem within RCA_0, we carefully formalize the usual proof so that the induction fits in $\mathrm{I}\Sigma^0_1$.

3 Weak Completeness Theorem in Second-Order Arithmetic

In this section, we will prove the weak completeness theorems for modal logic in RCA_0. For this purpose, we will formalize the proof of weak completeness theorem by Moss [2] in RCA_0. Unless otherwise specified, we will assume $\Sigma \subseteq \{T, B, 4, D\}$. First, for a given $\mathbf{K}\Sigma$-consistent formula, we aim to create the weak Kripke model in RCA_0.

For any $\varphi \in \mathrm{Fml}$, $sub(\varphi)$ is the set of subformulas of φ. A weak Kripke model for φ is a Kripke model with the valuation restricted to subformulas of φ.

Definition 3.1 (Weak Kripke model, RCA_0**).** Weak Kripke model of *φ is a tuple $M = (W, R, V)$ satisfying the following conditions:*

1. *$W \subseteq \mathbb{N}$ is a non-empty set,*
2. *R is a binary relation on W, i.e., $R \subseteq W \times W$,*
3. *$V : W \times sub(\varphi) \to \{0, 1\}$,*
4. *$V(w, \bot) = 0$ for any $w \in W$,*

5. $V(w, \psi \to \theta) = 1 - V(w, \psi)(1 - V(w, \theta))$ *for any* $w \in W$ *and any* $\psi, \theta \in sub(\varphi)$,
6. $V(w, \Box\psi) = 1 \iff \forall v \in W(wRv \to V(v, \psi) = 1)$ *for any* $w \in W$ *and any* $\psi \in sub(\varphi)$.

To construct the model, we make some preparations. Let $p_0, p_1, \ldots$ be an enumeration of Atm. We define the *height function* $ht : \mathrm{Fml} \to \mathbb{N}$ and the *order function* $ord : \mathrm{Fml} \to \mathbb{N}$ by primitive recursion:

- $ht(p_n) = 0$ for all n,
- $ht(\varphi \to \psi) = \max\{ht(\varphi), ht(\psi)\}$,
- $ht(\Box\varphi) = 1 + ht(\varphi)$.
- $ord(p_n) = n$ for all n,
- $ord(\varphi \to \psi) = \max\{ord(\varphi), ord(\psi)\}$,
- $ord(\Box\varphi) = ord(\varphi)$.

Definition 3.2 (RCA_0). *We define the following.*

- $\mathcal{L}_{h,n} = \{\varphi \in \mathrm{Fml} \mid ht(\varphi) \leq h \wedge ord(\varphi) \leq n\}$ *for all* h *and* n,
- $\hat{T} \equiv \bigwedge T \wedge \bigwedge_{p \in \{p_0,\ldots,p_n\} \setminus T} \neg p$ *for all* $T \subseteq \{p_0, \ldots, p_n\}$,
- $\bigoplus X \equiv \bigvee_{\alpha \in X}(\alpha \wedge \bigwedge_{\beta \in X \setminus \{\alpha\}} \neg\beta)$ *for all* $X \subseteq \mathrm{Fml}$.

For any n, *we define the finite set* $C_{h,n} \subseteq \mathrm{Fml}$ *recursively:*

- $C_{0,n} = \{\hat{T} \mid T \subseteq \{p_0, \ldots, p_n\}\}$,
- $C_{h+1,n} = \{\alpha_{S,T} \equiv (\bigwedge_{\psi \in S} \Diamond\psi) \wedge (\Box \bigvee S) \wedge \hat{T} \mid S \subseteq C_{h,n}, T \subseteq \{p_0, \ldots, p_n\}\}$.

The members of $C_{h,n}$ are called *canonical formulas*. They are of height $\leq h$ and built from the first n atomic formulas. Note that each $C_{h,n}$ exists as a finite set and they are uniformly available within RCA_0 by the primitive recursion. We see several basic lemmas.

Lemma 3.3 (RCA_0). *Fix* h *and* n. *For all* $\varphi \in \mathcal{L}_{h,n}$ *and for all* $\alpha \in C_{h,n}$, *either* $\mathrm{Pbl}_{\mathbf{K}\Sigma}(\emptyset, \alpha \to \varphi)$ *or* $\mathrm{Pbl}_{\mathbf{K}\Sigma}(\emptyset, \alpha \to \neg\varphi)$ *holds.*

Lemma 3.4 (RCA_0). *Let* $\varphi \in \mathcal{L}_{h,n}$ *and* $\psi \in \mathrm{Fml}$. *If* φ *is* $\mathbf{K}\Sigma$*-consistent, then there exists* $\alpha \in C_{h+1,n}$ *such that* α *is* $\mathbf{K}\Sigma$*-consistent and* $\mathrm{Pbl}_{\mathbf{K}\Sigma}(\emptyset, \alpha \to \varphi)$ *holds.*

Lemma 3.5 (Existence lemma, RCA_0**).** *Let* $\varphi \in \mathcal{L}_{h,n}$ *and* $\psi \in \mathrm{Fml}$. *If* $\psi \wedge \Diamond\varphi$ *is* $\mathbf{K}\Sigma$*-consistent, then there exists* $\alpha \in C_{h,n}$ *such that* $\psi \wedge \Diamond\alpha$ *is* $\mathbf{K}\Sigma$*-consistent and* $\mathrm{Pbl}_{\mathbf{K}\Sigma}(\emptyset, \alpha \to \varphi)$ *holds.*

Proof. Let $S = \{\alpha \in C_{h,n} \mid \mathrm{Pbl}_{\mathbf{K}\Sigma}(\emptyset, \alpha \to \varphi)\}$. Then, $\mathrm{Pbl}_{\mathbf{K}\Sigma}(\emptyset, \Diamond\varphi \leftrightarrow \Diamond(\bigvee S) \leftrightarrow (\bigvee\{\Diamond\alpha \mid \alpha \in S\}))$ holds. Thus,

$$\mathrm{Pbl}_{\mathbf{K}\Sigma}(\emptyset, (\psi \wedge \Diamond\varphi) \leftrightarrow (\psi \wedge \bigvee\{\Diamond\alpha \mid \alpha \in S\}) \leftrightarrow (\bigvee\{\psi \wedge \Diamond\alpha \mid \alpha \in S\})).$$

Since $\psi \wedge \Diamond\varphi$ is $\mathbf{K}\Sigma$-consistent, there exists $\alpha \in S$ such that $\psi \wedge \Diamond\alpha$ is $\mathbf{K}\Sigma$-consistent.

With the above preparation, we first show the weak completeness theorem for the case of $4 \notin \Sigma$ in RCA_0.

Definition 3.6 (RCA_0). *We assume that $\Sigma \subseteq \{T, B, D\}$. Let $\varphi \in \mathcal{L}_{h,n}$ be $\mathbf{K}\Sigma$-consistent. Then we define $H = (\tilde{W}, \xrightarrow{c}, \tilde{V})$ satisfying the following:*

- $\tilde{W} = \{\alpha \in C_{h+1,n} \mid \neg \mathrm{Pbl}_{\mathbf{K}\Sigma}(\alpha, \bot) \wedge \mathrm{Pbl}_{\mathbf{K}\Sigma}(\emptyset, \alpha \to \varphi)\}$,
- $\xrightarrow{c} = \{(\alpha, \beta) \in \tilde{W} \times \tilde{W} \mid \neg \mathrm{Pbl}_{\mathbf{K}\Sigma}(\alpha \wedge \Diamond\beta, \bot)\}$,
- *For all $\psi \in sub(\varphi)$ and $\alpha \in \tilde{W}$, $\tilde{V}(\alpha, \psi) = 1 \iff \mathrm{Pbl}_{\mathbf{K}\Sigma}(\emptyset, \alpha \to \psi)$.*

This H exists by Lemma 3.3 and $\mathrm{I}\Sigma^0_1$ (more precisely, we need the bounded Σ^0_1-comprehension to obtain $\xrightarrow{c}$). Moreover it is clear that $\tilde{V}(\alpha, \varphi) = 1$ for any $\alpha \in \tilde{W}$. By Lemma 3.4 and Lemma 3.5, $\tilde{W}$ and $\xrightarrow{c}$ are non-empty sets.

Lemma 3.7 (RCA_0). *The following hold:*

1. *if $T \in \Sigma$ then $\xrightarrow{c}$ is reflexive.*
2. *if $B \in \Sigma$ then $\xrightarrow{c}$ is symmetric.*
3. *if $D \in \Sigma$ then $\xrightarrow{c}$ is serial.*
4. *H is a weak Kripke model.*

Proposition 3.8 (RCA_0). *Let $\Sigma \subseteq \{T, B, D\}$. If $\varphi \in \mathrm{Fml}$ is $\mathbf{K}\Sigma$-consistent, then there exist a frame which is appropriate to Σ and a valuation V such that (F, V) is a weak model of φ.*

Proof. By Definition 3.6, we define $F = (\tilde{W}, \xrightarrow{c})$ and $V = \tilde{V}$. By Lemma 3.7, (F, V) is a weak model of φ.

We next aim to construct a weak Kripke model for the case $4 \in \Sigma$. In what follows, we will assume $4 \in \Sigma$ unless otherwise specified.

Lemma 3.9 (RCA_0). *For all $\mathbf{K}\Sigma$-consistent $\alpha \in C_{h+1,n}$, there is only one $\mathbf{K}\Sigma$-consistent formula $\alpha' \in C_{h,n}$ such that $\mathrm{Pbl}_{\mathbf{K}\Sigma}(\emptyset, \alpha \to \alpha')$ holds.*

Note that, for each $n \in \mathbb{N}$, we may consider the correspondence $\alpha \mapsto \alpha'$ in Lemma 3.9 as a function (with the finite domain) by $\mathrm{I}\Sigma^0_1$. Next we consider the following two models according to the conditions of Σ.

Definition 3.10 (RCA_0). *We assume that $4 \in \Sigma$ and $B \notin \Sigma$. Let $\varphi \in \mathcal{L}_{h,n}$ be $\mathbf{K}\Sigma$-consistent. Then we define $N = (\tilde{W}, \xrightarrow{m}, \tilde{V})$ satisfying the following:*

- $\tilde{W} = \{\alpha \in C_{h+1,n} \mid \neg \mathrm{Pbl}_{\mathbf{K}\Sigma}(\alpha, \bot) \wedge \mathrm{Pbl}_{\mathbf{K}\Sigma}(\emptyset, \alpha \to \varphi)\}$,
- $\xrightarrow{m} = \{(\alpha, \beta) \in \tilde{W} \times \tilde{W} \mid \mathrm{Pbl}_{\mathbf{K}\Sigma}(\emptyset, \alpha \to \Diamond\beta') \wedge \forall \xi \in C_{h,n}(\mathrm{Pbl}_{\mathbf{K}\Sigma}(\emptyset, \beta \to \Diamond\xi) \to \mathrm{Pbl}_{\mathbf{K}\Sigma}(\emptyset, \alpha \to \Diamond\xi))\}$,
- *For all $\psi \in sub(\varphi)$ and $\alpha \in \tilde{W}$, $\tilde{V}(\alpha, \psi) = 1 \iff \mathrm{Pbl}_{\mathbf{K}\Sigma}(\emptyset, \alpha \to \psi)$.*

This N exists by Lemma 3.3 and $\mathrm{I}\Sigma^0_1$.

Definition 3.11 (RCA$_0$). *We assume that* $4, B \in \Sigma$. *Let* $\varphi \in \mathcal{L}_{h,n}$ *be* **K**Σ-*consistent. Then we define* $M = (\tilde{W}, \xrightarrow{mm}, \tilde{V})$ *satisfying the following:*

- $\tilde{W} = \{\alpha \in C_{h+1,n} \mid \neg\mathrm{Pbl}_{\mathbf{K}\Sigma}(\alpha, \bot) \wedge \mathrm{Pbl}_{\mathbf{K}\Sigma}(\emptyset, \alpha \to \varphi)\}$,
- $\xrightarrow{mm} = \{(\alpha, \beta) \in \tilde{W} \times \tilde{W} \mid \alpha \xrightarrow{m} \beta \wedge \beta \xrightarrow{m} \alpha\}$,
- *For all* $\psi \in sub(\varphi)$ *and* $\alpha \in \tilde{W}$, $\tilde{V}(\alpha, \psi) = 1 \iff \mathrm{Pbl}_{\mathbf{K}\Sigma}(\emptyset, \alpha \to \psi)$.

This M exists by Lemma 3.3 and IΣ^0_1.

Proposition 3.12 (RCA$_0$). *Let* $\alpha, \beta \in \tilde{W}$ *and* $4 \in \Sigma$. *Then* $\alpha \xrightarrow{c} \beta$ *implies* $\alpha \xrightarrow{m} \beta$. *Moreover, if* $4, B \in \Sigma$, *then* $\alpha \xrightarrow{c} \beta$ *implies* $\alpha \xrightarrow{mm} \beta$.

Lemma 3.13 (RCA$_0$). *The following hold:*

1. $\xrightarrow{m}$ *is transitive.*
2. *if* $T \in \Sigma$ *then* $\xrightarrow{m}$ *is reflexive.*
3. *if* $D \in \Sigma$ *then* $\xrightarrow{m}$ *is serial.*

Lemma 3.14 (RCA$_0$). *The following hold:*

1. $\xrightarrow{mm}$ *is symmetric and transitive.*
2. *if* $T \in \Sigma$ *then* $\xrightarrow{mm}$ *is reflexive.*
3. *if* $D \in \Sigma$ *then* $\xrightarrow{mm}$ *is serial.*

Lemma 3.15 (Generalized truth lemma, RCA$_0$**).** *Let* $\varphi \in \mathcal{L}_{h,n}$ *be* **K**Σ-*consistent. Assume that* $R \subseteq \tilde{W} \times \tilde{W}$ *satisfies that* $\alpha \xrightarrow{c} \beta$ *implies* $\alpha R \beta$ *and* $\alpha R \beta$ *implies* $\mathrm{Pbl}_{\mathbf{K}\Sigma}(\emptyset, \alpha \to \Diamond\beta')$. *Then* $(\tilde{W}, R, \tilde{V})$ *is a weak Kripke model.*

From the above, we now see the weak completeness theorem for the case of $4 \in \Sigma$ in RCA$_0$.

Proposition 3.16 (RCA$_0$). *Let* $\Sigma \subseteq \{T, 4, B, D\}$ *and* $4 \in \Sigma$. *If* $\varphi \in \mathrm{Fml}$ *is* **K**Σ-*consistent, then there exist a frame* F *which is appropriate to* Σ *and a valuation* V *such that* (F, V) *is a weak model of* φ.

Proof. Assume that $\varphi \in \mathcal{L}_{h,n}$. By Definition 3.10 and Definition 3.11, we define

- Case1: If $B \notin \Sigma$, then we define $F = (\tilde{W}, \xrightarrow{m})$ and $V = \tilde{V}$.
- Case2: If $B \in \Sigma$, then we define $F = (\tilde{W}, \xrightarrow{mm})$ and $V = \tilde{V}$.

By Generalized truth lemma, (F, V) is a weak model of φ.

Summarizing, we have the following weak completeness theorem with the finite model property.

Theorem 3.17 (Weak completeness theorem, RCA$_0$**).** *Let* $\Sigma \subseteq \{T, 4, B, D\}$. *If* $\varphi \in \mathrm{Fml}$ *is* **K**Σ-*consistent, then there exist a finite frame* F *which is appropriate to* Σ *and a valuation* V *such that* (F, V) *is a weak model of* φ.

4 Strong Completeness Theorem Using Canonical Models in Second-Order Arithmetic

In this section, we will prove that the strong completeness theorem for modal logic using canonical models is equivalent to ACA_0 over RCA_0. In this section, we will assume $\Sigma \subseteq \{T, B, 4, 5, D, .2\}$ unless otherwise specified.

Basically, we still formalize the strong completeness proofs using canonical models. However, we cannot collect all completions of a given set of formulas within second-order arithmetic. Therefore, we use the well-known idea of the "low completion" from the computability theory, and make it possible to perform the construction in ACA_0.

In ACA_0, we can define some notions for computability theory with Turing jumps (see Simpson [3] for the details). Let $X, Y \subseteq \mathbb{N}$. Then, Y is said to be *low relative to* X if $\mathrm{TJ}(Y \oplus X) \leq_T \mathrm{TJ}(X)$ (here $\mathrm{TJ}(X)$ denotes the Turing jump of X and $Y \oplus X$ denotes the recursive join of Y and X as in [3]). If $Y \leq_T \mathrm{TJ}(X)$, we may consider a $\Delta_2^{0,X}$-index of Y (obtained as a $\mathrm{TJ}(X)$-recursive index of Y). Moreover, the statement "e is a $\mathrm{TJ}(X)$-recursive index of a set" is arithmetical, and in such a case, we write $\Delta_2^{0,X}[e]$ for the set indexed by e.

Definition 4.1 (ACA_0). *Let $X \subseteq \mathbb{N}$. We define the set $L(X)$ by the following:*

$$L(X) = \{e \in \mathbb{N} \mid e \textit{ is a } \Delta_2^{0,X}\textit{-index and } \Delta_2^{0,X}[e] \textit{ is a low set relative to } X.\}$$

Note that within ACA_0 the set $\Delta_2^{0,X}[e]$ always exists, and thus we usually identify the set $\Delta_2^{0,X}[e]$ and its index e.

Lemma 4.2 (Low Lindenbaum's Lemma, ACA_0**).** *If $\Gamma \subseteq \mathrm{Fml}$ is $\mathbf{K}\Sigma$-consistent, then there exists a low relative to Γ set $\Gamma^+ \supseteq \Gamma$ which is maximally $\mathbf{K}\Sigma$-consistent.*

Proof. By Low Basis Theorem and Lindenbaum's lemma. See Simpson [3].

Lemma 4.3 (ACA_0). *Let ψ be a formula, Γ be $\mathbf{K}\Sigma$-consistent, and E be the maximally $\mathbf{K}\Sigma$-consistent low relative to Γ set. If $\Box\psi \notin E$, then there exists a maximally $\mathbf{K}\Sigma$-consistent low relative to Γ set $\tilde{E}$ such that $\{\neg\psi\} \cup \{\xi | \Box\xi \in E\} \cup \Gamma \subseteq \tilde{E}$.*

Proof. Suppose $\Box\psi \notin E$. We show that $\{\neg\psi\} \cup \{\xi \mid \Box\xi \in E\} \cup \Gamma$ is $\mathbf{K}\Sigma$-consistent. Assume that $\{\neg\psi\} \cup \{\xi \mid \Box\xi \in E\} \cup \Gamma$ is $\mathbf{K}\Sigma$-inconsistent, then there exist some $\Box\xi_1, \dots \Box\xi_n \in E$ such that $\mathrm{Pbl}_{\mathbf{K}\Sigma}(\Gamma, (\xi_1 \wedge \cdots \wedge \xi_n \wedge \neg\psi) \to \bot)$, that is, $\mathrm{Pbl}_{\mathbf{K}\Sigma}(\Gamma, (\xi_1 \wedge \cdots \wedge \xi_n) \to \psi)$ holds. Thus $\mathrm{Pbl}_{\mathbf{K}\Sigma}(\Gamma, (\Box\xi_1 \wedge \cdots \wedge \Box\xi_n) \to \Box\psi)$ holds. Thus $\Box\psi \in E$, it is a contradiction. Therefore by Low Lindenbaum's Lemma, there exists a low relative to Γ set $\tilde{E}$ such that $\{\neg\psi\} \cup \{\xi | \Box\xi \in E\} \cup \Gamma \subseteq \tilde{E}$.

Definition 4.4 (ACA_0). *Let Γ be $\mathbf{K}\Sigma$-consistent. A $\mathbf{K}\Sigma$-model $M^{\mathbf{K}\Sigma}$ is a tuple $(W^{\mathbf{K}\Sigma}, R^{\mathbf{K}\Sigma}, V^{\mathbf{K}\Sigma})$ satisfying the following conditions:*

- $W^{\mathbf{K}\Sigma} = \{e \in L(\Gamma) \mid \Gamma \subseteq \Delta_2^{0,\Gamma}[e]$ *and* $\Delta_2^{0,\Gamma}[e]$ *is maximally* $\mathbf{K}\Sigma$*-consistent.*$\}$,
- $R^{\mathbf{K}\Sigma} = \{(e, \tilde{e}) \in W^{\mathbf{K}\Sigma} \times W^{\mathbf{K}\Sigma} \mid \forall \varphi \in \mathrm{Fml}(\Box\varphi \in \Delta_2^{0,\Gamma}[e] \to \varphi \in \Delta_2^{0,\Gamma}[\tilde{e}]\}$,
- $V^{\mathbf{K}\Sigma}(e, \varphi) = 1 \iff \varphi \in \Delta_2^{0,\Gamma}[e]$ *for any* $e \in W^{\mathbf{K}\Sigma}$ *and any* $\varphi \in \mathrm{Fml}$.

To simplify the notations, we may write $E, E', \tilde{E}, \ldots$ for the sets $\Delta_2^{0,\Gamma}[e]$, $\Delta_2^{0,\Gamma}[e']$, $\Delta_2^{0,\Gamma}[\tilde{e}], \ldots$.

Proposition 4.5 (RCA$_0$)**.** *Let* $\varphi, \psi \in \mathrm{Fml}$. *Let* E *be maximally* $\mathbf{K}\Sigma$*-consistent. Then the following holds:*

1. $\neg\varphi \in E \iff \varphi \notin E$,
2. $\varphi \to \psi \in E \iff (\varphi \notin E \vee \psi \in E)$,
3. $\varphi \wedge \psi \in E \iff \varphi \in E \wedge \psi \in E$.

Proof. It is clear since E is $\mathbf{K}\Sigma$-consistent.

Lemma 4.6 (ACA$_0$)**.** *Let* $W = W^{\mathbf{K}\Sigma}$ *and* $R = R^{\mathbf{K}\Sigma}$. *Then* (W, R) *is appropriate to* Σ.

Proof. Fix $\varphi \in \mathrm{Fml}$.

(T): Let $e \in W$. Assume that $\Box\varphi \in E$. Then by $T \in \Sigma$, that is, $T \in E$, $\Box\varphi \to \varphi \in E$, so $\varphi \in E$. Thus eRe holds.

(B): Let $eR\tilde{e}$. Assume that $\varphi \in E$, then by $B \in \Sigma$, that is, $B \in E$, $\varphi \to \Box\Diamond\varphi \in E$, so $\Box\Diamond\varphi \in E$. By $eR\tilde{e}$, $\Diamond\varphi \in \tilde{E}$. Thus $\tilde{e}Re$ holds.

(4): Let $e_0, e_1, e_2 \in W$ such that e_0Re_1 and e_1Re_2. Assume that $\Box\varphi \in E_0$, then by $4 \in \Sigma$, that is, $4 \in E_0$, so $\Box\Box\varphi \in E_0$. Then $\varphi \in E_2$. Thus e_0Re_2 holds.

(5): Let $e_0, e_1, e_2 \in W$ such that e_0Re_1 and e_0Re_2. Assume that $\varphi \in E_2$, then $\Diamond\varphi \in E_0$. By $5 \in \Sigma$, that is, $5 \in E_0$, $\Box\Diamond\varphi \in E_0$. Thus $\Diamond\varphi \in E_1$. Thus e_1Re_2 holds.

(D): Let $e \in W$. Note that $\top \in E$. Then $\Box\neg\bot \in E$. By $D \in \Sigma$, that is, $D \in E$, $\neg\Box\bot \in E$. Then $\Box\bot \notin E$. By Lemma 4.3, there exists $\tilde{e} \in W$ such that $\{\top\} \cup \{\xi \mid \Box\xi \in E\} \cup \Gamma \subseteq \tilde{E}$. Thus $eR\tilde{e}$ holds.

(.2): Let $e_0, e_1, e_2 \in W$ such that e_0Re_1 and e_0Re_2. We show that $\{\xi \mid \Box\xi \in E_1\} \cup \{\delta \mid \Box\delta \in E_2\}$ is $\mathbf{K}\Sigma$-consistent. Assume that $\{\xi \mid \Box\xi \in E_1\} \cup \{\delta \mid \Box\delta \in E_2\}$ is $\mathbf{K}\Sigma$-inconsistent, there exist some $\Box\xi_1, \ldots, \Box\xi_n \in E_1$ and $\Box\delta_1, \ldots, \Box\delta_m \in E_2$ such that $\mathrm{Pbl}_{\mathbf{K}\Sigma}(\emptyset, (\xi_1 \wedge \cdots \wedge \xi_n \wedge \delta_1 \wedge \cdots \wedge \delta_m) \to \bot)$ holds. Since $\Box(\xi_1 \wedge \cdots \wedge \xi_n) \in E_1$ and e_0Re_1, $\Diamond\Box(\xi_1 \wedge \cdots \wedge \xi_n) \in E_0$. On the other hand, $\mathrm{Pbl}_{\mathbf{K}\Sigma}(\emptyset, (\xi_1 \wedge \cdots \wedge \xi_n) \to \neg(\delta_1 \wedge \cdots \wedge \delta_m))$ holds, so $\mathrm{Pbl}_{\mathbf{K}\Sigma}(\emptyset, \Diamond\Box(\xi_1 \wedge \cdots \wedge \xi_n) \to \Diamond\Box\neg(\delta_1 \wedge \cdots \wedge \delta_m))$ holds. Thus $\mathrm{Pbl}_{\mathbf{K}\Sigma}(\emptyset, \Diamond\Box(\xi_1 \wedge \cdots \wedge \xi_n) \to \neg\Box\Diamond(\delta_1 \wedge \cdots \wedge \delta_m))$ holds, so we have $\neg\Box\Diamond(\delta_1 \wedge \cdots \wedge \delta_m) \in E_0$. However, by $.2 \in \Sigma$, that is, $.2 \in E_0$ and $\Diamond\Box(\delta_1 \wedge \cdots \wedge \delta_m) \in E_0$, $\Box\Diamond(\delta_1 \wedge \cdots \wedge \delta_m) \in E_0$, it is a contradiction that E_0 is $\mathbf{K}\Sigma$-consistent. Thus $\{\xi \mid \Box\xi \in E_1\} \cup \{\delta \mid \Box\delta \in E_2\}$ is $\mathbf{K}\Sigma$-consistent. By Lemma 4.3, there exists $e_3 \in W$ such that $\{\xi \mid \Box\xi \in E_1\} \cup \{\delta \mid \Box\delta \in E_2\} \cup \Gamma \subseteq E_3$. So e_1Re_3 and e_2Re_3 hold.

Lemma 4.7 (Truth Lemma, ACA$_0$)**.** *The truth valuation* $V^{\mathbf{K}\Sigma}$ *satisfies the following conditions:*

- $V^{\mathbf{K}\Sigma}(w, \bot) = 0$ *for any* $w \in W^{\mathbf{K}\Sigma}$,
- $V^{\mathbf{K}\Sigma}(w, \varphi \to \psi) = 1 - V^{\mathbf{K}\Sigma}(w, \varphi)(1 - V^{\mathbf{K}\Sigma}(w, \psi))$ *for any* $w \in W^{\mathbf{K}\Sigma}$ *and any* $\varphi, \psi \in \mathrm{Fml}$,
- $V^{\mathbf{K}\Sigma}(w, \Box\varphi) = 1 \iff \forall v \in W^{\mathbf{K}\Sigma}(wR^{\mathbf{K}\Sigma}v \to V^{\mathbf{K}\Sigma}(v, \varphi) = 1)$ *for any* $w \in W^{\mathbf{K}\Sigma}$ *and any* $\varphi \in \mathrm{Fml}$.

Proof. We show this lemma by induction on the length of φ using ACA_0. Fix $e \in W$. If $\varphi \equiv \bot$, it is clear.

Assume that $\varphi \equiv \psi \to \theta$. Then one of (1) $V^{\mathbf{K}\Sigma}(e, \psi) = 0 \iff \psi \notin E$ or (2) $V^{\mathbf{K}\Sigma}(e, \theta) = 1 \iff \theta \in E$ or (3) $(V^{\mathbf{K}\Sigma}(e, \psi) = 1 \land V^{\mathbf{K}\Sigma}(e, \theta) = 0) \iff (\psi \in E \land \theta \notin E)$ is satisfiable. If either (1) or (2) holds, then by Proposition 4.5, $\psi \to \theta \in E \iff V^{\mathbf{K}\Sigma}(e, \psi \to \theta) = 1$. If (3) holds, then by Proposition 4.5, $\psi \to \theta \notin E \Leftrightarrow V^{\mathbf{K}\Sigma}(e, \psi \to \theta) = 0$. Thus $V^{\mathbf{K}\Sigma}(w, \psi \to \theta) = 1 - V^{\mathbf{K}\Sigma}(w, \psi)(1 - V^{\mathbf{K}\Sigma}(w, \theta))$ holds.

Assume that $\varphi \equiv \Box\psi$. We have two cases. First, we assume that $V^{\mathbf{K}\Sigma}(e, \Box\psi) = 0$. Then $\Box\psi \notin E$. Then by Lemma 4.3, there exists $\tilde{e} \in W$ such that $\{\neg\psi\} \cup \{\xi | \Box\xi \in E\} \cup \Gamma \subseteq \tilde{E}$. Thus by the definition of $R^{\mathbf{K}\Sigma}$, $eR^{\mathbf{K}\Sigma}\tilde{e}$ and $\neg\psi \in \tilde{E}$, so $\psi \notin \tilde{E}$ holds. Thus $V^{\mathbf{K}\Sigma}(\tilde{e}, \psi) = 0$. The other case is when $V^{\mathbf{K}\Sigma}(e, \Box\psi) = 1$, that is, $\Box\psi \in E$ holds. Fix $\tilde{e} \in W$ with $eR^{\mathbf{K}\Sigma}\tilde{e}$. Then $\psi \in \tilde{E}$, that is, $V^{\mathbf{K}\Sigma}(\tilde{e}, \psi) = 1$.

From the above, we obtain the strong completeness theorem using a canonical model.

Theorem 4.8 (ACA_0). *We assume that* $\Sigma \subsetneq \{T, B, 4, 5, D, .2\}$. *Let* Γ *be a set of formulas. If* Γ *is* $\mathbf{K}\Sigma$*-consistent, then there exist a frame* $F = (W, R)$ *which is appropriate to* Σ *and a valuation* V *such that* $M = (F, V)$ *is a model of* Γ *and the following holds:*

$$\forall\varphi \in \mathrm{Fml}(\mathrm{Pbl}_{\mathbf{K}\Sigma}(\Gamma, \varphi) \leftrightarrow \forall e \in W(V(e, \varphi) = 1)).$$

Proof. Let $W = W^{\mathbf{K}\Sigma}$, $R = R^{\mathbf{K}\Sigma}$, and $V = V^{\mathbf{K}\Sigma}$ by Definition 4.4. This is the desired Kripke model by Truth Lemma. Moreover, it is clear that $\forall\varphi \in \Gamma\forall e \in W(V(e, \varphi) = 1)$. Let $\varphi \in \mathrm{Fml}$. We have the two cases. If $\mathrm{Pbl}_{\mathbf{K}\Sigma}(\Gamma, \varphi)$, then $\forall e \in W(V(e, \varphi) = 1)$ holds by the soundness theorem. If $\neg\mathrm{Pbl}_{\mathbf{K}\Sigma}(\Gamma, \varphi)$, then by Low Lindenbaum's Lemma, there exists $e \in W$ such that $\neg\varphi \in E$, that is, $\varphi \notin E$. Then $V(e, \varphi) = 0$.

Theorem 4.9. *The following statements are equivalent over* RCA_0*:*

1. ACA_0,
2. *Strong Completeness Theorem using canonical models.*

Proof. (1) implies (2) by Theorem 4.8. We show that (2) implies (1). It is enough to show Σ^0_1-CA. Let $\varphi(n) \equiv \exists m\theta(n, m)$ be Σ^0_1-formula ($\theta(n, m)$ is a Σ^0_0-formula). Let $\alpha_0, \alpha_1, \ldots$ be an enumeration of C. Then we define $\Gamma = \{\alpha_n \vee \cdots \vee \alpha_n \mid \exists m \leq \langle\alpha_n \vee \cdots \vee \alpha_n\rangle\theta(n, m)\}$. Any finite subset of Γ has a model, so by Soundness Theorem, any finite subset of Γ is $\mathbf{K}\Sigma$-consistent. Thus Γ is $\mathbf{K}\Sigma$-consistent.

Then by Strong Completeness Theorem using canonical models and Soundness Theorem, the following holds.

$$\varphi(n) \leftrightarrow \alpha_n \vee \cdots \vee \alpha_n \in \Gamma \leftrightarrow \mathrm{Pbl}_{\mathbf{K}\Sigma}(\Gamma, \alpha_n) \leftrightarrow \forall w \in W(V(w, \alpha_n) = 1)$$

Thus by Δ^0_1-CA, there exists a set X such that $\forall n(n \in X \leftrightarrow \varphi(n))$ holds. Thus Σ^0_1-CA holds.

5 Strong Completeness Theorem in Second-Order Arithmetic

We showed Theorem 4.8 by constructing a canonical model for a given $\mathbf{K}\Sigma$-consistent set Γ. On the other hand, we consider the following statement: for a given $\mathbf{K}\Sigma$-consistent set $\Gamma \cup \{\psi\}$, there exists a model for $\Gamma \cup \{\psi\}$. This is a weaker formulation of the strong completeness theorem. First of all, we give a natural first-order logic language $\mathcal{L}_{\mathcal{FO}}$ for the Kripke model of modal logic. Using the standard translation, we aim to obtain the Kripke model from the properties of first-order logic.

Definition 5.1 (RCA_0). *Let $\mathcal{L}_{\mathcal{FO}}$ be the first-order language (with equality) which has unary predicates $P_0, P_1, \ldots$ corresponding to the proposition letters $p_0, p_1, \ldots$, and 2-ary reation symbol r.*

Then, as in the case of modal logic, we can show the existence of the set of $\mathcal{L}_{\mathcal{FO}}$-formulas and axioms of first-order logic in RCA_0, i.e., we can show in RCA_0 that there exist sets $\mathrm{Trm}_{\mathcal{FO}}$, $\mathrm{Fml}_{\mathcal{FO}}$, $\mathrm{Snt}_{\mathcal{FO}}$, and $\mathrm{Axm}_{\mathcal{FO}}$ consisting of all $\mathcal{L}_{\mathcal{FO}}$-terms, $\mathcal{L}_{\mathcal{FO}}$-formulas, $\mathcal{L}_{\mathcal{FO}}$-sentences, and axioms of first-order logic in $\mathcal{L}_{\mathcal{FO}}$ respectively. Moreover, as in the case of modal logic, we can define the provability predicate of first-order logic in RCA_0. We assume that the only logical rule is modus ponens (See Simpson [3]).

Definition 5.2 (RCA_0). *Let $\Sigma \subseteq \{T, B, 4, D, 5, .2\}$. We define the function $A : \Sigma \to \mathrm{Snt}_{\mathcal{FO}}$ by the following:*

$$A(\psi) = \begin{cases} \forall x r(x,x) & \text{if } \psi \equiv T, \\ \forall x \forall y (r(x,y) \to r(y,x)) & \text{if } \psi \equiv B, \\ \forall x \forall y \forall z (r(x,y) \wedge r(y,z) \to r(x,z)) & \text{if } \psi \equiv 4, \\ \forall x \exists y r(x,y) & \text{if } \psi \equiv D, \\ \forall x \forall y \forall z (r(x,y) \wedge r(x,z) \to r(y,z)) & \text{if } \psi \equiv 5, \\ \forall x \forall y \forall z (r(x,y) \wedge r(x,z) \to \exists w (r(y,w) \wedge r(z,w))) & \text{if } \psi \equiv .2. \end{cases}$$

Moreover, we define $\Sigma_{\mathcal{FO}} = \{A(\psi) \mid \psi \in \Sigma\}$.

Definition 5.3 (RCA_0). *Let $X \subset \mathrm{Fml}_{\mathcal{FO}}$ and $\Sigma \subseteq \{T, B, 4, D, 5, .2\}$. We will define the following formulas:*

$$\begin{aligned}\mathrm{Prf}_{\mathcal{FO}(\Sigma)}(X,p) \equiv\ & p \in \mathrm{Seq} \wedge \forall k(k < lh(p) \to p(k) \in \mathrm{Fml}_{\mathcal{FO}}) \\ & \wedge \forall k\Big(k < lh(p) \to \Big(p(k) \in X \vee p(k) \in \mathrm{Axm}_{\mathcal{FO}} \vee p(k) \in \Sigma_{\mathcal{FO}} \\ & \vee (\exists i < k \exists j < k)(p(i) = p(j) \to p(k))\Big)\Big),\end{aligned}$$

$$\begin{aligned}\mathrm{Pbl}_{\mathcal{FO}(\Sigma)}(X,\varphi) \equiv & \exists \psi_1, \ldots, \psi_n \in X \exists p \\ & \big(\mathrm{Prf}_{\mathcal{FO}(\Sigma)}(X,p) \wedge (\exists i < lh(p))(p(i) = (\psi_1 \wedge \cdots \wedge \psi_n \to \varphi))\big).\end{aligned}$$

Furthermore, as in the case of modal logic, we can use RCA_0 to describe the consistency of the system and the satisfiability of the model in first-order logic.

Definition 5.4 (RCA_0). *Let M be a set. Let T_M and S_M be respectively the sets of closed terms and sentences of the expanded language $\mathcal{L}^M_{\mathcal{FO}} = \mathcal{L}_{\mathcal{FO}} \cup \{\overline{m} \mid m \in M\}$ with new constant symbols $\overline{m}$ for each element $m \in M$.*

$\mathcal{M} = (M, v)$ is a standard model on $\mathcal{L}_{\mathcal{FO}}$ if the function $v : T_M \cup S_M \to M \cup \{0,1\}$ satisfies the following conditions:

- $\forall t \in T_M(v(t) \in M)$,
- $\forall \sigma \in S_M(v(\sigma) \in \{0,1\})$,
- $\forall t_1, \forall t'_1, \ldots \forall t_n, \forall t'_n \in T_M(\forall i \leq n(v(t_i) = v(t'_i)) \to v(R(t_1, \ldots, t_n)) = v(R(t'_1, \ldots, t'_n)))$, *where R is a relation symbol,*
- $v(\neg\sigma) = 1 - v(\sigma)$,
- $v(\sigma_0 \wedge \sigma_1) = v(\sigma_0) \cdot v(\sigma_1)$,
- $v(\forall x \varphi(x)) = 1 \iff \forall a \in M(v(\varphi(\overline{a})) = 1)$,
- $v(\exists x \varphi(x)) = 1 \iff \exists a \in M(v(\varphi(\overline{a})) = 1)$.

Let $X \subseteq \mathrm{Snt}_{\mathcal{FO}}$. Then a standard model $\mathcal{M}$ is a model of X on $\mathcal{L}_{\mathcal{FO}}$ if $\forall \varphi \in X(v(\varphi) = 1)$ holds.

Definition 5.5 (RCA_0). *Let $\varphi \in \mathrm{Snt}_{\mathcal{FO}}$ and $X \subseteq \mathrm{Snt}_{\mathcal{FO}}$. Let $\mathcal{M} = (M, v)$ be a standard model on $\mathcal{L}_{\mathcal{FO}}$. We write $\mathcal{M} \vDash \varphi$ if M satisfies φ, i.e., $v(\varphi) = 1$, and we write $X \vDash \varphi$ if every model of X on $\mathcal{L}_{\mathcal{FO}}$ satisfies φ.*

Theorem 5.6 (Soundness Theorem for first-order logic, RCA_0[3]**).** *Let $\Sigma \subseteq \{T, B, 4, D, 5, .2\}$ and $\varphi \in \mathrm{Snt}_{\mathcal{FO}}$. If $\mathrm{Pbl}_{\mathcal{FO}(\Sigma)}(\emptyset, \varphi)$ holds, then $\Sigma_{\mathcal{FO}} \vDash \varphi$ holds.*

Theorem 5.7 (Gödel's Completeness Theorem, WKL_0 [3]**).** *Let $\varphi \in \mathrm{Snt}_{\mathcal{FO}}$ and $\Sigma \subseteq \{T, B, 4, D, 5, .2\}$. If $X \subseteq \mathrm{Snt}_{\mathcal{FO}}$ is consistent, then $\mathrm{Pbl}_{\mathcal{FO}(\Sigma)}(X, \varphi)$ holds if and only if $X \cup \Sigma_{\mathcal{FO}} \vDash \varphi$ holds.*

Theorem 5.8 ([3])**.** *The following statements are equivalent over* RCA_0*:*

1. WKL_0,
2. *Gödel's Completeness Theorem,*

3. The completeness theorem for propositional logic with countably many atoms.

The following is essentially a first-order reformulation of the modal satisfaction definition.

Definition 5.9 (RCA$_0$ [1]). *Let x be a first-order variable. The* standard translation ST_x *taking modal formulas to $\mathcal{L}_{\mathcal{FO}}$-formulas is defined as follows:*

- $ST_x(p_n) \equiv P_n(x)$,
- $ST_x(\bot) \equiv \bot$,
- $ST_x(\varphi \to \psi) \equiv ST_x(\varphi) \to ST_x(\psi)$,
- $ST_x(\Box\varphi) \equiv \forall y(r(x,y) \to ST_y(\varphi))$ *(y is fresh).*

Lemma 5.10 (RCA$_0$). *Let $\varphi \in$ Fml. For all $\mathcal{M} = (M, v)$ which is a standard model on $\mathcal{L}_{\mathcal{FO}}$, there exists a Kripke model $M^\star$ such that*

1. *for all $w \in M$, $M^\star, w \Vdash \varphi \iff \mathcal{M} \models ST_w(\varphi)$, and*
2. *$M^\star \Vdash \varphi \iff \mathcal{M} \models \forall x ST_x(\varphi)$.*

Proof. We define $M^\star = (W^\star, R^\star, V^\star)$ by the following:

- $W^\star = M$,
- $R^\star = r$,
- $V^\star(w,\varphi) = v(ST_w(\varphi))$ for all $w \in W^\star$ and $\varphi \in$ Fml.

We can show that $V^\star$ is a valuation with $F^\star = (W^\star, R^\star)$ by the definition of v and the standard translation. For example, for all $w \in W^\star$ and $\Box\varphi \in$ Fml,

$$\begin{aligned} V^\star(w,\Box\varphi) = 1 &\iff v(ST_w(\Box\varphi)) = 1 \\ &\iff v(\forall u(r(w,u) \to ST_u(\varphi))) = 1 \\ &\iff \forall u \in M(v(r(w,u) \to ST_u(\varphi)) = 1) \\ &\iff \forall u \in M(v(r(w,u)) \cdot (1 - v(ST_u(\varphi))) = 0) \\ &\iff \forall u \in M(v(r(w,u)) = 0 \vee v(ST_u(\varphi)) = 1) \\ &\iff \forall u \in W^\star(wRu \to V^\star(u,\varphi) = 1). \end{aligned}$$

Thus (1) holds by the definition of $V^\star$. (2) also holds from (1).

Lemma 5.11 (RCA$_0$). *Let $\Sigma \subseteq \{T, B, 4, D\}$ and $\Gamma \subseteq$ Fml be* **K**Σ*-consistent. Then 1 implies 2.*

1. $\mathrm{Pbl}_{\mathcal{FO}(\Sigma)}(\{\forall x ST_x(\theta) \mid \theta \in \Gamma\}, \forall x ST_x(\varphi))$,
2. $\mathrm{Pbl}_{\mathbf{K}\Sigma}(\Gamma, \varphi)$.

Proof. Let $\Gamma^* = \{\forall x ST_x(\theta) \mid \theta \in \Gamma\}$ and $\varphi^* \equiv \forall x ST_x(\varphi)$. Assume that $\neg\mathrm{Pbl}_{\mathbf{K}\Sigma}(\Gamma, \varphi)$ and $\mathrm{Pbl}_{\mathcal{FO}(\Sigma)}(\Gamma^*, \varphi^*)$ holds. Then there exists a finite set $\Gamma_0 \subseteq \Gamma$ such that $\mathrm{Pbl}_{\mathcal{FO}(\Sigma)}(\Gamma_0^*, \varphi^*)$ holds. By the assumption, $\neg\mathrm{Pbl}_{\mathbf{K}\Sigma}(\Gamma_0, \varphi)$ holds. Since $\bigwedge\Gamma_0 \wedge \neg\varphi$ is **K**Σ-consistent, by Weak Completeness Theorem, there exist a finite frame $F = (W, R)$ which is appropriate to Σ and a valuation V such that (F, V) is a weak model of $\bigwedge\Gamma_0 \wedge \neg\varphi$. Then we can define a model of $\bigwedge\Gamma_0^* \wedge \neg\varphi^*$ on $\mathcal{L}_{\mathcal{FO}}$ from (F, V). It is a contradiction by Soundness Theorem for first-order logic.

Theorem 5.12 (Strong Completeness Theorem, WKL_0**).** *Suppose* $\Sigma \subseteq \{T, B, 4, D\}$. *Let* Γ *be a set of formulas. Let* $\varphi \in \mathrm{Fml}$. *If* Γ *is* $\mathbf{K}\Sigma$*-consistent, then* $\mathrm{Pbl}_{\mathbf{K}\Sigma}(\Gamma, \varphi)$ *holds if and only if* $\Gamma \Vdash_{\mathbb{F}_\Sigma} \varphi$ *holds.*

Proof. By Soundness Theorem (for modal logic), it is enough to show that $\neg\mathrm{Pbl}_{\mathbf{K}\Sigma}(\Gamma, \varphi)$ implies $\Gamma \nVdash_{\mathbb{F}_\Sigma} \varphi$. Assume that $\neg\mathrm{Pbl}_{\mathbf{K}\Sigma}(\Gamma, \varphi)$. Let $\Gamma^* = \{\forall x ST_x(\theta) \mid \theta \in \Gamma\}$ and $\varphi^* \equiv \forall x ST_x(\varphi)$. By Lemma 5.11, $\neg\mathrm{Pbl}_{\mathcal{FO}(\Sigma)}(\Gamma^*, \varphi^*)$ holds. Then Γ^* is consistent since Γ is $\mathbf{K}\Sigma$-consistent. Thus by Gödel's Completeness Theorem, there exists a standard model $\mathcal{M} = (M, v)$ such that $\mathcal{M}$ is a model of $\Gamma^* \cup \{\neg\varphi^*\}$ on $\mathcal{L}_{\mathcal{FO}}$. Therefore by Lemma 5.10, there exists a Kripke model $M^\star = (F^\star, V^\star)$ such that $M^\star \Vdash \Gamma \cup \{\neg\varphi\}$. Moreover, for all $\Psi \in \Sigma_{\mathcal{FO}}$, $\mathcal{M} \models \Psi$, so $F^\star \Vdash \Sigma$. Thus $\Gamma \nVdash_{\mathbb{F}_\Sigma} \varphi$.

Theorem 5.13. *The following statements are equivalent over* RCA_0*:*

1. WKL_0,
2. *Strong Completeness Theorem.*

Proof. (1) implies (2) from Theorem 5.12. Clearly, Strong Completeness Theorem implies the completeness theorem for propositional logic with countably many atoms. Thus by Theorem 5.8, we have WKL_0.

6 Open Questions

Some questions remain open concerning the weak completeness theorem in second-order arithmetic. In particular, the completeness with respect to the modal logic cube is not yet fully known. We showed Weak Completeness Theorem for the case of $\Sigma \subseteq \{T, B, 4, D\}$ in RCA_0. However, we have not shown Weak Completeness Theorem for the case of $5 \in \Sigma$:

Question 6.1. *Is Weak Completeness Theorem for the case of* $5 \in \Sigma$ *provable in* RCA_0*?*

In fact, the following weak model $M = (W, \xrightarrow{s}, V)$ is the model for cases $\Sigma = \{4, 5\}$, $\Sigma = \{4, 5, D\}$.

- $W = \{\alpha \in C_{h+1,n} \mid \neg\mathrm{Pbl}_{\mathbf{K}\Sigma}(\alpha, \bot) \wedge \mathrm{Pbl}_{\mathbf{K}\Sigma}(\emptyset, \alpha \to \varphi)\}$,
- $\xrightarrow{s} = \{(\alpha, \beta) \in W \times W \mid \alpha \xrightarrow{m} \beta \wedge \forall \xi \in C_{h,n}(\mathrm{Pbl}_{\mathbf{K}\Sigma}(\emptyset, \alpha \to \Diamond\xi) \to \mathrm{Pbl}_{\mathbf{K}\Sigma}(\emptyset, \beta \to \Diamond\xi))\}$,
- For all $\psi \in sub(\varphi)$ and $\alpha \in W$, $V(\alpha, \psi) = 1 \iff \mathrm{Pbl}_{\mathbf{K}\Sigma}(\emptyset, \alpha \to \psi)$.

However, it is known that in the case of $5 \in \Sigma$ and $4 \notin \Sigma$, this model does not work for $h = 0$ and $n = 1$ by Moss [2]. We expect that the $\Sigma = \{5\}$ and $\Sigma = \{5, D\}$ cases will force us to construct different models.

Question 6.2. *Is Weak Completeness Theorem for the case of* $\Sigma = \{5\}$ *and* $\Sigma = \{5, D\}$ *provable in* RCA_0*?*

The modal logic **S4.2** is the smallest normal modal logic which contains $T, 4$, and .2. **S4.2** is the important object that appears in the analysis of modal logic using the forcing method.

Question 6.3. *Is Weak Completeness Theorem for* **S4.2** *provable in* RCA_0*?*

When weak completeness holds for a formula, which is important in the analysis of modal logic, we expect most of them to be provable in RCA_0.

Acknowledgements. Takeda is supported by JST SPRING, Grant Number JPMJSP2114. The third author is partially supported by JSPS KAKENHI grant numbers JP21KK0045 and JP23K03193.

References

1. Blackburn, P., de Rijke, M., Venema, Y.: Modal Logic. Cambridge Tracts in Theoretical Computer Science, vol. 53. Cambridge University Press, Cambridge (2001)
2. Moss, L.S.: Finite models constructed from canonical formulas. J. Philos. Logic **36**(6), 605–640 (2007)
3. Simpson, S.G.: Subsystems of second order arithmetic. Perspectives in Logic, 2nd edn. Cambridge University Press, Cambridge; Association for Symbolic Logic, Poughkeepsie (2009)
4. Yamazaki, T.: Reverse mathematics and completeness theorems for intuitionistic logic. Notre Dame J. Formal Logic **42**(3), 143–148 (2001)

Variants of Solovay Reducibility

Ivan Titov[1,2(✉)]

[1] Universität Heidelberg, Institut für Informatik, 69120 Heidelberg, Germany
me@ivantitov.de

[2] Université de Bordeaux, CNRS, Bordeaux INP, LaBRI, UMR 5800, 33400 Talence, France

Abstract. Solovay reducibility was introduced by Robert M. Solovay [6] in 1975 in terms of translation functions on rationals. It is a measure of relative approximation speed and thus also of relative randomness of reals. In the area of algorithmic randomness, Solovay reducibility has been intensively studied and several central results on left-c.e. reals have been obtained.

Outside of the left-c.e. reals, Solovay reducibility is considered to be behaved badly [2]. Proposals for variants of Solovay reducibility that are better suited for the investigation of arbitrary, not necessarily left-c.e. reals were made by Rettinger and Zheng [9], and, recently, by Titov [7] and by Kumabe and co-authors [3,4]. These variants all coincide with the original version of Solovay reducibility on the left-c.e. reals. Furthermore, they are all defined in terms of translation functions. The latter translate between computable approximations in the case of Rettinger and Zheng, are monotone in the case of Titov, and are functions between reals in the case of Kumabe et al.

In what follows, we derive new results on the mentioned variants and their relation to each other. In particular, we obtain that Solovay reducibility defined in terms of translation function on rationals implies Solovay reducibility defined in terms of translation functions on reals, and we show that the original version of Solovay reducibility is strictly weaker than its monotone variant.

Solovay reducibility and its variants mentioned so far have tight connections to Martin-Löf randomness, the strongest and most central notion of a random sequence. For the investigation of Schnorr randomness, total variants of Solovay reducibility have been introduced by Merkle and Titov [5] in 2022 and, independently, by Kumabe et al. [4] in 2024, the latter again via real-valued translation functions. In what follows, we show that total Solovay reducibility defined in terms of rational functions implies total Solovay reducibility defined in terms of real functions.

Keywords: relative randomness · Solovay reducibility · Martin-Löf randomness · Schnorr randomness · computable analysis

Supported in part by the ANR project FLITTLA ANR-21-CE48-0023.

© The Author(s), under exclusive license to Springer Nature Switzerland AG 2025
A. Beckmann et al. (Eds.): CiE 2025, LNCS 15764, pp. 467–481, 2025.
https://doi.org/10.1007/978-3-031-95908-0_33

1 Solovay Reducibility and its Variants

We start with reviewing the concept of Solovay reducibility introduced by Solovay [6] in 1975 as a measure of relative randomness. Its original definition uses the notion of a translation function on rationals, or a $\mathbb{Q}$-translation function, defined on the left cut of a real. Our notation is standard. All rationals and reals are supposed to be on the interval $[0,1)$ if not stated otherwise. A left-c.e. approximation is a strictly increasing computable approximation. Unexplained notation can be found in Downey and Hirschfeldt [2].

1.1 Solovay Reducibility

Definition 1 (Solovay, 1975). *The* LEFT CUT *of a real β is the set* $\mathrm{LC}(\beta)$ *of all rationals y where $0 \leq y < \beta$. A* $\mathbb{Q}$-TRANSLATION FUNCTION *from a real β to a real α is a partially computable function g from the set $\mathbb{Q} \cap [0,1)$ to itself such that for every $q \in \mathrm{LC}(\beta)$ the value $g(q)$ is defined and fulfills $g(q) < \alpha$, and it holds that*

$$\lim_{q \nearrow \beta} g(q) = \alpha, \tag{1}$$

where $\lim\limits_{q \nearrow \beta}$ denotes the left limit.

A real α is SOLOVAY REDUCIBLE *to a real β, also written as $\alpha \leq_{\mathrm{S}} \beta$, if there is a real constant c and a $\mathbb{Q}$-translation function g from β to α such that, for all $q < \beta$, it holds that*

$$\alpha - g(q) < c(\beta - q). \tag{2}$$

A $\mathbb{Q}$-translation function maps rationals to rationals—in contrast to $\mathbb{R}$-translation functions, to be introduced in the next section, which map reals to reals. We refer to inequality (2) as SOLOVAY CONDITION and to the constant c that occurs in it as SOLOVAY CONSTANT.

Recall that a function f is LIPSCHITZ CONTINUOUS on some interval I if there exists a constant d, called LIPSCHITZ CONSTANT, such that for every $p, q \in I$ in the domain of f it holds that $|f(q) - f(p)| < d|q - p|$.

The Solovay condition (2) resembles a localized version of the defining condition of Lipschitz continuity, where instead of arbitrary pairs of arguments, we consider only pairs with second component β. Accordingly, there are close relations between $\mathbb{Q}$-translation functions that are Lipschitz continuous and those that witness Solovay reducibility.

Proposition 1. *Let α and β be two reals.*

(a) If g is a Lipschitz continuous $\mathbb{Q}$-translation function from β to α, then $\alpha \leq_{\mathrm{S}} \beta$ via g.
(b) If α and β are left-c.e. and $\alpha \leq_{\mathrm{S}} \beta$, then $\alpha \leq_{\mathrm{S}} \beta$ via some Lipschitz continuous $\mathbb{Q}$-translation function. Moreover, the function can be chosen strictly increasing.

Proof. (a) The proof is standard and is left to the interested reader.

(b) By [2, Proposition 9.1.7], there exist a constant $d > c$ and two left-c.e. approximations $a_0, a_1, \ldots$ and $b_0, b_1, \ldots$ of α and β, respectively, that fulfill

$$a_{n+1} - a_n < d(b_{n+1} - b_n) \quad \text{for every } n. \tag{3}$$

Then, the function g from $\mathrm{LC}(\beta)$ to $\mathrm{LC}(\alpha)$ defined by

$$g(q) = a_n + \frac{a_{n+1} - a_n}{b_{n+1} - b_n}(q - b_n) \quad \text{for all rationals } q \in [b_n, b_{n+1})$$

and $g(q) = a_0$ for all rationals $q \in [0, a_0)$ is Lipschitz continuous with the Lipschitz constant d by (3). Further, g is strictly increasing since the sequences $a_0, a_1, \ldots$ and $b_0, b_1, \ldots$ are strictly increasing. Finally, g fulfills (1) because, for arbitrarily small distance $\beta - b_n$, we have $0 < \alpha - g(q) < d(\beta - b_n)$ for all $q \in [a_n, \alpha)$.

Therefore, g is a Lipschitz continuous strictly increasing $\mathbb{Q}$-translation function from β to α, hence, by the statement (a) of the current proposition, witnesses $\alpha \leq_{\mathrm{S}} \beta$.

□

In 2022, Merkle and Titov introduced [5] a version of Solovay reducibility via a totally defined $\mathbb{Q}$-translation function.

Definition 2 (Merkle, Titov, 2022). *A real α is* TOTAL SOLOVAY REDUCIBLE *to a real β, written $\alpha \leq_{\mathrm{S}}^{\mathrm{tot}} \beta$, if $\alpha \leq_{\mathrm{S}} \beta$ via a $\mathbb{Q}$-translation function g which is totally computable on $[0, 1)$.*

1.2 Monotone Solovay Reducibility

In 2024, Titov [7] proposed the following monotone variant of Solovay reducibility, which coincides with $\leq_{\mathrm{S}}$ on the set on left-c.e. reals.

Definition 3. *A real α is* MONOTONE SOLOVAY REDUCIBLE *to a real β, written $\alpha \leq_{\mathrm{S}}^{\mathrm{m}} \beta$, if α is* SOLOVAY REDUCIBLE *to β via a $\mathbb{Q}$-translation function that is nondecreasing.*

By definition, Solovay reducibility is implied by its monotone variant. In the remainder of this section, we demonstrate that the reverse implication does not hold in general, i.e., $\leq_{\mathrm{S}}^{\mathrm{m}}$ is strictly stronger than $\leq_{\mathrm{S}}$. For a start, we state somewhat technical observations about the behavior of monotone $\mathbb{Q}$-translation functions, their proofs are left to the interested reader.

Lemma 1. *Every monotone $\mathbb{Q}$-translation function g from a real β to a real α satisfies the following properties.*

(a) For every $q \geq \beta$ such that $g(q)$ is defined, we have $g(q) \geq \beta$.
(b) If $q_0, q_1, \ldots$ is a sequence of rationals in $[0, \beta)$ such that $\lim_{n\to\infty} g(q_n) = \alpha$, then $\lim_{n\to\infty} q_n = \beta$.

The next proposition shows that the monotone $\mathbb{Q}$-translation function can be only from a left-c.e. to a left-c.e. or from a nonleft-c.e. to a nonleft-c.e. real.

Proposition 2. *Let $\alpha, \beta \in \mathbb{R}$, where α is left-c.e. Then the following statements are equivalent:*

(i) β is left-c.e.;
(ii) there exists a nondecreasing $\mathbb{Q}$-translation function from α to β;
(iii) there exists a nondecreasing $\mathbb{Q}$-translation function from β to α.

Proof. Let $a_0, a_1, \ldots$ where $a_0 = 0$ be a left-c.e. approximation of α.

- *(i)* $\Longrightarrow$ *(ii)* and *(i)* $\Longrightarrow$ *(iii)*. In case β is left-c.e., let $b_0, b_1, \ldots$ where $b_0 = 0$ be a left-c.e. approximation of β. Then the functions

$$g(q) = a_{\min\{m: b_m \geq q\}} \quad \text{and} \quad h(q) = b_{\min\{m: a_m \geq q\}}$$

are nondecreasing translation functions from β to α and from α to β, respectively.
- *(ii)* $\Longrightarrow$ *(i)*. If there exists a nondecreasing $\mathbb{Q}$-translation function g from α to β, then the sequence $g(a_0), g(a_1), \ldots$ is a nondecreasing computable approximation of β, hence β is left-c.e.
- *(iii)* $\Longrightarrow$ *(i)*. If there exists a nondecreasing $\mathbb{Q}$-translation function g from β to α, then, for a fixed computable enumeration $q_0 = 0, q_1, \ldots$ of $\mathrm{dom}(g)$, we define $i_0, i_1, \ldots$ step-wise: at step 0, we set $i_0 = 0$; at step $n+1$, we search for an index i such that

$$q_{i_n} < q_i,\ g(q_i) > a_n,\ \text{and there is an index } j > i_n \text{ such that } g(q_i) < a_j \quad (4)$$

and set $i_{n+1} = i$.
Then every step terminates since, for every n, all rationals in the interval $(q_{i_n}, \beta) \cap (d_n, \beta)$, where d_n is some point in $(0, \beta)$ such that $g(d_n) \in (a_{i_n}, \alpha)$, fulfill (4). Therefore, the sequence $q_{i_0}, q_{i_1}, \ldots$ is infinite and computable. It is increasing by $q_{i_n} < q_{i_{n+1}}$. Further, for every j, we have $q_{i_n} < a_j < \alpha$ for some j, so, by Lemma 1(a), the sequence $q_{i_0}, q_{i_1}, \ldots$ lies in $[0, \beta)$. Finally, it fulfills $g(q_{i_{n+1}}) > a_n$ for every n, hence the sequence $g(q_{i_0}), g(q_{i_1}), \ldots$ tends to α; and thus, by Lemma 1(b), the sequence $q_{i_0}, q_{i_1}, \ldots$ tends to β. Therefore, it is a left-c.e. approximation of β, hence β is left-c.e.

□

Corollary 1. *The set of left-c.e. reals is $\leq_S^m$-closed downwards and upwards in $\mathbb{R}$.*

In contrast, the set of left-c.e. reals is not $\leq_S$-closed downwards, as we will see in the next proposition.

Proposition 3. *There exist a nonleft-c.e. real β such that $1 \leq_S \beta$.*

Proof. We fix an enumeration $q_1, q_2, \ldots$ of all rationals in the unit interval and define a computable test $I_n := [q_n, q_n + 2^{-2n}]$. Since $\sum_{n=1}^{\infty} \mu(I_n) = \frac{1}{3}$, while the set of all left-c.e. reals has the Lebesgue measure 0 because there are only countably many such reals, we can fix a nonleft-c.e. real $\beta \notin \bigcup_{i=1}^{\infty} I_n$.

On the one hand, the function g defined by

$$g(q_n) = 1 - 2^{-2n}$$

is a $\mathbb{Q}$-translation function from β to 1. Moreover, for every $q_n < \beta$, we easily obtain from $\beta \notin I_n$ that $\beta > q_n + \mu(I_n) = q_n + 2^{-2n}$, and thus

$$1 - g(q_n) = 1 - (1 - 2^{-2n}) = 2^{-2n} \in (0, \beta - q_n).$$

Therefore we obtain that $1 \leq_{\mathrm{S}} \beta$ with the constant 1 via g. □

On the other hand, the existence of a nondecreasing $\mathbb{Q}$-translation function from the real β defined in the latter proof to 1 would imply by Proposition 2 that β is left-c.e., which cannot be true. Therefore, Proposition 3 allows us to separate $\leq_{\mathrm{S}}^{\mathrm{m}}$ and $\leq_{\mathrm{S}}$ on $\mathbb{R}$.

Corollary 2. *There exist two reals α and β such that $\alpha \leq_{\mathrm{S}} \beta$ but $\alpha \not\leq_{\mathrm{S}}^{\mathrm{m}} \beta$.*

1.3 Computable Functions on Reals

Definition 4. *A sequence $q_0, q_1, \ldots$ of rationals is called* EFFECTIVE APPROXIMATION *if it fulfills $|q_n - q_{n+1}| < 2^{-n}$ for every n.*

Since every effective approximation $q_0, q_1, \ldots$ is a Cauchy sequence, it converges to some limit point $x \in \mathbb{R}$. The following properties of effective approximations and their limits can be obtained straightforwardly.

Lemma 2. *(a) If $q_0, q_1, \ldots$ is an effective approximation, then there exists a real x such that*

$$|x - q_n| < 2^{-(n-1)} \text{ for all } n \in \mathbb{N}. \tag{5}$$

In particular, $x = \lim_{n\to\infty} q_n$. In that case, we also say that $q_0, q_1, \ldots$ is an EFFECTIVE APPROXIMATION OF x.

(b) Let $q_0, q_1, \ldots$ be an approximation of a real x that satisfies

$$|q_{n+1} - q_n| < 2^{-n} \text{ for all } n \in \mathbb{N}. \tag{6}$$

Then $q_0, q_1, \ldots$ is an effective approximation of x.

Obviously, a real x is computable iff there exists a computable effective approximation of x.

The class of Turing machines that, using an infinite sequence of finite strings (in our case, encoded rationals) as an oracle, returns another sequence of finite strings was formally defined by Weihrauch [8, Chapter 2] under the name "Turing machines of Type 2". In what follows, we give a formal definition of a computable real function using a notion of a Turing machine of Type 2 specified for the sequences of rationals.

Definition 5 (Weihrauch, 2000). *A* TURING MACHINE M OF TYPE 2 *is an oracle Turing machine that, for every oracle $(p_0, p_1, \dots)$, where $p_0, p_1, \dots$ are (appropriately finitely encoded) rationals, produces either an infinite sequence of rationals $(q_0, q_1, \dots)$; in this case, we write $M^{(p_0,p_1,\dots)} \downarrow= (q_0, q_1, \dots)$ and say that M* RETURNS THE SEQUENCE $(q_0, q_1, \dots)$ FROM THE INPUT $(p_0, p_1, \dots)$*; or a finite set of rationals $(q_0, q_1, \dots, q_n)$; in the latter case, we say that $M^{(p_0,p_1,\dots)}$* IS UNDEFINED.

If, for an effective approximation $(p_0, p_1, \dots)$ given as an oracle, M returns at least $n+1$ first elements $(q_0, \dots, q_n)$ of some effective approximation $(q_0, q_1, \dots)$, we say that THE COMPUTATION OF $M^{(p_0,p_1,\dots)}$ WITH TOLERANCE $2^{-(n-1)}$ HALTS *and call p_n* RESULT OF $M^{(q_0,q_1,\dots)}$ WITH TOLERANCE $2^{-(n-1)}$.

A real function f from some subset of $\mathbb{R}$ to $\mathbb{R}$ is COMPUTABLE *on some set $X \subseteq \mathrm{dom}(f)$ if there exists an oracle Turing machine M such that, for every $x \in X$ and every effective approximation $p_0, p_1, \dots$ that converges to x, $M^{(p_0,p_1,\dots)}$ returns an effective approximation $q_0, q_1, \dots$ of $f(x)$.*

By [8, Theorem 4.3.1], computability of a real function implies its continuity.

Proposition 4. *If a real function f is computable on some interval (a, b), then f is continuous on (a, b).*

1.4 Variants of Solovay Reducibility Defined via Translation Function on Reals

According to Proposition 1(a), on the left-c.e. reals, the Solovay reducibility is equivalent to the Solovay reducibility via a Lipschitz continuous $\mathbb{Q}$-translation function. In 2020, Kumabe, Miyabe, Mizusawa, and Suzuki proposed [3, Definition 9] a new type of reducibility by replacing the Lipschitz continuous translation functions on rationals in the latter characterization by the translation functions on reals, without requiring for the reals α and β to be left-c.e.

In what follows, we give the formal definition of this reducibility denoted by Kumabe et al. as "L2" under the more intuitive name "real Solovay reducibility".

Definition 6 (Kumabe et al., 2020). *An* $\mathbb{R}$-TRANSLATION FUNCTION *from a real β to another real α is a real function which is computable on the interval $[0, \beta)$ and maps it to the interval $[0, \alpha)$ such that*

$$\lim_{x \nearrow \beta} f(x) = \alpha. \tag{7}$$

A real α is REAL SOLOVAY REDUCIBLE *to a real β, written $\alpha \leq_{\mathrm{S}}^{\mathbb{R}} \beta$, if there exists a Lipschitz continuous $\mathbb{R}$-translation function from β to α.*

Note that, between every two left-c.e. reals, there exist both an $\mathbb{R}$-translation function and a $\mathbb{Q}$-translation function.

Proposition 5. *If α and β are left-c.e. reals, then there exists an $\mathbb{R}$-translation function from β to α, whose restriction on $\mathbb{Q}$ is a $\mathbb{Q}$-translation function from β to α.*

Proof. Fixing two left-c.e. approximations $a_0, a_1, \dots \nearrow \alpha$ and $b_0, b_1, \dots \nearrow \beta$, we can construct an $\mathbb{R}$-translation function f from β to α by setting $f(x) = a_0$ for all $x < b_0$ and

$$f(x) = a_n + \frac{x - b_n}{b_{n+1} - b_n}(a_{n+1} - a_n) \quad \text{if there exists such } n \text{ that } b_n \leq x < b_{n+1}.$$

It is easy to see that, if x is rational, then $f(x)$ is rational too, so $f|_{\mathbb{Q}}$ is a $\mathbb{Q}$-translation function from β to α. □

In contrast to Solovay reducibility (see Corollary 2), the additional requirement for the $\mathbb{R}$-translation function in the latter definition to be nondecreasing (Kumabe et al. denoted [3, Definition 9] the resulting reducibility "L1") does not induce a strictly stronger reducibility on $\mathbb{R}$ than $\leq_{\mathrm{S}}^{\mathbb{R}}$, as we will see in the next proposition, which can be shown by essentially the same proof as [8, Corollary 6.2.5] (computability of the maximum operator).

Proposition 6. *If f is a $\mathbb{R}$-translation function from a real β to another real α, then*

$$\tilde{f}(x) = \max\{f(y) : y \in [0, x]\}$$

is a monotone nondecreasing $\mathbb{R}$-translation function from β to α. $\tilde{f}$ can be computed by a Machine M of Type 2 that satisfies the following property:

(A) For every real $x < \beta$, every rational $q > \beta$, every effective approximation $p_0, p_1, \dots$ of x, and every $n \in \mathbb{N}$, if the computation of $M^{(q,q,\dots)}$ with tolerance 2^{-n} halts, and its result is some rational r, then r fulfills the inequality $r > \tilde{f}^{(n)}(x) - 2^{-(n-1)}$, where $\tilde{f}^{(n)}(x)$ is the result of the computation of $M^{(p_0,p_1,\dots)}$ with the tolerance 2^{-n} (which obviously fulfills $|f(x) - \tilde{f}^{(n)}(x)| < 2^{-n}$; so, in particular, $r > f(x) - 2^{-(n-2)}$).

Moreover, if $\alpha \leq_{\mathrm{S}}^{\mathbb{R}} \beta$ via f, then $\alpha \leq_{\mathrm{S}}^{\mathbb{R}} \beta$ via $\tilde{f}$ with the same Lipschitz constant.

Corollary 3. *If $\alpha \leq_{\mathrm{S}}^{\mathbb{R}} \beta$ for two reals α and β, then $\alpha \leq_{\mathrm{S}}^{\mathbb{R}} \beta$ via a nondecreasing $\mathbb{R}$-translation function.*

The following property of monotone $\mathbb{R}$-translation functions, which is similar to Lemma 1(b) for $\mathbb{Q}$-translation functions, can be easily proved.

Lemma 3. *If f is a monotone $\mathbb{R}$-translation function from a real β to a real α and $x_0, x_1, \dots$ is a sequence of reals in $[0, \beta)$ such that $\lim_{n \to \infty} f(x_n) = \alpha$, then $\lim_{n \to \infty} x_n = \beta$.*

Similarly as for $\leq_{\mathrm{S}}^{\mathrm{m}}$, we can show the $\leq_{\mathrm{S}}^{\mathbb{R}}$-closure upwards and downwards of left-c.e. reals.

Proposition 7. *Let $\alpha, \beta \in \mathbb{R}$ where α is left-c.e. Then the following statements are equivalent:*

(i) β is left-c.e.;

(ii) there exists an $\mathbb{R}$-translation function from α to β;
(iii) there exists an $\mathbb{R}$-translation function from β to α.

Proof. Let $a_0, a_1, \ldots$ where $a_0 = 0$ be a left-c.e. approximation of α.

- *(i)* $\Longrightarrow$ *(ii)* and *(i)* $\Longrightarrow$ *(iii)*. Directly by Proposition 5.
- *(ii)* $\Longrightarrow$ *(i)*. If there exists an $\mathbb{R}$-translation function from α to β, then, by Proposition 6, we can fix a nondecreasing $\mathbb{R}$-translation function f from α to β. Then we compute a left-c.e. approximation of β step-wise: starting with $i_{-1} = -1$ and $m_{-1} = 0$, in every step $n \geq 0$, we approximate the values of $f(a_i)$ for all $i > i_{n-1}$ effectively until we find a natural $m > m_{n-1}$ and two indices i, j where $i_{n-1} < i < j$ such that, for the approximated with tolerance 2^{-m} values $\tilde{f}(a_i)$ and $\tilde{f}(a_j)$ of $f(a_i)$ and $f(a_j)$, respectively, the inequality

$$\tilde{f}(a_i) + 2^{-m} < \tilde{f}(a_j) - 2^{-m}$$

holds (so, by $\tilde{f}(a_j) < f(a_j) + 2^{-m} < \alpha + 2^{-m}$, we can guarantee that $\tilde{f}(a_i) < \alpha$), and set $i_n = i$, $m_n = m$, and $b_n = \tilde{f}(a_{i_n})$. Then the sequence $b_0, b_1, \ldots$ is a left-c.e. approximation of β.
- *(iii)* $\Longrightarrow$ *(i)*. The case $\beta \in \mathbb{Q}$ is trivial; so, in what follows, assume $\beta \notin \mathbb{Q}$. If there exists an $\mathbb{R}$-translation function from β to α, then, by Proposition 6, we can fix a nondecreasing $\mathbb{R}$-translation function f from β to α and a Turing machine M of Type 2 that computes f and fulfills (A). Then we compute a left-c.e. approximation of β step-wise: starting from some fixed $b_{-1} := 0 < \beta$ and $j_{-1} = -1$, in every step $n > 0$, we search for a rational $b > b_{n-1}$, a natural m, and an index $j > j_{n-1}$ such that the computation of $M^{(b,b,\ldots)}$ with tolerance 2^{-m} halts, and its result r (in case $b < \beta$, r we have $|f(b) - r| < 2^{-m}$ by definition) fulfills

$$a_{j_{n-1}} + 2^{-m} < r - 2^{-m} \quad \text{and} \quad r + 2^{-(m-2)} < a_j - 2^{-(m-2)}, \tag{8}$$

and set $b_n = b$, $m_n = m$, $r_n = r$, and $j_n = j$.
Step n terminates and yields a rational $b_n < \beta$ due to the following argumentation:
  - assuming by induction hypothesis that $b_{n-1} < \beta$, the first inequality in (8) is satisfied for all rationals $r = f^{(m_1)}(b)$ where $f^{(m_1)}(b)$ is the result of the computation of $M^{(b,b,\ldots)}$ with the tolerance 2^{-m_1} (which fulfills $|f(b) - \tilde{f}^{(m_1)}(b)| < 2^{-m_1}$ since $b < \beta$) for all rationals b in some left neighborhood of β (maybe intersected with the interval $[b_{n-1}, \beta)$, which is nonempty by induction hypothesis) and all sufficiently large m_1. For every such r, all but finitely many indices j fulfill $f(b) < a_j$, hence there exists a natural m_2 such that $r + 2^{-(m_2-2)} < a_j - 2^{-(m_2-2)}$. We conclude the termination proof by choosing $m > \max\{m_1, m_2\}$.
  - If, for some m and j, there exists a rational $b > \beta$ such that the computation of $M^{(b,b,\ldots)}$ with tolerance 2^{-m} halts, and its result r fulfills (8), then we obtain from the right part of (8) that $r + 2^{-(m-2)} < \alpha - 2^{-(m-2)}$, hence $r < \alpha - 2^{-(m-3)}$. By $\lim_{x \nearrow \beta} f(x) = \alpha$, there exists a real $x < \beta$ such

that $r < f(x) - 2^{-(m-2)}$, but this contradicts to the property (A) applied on x and b. Finally, from $b \in \mathbb{Q}$ and $\beta \notin \mathbb{Q}$, we obtain that $b \neq \beta$; thus, the only possible case is $b < \beta$.

Therefore, we have constructed an infinite sequence $b_0 < b_1 < \cdots < \beta$, hence, for every n, the definition of r_n and the construction of M imply that

$$|f(b_n) - r_n| < 2^{-m_n}. \tag{9}$$

From (8) and (9), we easily obtain that

$$a_{j_{n-1}} < f(b_n) < a_{j_n}. \tag{10}$$

Thus, the sequence $b_0, b_1, \ldots$ lies in $[0, \beta)$ by (9) and fulfills $\lim\limits_{n\to\infty} f(b_n) = \alpha$ by (10), hence, by Lemma 3, it converges to β. Since the sequence $b_0, b_1, \ldots$ is strictly increasing by (9), it is a left-c.e. approximation of β.

□

Corollary 4. *The set of left-c.e. reals is $\leq_{\mathrm{S}}^{\mathbb{R}}$-closed downwards and upwards.*

In 2024, Kumabe, Miyabe, and Suzuki introduced [4, Chapter 5] a slightly modified version of $\leq_{\mathrm{S}}^{\mathbb{R}}$ by omitting the requirement for the $\mathbb{R}$-translation function f in Definition 6 to fulfill $f(x) < \alpha$ for all $x < \beta$. Kumabe et al. also examined the totalized variant $\leq_{\mathrm{cL}}^{\mathrm{loc}}$ of the latter reducibility.

Definition 7 (Kumabe et al., 2024). *A* WEAKLY $\mathbb{R}$-TRANSLATION FUNCTION *from a real α to another real β is a real function which is computable on the interval $[0, \beta)$ and satisfies* (7).

A real α is COMPUTABLY LIPSCHITZ REDUCIBLE *to a real β* ON A C.E. OPEN INTERVAL, *written $\alpha \leq_{\mathrm{cL}}^{\mathrm{open}} \beta$, if there exists a Lipschitz continuous weakly $\mathbb{R}$-translation function f on reals from β to α.*

α is COMPUTABLY LIPSCHITZ REDUCIBLE *to a real β* LOCALLY, *written $\alpha \leq_{\mathrm{cL}}^{\mathrm{loc}} \beta$, if there exists a real $b > \beta$ and a Lipschitz continuous function f on reals, which is computable on $[0, b]$ and fulfills $f(\beta) = \alpha$.*

By [3, Theorem 1] and by [4, Observation 5.3], respectively, the reducibilities $\leq_{\mathrm{S}}^{\mathbb{R}}$ and $\leq_{\mathrm{cL}}^{\mathrm{open}}$ are equivalent to the Solovay reducibility $\leq_{\mathrm{S}}$ on the set of left-c.e. reals.

In Sect. 3, we prove that $\leq_{\mathrm{cL}}^{\mathrm{open}}$ is implied by $\leq_{\mathrm{S}}$ and $\leq_{\mathrm{cL}}^{\mathrm{loc}}$ is implied by $\leq_{\mathrm{S}}^{\mathrm{tot}}$ on $\mathbb{R}$ and that $\leq_{\mathrm{S}}^{\mathbb{R}}$ is implied by $\leq_{\mathrm{S}}$ on all but computable reals.

2 Reducibilities $\leq_{\mathrm{cL}}^{\mathrm{open}}$ and $\leq_{\mathrm{cL}}^{\mathrm{loc}}$ as Measures of Relative Martin-Löf and Schnorr Randomness, Respectively

On the one hand, it is easy to see that every computable real α is $\leq_{\mathrm{cL}}^{\mathrm{loc}}$-reducible and thus also $\leq_{\mathrm{cL}}^{\mathrm{open}}$-reducible to every further real β via the weakly $\mathbb{R}$-translation function $f(x) = \alpha$ which is computable (as a real function) and totally defined on the unit interval.

On the other hand, if a real α is $\leq_{\mathrm{cL}}^{\mathrm{open}}$-reducible to a computable real β, then we can easily compute α as well; the formal proof is left to the reader as an exercise. The latter two observations imply that the least $\leq_{\mathrm{cL}}^{\mathrm{open}}$- and $\leq_{\mathrm{cL}}^{\mathrm{loc}}$-degrees on $\mathbb{R}$ both contain exactly all computable reals.

Proposition 8. *The computable reals form the least degree on* $\mathbb{R}$ *relative to both* $\leq_{\mathrm{cL}}^{\mathrm{open}}$ *and* $\leq_{\mathrm{cL}}^{\mathrm{loc}}$.

The $\leq_{\mathrm{S}}$-closure upwards of Martin-Löf random reals has been proved by Solovay himself [6]; the $\leq_{\mathrm{S}}^{\mathrm{tot}}$-closure upwards of Schnorr random reals has been demonstrated by Merkle and Titov [5, Corollary 2.10].

In what follows, we prove the same closures for the reducibilities $\leq_{\mathrm{cL}}^{\mathrm{open}}$ and $\leq_{\mathrm{cL}}^{\mathrm{open}}$, respectively.

Proposition 9. *(a) The set of Martin-Löf random reals is* $\leq_{\mathrm{cL}}^{\mathrm{open}}$*-closed upwards.*

(b) The set of Schnorr random reals is $\leq_{\mathrm{cL}}^{\mathrm{loc}}$*-closed upwards.*

Proof. (a) Let α and β, where β is Martin-Löf nonrandom, be two reals that fulfill $\alpha \leq_{\mathrm{cL}}^{\mathrm{open}} \beta$ with a constant c via a real function f computed by a Turing machine M of Type 2. In particular, for every rational $q < \beta$, f satisfies

$$|\alpha - f(q) \downarrow | < c(\beta - q). \tag{11}$$

We define further a two-argument-function $g : \mathbb{Q}|_{[0,1)} \times \mathbb{N} \to \mathbb{Q}$ by setting

$$g(q, m) \text{ is the result of } M^{(q,q,\dots)} \text{ with tolerance } 2^{-m} \text{ if defined} \tag{12}$$

Then, for every rational $q < \beta$ and natural m, we know by the choice of M that $|f(q) \downarrow - g(q, m)| < 2^{-m}$, hence we obtain by (11) that

$$|\alpha - g(q, m) \downarrow | \leq c(\beta - q) + 2^{-m} \tag{13}$$

for every $q < \beta$ and $m \in \mathbb{N}$.

Let further $q_0, q_1, \dots$ be a standard enumeration of rationals on $[0, 1)$.

Since β is Martin-Löf nonrandom, we can further fix a Solovay test $S_0, S_1, \dots$ that fails on β, where we denote $S_n = [l_n, r_n]$ of length d_n for every n. In particular, this test has a finite measure:

$$L := \sum_{n \in \mathbb{N}} \mu(S_n) = \sum_{n \in \mathbb{N}} d_n < \infty. \tag{14}$$

Then we prove that α should be Martin-Löf nonrandom as well by constructing the following Solovay test: for every natural n, compute $g(l_n, n)$ and, if the computation halts, set

$$T_n = [g(l_n, n) - (cd_n + 2^{-n}), g(l_n, n) + (cd_n + 2^{-n})]. \tag{15}$$

Then, the test $T_0, T_1, \ldots$ that contains all defined T_n is computable and has a finite measure by

$$\sum_{n\in\mathbb{N}} \mu(T_n) \leq \sum_{n\in\mathbb{N}} (2^{-n+1} + 2cd_n) = \sum_{n\in\mathbb{N}} 2^{-n+1} + 2c \sum_{n\in\mathbb{N}} d_n = 4 + 2cL < \infty.$$

Further, for every of infinitely many n such that $\beta \in S_n$, we have $l_n < \beta < r_n$, hence, by (13), we have $g(l_n, n) \downarrow$ (hence T_n is defined) and

$$|\alpha - g(l_n, n)| \leq c(\beta - l_n) + 2^{-n} < c(r_n - l_n) + 2^{-n} = cd_n + 2^{-n}, \quad (16)$$

which implies that $\alpha \in T_n$. Therefore, the test $T_0, T_1, \ldots$ that contains all defined T_n is a Solovay test that fails on α, hence α is Martin-Löf nonrandom.

(b) If, in the latter proof, f is computable on some interval $[0, b]$ with $b > \beta$ (i.e. $\alpha \leq^{\text{loc}}_{\text{cL}} \beta$ via f), then we can fix some rational $\rho \in (\beta, b)$. Obviously, f is computable on $[0, \rho]$, and the function g defined as in (12) fulfills (13) for every $q < \rho$ and $m \in \mathbb{N}$.
If, additionally, $S_0, S_1, \ldots$ is a total Solovay test that fails on β (which witnesses the Schnorr nonrandomness of β by [1, Theorem 2.5]), then we can modify this test by replacing every S_n by $S_n \cap [0, \rho]$; the resulting test is, again, a total Solovay test that fails on β and has a computable measure:

$$L_{comp} := \sum_{n\in\mathbb{N}} \mu(S_n) = \sum_{n\in\mathbb{N}} d_n < \infty. \quad (17)$$

For every natural n the interval T_n as in (15) is well-defined since we know from $l_n < \rho$ that $g(l_n, n) \downarrow$. Hence, the test $T_0, T_1, \ldots$ has a measure exactly equal to $4 + 2cL_{comp}$, which is finite and computable. Moreover, for every of infinitely many n such that $\beta \in S_n$, it still holds by (16) that T_n is defined and contains α. Therefore, the test $T_0, T_1, \ldots$ is a total Solovay test that fails on α, hence α is Schnorr nonrandom.

□

3 Main Result

As a main result of this paper, we claim that the reducibilities via $\mathbb{Q}$-translation functions imply the corresponding reducibilities via $\mathbb{R}$-translation functions.

Theorem 1. *For arbitrary reals α and β, the following implications hold:*

$$\alpha \leq_{\text{S}} \beta \implies \alpha \leq^{\text{open}}_{\text{cL}} \beta \qquad \textit{and} \qquad \alpha \leq^{\text{tot}}_{\text{S}} \beta \implies \alpha \leq^{\text{loc}}_{\text{cL}} \beta. \quad (18)$$

Moreover, if α is not computable, then the following implication holds:

$$\alpha \leq_{\text{S}} \beta \implies \alpha \leq^{\mathbb{R}}_{\text{S}} \beta. \quad (19)$$

Proof. Let g be a $\mathbb{Q}$-translation function from β to α, and let α and β be two reals such that $\alpha \leq_S \beta$ with some constant c via g. If β is left-c.e., then α is left-c.e. as well since the set of left-c.e. reals is $\leq_S$-closed downwards, and the theorem statement holds by [3, Theorem 1]. So, in what follows, suppose that β is not left-c.e., i.e., $\mathrm{LC}(\beta)$ is not recursively enumerable.

Further, let b be the maximal real that fulfills the property

$$g(q)\downarrow \text{ for every } q < b. \tag{20}$$

It holds obviously that $b \geq \beta$ since g is a translation function from β to α, wherein, in case $\alpha \leq_S^{\mathrm{tot}} \beta$, we obtain $b = 1$.

In the scope of the proof, we set an enumeration $q_0, q_1, \ldots$ of the domain of g. If there exists a rational $b > \beta$ such that there is no $q_i \in (\beta, b)$, then $\mathrm{LC}(\beta) = \mathrm{dom}(g) \cap [0, \beta)$ is recursively enumerable, a contradiction. Therefore, there exists a subsequence of $q_0, q_1, \ldots$ which tends to β from above.

Construction of a Weakly $\mathbb{R}$-translation Function from β to α

Now, we construct a Lipschitz continuous weakly $\mathbb{R}$-translation function h from β to α in four steps: first, on the base of g, we construct another partial function $\tilde{g}$ on rationals that witnesses the Solovay reducibility $\alpha \leq_S \beta$ with the same constant c; second, on the base of $\tilde{g}$, we construct a technical computable two-argument partial function $f(\cdot,\cdot)$ on rationals; third, on the base of f, we construct another technical two-argument partial function $\tilde{f}(\cdot,\cdot)$ on rationals; and fourth, on the base of $\tilde{f}$, we finally construct the desired function h.

The lists of properties of the functions $\tilde{g}$ through h are combined in Claims 1–4, respectively. The proof of each claim can be easily obtained from the previous one through simple calculations.

First, we define a computable function $\tilde{g}$ from rationals to rationals by setting

$$Q_n = \{q_m : m \leq n \text{ and } q_m \leq q_n\} \quad \text{and} \quad \tilde{g}(q_n) = \max\{g(q) : q \in Q_n\} \text{ for all } n.$$

Claim 1. *$\tilde{g}$ satisfies the following properties:*

$$\mathrm{dom}(\tilde{g}) = \mathrm{dom}(g), \tag{21}$$

$$0 < \alpha - \tilde{g}(q) \downarrow \leq \alpha - g(q) < c(\beta - q) \textit{ for all } q \in \mathrm{LC}(\beta). \tag{22}$$

$$Q_m \subseteq Q_n \textit{ and } \tilde{g}(q_m) \leq \tilde{g}(q_n) \textit{ for all } q_m, q_n \in \mathrm{dom}(\tilde{g}) \textit{ where} \begin{cases} m < n, \\ q_m < q_n. \end{cases} \tag{23}$$

From (21), we obtain that, in particular, that $q_0, q_1, \ldots$ is at the same time a computable enumeration of $\mathrm{dom}(\tilde{g})$. The property (22) directly implies that the function g is a well-defined $\mathbb{Q}$-translation function that witnesses the Solovay reducibility of α to β with the same constant as g.

At the beginning of the second step, we fix a rational d and a natural $K > 1$ such that $2^K = d > c$ and define a computable two-argument partial function $f : \mathbb{Q} \times \mathbb{N} \to \mathbb{Q}$ as follows: given q and n, we enumerate $q_0 (= 0), q_1, \ldots$ and, in case if we meet after some enumeration step t a set $\{q_{i_0} (= 0), q_{i_1}, \ldots, q_{i_k}\}$ where $i_0, i_1, \ldots, i_k \leq t$ such that

$$0 < q_{i_n} - q_{i_{n-1}} < 2^{-n} \text{ for all } i \in \{1, \ldots, k\} \text{ and} \tag{24}$$
$$0 \leq q - q_{i_k} < 2^{-n}, \tag{25}$$

let $\{q_{i_0}, q_{i_1}, \ldots, q_{i_N}\}$ be the greatest (under inclusion) such set of indices not larger than t. Such set exists since, if two different sets $A = (q_{i_0}, \ldots, q_{i_a})$ and $B = (q_{j_0}, \ldots, q_{j_b})$, where $i_0, \ldots, i_a, j_0, \ldots, j_b \leq t$, fulfill the properties (24)–(25), then the same holds for the set $A \cup B$.

Next, we define the ordered tuple

$$P(q, n) = (q_{i_0}, \ldots, q_{i_k}) \quad \text{and} \quad f(q, n) = \min\left\{\tilde{g}(p) + d(q - p) : p \in P(q, n)\right\}.$$

Accordingly, we write $P(q, n) \uparrow$ and $f(q, n) \uparrow$ if the search for the set $P(q, n)$ never ends.

Further, we define the (in case $b = \beta$ coinciding) sets

$$D_\beta = \{(q, n) : n \in \mathbb{N} \text{ and } q \in [0, \beta + 2^{-n})\} \text{ and}$$
$$D_b = \{(q, n) : n \in \mathbb{N} \text{ and } q \in [0, b + 2^{-n})\},$$

that obviously fulfill $D_\beta \subseteq D_b$.

Claim 2. *f satisfies the following properties:*

$$P(q, n) \downarrow \text{ and } f(q, n) \downarrow \text{ for all naturals } n \text{ and rationals } q \in [0, b), \tag{26}$$
$$P(q, n) \downarrow \text{ and } f(q, n) \downarrow \text{ for all } (q, n) \in D_b, \tag{27}$$
$$f(q, n) \downarrow < \alpha + 2^{-(n-K-1)} \text{ for all } (q, n) \in D_\beta, \tag{28}$$
$$P(p, n) \downarrow \subseteq P(q, n) \text{ for all } (q, n) \in \mathrm{dom}(f) \text{ and } p < q, \tag{29}$$
$$\begin{cases} P(q, m) \downarrow \subseteq P(q, n) \\ 0 \leq f(q, m) - f(q, n) \downarrow < 2^{-(m-K)} \end{cases} \text{for all } (q, n) \in \mathrm{dom}(f) \text{ and } m < n, \tag{30}$$
$$f(q, n) - f(p, n) \downarrow < d(q - p) \text{ for all } (q, n) \in \mathrm{dom}(f) \text{ and } p < q, \tag{31}$$
$$\alpha - f(q, n) \downarrow < c(\beta - q) \text{ for every } n \in \mathbb{N} \text{ and } q \in \mathrm{LC}(\beta). \tag{32}$$

In the second step, we define a computable two-argument partial function $\tilde{f} : \mathbb{Q} \times \mathbb{N} \to \mathbb{Q}$ by setting

$$\tilde{f}(q, n) = \max\left(\{f(q', n) : q' \in P(q, n)\} \cup \{f(q, n)\}\right)$$

for all $(q, n) \in \mathrm{dom}(f)$.

Claim 3. *$\tilde{f}$ satisfies the following properties:*

$$\text{dom}(\tilde{f}) = \text{dom}(f), \tag{33}$$

$$\tilde{f}(q,n) \geq f(q,n) \textit{ for all } (q,n) \in \text{dom}(\tilde{f}), \tag{34}$$

$$\tilde{f}(q,n) \downarrow < \alpha + 2^{-(n-K-1)} \textit{ for all } (q,n) \in D_{\beta}, \tag{35}$$

$$|\tilde{f}(q,n) - \tilde{f}(q,m)| < 2^{-(m-K-1)} \textit{ for all } (q,m) \in \text{dom}(f) \textit{ and } m < n, \tag{36}$$

$$|\tilde{f}(q,n) - \tilde{f}(p,n) \downarrow| < 2^{-(n-K)} + 2^{K}|q-p| \textit{ for all } (q,n) \in \text{dom}(f) \textit{ and } p < q, \tag{37}$$

$$\alpha - \tilde{f}(q,n) \downarrow < c(\beta - q) \textit{ for every } n \in \mathbb{N} \textit{ and } q \in \text{LC}(\beta). \tag{38}$$

In the third step, we define a real function

$$h(x) = \lim_{n\to\infty} f(q_n, n) \text{ for every effective approximation } q_n \underset{n\to\infty}{\longrightarrow} x.$$

Claim 4. *h satisfies the following properties:*

$$[0,b) \in \text{dom}(h); \tag{39}$$

$$h \textit{ is Lipschitz continuous}; \tag{40}$$

$$h(x) \leq \alpha \textit{ for every } x < \beta; \tag{41}$$

$$\lim_{\substack{x\to\beta \\ x\in[0,b)}} h(x) = \alpha; \tag{42}$$

$$h \textit{ is (totally Type 2) computable on } (0,b). \tag{43}$$

The Constructed Weakly $\mathbb{R}$-translation Function Demonstrates the Theorem Statement

By (39) through (43), we obtain that h witnesses $\alpha \leq_{\text{cL}}^{\text{open}} \beta$ and, in case $b = 1$, even $\alpha \leq_{\text{cL}}^{\text{loc}} \beta$.

Now, in case α is not computable, we even obtain that $h(x) < \alpha$ for every $x < \beta$ by contradiction: assuming $h(\tilde{x}) = \alpha$ for some real $\tilde{x} < \beta$, we can fix two rationals $\tilde{p}$ and $\tilde{q}$ such that $\tilde{x} < \tilde{p} < \tilde{q} < \beta$. Then function $\tilde{h}$ defined on the compact interval $[\tilde{p}, \tilde{q}]$ by

$$\tilde{h}(x) = \max\{h(y) : y \in [0,x]\}$$

is computable on the whole interval $[\tilde{p}, \tilde{q}]$ by [8, Corollary 6.2.5]. On the other hand, by (41), we have $h(y) \leq \alpha$ for every $y \in [0, \beta)$, hence our assumption implies for every $x \in [\tilde{p}, \tilde{q}]$ that $\tilde{h}(x) = \max\{h(y) : y \in [0,x]\} = h(\tilde{x}) = \alpha$; thus, the function $\tilde{h}$ is a constant function defined on $[\tilde{p}, \tilde{q}]$ that returns α for every input. Therefore, α should be computable as a limit point of an effective approximation returned by a Turing machine that computes $\tilde{h}$ from the (computable) input $(\frac{1}{2}(\tilde{p}+\tilde{q}), \frac{1}{2}(\tilde{p}+\tilde{q}), \dots)$. A contradiction.

Thus, h is an $\mathbb{R}$-translation function from β to α, and we obtain from its Lipschitz continuity that $\alpha \leq_{\text{S}}^{\mathbb{R}} \beta$. □

On the other hand, by Proposition 3, there exists a nonleft-c.e. real β such that 1 is Solovay reducible to it, while, by Corollary 4, as a computable real, 1 cannot be real Solovay reducible to β. These two observations imply together the following result.

Proposition 10. *There exists a computable real α and a nonleft-c.e. real β such that $\alpha \leq_{\mathrm{S}} \beta$ and $\alpha \not\leq^{\mathbb{R}}_{\mathrm{S}} \beta$.*

The implication $\leq_{\mathrm{S}} \implies \leq^{\mathrm{open}}_{\mathrm{cL}}$ is strict on $\mathbb{R}$ since computable reals form the least degree on $\mathbb{R}$ relative to $\leq^{\mathrm{open}}_{\mathrm{cL}}$ by Proposition 8 but not relative to $\leq_{\mathrm{S}}$ by [2, Proposition 9.6.1]. We still do not know whether the implication $\leq_{\mathrm{S}} \implies \leq^{\mathbb{R}}_{\mathrm{S}}$ is strict on noncomputable reals.

The implication $\leq^{\mathrm{tot}}_{\mathrm{S}} \implies \leq^{\mathrm{loc}}_{\mathrm{cL}}$ is strict since computable reals form the least degree on $\mathbb{R}$ (and thus also on the set of left-c.e. reals) relative to $\leq^{\mathrm{loc}}_{\mathrm{cL}}$ by Proposition 8 but not relative to $\leq^{\mathrm{tot}}_{\mathrm{S}}$ by [5, Proposition 2.5].

Acknowledgments. I would like to thank Peter Hertling, Wolfgang Merkle, and Laurent Bienvenu for meticulous proofreading of this article, as well as Subin Pulari, Andrei Romashchenko, and Alexander Shen for attending an unofficial presentation and fixing some logical errors.

References

1. Downey, R., Griffiths, E.: Schnorr randomness. J. Symb. Logic **69**, 533–554 (2004)
2. Downey, R., Hirschfeldt, D.: Algorithmic Randomness and Complexity. Springer, Berlin (2010)
3. Kumabe, M., Miyabe, K., Mizusawa, Y., Suzuki, T.: Solovay reducibility and continuity. J. Log. Anal. **12**(2), 1–22 (2020)
4. Kumabe, M., Miyabe, K., Suzuki, T.: Solovay reducibility via Lipschitz functions and signed-digit representation. Computability 1–27 (2024)
5. Merkle, W., Titov, I.: A total Solovay reducibility and totalizing of the notion of speedability. In: 13th Panhellenic Logic Symposium, University of Thessaly, Volos, Greece, 6–10 July 2022, Proceedings, vol. II, pp. 68–78 (2022)
6. Solovay, R.: Draft of paper (or series of papers) on Chaitin's work. Unpublished notes, 215 p (1975)
7. Titov, I.: Solovay reducibility and speedability outside of left-c.e. reals. Doctoral dissertation, University of Heidelberg, Department of Mathematics and Computer Science, heiDOK (2024)
8. Weihrauch, K.: Computable Analysis: An Introduction. Springer, Heidelberg (2000)
9. Zheng, X., Rettinger, R.: On the extensions of Solovay-Reducibility. In: Chwa, K.-Y., Munro, J. (eds.) COCOON 2004. LNCS, vol. 3106, pp. 360–369. Springer, Heidelberg (2004). https://doi.org/10.1007/978-3-540-27798-9_39

Constructive Analysis of Maximal Ideals in $\mathbb{Z}[X]$ by the Material Interpretation

Franziskus Wiesnet(✉)

Technische Universität Wien, Wiedner Hauptstraße 8-10, 1050 Wien, Austria
franziskus.wiesnet@tuwien.ac.at

Abstract. This article presents the concept of material interpretation as a method to transform classical proofs into constructive ones. Using the case study of maximal ideals in $\mathbb{Z}[X]$, it demonstrates how a classical implication $A \to B$ can be rephrased as a constructive disjunction $\neg\!\!\!\neg A \vee B$, with $\neg\!\!\!\neg A$ representing a strong form of negation. The approach is based on Gödel's Dialectica interpretation, the strong negation, and potentially Herbrand disjunctions.

The classical proof that every maximal ideal in $\mathbb{Z}[X]$ contains a prime number is revisited, highlighting its reliance on non-constructive principles such as the law of excluded middle. A constructive proof is then developed, replacing abstract constructs with explicit case distinctions and direct computations in $\mathbb{Z}[X]$. This proof clarifies the logical structure and reveals computational content.

The article discusses broader applications, such as Zariski's Lemma, Hilbert's Nullstellensatz, and the Universal Krull-Lindenbaum Lemma, with an emphasis on practical implementation using tools such as Python and proof assistants. The material interpretation offers a promising framework for bridging classical and constructive mathematics, enabling algorithmic implementations.

Keywords: material interpretation · constructive algebra · program extraction · strong negation · proof interpretation

1 Motivation: Material Interpretation

Based on the case study of maximal ideals in $\mathbb{Z}[X]$, our aim is to illustrate the concept of material interpretation. The approach begins with a (possibly classical) proof of a statement of the form $A \to B$ and aims to transform it as far as possible into a proof of the form $\neg\!\!\!\neg A \vee B$. Here, $\neg\!\!\!\neg A$ should represent a stronger form of the negation of A which is quite similar to what is known as *strong negation* [9,12,16]. For our purposes, it is sufficient to know that the strong negation of A is a statement with constructive content that implies the ordinary negation $\neg A$ and in particular

$$\neg\!\!\!\neg(\forall_x(A_1 \wedge \cdots \wedge A_k)) \quad \Leftrightarrow \quad \exists_x(\neg\!\!\!\neg A_1 \vee \cdots \vee \neg\!\!\!\neg A_k). \tag{1}$$

© The Author(s), under exclusive license to Springer Nature Switzerland AG 2025
A. Beckmann et al. (Eds.): CiE 2025, LNCS 15764, pp. 482–495, 2025.
https://doi.org/10.1007/978-3-031-95908-0_34

In our case, the strong negation for prime formulas is simply the ordinary negation. Many algebraic definitions, in particular that of a maximal ideal, have exactly the form $\forall_x(A_1 \wedge \cdots \wedge A_k)$. Therefore, we have a way to redefine their negation in a constructive stronger version.

A formula of the form $\tilde{\neg}A \vee B$ can often be found using Gödel's Dialectica interpretation [10], as Gödel's interpretation was used in [18,19] to motivate a similar constructive approach. For example, if in Formula (1) the A_i are atomic and decidable formulas and $\tilde{\neg}$ denotes the standard negation, then the right-hand side is equivalent to Gödel's Dialectica interpretation of the left-hand side. Herbrand disjunctions could also play a role in this context, as they have a similar form and can be derived from the Dialectica interpretation as well [8].

Since a classical proof of $A \to B$ typically relies on some non-constructive axioms, it will generally not be possible to transform it directly into a constructive proof. Therefore, the idea is to slightly modify A and B to $\tilde{A}$ and $\tilde{B}$ in advance to obtain a constructive proof of $\tilde{\neg}\tilde{A} \vee \tilde{B}$. The new statements $\tilde{A}$ and $\tilde{B}$ should remain as close as possible to A and B and the modifications should be derivable from the classical proof. At first glance, this may seem very abstract, but let us now examine it in a more concrete way using our case study on maximal ideals in $\mathbb{Z}[X]$:

2 A Case Study: Maximal Ideal in $\mathbb{Z}[X]$

For clarity, we will present the classical definition of ideal variants in a ring. Note that in this article, every ring is commutative. A subset $I \subseteq A$ of a ring A is called an *ideal* if

- $0 \in I$,
- $\forall_{x,y \in I}\ x + y \in I$,
- $\forall_{\lambda \in A, x \in I}\ \lambda x \in I$.

An ideal P is called a *prime ideal* if $1 \notin P$ and $\forall_{x,y \in A}(xy \in P \to x \in P \vee y \in I)$. An ideal M is called a *maximal ideal* if $1 \notin M$ and there is no ideal J with $M \subsetneq J$ and $1 \notin J$. Classically (but not constructively) equivalent to the last property is that for all $x \notin M$ there is $\lambda \in A$ with $\lambda x - 1 \in M$.

With these definitions in place, we now illustrate the concept of material interpretation through the case study that every maximal ideal in $\mathbb{Z}[X]$ contains a prime number. As is well known, this theorem can even be extended to the statement that every maximal ideal is of the form

$$\langle p, f \rangle := \{\lambda p + \mu f \mid \lambda, \mu \in A\},$$

where p is a prime ideal and $f \in \mathbb{Z}[X]$ is a polynomial such that its canonical projection $\overline{f} \in (\mathbb{Z}/p\mathbb{Z})[X]$ is irreducible (see [4, Theorem 5.3]). However, the main task is to prove the existence of the prime number p. The extended statement follows quite directly and is not of particular interest for our purposes.

In the spirit of what we discussed in the first section, particularly Formula (1), our aim is to transform the statement

Let $M \subseteq \mathbb{Z}[X]$ be a maximal ideal, then M contains a prime number.

along with a classical proof, into a constructive proof of the statement:

Let $M \subseteq \mathbb{Z}[X]$ be any subset, then either M contains a prime number or there is concrete evidence that M is not a maximal ideal.

We will also work out exactly what is meant by "concrete evidence that M is not a maximal ideal".

2.1 Examination of the Classical Proof

We begin by presenting the classical proof that every maximal ideal in $\mathbb{Z}[X]$ contains a prime number. In doing so, we also present all lemmas whose proofs are necessary to our constructive proof of Theorem 2. The statement is often viewed as an application of Zariski's Lemma, which can be used to prove Hilbert's Nullstellensatz [14,24]. Indeed, our proof does not directly use Zariski's Lemma but does adopt some of its ideas, as Lemma 2 and Lemma 3 can also be used in the proof of Zariski's Lemma. We will return to a constructive treatment of Zariski's Lemma in Sect. 3.1.

Lemma 1. *$\mathbb{Z}[d^{-1}]$ is not a field for $d \in \mathbb{Z} \setminus \{0\}$.*

We present the proof in detail here so that we can later see the parallels in the constructive part. In many articles, this fact – if mentioned at all – is proved in only two lines, for example the last two lines in the proof of Theorem 5.1 in [4].

Proof. Let q be a prime number that does not divide d, and assume that $\mathbb{Z}[d^{-1}]$ is a field. Then there exist $a_0, \dots, a_n \in \mathbb{Z}$ such that $q^{-1} = \sum_{i=0}^{n} a_i d^{-i}$. Multiplying this equation by qd^n gives

$$d^n = q \sum_{i=0}^{n} a_i d^{n-i},$$

where $\sum_{i=0}^{n} a_i d^{n-i} \in \mathbb{Z}$. Hence, q divides d^n, and since q is prime, q divides d, which leads to a contradiction. □

For the classical proof, we need the definition of an integral ring extension and some properties of it. For a ring extension $A \subseteq B$, an element $x \in B$ is called *integral* over A if there are $a_0, \dots, a_{n-1} \in A$ with

$$x^n + a_{n-1}x^{n-1} + \cdots + a_1 x + a_0 = 0.$$

The ring extension $A \subseteq B$ is called an *integral ring extension* if all $x \in B$ are integral over A.

The following two lemmas provide the necessary properties of integral ring extensions we need. Even though the proof of particularly the next lemma is straightforward and well known, for example in [11, Lemma 2.1.7]. Again, we present it here in full, as the reader will encounter this proof idea in the constructive proof.

Lemma 2. *Let $A \subseteq B$ be an integral ring extension. If B is a field, so is A.*

Proof. Let $x \in A \setminus \{0\}$ be arbitrary. As B is a field, there is $x^{-1} \in B$. We show $x^{-1} \in A$:
The field extension $A \subseteq B$ is integral, hence there are $a_0, \ldots, a_{n-1}$ with

$$(x^{-1})^n + a_{n-1}(x^{-1})^{n-1} + \cdots + a_1 x^{-1} + a_0 = 0.$$

Multiplying this equation by x^{n-1} and isolating x^{-1} leads to

$$x^{-1} = -a_{n-1} - a_{n-2}x^1 - \cdots - a_0 x^{n-1} \in A.$$

□

Lemma 3. *Let $A \subseteq B$ be a ring extension generated by an element $x \in B$, i.e. $B = A[x]$. If x is integral over A, also the ring extension $A \subseteq B$ is integral.*

The proof is a combination of Proposition 2.4 and Proposition 5.1 in [2]:

Proof. As x is integral over A, there are $a_0, \ldots, a_{n-1} \in A$ with

$$x^n + a_{n-1}x^{n-1} + \cdots + a_1 x + a_0 = 0. \tag{2}$$

Given $y \in B = A[x]$ there are $b_0, \ldots, b_m \in A$ with $y = b_m x^m + \cdots + b_1 x + b_0$. By Eq. (2) we can reduce this to

$$y = c_{n-1}x^{n-1} + \cdots + c_1 x + c_0$$

for some $c_0, \ldots, c_{n-1} \in A$. Hence, all elements in B can be represented as a linear combination of $1, x, x^2, \ldots, x^{n-1}$ in A. Therefore, B is a finitely generated A-Module.

With these preliminaries in place, let $z \in B$ be given and we show that z is integral over A:

For $i \in \{0, \ldots, n-1\}$ we get $a_{ij} \in A$ such that $zx^i = \sum_{j=0}^{n-1} a_{ij}x^j$ by the first part of this proof. Using the Kronecker delta, we obtain

$$\sum_{j=0}^{n-1} (z\delta_{ij} - a_{ij})x^j = 0.$$

Let M be the matrix

$$M := (z\delta_{ij} - a_{ij})_{ij}.$$

In matrix representation, we obtain

$$M \begin{pmatrix} 1 \\ x \\ \vdots \\ x^{n-1} \end{pmatrix} = 0.$$

Multiplying on the left by the adjoint matrix $\hat{M}$ of M and using $\hat{M}M = \det(M)E$ with E being the unit matrix, we get $\det(M) = 0$ on the first line. Expanding the definition of M and the determinant, we have an integral equation of z. □

With this preparation, the proof of the classical statement is now fairly brief:

Theorem 1. *Let $M \subseteq \mathbb{Z}[X]$ be a maximal ideal. Then there is a prime number p such that $p \in M$.*

Proof. If $X \notin M$, there exists some $g \in \mathbb{Z}[X]$ with $gX - 1 \in M$ because M is a maximal ideal. $gX - 1$ is not constant since its constant coefficient is -1 and $g \neq 0$ because otherwise $1 \in M$. Hence, in both cases ($X \in M$ and $X \notin M$) there is some non-constant $f \in M$. Let d be the leading coefficient of f.

We now assume that there is no prime number p with $p \in M$. As M is maximal and therefore a prime ideal, it follows $M \cap \mathbb{Z} = \{0\}$. Hence, the canonical homomorphism $\mathbb{Z} \to \mathbb{Z}[X]/M$ is injective and induces a ring extension $\mathbb{Z}[d^{-1}] \to \mathbb{Z}[X]/M$, where $d^{-1}f$ is an integral polynomial for $X \in \mathbb{Z}[X]/M$. As $f \in M$ we have $d^{-1}f = 0$ in $\mathbb{Z}[X]/M$. By Lemma 3 the ring extension $\mathbb{Z}[d^{-1}] \to \mathbb{Z}[X]/M$ is integral. As $\mathbb{Z}[X]/M$ is a field, $\mathbb{Z}[d^{-1}]$ must also be a field by Lemma 2, which is a contradiction to Lemma 1. □

2.2 Developing a Constructive Proof

What stands out in the classical proof is that it often uses case distinctions based on whether an element belongs to the maximal ideal M or not. If we now want to transform the classical proof into a constructive proof that stays as close as possible to the classical proof, we will need to retain this principle. This means that we must assume that membership in M is decidable. To make this decidability explicit, we will not view M as a subset of a ring R (in our case $R = \mathbb{Z}[X]$) in the constructive proof, but rather as a function from R to the Boolean set $\mathbb{B} := \{\mathrm{tt}, \mathrm{ff}\}$. Here, we always consider functions to be total, i.e. $M(f)$ is either tt or ff for all $f \in \mathbb{Z}[X]$. However, we still use the notation $x \in M$ and $x \notin M$, where we mean $M(x) = \mathrm{tt}$ and $M(x) = \mathrm{ff}$, respectively.

We also need to refine the notion of a maximal ideal here, as there are several variants of maximal ideals that are classically equivalent. In the proof, maximality was used in the sense that if an element a is not in M, then there is an element λ with $\lambda a - 1 \in M$. This leads us to the notion of an explicit maximal ideal, which we even define for any ring R.

Definition 1. *Let R be a ring. For a subset $M : R \to \mathbb{B}$ and a function $\nu : R \to R$, we say that (M, ν) is an explicit maximal ideal if M is an ideal, $1 \notin M$ and $a\nu(a) - 1 \in M$ for all $a \in R \setminus M$.*

Furthermore, we say that there is evidence that (M, ν) is not an explicit maximal ideal if one of the following cases holds:

1) $0 \notin M$,
2) *there are* $a, b \in M$ *with* $a + b \notin M$,
3) *there are* $\lambda \in R$ *and* $a \in M$ *with* $\lambda a \notin M$,
4) $1 \in M$, *or*
5) *there is* $a \in R \setminus M$ *with* $a\nu(a) - 1 \notin M$.

Remark. The evidence that (M, ν) is not an explicit maximal ideal constitutes a stronger version of the statement that M is not a maximal ideal.

Note that if there is evidence that (M, ν) is not an explicit maximal ideal, we want to be able to explicitly specify which of the statements 1–5 is satisfied. In the case of statements 2, 3, or 5, we also want the elements a, b and λ to be explicitly provided.

The following lemma is the constructive variant of the classical statement that every maximal ideal is also a prime ideal. Classically, this statement is straightforward to prove. Constructively, however, we need to explicitly identify the specific factor of a product in M that also lies in M. In short terms: each explicit maximal ideal is an explicit prime ideal.

In our case, however, we even need a stronger form of this statement, which is intended to precisely capture the material interpretation. Therefore, we will formulate this statement and provide a proof as follows:

Lemma 4. *Let R be a ring, $M : R \to \mathbb{B}$, $\nu : R \to R$ and $a_1, \ldots, a_n \in R$ be given such that $a_1 \cdot \ldots \cdot a_n \in M$. Then, either some a_i is in M, or there is evidence that (M, ν) is not an explicit maximal ideal.*

Proof. Induction over n. For $n = 0$ it follows that $a_1 \cdot \ldots \cdot a_n := 1 \in M$, which is evidence that (M, ν) is not an explicit maximal ideal. For the induction step, let $a_0 \cdot \ldots \cdot a_n \in M$. If $a_0 \in M$, the proof is complete. Otherwise, $a_0\nu(a_0) - 1 \in M$ or there is evidence that (M, ν) is not an explicit maximal ideal.

Moreover, $\nu(a_0) \cdot a_0 \cdot \ldots \cdot a_n \in M$ or there is evidence that (M, ν) is not an explicit maximal ideal. Hence, we can assume that $a_0\nu(a_0) - 1 \in M$ and $\nu(a_0) \cdot a_0 \cdot \ldots \cdot a_n \in M$. This implies $(1 - a_0\nu(a_0)) \cdot a_1 \cdot \ldots \cdot a_n \in M$ and $\nu(a_0) \cdot a_0 \cdot \ldots \cdot a_n \in M$ or there is evidence that (M, ν) is not an explicit maximal ideal. It follows that $a_1 \cdot \ldots \cdot a_n \in M$, or there is evidence that (M, ν) is not an explicit maximal ideal. Applying the induction hypothesis to $a_1 \cdot \ldots \cdot a_n \in M$ completes the proof. □

The following lemma provides a more general form of polynomial division, which takes into account that the divisor is not necessarily monic.

Lemma 5. *Let $f, g \in \mathbb{Z}[X]$ and $d \neq 0$ be the leading coefficient of f. Then there is $k \in \mathbb{N}$ and $h \in \mathbb{Z}[X]$ such that* $\deg(d^k g + hf) < \deg(f)$.

Proof. Let $m := \deg(f)$ and $n := \deg(g)$. For fixed m, we use induction on n. If $n < m$, we take $k := 0$ and $h := 0$. Otherwise, let c be the leading coefficient of g. Then $\deg(dg - cx^{n-m}f) < n$, thus we obtain k' and h' such that $\deg(d^{k'}(dg - cx^{n-m}f) + h'f) < m$. Hence, $k := k' + 1$ and $h := h' - d^{k'}cx^{n-m}$ fulfil the requirement. □

With these two lemmas, we can now prove the material interpretation of Theorem 1. Note that we do not need a material interpretation of Lemmas 1–3, as they are implicitly incorporated into the proof. As a result, the proof itself may appear longer than the classical proof. However, this is not the case, as the corresponding lemmas are already integrated into the proof and even simplified.

Theorem 2. *Let $M \subseteq \mathbb{Z}[X]$ and $\nu : \mathbb{Z}[X] \to \mathbb{Z}[X]$ be an arbitrary map. Then, either there exists a prime number $p \in M$, or there is evidence that (M, ν) is not an explicit maximal ideal in $\mathbb{Z}[X]$.*

Proof. First, we construct some non-constant $f \in M$: If $X \in M$, we are done. Otherwise, $X \notin M$, and therefore $X\nu(X) - 1 \in M$ or there is evidence that (M, ν) is not an explicit maximal ideal. Let d be the leading coefficient of f and $n := \deg(f)$. If $n = 0$ we have $-1 = f \in M$ and there is witness that (M, ν) is not an explicit maximal ideal. Thus, we can assume that f is non-constant and $n > 0$.

We take some prime number q which is not a divisor of d and consider $\nu(q) \in \mathbb{Z}[X]$. We check if $q \in M$ or $m := q\nu(q) - 1 \notin M$, if yes, the proof is complete. Otherwise, we continue: For each $i \in \{0, \ldots, n-1\}$ we apply $\nu(q)x^i$ to Lemma 5 and get some $k_i \in \mathbb{N}$, $h_i \in \mathbb{Z}[X]$ and $(a_{ij})_{j\in\{0,\ldots,n-1\}} \in \mathbb{Z}^n$ with

$$d^{k_i}\nu(q)x^i + h_i f = \sum_{j=0}^{n-1} a_{ij}x^j. \tag{3}$$

Using the Kronecker delta $(\delta_{ij})_{ij}$ we get

$$\sum_{j=0}^{n-1}(d^{k_i}\nu(q)\delta_{ij} - a_{ij})x^j = -h_i f. \tag{4}$$

Let A be the matrix $(d^{k_i}\nu(q)\delta_{ij} - a_{ij})_{i,j\in\{0,\ldots,n-1\}}$ then we have

$$A\begin{pmatrix} 1 \\ x \\ \vdots \\ x^{n-1} \end{pmatrix} = \begin{pmatrix} -h_0 f \\ -h_1 f \\ \vdots \\ -h_{n-1} f \end{pmatrix} \tag{5}$$

Multiplying both sides by the adjugate matrix $\hat{A}$ of A and using $\hat{A}A = \det(A)I$ leads to

$$\begin{pmatrix} \det(A) \\ \det(A)x \\ \vdots \\ \det(A)x^{n-1} \end{pmatrix} = \hat{A}\begin{pmatrix} -h_0 f \\ -h_1 f \\ \vdots \\ -h_{n-1} f \end{pmatrix} \tag{6}$$

In particular, the first line is $\det(A) = -\sum_{j=0}^{n-1} \hat{A}_{0j}h_j f$. Looking at the definition of A, we have $\det(A) = d^K\nu(q)^n + b_{n-1}\nu(q)^{n-1} + \cdots + b_1\nu(q) + b_0$ for some $b_0, \ldots, b_{n-1} \in \mathbb{Z}$ and $K := \sum k_i$. Hence,

$$d^K\nu(q)^n + b_{n-1}\nu(q)^{n-1} + \cdots + b_1\nu(q) + b_0 = \sum_{j=0}^{n-1}(-\hat{A}_{0j}h_j)f. \tag{7}$$

Multiplying both sides by q^n leads to

$$d^K(q\nu(q))^n + b_{n-1}q(q\nu(q))^{n-1} + \cdots + b_1q^{n-1}(q\nu(q)) + b_0q^n = \sum_{j=0}^{n-1}(-q^n\hat{A}_{0j}h_j)f \tag{8}$$

We define $m := q\nu(q) - 1$, which is equivalent to $q\nu(q) = m + 1$. For each $i \in \{1, \ldots, n\}$ one can easily compute some polynomial g_i with $(m+1)^i = 1+mg_i$. This leads to

$$\begin{aligned} &d^K + b_{n-1}q + \cdots + b_1q^{n-1} + b_0q^n = \\ &\sum_{j=0}^{n-1}(-q^n\hat{A}_{0j}h_j)f + (-d^Kg_n - b_{n-1}qg_{n-1} - \cdots - b_1q^{n-1}g_1)m \end{aligned} \tag{9}$$

Since the first part of the equation lies in $\mathbb{Z}$, the second part must also lie in $\mathbb{Z}$. Furthermore, the first part cannot be zero because otherwise $q \mid d$ (or $q \mid 1$ if $K = 0$) and the second part is a linear combination of the elements f und m, which are both in M. By Lemma 4, one prime factor is in M or there is evidence that (M, ν) is not an explicit maximal ideal. □

2.3 Translation from the Classical to the Constructive Proof

First, we would like to review and summarize how we come from the classical statement to the constructive one. As stated in Sect. 1, we wanted to translate a classical proof of $A \to B$ into a classical proof of $\neg\tilde{A} \vee \tilde{B}$. Here, A was the statement "M is maximal ideal" and B was the statement "M contains a prime number".

When introducing classical maximal ideals, we presented two variants. As a result, the statement "A is a maximal ideal" is, in fact, a classical statement. Constructively, this statement must be made more specific. Already the first sentence of the classical proof makes clear which version applies: Here, we concluded from $X \notin M$ that there exists a g such that $gX - 1 \in M$. For this reason we had to use the classical stronger version of maximality, namely for all $f \notin M$ there is g with $gf - 1 \in M$. Definition 1 is then the strong negation of this version.

A similar situation occurred with the type of M: in the classical proof, M was a subset, but the proof sometimes required knowing whether a given element was in M or not. Therefore, we replaced the vague notion of a set with the precise statement that M is a total function into $\mathbb{B}$. Moreover, we also made frequent use of this property in the constructive proof, which now allows us to discuss how we transformed the classical proof into a constructive one:

What immediately stands out is that, in the constructive proof, instances of modus ponens from the classical proof are often replaced by a case distinction. This is particularly the case when statements of the form $a \in M$ are considered. For example, a statement such as "$a \in M$ because M is a maximal ideal" is replaced by the statement "$a \in M$ or we have evidence that M is not an explicitly maximal ideal". This is another reason why the assumption that $a \in M$ is decidable was indeed necessary.

Apart from this transformation, the first part of the proof, where the non-constant $f \in M$ is obtained, is quite similar. However, in the classical proof, this f is only used as a witness that $\mathbb{Z}[d^{-1}] \to \mathbb{Z}[X]/M$ is an integral ring extension. In the constructive proof, on the other hand, the notion of an integral ring extension and the embedding $\mathbb{Z}[d^{-1}] \to \mathbb{Z}[X]/M$ do not explicitly appear but are incorporated into the proof itself. For this reason, f is used further; in particular, it is applied to Lemma 5, and appears in several subsequent equations, such as Eqs. (3) to (9). The same holds for the prime number q, which is coprime to the leading coefficient d of f. The classical proof only uses it in Lemma 1, whereas the constructive proof defines it directly after the construction of f and then uses q and $\nu(q)$ throughout the proof. One can see that the details of the proof become much clearer in the constructive proof, whereas in the classical proof, they are less explicit.

Note also that the constructive proof consistently argues within $\mathbb{Z}[X]$ or $\mathbb{Z}$. As $\mathbb{Z}$ and $\mathbb{Z}[X]$ are concrete objects, we do not need to make case distinctions but can argue directly. Which we did to use statements such as $\hat{A}A = \det(A)I$ and "the first part can not be zero as otherwise $q \mid d$ (or $q \mid 1$ if $K = 0$)".

2.4 The Constructive Content

An Python implementation of the algorithm, derived from the constructive proof in Sect. 2.2, is provided at the author's GitHub page (https://github.com/FranziskusWiesnet/maxzx) using the built-in computer algebra system *SymPy* [15]. The main algorithm is referred to as `MaxZX`. In addition, there is a simple algorithm `unbounded_search` that takes any $M : \mathbb{Z}[X] \to \mathbb{B}$ and performs an unbounded search over all prime numbers and tests which prime lies in M. There are two examples: one is the ideal generated by $X^2 + 1$ and 3, and the other is the ideal generated by X and 1019.

While the unbounded search finds a prime (in this case, 3) in M significantly faster than `MaxZX` in the first example, the situation is clearly the opposite in the second case. The larger the prime number and the higher the runtime of M, the longer the unbounded search takes. In contrast, the runtime of `MaxZX` scales with the runtime of M and ν.

`MaxZX` also has the advantage that it allows the use of arbitrary subsets and functions. Moreover, it provides a witness that explains why the conditions are not satisfied. For example, if we pass the constant false function ($\lambda\_$ False) for M and the zero function (λ_1) for ν to `MaxZX`, it returns (False, $3, x$). This indicates that (M, ν) is not an explicit maximal ideal, as witnessed by $X\nu(X) - 1 \notin M$.

One can now proceed, for example, by setting $\nu(x) = 2$ and $M(2x-1) =$ true. Then `MaxZX` will produce a different witness showing why (M, ν) is not a maximal ideal. Then one can further modify M and ν so that this case also does not lead to a counterexample, and `MaxZX` outputs a different counterexample or a prime number. This process can be continued as far as one wishes. This is illustrated in the Python file with the example `M_2` and `nu_2` in the Python code. In [18,19], we followed a similar procedure in the context of general maximal objects. This demonstrates that even counterexamples or failure evidence can be informative in constructive mathematics.

3 Outlook

3.1 Further Applications of Material Interpretation

The material interpretation offers a versatile framework with potential applications in various areas of mathematics and logic. The example above served as a case study demonstrating the underlying idea. There are many cases where material interpretation can be of substantial benefit. The aim of future research will be to explore these examples further. We highlight the following two applications, where some preliminary work has already been carried out:

Zariski's Lemma and Hilbert's Nullstellensatz. Zariski's Lemma can be stated as follows:

> Let K be a field and R a K-algebra, which is also a field. Suppose that $R = K[x_1, \ldots, x_n]$ for some $x_1, \ldots, x_n \in R$. Then R is algebraic over K, i.e., there are non-zero $f_1, \ldots, f_n \in K[X]$ such that $f_i(x_i) = 0$ for all i.

An algorithmic version of Zariski's Lemma is considered in [22]. There, a constructive proof of Zariski's Lemma is provided, from which an algorithm is then developed, and the correctness of the algorithm is proved. The algorithmic version is divided into several lemmas, but they all share the structure in which an algorithm is first presented, followed by a theorem about the algorithm. These theorems typically state that if the objects in the algorithm satisfy certain axioms (e.g. field or ring axioms), then the algorithm's output satisfies certain properties. Thus, a material interpretation means that either the output satisfies the desired properties, or there exists a witness showing that one of the axioms about the objects in the algorithm does not hold. A proof sketch was already provided in Section 4.5.1 of [23] and will be the subject of further research. In Section 4.5.2 of [23], it was shown that applying the material interpretation to Zariski's Lemma yields an algorithmic version of Hilbert's Nullstellensatz, where Hilbert's Nullstellensatz states the following:

> Let K be an algebraically closed field, and let $f_1, \ldots, f_m \in K[X_1, \ldots, X_n]$ be given. Then either there are $g_1, \ldots, g_m \in K[X_1, \ldots, X_n]$ with $g_1 f_1 + \cdots + g_m f_m = 1$ or there are $x_1, \ldots, x_n \in K$ with $f_i(x_1, \ldots, x_n) = 0$ for all i.

What remains is to fully prove the material interpretation of Zariski's Lemma and to isolate its constructive content, preferably as a computer program. As previously mentioned, the case study in Sect. 2 of this article is a special case of

Zariski's Lemma. More precisely, it follows from Zariski's Lemma that, considering the embedding $\mathbb{Q} \to \mathbb{Z}[X]/M$, the element X is algebraic over $\mathbb{Q}$, i.e. there is some $g \in \mathbb{Q}[X]$ with $g(X) = 0 \in \mathbb{Z}[X]/M$. If we now multiply g by the least common denominator of all its coefficients, we indeed obtain f, which is constructed in the first part of the proof. Thus, the first part of the proof was essentially the base case of the inductive proof of Zariski's Lemma. Interestingly, Lemma 2 and Lemma 3 are also applied in the induction step of Zariski's Lemma [22, Lemma 1–2]. Therefore, studying it is particularly well-suited as a stepping stone to understanding the material interpretation of Zariski's Lemma, and consequently, of Hilbert's Nullstellensatz.

The Universal Krull-Lindenbaum Lemma. In [20], Schuster and Rinaldi gave a constructive version of the Universal Krull-Lindenbaum lemma (UKLL), and in [18,19] an algorithm realising UKLL under some general assumptions is developed.

Ring Theory. In the context of ring theory the universal Krull-Lindenbaum lemma says that the intersection of all prime ideals of a commutative ring $R \neq \{0\}$ is equal to the nilradical:

$$\bigcap \{P \subseteq R \mid P \text{ is a prime ideal in } R\} = \sqrt{\langle 0 \rangle}$$

For a fixed $r \in R$, the construed algorithm, takes a so-called *Krull functional* as input. In the context of ring theory a Krull functional for a subset takes an arbitrary subset $P \subseteq R$ and returns a witness that either $r \in P$ or that P is not a prime ideal. This can be seen as a material interpretation of the statement "r is in every prime ideal". The output of the constructed algorithm in the context of ring theory is indeed an $e \in \mathbb{N}$ such that $r^e = 0$. Through the material interpretation of general proofs of the following statements, which use UKLL in the version above, we were indeed able to give constructive variants of the following theorems:

> Let A be a ring and $f = \sum_{i=0}^{n} a_i X^i \in A[X]$ be a unit in $A[X]$, then a_i is nilpotent for each $i > 0$.

> Let A be a ring and $f = \sum_{i=0}^{n} a_i X^i, g = \sum_{i=0}^{m} b_i X^i \in A[X]$ be two polynomials with $fg = \sum_{i=0}^{n+m} c_i X^i$, then
>
> $$a_i b_j \in \sqrt{\langle c_0, \ldots, c_{i+j} \rangle}\ .$$

The first statement is a typical example and has already been studied, among others, in [17,21]. The second statement is known as theorem by Gauß-Joyal and its constructive meaning was also studied as a standard example in many articles such as [3,5].

Hence, in the case of ring theory and on test examples, the material interpretation has already shown success, partly due to the simplicity of the test examples. In the context of ordered fields, however, the proofs are significantly more complex:

Ordered Fields. For a field K a total order $\leq$ on K is called a *field order*, if for all $a, b, c \in K$: $a \leq b \to a + c \leq b + c$ and $0 \leq a \wedge 0 \leq b \to 0 \leq a \cdot b$. We call a subset $U \subseteq K$ an *order* if there is a field order $\leq$ on K such that $U = \{x \in K \mid 0 \leq x\}$. In this setting the universal Krull-Lindenbaum-Lemma coincides with the Artin-Schreier Theorem and says

$$\bigcap \{U \subseteq K \mid U \text{ is an order of } K\} = \left\{ \sum_{i=0}^{n} x_i^2 \mid n \in \mathbb{N},\ x_0, \ldots, x_n \in K \right\}.$$

This statement is only non-trivial if -1 is not a sum of squares in K which implies $\text{char}(K) = 0$. The fields $\mathbb{R}$ and $\mathbb{Q}$ and all function fields over them fulfil this property.

Artin proved the Artin-Schreier Theorem and used it to prove Hilbert's 17th Problem [1]:

> Let $f \in \mathbb{Q}[X_1, \ldots, X_n]$ be a polynomial in $n \in \mathbb{N}$ variables such that $f(x) \geq 0$ for all $x \in \mathbb{Q}^n$. Then f is the sum of squares in $\mathbb{Q}(X_1, \ldots, X_n)$.

Artin's proof of Hilbert's 17th problem, however, is very complex and involves, in addition to the Artin-Schreier theorem, Sturm's theorem [7, Section 5.8.2] and the fact that every ordered field has a real closure. A constructive consideration of the real closure was given by Lombardi and Roy [13] and could be used to construct a Krull functional.

There is already a constructive proof of Hilbert's 17th problem by Delzell [6], and it would be very interesting to compare the computational content of Delzell's proof with that of the proof using a Krull functional. However, since Artin's proof is very complex, it will hardly be possible to trace the computational content using only pen and paper. Arguably, the same applies to the material interpretation of Zariski's Lemma for the section above, especially when aiming to extract its precise constructive content. Therefore, computer assistance will be absolutely necessary.

3.2 Implementation of the Material Interpretation

In Sect. 2.4, we discussed an implementation of the algorithm in Python. The algorithm was created manually. However, it is of interest to explore whether this algorithm could be generated automatically. This applies not only to the algorithm itself, but also to the entire material interpretation as well. An implementation in a proof assistant such as Agda, Coq, or Lean would be highly valuable.

The ultimate goal would be to at least partially automate the material interpretation. As we have seen in Sect. 2.3, applications of modus ponens are often replaced by a case distinction. Such transformations could be implemented in a proof assistant. Furthermore, in the classical proof, a case distinction based on $x \in M$ was performed and the property that M is a maximal ideal was utilized to imply that for $x \notin M$, there exists an a with $ax - 1 \in M$. In the constructive setting, this motivates defining M as a total Boolean-valued function and the definition of an explicit maximal ideal. Thus, it can be observed that the modified statement can, in fact, be extracted from the classical proof. This suggests that automation might be possible here as well.

Acknowledgements. This research was funded in whole by the Austrian Science Fund (FWF) under Grant 10.55776/ESP576.

Great thanks also go to the anonymous reviewers, whose comments and suggestions have substantially enhanced the quality of this article.

Disclosure of Interests. The author has no competing interests to declare that are relevant to the content of this article.

References

1. Artin, E.: Über die Zerlegung definiter Funktionen in Quadrate. Abh. Math. Semin. Univ. Hambg. **5**(1), 100–115 (1927). https://doi.org/10.1007/bf02952513
2. Atiyah, M.F., Macdonald, I.G.: Introduction to Commutative Algebra. The Advanced Book Program, CRC Press, Taylor & Francis Group, Boca Raton (2018)
3. Banaschewski, B., Vermeulen, J.: Polynomials and radical ideals. J. Pure Appl. Algebra **113**(3), 219 (1996)
4. Conrad, K.: Maximal Ideals in Polynomial Rings. https://kconrad.math.uconn.edu/blurbs/ringtheory/maxideal-polyring.pdf. Accessed 10 Apr 2025
5. Coquand, T.: Space of valuations. Ann. Pure Appl. Logic **157**(2–3), 97–109 (2009). https://doi.org/10.1016/j.apal.2008.09.003
6. Delzell, C.N.: A continuous, constructive solution to Hilbert's 17th problem. Invent. Math. **76**(3), 365–384 (1984). https://doi.org/10.1007/bf01388465
7. Freund, R., Hoppe, R.: Stoer/Bulirsch: Numerische Mathematik 1, 10th edn. Springer, Heidelberg (2007)
8. Gerhardy, P., Kohlenbach, U.: Extracting Herbrand disjunctions by functional interpretation. Arch. Math. Logic **44**(5), 633–644 (2005). https://doi.org/10.1007/s00153-005-0275-1
9. Gurevich, Y.: Intuitionistic logic with strong negation. Studia Logica Int. J. Symb. Logic **36**(1/2), 49–59 (1977). http://www.jstor.org/stable/20014839
10. Gödel, K.: Über eine bisher noch nicht benützte Erweiterung des finiten Standpunktes. Dialectica **12**(3–4), 280–287 (1958). https://doi.org/10.1111/j.1746-8361.1958.tb01464.x
11. Huneke, C., Swanson, I.: Integral Closure of Ideals, Rings, and Modules. Cambridge University Press, Cambridge (2006). Includes bibliographical references (p. [405]-421) and index
12. Köpp, N., Petrakis, I.: Strong negation in the theory of computable functionals TCF. arXiv:2210.05491 (2024). https://doi.org/10.48550/ARXIV.2210.05491
13. Lombardi, H., Roy, M.F.: Elementary constructive theory of ordered fields, pp. 249–262. Birkhäuser Boston (1991). https://doi.org/10.1007/978-1-4612-0441-1_17

14. McCabe, J.: A note on Zariski's lemma. Am. Math. Mon. **83**(7), 560–561 (1976). https://doi.org/10.1080/00029890.1976.11994170
15. Meurer, A., Čertík, O., Smith, C.P.: SymPy. https://www.sympy.org/en/index.html. Accessed 10 Apr 2025
16. Nelson, D.: Constructible falsity. J. Symb. Log. **14**(1), 16–26 (1949). https://doi.org/10.2307/2266838
17. Persson, H.: An Application of the Constructive Spectrum of a Ring. Type Theory and the Integrated Logic of Programs. Chalmers University and University of Göteborg **1**(1), 19 (1999)
18. Powell, T., Schuster, P., Wiesnet, F.: An algorithmic approach to the existence of ideal objects in commutative algebra. In: Iemhoff, R., Moortgat, M., de Queiroz, R. (eds.) WoLLIC 2019. LNCS, vol. 11541, pp. 533–549. Springer, Heidelberg (2019). https://doi.org/10.1007/978-3-662-59533-6_32
19. Powell, T., Schuster, P., Wiesnet, F.: A universal algorithm for Krull's theorem. Inf. Comput. **287**, 104761 (2022). https://doi.org/10.1016/j.ic.2021.104761
20. Rinaldi, D., Schuster, P.: A universal Krull-Lindenbaum theorem. J. Pure Appl. Algebra **220**(9), 3207–3232 (2016). https://doi.org/10.1016/j.jpaa.2016.02.011
21. Schuster, P.: Induction in algebra: a first case study. Log. Methods Comput. Sci. **9**(3) (2013). https://doi.org/10.2168/lmcs-9(3:20)2013
22. Wiesnet, F.: An algorithmic version of Zariski's lemma. In: De Mol, L., Weiermann, A., Manea, F., Fernández-Duque, D. (eds.) CiE 2021. LNCS, vol. 12813, pp. 469–482. Springer, Cham (2021). https://doi.org/10.1007/978-3-030-80049-9_46
23. Wiesnet, F.: The Computational Content of Abstract Algebra and Analysis. Ph.D. thesis, Ludwig-Maximilians-Universität München, Università degli Studi di Verona, Università degli Studi di Trento (2021). https://doi.org/10.5282/EDOC.28929
24. Zariski, O.: A new proof of Hilbert's Nullstellensatz. Bull. Am. Math. Soc. **53**, 362–368 (1947). https://doi.org/10.1090/S0002-9904-1947-08801-7

Alternation-Bounded Semi-unbounded Fan-in Cascading Circuits and the Complementation Closure Property

Tomoyuki Yamakami(✉)

Faculty of Engineering, University of Fukui, 3-9-1 Bunkyo, Fukui 910-8507, Japan
tomoyukiyamakami@gmail.com

Abstract. Semi-unbounded fan-in circuit families have played an important role in characterizing languages accepted by nondeterministic auxiliary pushdown automata. Such families of polynomial-size circuits of depth $O(\log^k n)$ induce the complexity class SAC^k. As a natural extension of these circuit families, another model of semi-unbounded fan-in cascading circuits was introduced in 2022 with the use of additional gates, called $\mathrm{AND}_{(\omega)}$ gates, of unbounded fan-out. The nondeterminism-vs-co-nondeterminism question (or the complementation closure property question) has been centering in computational complexity theory. The complementation closure property is still unknown for NP and NEXP, whereas NL and SAC^k are well-known to be closed under complementation. We particularly focus on the languages accepted by uniform families of polynomial-size semi-unbounded fan-in k-cascading circuits of $O(\log^r n)$ alternations. We prove that the collection of those languages indeed enjoys the complementation closure property. This result significantly improve the known fact of $\mathrm{SAC}^k = \text{co-}\mathrm{SAC}^k$.

Keywords: cascading circuit · complementation closure · inductive counting · storage automata · semi-unbounded fan-in · SAC^k

1 A Historical Account and an Overview of Contributions

1.1 Historical Background and New Challenges

The notions of *nondeterminism* and *co-nondeterminism* have been a centerpiece of modern theoretical research in computer science for decades, where "co-nondeterminism" is the complementary notion of "nondeterminism". In the form of the *nondeterminism-vs-co-nondeterminism question*, one often asks whether or not these two notions are computationally equivalent. For most runtime-bounded computations, nondeterminism and co-nondeterminism seem to be computationally opposite notions. For example, polynomial-time nondeterministic and co-nondeterministic computations respectively give rise to the well-known complexity classes NP and co-NP, and they are believed to be different (which further implies $\mathrm{P} \neq \mathrm{NP}$) by many researchers. For space-bounded computation, on

© The Author(s), under exclusive license to Springer Nature Switzerland AG 2025
A. Beckmann et al. (Eds.): CiE 2025, LNCS 15764, pp. 496–510, 2025.
https://doi.org/10.1007/978-3-031-95908-0_35

the contrary, no difference is often observed between nondeterminism and co-nondeterminism. As a typical example, Immerman [3] and Szelepcsényi [5] independently proved that NL is closed under complementation; namely, the *complementation closure property* holds for NL. In other words, nondeterminism and co-nondeterminism computationally coincide with each other for logarithmic-space (or log-space, for short) computations. Furthermore, Borodin, Cook, Dymond, Ruzzo, and Tompa [1] succeeded in demonstrating that LOGCFL is also closed under complementation, where LOGCFL is the closure of CFL (context-free language class) under log-space many-one reductions [2]. Their proof heavily depends on Venkateswaran's characterization of LOGCFL in terms of families of polynomial-size semi-unbounded fan-in Boolean circuits of logarithmic depth [6]. These special circuit families form the complexity class SAC^1 of decision problems (or equivalently, languages). Actually, Borodin et al. proved the equality between SAC^1 and co-SAC^1 (class of complementary languages of SAC^1), which therefore leads to LOGCFL = co-LOGCFL. In general, circuit-based proofs seem to be much easier and more intuitive than machine-based ones. It had been expected to improve their proof techniques for other complexity classes and to show their complementary closure property. A more general statement, SAC^k = co-SAC^k for all numbers $k \geq 2$, was also given in [1]. See also a textbook by, e.g., Vollmer [7]. Naturally, we wonder if there exist any other complexity classes $\mathcal{C}$, which can be proven to be closed under complementation (i.e., $\mathcal{C}$ = co-$\mathcal{C}$) without any unproven assumptions.

Although SAC^1 = co-SAC^1, it is widely believed that NP $\neq$ co-NP. Since $\mathrm{SAC}^1 \subseteq$ NP, it is natural to ask which complexity class $\mathcal{C}$ between SAC^1 and NP is closed under complementation. Concerning this question, how far can we go beyond SAC^1?

Open Problem 1. *Between* SAC^1 *and* NP, *is there any natural complexity class that also enjoys the complementation closure property?*

The purpose of this work is to make a further contribution in this line of study on the nondeterminism-vs-co-nondeterminism question. As a natural extension of semi-unbounded fan-in Boolean circuits, we wish to look into a non-standard computational model of *semi-unbounded fan-in* k*-cascading circuits*, which was lately introduced in [9,10]. We are particularly interested in decision problems solvable by uniform families of polynomial-size semi-unbounded fan-in k-cascading circuits of $O(\log n)$ alternations (see [11]). For notational convenience, we use the notation $k\mathrm{SCC}^1$ to denote the collection of all such decision problems. This circuit model was originally sought to characterize the machine model of *nondeterministic auxiliary depth-*k *storage automata* (or aux-k-sna's, for short) [9,10], which are equipped with read-only input tapes, auxiliary (work) tapes, and storage tapes, where a storage tape means a two-way semi-infinite rewritable work tape with the specific restriction that, after each tape cell is visited by its tape head k times, the tape cell becomes non-rewritable (or frozen) blank forever. Languages recognized by polynomial-time aux-k-sna's using only $O(\log n)$ auxiliary (work) space naturally expand LOGCFL. These languages are proven to be contained in NSC^k (i.e. the nondeterministic variant of the kth level of Steve's class SC) [9,10].

Therefore, $k\mathrm{SCC}^1$ seems to be a perfect candidate for Open Problem 1. Namely, we ask as to whether $k\mathrm{SCC}^1$ enjoys the complementary closure property.

Open Problem 2. *For each fixed integer $k \geq 1$, is $k\mathrm{SCC}^1$ closed under complementation?*

1.2 Our Contributions

This work intends to tackle Open Problem 2, which is the nondeterminism-vs-co-nondeterminism question regarding $k\mathrm{SCC}^1$. As a clear affirmative answer to this question, we intend to prove in Theorem 8 that $k\mathrm{SCC}^1$ equals its "complementary" class co-$k\mathrm{SCC}^1$. As a result, this work significantly expands the aforementioned well-known complementation closure property of NL and SAC^1 (or equivalently, LOGCFL) to the cases of cascading circuit families.

Our initial strategy to prove Theorem 8 is the heavy use of circuit simulations. As noted in Sect. 1.1, the proof of $\mathrm{SAC}^1 = \text{co-}\mathrm{SAC}^1$ in [1] exploits a fundamental technique of inductive counting, initially developed in [3,5], for space-bounded Turing machines. We attempt to take a similar proof strategy but expand it to fit into the proof of our main result, Theorem 8. More precisely, we provide how to simulate, using log space, the negation of an alternation-bounded semi-unbounded fan-in cascading circuit by another alternation-bounded semi-unbounded fan-in cascading circuit of proper size and alternation. Furthermore, we expand our proof into $O(\log^r n)$-alternation k-cascading circuits as in Proposition 10 for an arbitrary number $r \geq 1$. Since families of such circuits easily compute languages in SAC^r, our result truly expands the scope of the nondeterminism-vs-co-nondeterminism question.

The result of this work can be also seen as an extension of a circuit-based inductive counting technique of [1] to a wider rage of circuit families.

2 Fundamental Notions and Notation

We briefly explain basic terminology relevant to the content of this work.

2.1 Numbers, Alphabets, and Languages

We use the standard notations $\mathbb{N}$ and $\mathbb{N}^+$ to denote the sets of all *natural numbers* including 0 and of all positive natural numbers, respectively. For two integers m and n with $m \leq n$, we write $[m,n]_{\mathbb{Z}}$ for the *integer interval* $\{m, m+1, m+2, \ldots, n\}$ and abbreviate $[1,n]_{\mathbb{Z}}$ for each $n \in \mathbb{N}^+$ as $[n]$. In this work, all *polynomials* have coefficients of only natural numbers and all *logarithms* are taken to the base 2.

An *alphabet* is a finite nonempty set of "letters" or "symbols" and a *language* over alphabet Σ is a subset of Σ^*. For a language L over alphabet Σ, its *complement* denotes the language $\Sigma^* - L$. We often abbreviate it as $\overline{L}$ whenever Σ

is clear from the context. Given a complexity class $\mathcal{C}$ of languages, the notation co-$\mathcal{C}$ denotes the collection of all complements $\overline{L}$ of any languages L in $\mathcal{C}$. We freely identify "decision problems" with the corresponding "languages".

The collection of all context-free languages is denoted CFL in this work. Let FL denote the set of all functions computable by polynomial-time log-space deterministic Turing machines (or DTMs, for short). Given a complexity class $\mathcal{C}$ of languages, the notation LOG($\mathcal{C}$) denotes the collection of all languages L for which there exist another language $A \in \mathcal{C}$ and a function $f \in \mathrm{FL}$ such that, for any string x, $x \in L$ holds exactly when $f(x) \in A$ holds. As customary, we abbreviate LOG(CFL) as LOGCFL.

Since our main target is cascading circuits, we only briefly mention the machine model of auxiliary depth-k storage automata and its correspondence to LOGCFL in the special case of $k = 2$. This correspondence shows that auxiliary depth-k storage automata truly expand auxiliary pushdown automata. A *storage tape* refers to a two-way semi-infinite rewritable work tape with the left endmarker used for an underlying machine as a memory device, and it is said to be of *depth k* if any tape cell in the tape becomes *frozen blank* (with the designated frozen blank symbol B) and cannot be revised forever after its tape head visits this tape cell k times. The machine model of *deterministic auxiliary depth-k storage automata* (or aux-k-sda's, for short) was introduced in [8] and its nondeterministic variant, called *nondeterministic auxiliary depth-k storage automata* (or aux-k-sna's), was discussed in [9,10]. An aux-k-sna is a two-way 3-tape nondeterministic Turing machine with a read-only input tape, a rewritable auxiliary (work) tape, and a depth-k storage tape. An aux-k-sna is said to be *frozen blank sensitive* if, while reading B on its storage tape, its input and auxiliary tape heads never move to adjacent tape cells.

The notation AuxSNAdepth,space,time($k, s(n), t(n)$) denotes the class of all languages recognized by aux-k-sna's running in time $t(n)$ using work space $s(n)$.

Fact 3. [10] LOGCFL $=$ AuxSNAdepth,space,time($2, O(\log n), n^{O(1)}$), *provided that all aux-2-sna's are frozen blank sensitive.*

2.2 k-Cascading Circuits

Let us explain the central concept of k-cascading circuits, which were introduced in [9,10]. These special circuits are basically composed of the Boolean gates of AND ($\wedge$), OR ($\vee$), and negation ($\overline{x}$) as well as input and output gates. As customary in circuit complexity theory, we assume that the negation gate is applied only to input variables so that we can take x and $\overline{x}$ as inputs given to the target circuit. We view a circuit as a directed acyclic graph whose vertices are gates and edges are wires (between gates). The direction of each edge indicates the flow of truth/falsehood information from input gates to output gates of the circuit. Given a gate, its *fan-in* (resp., *fan-out*) is the number of incoming wires connecting to (resp., outgoing wires connecting from) this gate. We consider only families of circuits C_n indexed by natural numbers n. In such a circuit family $\{C_n\}_{n\in\mathbb{N}}$, a gate in each circuit is said to have *bounded fan-in* if the gate's fan-in

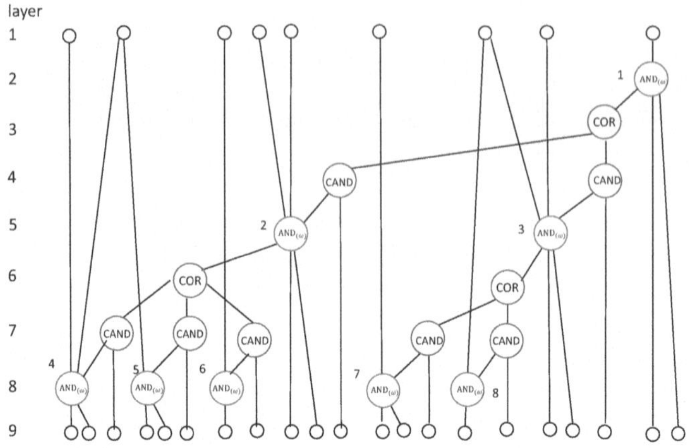

Fig. 1. A cascading block composed of $\text{AND}_{(\omega)}$, CAND, and COR gates marked as large circles. The small circles on layer 1 and layer 9 represent respectively output gates and input gates of this cascading block. The numbers adjacent to gates are gate labels. The link length of this subcircuit is 3.

is upper bounded by a fixed constant, independent of the choice of the circuit in the family. Otherwise, the gate has *unbounded fan-in*. A similar notion is introduced for "fan-out". We use AND gates that have fan-in 2 and fan-out 1; in contrast, OR gates always have unbounded fan-in but fan-out 1. As in [9,10], we also consider a special type of "AND" gates, which have bounded fan-in and unbounded fan-out, called $\text{AND}_{(\omega)}$, distinguished from ordinary AND gate whose fan-out is always 1. Even though AND can be seen as a special case of $\text{AND}_{(\omega)}$, we generally "distinguish" them according to their distinctive roles in cascading circuits.

In what follows, we assume that a circuit is *layered* in the sense that (i) any gate at depth i takes inputs from gates at depth less than i (except for the case of $i = 1$) and (ii) all gates at the same depth are placed in the same layer and are indexed from left to right. We should differentiate "layered" circuits from "leveled" circuits, in which all gates at level i take inputs from only gates at level $i - 1$. Remember that layers are numbered from outputs to inputs. See an illustration of a layered subcircuit given in Fig. 1.

We first define a *cascading block*, which will be embedded into a much larger subcircuit called a *cascading semi-circuit*. It is useful to distinguish between OR gates that appear in certain cascading blocks and OR gates not in any cascading blocks. For clarity reason, we intend to call the former OR gates *COR gates*, where the prefix "C" indicates "cascading" blocks. In a similar fashion, we also distinguish between AND gates in cascading blocks and AND gates not in any cascading blocks. The former AND gates are distinctly called *CAND gates*.

Definition 4. *A* cascading block D *is a special layered subcircuit composed of* $\text{AND}_{(\omega)}$*, CAND, and COR gates (explained above) with numerous input and* *output gates satisfying the following seven conditions. Let i denote any natural number.*

(1) Gates on the first and the last layers are respectively used as outputs and inputs of this block. These gates should be replaced by appropriate gates when the block is inserted into a larger circuit.
(2) On layer 2, there is only one $AND_{(\omega)}$ gate and its fan-out is 1.
(3) The subcircuit D contains $AND_{(\omega)}$ gates only on layer $3i+2$, COR gates on layers $3i+3$, and CAND gates on layer $3i+4$.
(4) Each $AND_{(\omega)}$ gate on layer $3i+5$ is directly connected to (possibly) input gates at the first layer and exactly one CAND gate on layer $3i+4$.
(5) Each $AND_{(\omega)}$ gate on layer $3i+2$ is directly connected from (possibly) output gates on the last layer and exactly one COR gate on layer $3i+3$.
(6) Each COR gate on layer $3i+3$ is directly connected only from CAND gates on layer $3i+4$.
(7) For any two distinct $AND_{(\omega)}$ gates, there is no more than one directed path between them.

Two distinct $\mathrm{AND}_{(\omega)}$ gates in a cascading block are said to be *linked* if there is a path between them. Such a path is referred to as a *link*. The *link length* of a cascading block D is the maximum number of $\mathrm{AND}_{(\omega)}$ gates (not including CAND gates) that appear in any path from an input gate to an output gate of D. Figure 1 illustrates a cascading block of link length 3.

Cascading blocks are further embedded into a much larger circuit. In such a larger circuit G containing two cascading blocks C_1 and C_2, C_1 is called an *ancestor* of C_2 (or C_2 is a *descendant* of C_1) in G if the following condition holds: for any two $\mathrm{AND}_{(\omega)}$ gates g_1 and g_2 in C_1 and C_2, respectively, if g_1 and g_2 are connected by a path of G, then g_1 must be an ancestor of g_2. A *common ancestor* of two $\mathrm{AND}_{(\omega)}$ gates g_1 and g_2 in a circuit G is a gate that is ancestors of both g_1 and g_2. The *minimum common ancestor* of g_1 and g_2 is the first gate that is present on "all" paths from g_1 to the root as well as from g_2 to the root. Assuming that C_1 and C_2 have link lengths k_1 and k_2, respectively, we say that C_1 and C_2 are *orderly* if, for any number $i \in [\min\{k_1, k_2\}]$, the ith $\mathrm{AND}_{(\omega)}$ gate of C_1 is an ancestor of the ith $\mathrm{AND}_{(\omega)}$ gate of C_2, provided that all $\mathrm{AND}_{(\omega)}$ gates are enumerated in a cascading block according to their layer numbers.

When a cascading block D is embedded into a larger circuit G, the *block-top layer number* of D in G refers to the layer number of the top $\mathrm{AND}_{(\omega)}$ gate of D in G, and the *block-bottom layer number* of D in G is the layer number of the bottom $\mathrm{AND}_{(\omega)}$ gate of D in G.

A *decisive fragment* of a circuit G is obtained from G by deleting all but one child (and all the subcircuits rooted at them) of each OR gate (including COR gate). The *unlinked version* of G is the subcircuit obtained from G by deleting all CAND and COR gates and all wires associated with them. Moreover, the *leaf-free version* of G is obtained from G by deleting all input gates of G.

We next introduce *cascading semi-circuits*, each of which is composed of cascading blocks of a certain restricted form.

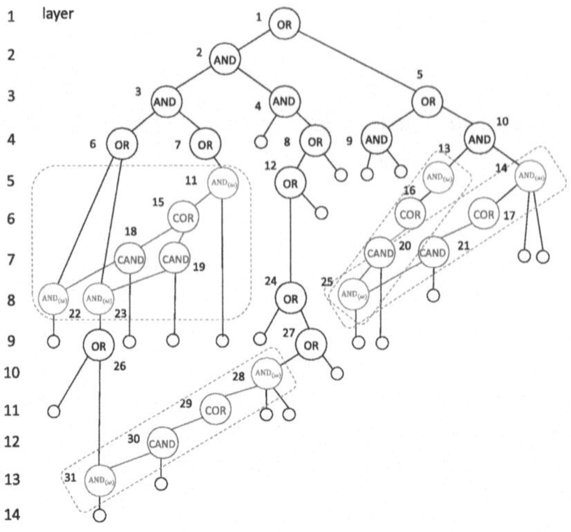

Fig. 2. A cascading circuit made up from two cascading semi-circuits. Four cascading blocks (excluding their input and output gates) are marked by green dotted boxes. The numbers adjacent to circles are gate labels. In the two left cascading blocks, gate 3 is the minimum common ancestor of gates 11 and 23, and gate 2 is the minimum common ancestor of gates 28 and 31. The subcircuit rooted at gate 2 is a cascading semi-circuit, and the subcircuit rooted at gate 5 is another cascading semi-circuit composed of two cascading blocks. The basis subcircuit is made up of gates 1, 2, and 5. (Color figure online)

Definition 5. *A* cascading semi-circuit *is a circuit such that every decisive fragment C of it satisfies the following ten conditions.*

(1) The unlinked version of C must be a tree.
(2) On each layer of C, there is at most one cascading block of length ≥ 2 and, if any, it must contain the leftmost node on the layer.
(3) In the unlinked, leaf-free version of C, every path from each bottom node to the root must pass through at least one cascading block.
(4) The root of C is the minimum common ancestor of two $AND_{(\omega)}$ gates of a certain cascading block of C.
(5) For any two linked $AND_{(\omega)}$ gates in C, their minimum common ancestor must be either an AND gate or an $AND_{(\omega)}$ gate.
(6) In any cascading block D of C, all $AND_{(\omega)}$ gates of D have leafless fan-in at most 2.
(7) Every $AND_{(\omega)}$ gate g in C has at most two outputs. If g has exactly two outputs, then one of them must be connected to a CAND gate in a certain cascading block.
(8) For any two cascading blocks A and B of C, if A is an ancestor of B, then A and B must be orderly.

(9) *For any two cascading blocks A and B of C, if A is an ancestor of B, then the block-bottom layer number of A is larger than the block-top layer number of B.*

(10) *On any layer ℓ between the block-top layer number and the block-bottom layer number of a cascading block D of C having cascading length ≥ 2, any gate that is located on the right side of D on layer ℓ must have leafless fan-in* 1.

Given a circuit G, the *leafless fan-in* of an AND_ω gate of G is the number of inputs wired directly to this gate except for the inputs wired directly from input gates. Given a cascading semi-circuit G and a cascading block D in G, the *cascading length* of D is one plus the sum of the leafless fan-ins of any $\mathrm{AND}_{(\omega)}$ gates of D except for the top $\mathrm{AND}_{(\omega)}$ gate. A cascading semi-circuit whose cascading length is at most k is succinctly called a *k-cascading semi-circuit.*

Now, we combine various cascading semi-circuits to build a (full-fledged) cascading circuit.

Definition 6. *A* basis subcircuit *refers to a tree-form subcircuit composed only of OR and AND gates. A* cascading circuit *is built from a basis subcircuit by connecting its leaves to cascading semi-circuits. A* k-cascading circuit *refers to a cascading circuit whose cascading semi-circuits all have cascading lengths at most k.*

In a cascading circuit depicted in Fig. 2, the basis subcircuit is composed only of gates 1, 2, and 5. The two cascading semi-circuits of this circuit are the subcircuits rooted at gates 2 and 5.

2.3 Known Facts on Semi-unbounded Fan-in k-Cascading Circuits

In what follows, we are interested only in families of cascading circuits of the following form. Any family of circuits is said to have *semi-unbounded fan-in* if there exists a constant $c \in \mathbb{N}^+$, independent of the choice of any circuit in the family, for which any path from the root to a leaf of any circuit has at most c consecutive gates of the types AND and $\mathrm{AND}_{(\omega)}$. The *alternation* of such a circuit is the maximum number of times when the gate types switch among AND, $\mathrm{AND}_{(\omega)}$, and OR along any path from the output gate to an input gate. The *size* of a circuit is the total number of gates and wires in it.

A family $\mathcal{C} = \{C_n\}_{n\in\mathbb{N}}$ of Boolean circuits is *logspace uniform* if there exists a DTM M equipped with a read-only input tape, a rewritable index tape, multiple work tapes, and a write-once output tape such that M takes inputs of the form 1^n and produces an appropriately-defined encoding $\langle C_n\rangle$ of the nth circuit C_n in time $n^{O(1)}$ and using space $O(\log n)$.

Now, it is convenient to introduce the general notation CCIRcasc,alt,size($k, d(n), s(n)$) to express the collection of all languages computed by logspace-uniform families of semi-unbounded fan-in k-cascading circuits of alternation at most $d(n)$ and size at most $s(n)$.

Yamakami [9–11] demonstrated that logspace-uniform $O(\log n)$-alternation $n^{O(1)}$-size semi-unbounded fan-in k-cascading circuit families are as powerful as $n^{O(1)}$-time $O(\log n)$-space aux-k-sna's, provided that the aux-k-sna's satisfy the *frozen blank sensitivity* (see [10]). In other words, we obtain the following. Hereafter, we wish to abbreviate CCIRcasc,alt,size$(k, O(\log^r n), n^{O(1)})$ as kSCC^r.

Fact 7 [9–11]. *Let k be any positive integer. The complexity class kSCC^1 coincides with* AuxSNAdepth,space,time$(2k, O(\log n), n^{O(1)})$, *provided that all aux-$2k$-sna's are frozen blank sensitive.*

By Facts 3 and 7, we instantly conclude that SAC^1 equals 1SCC^1 because SAC^1 coincides with LOGCFL [6].

In the case of SAC^1, Borodin et al. [1] demonstrated that SAC^1 coincides with its complementary class co-SAC^1. This instantly leads to the conclusion that 1SCC^1 coincides with co-1SCC^1. As the main contribution of this work, we prove the following generalized statement on k-cascading circuit families.

Theorem 8. *For any constant $k \in \mathbb{N}^+$, the complexity class kSCC^1 coincides with the class* co-kSCC^1 *of complementary languages of* kSCC^1.

The proof of Theorem 8 will be given in Sect. 3. Fact 7 directly yields the following statement on auxiliary depth-k storage automata.

Corollary 9. AuxSNAdepth,space,time$(2k, O(\log n), n^{O(1)})$ *is closed under complementation, provided that all aux-k-sna's are frozen blank sensitive.*

Theorem 8 will be further expanded in Sect. 3.6 to kSCC^r = co-kSCC^r for any $r \geq 1$.

3 Proof of the Main Theorem

3.1 Underlying Proof Ideas

In the subsequent subsections provides the detailed proof of Theorem 8. We begin with fixing an integer k with $k \geq 2$ arbitrarily throughout the following argument because the case of $k = 1$ follows immediately from SAC^1 = 1SCC^1.

Given a Boolean circuit C with n inputs, its *complementary circuit* generally refers to another circuit with n inputs that always produces the negation of C on the same inputs. Thus, we need to construct from a given cascading circuit C another cascading circuit that computes the complementary circuit of C. A standard method of obtaining the complementary circuit is to inductively apply de Morgan's law to swap between AND and OR from an output gate to input gates (assuming that the negation is applied only to input variables). Unfortunately, the circuit obtained from a given semi-unbounded fan-in circuit by this method is no longer semi-unbounded fan-in. Moreover, when the given circuit contains a number of cascading blocks, these blocks are completely destroyed, and thus the circuit is no longer a cascading circuit. These difficulties therefore need to be overcome. A basic idea behind the proof of SAC^k = co-SAC^k in [1] is to count the total number of gates on each "level" of the circuit and to use this number to decide whether or not the computation leading to it is indeed a complementary subcircuit. For this purpose, we utilize the knowledge on the number of AND gates on the same layer that outputs 1.

A key to the construction of complementary circuits presented in [1] is the free use of *threshold gates* T^n_m, by which we can check whether the number of 1s among n inputs is more than or equal to m. It is known that T^n_m can be implemented by an appropriate semi-unbounded fan-in Boolean circuit of $n^{O(1)}$ size and $O(\log n)$ alternation. See also a textbook by, e.g., Vollmer [7] for the proof.

3.2 Starting Point

Let us take an arbitrary non-empty language L over the binary alphabet $\Sigma = \{0,1\}$ in $k\mathrm{SCC}^1$. There exists a logspace-uniform family $\mathcal{C} = \{C_n\}_{n\in\mathbb{N}}$ of polynomial-size semi-unbounded fan-in k-cascading circuits of $O(\log n)$-alternation solving L. From this family $\mathcal{C}$, we will construct another logspace-uniform family $\overline{\mathcal{C}} = \{\overline{C}_n\}_{n\in\mathbb{N}}$ of polynomial-size semi-unbounded fan-in k-cascading circuits of $O(\log n)$-alternation layer by layer and then show that $\overline{\mathcal{C}}$ indeed solves $\overline{L}$. This demonstrates that $\overline{L}$ also belongs to $k\mathrm{SCC}^1$, yielding the theorem.

For the subsequent argument, we now fix n arbitrarily and consider the nth circuit C_n of $\mathcal{C}$. This k-cascading circuit C_n must be made up of a basic subcircuit, say, A_n and a number of k-cascading semi-circuits, say, $S_1, S_2, \ldots, S_t$ ($t \in \mathbb{N}^+$), each of which is attached to an appropriate leaf (i.e., an input gate) of A_n. We wish to construct the desired circuit $\overline{C}_n$ inductively layer by layer from the original circuit C_n. We split our construction into two independent parts targeting A_n and S_i's.

It is important to note that, if there is no cascading semi-circuit (i.e., $t = 0$), the construction is the same as that of Borodin et al. [1]. For simplicity, in what follows, we assume that $t \geq 1$ and $s > n$.

3.3 A Basis Subcircuit

Let $\overline{A}_n$ denote the complementary circuit of A_n.

Since the basis subcircuit A_n has no $\mathrm{AND}_{(\omega)}$ gate, we can apply the construction of Borodin et al. to obtain its complementary circuit $\overline{A}_n$. Unfortunately, A_n may not be a "leveled" circuit in general. Hence, we first modify it so that it becomes leveled. For each path from a leaf to the root of A_n, we insert a number of extra OR gates of fan-in 1 between two consecutive gates lying on this path so that every layer should contain a certain gate. Moreover, we further add OR gates to shorter paths so that all paths from leaves to the root in A_n have exactly the same length. This makes it possible to turn A_n into a "leveled" circuit. By slightly abusing the notation, A_n also refers to this rearranged subcircuit. Under the above rearrangement, we briefly describe how to define the complementary subcircuit $\overline{A}_n$ from A_n.

Let d denote the highest layer number of A_n. Let s_m be the total number of gates on layer m in A_n and set $s_{max} = \max\{s_m \mid m \in [0,d]_{\mathbb{Z}}\}$.

For each bit $b \in \{0,1\}$, we set $\tau_m(b)$ to be the total number of gates of A_n lying on layer m that output b. For convenience, a unique output gate of A_n is located on the lowest layer 1 and all input gates (i.e., leaves) are placed on the highest layer d. Remember that, for any index $m \in [d]$, $\tau_m(0) + \tau_m(1)$ equals s_m.

Firstly, we consider a subcircuit C whose top gate is located on layer m, where $m \in [d]$. We then define $\overline{C} \equiv \bigvee_{r=0}^{s_{max}} (H_m^{(r)} \wedge C^{(r)})$, where two subcircuits $H_m^{(r)}$ and $C^{(r)}$ for each number $r \in [0, s_{max}]_{\mathbb{Z}}$ are inductively defined from A_n and C in a way similar to [1,7].

When C is A_n, a similar to the construction in [1,7] yields $\overline{A}_n$. Since this subcircuit $\overline{A}_n$ has no cascading block, $\overline{A}_n$ also forms a basis subcircuit. The semi-unboundedness of $\overline{A}_n$ follows from the fact that $H_m^{(r)}$ is a semi-unbounded fan-in subcircuit.

3.4 Cascading Blocks

We have already dealt with the basis subcircuit A_n in Sect. 3.3. The remaining task is to handle the cascading semi-circuits $S_1, S_2, \ldots, S_t$, which are attached to the leaves of A_n. Note that each S_i contains at least one cascading block. Let us consider a k-cascading semi-circuit $S \in \{S_1, S_2, \ldots, S_t\}$. This subcircuit S constitutes two different categories of gates. Gates of the first category appear in certain cascading blocks and gates of the second category do not appear in any cascading blocks. For convenience, we intend to split the subsequent construction process into two according to these categories.

In what follows, we consider gates of the first category and, meanwhile, we focus on how to simulate the complementary computation of a single cascading block in S. Now, we pick up an arbitrary cascading block of S and call it by D. We explain how to compute its complement block $\overline{D}$.

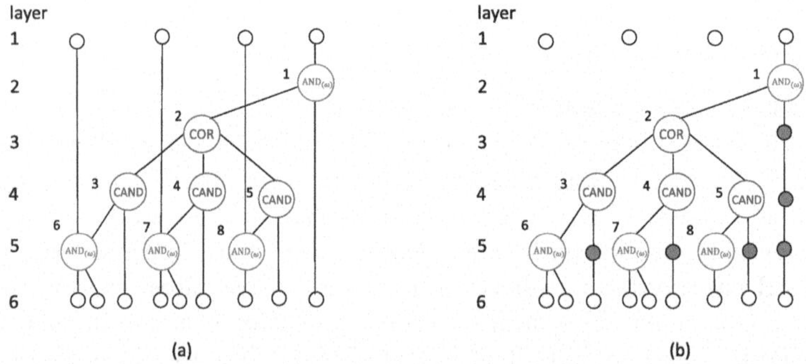

Fig. 3. (a) A cascading block. (b) A rearranged block of (a). The green small circles represent newly inserted OR gates of fan-in 1. All wires from $\mathrm{AND}_{(\omega)}$ gates labeled 6, 7, and 8 to the output gates on layer 1 are deleted. (Color figure online)

The layers of D are temporarily called $1, 2, \ldots, d$ in order from the block top-layer number to the block bottom-layer number. Input gates of D, say, $h_1, h_2, \ldots, h_t$ are laid on layer d and D is composed of only output gates, say, $g_1, g_2, \ldots, g_{t'}$ on layer 1. By the definition of cascading blocks, the COR gates are on layers $3l$ for any integer l with $1 \leq l < d/3$.

To simplify our construction process, we make the first rearrangement of D as follows. Since there may be output gates of fan-in more than one, we insert extra OR gates of fan-in 1 between each output gate and its directly connecting gate of D, we exclude all the original output gates of D, and we then reset the newly inserted OR gates to be the new output gates of D.

As the second rearrangement, we also insert extra OR gates of fan-in 1 to further modify D. Note that only $\mathrm{AND}_{(\omega)}$ and CAND can be attached directly to input gates of D. If a gate, say, p is attached to such an input gate, say, h, then we insert OR gates of fan-in 1 between p and h so that h is located on layer d of D and that the inserted OR gates occupy every layer on the path between p and h.

Unlike the basis subcircuit A_n, D contains at least one $\mathrm{AND}_{(\omega)}$ gate. Note that the complementary gate of $\mathrm{AND}_{(\omega)}$ is not even an admissible gate. As the third rearrangement, we target $\mathrm{AND}_{(\omega)}$ gates of fan-out at least 2 and turn them into $\mathrm{AND}_{(\omega)}$ of fan-out exactly 1. Corresponding to the output gates $g_1, g_2, \ldots, g_{t'}$ ($t' \in \mathbb{N}^+$) of D on layer 1, there are unique $\mathrm{AND}_{(\omega)}$ gates directly connected to them. We write p_i for such a gate leading to g_i. We then construct t' cascading blocks $D_1, D_2, \ldots, D_{t'}$ as follows. For each index j, we first delete all wires between p_i and g_i for all i except for j. We then delete all gates and wires that are not lying between g_j and some of the input gates $h_1, h_2, \ldots, h_t$. We call the resulting blocks *rearranged blocks*. See an illustration in Fig. 3.

We remark that the rearranged blocks are "independent" in the sense that any further change of the design of D_i does not affect the outcome of D_j if $i \neq j$ (unless adding any other input gates). This makes it possible to treat each D_i separately. In what follows, we pick up an arbitrary block out of $D_1, D_2, \ldots, D_t$ and write it as D'. Notice that D' is also a cascading block of cascading length at most k.

We re-number the layers of D' so that its input gates of D' lie on layer 1 and its output gates lie on layer, say, d'. Let $\overline{s}_m$ denote the total number of gates of D' on layer $m \in [d']$. Let $3l_0$ denote the highest layer that contains at least one COR gate of D'. For each number l with $1 \leq l \leq l_0$, we assume that layer $i+1$ with $i < 3l$ has been already handled and we then consider a subcircuit C whose top gate is located on layer $i \geq 1$. For this C, we define its complementary subcircuit $\overline{C}$ as $\bigvee_{r=0}^{\overline{s}_{max}} (\hat{H}_i^{(r)} \wedge C^{(r)})$, where $\overline{s}_{max} = \max\{\overline{s}_m \mid m \in [d']\}$, and $\hat{H}^{(r)}$ and $C^{(r)}$ are introduced in the following way.

(1) Hereafter, we intend to define $C^{(r)}$ from C inductively.
(a) If C consists of an input gate h on layer d', then we set $C^{(r)}$ as the complement of h for any r. (b) If C consists of a newly inserted OR gate of the form $OR(E)$ of fan-in 1 for a certain subcircuit E, then we set $C^{(r)} \equiv OR(E^{(r)})$. (c) If C is of the form $CAND(C_1, C_2)$ for subcircuits C_1 and C_2 whose top

gates lie on layer $i+1$, then we set $C^{(r)} \equiv C_1^{(r)} \vee C_2^{(r)}$. (d) Now, assume that C's top gate is $AND_{(\omega)}$. Since C has fan-out 1, let C be of the form $C_1 \wedge C_2$ for two subcircuits C_1 and C_2. We then set $C^{(r)} \equiv C_1^{(r)} \vee C_2^{(r)}$. Note that, in the cases (a)–(d), if $C_1^{(r)}$ and $C_2^{(r)}$ respectively compute the negation of the outcomes of C_1 and C_2, then $C^{(r)}$ also computes the negation of C.
(e) Next, we consider the case of $i = 3l$. We focus on C whose top gate is an COR gate. We assume that C is of the form $COR(C_{j_1}, C_{j_2}, \ldots, C_{j_c})$, where the top gate of each C_{j_e} is an CAND gate. Let $\overline{\tau}(1)$ and $\overline{\tau}(0)$ respectively denote the number of 1s and the number of 0s of the outcomes of subcircuits $C_1, C_2, \ldots, C_{s'}$, where $s' = \overline{s}_{i+1}$. Here, we do not define $C^{(r)}$ to be $T_c^c(C_{j_1}^{(r)}, \cdots, C_{j_c}^{(r)})$. Instead, we set $C^{(r)}$ to be $T_r^{s'}(C'_1, \ldots, C'_{s'-c})$, where $C'_1, \ldots, C'_{s'-c}$ are all subcircuits in $\{C_1, C_2, \ldots, C_{s'}\} - \{C_{j_1}, C_{j_2}, \ldots, C_{j_c}\}$. It follows that C outputs 0 iff all gates C_{j_e} output 0 iff $r = \overline{\tau}(1)$ and $C^{(r)}$ outputs 1.
It is important to note by the construction of $C^{(r)}$ that, since $C^{(r)}$ contains no $\text{AND}_{(\omega)}$ gate, $C^{(r)}$ does not form any part of the cascading block. This property is useful in (2).

(2) How can we define $\hat{H}_m^{(r)}$ from D'? When $m = d'$, we set $\hat{H}_m^{(\tau(1))} = 1$ and $\hat{H}_m^{(j)} = 0$ for all $j \neq \tau(1)$, where $\tau(1)$ denotes the total number of 1s of (the outcomes of) the input gates of D'. Assuming that $\hat{H}_{m+1}^{(r)}$ is already defined, we define $\hat{H}_m^{(r)}$ as $\bigvee_{j=0}^{s'}(\hat{H}_{m+1}^{(j)} \wedge \hat{P}_m^{(s',r)} \wedge \hat{N}_m^{(s'-r,j)})$. The subcircuit $\hat{P}_m^{(s',r)}$ is $T_r^{s'}(C_1, C_2, \ldots, C_{s'})$ and $\hat{N}_m^{(s'-r,j)}$ is $T_{s'-r}^{s'}(C_1^{(j)}, C_2^{(j)}, \ldots, C_{s'}^{(j)})$, where the top gate of each C_i lies on layer $m+1$. Since $C_1, C_2, \ldots, C_{s'}$ are all original subcircuits of D, $\hat{P}_m^{(s',r)}$ contains the cascading blocks of $C_1, C_2, \ldots, C_{s'}$; on the contrary, $\hat{N}_m^{(s'-r,j)}$ has no cascading block. Although D' is a cascading block, $\hat{H}_m^{(r)}$ may be no longer a cascading block by itself. Rather, $\hat{H}_m^{(r)}$ forms a cascading semi-circuit, which is induced from D'.

3.5 Cascading Semi-circuits

In Sect. 3.4, we have dealt with gates of the first category and have shown how to compute the complement of a cascading block D. To construct the desired complementary subcircuit $\overline{S}$ of a given cascading semi-circuit S, we still need to deal with gates of the second category. Notice that these gates do not belong to any cascading block. We convert the original cascading semi-circuit S in three stages and produce a new complementary subcircuit $\overline{S}$ that correctly computes the complement of S; that is, $S = 0$ iff $\overline{S} = 1$.

Assume that S consists of cascading blocks $D_1, D_2, \ldots, D_m$, where D_i is an immediate descendent of D_{i+1} for any index $i \in [m-1]$. Here, the output gates of each D_i have fan-in 1. As done in Sect. 3.4, we can obtain its complementary block $\overline{D}_i$. We make a further rearrangement of D_i's as follows.

By the definition of S, we wish to distinguish between two types of gates: (1) all gates appearing on paths between two consecutive cascading blocks and (2)

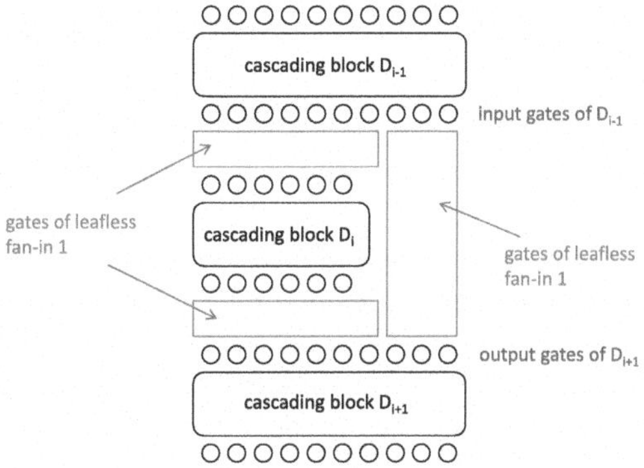

Fig. 4. An illustration of a part of cascading semi-circuit, consisting of three consecutive cascading blocks D_{i-1}, D_i, D_{i+1}.

any gate located in the right side of D_i on layers between the block-top layer and the block-bottom layer of D_i. It is important to note that gates in (1) and (2) have leafless fan-in 1.

(1) Let us consider the gates g that are used as output gates of D_{i+1} and the gates h that are used as input gates of D_i. For each gate g, we consider the subcircuit C whose top gate is exactly g. By the construction given in Sect. 3.4, we obtain the complementary circuits $\overline{C}$ from C. For each h, we consider the subcircuit C' whose top gate is h. (a) If E is a subcircuit of C' whose top gate is an OR of fan-in 1 of the form $OR(E')$ for a certain subcircuit E', then we set $\overline{E} \equiv OR(\overline{E'})$. (b) If E is of the form $AND(E_1, E_2)$ for two subcircuits E_1 and E_2, then we set $\overline{E} \equiv OR(\overline{E_1}, \overline{E_2})$. (c) If E is of the form $OR(E_1, E_2)$, then we set $\overline{E} \equiv AND(\overline{E_1}, \overline{E_2})$.
(2) Next, we consider the gates located in the right side of D_i. We assume that these gates connect two non-consecutive cascading blocks, say, D_{i+1} and D_{i-1}. See Fig. 4 for this situation. Since all those gates have leafless fan-in 1, they can be handled similarly to (1).

Finally, we argue that the constructed circuit correctly simulates the negation of the outcomes of the original circuit C_n. Moreover, the obtained circuit is semi-unbounded fan-in because of the use of counting gates (which are semi-unbounded fan-in). The obtained circuit is therefore a k-cascading circuit. The logspace-uniformity of the obtained circuit family comes from the fact that the construction procedure given in Sects. 3.3–3.5 can be carried out using only space logarithmic in n.

3.6 Extension to $O(\log^r n)$-Alternations

We can generalize Theorem 8 to cascading circuits of $O(\log^r n)$-alternations for any constant $r \in \mathbb{N}^+$.

Proposition 10. *For any constants $k, r \in \mathbb{N}^+$, $k\mathrm{SCC}^r$ is closed under complementation.*

4 A Brief Discussion and Open Problems

In this work, we have proven that $k\mathrm{SCC}^1$ is closed under complementation (Theorem 8) by exploiting a fundamental idea of circuit-based inductive counting [1,7]. This result is further expanded to $k\mathrm{SCC}^r$ for any integer $r \geq 1$ (Proposition 10). A crucial point of our proof is how to modify the existing proof technique of inductive counting used for semi-unbounded fan-in Boolean circuit families in order to make it applicable to the computational model of cascading circuits.

It is interesting to know how far we can expand the proof technique of inductive counting to verify the complementation closure property for other classes of language families of higher complexity.

References

1. Borodin, A., Cook, S.A., Dymond, P.W., Ruzzo, W.L., Tompa, M.: Two applications of inductive counting for complementation problems. SIAM J. Comput. **18**, 559–578 (1989)
2. Cook, S.A.: Characterizations of pushdown machines in terms of time-bounded computers. J. ACM **18**, 4–18 (1971)
3. Immerman, N.: Nondeterministic space is closed under complementation. SIAM J. Comput. **17**, 935–938 (1988)
4. Sudborough, I.H.: On the tape complexity of deterministic context-free languages. J. ACM **25**, 405–414 (1978)
5. Szelepcsényi, R.: The method of forced enumeration for nondeterministic automata. Acta Inform. **26**, 279–284 (1988)
6. Venkateswaran, H.: Properties that characterize LOGCFL. J. Comput. Syst. Sci. **43**, 380–404 (1991)
7. Vollmer, H.: Introduction to Circuit Complexity. Springer (1999)
8. Yamakami, T.: Between SC and LOGDCFL: families of languages accepted by logarithmic-space deterministic auxiliary depth-k storage automata. Int. J. Comput. Math. Comput. Syst. Theory **8**, 1–31 (2023)
9. Yamakami, T.: Nondeterministic auxiliary depth-bounded storage automata and semi-unbounded fan-in cascading circuits (extended abstract). In: Proceedings of the 28th International Conference on Computing and Combinatorics (COCOON 2022), Lecture Notes in Computer Science, vol. 13595, pp. 61–69, Springer (2022)
10. Yamakami, T.: Nondeterministic auxiliary depth-bounded storage automata and semi-unbounded fan-in cascading circuits. This is a complete, corrected version of [9]. arXiv:2412.09186 (2024)
11. Yamakami, T.: Polynomial-size semi-unbounded fan-in cascading circuit families. To be posted at arXiv.org (2025)

Author Index

© The Editor(s) (if applicable) and The Author(s), under exclusive license
to Springer Nature Switzerland AG 2025

A. Beckmann et al. (Eds.): CiE 2025, LNCS 15764, pp. 511–512, 2025.
https://doi.org/10.1007/978-3-031-95908-0

The manufacturer's authorised representative in the EU is Springer Nature Customer Service Centre GmbH, Europaplatz 3, 69115 Heidelberg, Germany. If you have any concerns regarding our products, please contact ProductSafety@springernature.com

Printed and bound by CPI Group (UK) Ltd, Croydon, CR0 4YY
30/09/2026
02225560-0018